Communications in Computer and Information Science 3095

Rationale

The CCIS series is devoted to the publication of proceedings of computer science conferences. Its aim is to efficiently disseminate original research results in informatics in printed and electronic form. While the focus is on publication of peer-reviewed full papers presenting mature work, inclusion of reviewed short papers reporting on work in progress is welcome, too. Besides globally relevant meetings with internationally representative program committees guaranteeing a strict peer-reviewing and paper selection process, conferences run by societies or of high regional or national relevance are also considered for publication.

Topics

The topical scope of CCIS spans the entire spectrum of informatics ranging from foundational topics in the theory of computing to information and communications science and technology and a broad variety of interdisciplinary application fields.

Information for Volume Editors and Authors

Publication in CCIS is free of charge. No royalties are paid, however, we offer registered conference participants temporary free access to the online version of the conference proceedings on SpringerLink (http://link.springer.com) by means of an http referrer from the conference website and/or a number of complimentary printed copies, as specified in the official acceptance email of the event.

CCIS proceedings can be published in time for distribution at conferences or as post-proceedings, and delivered in the form of printed books and/or electronically as USBs and/or e-content licenses for accessing proceedings at SpringerLink. Furthermore, CCIS proceedings are included in the CCIS electronic book series hosted in the SpringerLink digital library at http://link.springer.com/bookseries/7899. Conferences publishing in CCIS are allowed to use our online conference service (Meteor) for managing the whole proceedings lifecycle (from submission and reviewing to preparing for publication) free of charge.

Publication process

The language of publication is exclusively English. Authors publishing in CCIS have to sign the Springer CCIS copyright transfer form, however, they are free to use their material published in CCIS for substantially changed, more elaborate subsequent publications elsewhere. For the preparation of the camera-ready papers/files, authors have to strictly adhere to the Springer CCIS Authors' Instructions and are strongly encouraged to use the CCIS LaTeX style files or templates.

Abstracting/Indexing

CCIS is abstracted/indexed in DBLP, Google Scholar, EI-Compendex, Mathematical Reviews, SCImago, Scopus. CCIS volumes are also submitted for the inclusion in ISI Proceedings.

How to start

To start the evaluation of your proposal for inclusion in the CCIS series, please send an e-mail to ccis@springer.com

Hiram Ponce · Hugo Jair Escalante ·
Roberto A. Vazquez
Editors

Artificial Intelligence – COMIA 2026

18th Mexican Congress
Cuernavaca, Mexico, May 25–29, 2026
Proceedings, Part I

Springer

Editors
Hiram Ponce
Universidad Panamericana
Ciudad de México, Mexico

Hugo Jair Escalante
University of Texas at El Paso and INAOE
El Paso, TX, USA

Roberto A. Vazquez
Universidad La Salle México
Ciudad de México, Mexico

ISSN 1865-0929 ISSN 1865-0937 (electronic)
Communications in Computer and Information Science
ISBN 978-3-032-37437-0 ISBN 978-3-032-37438-7 (eBook)
https://doi.org/10.1007/978-3-032-37438-7

This Springer imprint is published by the registered company Springer Nature Switzerland AG
The registered company address is: Gewerbestrasse 11, 6330 Cham, Switzerland

Preface

The Mexican Conference on Artificial Intelligence (COMIA) is an annual academic event organized by the Mexican Society for Artificial Intelligence (SMIA, Sociedad Mexicana de Inteligencia Artificial) since 2008. COMIA serves as the main national forum in the field of artificial intelligence (AI) and represents a key meeting point for the country's academic and scientific community dedicated to AI.

This year, COMIA 2026 was organized by the SMIA in collaboration with Universidad Autónoma del Estado de México and Instituto Nacional de Electricidad y Energías Limpias; and was held from May 25 to 29, 2025, in Cuernavaca, Morelos, Mexico.

The COMIA 2026 website is https://comia.com.mx/. The website of the Mexican Society for Artificial Intelligence, SMIA, is www.SMIA.mx. Contact options and additional information can be found on these websites.

The conference showcased a wide range of research areas and topics of AI. The program included keynote lectures by international experts, government and industrial panels, technical paper presentations, hands-on tutorials, and specialized workshops that enriched the academic experience. These activities developed a dynamic and engaging environment designed to appeal to a diverse audience interested in the latest developments and applications of AI. In addition, the historic and culturally rich town of Cuernavaca was our chosen destination.

COMIA conferences publish peer-reviewed papers in all areas of AI and its applications. Since its foundation in 2008, the conference has grown in popularity and improved in quality. This year is the second time that the proceedings have been published in Springer.

The proceedings of COMIA 2026 are published in three volumes.

The first volume contains 63 papers structured into two sections:

- Applications of Artificial Intelligence
- Medical Applications

The second volume contains 61 papers structured into three sections:

- Machine Learning and Pattern Recognition
- Artificial Vision and Robotics
- Computational Intelligence and Data Mining

The third volume contains 59 papers structured into three sections:

- Natural Language Processing
- Intelligent Systems
- Deep Learning

The three-volume set will be of interest for researchers in all fields of artificial intelligence, students specializing in related topics, and the general public interested in recent developments in AI.

The conference received 312 submissions from six countries: Bolivia, Colombia, Mexico, Peru, Russia, and the USA. Given that COMIA was founded as a key meeting point for the Mexican AI community, national representation remains essential. This year, we received submissions from 26 Mexican states, demonstrating widespread engagement across the country: Aguascalientes, Baja California, Chiapas, Chihuahua, Ciudad de México, Estado de México, Guanajuato, Guerrero, Hidalgo, Jalisco, Michoacán, Morelos, Nayarit, Nuevo León, Oaxaca, Puebla, Querétaro, San Luis Potosí, Sinaloa, Sonora, Tabasco, Tamaulipas, Tlaxcala, Veracruz, Yucatán, and Zacatecas. Of these submissions, 183 papers were selected for publication in these three volumes after at least three reviews per submission in a double-blind peer review process conducted by the international Program Committee. The acceptance rate was 59%.

The international Program Committee consisted of 254 experts from 10 countries: Austria, Brazil, Colombia, Cuba, Germany, Mexico, Spain, Switzerland, the UK, and the USA.

A workshop was held in conjunction with the conference

– Intelligent Industry Applications

We express our gratitude to everyone who contributed to the organization of this conference: the authors of the papers included in these three volumes, whose research gives true value to these proceedings, and the organizing team for their dedication. We also extend our thanks to the reviewers for the time and effort they devoted to evaluating the submissions, and to the members of the Program and Organizing Committees.

The entire submission, reviewing, and selection process, as well as preparation of the proceedings, was supported by Microsoft's Conference Management Toolkit (https://cmt3.research.microsoft.com/).

Last but not least, we are grateful to Springer for their patience and help in the preparation of these volumes.

In summary, COMIA 2026 was more than just a conference —it was a meeting of minds, a reflection of the AI community's unwavering dedication, and a hopeful signal for the future of the field. We hope these proceedings not only inform but also inspire new ideas, collaborations, and progress along the journey in AI.

May 2026

Hiram Ponce
Hugo Jair Escalante
Roberto A. Vazquez

Organization

Conference Committee

General Chair

Hiram Ponce	Universidad Panamericana, Mexico

Program Chairs

Hugo Escalante	University of Texas at El Paso, USA
Roberto A. Vazquez	Universidad La Salle, Mexico

Keynote Talks Committee

Gilberto Ochoa	Tecnológico de Monterrey, Mexico
Bella Martínez	Instituto Politécnico Nacional, Mexico

Workshop Committee

Lourdes Martínez	Universidad Panamericana, Mexico
David Pinto	Benemérita Universidad Autónoma de Puebla, Mexico

Tutorials Committee

Joanna Alvarado	Tecnológico de Monterrey, Mexico
Claudia González	Instituto Tecnológico de Tijuana, Mexico

Panel Committee

Miguel Gonzalez	Tecnológico de Monterrey, Mexico
Obdulia Pichardo	Instituto Politécnico Nacional, Mexico

Logistic Committee

Iris Méndez	Universidad Autónoma de Ciudad Juárez, Mexico
Ari Barrera	Universidad Panamericana, Mexico

Volunteer Committee

Ari Barrera	Universidad Panamericana, Mexico
Omar Saldivar	Barbarosos, Mexico

Communication and Design Committee

Omar Saldivar	Barbarosos, Mexico
Luis Gerardo Téllez	Barbarosos, Mexico
Josue Ruiz	CENIDET, Mexico
Noé Castro	CENIDET, Mexico
Iván Olvera	Grupo SAC, Mexico

Sponsorship Committee

Leobardo Morales	IBM, Mexico
Ismael Medina	Databricks, Mexico

Local Organizing Committee

Jose Hernández	Universidad Autónoma del Estado de Morelos, Mexico
Gustavo Arroyo	Instituto Nacional de Electricidad y Energías Limpiaz, Mexico

Program Committee

Abdiel Reyes Vera	National Polytechnic Institute, Mexico
Alberto Aguilar	Tecnológico de Monterrey, Mexico
Alberto Ochoa-Zezzatti	Universidad Autónoma de Ciudad Juárez, Mexico
Alberto Espinosa	Centro de Investigación en Computación, Mexico
Alberto Alfonso Aguilar-Lasserre	Tecnológico Nacional de México/Instituto Tecnológico de Orizaba, Mexico
Aldo Márquez-Grajales	Universidad Autónoma del Estado de Hidalgo, Mexico
Alejandro Rodríguez-Sánchez	Universidad Panamericana, Mexico
Alejandro Israel Barranco Gutiérrez	Tecnológico Nacional de México en Celaya, Mexico
Alma Rosa Mena-Martinez	Tecnológico de Monterrey, Mexico

Alvaro Sánchez Márquez	Universidad Autónoma de Tlaxcala, Campus Calpulalpan, Mexico
Ana Laura Lezama-Sánchez	SECIHTI, Benemérita Universidad Autónoma de Puebla, Mexico
André Araújo	Federal University of Alagoas, Brazil
Andrea Magadán-Salazar	TecNM/CENIDET, Mexico
Andres Garcia Floriano	Instituto Politécnico Nacional, Mexico
Andrés Espinal	Universidad de Guanajuato, Mexico
Angel Sánchez García	Universidad Veracruzana, Mexico
Angel Vergara Betancourt	TecNM - ITS Zacapoaxtla, Mexico
Ángeles Quezada	Instituto Tecnológico de Tijuana, Mexico
Angeles Belém Priego Sánchez	Universidad Autónoma Metropolitana Unidad Azcapotzalco, Mexico
Angélica Guzmán	Universidad Autónoma del Estado de México, Mexico
Anilu Franco	Universidad Autónoma del Estado de Hidalgo, Mexico
Antonio Marin Hernandez	University of Veracruz, Mexico
Antonio Neme	Universidad Nacional Autónoma de México, Mexico
Ari Barrera Animas	Universidad Panamericana, Mexico
Armando Pérez Aguilar	Instituto Tecnológico Superior de Villa la Venta, Mexico
Armando Levid Rodríguez Santiago	Universidad Tecnológica de los Valles Centrales de Oaxaca, Mexico
Asdrúbal López Chau	Universidad Autónoma del Estado de México, Mexico
Atoany Fierro-Radilla	Tecnológico de Monterrey, Mexico
Beatriz González-Beltrán	Universidad Autónoma Metropolitana, Mexico
Bell Manrique Losada	Universidad de Medellín, Colombia
Bella Martinez Seis	IPN, Mexico
Bernardo Corona Domínguez	Universidad Nacional Rosario Castellanos, Mexico
Betania Hernandez-Ocaña	Universidad Juárez Autónoma de Tabasco, Mexico
Brian Sarmina	Centro de Investigación en Computación IPN, Mexico
Carlos Aguilar	CICESE-UT3, Mexico
Carlos Reyes	Instituto Nacional de Astrofísica, Óptica y Electrónica, Mexico
Carlos Lastre Domínguez	TecNM/ITOaxaca, Mexico
Carlos Saldaña	Instituto Tecnológico de Estudios Superiores de Monterrey, Mexico

Carlos Manuel Ruano González	Universidad de Camagüey, Cuba
Carolina Del-Valle-Soto	Universidad Panamericana, Mexico
Cecilia Reyes-Peña	UPIIT-IPN, Mexico
Cecilia-Irene Loeza-Mejía	Tecnológico Nacional de México/Instituto Tecnológico Superior de Misantla, Mexico
Cesar Macias	Centro de Investigación en Computación, Mexico
Cesar Puente Montejano	Universidad Autónoma de San Luis Potosí, Mexico
César Jesús Núñez Prado	Instituto Politécnico Nacional, Mexico
Cinthia Peraza Ramirez	Tijuana Institute of Technology, Mexico
Claudia Gómez	Instituto Tecnológico de Ciudad Madero, Mexico
Claudia I. Gonzalez	TecNM - InstitutoTecnológico de Tijuana, Mexico
Cosijopii Garcia-Garcia	Universidad del Istmo, Mexico
Daniel Sánchez-Ruiz	UPIIT-IPN, Mexico
Daniel Gonzalez Arriaga	Benemérita Universidad Autónoma de Puebla, Mexico
Daniel Canton-Enriquez	Universidad Autónoma de Querétaro, Mexico
Daniela Alejandra Ochoa	CentroGeo, Mexico
Dante Mújica-Vargas	CENIDET, Mexico
David Pinto	Benemérita Universidad Autónoma de Puebla, Mexico
David Tinoco-Varela	Universidad Nacional Autónoma de México, Mexico
Diana Jacobo	Instituto Tecnológico Nacional de México, Mexico
Diana Aurora Cruz	Universidad Autónoma de la Ciudad de México, Mexico
Diana-Margarita Córdova-Esparza	Universidad Autónoma de Querétaro, Mexico
Diego Uribe	TecNM ITL, Mexico
Eddie Clemente	CENIDET, Mexico
Eddy Sánchez-DelaCruz	Tecnológico Nacional de México - Campus Misantla, Mexico
Edgar Avalos Gauna	Rice University, USA
Edmundo Bonilla	TecNM, Mexico
Eduardo Espinosa-Avila	Universidad Nacional Autónoma de México, Mexico
Eduardo Valdez	IPN, Mexico
Efrén Mezura-Montes	University of Veracruz, Mexico
Elisabetta Crescio	Tecnológico de Monterrey, Mexico
Eloísa García-Canseco	Universidad Autónoma de Baja California, Mexico
Elva Lilia Reynoso Jardon	Universidad Autónoma de Ciudad Juárez, Mexico
Emmanuel Santos Rodriguez	Nu Bank, Mexico

Erik Carbajal-Degante	Universidad Nacional Autónoma de México, Mexico
Ernesto Moya-Albor	Universidad Panamericana, Mexico
Esperanza Aleman-Aguilar	Medica Sur, Mexico
Esther Vargas	Universidad La Salle, Mexico
Federico Alonso Pecina	Universidad Autónoma del Estado de Morelos, Mexico
Félix Castro Espinoza	Autonomous University for the State of Hidalgo, Mexico
Fernando Camarena	Tecnológico de Monterrey, Mexico
Fernando Gudino	National Autonomous University of Mexico, Mexico
Fernando Elizalde-Ramírez	Instituto Tecnológico y de Estudios Superiores de Monterrey, Mexico
Fevrier Valdez	Tijuana Institute of Technology/TecNM, Mexico
Francisco Lopez Orozco	Universidad Autónoma de Ciudad Juárez, Mexico
Gabriel Aguilera	Tecnológico Nacional de México campus Sur de Guanajuato, Mexico
Gabriel González-Serna	TecNM/CENIDET
Gabriela Ramirez-de-la-Rosa	Universidad Autónoma Metropolitana, Mexico
Garibaldi Pineda Garcia	Applied AGI, UK
Gemma Bel Enguix	Universidad Nacional Autónoma de México, Mexico
Gerardo Altamirano-Gomez	Instituto Tecnológico y de Estudios Superiores de Monterrey, Mexico
Gerardo Laguna-Sanchez	Universidad Autónoma Metropolitana Lerma, Mexico
German Preciat	Universidad de Guadalajara, Mexico
Gilberto Rivera	Universidad Autónoma de Ciudad Juárez, Mexico
Gildardo Sanchez-Ante	Tecnológico de Monterrey, Mexico
Giner Alor-Hernandez	Tecnológico Nacional de México/I.T. Orizaba, Mexico
Grissel Rodríguez Roldán	Universidad Nacional Autónoma de México, Mexico
Guadalupe Carmona Arroyo	Universidad Veracruzana, Mexico
Guillermo Santamaría-Bonfil	BBVA México, Mexico
Gustavo Arroyo	Instituto Nacional de Electricidad y Energías Limpias, Mexico
Hansel Lázaro Valdés Pérez	Benemérita Universidad Autónoma de Puebla, Mexico
Héctor Maravillo Gómez	Autonomous University of Mexico City, Mexico
Hipólito Aguilar Sierra	Universidad La Salle México, Mexico

Hiram Calvo	Center for Computing Research - Instituto Politécnico Nacional, Mexico
Hiram Ponce	Universidad Panamericana, Mexico
Hugo Jimenez	Universidad Autónoma de Querétaro, Mexico
Hugo Jair Escalante	Instituto Nacional de Astrofísica, Óptica y Electrónica, Mexico
Humberto Sossa	National Polytechnic Institute, Mexico
Iclia Villordo-Jimenez	Instituto Politécnico Nacional, Mexico
Irahan Jose-Guzman	Tecnológico Nacional de Mexico/ITS de Misantla, Mexico
Iris Iddaly Méndez-Gurrola	Universidad Autónoma de Ciudad Juárez, Mexico
Irvin Hussein Lopez-Nava	Centro de Investigación Científica y de Educación Superior de Ensenada, Mexico
Isidro Gómez Vargas	Instituto de Astrofísica de Andalucía, Mexico
Ismael Osuna-Galán	Universidad Autónoma del Estado de Quintana Roo, Mexico
Israel Tabarez	Universidad Autónoma del Estado de México, Mexico
Ivan Vladimir Meza Ruiz	Universidad Nacional Autónoma de México, Mexico
Ivette Miramontes	Tijuana Institute of Technology, Mexico
J. Carlos Urteaga Reyesvera	Globant, Mexico
J. Octavio Gutierrez-Garcia	Instituto Tecnológico Autónomo de México, Mexico
J.-Martin Castro-Manzano	Universidad Popular Autónoma del Estado de Puebla, Mexico
Jacobo Sandoval-Gutierrez	Universidad Autónoma Metropolitana, Mexico
Jaime Cerda	Universidad Michoacana de San Nicolás de Hidalgo, Mexico
Javad Khodadoust	Tecnológico de Monterrey, Mexico
Javier Ortiz-Hernandez	TecNM/CENIDET
Jeremy Horne	Newlane University, USA
Jesus Solano Noriega	Universidad Autónoma de Occidente, Mexico
Jesús García-Ramírez	UPIIT-IPN, Mexico
Jhacer Kharen Ruiz	Instituto Tecnológico de Zitácuaro, Mexico
Joanna Alvarado Uribe	Tecnológico de Monterrey, Mexico
Jonathan Rojas-Simón	Autonomous University of the State of Mexico, Mexico
Jorge de la Calleja	Universidad Politécnica de Puebla, Mexico
Jorge Reyes-Magaña	Universidad Autónoma de Yucatán, Mexico
Jorge Ciprián-Sánchez	Hasso Plattner Institut, Germany
Jorge Hermosillo Valadez	Universidad Autónoma del Estado de Morelos, Mexico

Jorge Perez Ramos	Universidad Autónoma de Querétaro, Mexico
Jorge Armando Ramirez Diaz	Universidad Autónoma de Querétaro, Mexico
Jorge Isaac Chairez Oria	Tecnológico de Monterrey, Mexico
Jose Alanis	Universidad Tecnológica de Puebla, Mexico
Jose Sosa	Hydrosat, USA
Jose Cruz Mora	Instituto Politécnico Nacional, Mexico
José A. Reyes-Ortiz	Universidad Autónoma Metropolitana, Mexico
Jose Alberto Hernandez-Aguilar	Universidad Autónoma del Estado de Morelos, Mexico
Jose Antonio Leon Borges	Universidad Autónoma del Estado de Quintana Roo, Mexico
Jose Carlos Ortiz-Bayliss	Universidad Nacional Autónoma de México, Mexico
José Carlos Ortiz-Bayliss	Tecnológico de Monterrey, Mexico
Jose Luis López-Ramírez	Tecnológico Nacional de México/Instituto Tecnológico Superior de Purisima del Rincón, Mexico
José Luis Hernández González	TecNM/ITApizaco, Mexico
José Manuel Gámez	Universidad Autónoma de Zacatecas, Mexico
Jose Sergio Ruiz Castilla	Universidad Autónoma del Estado de México, Mexico
Jose-Agustin Almaraz-Damian	CICESE UAT, Mexico
Jose-Federico Ramirez-Cruz	TecNM - Instituto Tecnológico de Apizaco, Mexico
Juan Acosta	Universidad Autónoma de Ciudad Juárez and Philosophic Systems Institute, Mexico
Juan Martínez-Miranda	CICESE-UT3, Mexico
Juan Villegas-Cortez	Universidad Autónoma Metropolitana-Azcapotzalco, Mexico
Juan Carlos Olivares Rojas	Tecnológico Nacional de México/Instituto Tecnológico de Morelia
Juan Carlos Conde Ramírez	Benemérita Universidad Autónoma de Puebla, Mexico
Juan M. Fortuna-Cervantes	Universidad Autónoma de San Luis Potosí, Mexico
Juan Moisés Arredondo Velázquez	Centro de Investigación y Estudios Avanzados Unidad Tamaulipas, Mexico
Juana Canul-Reich	Juárez Autónoma de Tabasco, Mexico
Julia Melo Morín	TecNM - Instituto Tecnológico Superior de Pánuco, Mexico
Julio Galván	UPIITA-IPN, Mexico
Karina Figueroa Mora	Universidad Michoacana de San Nicolás de Hidalgo, Mexico

Karla Valencia	Instituto Nacional de Astrofísica, Óptica y Electrónica, Mexico
Katya Rodriguez-Vazquez	Universidad Nacional Autónoma de México, Mexico
Kay García Sánchez	Universidad Autónoma de Querétaro, Mexico
Lazaro Bustio	Universidad Iberoamericana, Mexico
Leandro Pazmiño Ortiz	Escuela Politécnica Nacional, Mexico
Leonardo Glasserman-Morales	Tecnológico de Monterrey, Mexico
Leopoldo Altamirano Robles	Instituto Nacional de Astrofísica, Óptica y Electrónica, Mexico
Leticia Flores Pulido	Universidad Autónoma de Tlaxcala, Mexico
Liliana Barbosa-Santillan	Universidad de Guadalajara, Mexico
Lorena Diaz Gonzalez	Universidad Autónoma del Estado de Morelos, Mexico
Lourdes Martinez-Villaseñor	Universidad Panamericana, Mexico
Luis Torres-Treviño	Universidad Autónoma de Nuevo León, Mexico
Luis Luevano	Idiap Research Institute, Switzerland
Luis Guarneros-Nolasco	Universidad Tecnológica del Centro de Veracruz, Mexico
Luis Arellano	Instituto Nacional de Astrofísica, Óptica y Electrónica, Mexico
Manuel Uche Soria	Universidad Politécnica de Madrid, Spain
Marco Aurelio Sotelo Figueroa	Universidad de Guanajuato, Mexico
Maria Auxilio Medina Nieto	Universidad Politécnica de Puebla, Mexico
Maria de la Concepción Pérez de Celis Herrero	Benemérita Universidad Autónoma de Puebla, Mexico
María Lucia Barron Estrada	TecNM Instituto Tecnológico de Culiacán, Mexico
Maria Luisa Davila Garcia	Sheffield Hallam University, UK
Maricela Quintana	UAEMéx Valle de México, Mexico
Mario Rossainz López	Benemérita Universidad Autónoma de Puebla, Mexico
Martha Pulido	Instituto Tecnológico de Tijuana, Mexico
Martin Montes Rivera	Universidad Politécnica de Aguascalientes, Mexico
Martín Contreras	Instituto Tecnológico Superior de Zongolica, Mexico
Mauricio Gonzalez-Soto	Tecnológico de Monterrey, Mexico
Mauricio Arturo Ibarra Corona	Universidad Autónoma de Querétaro, Mexico
Mayra HernándezOramas	TecNM, Mexico
Meliza Contreras González	Benemérita Universidad Autónoma de Puebla, Mexico
Miguel Hidalgo-Reyes	Instituto Tecnológico Superior de Xalapa, Mexico

Miguel Sanchez-Brito	Instituto Politécnico Nacional, Mexico
Miguel Soto	Centro de Investigación en Computación, Mexico
Miguel Gonzalez-Mendoza	Tecnológico de Monterrey, Mexico
Miguel Mora-Gonzalez	University of Guadalajara, Mexico
Miguel Alberto Domínguez Gurría	Universidad Tecnológica de la Mixteca, Mexico
Miguel Angel Rios Gaona	University of Vienna, Austria
Mireya Tovar Vidal	Benemérita Universidad Autónoma de Puebla, Mexico
Miroslava Cano	TecNM/ITS de Irapuato, Mexico
Mónica Acosta Montes de Oca	Universidad Autónoma de Tamaulipas, Mexico
Nelson Rangel Valdez	TecNM - Instituto Tecnológico de Cuidad Madero, Mexico
Néstor Leyva López	TecNM - Instituto Tecnológico de Culiacán, Mexico
Nicasio Hernández Flores	Instituto Nacional de Electricidad y Energías Limpias, Mexico
Noé-Alejandro Castro-Sánchez	TecNM-CENIDET, Mexico
Noel E. Rodríguez-Maya	TecNM campus Zitácuaro, Mexico
Obdulia Pichardo-Lagunas	IPN, Mexico
Omar Jehovani López Orozco	Tecnológico Nacional de México Campus Apatzingán, Mexico
Omar-Arturo Domínguez-Ramírez	Universidad Autónoma del Estado de Hidalgo, Mexico
Oscar Chávez-Bosquez	Universidad Juárez Autónoma de Tabasco, Mexico
Pablo Sánchez	Universidad Autónoma del Estado de Morelos, Mexico
Pablo Pancardo	Juárez Autonomous University of Tabasco, Mexico
Pedro Bello López	Benemérita Universidad Autónoma de Puebla, Mexico
Rafael Guzman-Cabrera	Universidad de Guanajuato, Mexico
Rafael Salas Zarate	TecNM - Instituto Tecnológico de Zitacuaro, Mexico
Rafael Rojas	Universidad Autónoma del Estado de México
Raghavender Puchhakayala	JPMorgan Chase & Co., USA
Ramon Brena	Instituto Tecnológico de Sonora, Mexico
Ramón Zatarain Cabada	TecNM- Instituto Tecnológico de Culiacán, Mexico
Ramón Iván Barraza-Castillo	Universidad Autónoma de Ciudad Juárez, Mexico
Raquel Diaz-Hernandez	National Institute of Astrophysics, Optics and Electronics, Mexico

Rasikh Tariq	Tecnológico de Monterrey, Institute for the Future of Education, Mexico
Raul Jimenez Cruz	Universidad La Salle, Mexico
Raúl Monroy Borja	Tecnológico de Monterrey, Mexico
René García Reyes	Instituto Nacional Electoral, Mexico
Ricardo Calderon	Universidad Autónoma del Estado de Hidalgo, Mexico
Roberto Melendez-Armenta	Instituto Tecnológico Superior de Misantla, Mexico
Roberto Ruiz	TecNM/Zongolica, Mexico
Roberto Antonio Vazquez	Universidad La Salle México, Mexico
Rocio Ochoa-Montiel	Universidad Autónoma de Tlaxcala
Rodrigo Villalvazo-Covián	TecNM/IT Ensenada, Mexico
Rosa Ortega-Mendoza	Universidad Autónoma del Estado de Hidalgo, Mexico
Rosa María Valdovinos	Universidad Autónoma del Estado de México, Mexico
Sabino Miranda	Universidad Autónoma de la Ciudad de México
Saiph Savage	Northeastern University, USA
Salvador Hinojosa	Tecnológico de Monterrey, Mexico
Samuel Viñas	Tecnológico Nacional de México campus Zitácuaro, Mexico
Sara Edith Pinzón Pineda	Instituto Nacional de Electricidad y Energías Limpias, Mexico
Saturnino Job Morales	Universidad Autónoma del Estado de México, Mexico
Saúl Zapotecas-Martínez	Instituto Nacional de Astrofísica, Óptica y Electrónica, Mexico
Selene Ramirez Rosales	Universidad Autónoma de Querétaro, Mexico
Sofia Galicia Haro	Universidad Nacional Autónoma de México, Mexico
Sravan Kumar Vazrapu	Salesforce, Mexico
Susana De León	Cenidet, Mexico
Tania Ramirez-delReal	CONACyT-CentroGeo, Mexico
Teth Azrael Cortes Aguilar	Tecnológico Nacional de México: TJMMPyH, Mexico
Valeria Maeda-Gutiérrez	Universidad Autónoma de Zacatecas, Mexico
Veronica Neri-Mendoza	Universidad Autónoma del Estado de México, Mexico
Verónica Rodríguez López	Universidad Tecnológica de la Mixteca, Mexico
Vicente Garcia Jimenez	Universidad Autónoma de Ciudad Juárez, Mexico
Victor Lomas-Barrie	Universidad Nacional Autónoma de México, Mexico

Victor Arias Peregrino	TecNM, Mexico
Victor Maldonado Benitez	Instituto Politécnico Nacional, Mexico
Víctor Ruiz	TecNM/CENIDET, Mexico
Víctor Pacheco-Valencia	Autonomous University of the State of Morelos, Mexico
Vidblain Amaro Ortega	Instituto Tecnológico de Mexicali, Mexico
Virginia Lagunes Barradas	TecNM/Instituto Tecnológico Superior de Xalapa, Mexico
Yaxk'in Coronado	Universidad LaSalle, Mexico
Yesenia Gonzalez-Navarro	Instituto Politécnico Nacional, Mexico
Yessica Calderon Segura	Universidad Autónoma del Estado de México, Mexico
Yoqsan Angeles	Instituto Politecnico Nacional, Mexico

Contents

Medical Applications

Applications of Artificial Intelligence

Virtue Artificial Intelligence With Examples

J.-Martín Castro-Manzano(✉)

Secretaría de Ciencia, Humanidades, Tecnología e Innovación del Estado de Puebla, Faculty of Philosophy, UPAEP University, 21 sur 1103, Puebla, México
josemartin.castro@upaep.mx

Abstract. In this contribution, we briefly present the model of Virtue Artificial Intelligence, a normative framework based upon virtue ethics that coordinates the standard models of Explainable, Responsible, and Trustworthy Artificial Intelligence. We then offer three concrete examples of how this normative model can be implemented.

Keywords: Virtue ethics · Ethics of artificial intelligence · Implementation

1 Introduction

Elsewhere we have presented Virtue Artificial Intelligence (VAI) as a normative model based upon virtue ethics that coordinates the standard models of Explainable, Responsible, and Trustworthy Artificial Intelligence. While our model offered a teleological critique of these approaches and organized them neatly, it lacked precise examples in the face of implementation. Given this methodological gap, in this contribution we briefly explain what our model consists of and then we provide three examples of how we can implement it, thus closing said breach.

2 Virtue Artificial Intelligence

Current approaches to the ethics of artificial intelligence (AI) largely focus on the relationships between explanation, trust, and accountability, giving rise to projects on Explainable Artificial Intelligence (XAI) [2,6], Trustworthy Artificial Intelligence (TAI) [3,4], and Responsible Artificial Intelligence (RAI) [1,5].

Broadly construed, XAI is a paradigm that requires that the actions, decisions, or outcomes produced by AI systems be analyzable and intelligible in accordance with our social right to explanation. Consequently, core concepts in XAI include justification, causality, accuracy, and transparency. TAI, in contrast, deals with the design of systems that deserve our trust, and its core concepts include reliability, robustness, competence, and data privacy. Finally, RAI refers

H. Ponce et al. (Eds.): COMIA 2026, CCIS 3095, pp. 3–12, 2027.
https://doi.org/10.1007/978-3-032-37438-7_1

to a set of principles intended to guide the design, development, and deployment of AI systems, emphasizing values such as fairness, accountability, explainability, transparency, and privacy.

These models are valuable and well-motivated on their own; however, given their conceptual structure, they are not exempt from what we call an "ergon situation"—borrowing the terms from Aristotle. By an ergon situation we mean a methodological situation in which the core concepts of a model are instrumentally reciprocal in the sense that each serves as a means to the others, such that none can be considered an end in itself. In other words, these models are in an ergon situation because they assign functions to AI systems that serve various purposes, but these purposes are intermediate rather than final: to the extent that explanation, trusst, and responsibility are invoked in a circular fashion to mutually assure each other, none of them fulfills the role of a genuine end.

Indeed, in some cases we demand explanations to foster trust; in others, we demand trust as a precondition for accepting an explanation. Sometimes we invoke responsibility to ensure trust, while in other cases we require explanations to assign responsibilities. And so on. The result, then, is a network of interdependent so-called ends—which are rather means—that lacks a defined teleological center. We can summarize this situation in Fig. 1.

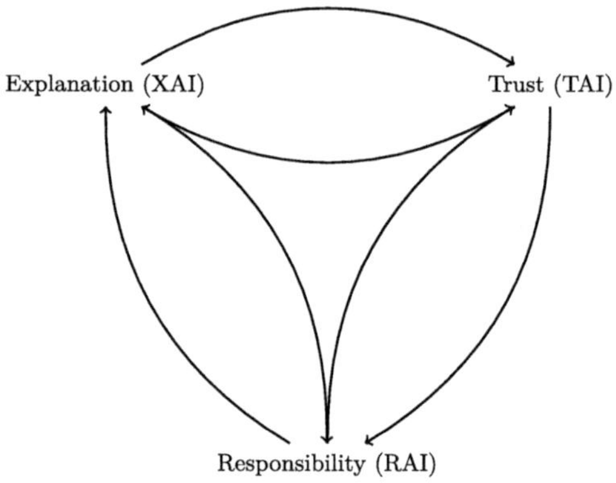

Fig. 1. An ergon situation.

From an Aristotelian perspective, situations like these are conceptually problematic. Aristotle argued, and not without good reason, that if every end is pursued for the sake of another, our practical reasoning collapses into an infinite regress (*EN* I, 1094a18-22), in which case none of the functions involved can count as a genuine ergon, since an ergon is precisely defined by reference to a final cause.

This perspective allows us to offer a teleological critique of contemporary AI ethics: while XAI, TAI, and RAI focus on means, VAI introduces final causes to focus on ends. In other words, current ethical models lack a substantive account of what AI systems ought to be for. And so, just as virtues resolve the ergon situation in an ethical theory, VAI can help resolve this problem between these AI models. We can represent this critique and our proposal in Fig. 2—which, incidentally, reminds us of the *Scutum Fidei.*

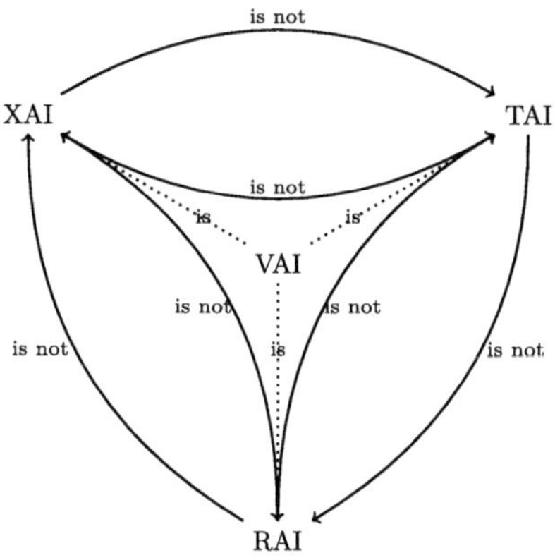

Fig. 2. The shield of VAI.

Given this motivation, we can define VAI as the normative model that prescribes the study, design, and implementation of artificial intelligence in terms of virtues. Formally, it is a triple VAI $= \langle \mathcal{M}_{ai}, \mathcal{V}, \varphi \rangle$, where $\mathcal{M}_{ai}$ is a Markov decision process and $\mathcal{V}$ is a set of virtue evaluators along with a phronesis function φ. To understand this formal definition, let us define a concrete artificial intelligence system, call it an *ai*, as a Markov decision process:

$$\mathcal{M}_{ai} = \langle \mathcal{S}, \mathcal{A}, P, r, \gamma \rangle, \tag{1}$$

where $s \in \mathcal{S}$ are states, $a \in \mathcal{A}$ are actions, $P(s' \mid s, a)$ is a transition probability function, $r(s, a, s') \in \mathbb{R}$ is a reward function, and $\gamma \in (0, 1)$ is a discount factor. Then, we define a policy $\pi(a \mid s)$ that induces a value function:

$$V^{\pi}(s) = \mathbb{E}_{\pi} \left[\sum_{t=0}^{\infty} \gamma^t \, r(s_t, a_t, s_{t+1}) \,\middle|\, s_0 = s \right]. \tag{2}$$

This Markov decision process captures a function maximization criteria presupposed in every *ai*; and now we augment this model by defining virtues as dispositional regulators—not as goals to be maximized. Let $\mathcal{V} = \{v_1, \ldots, v_k\}$ be a finite set of virtues, then each virtue is modeled as an evaluative functional:

$$v_i : \mathcal{S} \times \mathcal{A} \times \mathcal{S} \to \mathbb{R}, \tag{3}$$

that assigns positive values to transitions that express a virtue and negative values to those that express a vice. Thus, in the VAI model, virtues are not independent objectives to be maximized, but rather dispositional regulators that shape how an *ai* performs its task. Consequently, we can redefine a virtue-augmented reward function like so:

$$r_V(s, a, s') = r(s, a, s') + \sum_{i=1}^{k} \lambda_i(s)\, v_i(s, a, s'), \tag{4}$$

where $\lambda_i(s) \geq 0$ are context-sensitive weights. The base reward r encodes the task-relative goal of an *ai*, while the virtue terms modulate learning and action selection without replacing the underlying original goal.

From a purely formal point of view, this virtue-based modeling departs from the standard utility-maximization model that is usually assumed in the development of *ais*; however, such a formulation treats virtues as uniformly weighted and static regulators. But since the Aristotelian concept of virtue is context-sensitive—insofar as the same virtue may require different expressions depending on the circumstances—we need to add the role of *phronesis* (prudence) to mediate its application. Consequently, while phronesis is not necessary for a minimal model inspired by a virtue theory, it is essential for a more Aristotelian characterization of AI. We therefore add phronesis as a higher-order control mechanism that selects the relative importance of certain virtues according to the context. Formally, let:

$$\varphi : \mathcal{S} \to \mathbb{R}^k_{\geq 0} \tag{5}$$

be a phronesis or prudential function that assigns virtue weights $\lambda(s) = (\lambda_1(s), \ldots, \lambda_k(s))$ to each state. The behavior of an *ai* is thus governed by a two-level structure: an object-level policy $\pi(a \mid s; \lambda(s))$ that selects actions, and a meta-level controller φ that determines how virtues shape action in context. This reflects the Aristotelian idea that virtues do not operate uniformly, but are guided by situational judgment; and since some virtues function less as graded evaluators and more as normative limits, let

$$c_j : \mathcal{S} \times \mathcal{A} \times \mathcal{S} \to \mathbb{R} \tag{6}$$

be a family of constraint functions that encode requirements such as privacy, non-deception, or security. Therefore, a particular *ai* is required to satisfy:

$$\mathbb{E}_\pi \left[\sum_{t=0}^{\infty} \gamma^t c_j(s_t, a_t, s_{t+1}) \right] \leq \epsilon_j. \tag{7}$$

Thus, in VAI, virtues appear both as habituated regulations (reward modeling) and moral boundaries (constraints). Consequently, this framework allows virtues to give teleological structure to typical AI models without treating them as ultimate ends in themselves (organizing XAI, TAI, and RAI), it is consistent with an Aristotelian conception of virtue (which has philosophical pedigree), and can also be properly formalized.

3 Three Examples

At this moment, however, someone might object that the VAI model may sound interesting in theory, but without concrete implementation, its value is reduced to wishful thinking. And they would have a point. Therefore, what we will do now is to show how we could implement this normative AI model through three examples.

3.1 Computer Vision

Consider a computer vision system based on a deep neural network trained to classify images, for example, for facial recognition. The system receives an image x and produces a prediction $\hat{y}$, and its base reward reflects standard task performance, such as accuracy. Let the state s encode the input image x and the relevant model features. The action a corresponds to the prediction $\hat{y}$, and the base reward is defined as:

$$r(s, a, s') = -\ell(\hat{y}, y), \tag{8}$$

where ℓ is a standard loss function. Let us now introduce the virtue of justice, understood as a disposition to avoid structural and unjustified disparities in the system's performance between different relevant groups, and suppose we have a measure of error disparity between groups $g \in G$, for example:

$$\mathsf{disp}(s) = \max_{g,g' \in G} |\mathrm{err}_g - \mathrm{err}_{g'}|, \tag{9}$$

where err_g is the system error rate in group g. We then define the virtue evaluator:

$$v_{\mathrm{jus}}(s, a, s') = \delta - \mathsf{disp}(s), \tag{10}$$

such that δ is a maximum tolerable threshold of disparity. Thus, virtue penalizes system configurations that generate excessive inequalities, even when the overall system performance is high. The virtue-enhanced reward is then defined as follows:

$$r_V(s, a, s') = r(s, a, s') + \lambda(s)\, v_{\mathrm{jus}}(s, a, s'), \tag{11}$$

where the weight $\lambda(s)$ can be increased in socially sensitive contexts by a prudential controller. Thus, we can note that justice does not replace the predictive objective, but rather regulates the way in which it is achieved, introducing normative limits by design, as in Algorithm 1.

Algorithm 1: VAI: fair facial recognition

Input: Dataset $D = \{(x, y, g)\}$ with relevant group $g \in G$; vision model f_θ; base loss $\ell(\hat{y}, y)$; disparity measure $\mathsf{disp}(\cdot)$; justice threshold δ; phronesis control φ.
Output: Parameters θ trained with justice.

foreach *training step* **do**
  sample $B \subset D$;
  // Prediction and base reward
  foreach $(x, y, g) \in B$ **do**
    $\hat{y} \leftarrow f_\theta(x)$;
    $r \leftarrow -\ell(\hat{y}, y)$;
  end
  // Measure of intergroup disparity
  $\mathsf{disp} \leftarrow \max_{g,g' \in G} \left|\mathrm{err}_g(B) - \mathrm{err}_{g'}(B)\right|$;
  // Justice as a regulator
  $v_{\mathrm{jus}} \leftarrow \delta - \mathsf{disp}$;
  // Context-sensitive prudence
  $\lambda \leftarrow \varphi(\mathrm{ctx})$;
  // Virtue-augmented reward
  $r_V \leftarrow \mathbb{E}_{(x,y,g)\in B}[r] + \lambda \cdot v_{\mathrm{jus}}$;
  // Model update
  $\theta \leftarrow \theta + \alpha \nabla_\theta r_V$;
end

3.2 Generative Models

Let us now consider a large-scale generative language model that receives a prompt x and generates a textual response y. The state s includes the prompt, the dialogue history, and internal model representations; the action a consists of the generated sequence. The base reward can capture fluency or coherence:

$$r(s, a, s') = \mathsf{qual}(y), \tag{12}$$

where qual is a metric of linguistic quality. Let us now introduce the epistemic virtue of skepticism, understood as the willingness to modulate the degree of assertion according to the level of available evidence. Suppose the system can estimate a degree of epistemic support $\sigma(y \mid s) \in [0, 1]$, which measures how well-supported a claim is given the available information. We then define the virtue evaluator:

$$v_{\mathrm{ske}}(s, a, s') = \sigma(y \mid s) - \kappa(y), \tag{13}$$

where $\kappa(y)$ measures the degree of assertive strength of the generated content (e.g., categorical statements versus conditional formulations). The virtue-enhanced reward is:

$$r_V(s, a, s') = r(s, a, s') + \lambda(s)\, v_{\mathrm{ske}}(s, a, s'), \tag{14}$$

so that the system is incentivized to adjust its level of expressive security to the available epistemic support. Thus, skepticism does not prohibit the generation of content, but rather normatively structures its form and scope, penalizing both unfounded overstatement and unjustified evasion: Algorithm 2.

Algorithm 2: VAI: skeptic generation

Input: Prompt s; generator $p_\theta(y \mid s)$; measure of quality $\mathsf{qual}(y)$; epistemic estimator $\sigma(y \mid s) \in [0,1]$; measure of assertive strength $\kappa(y) \in [0,1]$; phronesis control φ.
Output: Answer y skeptically regulated.

// Candidate generation
sample $y \sim p_\theta(\cdot \mid s)$;
// Base reward
$r \leftarrow \mathsf{qual}(y)$;
// Skepticism
$\sigma \leftarrow \sigma(y \mid s)$;
$\kappa \leftarrow \kappa(y)$;
$v_{\text{ske}} \leftarrow \sigma - \kappa$;
// Prudence: increases weigth in high risk domains
$\lambda \leftarrow \varphi(\text{ctx}(s))$;
// Virtue-augmented reward
$r_V \leftarrow r + \lambda \cdot v_{\text{ske}}$;
// Decision: accept, revise, or mitigate assertive strength
if r_V *is sufficient* **then**
  return y;
else
  // Adjust: cautious re-statement / ask for evidence / admit uncertainty
  $y' \leftarrow \textsc{Adjust}(y)$;
  return y';
end

Algorithm 3: VAI: tempered education

Input: Student status s; repertoire of aids $\mathcal{A}$; base reward $r_{\text{prog}}(s, a, s')$; dependency estimator $\mathsf{dep}(s') \in [0,1]$; temperance threshold η; phronesis control φ.
Output: Temperate pedagogical action.

// Select an intervention
choose $a \in \mathcal{A}$ (e.g., the minimum effective aid);
// Apply intervention and observe transition
execute a and observe s' ;
// Base reward: immediate learning
$r \leftarrow r_{\text{prog}}(s, a, s')$;
// Virtue: temperance
$d \leftarrow \mathsf{dep}(s')$;
$v_{\text{temp}} \leftarrow \eta - d$;
// Prudence: weighs temperance according to context
$\lambda \leftarrow \varphi(\text{ctx}(s))$;
// Reward augmented by virtue
$r_V \leftarrow r + \lambda \cdot v_{\text{temp}}$;
// If the intervention induces excessive dependence, reduce
while r_V *is low* ***and*** *exists* $a_{minus} \in \mathcal{A}$ **do**
  $a \leftarrow a_{\text{less}}$ execute a and observe new s';
  $r \leftarrow r_{\text{prog}}(s, a, s')$;
  $d \leftarrow \mathsf{dep}(s')$;
  $v_{\text{temp}} \leftarrow \eta - d$;
  $\lambda \leftarrow \varphi(\text{ctx}(s))$;
  $r_V \leftarrow r + \lambda \cdot v_{\text{temp}}$;
end
return a;

3.3 Tutoring Systems

Finally, let us consider a tutoring system designed to assist students in problem-solving. The state s includes the student's level of competence, their interaction

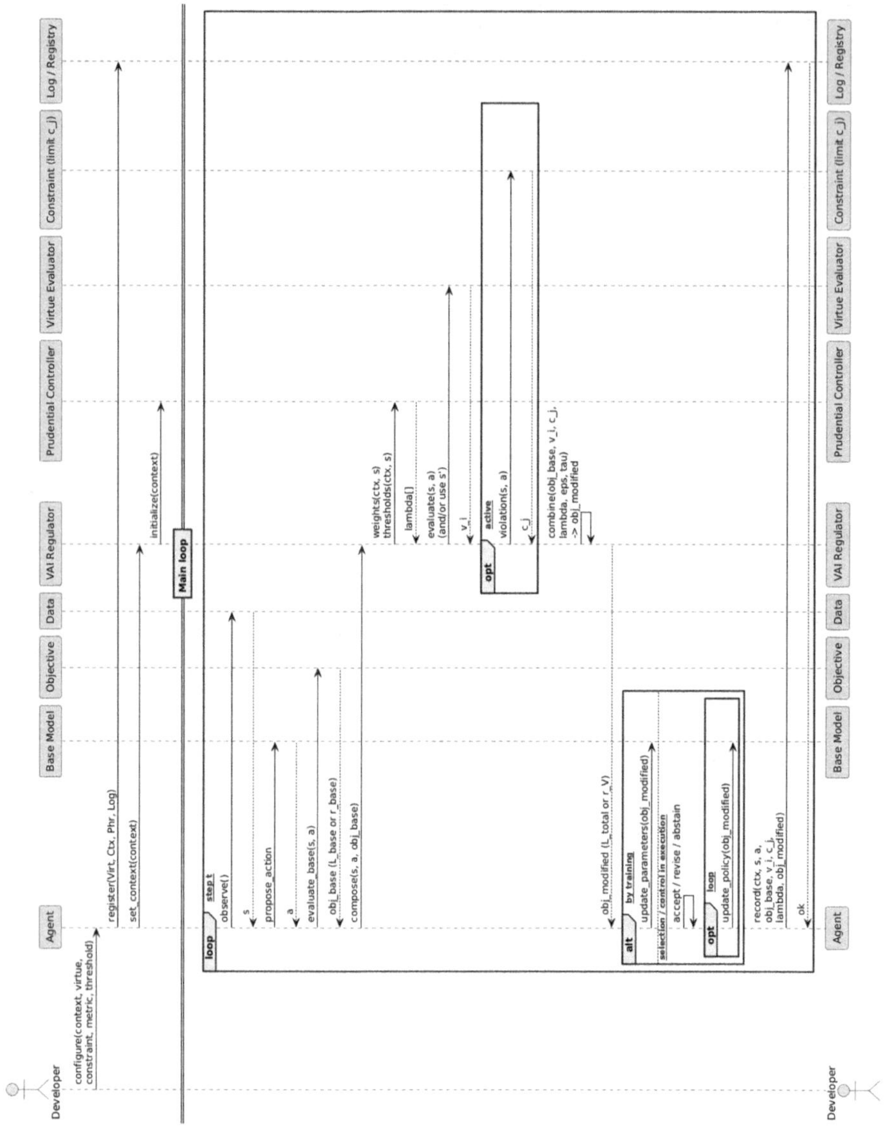

Fig. 3. An implementation model of VAI.

history, and the task context; the action a corresponds to the type and degree of assistance provided by the system:

$$r(s, a, s') = \mathbf{1}[\text{task completed}], \tag{15}$$

Let us now introduce the virtue of temperance, understood as a disposition to avoid over-assistance that undermines autonomous development. Let us define an induced dependency measure $\mathsf{dep}(s')$ that estimates the degree to which the student's progress depends on system intervention. We define the virtue evaluator:

$$v_{\text{temp}}(s, a, s') = \eta - \mathsf{dep}(s'), \tag{16}$$

where η is a maximum acceptable dependency threshold. The augmented reward then becomes:

$$r_V(s, a, s') = r(s, a, s') + \lambda(s)\, v_{\text{temp}}(s, a, s'), \tag{17}$$

so that the system is incentivized to offer graduated aids to encourage active reflection and withdraw support when appropriate. See Algorithm 3.

Finally, to conclude this section of examples, let us consider a diagram that allows us to summarize how these (and other) VAI-based implementations can be carried out (Fig. 3).

4 Conclusion

After briefly presenting the VAI model, we have offered three examples of how we can implement it in order to fill a methodological gap and show that the model is feasible. We believe, therefore, that our goal has been achieved; however, at this point, some further objections may arise. We pinpoint just three.

Objection 1. The lack of novelty. The VAI model is nothing more than a reformulation of existing approaches in AI ethics, such as explainability, fairness, or accountability—*nihil novum sub sole. Answer:* This objection overlooks three important points. First, VAI is conceptually distinct from those models, as it is organized around specific ethical criteria of virtues, in contrast to duties, values, rights, or utilities. Second, VAI is explicitly normative and teleological: it introduces final causes into the design of AI systems and offers a criterion for what counts as good performance, beyond the mere maximization of functions. Finally, the proposal boasts a clear philosophical pedigree supported by established traditions, as well as the Aristotelian conception of habits as acquired dispositions. For these reasons, VAI is not a mere rebranding, but a genuine normative proposal.

Objection 2. The lack of implementation. The VAI model may be a genuine normative proposal, but these philosophical ideas cannot be implemented. *Answer:* We believe this objection is easily answered by examining the algorithms presented in this work, as they demonstrate how to integrate virtues as internal regulators in predictive, generative, and educational systems. In these algorithms, virtues do not operate as external or *ex post* evaluations, but rather as components that modulate action selection in a context-sensitive manner.

Objection 3. The impossibility of programming. The VAI model can be implemented in theory, but not in practice because virtues cannot be programmed ultimately. *Answer:* This objection rests on a false dichotomy. While it may be true that virtues cannot be exhaustively captured by formal rules, it does not follow that they cannot be operationally implemented. From an Aristotelian perspective, virtues are habits: stable, acquired, and goal-oriented patterns that regulate action. In this operational sense, virtues are habits, habits are algorithms, and algorithms can be programmed. Therefore, although VAI does not reduce virtue to a calculus, it does implement it as a programmable pattern.

Finally, to close this contribution, we would like to mention two areas of research: *i)* we need to implement concrete systems with VAI and *ii)* compare their viability against equivalent systems or standard approaches. In the meantime, we will continue working on the generation and development of technological models that take our common goods into account.

Acknowledgments. This study was funded by an UPAEP grant number w/no.

Disclosure of Interests. The authors have no competing interests to declare that are relevant to the content of this article.

References

1. Dignum, V.: Responsible Artificial Intelligence: How to Develop and Use AI in a Responsible Way. Springer, Heidelberg (2019). https://doi.org/10.1007/978-3-030-30371-6
2. Doshi-Velez, F., Kim, B.: Towards a rigorous science of interpretable machine learning. arXiv preprint arXiv:1702.08608 (2017)
3. European Commission High-Level Expert Group on Artificial Intelligence: Ethics guidelines for trustworthy artificial intelligence (2019). https://digital-strategy.ec.europa.eu/en/library/ethics-guidelines-trustworthy-ai
4. Floridi, L., et al.: Ai4people—an ethical framework for a good ai society. Mind. Mach. **28**(4), 689–707 (2018). https://doi.org/10.1007/s11023-018-9482-5
5. IEEE: Ethically aligned design: a vision for prioritizing human well-being with autonomous and intelligent systems (2019). https://standards.ieee.org/industry-connections/ec/autonomous-systems.html
6. Miller, T.: Explanation in artificial intelligence: insights from the social sciences. Artif. Intell. **267**, 1–38 (2019). https://doi.org/10.1016/j.artint.2018.07.007

Large Language Models for Mexican Civil Law: A Comparative Evaluation of Specialized Fine-Tuning and RAG Approaches

Bogdan Kaleb García Rivera(✉), Ignacio Arroyo Fernández,
Cesar Felipe Martínez Cisneros, Jesús Ulises Quiroz Bautista,
and Claudia Anahí Guzmán Solano

Universidad Tecnológica de la Mixteca, Huajuapan de León, Oaxaca, Mexico
{garb981202,iaf}@gs.utm.mx

Abstract. Access to legal guidance in Mexico is often hindered by the complexity of civil jurisprudence and the high cost of professional counsel. This paper presents a specialized approach to optimizing Large Language Models (LLMs) for the Mexican civil law domain, focusing on providing accurate responses to basic legal inquiries. We detail the construction of a novel Legal Skills Dataset and the implementation of a fine-tuning pipeline that incorporates LLM-assisted semantic cleaning and a comparative study of training hyperparameters. Our methodology evaluates the impact of learning rates and quantization on model robustness and consistency. Preliminary results demonstrate that our specialized fine-tuning strategy significantly enhances the model's ability to handle domain-specific terminology and reduces hallucination rates compared to general-purpose models. This work contributes to the advancement of LegalTech in Latin America by providing a framework for developing reliable, accessible, and automated legal assistance systems.

Keywords: Large Language Models · Mexican Civil Law · fine-tuning · RAG · LegalTech · Natural Language Processing

1 Introduction

The rapid evolution of LLMs has opened new frontiers for automated legal assistance [1]. However, their application in the legal domain faces significant hurdles due to the inherent ambiguity and complexity of legal terminology. In the specific context of Mexican Civil Law, the multiplicity of meanings and the high degree of technicality often lead general-purpose LLMs to generate inaccurate or entirely fabricated information, a phenomenon known as hallucination [2]. These errors are particularly critical in legal settings, where misinformation can result in severe real-world consequences.

H. Ponce et al. (Eds.): COMIA 2026, CCIS 3095, pp. 13–26, 2027.
https://doi.org/10.1007/978-3-032-37438-7_2

1.1 Problem Statement

Current LLMs struggle with the high specificity required by the Mexican legal framework. General models lack the specialized "common sense" of civil jurisprudence, leading to imprecise semantic retrieval and response generation [3]. Furthermore, the lack of transparency in how these models cite sources makes them unreliable for professional or public use [4]. There is a clear need for a specialized system that not only understands the nuances of Mexican civil statutes but also provides a verifiable bridge between AI-generated advice and official legal texts.

1.2 Proposed Solution and Contributions

To address these challenges, this paper introduces a specialized web-based chatbot prototype focused on basic Mexican civil law assistance. Unlike general-purpose assistants, our system integrates a specialized fine-tuning pipeline and a Retrieval-Augmented Generation (RAG) architecture. The main contributions of this work are:

- **Information Extraction and Semantic Search:** Implementation of a vector database to identify and retrieve key legal obligations and rights from civil codes.
- **Verifiable Question-Answering:** A logic-handling framework that enables the system to reference and display specific legal fragments used to generate its responses, ensuring traceability.
- **User-Centric Personalization:** Support for registered users to integrate their own legal data via personalized RAG or fine-tuning, adapting the model to specific case requirements.
- **Robust Evaluation Framework:** A multi-metric approach to quantify performance, focusing on precision, coherence, and relevance, alongside specialized techniques to detect and mitigate hallucinations using BERTScore and internal consistency checks.

2 Related Work

The application of Natural Language Processing (NLP) in the legal domain has transitioned from simple classification to specialized benchmarks and predictive models. Chalkidis et al. [5] introduced LexGLUE to evaluate legal NLU (Natural Language Understanding). They showed legal-specific models like Legal-BERT outperform generic ones by 2–4% in Macro-F1 due to legal language's unique characteristics.

Further research by Chalkidis et al. [6] explored automatic judicial prediction using neural networks on European Court of Human Rights (ECHR) cases. Their study utilized architectures like Hier-BERT to handle long texts, achieving an F1-score of 82% in violation classification. Regionally, Martinez Pinilla et al. [7] addressed Colombian judicial congestion via semantic search, inspiring our pipeline. Their approach, illustrated in Fig. 1, underscores a structured sequence involving judicial text preprocessing, embedding generation with models like distiluse-base-multilingual-cased-v1, and efficient vector indexing. By adopting a similar logic for the Mexican civil law domain, this research builds upon these retrieval foundations to implement advanced RAG and fine-tuning techniques. This ensures that the system not only retrieves information but also addresses domain-specific challenges such as the detection of hallucinations and semantic verification in legal outputs.

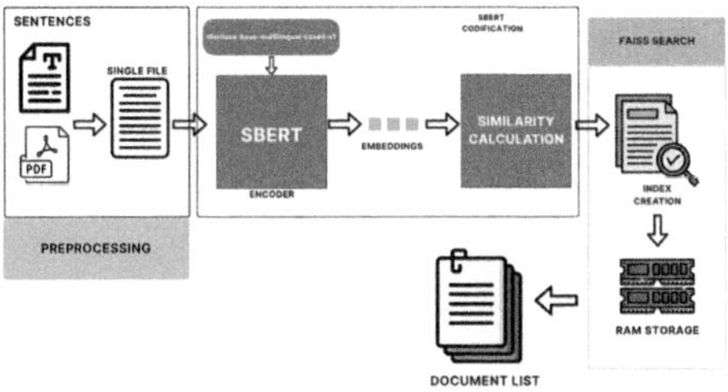

Fig. 1. General process diagram of the judicial search system proposed by Martinez Pinilla et al. [7], illustrating the pipeline from text preprocessing to RAM-based index storage. Source: Adapted from [7].

In Mexico, AI has been integrated into fiscal auditing by the SAT to identify tax evasion [8,9]. Research by Cáceres [10] laid the groundwork for judicial decision-making systems based on legal constructivism, while Hernandez M [11] evaluated generative AI for legal education in customs and international trade. Academic efforts, such as the 2019 congress organized by SMIA and ITAM, continue to explore NLP applications for discovering relevant legal information in the Mexican context [12,13].

3 Theoretical Framework

3.1 LLMs and Hallucinations

LLMs based on deep neural networks have revolutionized the processing of unstructured information [14]. However, a significant challenge remains: "hallucinations," where models generate coherent but factually unsupported responses. This is particularly critical in legal contexts where accuracy is non-negotiable.

3.2 Transformer Architecture and Long-Context Challenges

Modern LLMs predominantly utilize the decoder-only Transformer architecture [15]. By removing encoder and cross-attention layers, this structure relies on *causal self-attention*, where a look-ahead mask ensures tokens only attend to preceding positions for autoregressive generation. Long-context processing faces challenges from $O(L^2)$ self-attention complexity and KV cache constraints. Key mitigations include efficient attention, extrapolative encodings (e.g., RoPE), and Mixture-of-Experts scaling [15]. Llama 3.1 [16] and Phi-4 [17] adopt this architecture.

3.3 Vector Databases and RAG

To optimize searches in large systems, Vector Database Management Systems (VDBMS) [18] store data as high-dimensional vectors [19]. ElasticSearch [20] is utilized as a non-relational vector database, using inverted indices to achieve latencies of less than one second for thousands of requests. RAG combines text generation with information retrieval to extract knowledge from a specific corpus [21]. This process ensures that the system can cite the specific legal sources used to answer a query [22].

3.4 Model Optimization via Quantization

Quantization is a compression technique that reduces the bit-precision of model parameters to decrease memory usage and accelerate inference [23]. This methodology achieves the compression of weight matrices (W) through a scale factor (Δ). The quantization process is formally defined in Eq. 1:

$$W_Q = \left\lfloor \frac{W}{\Delta} \right\rceil, \quad \Delta = \frac{\max(|W|)}{2^N - 1} \tag{1}$$

This process allows models to run on limited GPU resources without significant accuracy loss [24].

3.5 Evaluation Metrics and LLM-as-a-Judge

Beyond traditional metrics like BLEU [25], ROUGE [26], and BERTScore [27], this work adopts the LLM-as-a-judge approach [28]. Using high-performance models as evaluators allows for Pairwise Comparison and Reference-guided Grading, providing a scalable and cost-effective alternative to human evaluation that aligns with human preferences in over 80% of cases [29].

4 Methodology and Experimentation

4.1 Experimental Setup

All experiments were conducted in a controlled computational environment. Local fine-tuning utilized NVIDIA A100 (40 GB) and RTX 6000 Ada (48 GB) GPUs, leveraging PyTorch 2.1.0, CUDA 11.8, and the Unsloth framework for optimized QLoRA (4-bit NF4 quantization). For the evaluation of large-scale commercial models, inference was performed via the NVIDIA API. The legal corpus comprises 129,739 indexed articles from the 32 Mexican state civil codes and federal legislation, stored in an Elasticsearch 8.15 vector database with a hybrid retrieval strategy combining metadata filtering and dense semantic search (*paraphrase-multilingual-MiniLM-L12-v2*).

4.2 Overview of the Methodological Pipeline

To enhance clarity and reproducibility, Fig. 2 illustrates the six-stage workflow of our proposal, showing both the isolated function of each component and their integration into the final system.

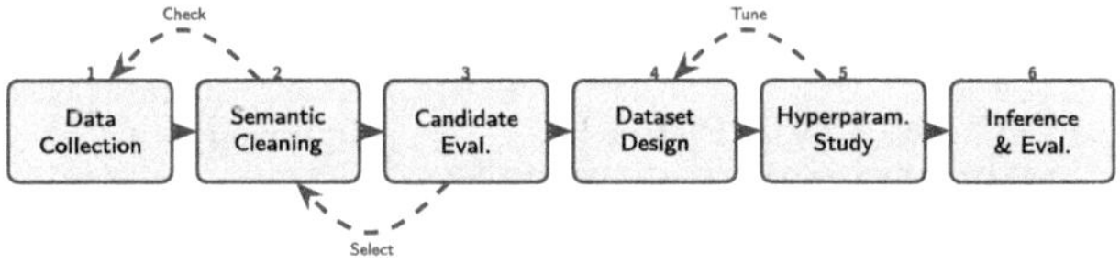

Fig. 2. High-level overview of the proposed methodology. Each numbered block corresponds to a subsection in Sect. 4: (1) Data collection and normalization, (2) LLM-assisted semantic cleaning, (3) Candidate model evaluation, (4) Legal Skills Dataset design, (5) Hyperparameter study, and (6) Inference and models evaluation. Dashed arrows indicate iterative feedback loops for refinement. Source: Own elaboration.

Stages 1–2 (Data Preparation): Legal PDFs are ingested and cleaned via regex pipelines, followed by LLM-assisted semantic refinement to produce a normalized, structured corpus (Sects. 4.3–4.4).

Stage 3 (Candidate Selection): Six models are initially screened on Oaxaca civil laws (Sect. 4.5) to identify the most promising bases for specialization.

Stages 4–5 (Specialization): A synthetic Legal Skills Dataset is constructed via instruction tuning; selected models are fine-tuned by optimizing learning rates, decoding strategies, and RAG integration (Sects. 4.6–4.7).

Stage 6 (Final Evaluation): The selected top-performing models are contrasted against commercial API models under the comprehensive factorial framework detailed in Sect. 4.8 to determine the production-ready configuration.
This modular design ensures that each component can be validated independently while maintaining full traceability to the end-to-end system performance.

4.3 Data Collection and Normalization

The corpus was established through expert consultation, comprising 32 federal elements and specialized state civil codes retrieved from official government portals. To prevent embedding degradation caused by PDF noise—such as pagination, institutional headers, and non-canonical metadata—we developed a Python-based Regex pipeline to isolate the normative text. Table 1 illustrates this transformation process.

Table 1. Legal text normalization: regex-based noise removal. Source: Own elaboration.

Component	Raw PDF Segment (Before)	Normalized Text (After)
Pagination	1	[Removed]
Header	CÓDIGO CIVIL PARA EL ESTADO LIBRE Y SOBERANO DE BAJA CALIFORNIA SUR	[Removed]
Metadata	Última Reforma BOGE. 51 Ext. 11-Julio-2024	[Removed]
Legal Content	Artículo 14.- Los bienes inmuebles sitos en el Estado...	**Artículo 14.- Los bienes inmuebles sitos en el Estado...**

To preserve legal hierarchies in a NoSQL environment, we developed a metadata-driven schema in Elasticsearch. Each article or fraction is indexed as a denormalized document enriched with specific identifiers (`state`, `law_id`, `parent_id`, `type_of_law`). The system employs a hybrid retrieval strategy:

- **Semantic Layer**: Uses *paraphrase-multilingual-MiniLM-L12-v2* for vector generation.
- **Deterministic Layer**: Implements metadata filtering before vector similarity calculations.

By filtering for a specific state or law type (Civil or Procedural), the system eliminates "semantic noise" from unrelated jurisdictions. This ensures high precision and low latency across the extensive collection of legal texts, combining the speed of an inverted index with the flexibility of semantic understanding.

4.4 LLM-Assisted Semantic Cleaning

Despite the efficiency of Regular Expressions, legal texts often present complex fragmentation and OCR noise (e.g., "a rticulo" or "ti en. en") that rule-based methods cannot resolve. To address this, we integrated remote LLM APIs via NVIDIA Build, utilizing models such as Llama-3.3-Nemotron, GPT-OSS, and DeepSeek-V3. Following the CheckList methodology [30], we evaluated model robustness across 13 structural complexity categories, including legal format preservation (Roman numerals, "BIS" suffixes) and article status detection (reformed or derogated). The evaluation relies on two primary metrics:

- Exact Match: Identity between the processed output and the ground truth.
- Semantic Similarity: Meaning preservation between the generated response and the ideal result.

To ensure deterministic and consistent results, inference parameters were fixed at temperature=0 and top-p=1. We developed a specialized prompt designed to correct typographical noise while strictly preserving key legal terminology (e.g., "emplazamiento", "orden público") to prevent hallucinations that could alter judicial meaning.

4.5 Candidate Model Evaluation and Selection

The selection of a base model was performed through a comparative analysis of six models: Gemma-3-27B, GPT-OSS-20B, Llama-3.1-8B-Bárcenas, Llama-3.3-70B, Phi-4-14B, and Qwen-3-14B. To establish a controlled experimental baseline, we initially focused the training and evaluation exclusively on the Civil and Procedural laws of the State of Oaxaca. Evaluation utilized a split of 14,773 training and 1,642 test records (90/10). Based on a preliminary assessment balancing domain adaptation potential, computational efficiency, and baseline response consistency, `Llama-3.1-8B-Bárcenas` and `Phi-4-14B` were identified as the most promising candidates for subsequent specialized fine-tuning. The initial fine-tuning parameters applied during this phase are summarized in Table 2:

Table 2. Initial fine-tuning hyperparameters. Source: Own elaboration.

Parameter	Value	Parameter	Value
Epochs	2	Learning Rate	2e-5
Batch Size	2	LR Scheduler	Linear
Grad. Accumulation	4	Optimizer	AdamW 8-bit
Max Seq. Length	2048*	Weight Decay	0.01

*1024 for Llama-3.3-70B

4.6 Design of the Legal Skills Dataset

To mitigate hallucinations and enhance instruction following, we developed a synthetic Legal Skills Dataset based on the Instruction Tuning paradigm. We employed a Teacher-Student Distillation technique using Qwen3-235B via the NVIDIA API, selected for its high semantic similarity and technical precision. The dataset (5,000 records) is structured around three core pillars, operationalized into five modular skills (H1–H5) as summarized in Table 3. This multifaceted approach ensures the model can prioritize retrieved context, detect information gaps, rectify false premises, and maintain a professional, traceable persona via JSON-structured outputs.

Table 3. Modular skills and technical objectives for the Legal Skills Dataset. Source: Own elaboration.

ID	Skill	Technical Objective
H1	Direct RAG	Literal/paraphrased extraction from context
H2	Negation	Hallucination avoidance via information gap detection
H3	ELI5	Simplification of complex terms for non-experts
H4	Correction	Identification and rectification of false premises
H5	Warning	Inclusion of legal disclaimers

4.7 Training Hyperparameter Study

With the dataset structure established, we optimized the fine-tuning process to balance domain adaptation and generalization. The training protocol focused on key hyperparameters including learning rate (2×10^{-5} to 4×10^{-4}), batch sizes (2–8), and LoRA ranks ($r \in \{8, 16, 32\}$). Early stopping was applied if validation loss plateaued for three consecutive epochs, ensuring computational efficiency while preventing overfitting on the specialized legal corpus.

4.8 Inference and Models Evaluation

This phase identifies the optimal configuration in terms of legal precision and semantic coherence. We contrast the top-performing fine-tuned local models (Llama-3.1-8B-Bárcenas and Phi-4-14B) against massive general-purpose models via the NVIDIA API, including Llama-4-Maverick-17B, Mixtral-8x22B, and DeepSeek-V3.2.

To systematically assess performance, we designed a factorial experimental framework (Fig. 3) that explores three key dimensions: (1) model state (pre-trained vs. fine-tuned), (2) context injection (RAG enabled vs. disabled), and (3) decoding hyperparameters varied across three levels each. This structure yields $2 \times 2 \times 3^3 = 108$ unique configurations per model, enabling a granular analysis of interaction effects while maintaining computational feasibility.

At the HP level, we evaluated decoding strategies through a structured protocol:

- **Sampling:** Variation of Top-p ($p \in \{0, 0.5, 1\}$) to balance diversity and repetition.
- **Temperature:** Testing deterministic reproducibility ($T = 0$) vs. controlled variability ($T \in \{0.5, 1\}$).
- **Length:** Impact of generation limits (512, 1024, and 2048 tokens) on conciseness and redundancy.

These experiments facilitate a direct comparison between fine-tuned specificity and the "raw power" of massive vanilla models under varied RAG contexts, ensuring the selection of a production-ready candidate for the Mexican civil law domain.

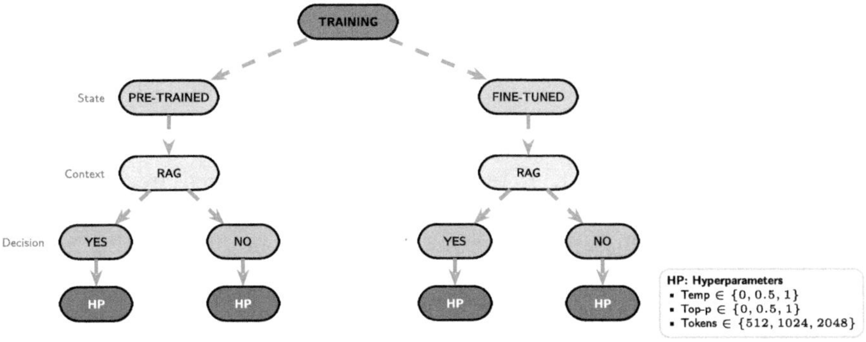

Fig. 3. Experimental design tree illustrating the factorial combination of model state (pre-trained vs. fine-tuned), context injection (RAG on/off), and decoding hyperparameters (temperature, top-p, max tokens). Each leaf node represents a unique experimental configuration evaluated in this study. Source: Own elaboration.

Model evaluation employed a dual-metric protocol: (1)Deterministic metrics including ROUGE-L, BLEU-4, BERTScore (F1), and Cosine Similarity, computed against ground-truth legal articles; and (2) Qualitative assessment via an LLM-as-a-judge framework using Llama-3.3-Nemotron-49B, scoring responses on a 1–10 scale across legal accuracy, contextual grounding, and clarity. Robustness was further validated through 16 structured queries, including out-of-domain and adversarial "trick" questions, following the CheckList methodology [30].

4.9 System Integration and End-to-End Model

The components from Sects. 4.3–4.8 are orchestrated into a unified RAG pipeline. Figure 4 illustrates the end-to-end flow from user query to grounded response.

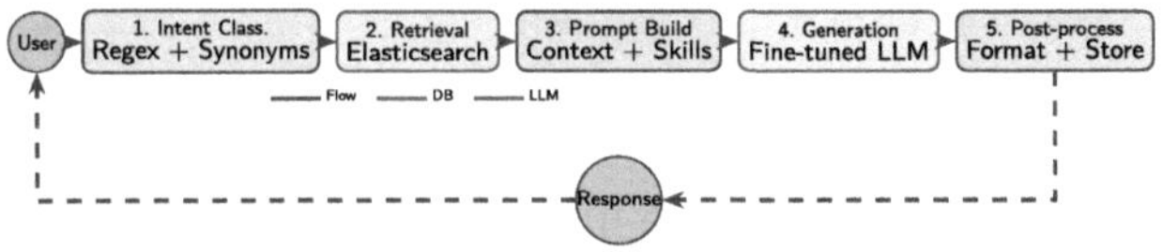

Fig. 4. End-to-end architecture of the Mexican civil law assistant. The FastAPI backend orchestrates intent classification, hybrid retrieval from Elasticsearch, prompt construction, and LLM generation, returning a response that cites relevant legal articles. Source: Own elaboration.

The pipeline operates in five sequential stages:

1. **Intent Classification.** Casual queries (greetings) are detected via regex and bypass retrieval. Legal queries undergo synonym normalization (e.g., "casero"→"arrendador") to improve semantic search recall.

2. **Hybrid Retrieval.** The normalized query is embedded (*paraphrase-multilingual-MiniLM-L12-v2*) and searched in Elasticsearch using: (i) deterministic metadata filtering (`state`, `law_id`) to eliminate cross-jurisdiction noise, and (ii) dense vector similarity over 129,739 indexed articles. Top-k fragments (k=4–6) are retrieved with traceable identifiers.
3. **Prompt Construction.** Retrieved articles are injected into a template instructing the fine-tuned LLM to: (i) ground answers exclusively on provided context (skill `H2_Negation`), (ii) explain concepts in plain language (skill `H4_ELI5`), (iii) cite specific articles, and (iv) include jurisdictional disclaimers. Listing 2.1 shows the prompt structure.
4. **Response Generation.** The prompt is fed to the selected fine-tuned model (`Llama-3.1-8B-Bárcenas` or `Phi-4-14B`) with deterministic temperature (T=0.3). Internal reasoning tokens are stripped post-generation.
5. **Post-processing.** The response is formatted for display (with hyperlinked citations) and persisted in PostgreSQL alongside the user query, enabling conversation history review.

Listing 2.1. Simplified prompt template for inference. The LLM is instructed to ground its answer in retrieved context and cite relevant articles. Source: Own elaboration.

```
You are a legal assistant specialized in the Mexican Civil
    Code.
Respond ONLY based on the provided legal context.
If the information is not in the context, clearly state so.

Legal context:
{retrieved_articles}

User query:
{query}

Response (cite relevant articles):
```

This integration ensures that specialized knowledge from fine-tuning (Sect. 4.6) is paired with current, jurisdictionally correct legal texts, minimizing hallucination and guaranteeing full traceability.

5 Results and Discussion

5.1 Semantic Cleaning Performance

We evaluated three LLMs for corpus refinement via NVIDIA API, measuring Exact Match, Semantic Similarity (all-MiniLM-L6-v2), and Response Latency. The comparative results, summarized in Table 4, indicate significant performance variances. The NVIDIA/Llama-3.3-Nemotron-49B model outperformed competitors across all metrics, achieving an EM average of 0.8462 and a perfect 1.0000 semantic similarity score. In contrast, the OpenAI/GPT-OSS-120B

Table 4. Summary of semantic cleaning metrics across LLMs. Source: Own elaboration.

Model	Exact Match	Semantic Sim.	Avg. Latency (s)
DeepSeek-V3.1	0.5385	0.9941	229.16
Llama-3.3-Nemotron	**0.8462**	**1.0000**	**42.40**
GPT-OSS-120B*	0.1818	0.9663	243.64

*Exhibited API failure rates (NaN) during testing

model exhibited reliability issues, frequently returning "NaN" values due to API failures, which resulted in the lowest EM score of 0.1818.

Efficiency analysis further distinguishes the models. Llama-3.3-Nemotron processed queries significantly faster (42.40 s) than both DeepSeek and OpenAI, which exceeded 220 s on average. These delays are likely attributable to model scale and API optimization differences. Consequently, Llama-3.3-Nemotron was selected as the primary engine for the legal corpus cleaning phase due to its superior balance of deterministic accuracy and operational efficiency.

5.2 Experiments Results

To evaluate robustness, we executed a comprehensive experimental framework. This phase focused on measuring the interaction between stochastic hyperparameters (temperature, `top-p`, and `max_tokens`), specialized fine-tuning (FT), and RAG. By comparing local models with commercial APIs, we identified the performance trade-offs between lexical precision and stylistic fluency.

Performance Summary of Local Models Table 5 confirms FT + RAG supremacy: the top four positions are RAG-enabled models, proving external legal context outweighs internal parametric knowledge. A significant finding is the discrepancy between automated metrics and the LLM-as-a-judge scores. While *Phi-4-14B (fine-tuned, civil)* achieved the highest Judge rating (9.10), the *Llama-3.1-8B-Bárcenas (fine-tuned, civil)* configuration consistently outperformed all models in deterministic metrics, such as BERT F1 (0.848) and ROUGE-L (0.545). This indicates that the fine-tuned LLaMA variant maintains a higher level of factual adherence to the legal source material, even when the judge-evaluator favors the stylistic delivery of the Phi-4 family.

Benchmarking API Models To establish a performance ceiling, we benchmarked commercial models consumed via the NVIDIA API (Table 6). **Llama-4-Maverick-17B** proved to be the most robust model for this legal task when combined with RAG, showing a balanced performance across all metrics. Interestingly, *DeepSeek-V3.2* exhibited a behavior similar to Phi-4: it received the highest Judge ratings (9.45) despite having significantly lower lexical overlap (ROUGE-L) and semantic similarity scores compared to the Llama 4 family.

Table 5. Global ranking of local model configurations (averaged across all hyperparameters). *Judge Avg.*: LLM-as-a-judge scores on a 1–10 scale (10 = perfect legal alignment). *FT*: fine-tuned; *RAG*: retrieval-augmented generation enabled. Source: Own elaboration.

Rank	Model Configuration	BERT F1	Cosine	ROUGE-L	Judge Avg.
1	Phi-4-14B (FT + RAG)	0.8406	0.8440	0.5220	9.1075
2	**Llama-3.1-8B-Bárcenas (FT + RAG)**	**0.8483**	**0.8532**	**0.5458**	**9.0866**
3	Phi-4-14B (Base + RAG)	0.7098	0.6470	0.1632	9.0759
4	Llama-3.1-8B-Bárcenas (Base + RAG)	0.7949	0.7418	0.3894	8.0862
5	Phi-4-14B (Base)	0.6693	0.5089	0.1001	8.0428
6	Phi-4-14B (FT)	0.7346	0.6391	0.2369	7.2619
7	Llama-3.1-8B-Bárcenas (FT)	0.7434	0.6382	0.2540	7.0297
8	Llama-3.1-8B-Bárcenas (Base)	0.7400	0.6235	0.2520	6.1868

Table 6. Global ranking of commercial API model configurations (averaged across all hyperparameters). *Judge Avg.*: LLM-as-a-judge scores on a 1–10 scale (10 = perfect legal alignment). Source: Own elaboration.

Rank	Model Configuration	BERT F1	Cosine	ROUGE-L	Judge Avg.
1	**Llama-4-Maverick-17B (Base + RAG)**	**0.7665**	**0.5018**	**0.2671**	**9.2971**
2	DeepSeek-V3.2 (Base + RAG)	0.6772	0.4798	0.1367	9.4564
3	Llama-4-Maverick-17B (Base)	0.6574	0.3983	0.1257	8.8286
4	DeepSeek-V3.2 (Base)	0.6363	0.3894	0.0854	9.1429
5	Mixtral-8x22B (Base + RAG)	0.5873	0.0000	0.0002	1.0000
6	Mixtral-8x22B (Base)	0.5837	0.0032	0.0013	1.0589

6 Conclusions

This study systematically evaluated the efficacy of LLMs for a Mexican Civil Law assistant. Our findings confirm that RAG is the decisive factor for semantic accuracy, significantly outweighing the benefits of standalone fine-tuning when navigating the vast complexity of the 32 Mexican state codes and federal laws. Notably, while proprietary evaluators exhibit a stylistic bias toward verbose models, deterministic metrics confirm that the Llama-3.1-8B-Bárcenas (fine-tuned) model, integrated with RAG and configured at a balanced temperature of 0.5, offers the optimal solution. This configuration achieves the necessary equilibrium between juridical precision, factual fidelity to the civil code, and generative fluency, establishing a robust framework for AI-driven legal assistance in the Mexican context.

6.1 Future Research Directions

Given the preliminary nature of this study, several critical avenues remain for future investigation:

- **Continuous Legal Corpus Updating:** Implementation of automated pipelines to ingest, clean, and index newly published judicial reforms, official gazettes, and court rulings, ensuring the RAG knowledge base remains legally current.
- **Cross-Domain and Multi-Jurisdictional Expansion:** Extending the fine-tuning and evaluation framework to criminal, labor, and administrative law, alongside comparative analyses across Latin American civil law systems.
- **Human-in-the-Loop Validation:** Integrating licensed legal practitioners into the evaluation pipeline to refine the LLM-as-a-judge rubrics, establish expert-grounded benchmarks, and mitigate the stylistic bias observed in automated evaluators.
- **Production Deployment and Usability Studies:** Conducting controlled field trials with law students, paralegals, and end-users to measure task completion rates, error recovery, trust calibration, and latency in real-world legal assistance scenarios.

The CiviLIA chatbot prototype described in this work is publicly accessible for research and evaluation purposes at https://legaliautm.short.gy/vXbMBF.

Acknowledgements. We thank Secihti project CF-2023-I-2854 for the resources provided to conduct this work.

References

1. Hu, Y., Gan, L., Xiao, W., Kuang, K., Wu, F.: Fine-tuning large language models for improving factuality in legal question answering. In: Proceedings of COLING 2025, pp. 4410–4427 (2025)
2. Hou, A.B., Jurayj, W., Holzenberger, N., Blair-Stanek, A., Van Durme, B.: Gaps or hallucinations? Gazing into machine-generated legal analysis for fine-grained text evaluations. arXiv preprint arXiv:2409.09947 (2024)
3. Bashir, A.H., et al.: Domain-adaptation through synthetic data: fine-tuning large language models for German law. arXiv preprint arXiv:2601.14160 (2026)
4. Dahl, M., Magesh, V., Suzgun, M., Ho, D.E.: Large legal fictions: profiling legal hallucinations in large language models. J. Legal Anal. **16**(1), 64–93 (2024)
5. Chalkidis, I., et al.: LexGLUE: a benchmark dataset for legal language understanding in English. arXiv preprint arXiv:2110.00976 (2021)
6. Chalkidis, I., Androutsopoulos, I., Aletras, N.: Neural legal judgment prediction in English. In: Korhonen, A., Traum, D., Màrquez, L. (eds.) Proceedings of the 57th Annual Meeting of the Association for Computational Linguistics, pp. 4317–4323. Association for Computational Linguistics, Florence (2019)
7. Martínez Pinilla, M.F., et al.: Búsqueda semántica de sentencias judiciales usando técnicas actuales del NLP y métodos de recuperación de información. Pontificia Universidad Javeriana, Bogotá (2021)
8. Gillis, T., McStocker, A., Percival, A.: Indirect tax compliance in an era of big data. Tax Plan. Int. Indirect Taxes **13**(3), 1–6 (2015)
9. Faúndez-Ugalde, A., Mellado-Silva, R., Aldunate-Lizana, E.: Use of artificial intelligence by tax administrations: an analysis regarding taxpayers' rights in Latin American countries. Comput. Law Secur. Rev. **38**, 105441 (2020)

10. Cáceres, E.: Inteligencia artificial, derecho y e-justice (el proyecto iij-conacyt). Boletín mexicano de derecho comparado **39**(116), 593–611 (2006)
11. Hernández, J.M.M.: Aplicaciones de la inteligencia artificial generativa en la enseñanza del derecho aduanero e internacional. Región Científica **4**(1), 2024432 (2025)
12. Dirección General de Comunicación Social, UNAM: La UNAM impulsa proyectos de inteligencia artificial para beneficio social. Boletín UNAM, núm. 618 (2022). https://www.dgcs.unam.mx/boletin/bdboletin/2022_618.html. Accessed 15 Nov 2024
13. Nakamura, L.A.C., Madrigal, J.A.G.: La perspectiva ética y jurídica de la Inteligencia Artificial en México. Misión Jurídica **16**(25), 199–214 (2023)
14. Palomino Bravo, M.: Los Grandes Modelos del Lenguaje basados en Transformers: revisión y aplicación práctica con ChatGPT. Universidad Nacional Mayor de San Marcos, Lima, Tesis de grado (2023)
15. Huang, Y., Xu, J., Lai, Z., Jiang, T., Chen, T.: Advancing transformer architecture in long-context large language models: a comprehensive survey. arXiv preprint arXiv:2311.12351 (2023)
16. Roque, L.: The evolution of llama: From llama 1 to llama 3.1. arXiv preprint (2025)
17. Abouelenin, A., et al.: Phi-4-mini technical report: compact yet powerful multimodal language models via mixture-of-loras. arXiv preprint arXiv:2503.01743 (2025)
18. Zhao, W.X., et al.: A survey of large language models. arXiv preprint arXiv:2303.18223 (2024)
19. Heflin, J.J.A.: AI-generated literature and the vectorized word, Cambridge. Ph.D. thesis (2020)
20. Voit, A., Stankus, A., Magomedov, S., Ivanova, I.: Big data processing for full-text search and visualization with Elasticsearch. Int. J. Adv. Comput. Sci. Appl. **8**(12) (2017)
21. Calahorro Jiménez, P.: UJA-GPT: Un chatbot con generación aumentada por recuperación para la comunidad universitaria. Universidad de Jaén, Jaén, Trabajo de Fin de Grado (2024)
22. Gao, Y., et al.: Retrieval-augmented generation for large language models: a survey. arXiv preprint arXiv:2312.10997 (2023)
23. Marchisio, K., et al.: How does quantization affect multilingual LLMs? arXiv preprint arXiv:2407.03211 (2024)
24. Yang, J., et al.: Quantization networks. In: Proceedings of the IEEE/CVF Conference on Computer Vision and Pattern Recognition, pp. 7308–7316 (2019)
25. Reiter, E.: A structured review of the validity of BLEU. Comput. Linguist. **44**(3), 393–401 (2018)
26. Lin, C.Y.: Rouge: a package for automatic evaluation of summaries. In: Text Summarization Branches Out, pp. 74–81 (2004)
27. Zhang, T., Kishore, V., Wu, F., Weinberger, K.Q., Artzi, Y.: Bertscore: evaluating text generation with bert. arXiv preprint arXiv:1904.09675 (2019)
28. Li, D., et al.: From generation to judgment: opportunities and challenges of LLM-as-a-judge. In: Proceedings of the 2025 Conference on Empirical Methods in Natural Language Processing, pp. 2757–2791 (2025)
29. Zheng, L., et al.: Judging LLM-as-a-judge with MT-Bench and Chatbot Arena. Adv. Neural. Inf. Process. Syst. **36**, 46595–46623 (2023)
30. Ribeiro, M.T., Wu, T., Guestrin, C., Singh, S.: Beyond accuracy: behavioral testing of NLP models with CheckList. In: Proceedings of the 58th Annual Meeting of the Association for Computational Linguistics, pp. 6326–6341 (2020)

Factual Reliability Analysis of Large Language Models Using a Semantic Stability Approach

Vitali Herrera-Semenets[1], Lázaro Bustio-Martínez[2](✉), and Jan van den Berg[3]

[1] OnTheDime S.r.l., Ancona, Italy
vitali.herrera@onthedime.it
[2] Universidad Iberoamericana, Mexico City, Mexico
lazaro.bustio@ibero.mx
[3] Delft University of Technology (TU Delft), Delft, The Netherlands
j.vandenBerg@tudelft.nl

Abstract. Assessing factual reliability in large language models (LLMs) remains an unresolved methodological challenge. Most evaluation strategies rely on external ground-truth comparison or retrieval augmentation, which limits scalability and domain transferability. This work introduces a computational approach for factual reliability assessment based on semantic stability analysis. The central assumption is that the consistency of independently generated responses to the same factual claim, measured as semantic similarity in embedding space, reflects the degree of certainty encoded in the model's conditional distribution. The proposed framework is domain-agnostic and does not require external verification during inference. As a case study, the method is applied to misinformation detection using claims derived from the LIAR-PLUS benchmark, and evaluated on a representative instruction-tuned LLM instance deployed locally. For each claim, multiple independent generations are sampled and their pairwise semantic similarity is computed to quantify response stability. The results reveal a systematic asymmetry between false and true claims: false statements produce higher semantic stability, indicating that the model responds with greater uniformity when rejecting incorrect claims than when explaining correct ones. The contribution of this work lies in reframing factual reliability assessment as a computational semantic stability analysis problem. The proposed strategy complements evidence-based fact-checking by providing a quantitative diagnostic signal of generative behavior, with potential applicability across high-stakes domains where reliability auditing is required.

Keywords: factual reliability · Large Language Models · semantic stability · self-consistency · misinformation detection

H. Ponce et al. (Eds.): COMIA 2026, CCIS 3095, pp. 27–38, 2027.
https://doi.org/10.1007/978-3-032-37438-7_3

1 Introduction

Large language models (LLMs) have demonstrated notable performance across a wide range of language understanding and generation tasks [2]. Despite these advances, factual reliability remains a structural limitation. Models frequently produce statements that are syntactically coherent and semantically plausible yet factually incorrect. This phenomenon, commonly referred to as hallucination [4], represents a fundamental challenge for deploying generative systems in high-stakes environments such as journalism, healthcare, law, and public information systems.

Most existing approaches to factual correctness evaluation rely on external verification [5,9]. In automated fact-checking, model outputs are compared against curated knowledge bases or retrieved documentary evidence [3,7]. In benchmark-based evaluations, generated responses are assessed against ground-truth references or human annotations [4,7,9]. While effective in controlled settings, these strategies exhibit two important limitations. First, they require domain-specific supervision or curated corpora. Second, they do not scale easily to arbitrary domains where ground truth may be incomplete, evolving, or expensive to obtain [5].

Rather than measuring correctness against external truth, this work evaluates the semantic stability of a generative model's responses. Semantic stability, in this context, refers to the degree of similarity among independently generated responses to the same factual claim, measured via pairwise cosine similarity in a sentence embedding space. When a model encodes a well-defined representation of a claim, independently generated responses are expected to exhibit high pairwise similarity. When that representation is uncertain or underspecified, repeated responses to the same claim are more likely to diverge in content, emphasis, or argumentative structure. Semantic dispersion across samples thus provides a measurable signal of how decisively the model represents a given claim.

Under this perspective, semantic stability does not necessarily imply factual correctness. Throughout this paper, the term *factual correctness* (or simply *factuality*) refers to the correspondence between a generated statement and verifiable real-world facts. A model may consistently reject a false claim, producing highly similar refutations across samples. In contrast, when responding to a true claim, the model may generate multiple valid explanations, paraphrases, or contextualizations, increasing semantic variability. Therefore, semantic stability reflects decisiveness of representation rather than correctness.

The primary contribution of this paper is a computational evaluation strategy that measures semantic stability in LLM responses as a proxy for characterizing the model's generative behavior on factual claims. The framework is validated through a case study on misinformation detection using claims from an existing benchmark to analyze the relationship between semantic stability and claim veracity. The proposed metrics do not replace external verification but provide an additional quantitative signal that characterizes the generative behavior of language models when confronted with factual claims.

The remainder of this paper is organized as follows. Section 2 reviews related work in automated fact-checking, hallucination detection, and sampling-based reasoning, highlighting the methodological gaps addressed by our approach. Section 3 formalizes the proposed computational semantic stability assessment strategy, detailing the dispersion metrics and the interpretational hypothesis. Section 4 presents the experimental protocol and statistical results of the case study on misinformation claims, followed by a discussion of its practical implications. Finally, Sect. 5 concludes the paper.

2 Related Work

Factual reliability in natural language systems has evolved through three main research directions: (i) automated fact-checking [1,7,9], (ii) hallucination analysis in generative systems [3–5,8], and (iii) sampling-based self-consistency methods [10]. These paradigms differ in their objectives, supervision requirements, and evaluation assumptions. The following discussion reviews these approaches in order to identify the methodological gap addressed in this work.

Early approaches to factual verification framed the problem as supervised classification or textual entailment grounded in curated knowledge bases. The FEVER benchmark [7] formalized fact-checking as textual entailment supported by Wikipedia evidence, establishing a standard pipeline consisting of claim detection, evidence retrieval, and veracity prediction. These systems evaluate the truthfulness of a claim by comparing it against supporting evidence retrieved from external corpora. While this paradigm established the foundations of computational fact-checking, its effectiveness depends heavily on the availability and reliability of external knowledge sources.

With the emergence of Large Language Models (LLMs), research attention shifted toward hallucinations in generative outputs [4]. Instead of relying solely on static evaluation datasets, more recent approaches attempt to reduce hallucinations by integrating external information during inference. Retrieval-Augmented Generation (RAG) [3] augments the generation process with documents retrieved from external corpora, allowing the model to condition its output on supporting evidence. However, recent surveys show that hallucinations remain a persistent phenomenon even when external evidence is provided [5,8]. Models may still produce inconsistent or misleading responses, reflecting the probabilistic nature of generative language modeling. External retrieval improves grounding but does not provide insight into the reliability of the model's generative behavior.

A different research direction exploits the stochastic nature of language model generation itself. Sampling-based self-consistency was introduced as a decoding strategy for improving multi-step reasoning [10]. In this approach, multiple reasoning paths are sampled and the most frequent answer is selected. Empirical results show that agreement among sampled outputs can improve reasoning accuracy. However, existing self-consistency methods implicitly assume that agreement among generated outputs correlates with correctness. In practice, consistency is used primarily as an optimization mechanism to select a final answer rather than as an object of analysis. Importantly, high agreement among generated responses does not imply factual correctness. This observation raises the question of whether the variability of model outputs across repeated generations may itself serve as a diagnostic signal, reflecting how decisively the model represents a given claim.

Table 1. Conceptual comparison of factuality evaluation paradigms

Approach	Evaluation Perspective
Fact-checking systems (FEVER) [7]	Verification of claims against external evidence
Truthfulness benchmarks (TruthfulQA) [4]	Correctness of generated answers relative to reference responses
Retrieval-Augmented Generation [3]	Grounded generation using retrieved documents during inference
Self-consistency decoding [10]	Agreement among sampled outputs to select a final answer
Proposed strategy	Semantic stability of independently generated model responses

Table 1 summarizes the conceptual focus of these paradigms in comparison with the strategy proposed in this work.

Existing approaches emphasize either external grounding or output agreement as a mechanism for selecting answers. The possibility that the variability of model outputs itself may serve as a diagnostic signal of internal representational stability has received comparatively little attention. This work investigates that possibility by analyzing semantic dispersion across independently generated responses to factual claims.

3 Proposed Computational Semantic Stability Assessment Strategy

The proposed framework evaluates factual reliability through semantic stability analysis. Instead of verifying outputs against external evidence, the method estimates the dispersion of independently sampled responses conditioned on the same factual claim.

Let c_i denote the i-th factual claim, that is, a natural-language statement whose truth value can in principle be verified. Let $r_{i,j}$ denote the j-th textual response generated for claim c_i, and let θ denote the parameters of the language model. For a given claim c_i, the model induces a conditional probability distribution over possible textual responses:

$$p_\theta(r \mid c_i), \tag{1}$$

where r denotes a generic textual response. Under stochastic decoding with fixed hyperparameters, generated responses for claim c_i can be modeled as samples from this conditional distribution. For each claim c_i, n such responses are generated:

$$r_{i,1}, r_{i,2}, \ldots, r_{i,n} \sim p_\theta(r \mid c_i). \tag{2}$$

In the empirical instantiation presented in this study, variability is induced through semantically equivalent paraphrased formulations of the same claim rather than repeated stochastic sampling of an identical prompt. This design evaluates the robustness of the conditional response distribution under surface-level linguistic perturbations.

Low-temperature sampling (i.e., setting the temperature parameter T close to zero in the softmax distribution over the vocabulary) is used to constrain randomness while preserving minor stochastic fluctuations [2]. The objective is not to maximize diversity but to expose the variability inherent in the model's representation of the claim. If the model encodes a stable representation of the claim, the generated responses are expected to exhibit low dispersion in semantic space.

A pretrained encoder $E(\cdot)$ is used to map textual responses into a continuous semantic space:

$$E : \mathcal{R} \to \mathbb{R}^d. \tag{3}$$

Each sampled response is then embedded as:

$$X_{i,j} = E(r_{i,j}), \tag{4}$$

where the encoder maps each full response to a fixed-dimensional vector via mean pooling over token representations, as implemented in Sentence-BERT [6]. The vector $X_{i,j}$ can be interpreted as a realization of a random variable induced in embedding space by $p_\theta(r \mid c_i)$ through $E(\cdot)$.

The methodology computes pairwise semantic similarity within this induced distribution using cosine similarity, consistent with the native representation space of siamese sentence encoders [6]:

$$S_{i,jk} = \frac{X_{i,j} \cdot X_{i,k}}{\|X_{i,j}\| \|X_{i,k}\|}, \quad j \neq k. \tag{5}$$

For readability, the claim index i may be omitted when no ambiguity arises. Let $\mathcal{S}_i = \{S_{i,jk} \mid 1 \leq j < k \leq n\}$ denote the set of all pairwise similarities for claim c_i, with cardinality $|\mathcal{S}_i| = \frac{n(n-1)}{2}$. The empirical mean and variance over this set are defined as:

$$\mathbb{E}_{emp}[S_{i,jk}] = \frac{1}{|\mathcal{S}_i|} \sum_{j<k} S_{i,jk}, \tag{6}$$

$$\mathbb{V}_{emp}[S_{i,jk}] = \frac{1}{|S_i|}\sum_{j<k}(S_{i,jk} - \mathbb{E}_{emp}[S_{i,jk}])^2. \tag{7}$$

From this similarity structure, the following stability metrics are computed:

$$\mu_i = \mathbb{E}_{emp}[S_{i,jk}], \quad \sigma_i = \sqrt{\mathbb{V}_{emp}[S_{i,jk}]}, \quad m_i = \min_{j<k} S_{i,jk}. \tag{8}$$

Here, μ_i serves as an empirical estimator of central semantic concentration, and σ_i captures the overall semantic dispersion. Furthermore, m_i isolates the minimum pairwise similarity, which mathematically bounds the maximal semantic divergence within the sampled responses.

Under this formulation, high semantic stability (i.e., self-consistency across generated responses) corresponds to low empirical variance of the embedding distribution conditioned on c_i. Conversely, high dispersion suggests that the conditional distribution $p_\theta(r \mid c_i)$ is structurally diffuse.

No systematic positive correlation between consistency and correctness is assumed. Instead, dispersion is treated as a behavioral property of the generative model. High semantic concentration indicates decisiveness in representation, whereas low concentration indicates representational flexibility or ambiguity.

This probabilistic view clarifies that the proposed metric estimates properties of the induced response distribution rather than ground-truth veracity. Labels, when available, are used only to analyze correlations between behavioral stability and factual status. The overall pipeline of the proposed methodology is illustrated in Fig. 1.

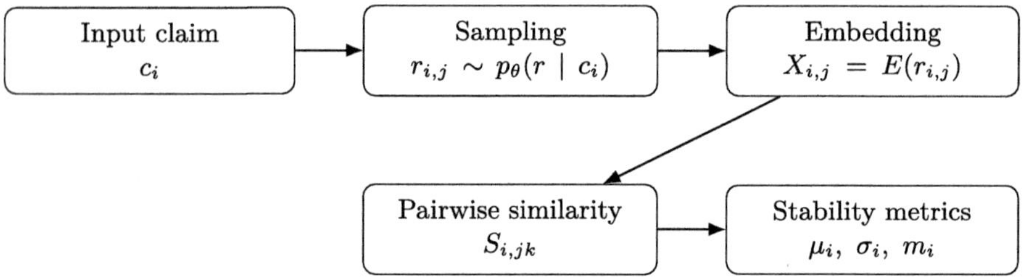

Fig. 1. Pipeline of the proposed computational semantic stability assessment strategy.

While the proposed framework removes the need for external retrieval and supervised verification models, it introduces a necessary inference overhead. Specifically, generating n independent responses increases the decoding latency by a factor of n compared to standard greedy decoding. However, the subsequent semantic dispersion estimation operates exclusively on dense vectors in $\mathbb{R}^d$ and requires only $n(n-1)/2$ pairwise cosine similarity computations.

For practical diagnostic values (*e.g.*, $n \leq 10$), the mathematical overhead of calculating dispersion is computationally negligible. The fundamental trade-off of this methodology is deliberate: inference latency is exchanged for a robust diagnostic signal of factual reliability that requires no external resources.

4 Evaluating Semantic Stability on Misinformation Claims

This section evaluates the proposed semantic stability framework on a misinformation-oriented claim dataset. The objective is to examine whether internal semantic consistency varies systematically with claim veracity, using labeled statements as an empirical testbed rather than as a benchmark contribution.

The factual claims in this study were drawn from the LIAR-PLUS dataset [1], an extended version of the LIAR benchmark originally introduced for fake news and fact-checking research [9]. The LIAR dataset consists of 12,836 short statements manually labeled for veracity across multiple categories, collected from http://politifact.com and extensively used in automated fact-checking evaluations. LIAR-PLUS further augments these claims with evidence extracted from original fact-checking reports, providing richer contextual information for each statement.

Claims in this case study are categorized as either false (Class 0) or true (Class 1). To instantiate the multi-sample protocol without incurring prohibitive computational overhead, the experimental design restricts the sample size to $n = 10$ independent responses per claim. Specifically, the framework elicits these responses using ten manually constructed, semantically equivalent reformulations of each original statement. This sample size provides a sufficient empirical basis to characterize response variability under semantically equivalent prompt reformulations while maintaining feasible inference latency.

The following subsections detail the experimental protocol and results.

4.1 Experimental Protocol

The experimental subset consisted of 50 claims randomly sampled from the LIAR-PLUS corpus, balanced across true and false categories. Given the exploratory nature of this study, this subset is intentionally limited in scope and should be interpreted as a pilot evaluation rather than a definitive empirical assessment. For each claim, 10 semantically equivalent reformulations were manually constructed, yielding a total of 500 distinct query instances. This design evaluates robustness of model behavior under paraphrastic invariance rather than repeated stochastic sampling of an identical prompt.

All experiments were conducted using a locally deployed instruction-tuned language model via `llama.cpp`. The evaluated model was `Phi-4-mini-instruct-Q4_K_M.gguf`. Each reformulated prompt was submitted once under fixed decoding parameters (context window $n_{ctx} = 4096$, maximum generation length 256 tokens, temperature $T = 0.1$, repetition penalty 1.1). A low temperature was selected to minimize stochastic noise while preserving sensitivity to semantic perturbations.

The selection of a single model instance was intentional and methodological rather than comparative. The objective of this study is not to benchmark architectures, but to examine whether semantic self-consistency constitutes a meaningful behavioral signal within a generative model. By isolating a single architecture, the analysis focuses on internal response stability without introducing cross-model variability as a confounding factor.

Semantic representations were obtained using Sentence-BERT embeddings [6], enabling pairwise cosine similarity estimation in a fixed embedding space.

4.2 Statistical Results

To examine whether semantic stability differs systematically between false and true claims, the aggregate statistics summarized in Table 2 are examined. Given the exploratory scale of this study, formal inferential testing is deferred to future work; the analysis focuses on the magnitude and directional consistency of the observed separation.

Table 2. Aggregate semantic stability metrics by claim class

Class	Mean Similarity (μ)	Std. Dev. (σ)	Min. Similarity (m)
False (0)	0.6854	0.1398	0.3785
True (1)	0.6296	0.1709	0.2753

The mean semantic similarity for false claims exceeds that of true claims by $\Delta = 0.0558$. The relative increase in central stability for false claims corresponds to an approximate 8.9% gain over the true-claim baseline. At the same time, dispersion (σ) is lower for false claims, and the minimum similarity value is substantially higher.

This pattern suggests that responses to false statements may cluster more tightly in embedding space, whereas responses to true claims may exhibit broader semantic spread and a higher likelihood of outlier or divergent samples.

Given the exploratory scale of this study, no formal hypothesis testing is performed. Consequently, the observed differences should be interpreted as descriptive trends rather than statistically validated effects. Future work should include formal nonparametric testing (*e.g.*, the Mann–Whitney U test) together with effect-size estimation to quantify the statistical significance and magnitude of this separation.

4.3 Qualitative Example

Table 3 shows an illustrative example of how paraphrased prompts targeting the same underlying claim can elicit different response framings. The goal is not to claim a new benchmark result, but to provide an interpretable qualitative counterpart to the stability metrics.

Table 3. Illustrative prompts and condensed responses for the same underlying claim. Responses are paraphrased summaries to preserve space and avoid verbatim reproduction of model outputs.

Prompt (paraphrase variant)	Model response (condensed)
Underlying claim (False): "COVID-19 vaccines contain microchips for tracking"	
Is it true that COVID-19 vaccines include tracking microchips?	Rejects the claim; states there is no credible evidence; frames it as a conspiracy narrative
Do coronavirus vaccines secretly contain microchip devices?	Rejects the claim; emphasizes scientific/medical scrutiny and manufacturing transparency; no microchips
Are governments using vaccines to implant tracking technology?	Rejects the claim; notes absence of verified sources; highlights implausibility at scale
Underlying claim (True): "The Earth revolves around the Sun"	
Does the Earth orbit the Sun?	Confirms; briefly states heliocentric orbit as basic astronomy
Is it correct that the Sun is at the center of the solar system?	Confirms with nuance; notes barycenter and that the Sun is the dominant mass, not a strict fixed center
Does our planet move around the Sun?	Confirms; expands with orbital period and elliptical orbit description

The table highlights that false claims tend to elicit semantically convergent refutations, whereas true claims admit multiple correct explanatory framings, increasing semantic variability without implying incorrectness.

4.4 Behavioral Interpretation

The results challenge the implicit assumption that internal agreement correlates positively with correctness. Instead, high semantic concentration appears associated with decisive rejection behavior. When confronted with false claims, the model consistently produces similar refutations. In contrast, true claims admit multiple valid explanatory formulations, increasing representational dispersion.

This asymmetry is not paradoxical. In probabilistic terms, rejecting a false claim may correspond to a concentrated mode in the conditional distribution $p_\theta(r \mid c_i)$, whereas explaining a true claim may involve a wider manifold of semantically acceptable responses.

4.5 Discussion

The case study indicates that, for the specific model and configuration evaluated, semantic self-consistency captures a behavioral property that is relevant to misinformation settings: the model exhibits higher semantic concentration when responding to false claims than when responding to true claims. In operational terms, this suggests that the model tends to converge toward a relatively stable refutation pattern when a claim is incorrect. For misinformation-sensitive workflows, such stability can be valuable because corrective messaging benefits from predictability. If repeated interactions yield substantially different refutations, trust deteriorates and downstream moderation decisions become less reproducible.

At the same time, the observed asymmetry clarifies a crucial point: semantic stability is not equivalent to factual correctness. A model may be internally consistent and still be wrong, particularly when it produces confident but systematically biased explanations. Therefore, the proposed metrics should be interpreted as indicators of representational decisiveness rather than truth estimators. In practice, this means self-consistency can support pre-deployment diagnostics and monitoring, but it should not replace evidence-based verification. A stable refutation behavior may be desirable, yet external grounding remains necessary to ensure epistemic alignment with reality.

The broader value of the strategy lies in its domain-agnostic formulation. However, the present empirical evaluation is limited to a small subset of 50 claims derived from a single dataset and to a single model instance (Phi-4-mini). The observed patterns may therefore be influenced both by the specific characteristics of misinformation-related statements in this corpus and by the architectural and training properties of the evaluated model. Evaluating the proposed framework across a broader range of models, including large-scale and commercially deployed systems, would be necessary to determine whether the observed stability patterns generalize across architectures. Expanding the evaluation to larger and more diverse datasets, including domains such as scientific, medical, or legal claims, would be necessary to determine whether the observed stability patterns persist across different knowledge contexts. Within the scope of the present evaluation, the proposed strategy may provide a useful diagnostic signal in applications involving factual assertions, especially when ground truth is incomplete, expensive to obtain, or rapidly evolving. In clinical, legal, or scientific assistance scenarios, high dispersion across repeated samples can signal internal ambiguity, incomplete knowledge encoding, or heightened sensitivity to prompt phrasing. In these environments, dispersion monitoring can function as an early warning mechanism: it flags claims or topics for which the model's internal behavior is unstable and therefore less reliable, even before external verification is performed.

These observations also expose the limits of self-consistency as a standalone reliability signal. Semantic dispersion measures the geometry of the induced response distribution in embedding space; it does not directly measure correspondence to external facts. Consequently, the most realistic role for self-consistency is as a complementary signal within multi-component reliability assessment pipelines, where internal stability can be combined with retrieval-based grounding, calibration analysis, and cross-model agreement to obtain more robust trust estimates.

5 Conclusion

This work introduces a model-centric strategy for assessing factual reliability in large language models based on semantic self-consistency. Instead of relying on external evidence retrieval or supervised verification, the framework analyzes the dispersion of independently generated responses conditioned on the same claim. By interpreting stability as an empirical property of the induced response distribution, the method reframes factuality assessment as a behavioral analysis problem.

The case study on misinformation claims revealed a systematic asymmetry: false claims produced higher semantic concentration than true claims. This finding challenges the common assumption that internal agreement necessarily correlates with correctness. Instead, semantic stability appears to reflect decisiveness in representation rather than epistemic alignment with reality. The results indicate that self-consistency captures structural characteristics of generative behavior that are not observable under single-sample evaluation.

Conceptually, the contribution lies in reframing factuality assessment from evidence-centric validation to distributional stability analysis of generative behavior. Methodologically, the framework is domain-agnostic and requires no modification of the underlying model architecture. Practically, it provides an auxiliary diagnostic signal that can support pre-deployment evaluation, behavioral monitoring, and reliability auditing in misinformation-sensitive environments and other high-stakes domains.

The approach does not replace fact-checking or retrieval-based grounding. Instead, it complements them by quantifying internal representational coherence. Future work should formalize statistical testing across larger datasets, evaluate cross-model robustness, analyze sensitivity to decoding parameters, and integrate stability metrics with external grounding signals to construct multi-dimensional reliability assessment frameworks.

Declaration on Generative AI

During the preparation of this work, the authors used ChatGPT to improve the grammar and clarity of the manuscript. Following the use of this tool, the authors thoroughly reviewed and edited the content as needed, and take full responsibility for the final version of the publication. This practice is fully aligned with the critical perspective developed throughout this study.

Acknowledgments. This work was supported by the "Instituto de Investigación Aplicada y Tecnología" and the "Universidad Iberoamericana, Ciudad de México" through the research project "Detecting Disinformation in Texts through Psychological and Linguistic Patterns with Explainable AI".

References

1. Alhindi, T., Petridis, S., Muresan, S.: Where is your evidence: improving fact-checking by justification modeling. In: Proceedings of the First Workshop on Fact Extraction and VERification (FEVER), pp. 85–90. Association for Computational Linguistics, Brussels (2018). https://doi.org/10.18653/v1/W18-5513
2. Kumar, P.: Large language models (LLMs): survey, technical frameworks, and future challenges. Artif. Intell. Rev. **57**(260) (2024). https://doi.org/10.1007/s10462-024-10888-y
3. Lewis, P., et al.: Retrieval-augmented generation for knowledge-intensive NLP tasks. In: Advances in Neural Information Processing Systems, vol. 33, pp. 9459–9474. Curran Associates, Inc. (2020)
4. Lin, S., Hilton, J., Evans, O.: TruthfulQA: measuring how models mimic human falsehoods. In: Proceedings of the 60th Annual Meeting of the Association for Computational Linguistics, vol. 1: Long Papers, pp. 3214–3252. Association for Computational Linguistics, Dublin (2022). https://doi.org/10.18653/v1/2022.acl-long.229
5. Rahman, S.S., et al.: Hallucination to truth: a review of fact-checking and factuality evaluation in large language models. Artif. Intell. Rev. **59**(2), 70 (2026). https://doi.org/10.1007/s10462-025-11454-w
6. Reimers, N., Gurevych, I.: Sentence-BERT: sentence embeddings using siamese BERT-networks. In: Proceedings of the 2019 Conference on Empirical Methods in Natural Language Processing and the 9th International Joint Conference on Natural Language Processing (EMNLP-IJCNLP), pp. 3982–3992. Association for Computational Linguistics, Hong Kong (2019). https://doi.org/10.18653/v1/D19-1410
7. Thorne, J., Vlachos, A., Christodoulopoulos, C., Mittal, A.: FEVER: a large-scale dataset for fact extraction and VERification. In: Proceedings of the 2018 Conference of the North American Chapter of the Association for Computational Linguistics: Human Language Technologies, vol. 1 (Long Papers), pp. 809–819. Association for Computational Linguistics, New Orleans (2018). https://doi.org/10.18653/v1/N18-1074
8. Wang, C., et al.: Survey on factuality in large language models. ACM Comput. Surv. **58**(1) (2025). https://doi.org/10.1145/3742420
9. Wang, W.Y.: "Liar, Liar Pants on Fire": a new benchmark dataset for fake news detection. In: Proceedings of the 55th Annual Meeting of the Association for Computational Linguistics, vol. 2: Short Papers, pp. 422–426. Association for Computational Linguistics, Vancouver (2017). https://doi.org/10.18653/v1/P17-2067
10. Wang, X., et al.: Self-consistency improves chain of thought reasoning in language models. In: The Eleventh International Conference on Learning Representations (2023). https://openreview.net/forum?id=1PL1NIMMrw

Use and Perception of Artificial Intelligence in Higher Education Students in Creative Disciplines: Pedagogical, Creative, and Ethical Implications

Silveyra Rosales Mariana Teresa[1,2](✉)

[1] Universidad Autónoma del Estado de Morelos, Cuernavaca, Mexico
mariana.silveyra@uaem.mx
[2] Universidad La Salle Cuernavaca, Cuernavaca, Morelos, Mexico

Abstract. This paper analyzes the use and perception of artificial intelligence (AI) among higher education students aged 18 to 22 in creative disciplines at La Salle University Cuernavaca, integrating a theoretical approach grounded in education, ethics, and creativity studies. Based on a mixed-method study, patterns of use, impacts on design processes, and emerging ethical tensions are examined. The results reveal a widespread adoption of AI as a cognitive infrastructure, particularly in the stages of ideation and representation, as well as risks associated with dependency, superficial learning, and regulatory ambiguity. The study proposes advancing toward critical AI literacy and the development of pedagogical and institutional frameworks that redefine authorship and creativity in algorithm-mediated contexts.

Keywords: artificial intelligence · creativity · higher education · ethics · design · learning

1 Introduction

The recent expansion of generative artificial intelligence has profoundly transformed knowledge production processes in higher education. In creative disciplines such as architecture, design, and communication, this transformation is particularly significant due to the project-based, iterative, and exploration nature of their processes.

Unlike previous digital tools, AI not only automates tasks but actively participates in the generation of ideas, texts, and images, positioning itself as an agent within the creative process. This implies a reconfiguration of fundamental concepts such as authorship, originality, and learning.

Contemporary literature has highlighted that AI constitutes a new form of cognitive mediation that redefines the relationship between individuals, tools, and knowledge [5]. In this context, organizations such as UNESCO emphasize that its integration must be accompanied by clear ethical principles, focused on transparency, accountability, and human oversight [1].

H. Ponce et al. (Eds.): COMIA 2026, CCIS 3095, pp. 39–50, 2027.
https://doi.org/10.1007/978-3-032-37438-7_4

This study analyzes the use, perception, and implications of AI among students in creative disciplines, addressing both its potential and the tensions it generates in educational processes.

2 Theoretical Framework

Artificial intelligence in education has been conceptualized as a system capable of expanding human cognitive capacities through processes of automation and content generation [2]. However, with the emergence of generative models, this relationship has evolved toward forms of knowledge co-production.

From the perspective of creativity theory, Csikszentmihalyi argues that creativity is not an individual attribute, but a system involving the interaction between the individual, the domain, and the social field. In this sense, AI can be understood as a new actor within the creative domain, capable of transforming the rules of symbolic production (Fig. 1).

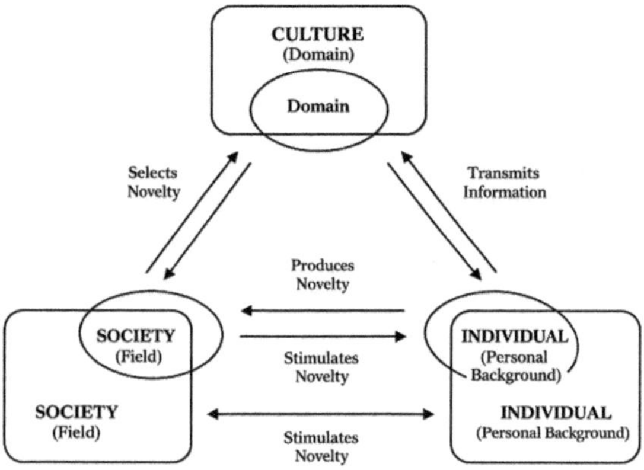

Fig. 1. Model of the theory of creativity developed by Csikszentmihalyi

The diagram above represents the systemic model of creativity proposed by Mihaly Csikszentmihalyi, in which creativity is not understood as an isolated individual capacity, but as the result of the dynamic interaction between three components: the individual, the domain (culture), and the field (society). The individual, from their experience and personal context, generates novel proposals or ideas; however, these only become "creative" if they are recognized and validated by the field, that is, by the community or agents that legitimize knowledge (academics, institutions, experts). In turn, the domain contains the set of existing knowledge, rules, and cultural languages that are transmitted to the individual and upon which they operate. The process is circular: the individual produces novelty, the field selects it, and if it is accepted, it is incorporated into the domain, thereby transforming culture.

In a contemporary reading, particularly relevant in the context of artificial intelligence, this model allows us to understand that creativity is no longer exclusively human but is mediated by technological systems that intervene in the domain, for example, databases and generative models, and in individual production using AI tools. AI expands the individual's capacity to generate variations but does not replace the role of the field, which remains responsible for validating what is relevant, valuable, or innovative. In this sense, creativity in the age of AI shifts toward a logic of co-production, where what matters is not only generating ideas, but also knowing how to select, interpret, and legitimize them within a specific cultural system (Fig. 2).

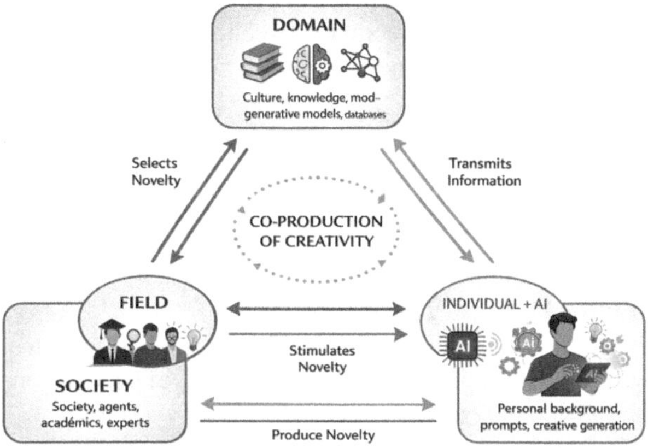

Fig. 2. Model of the theory of creativity integrating Artificial Intelligence as a co-producer

In the field of design, Nigel Cross argues that design thinking is based on abductive and iterative processes, where knowledge is constructed through the exploration of possible solutions. AI enhances these divergent phases by expanding the capacity to generate alternatives.

For his part, Schön introduces the concept of reflection-in-action, where learning occurs during the design process. Excessive use of AI may weaken this reflective dimension if it replaces cognitive processes rather than complementing them.

From an ethical perspective, UNESCO [1] and Floridi et al. [4] emphasize the need to establish principles of transparency, accountability, and governance, particularly in educational contexts where AI may affect academic integrity.

3 Methodology

A mixed-method approach was adopted, with a descriptive-interpretive emphasis, aimed at analyzing both quantifiable patterns and qualitative perceptions regarding the use of artificial intelligence among students in creative disciplines at the School of Architecture, Design, and Communication (EADyC) at La Salle University Cuernavaca. To this end, a survey-based instrument was designed and administered to a sample of 68 higher education students, primarily from the Architecture program, followed by Communication and Design.

The selection of this approach responds to the need to understand the relationship between technology and creativity not only in terms of frequency of use, but also in its symbolic, cognitive, and ethical dimensions. In this sense, the instrument was structured around five analytical dimensions that allow the phenomenon to be addressed from a comprehensive perspective, in dialogue with the systemic model of creativity (Csikszentmihalyi, 1998) and with contemporary approaches to digital literacy and technological culture.

The selected dimensions derive from the articulation between the theoretical framework of creativity, studies on technological integration in education, and emerging issues in the use of artificial intelligence in educational contexts. Taking together, these dimensions enable the analysis of the interaction between the individual, technological tools, and the academic environment, understanding creativity as a situated and mediated process.

a) **Artificial Intelligence Literacy**

This dimension evaluates the level of familiarity, knowledge, and basic competencies that students possess regarding the use of AI tools. Its inclusion is justified insofar as technological literacy constitutes a prerequisite for the meaningful appropriation of these tools (Area, 2012; Buckingham, 2008).

Additionally, from Csikszentmihalyi's model, access to the domain (in this case, AI systems as part of contemporary technical knowledge) depends on the individual's ability to understand and operate their codes and logics.

Associated questions in the instrument:

Have you used artificial intelligence tools before entering university?
What level of knowledge do you consider you have about AI? (Likert scale)
What AI tools do you know or have you used?

b) **Types of AI Use**

This dimension seeks to identify how students integrate AI into their academic and creative processes, distinguishing between instrumental, exploratory, and generative uses. It is based on the need to understand not only access to technology, but also the specific forms of appropriation and practice (Selwyn, 2016).

It also allows the role of AI within the creative process to be located, that is, whether it functions as a support tool, as a generator of ideas, or as a partial substitute for certain tasks.

Associated questions:

Which AI tools do you use most frequently?
For what activities do you use AI? (research, image generation, writing, etc.)
At which stage of the creative process do you consider AI most useful? (ideation, development, presentation)

c) **Impact on Learning and Creativity**

This dimension analyzes students' perceptions of how AI influences their learning, creative capacity, and cognitive processes. It is grounded in contemporary debates on the externalization of thought, distributed cognition, and potential technological dependency (Carr, 2010; Clark & Chalmers, 1998).

From the systemic model, this dimension makes it possible to observe how the individual reconfigures their relationship with the domain through tools that expand (or limit) their creative agency.

Associated questions:

Do you consider that AI improves your creativity?
Does AI facilitate your learning?
Do you think that the use of AI can generate dependency?
Briefly describe how AI influences your creative process.

d) **Ethics and Responsible Use**

The inclusion of this dimension responds to the growing concern about ethical dilemmas associated with the use of artificial intelligence in educational contexts, such as plagiarism, authorship, information reliability, and transparency in its use (UNESCO, 2021).

In this sense, it is not enough to analyze the use of technology; it is also necessary to understand the normative frameworks and ethical perceptions that guide its implementation.

Associated questions:

Do you consider the use of AI in academic activities to be ethical?
In which cases do you consider its use inappropriate?
Do you think AI use should be regulated at university?

e) **Training Needs**

This dimension explores students' demands, interests, and expectations regarding training in artificial intelligence. Its inclusion is justified from a prospective perspective, aimed at identifying training gaps and opportunities for curricular integration.

It also allows linking the field (educational institution) with the individual, in terms of how conditions are structured for the critical appropriation of these technologies.

Associated questions:

Would you like to receive training on the use of AI in your discipline?
Which topics do you consider most relevant to learning about AI?
In which format would you prefer this training? (workshops, courses, subjects) (Fig. 3)

4 Results

As mentioned previously, the instrument was administered to 68 students through a virtual form, presented as part of an academic study aimed at understanding how students from the School of Architecture, Design, and Communication use and perceive Artificial Intelligence (AI) tools in their learning and academic creation processes.

Regarding their fields of study, 33 students are enrolled in architecture, 18 in design, and 17 in communication. In response to the question, "Have you used AI tools before entering university?", 55% answered yes, while 45% answered no.

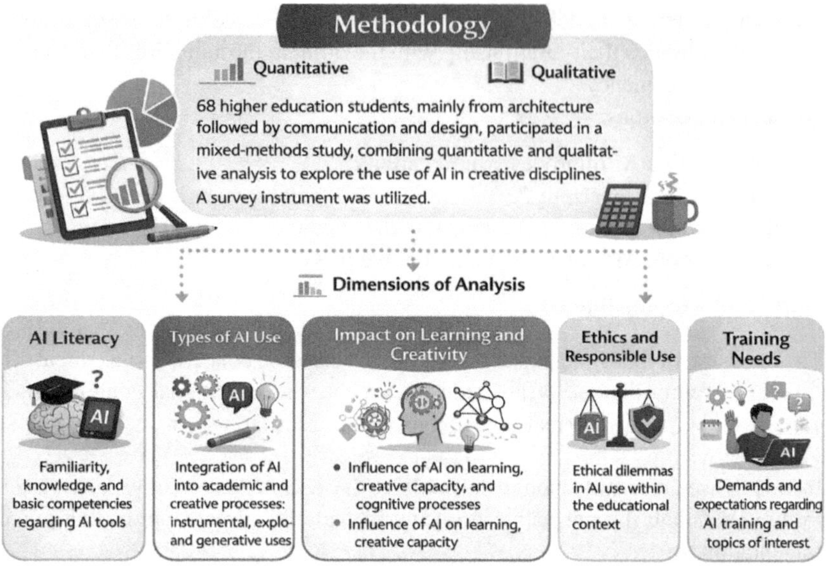

Fig. 3. Proposed methodological model based on five dimensions of analysis

The results show a widespread adoption of AI tools. In terms of frequency of use, the majority, specifically 45% use AI two to three times per week, followed by 24% who use it daily. There is a predominance of use in text-generation platforms. AI is primarily integrated into the ideation and representation stages, functioning as a cognitive catalyst in the early phases of the creative process.

In response to the question about current frequency of AI use, participants answered as follows (Fig. 4):

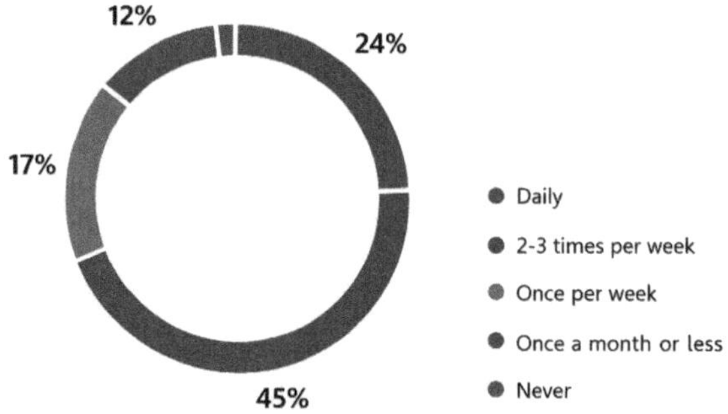

Fig. 4. Responses to the question on current frequency of AI use

The most used AI tools identified are ChatGPT, Claude, and Gemini, with a usage rate of 98%, followed by mobile AI applications at 41%. In response to the question, “Which AI tools do you use? (Select all that apply),” participants answered as follows (Fig. 5):

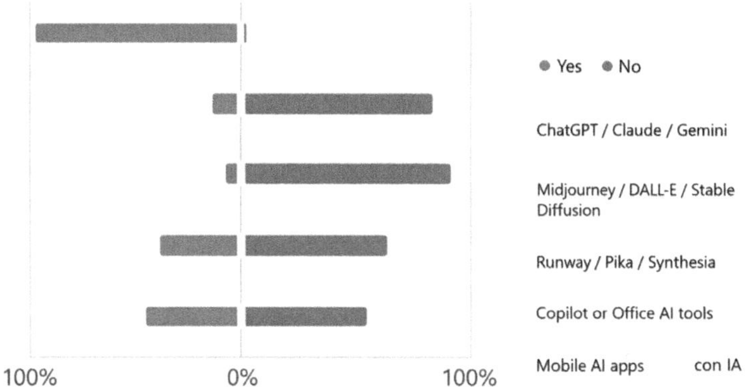

Fig. 5. Responses to the question "Which AI tools do you use"

Students recognize benefits in terms of efficiency and the expansion of creative possibilities. When asked what they use AI for in their studies, the majority (83%) indicated searching for ideas or references, followed by writing or text correction (72%), and analyzing or synthesizing information (67%) (Fig. 6).

Fig. 6. Responses to the question "What do you use AI for in your studies"

In addition to these benefits, students also identify risks associated with technological dependency, such as the loss of skills and superficial learning.

From an ethical perspective, a diversity of positions is observed regarding the use of AI without explicit disclosure, reflecting the absence of clear institutional guidelines. Furthermore, there is a high demand for training, particularly in the critical and professional use of these tools. In response to the question, "Would you like to receive training in AI applied to your discipline?", 60% indicated that they would like to receive practical workshops, followed by 16% who expressed interest in both practical workshops and guidelines for responsible use (Fig. 7).

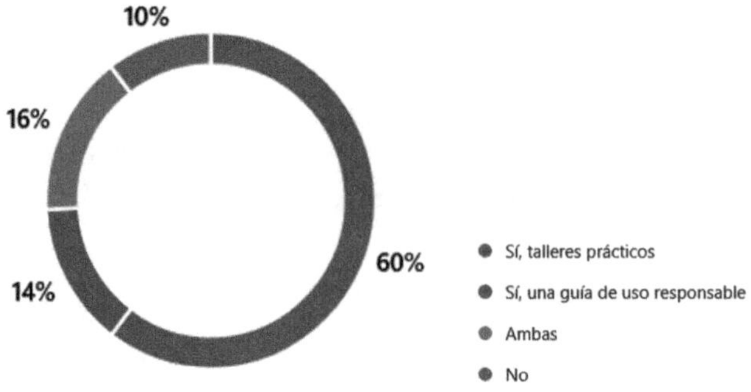

Fig. 7. Responses to the question on training interests related to AI

5 Discussion

The results obtained support the idea that artificial intelligence is not being incorporated by students merely as an auxiliary or external tool, but as a cognitive infrastructure that directly intervenes in processes of searching, organizing, producing, and validating knowledge. This idea is significant because it shifts the understanding of AI from a simple technical resource to a component of the environment from which thinking, designing, and learning currently take place in higher education. In creative disciplines, this transformation is even more significant, as academic work is not limited to the repetition of content but involves ideation, formal exploration, representation, and decision-making. In this sense, AI becomes part of the system of mediations through which students construct responses, test alternatives, and formulate solutions.

From the perspective of Luckin et al. [2] and Holmes et al. [5], AI in education should not be understood exclusively as automation, but as an environment that modifies the conditions of learning and redistributes certain cognitive functions. Likewise, from Csikszentmihalyi's systemic theory of creativity, creative production does not depend on an isolated individual but on the interaction between the subject, the domain, and the field. If we translate this model to the current context, AI appears as a mediation that transforms the relationship between the individual and the domain by expanding access to information, references, languages, and forms of symbolic combination. At the same time, it indirectly modifies the relationship with the field, as it forces a reconsideration of the criteria by which originality, value, and legitimacy are validated within the academic space.

The proposed analytical dimension related to AI literacy showed that a significant portion of students arrive at university with prior experience using AI, suggesting that their relationship with AI does not begin in higher education but precedes it. However, this initial familiarity does not guarantee a deep understanding of how these tools work, their scope, or their limitations. Here, the first central tension of the study becomes visible: the gap between instrumental use and critical understanding.

This gap is reflected in the fact that students frequently use AI, are familiar with its most common applications, and integrate it into specific tasks; however, this does not necessarily imply that they understand how results are produced, what biases these systems operate under, or the epistemological implications of relying on algorithmically generated responses. Questions associated with the dimensions of literacy and types of use allowed for the identification of this difference between operational familiarity and reflective mastery. Knowing how to use a tool is not equivalent to knowing how to interpret it critically. From a pedagogical perspective, this distinction is crucial, as superficial literacy may enhance immediate efficiency but does not ensure solid learning processes or a conscious appropriation of technology.

This tension can be analyzed from Schön's perspective and his notion of reflection-in-action. If, in design thinking, learning occurs while the student experiments, corrects, interprets, and revisits their own decisions, then the excessive automation of certain phases may weaken that reflective moment. Not because AI eliminates creativity, but because it may reduce the time for elaboration, doubt, and contrast that are themselves formative. At this point, the study does not suggest an opposition between AI and learning, but rather the need to distinguish between a use that complements reflection and one that replaces it.

The second identified tension is the ambivalence between creative enhancement and technological dependency. The results show that students perceive AI as particularly useful in stages of ideation, conceptual development, and representation, which was captured in questions such as "At what stage of the creative process do you consider AI most useful?" and in items related to the impact on creativity and learning. This confirms that, in the context of creative disciplines, AI functions as a device for expanding possibilities: it allows the generation of variations, the testing of formulations, the production of visual or textual drafts, and the acceleration of exploratory tasks. From Nigel Cross's perspective, this is consistent with the iterative nature of design thinking: AI primarily strengthens the divergent phases of the process, where the goal is to open up options rather than close them prematurely.

However, this enhancement is not without risks. Responses also reveal concerns about dependency, the loss of individual skills, and the potential superficiality of outcomes. Here, ambivalence emerges: the same tool that expands ideation capacity can, if used without a critical perspective, displace cognitive and creative processes that previously required greater personal elaboration. The dimension of impact on learning and creativity, addressed through questions such as "Do you think AI improves your creativity?", "Does AI facilitate your learning?", and "Do you think the use of AI can generate dependency?", allowed this dual evaluation to be captured. AI is perceived simultaneously as support and as risk. This ambivalence should not be understood as a contradiction, but as a constitutive feature of the phenomenon: technology enhances but also reorganizes the conditions under which thinking occurs.

From Csikszentmihalyi's theory, this can be interpreted as a transformation in the way the individual interacts with the domain. AI does not replace the creative subject, but it does alter the rules of access, combination, and reformulation of cultural elements. Consequently, creativity no longer depends exclusively on the ability to generate something from scratch, but also involves selection, evaluation, curation, and interpretation. In other words, in the current context, creativity is not reduced to producing novelty, but to knowing how to work critically with systems that generate variations at speed. This idea is central to educational discussion: the issue is not that AI generates content, but what kind of skills we aim to develop when generation is no longer an exclusively human activity.

The third identified tension is the disconnection between student practices and institutional regulation. This dimension was mainly examined through questions on ethics and responsible use. The results suggest that students already incorporate AI into their daily work but do so in a context where rules are diffused, uneven, or nonexistent. This generates a regulatory vacuum: practices evolve faster than institutional guidelines capable of guiding them.

This disconnection is particularly significant when considered in relation to UNESCO guidelines [1], which emphasize the need for transparency, accountability, human oversight, and the protection of academic integrity. If institutions do not define clear criteria regarding when, how, and under what conditions AI can be used, regulation shifts to the level of individual interpretation or instructor discretion. This leads to uncertainty: what may be considered legitimate support in one course may be interpreted as unethical in another. The problem, therefore, lies not only in the use of AI, but in the absence of shared frameworks to interpret it.

It is not enough to know that students use AI; it is also necessary to understand how they justify its use, what they consider acceptable, and what they perceive as problematic. The findings show that the boundaries between assistance, co-authorship, technical support, and plagiarism are not fully resolved in students' experiences. This confirms that the discussion of AI in education cannot be reduced to a technical issue but must be framed as an educational and institutional debate.

The dimension of training needs suggests that students not only use these tools but also recognize the need for more structured training to use them appropriately. This finding reinforces the idea that universities cannot simply prohibit or tolerate AI use; they must take on a role in developing critical competencies. At this point, the three identified tensions converge: the lack of deep understanding, the ambivalence between support and dependency, and the absence of clear regulation all point toward the same need the development of educational models capable of integrating AI from a critical, ethical, and reflective perspective.

The results confirm that AI reconfigures both production processes and forms of learning in creative higher education. However, this reconfiguration is not neutral. It affects students' relationship with knowledge, modifies the timing and modes of creativity, and challenges traditional categories such as authorship, originality, and evaluation. Therefore, in line with international guidelines [1, 5], the educational response should move toward an AI literacy that goes beyond technical use and includes critical understanding, ethical reflection, and clear criteria for academic use, in order to integrate these technologies without undermining the formative processes aimed at strengthening the capacity to think, create, and make decisions in complex contexts.

6 Conclusions

Artificial intelligence emerges in the case study as a key component within the educational environment. Its integration into creative disciplines not only transforms the available tools but also directly impacts on the epistemological foundations of learning by reconfiguring the relationship between students, knowledge, and creative production processes. In this sense, AI should not be understood merely as a technical resource, but rather as a form of mediation that reshapes the conditions under which learning, creation, and evaluation take place within the School of Architecture, Design, and Communication.

The results confirm that students have widely incorporated AI into their academic practices, particularly in the stages of ideation, conceptual development, and representation. However, this appropriation is predominantly instrumental in nature. This finding highlights the need to strengthen critical literacy processes, enabling students not only to use these tools, but also to understand their underlying logics, biases, and implications for knowledge construction.

Within this framework, the educational challenge does not lie in restricting the use of AI, but in developing pedagogical models that foster a reflective, informed, and ethical engagement with these technologies. This involves, on the one hand, the design of teaching strategies that integrate AI into learning processes, and, on the other, the establishment of clear institutional guidelines that regulate its use, promote transparency, and define the boundaries between technological assistance, co-authorship, and academic misconduct.

This study presents several limitations that should be acknowledged. First, the sample size and composition, focused on a single institution and predominantly on architecture students, limit the generalizability of the findings to other disciplinary and geographical contexts. Second, the use of a survey-based instrument implies reliance on self-reported data, which may be affected by social desirability bias and subjective interpretation. Furthermore, the study is situated within a rapidly evolving technological context, meaning that the patterns of use and perception identified may change in the short term as new tools and practices emerge.

Considering these limitations, several avenues for future research can be identified. First, comparative studies across different disciplines would allow for a more nuanced understanding of how AI is appropriated in diverse educational contexts. Second, the incorporation of qualitative methods, such as interviews and focus groups, would enable a deeper exploration of actual practices and cognitive strategies. Third, further research is needed on the implications of AI for academic assessment, particularly regarding authorship, originality, and evaluation criteria.

Finally, a priority line of work lies in the design and implementation of pedagogical approaches that integrate AI from a critical and situated perspective. The aim is to assess its impact on the development of creative, reflective, and ethical competencies. In this regard, the challenge is not merely to adapt to the presence of AI, but to develop educational models capable of redefining creativity and learning in contexts increasingly mediated by intelligent systems.

Acknowledgments. This research has been funded by the Research Department of La Salle University Cuernavaca because of a research call approved by the Scientific and Advisory Research Council.

Disclosure of Interests The author declares that they have no competing interests relevant to the content of this article.

References

1. UNESCO: Recommendation on the Ethics of Artificial Intelligence. UNESCO, Paris (2021)
2. Luckin, R., Holmes, W., Griffiths, M., Forcier, L.B.: Intelligence Unleashed: an Argument for AI in Education. Pearson (2016)
3. Holmes, W., Bialik, M., Fadel, C.: Artificial Intelligence in Education: Promises and Implications for Teaching and Learning. Center for Curriculum Redesign (2019)
4. Csikszentmihalyi, M.: Creativity: flow and the psychology of discovery and invention. HarperCollins (1998)
5. Holmes, W., Persson, J., Chounta, I.A., Wasson, B., Dimitrova, V.: Artificial Intelligence and Education: a critical view through the lens of human rights. Democracy and the Rule of Law, Council of Europe (2022)
6. Area, M.: Digital literacy and the formation of 21st-century citizenship. Revista de Education a Distancia (2012)
7. Cross, N.: Design Thinking: understanding how designers think and work, Berg (2011)
8. Clark, A., Chalmers, D.: The extended mind. Analysis. **58**(1), 7–19 (1998)

Detection of *Struthanthus Interruptus* Mistletoe in Trees in Mexico City Using Multispectral UAV Images and Convolutional Neural Networks

Angel Gerardo Luna-Romo[1](✉), León Felipe Dozal-García[2], Paola Andrea Mejía-Zuluaga[3], Juan C. Valdiviezo-Navarro[4], Francisco Javier Álvarez-Rodríguez[1], and Mauricio Gabriel Orozco-del-Castillo[5]

[1] Universidad Autónoma de Aguascalientes, Av. Universidad 940, Aguascalientes, Mexico
al322828@edu.uaa.mx

[2] Centro de Investigación en Ciencias de Información Geoespacial, Circuito Tecnopolo Norte, No.107, Aguascalientes, Mexico
leon.dozal@gmail.com

[3] Centro de Investigación en Ciencias de Información Geoespacial, C. Contoy 137, Tlalpan, Ciudad de México, Mexico

[4] SECIHTI - Centro de Investigación en Ciencias de Información Geoespacial, Mérida 97302, Yucatán, Mexico
jvaldiviezo@centrogeo.edu.mx

[5] Tecnológico Nacional de México/IT de Mérida, Av. Tecnológico km. 4.5 S/N, 97118 Mérida, Yucatán, Mexico

Abstract. The excessive proliferation of mistletoe on trees in the urban area of Mexico City is becoming a significant environmental problem. This study presents a deep learning approach for the automatic detection of mistletoe using multispectral aerial imagery. A review of related work revealed several studies on automated mistletoe detection, most of them using classical machine learning techniques on RGB images of forested areas. Four well-known CNN architectures were implemented and tested. The results showed that models such as U-Net and SegNet correctly classified more pixels as mistletoe, while PSPNet and DeepLabv3+ showed fewer false positive detections, likely due to the benefits of the pretrained Resnet-50 architecture and its more advanced feature extraction mechanisms. Furthermore, visual analysis revealed that all four models consistently identified regions not labeled as mistletoe in the original truth mask, this discrepancy may indicate the presence of annotation inaccuracies within the dataset. On the other hand, U-Net and SegNet exhibited similar errors when classifying yellow patterns as mistletoe. Furthermore, they achieved greater accuracy compared to the other models. However, the results of all four models were comparable to those obtained by human experts.

H. Ponce et al. (Eds.): COMIA 2026, CCIS 3095, pp. 51–63, 2027.
https://doi.org/10.1007/978-3-032-37438-7_5

Keywords: mistletoe · *Struthanthus interruptus* · CNN · semantic segmentation · UAV · multispectral images

1 Introduction

Mistletoe is the common name for a hemiparasitic plant that infests trees and shrubs, establishing a parasitic relationship with them. Mistletoe extracts water and minerals directly from the host tree through structures called haustoria [19]. As it grows, mistletoe takes over the space occupied by the host's foliage and increases its susceptibility to fungal and bacterial attacks [3]. Other effects on the host tree include slow growth, water stress, reduced flowering and fruiting, decline and death [5]. Although mistletoe plays an important ecological role by contributing to biodiversity and providing food resources for various species, its excessive proliferation poses a significant problem for tree health, which in urban environments already faces stress caused by pollution, soil compaction, and water scarcity [18]. At the present, the distribution and impact of mistletoe infestation in urban trees in Mexico City are unknown. This information could help environmental management authorities plan and make decisions about urban tree maintenance. However, several studies over time have revealed that the infestation has increased in various locations in Mexico City, as have the number of host tree species infested with mistletoe [1,11,21].

It is clear that, to monitor mistletoe infestation, determine the affected area, plan phytosanitary control, and measure the effectiveness of phytosanitary methods, such as herbicides and pruning, it is necessary to identify and locate mistletoe first. Environmental authorities face significant limitations in addressing this situation, given that the most common phytosanitary treatment, manual pruning, is time-consuming, requires trained personnel, and generates high operating costs, preventing efficient large-scale intervention [6,12]. Monitoring the phytosanitary status of urban trees provides valuable information for government agencies responsible for environmental management in the city. However, it is a complex task that must be performed by people capable of identifying pests or signs of stress in trees. The traditional method is visual identification *in situ*, although ground surveys are time-consuming and expensive. A much less used method among environmental authorities is the visual interpretation of aerial images obtained by unmanned aerial vehicles (UAVs). This contrasts with the results of a recent review of 16 academic studies on mistletoe detection using airborne remote sensing [14] published between 2007 and 2023, revealing relevant findings, such as nine studies (56%) used UAVs; seven (44%) used the visual interpretation of aerial RGB images as an analysis method and also seven used image classification algorithms, including those to automatically detect mistletoe in aerial RGB images.

In 2020, Sabrina et al. [17] reviewed existing research on the detection of mistletoe and other infestations through Machine Learning (ML) techniques, and highlighted the need for an automated mistletoe detection method. In recent years, various ML approaches have been used for the automatic detection of mistletoe in UAV-acquired RGB images of trees in forests areas. In 2021, Mojdeh Miraki et al. [13] calculated several vegetation spectral indices (VI) using the R, G, and B spectral bands from aerial images acquired in winter and summer 2020 in a forest in Iran. Then, they used the Random Forest approach, together with GINI to obtain the importance of input variables, to delineate tree crowns and to detect mistletoe (*Viscum album*) infected trees. They obtained a 76% overall accuracy (OA) and increased to 87% OA when trees crowns were manually delineated. They found that VIs were more effective than raw RGB bands, specifically Woebbecke Index (WI) [22] and Excess Green Index (EXG) [20]. One year later, Mejia-Zuluaga et al. [9] proposed a Genetic Programming (GP) approach to find an arithmetic combination of five spectral bands of aerial images, RGB, red edge and near-infrared, of a forestry conservation area to highlight *Phoradendron velutinum* they obtained a highly non-linear algorithm depending on three bands, with 96% of overall accuracy (OA) detecting mistletoe, overfitting support vector machine (SVM).

Surprisingly, after a literature review, we were unable to find any studies using Convolutional Neural Networks (CNN) to detect mistletoe in optical images acquired by a UAV. However, CNNs have revolutionized vegetation remote sensing by enabling efficient, accurate and flexible analysis of spatial patterns in high-resolution imagery. Furthermore, the modularity of common CNN architectures allows adaptation to various remote sensing problems involving complex vegetation structures [7]. For these reasons, we conducted a comparative study of four deep CNN architectures evaluated in the semantic segmentation of *Struthanthus interruptus* mistletoe using multispectral images, composed of red (R), green (G), blue (B), red edge (RE) and near infrared (NIR) spectral bands, acquired by a UAV in Mexico City.

The main contributions of our study are the following. Our study is a pioneer in mistletoe detection using deep CNN on aerial images acquired by UAVs. Also, according to [14] most of the studies on airborne remote sensing for mistletoe detection used RGB and hyperspectral images, instead we used multispectral images to explore their potential. In addition, whereas most mistletoe detection studies have been conducted in forest environments, our study focuses on an urban scenario, which is important for the ecological sustainability of cities but remains largely neglected.

2 Materials and Methods

This section describes the materials and methods used in both the development and evaluation of the convolutional neural network models used for mistletoe detection.

2.1 Dataset

The database used in this work is composed of high-resolution multispectral aerial images acquired in green areas of Mexico City, where the presence of the mistletoe *Struthanthus interruptus* is observed. Data acquisition was carried out independently in different years in order to capture different phenological conditions. Image capture was performed using a P4 Multispectral drone equipped with a five-band sensor: B (450 nm ± 16nm), G (560 nm ± 16nm), R (650 nm ± 16nm), RE (730 nm ± 16nm), NIR (840 nm ± 26 nm).

The RE and NIR bands are particularly relevant for vegetation analysis, as they are sensitive to chlorophyll content and plant physiological state. This allows for better discrimination between healthy and mistletoe-infested tree regions, which tend to exhibit distinct spectral signatures due to differences in water content, pigmentation, and canopy structure. The database, at the time of this study, is composed of 1300 multispectral tiles with their corresponding binary mask and RGB image for each tile. The collection is actively maintained and intended to expand as new data becomes available [10].

Geometric and radiometric corrections were applied to correct lens distortion and ensure spatial correspondence between pixels of five spectral bands, as well as to reduce illumination variability and improve radiometric consistency. After that, the images were divided into tiles of 250 × 250 pixels to facilitate the assessment of ML models. The database includes pixel-level segmentation masks indicating the presence or absence of *Struthanthus interruptus* mistletoe in each tile. These masks were generated through manual annotation based on visual interpretation of multispectral information. The resulting set of co-registered multispectral tiles and corresponding segmentation masks enables supervised training of detection models under varying conditions. Figure 1 shows four examples of images and their respective masks.

2.2 Image Preprocessing

The images were divided into 5 groups of 260 images to perform a 5-fold cross-validation, using the same groups across all experiments. Then, all images were resized from 250 × 250 to 256 × 256 pixels to ensure compatibility with CNN architectures during the down-sampling operation. The resize process was performed using OpenCV 4.12.0, applying bilinear interpolation to the multispectral images and nearest-neighbor interpolation to the segmentation masks in order to preserve discrete class labels and avoid the introduction of intermediate values, having a total of 18,952,839 pixels labeled mistletoe and 65,118,045 pixels labeled as background, corresponding to a class imbalance ratio of approximately 0.297. For each iteration, four folds were used as the training set and the remaining fold as the external test set. Within the training folds, a random internal split of 80% for training and 20% for validation, generated once and kept constant throughout the experiments, was applied to enable early stopping and hyperparameter tuning. This division resulted in 832 images for training, 208 images for validation, and 260 for testing.

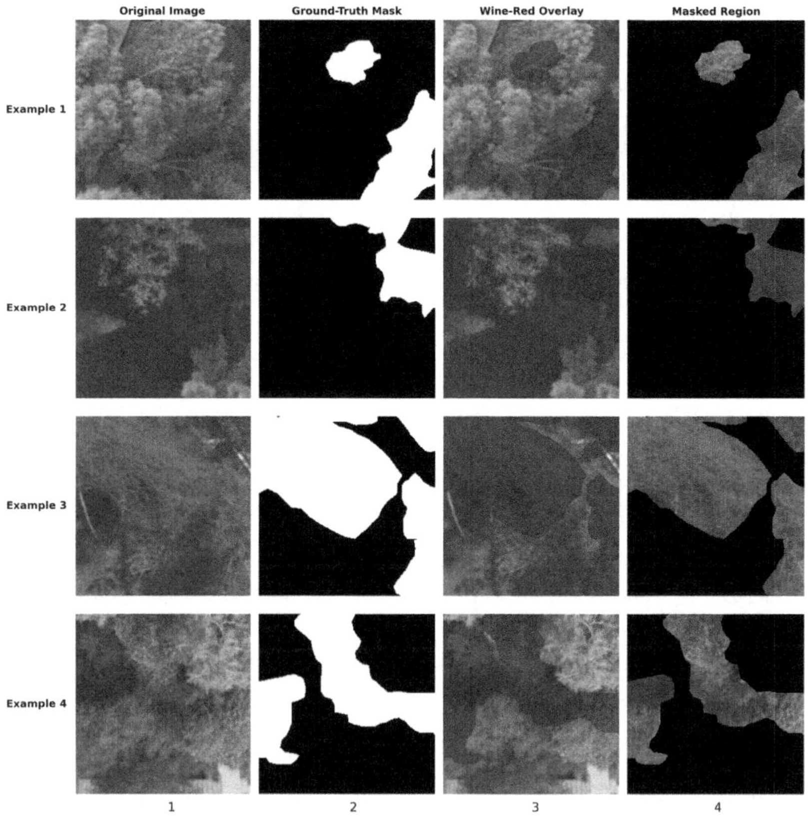

Fig. 1. Four examples from the dataset with ground-truth segmentation masks and their corresponding overlays. Each columns presents: (1) the original aerial image tile, (2) the binary ground-truth mask where white regions indicate the mistletoe in the image, (3) a wine-red overlay on the original image to illustrate the segmented region, and (4) the isolated masked region displayed against a black background. (Color figure online)

2.3 Deep CNN Architectures

In this study, four CNN encoder-decoder architectures including U-Net, SegNet, PSPNet, and DeepLabv3+ with a ResNet-50 backbone were evaluated for mistletoe semantic segmentation, that is, the detection of mistletoe (*Struthanthus interruptus*) in Mexico City. We chose U-Net and SegNet architectures because they are frequently applied for semantic segmentation of vegetation in remote sensed data [7]. U-Net follows two paths in a U-shaped design, first, the encoder contracts the image using a set of convolution and pooling operations; second, the decoder expands the image map to its original resolution using transposed convolutions. We also used the skip connections option between the encoder and decoder, which helps to preserve fine spatial details and improves boundary localization [16]. Similarly, SegNet uses convolutions and max-pooling operators in the encoder and un-pooling operations in the decoder, but SegNet stores the indices of pooling operations to use them to locate the features in the upsampled image map, enabling efficient reconstruction of object structures while reducing memory demand [2]. Table 1 shows the details of the U-Net and SegNet architectures used in this study, which had 17.2M and 29.4M parameters, respectively.

Table 1. Details of U-Net and SegNet architectures.

Part	U-Net (bilinear)		SegNet	
	Layer	Kernel No.	Layer	Kernel No.
Encoder	2 × CB	64	2 × CB + MP	64
	MP + 2 × CB	128	2 × CB + MP	128
	MP + 2 × CB	256	3 × CB + MP	256
	MP + 2 × CB	512	3 × CB + MP	512
	MP + 2 × CB	512	3 × CB + MP	512
Decoder	BU + Concat + 2 × CB	256	Up + 3 × CB	512
	BU + Concat + 2 × CB	128	Up + 3 × CB	256
	BU + Concat + 2 × CB	64	Up + 3 × CB	128
	BU + Concat + 2 × CB	64	Up + 2 × CB	64
	Conv(1 × 1)	1	Up + CB + Conv(3 × 3)	1

For both architectures, **CB** denotes a convolutional block (Conv 3×3 + Batch Normalization + ReLU), and **MP** denotes max pooling with a kernel size and stride of 2. For U-Net, **BU** denotes bilinear upsampling with a scale factor of 2, and **Concat** denotes the skip connection concatenation. For SegNet, **Up** denotes max unpooling with a kernel size and stride of 2.

We also chose PSPNet and DeepLabv3+ with pretrained ResNet-50 backbone with the aim of obtaining performance data to contrast with. PSPNet introduces pyramid pooling to aggregate multi-scale contextual information, improving robustness in scenes with heterogeneous vegetation patterns [23]. DeepLabV3+ combines spatial pyramid pooling with a decoder module to capture both global context and fine spatial detail, making it effective for complex vegetation segmentation tasks [4]. Both architectures are bigger because of the backbone, they have 46.5M and 39.2M parameters, respectively.

The experiments were conducted on a Lenovo IdeaPad Gaming 3 laptop equipped with an NVIDIA GeForce RTX 4050 Laptop GPU (6 GB GDDR6 VRAM), an AMD Ryzen 7 processor, 16 GB of DDR5 system RAM, and 1.5 TB of NVMe SSD storage. The proposed architectures were implemented in PyTorch [15] 2.9.0, running on Python 3.13.5 with CUDA 12.6 support. Image preprocessing and data handling were performed using OpenCV 4.12.0.

All CNNs were trained using Binary Cross Entropy as the loss function, with a `pos_weight` of 2 to account for class imbalance. Optimization was performed using Adam with an initial learning rate of 4×10^{-4}, determined empirically, and adjusted during training via a `ReduceLROnPlateau` scheduler with a reduction factor of 0.5 and a patience of 8 epochs for U-Net and 5 epochs for the others. Early stopping was applied when validation performance did not improve over 20 consecutive epochs, with a maximum of 150 epochs, to prevent overfitting, and the model checkpoint with the highest IoU on the validation set was selected for the final evaluation. All models were trained with a batch size of 10 images of 256×256 pixels. For the final network output, a sigmoid activation function was applied and the resulting probability map was thresholded at 0.5, classifying pixels with values greater or equal than this threshold as mistletoe (1) and those below as background (0). PSPNet and DeepLabv3+ used a ResNet-50 backbone initialized with `ResNet50_Weights.IMAGENET1K_V2` pretrained weights, while U-Net and SegNet were trained from scratch. Table 2 shows the hyperparameters used during the experiments. After training of the CNNs, we obtained four models for automatic detection of mistletoe, then we assessed their performance using six standard and well-known metrics: OA, precision, recall, F1-score and Intersection over Union (IoU) as defined in [8], and Matthews Correlation Coefficient (MCC), using the values of the binary confusion matrix generated by comparing the model predictions against the actual masks.

Table 2. Hyperparameters used during the experiments.

Hyperparameters	Values
Loss function	BCEWithLogits(pos_weight = 2)
Optimizer	Adam
Learning rate scheduler	ReduceLROnPlateau
Threshold	0.5
Limit epoch	150
Epoch for early stopping	20
Batch size	10
Image size	256×256
Evaluation metrics	Loss, Accuracy, Precision, Recall, F1-score, IoU, MCC
Initial learning rate	4×10^{-4}

The performance of the four implemented architectures: U-Net, SegNet, PSPNet, and DeepLabv3+, was evaluated using metrics derived from the binary confusion matrix, computed by comparing the ground-truth segmentation masks against the model predictions. Specifically, the following metrics were considered for the evaluation: OA, precision, recall, F1-score, MCC and IoU, the latter being the primary metric for model selection, as it provides a robust measure of segmentation quality by accounting for both false positives and false negatives and its relevance to semantic segmentation problems.

3 Results and Discussion

The four models showed similar performance as shown in Table 3, with DeepLabv3+ achieving the highest scores across most metrics including OA (85.41%), precision (67.31%), F1-score (68.66%), MCC (0.5918), and IoU (52.33%), followed closely by PSPNet with an IoU of 51.91% and an F1-score of 68.31%. U-Net and SegNet showed competitive but slightly lower performance, with IoU values of 51.31% and 49.95%, respectively.

Regarding recall, U-Net obtained the highest value (72.68%), suggesting a greater ability to detect positive instances compared to the ResNet-based models, which achieved slightly lower recall despite their higher precision. This precision-recall trade-off indicates that U-Net and SegNet exhibit a higher sensitivity to the target class than PSPNet and DeepLabv3+ that have better precision.

Table 3. Mean of the metrics in the 5-fold for all architectures.

Model	Overall accuracy (%)	Precision (%)	Recall (%)	F1-Score (%)	MCC	IoU (%)
U-Net	84.23	63.56	72.68	67.77	0.5765	51.31
SegNet	83.68	62.78	71.07	66.60	0.5609	49.95
PSPNet (ResNet50)	85.10	66.42	70.38	68.31	0.5863	51.91
DeepLabv3+ (ResNet50)	85.41	67.31	70.13	68.66	0.5918	52.33

To assess the computational efficiency of the models, the training and inference times were measured across all five folds, with the average values reported in Table 4. Training time includes both training and validation phases over the four training folds per run, while inference time corresponds to the average over the five held-out test folds. As shown, U-Net and SegNet were slower in both training and inference time, as they were trained from scratch and required more epochs to converge. In contrast, PSPNet and DeepLabv3+ benefited from a pretrained ResNet-50 backbone, which not only accelerated convergence but also resulted in faster inference times, likely due to the highly optimized feature extraction of the ResNet architecture.

Table 4. Average processing time for each CNN architecture.

Architecture	Training time (mm:ss)	Inference time (s)
U-Net	48:02	4.73
SegNet	41:09	4.18
PSPNet (ResNet50)	08:30	1.38
DeepLabv3+ (ResNet50)	09:16	1.49

Regarding computational efficiency, training and inference times were recorded for all five folds, and the averages are summarized in Table 4. U-Net and SegNet exhibited the longest training and inference times, which can be attributed to the fact that both models were trained from scratch, requiring a greater number of epochs to converge. They were modified to use 5 multispectral bands and perform more complex operations in upsampling or concatenation stages. In contrast, PSPNet and DeepLabv3+ used a ResNet-50 backbone architecture for both with pretraining, which result in accelerated convergence during training and contributed to faster inference times, furthermore, the fact that this backbone architecture is specially optimized for modern GPUs may help explain why these models achieved shorter processing times despite having a greater number of parameters.

We used U-Net and SegNet because they are well suited for training from scratch on small datasets, as their architectures were originally designed for scenarios with limited annotated data. Unlike many modern segmentation models that rely on a pretrained backbone, U-Net and SegNet feature a fully integrated encoder-decoder design, where both components are trained jointly from scratch. In contrast, PSPNet and DeepLabv3+ rely on a CNN backbone architecture (in this case, ResNet50) that requires pretraining to achieve competitive performance; they were included to enable a comparison between these two approaches. The preliminary experiments conducted in this study confirmed that the training of these models from scratch yielded results inferior to those of U-Net and SegNet, suggesting that without pretrained weights, ResNet-based architectures do not generalize effectively on small datasets.

Visual analysis revealed notable differences in the behavior of the evaluated models. As shown in the second and third rows of Fig. 2, both U-Net and SegNet exhibited similar error patterns, frequently detecting objects that were clearly not mistletoe. In several cases, these models appeared to confuse background elements with yellowish tones with mistletoe regions, leading to false positive detections. In contrast, such errors were less frequent in the predictions generated by PSPNet and DeepLabv3+. This difference may be attributed to the higher representational capacity of these architectures and the use of transfer learning from ImageNet, which likely provided more robust feature representations and improved discrimination between mistletoe and visually similar background elements. In other cases, such as the first row of Fig. 2, the four models consistently identify the same regions as mistletoe. However, these regions are not labeled as mistletoe in the original ground-truth mask. This discrepancy may indicate the presence of annotation inaccuracies within the dataset. Alternatively, it may suggest that additional data augmentation or improved labeling strategies could help the models better generalize and align predictions with the ground-truth annotations. The fourth row of Fig. 2 further confirms that SegNet, and particularly U-Net, achieve higher recall compared to the other models, as they detect a larger number of mistletoe regions within the image. Finally, as illustrated in the fifth row of Fig. 2, the models are capable of producing segmentation results that are visually comparable to those obtained by a human expert.

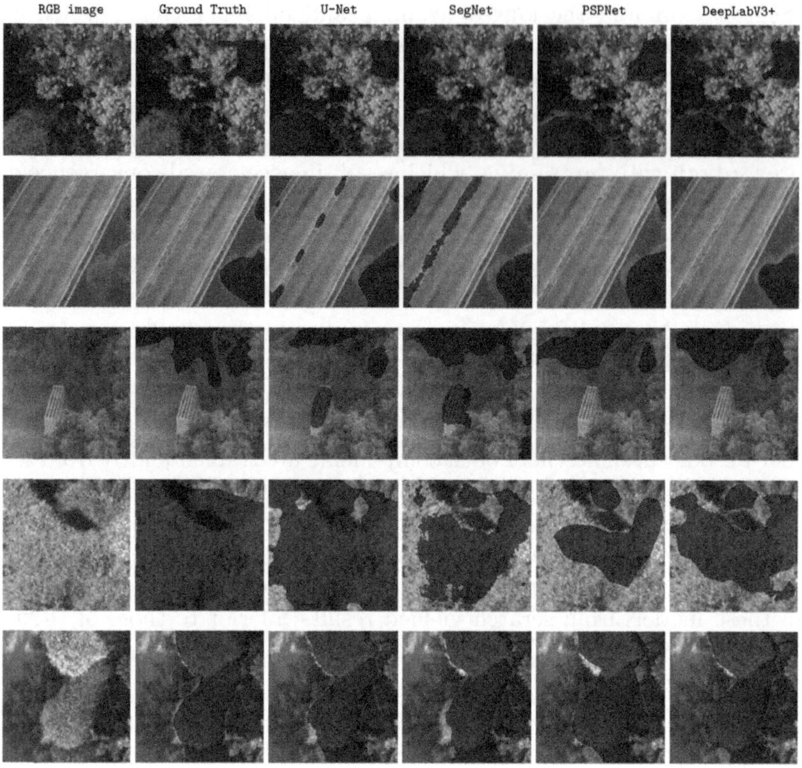

Fig. 2. Illustrative examples of the prediction results for the models.

4 Conclusions

This study evaluated the effectiveness of CNN architectures for the semantic segmentation of mistletoe in multispectral UAV imagery of urban trees in Mexico City. The results demonstrate that deep learningbased models can accurately identify mistletoe infested regions, even in images that appear challenging to humans, highlighting the potential of semantic segmentation techniques to support urban vegetation monitoring. During the conduct of this research, we identified a lack of information on automation for mistletoe detection and a limited application of artificial intelligence to this problem; therefore, we believe that this work is particularly significant for expanding knowledge in this area.

Among the evaluated architectures, differences in performance were observed in terms of recall, precision, and specifically IoU, showing that models such as U-Net and SegNet tended to detect a greater number of pixels classified as mistletoe, while PSPNet and DeepLabv3+ showed fewer false positive detections, likely due to the benefits of transfer learning from ImageNet and their more advanced feature extraction mechanisms.

A limitation of this study is the absence of a comparison with traditional machine learning methods, such as SVM or Random Forest, which have been applied to vegetation mapping, remote sensing tasks, and even mistletoe detection in other species, achieving in some cases competitive performance relative to deep learning approaches. However, studies specifically addressing the detection of *Struthanthus interruptus* using such methods remain scarce. As evidenced by the false positive rates observed in this study, this species poses notable detection challenges, likely due to its spectral similarity to surrounding vegetation, which makes a direct comparison with existing ML benchmarks difficult and underscores the relevance of this line of inquiry as a direction for future work. Including such comparisons in subsequent studies would provide a more comprehensive evaluation framework and help establish the relative advantages of deep learning methods for this task. Beyond traditional ML methods, future work could also explore alternative neural network architectures beyond CNNs, such as transformer-based models, which have demonstrated strong performance in semantic segmentation tasks due to their ability to capture global contextual information, a capability that could prove particularly beneficial for detecting spatially distributed mistletoe infestations.

To address false positive detections caused by yellowish background elements and improve upon the results obtained here, several data augmentation strategies could be explored. On one hand, geometric and photometric transformations (such as rotation, flipping, noise injection, and brightness adjustment) could be applied to increase sample diversity. On the other hand, explicitly incorporating examples of non-mistletoe vegetation with similar spectral characteristics, as well as images of uninfested tree regions, would help the model better discriminate between mistletoe and visually confusable elements. Collecting and annotating images across different seasons would further allow for the construction of a more phenologically diverse dataset, improving the model's ability to distinguish mistletoe from similar-looking foliage under varying lighting and vegetation conditions. Finally, expanding the dataset to include a greater variety of host tree species could enhance model generalizability, given that canopy structure, leaf morphology, and spectral response may vary considerably across species and directly affect detection performance.

This work has considerable room for improvement. Testing alternative CNN architectures such as UNet++, employing different backbone networks such as ResNet101 or MobileNet, or experimenting with different hyperparameter configurations for each model could enhance the segmentation metrics obtained here. Additionally, incorporating the methodological directions discussed above including transformer-based architectures, traditional ML comparisons, and a more diverse and seasonally representative dataset, could collectively advance the development of more robust and reliable automated tools for monitoring and controlling mistletoe infestations in urban environments.

Acknowledgments. The authors would like to thank to *Secretaría de Educación, Ciencia, Tecnología e Innovación* (SECTEI) of Mexico City for providing the founds to develop and publish the present study as part of the activities of the project "Monitoreo de especies de Muérdago en Parques Urbanos Usando Imágenes Aéreas e Inteligencia Artificial" CAR SECTEI/156/2024. We also thank to *Secretaría del Medio Ambiente* (SEDEMA) of Mexico City for the valuable support during the image acquisition process of mistletoe species.

References

1. Alvarado-Rosales, D., Saavedra-Romero, L.d.L.: Tree damage and mistletoe impact on urban green areas. Revista Árvore **49**(1) (2021). https://doi.org/10.1590/1806-908820210000030
2. Badrinarayanan, V., Kendall, A., Cipolla, R.: Segnet: a deep convolutional encoder-decoder architecture for image segmentation. IEEE Trans. Pattern Anal. Mach. Intell. **39**(12), 2481–2495 (2017). https://doi.org/10.1109/TPAMI.2016.2644615
3. Boussim, I., Guinko, S., Tuquet, C., Sallé, G.: Mistletoes of the agroforestry parklands of burkina faso. Agrofor. Syst. **60**, 39–49 (2004). https://doi.org/10.1023/B:AGFO.0000009403.36419.20
4. Chen, L., Zhu, Y., Papandreou, G., Schroff, F., Adam, H.: Encoder-decoder with atrous separable convolution for semantic image segmentation. CoRR arxiv:1802.02611 (2018)
5. Coder, K.D.: American mistletoes: kissing under a parasite. Arborist News **13**(6), 37–44 (2004)
6. Espinoza-Zúñiga, P., Lara-Vázquez, F., Cibrián-Tovar, D., Ruiz-Orta, A., Galacho-Jiménez, F.B., Ramírez Dávila, J.F.: Variabilidad espacial de muérdago (loranthaceae) en áreas verdes de la alcaldía coyoacán, ciudad de méxico. Revista Mexicana de Ciencias Forestales **15**(85) (2024). https://doi.org/10.29298/rmcf.v15i85.1469
7. Kattenborn, T., Leitloff, J., Schiefer, F., Hinz, S.: Review on convolutional neural networks (cnn) in vegetation remote sensing. ISPRS J. Photogramm. Remote. Sens. **173**, 24–49 (2021). https://doi.org/10.1016/j.isprsjprs.2020.12.010
8. Lobo Torres, D., et al.: Applying fully convolutional architectures for semantic segmentation of a single tree species in urban environment on high resolution uav optical imagery. Sensors **20**(2) (2020). https://doi.org/10.3390/s20020563
9. Mejia-Zuluaga, P.A., Dozal, L., Valdiviezo-N., J.C.: Genetic programming approach for the detection of mistletoe based on uav multispectral imagery in the conservation area of mexico city. Remote Sensing **14**(3) (2022).https://doi.org/10.3390/rs14030801
10. Mejia Zuluaga, P.A., Valdiviezo Navarro, J., Dozal, L., Flores-Gallegos, E., Olvera-González, E.: UAV multispectral image dataset for urban mistletoe (*struthanthus interruptus*) detection — Mexico city (2026). https://doi.org/10.6084/m9.figshare.31245229.v1
11. Mendoza-Ruíz, E.E., Dozal-García, L.F., Mejía-Zuluaga, P.A., Escalante-García, N., Orozco-del Castillo, M.G., Valdiviezo-Navarro, J.C.: Evaluación y caracterización de la plaga de muérdago en el arbolado de parques urbanos en la ciudad de méxico. Revista Chapingo Serie Ciencias Forestales y del Ambiente (2026). in press
12. Meza Aguilar, M.D.C.: Los árboles de la ciudad de méxico. guardianes de su imagen y calidad ambiental. Bitácora Arquitectura (31), 96–102 (2015). https://doi.org/10.22201/fa.14058901p.2015.31.56652

13. Miraki, M., Sohrabi, H., Fatehi, P., Kneubuehler, M.: Detection of mistletoe infected trees using uav high spatial resolution images. J. Plant Dis. Protect. **128**, 1679–1689 (2021). https://api.semanticscholar.org/CorpusID:237644024
14. Missarov, A., et al.: Vertical botany: airborne remote sensing as an emerging tool for mistletoe research. Botany **102**(3), 58–71 (2024). https://doi.org/10.1139/cjb-2023-0049
15. Paszke, A., et al.: Pytorch: an imperative style, high-performance deep learning library. CoRR arxiv:1912.01703 (2019)
16. Ronneberger, O., Fischer, P., Brox, T.: U-net: convolutional networks for biomedical image segmentation. In: Navab, N., Hornegger, J., Wells, W.M., Frangi, A.F. (eds.) MICCAI 2015. LNCS, vol. 9351, pp. 234–241. Springer, Cham (2015). https://doi.org/10.1007/978-3-319-24574-4_28
17. Sabrina, F., Sohail, S., Thakur, S., Azad, S., Wasimi, S.: Use of deep learning approach on uav imagery to detect mistletoe infestation. In: 2020 IEEE Region 10 Symposium (TENSYMP), pp. 556–559 (2020).https://doi.org/10.1109/TENSYMP50017.2020.9230971
18. Skrypnik, L., Maslennikov, P., Feduraev, P., Pungin, A., Belov, N.: Specific features of the response of the antioxidant system of urban trees to mistletoe infection. E3S Web Conf. **291**, 02013 (2021). https://doi.org/10.1051/e3sconf/202129102013
19. Tennakoon, K., Pate, J.: Effects of parasitism by a mistletoe on the structure and functioning of branches of its host. Plant Cell Environ. **19**, 517–528 (1996). https://doi.org/10.1111/j.1365-3040.1996.tb00385.x
20. Torres-Sánchez, J., Peña, J., de Castro, A., López-Granados, F.: Multi-temporal mapping of the vegetation fraction in early-season wheat fields using images from uav. Comput. Electron. Agric. **103**, 104–113 (2014). https://doi.org/10.1016/j.compag.2014.02.009
21. Valencia, D.M.: El muérdago en la ciudad de méxico (2009)
22. Woebbecke, D., Meyer, G., Von Bargen, K., Mortensen, D.: Shape features for identifying young weeds using image analysis. Trans. Am. Soc. Agric. Eng. **38**(1), 271–281 (1995). https://doi.org/10.13031/2013.27839
23. Zhao, H., Shi, J., Qi, X., Wang, X., Jia, J.: Pyramid scene parsing network. CoRR arxiv:1612.01105 (2016)

From Graphite to Prompt: Literacy, Workflow, and Perceptions Among Industrial Design Students

Angel Gómez-Navarro[1], Juan Villegas Cortez[2](✉), Marco-Vinicio Ferruzca-Navarro[3], and Cayetano-José Cruz-García[4]

[1] Department of Medio Ambiente, Universidad Autónoma Metropolitana, Azcapotzalco Unit, Mexico City, Mexico
angona@azc.uam.mx

[2] Department of Systems, Universidad Autónoma Metropolitana, Azcapotzalco Unit, Mexico City, Mexico
juanvc@azc.uam.mx

[3] Departamento de Investigación y Conocimiento del Diseño, Universidad Autónoma Metropolitana, Azcapotzalco Unit, Mexico City, Mexico
mvfn@azc.uam.mx

[4] Department of Music, Visual Arts, and Movement Education, Drawing Section, University of Extremadura, C. Santa Teresa de Jornet, 38, 06800 Mérida, Spain
ccruz@unex.es

Abstract. This exploratory mixed-methods study analyzed the technical and ethical adoption of Generative Artificial Intelligence (GAI) among 66 industrial design Mexican students. The results evidence a computational literacy gap: although adoption rates are high, only 20% demonstrate mastery of advanced prompt engineering, with basic instructions remaining predominant. Regarding design praxis, a preference for a hybrid workflow was observed, utilizing GAI for rapid creative divergence while reserving manual control for convergence and technical postproduction. However, a marked affective dissonance was identified, characterized by an authorship crisis and lower personal satisfaction compared to traditional, handcrafted methods. The study concludes that - The education of undergraduate design students should urgently transition from traditional graphic representation toward text-to-image literacy, with the ethical dimension integrated as a central pillar of contemporary design practice.

Keywords: Generative Artificial Intelligence (GAI) · Prompt engineering · Industrial design

1 Introduction

The emergence of Generative Artificial Intelligence (GAI), with interfaces such as ChatGPT, MidJourney, Adobe Firefly, and others, is disrupting traditional industrial design education. Historically, conceptualization was based on manual sketching and physical prototyping, which ensured knowledge regarding form,

H. Ponce et al. (Eds.): COMIA 2026, CCIS 3095, pp. 64–75, 2027.
https://doi.org/10.1007/978-3-032-37438-7_6

function, and ergonomics. Today, GAI has transformed the barrier between idea and representation, posing an unprecedented pedagogical challenge: *How should the validity and ethical implications of projects generated by these tools be evaluated?*

This essential change redefines the role of the student, forcing education to evolve when conceptualization is reduced to a textual indication (prompt). This curricular adaptation is justified within the framework of the UN 2030 Agenda, responding to Sustainable Development Goal (SDG) 4 by increasing technical and professional competencies, and SDG 9 by promoting responsible innovation.

The objective of this work is to analyze the implications of GAI in the training of 66 Mexican undergraduate Industrial Design students from a Mexican university. A 12-item survey was applied to obtain information on three main areas: their technical competence in prompt development, the transition toward hybrid work processes, and the perception of authorship and ethics in their practice.

2 State of the Art

The state of the art reveals a significant evolution of industrial design, moving from its artisanal roots to becoming a complex and interdisciplinary discipline consolidated after the Industrial Revolution and key movements such as the Bauhaus, among others. In the contemporary context, Rodríguez Morales's "General Model of the Design Process" [9] is established as the fundamental methodological structure. This model organizes the design workflow into critical phases—Case, Problem, Hypothesis, Project, and Realization—providing a flexible framework that allows for the operational integration of emerging technologies, especially during the ideation and hypothesis validation phase.

Furthermore, Artificial Intelligence (AI) has been redefined in this scenario, evolving from an automation tool to a "co-creative" agent. The evolution from rule-based systems toward Deep Learning and neural networks has enabled the emergence of generative design. Under the definition of the OECD, which integrates data, algorithms, and hardware, current AI possesses the capacity to recognize complex patterns and optimize design solutions, allowing designers to explore massive solution spaces that exceed traditional human processing capacity [1].

The convergence of AI and Industrial Design aligns with "Generative Design". In Fig. 1 (see Fig. 1), we can see the timeline with important milestones in the evolution of this emerging technology. At this intersection, AI acts as the logical engine that generates and optimizes proposals [5].

This symbiosis is enhanced by object recognition and natural language processing capabilities, enabling more intuitive interactions and dynamic adaptation to user needs. Figure 2 presents the conceptual map that synthesizes the structure of 21st-century industrial design, showing how AI can support the foundations and methodology of the "General Design Process Model" and how it must be framed within ethical and managerial considerations. At this intersection, AI acts as the logical engine that generates and optimizes proposals. [1].

Finally, research underscores that this technological integration must be governed by solid professional ethics and structured management frameworks. Crucial challenges are identified, such as algorithmic bias, data privacy, and sustainability. To manage the complexity of these hybrid projects, the creation of a framework is proposed to help manage the use of these technologies and their integration with industrial design, helping ensure that technological innovation is implemented in a responsible and efficient manner, aligned with the human values of design [1].

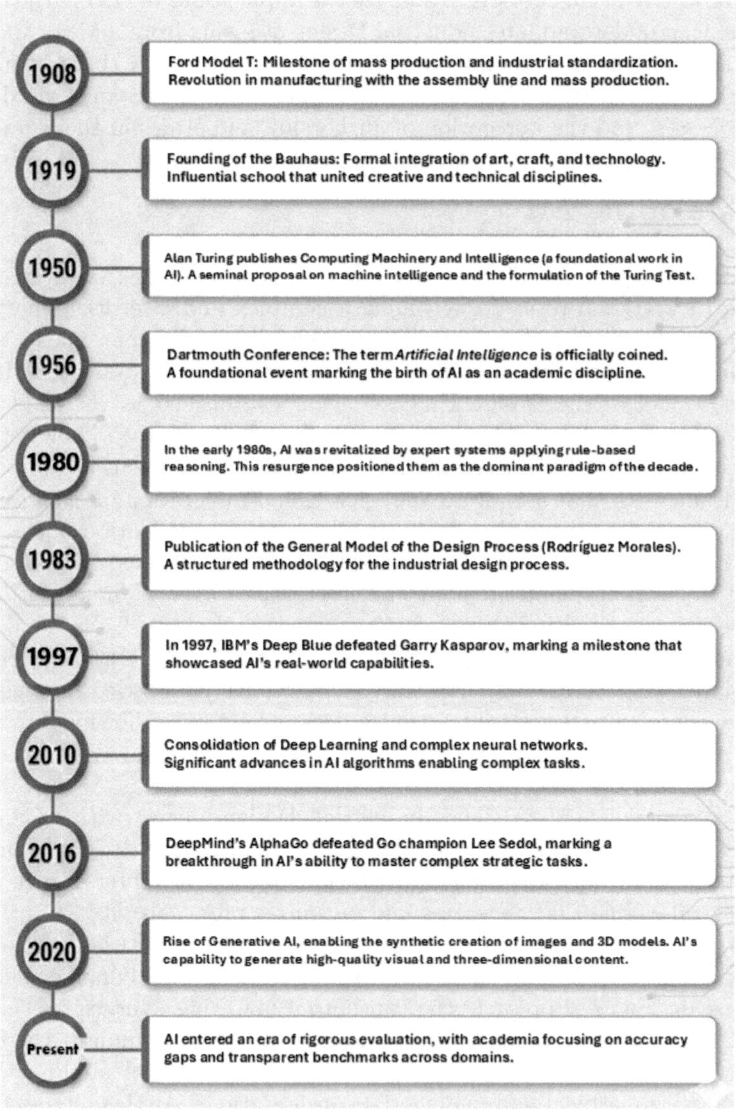

Fig. 1. Milestones in design and AI.

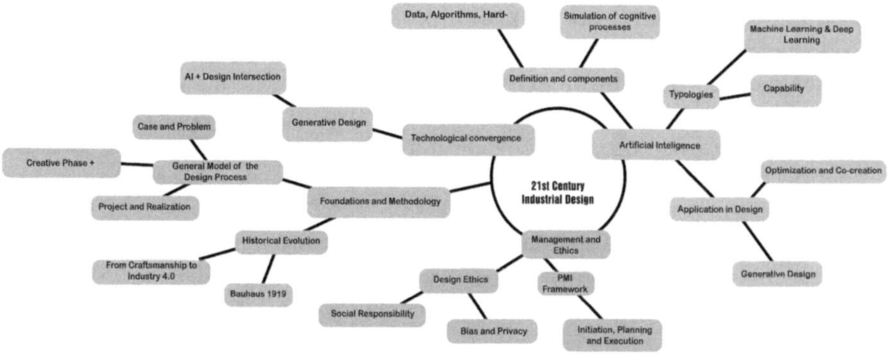

Fig. 2. Conceptual map of the integration of Industrial Design and AI.

3 Methodology

To fulfill the study's objectives, a mixed-methods methodology (qualitative-quantitative) was used with a descriptive and exploratory scope. This design allowed not only for the quantification of the frequency of use of GAI tools but also for the analysis of the semantic depth of instructions (prompts) and the phenomenological implications (feelings and perceptions) experienced by the participants during the creative process. [3].

Participants and context. The participants were 66 Mexican undergraduate students enrolled in Industrial Design at a Mexican university. The collaborators were active design participants, included solely based on prior experience with digital visualization applications.

Instrument. Data collection was conducted through an ad hoc structured digital questionnaire hosted on the Google Forms platform, divided into four analytical dimensions:

1. **Technical profile and digital literacy:** Multiple-choice questions about the platforms used (Gemini[1], ChatGPT[2], Copilot[3], Meta AI[4], Adobe Firefly[5], Midjourney[6], Stable Diffusion[7], etc.) and a 5-point Likert scale (1 = very uncomfortable, 5 = very comfortable) to measure perceived self-efficacy in directing AI toward specific aesthetic outcomes.
2. **Workflows and correction:** Situational aspects to identify problem-solving strategies (e.g., "What do you do in the face of a rendering error: regenerate or post-produce manually?").

---

[1] https://gemini.google.com.
[2] https://chat.openai.com.
[3] https://copilot.microsoft.com.
[4] https://ai.meta.com.
[5] https://firefly.adobe.com.
[6] https://www.midjourney.com.
[7] https://stability.ai.

3. **Ethical perception and authorship:** Items regarding intellectual property and the feeling of personal gratification compared to manual methods.
4. **Practical prompt engineering evaluation:** Participants were given a controlled practical task to write a prompt for creating a "smart water dispenser with a biomimetic aesthetic". This item served as a performance measure to objectively assess the use of technical vocabulary, design constraints, and command syntax.
   To ensure research transparency and support study replicability, the full 12-item structured questionnaire developed for this research has been deposited in the Zenodo open-access repository (available at: DOI 10.5281/zenodo.20087704).

The instrument validation process employed a structured expert judgment methodology to establish content validity for the 12-item questionnaire, addressing the exploratory nature of the study. A panel of specialists in Industrial Design and Educational Technologies evaluated each item against criteria of clarity, relevance, coherence, and sufficiency. This rigorous review specifically assessed the pertinence of the four analytical dimensions: technical profile and digital literacy, AI workflow and error correction strategies, ethical and authorship perception, and practical evaluation of syntax and prompt engineering commands.

Following the expert review, minor refinements were made to the technical terminology within Dimension 4 to ensure alignment between design vocabulary and command syntax. These adjustments were critical for the objective measurement of prompt literacy levels across Basic, Aesthetic, and Pro categories. The process concluded with full consensus among panelists, confirming the instrument's suitability and robustness for achieving the specific research objectives within the interdisciplinary framework of computational sciences and industrial design.

Procedure and Data Analysis. Data were collected asynchronously and consolidated into a master spreadsheet for cleaning. The analysis was divided into two parts [3]:

- **Quantitative Phase:** This phase involved a **descriptive statistical analysis** of frequencies and percentages of technological adoption. Additionally, **contingency tables** were created to relate technical comfort with image refinement methods.
- **Qualitative Phase: Thematic content analysis** was used to analyze open-ended responses related to feelings of authorship. To evaluate the prompts from the practical exercise, a **skill-level rubric** was prepared according to the following three levels: **Level 1** (Basic linguistic instruction), **Level 2** (Use of aesthetic descriptors), and **Level 3** (Use of technical parameters and roleplay).

4 Results

4.1 Quantitative Results

Analysis of the collected data revealed a high adoption rate of GAI (85%) among participants, while 15% stated they had not used AI for design purposes. However, significant variations were found regarding their technical depth and operational confidence. The statistical results are presented across three dimensions: platform dominance, technical self-efficacy, and workflow strategies.

1. **Hegemony of tools and closed (private) ecosystems:** Despite the proliferation of open-source models, the sample showed a strong dependency on commercial tools integrated into pre-existing workflows (see Fig. 3). ChatGPT positioned itself as the predominant tool, serving as the main vector of adoption in the university environment and doubling the usage of other image generation platforms. Gemini and Copilot remained secondary tools, used primarily for quick consultation and exploration.

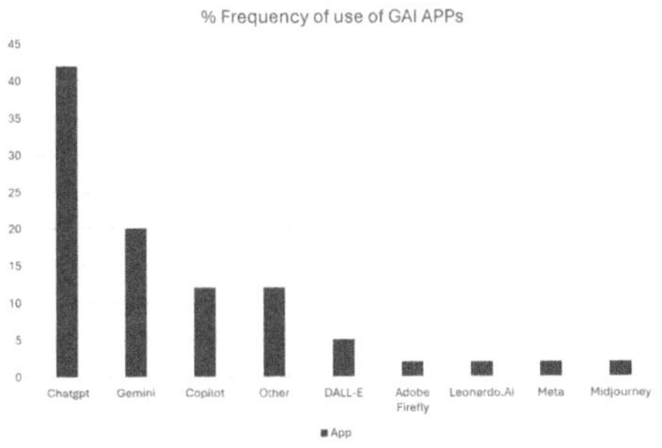

Fig. 3. Percentage of GAI App usage.

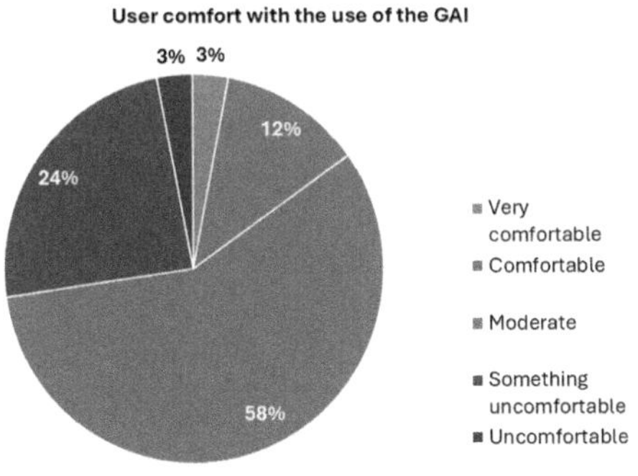

Fig. 4. User comfort with the use of the GAI.

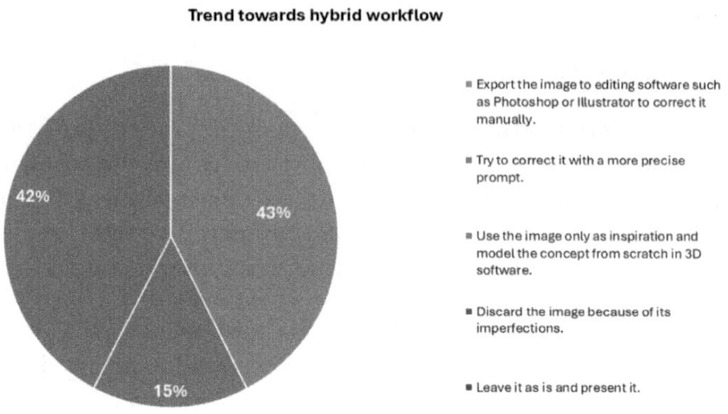

Fig. 5. Trend towards hybrid workflow.

Fig. 6. Trend towards hybrid workflows with the use of GAI.

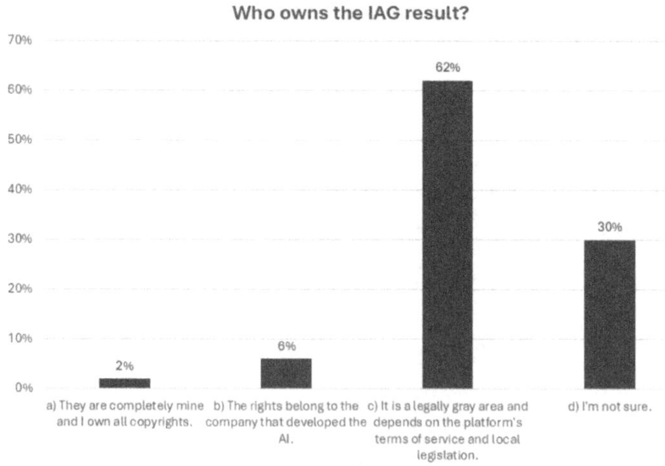

Fig. 7. Trend towards hybrid workflows with the use of GAI.

2. **The technical comfort curve:** Participants rated their comfort in guiding AI toward specific aesthetic outcomes using a Likert scale ranging from 1 to 5. The results showed a normal distribution slightly shifted toward neutrality, with an average score of 2.88 out of 5. Only 3% of participants placed themselves at level 5 (mastery), while another 3% reported very low knowledge levels, expressing frustration when trying to translate creative intent into effective text commands (see Fig. 4).
3. **Workflows and error correction:** One of the study's most relevant findings quantified how designers respond to errors, such as AI hallucinations or geometric failures (see Fig. 5). The data confirmed a clear preference for a hybrid workflow:
   - 43% preferred to export the image for **manual correction**.
   - 42% chose to use the image only as **inspiration** to model the concept from scratch in 3D software.
   - Only 15% opted to persist with AI refinement techniques by modifying the prompt.
4. **Perception of bias and intellectual property:** Regarding critical awareness of these tools: Bias: 82% of students claimed to be aware of cultural or aesthetic biases in GAI models, though 49% admitted they did not know how to mitigate them (see Fig. 6).
5. **Intellectual Property:** There is widespread uncertainty; 62% of the sample categorized the authorship of their generated works as a "legal gray area", and only 2% believed the images belonged to them entirely (see Fig. 7). Students' discomfort with AI use is closely tied to unclear legal and copyright issues. The current regulatory framework is seen as ambiguous, creating uncertainty about authorship and intellectual property. This lack of definition fosters emotional distancing and a perceived loss of creative merit.

4.2 Qualitative Results

Beyond usage metrics, the qualitative analysis of open-ended responses and the practical prompting exercise revealed a **profound transformation in the design student's identity**. Three emerging categories were identified: the taxonomy of technical competence, affective dissonance regarding authorship, and the reconfiguration of the creative role.

1. **Prompt taxonomy: Levels of technical competence.** Semantic analysis of the instructions generated for the "Fire lamp" exercise allowed for the classification of the student population into **three levels of GAI literacy** [3,10]:
   - **Level 1: Basic linguistic instruction** (approx. 50%). Participants used natural language without technical structure, resembling a Google search and lacking parameters for lighting, style, or rendering engines. *Example: "Create a fire lamp".*

- **Level 2: The visual aesthete** (approx. 30%). **Stylistic adjectives** and references to specific materials were incorporated, demonstrating an intent for art direction, although technically limited. *Example: "Create a cilindric lamp, with fire inside, organic shapes, green and gray tones, rock and metal texture, minimalist style".*
- **Level 3: Prompt engineering and roleplay** (approx. 20%). A significant minority used **advanced commands**, defining the camera, lighting type, rendering style, and even assigning a **role to the AI** to improve the result [2]. *Example: "...act as an award-winning industrial designer".*

This stratification confirms that, while access to AI is democratic, the ability to obtain professional results is a **specialized skill**, where only 20% demonstrate real technical mastery, evidencing the literacy gap.

2. The craft satisfaction gap When questioning participants about their experience, a pattern of **"emotional distancing"** emerged [2]. The data revealed a clear distinction between **"satisfaction through efficiency"** and **"satisfaction through creation"**. While the tool eliminated technical friction, it also seemed to eliminate the **sense of achievement** associated with manual mastery. One student noted: *"the satisfaction response the AI gives me is not the same... it is a different satisfaction, more professional than sentimental to my pride"*. Others expressed a sense of **"fraud" or lack of merit**, feeling as though they were "stealing a design".
3. From author to curator: The hybrid designer analysis indicated a shift in the cognitive role: students no longer perceive themselves solely as "form generators", but as **"editors of possibilities"**. AI was highly valued during the **hypothesis creation phase** (brainstorming and sketching) but lost relevance during **convergence** (final detailing), where participants preferred to regain manual control. This contradicts the narrative of total replacement; instead, participants use AI to overcome the **"fear of the blank page"** but reclaim final authorship through manual intervention, validating the preference for a **hybrid workflow**.

5 Discussion

The integration of GAI into the design model is not merely an instrumental update, but a structural reconfiguration of design methodology. The theoretical and practical implications of this paradigm shift are discussed below:

1. **Redefining rationality:** From cognitive limits to generative expansion. This axis addresses the fundamental tension in the **"Hypothesis" phase** of the Rodriguez Morales model. Traditionally, generating alternatives was restricted by **"bounded rationality"**, where a designer's ability to process information and generate variants was finite and sequential.
   Quantitative leap and disciplinary resilience: The introduction of AI breaks this barrier, transforming the hypothesis phase from manual selection

to a **massive exploration of the solution space**. However, it is crucial to qualify this with the concept of **disciplinary resilience** [1].
   - **Divergence vs. Convergence:** AI acts as a motor of **infinite divergence** (speed and randomness) but lacks **intentional criteria** [6].
   - **Hybrid Flow:** Contrary to human obsolescence, data validates a workflow where the student **"demotes" the AI from author to "generator of raw material"** [8].
   - **Synthesis:** Rationality no longer resides solely in the subject's mind, but in the **hybrid system**. The designer provides **technical coherence (convergence)** over the machine's creative chaos [10].
2. **Temporal shift of validation: The "Do-Evaluate" cognitive fracture.** The most disruptive impact of technology is observed in the chronology of the process and the associated cognitive load [8].
   From "Project" to "Hypothesis": In the traditional model, rigorous validation occurred in late stages, raising the cost of error. AI allow for **virtual prototyping**, bringing validation to early stages, which reduces uncertainty and economic risk.
   Affective Dissonance and the Inversion of the Pyramid: This shift creates a critical pedagogical friction: the **fracture between creating and evaluating** [7].
   - **Automation of "Technique":** By automating manual skill, the student obtains final visual results without the traditional physical effort [7].
   - **Premature Evaluation:** This forces the student into the **evaluation phase without having experienced the "doing"** [8].
   - **Resulting in a loss of sense of merit** because the traditional link between physical effort and cognitive reward is broken [8].
3. **From Solitary Authorship to Semantic Orchestration** The transition from a single-author model to **symbiotic co-creation** redefines the industrial designer's profile [11].
   Inversion of the cognitive pyramid:
   Analyzing findings under **Bloom's taxonomy**, design education can no longer be based on a linear sequence of observing, remembering, understanding, and applying [2]. GAI forces students to operate immediately at **higher levels of analysis and evaluation**.
   The New digital divide (semantic competence): The designer evolves from a "form creator" to an **"orchestra conductor"** or systems manager. This reveals a new academic inequality based on **linguistic competence** [1]:
   - Basic/intermediate levels produce generic results [4].
   - Advanced levels use specific technical nomenclature [4].

   Critical finding: The ability to prompt (semantic articulation) becomes a design skill as decisive as drawing [4]. Those without technical vocabulary cannot direct the algorithmic orchestra.
4. **Ethics, viability, and management of uncertainty**

Generative AI introduces ethical frictions in industrial design, requiring internationally recognized governance frameworks. This study aligns with the OECD definition of AI as systems optimizing solutions across vast design spaces, where ethics must translate into responsible management to mitigate risks such as bias and intellectual property opacity.

Curricular adaptation responds to the UN 2030 Agenda, advancing SDG 4 (Quality Education) through technical competencies and SDG 9 (Industry, Innovation, Infrastructure) via responsible innovation. To address student-reported uncertainty, management frameworks are proposed to structure AI-assisted projects under social responsibility and professional ethics [7].

Risks and Management: The new paradigm faces

- **Algorithmic bias** (perpetuating historical design prejudices) [1].
- **Intellectual property** issues (legal definition of authorship) [1].
  Imperative Action: To mitigate the technical complexity of integrating AI, it is imperative to adopt standardized management frameworks to control technological uncertainty and ensure the productive viability of the design.

6 Conclusions

The instrument validation process employed a structured expert judgment methodology to establish content validity for the 12-item questionnaire, addressing the exploratory nature of the study. A panel of specialists in Industrial Design and Educational Technologies evaluated each item against criteria of clarity, relevance, coherence, and sufficiency. This rigorous review specifically assessed the pertinence of the four analytical dimensions: technical profile and digital literacy, AI workflow and error correction strategies, ethical and authorship perception, and practical evaluation of syntax and prompt engineering commands.

Following the expert review, minor refinements were made to the technical terminology within Dimension 4 to ensure alignment between design vocabulary and command syntax. These adjustments were critical for the objective measurement of prompt literacy levels across Basic, Aesthetic, and Pro categories. The process concluded with full consensus among panelists, confirming the instrument's suitability and robustness for achieving the specific research objectives within the interdisciplinary framework of computational sciences and industrial design.

Acknowledgments. This publication was 85% co-financed by the European Union, the European Regional Development Fund, and the Regional Government of Extremadura. Managing Authority: Ministry of Finance. IB24052. "ERDF A way of making Europe"

References

1. Barranco, C.A.A., Moreno, G.M., Morales, H.N.: La integración de la IA en el diseño industrial y de productos: transformación, impactos y perspectivas futuras. Economía Creativa (22) (2025). https://doi.org/10.46840/ec.2025.22.751, https://ciecpress.centro.edu.mx/ojs/index.php/CentroTI/article/view/751
2. Chengyuan Zhang, Suzhe Xu: Aesthetic Experience and Educational Value in Co-creating Art with Generative AI: Evidence from a Survey of Young Learners (2025). https://arxiv.org/html/2509.10576v1
3. Gulzar, S., Harun, J., Yahaya, N.: Student Engagement in the Age of GenAI: A Mixed-Method Study on Integration of Generative AI in Visual Communication Design Education Classroom. Int. J. Acad. Res. Bus. Soc. Sci. **15**(10), 1669–1682 (2025). https://doi.org/10.6007/IJARBSS/v15-i10/26866
4. Jin, Y., Martinez-Maldonado, R., Gašević, D., Yan, L.: GLAT: The generative AI literacy assessment test. Comput. Edu. Artif. Intell. **9**, 100436 (2025). https://doi.org/10.1016/j.caeai.2025.100436, https://www.sciencedirect.com/science/article/pii/S2666920X25000761
5. Keshavarzi, M., Bidgoli, A., Kellner, H.: V-Dream: Immersive Exploration of Generative Design Solution Space, pp. 477–494 (2020). https://doi.org/10.1007/978-3-030-60117-1_35, https://consensus.app/papers/vdream-immersive-exploration-of-generative-design-keshavarzi-bidgoli/172d8e4ee7985656bdfe65fde68dc08f/
6. Lev Manovich: Artificial Aesthetics: Generative AI, Art and Visual Media (2024). https://manovich.net/index.php/projects/artificial-aesthetics
7. Yahaya, N., Gulzar, S., Harun, J.: Integration of Generative Artificial Intelligence in Design Education: Evidence Review (2020–2024). ResearchGate (2025). https://doi.org/10.6007/IJARBSS/v15-i11/26903, http://dx.doi.org/10.6007/IJARBSS/v15-i11/26903
8. Pallant, J.L., Blijlevens, J., Campbell, A., Jopp, R.: Mastering knowledge: the impact of generative AI on student learning outcomes. Stud. High. Educ. **0**(0) (2025). https://doi.org/10.1080/03075079.2025.2487570
9. Rodriguez Morales, G.: Manual de diseño industrial : curso básico (1983). http://ilitia.cua.uam.mx:8080/jspui/handle/123456789/953, Accepted 15 Oct 2021
10. Schulhoff, S., et al.: The Prompt Report: A Systematic Survey of Prompting Techniques (2024). https://doi.org/10.48550/arXiv.2406.06608
11. Soto Ayala, M.F.: Impacto ético y creativo de la inteligencia artificial generativa en la formación en diseño gráfico. Latam: revista latinoamericana de Ciencias Sociales y Humanidades **6**(3), 21 (2025). https://dialnet.unirioja.es/servlet/articulo?codigo=10301334, publisher: REDILAT: Red de Investigadores Latinoamericanos Section: Latam: revista latinoamericana de Ciencias Sociales y Humanidades

A Real-Time Emotional State Management System Using Physiological Signals

Luis Mario Martinez Rueda(✉) and Angela Mayhua-Quispe

Laboratorio de Inteligencia Artificial, Facultad de Ingeniería, Universidad de Lima, Lima, Peru

20211628@aloe.ulima.edu.pe, amayhua@ulima.edu.pe

Abstract. Monitoring students' emotional states during learning activities is an important challenge in educational environments, as emotional factors such as stress or engagement can significantly influence academic performance and well-being. This paper presents an intelligent system for real-time emotional state management in educational environments using physiological signals acquired through non-invasive wearable devices. The approach focuses on temporal patterns extracted from electrodermal activity, blood volume pulse, and skin temperature to objectively identify emotional states of students. Feature extraction is performed using fixed time windows, and several machine learning models are trained and evaluated on the WESAD dataset. The experimental results show that the proposed system achieves an accurate emotional classification, XGBoost achieving the highest performance at 91.2% accuracy and a 0.9127 F1-score. In addition, usability evaluation using the System Usability Scale (SUS) yielded a mean score of 88.0, indicating high perceived usability. The system enables real-time monitoring, visualization of emotional trends, supporting informed pedagogical decision-making.

Keywords: Emotion recognition · Physiological signals · Machine learning · Wearable sensing · Student well-being

1 Introduction

Wearable technology has gained increasing relevance in physiological monitoring due to its ability to continuously and non-invasively acquire biometric signals, enabling its integration into intelligent systems for emotional and educational well-being. This need is reinforced by the high prevalence of emotional problems in academic contexts: 73.2% of students report moderate to high stress [15], between 12% and 46% experience psychological distress [2], and more than 25% report recurrent mental fatigue [14,18]. Since prolonged emotional distress has been associated with impairments in attention, working memory, and executive

H. Ponce et al. (Eds.): COMIA 2026, CCIS 3095, pp. 76–89, 2027.
https://doi.org/10.1007/978-3-032-37438-7_7

functioning [3], the identification of emotional states in educational environments has become an important research challenge.

Conventional approaches, including surveys and interviews, mainly rely on subjective self-reports and are limited in their ability to capture continuous emotional fluctuations. Although electroencephalography (EEG) provides high precision, its hardware complexity and operational constraints reduce its applicability in real educational settings. Consequently, there is a need for practical systems capable of supporting real-time emotional awareness and assisting educators in identifying potentially critical situations without requiring laboratory conditions.

Recent studies have demonstrated that physiological signals combined with machine learning can achieve high accuracy in emotion recognition. Zhu et al. [22] reported 92.9% accuracy using electrodermal activity (EDA) and SVM, while Fang et al. [8] achieved over 95% F1-score using ROCKET, and Wu et al. [21] reported AUC values greater than 90% with Transformer-based multimodal approaches. In educational contexts, systems such as SensEmo [6] have also demonstrated the feasibility of wearable-based emotional monitoring in real time.

Motivated by these advances, this work proposes a real-time emotional state monitoring system for higher education environments based on non-invasive physiological sensing. The proposed approach uses EDA, blood volume pulse (BVP), and skin temperature (SKT) acquired through the Shimmer3 GSR+ device as input data for a machine learning pipeline trained on the WESAD dataset [17], evaluating models such as SVM, Random Forest, KNN, and XGBoost. Beyond emotion classification, the system incorporates real-time inference, visualization of individual and group emotional trends, and automated alerts to support practical deployment in real-world educational environments. Consequently, the central hypothesis of this study is that machine learning models trained with wearable-based multimodal physiological signals can classify emotional states in educational settings, enabling continuous monitoring without the constraints of specialized laboratory infrastructure.

The remainder of this paper is organized as follows. Section 2 reviews related work, Sect. 3 describes the proposed system, Sect. 4 presents the results, Sect. 5 discusses the findings, and Sect. 6 concludes the paper.

2 Related Work

This section reviews the most relevant studies on emotion recognition using physiological signals, machine learning methods, and educational monitoring systems.

2.1 Physiological Signals and Emotional States

Physiological signals provide an objective and non-invasive basis for emotion recognition, as they capture autonomic nervous system responses that are difficult to consciously control. Among the most commonly used are EDA, BVP, and

SKT, which have demonstrated consistent relationships with stress, relaxation, and engagement.

EDA has been extensively used as a primary indicator of emotional arousal because it reflects changes in skin conductance associated with the activity of the sympathetic nervous system, making it highly sensitive to emotional fluctuations. Lower values are associated with low-arousal states such as sadness, whereas higher values are associated with stress. In addition, phasic responses tend to increase in the presence of emotional stimuli, confirming the sensitivity of EDA to emotional changes [4,20]. Complementing EDA, BVP provides additional information on cardiovascular responses associated with emotional and cognitive processes. Obtained through photoplethysmography, BVP enables the extraction of heart rate and heart rate variability (HRV), which have been strongly correlated with states such as stress, engagement, boredom, and cognitive load [7,11]. On the other hand, SKT serves as a complementary signal that reflects peripheral physiological regulation. The peripheral temperature typically decreases during high-arousal states due to vasoconstriction and increases during relaxation, showing an inverse relation with emotional activation [13]. Although less reactive than EDA, temperature provides valuable contextual information that enhances the interpretation of emotional states over time.

Taken together, these signals offer a multimodal representation of emotional responses, where each modality contributes complementary information about different physiological mechanisms. This integration not only improves robustness against noise or missing data, but also enables a more stable and accurate characterization of emotional states, particularly in real-world wearable sensing environments.

2.2 Machine Learning for Emotion Classification

Machine learning has advanced emotion recognition by enabling the extraction of complex patterns from physiological data. Traditional models such as SVM and ensemble methods (e.g., XGBoost) have demonstrated strong and stable performance on different datasets [4,11]. More recently, deep learning approaches, including convolutional and recurrent architectures (e.g., LSTM), have improved the modeling of temporal dependencies in affective prediction [23]. In addition, transformer-based models and techniques such as ROCKET have further enhanced performance in sequential and multimodal emotion classification tasks [8,9,21].

The WESAD dataset [17] has become a standard benchmark in this domain due to its multimodal physiological recordings collected under controlled conditions. Given the class imbalance often present in emotion datasets, evaluation metrics such as F1-score and AUC are commonly preferred over accuracy, as they provide a more balanced assessment of model performance [5].

2.3 Emotion Recognition Systems in Educational Contexts

Several systems have explored real-time emotion recognition in educational settings using wearable sensors and machine learning. For example, SensEmo [6] leverages smartwatch signals such as heart rate (HR) and EDA to monitor student engagement, while M3Sense [16] employs multimodal physiological data for real-time detection. Other approaches incorporate additional modalities, including computer vision, speech, and text, to better capture the affective dynamics of the classroom [12, 19].

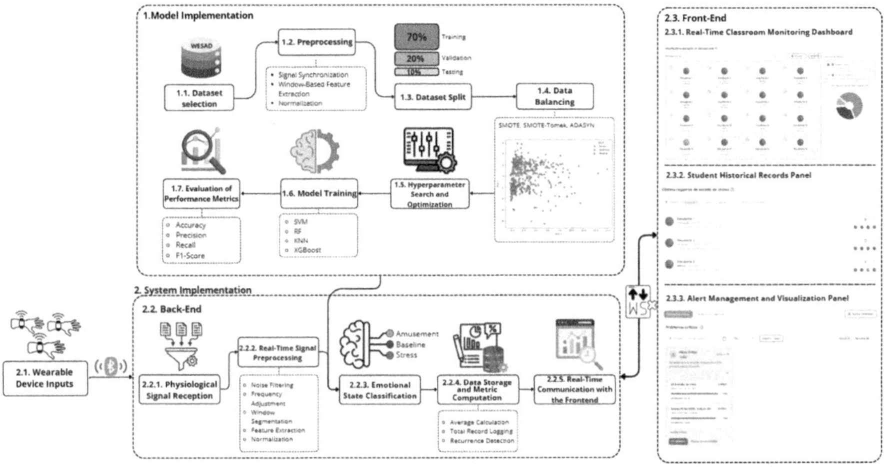

Fig. 1. Structure and workflow of the proposed system.

However, most existing systems focus primarily on instantaneous classification and provide limited support for longitudinal analysis. In educational contexts, this restricts the understanding of how emotional states evolve over time. Moreover, system-level integration, real-time deployment constraints, and comprehensive usability validation are not consistently addressed across studies, highlighting ongoing challenges in translating affective computing models into practical classroom applications.

3 Methodology

This section presents the methodology used for the development and implementation of the emotional state recognition system. Figure 1 outlines the workflow from model training to system integration.

3.1 Emotion Recognition Based on Physiological Signals

Dataset Selection. The WESAD dataset [17] was used in this study due to its availability as a public benchmark for physiological-based emotion classification. It includes multimodal signals, specifically accelerometer data, BVP, EDA, and SKT, recorded at different sampling rates, including 64 Hz for BVP and 4 Hz for EDA and SKT. These signals are associated with neutral, stress, and amusement states induced through standardized laboratory protocols. Because the induction periods differ in duration, the dataset presents an inherent class imbalance, which is addressed during preprocessing to reduce potential bias in the machine learning pipeline.

Furthermore, the structured design of WESAD also enables comparison with previous studies and provides a baseline for evaluating model performance. In addition, its use in education-oriented systems [16] provides a relevant reference for the signals and emotional states considered in this work, despite differences from real classroom conditions.

Table 1. Extracted features from physiological signals

Signal	Extracted features
EDA	Phasic mean, phasic standard deviation (std), phasic minimum, phasic maximum, SMNA mean, SMNA std, SMNA minimum, SMNA maximum, tonic mean, tonic std, tonic minimum, tonic maximum
BVP	Mean, std, minimum, maximum, peak frequency
SKT	Mean, std, minimum, maximum, slope
Label	Emotional category label: 1 = baseline, 2 = stress, 3 = amusement

Preprocessing and Feature Extraction. Due to heterogeneous sampling rates, preprocessing was required to synchronize signals and labels. Timestamps were aligned using a 700 Hz reference, enabling integration of EDA and SKT (4 Hz) with BVP (64 Hz), while removing missing values and incomplete segments. EDA was filtered and decomposed using cvxEDA into tonic, phasic, and SMNA components. Signals were segmented using sliding windows of 10 s (7000 samples at 700 Hz) with a 2-s step size (80% overlap), allowing emotional state updates every 2 s while preserving sufficient temporal context for stable physiological characterization.

From each window, statistical features (mean, standard deviation, minimum, maximum), BVP peak frequency, and temperature slope were extracted and normalized using Z-score. This configuration balances temporal responsiveness and robustness, supporting near real-time inference without sacrificing classification stability. The final set (see Table 1) includes 22 features per window for model training and evaluation.

Data Balancing and Model Training. The class imbalance may bias the model toward majority classes and reduce its ability to identify minority emotional states. To address this issue, three synthetic oversampling techniques – SMOTE, SMOTE-Tomek, and ADASYN – were applied exclusively to the training set, while the test set remained unchanged. Their impact was assessed using precision, recall, F1-score, and AUC.

Then multiple classifiers were trained and optimized, including Random Forest (RF), Support Vector Machine (SVM), K-Nearest Neighbors (KNN), and XGBoost. The hyperparameters were tuned through a systematic search and the resulting models were evaluated using cross-validation. The best-performing model was selected for integration into the real-time monitoring system.

Model Validation Using Self-Reported Ground Truth. To evaluate the performance of the trained model, a controlled experimental protocol was implemented based on the WESAD methodology [17]. This setup integrated real-time physiological acquisition with participant self-reports to align model predictions with perceived emotional states. To assess the robustness of the system across different hardware configurations, a Shimmer3 GSR+ device was utilized instead of the original sensing equipment.

Participation was voluntary and conducted under informed consent. Prior to the sessions, participants were briefed on the study, and demographic data were collected. Following institutional ethical guidelines, all data were anonymized before analysis, with no personally identifiable information stored within the system.

The protocol combined continuous signal recording with subjective assessments using the Positive and Negative Affect Schedule (PANAS) and the Self-Assessment Manikin (SAM). Participants underwent a baseline phase followed by emotion induction through validated video stimuli [1], targeting neutral, stress, and amusement states. Each experimental block consisted of a pre-assessment, a 10-minute induction phase, and a post-assessment. A five-minute rest period was introduced between blocks to minimize carry-over effects while physiological signals were continuously processed. Model performance was evaluated by comparing predicted emotional states with both experimental conditions and self-reported measures, assessing the consistency between model outputs and subjective experience.

3.2 Emotional State Management System

This section presents the proposed real-time emotional monitoring system, where wearable physiological signals are processed and classified on a server, with results visualized as states, trends, and alerts.

Wearable Device Inputs. The system connects to a Shimmer3 GSR+ device, enabling real-time acquisition of the galvanic skin response (GSR), photoplethysmography (PPG) and SKT. GSR is used as EDA, PPG is processed to derive

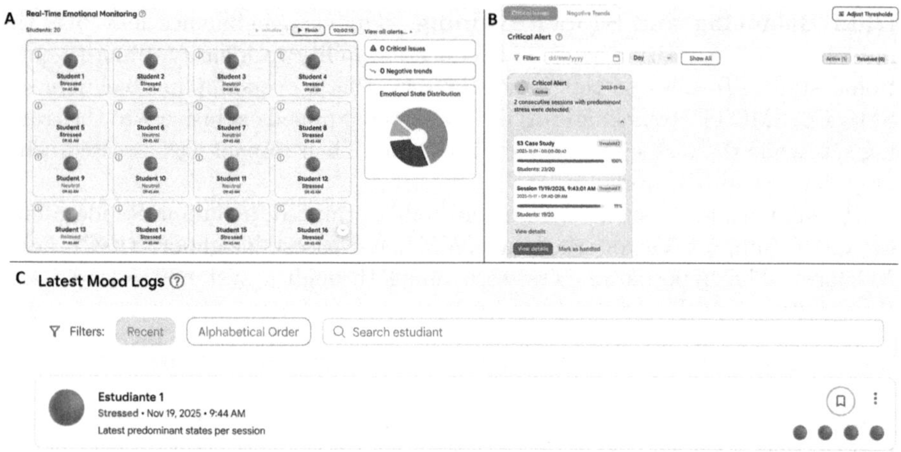

Fig. 2. User interface of the emotional monitoring system: (A) real-time classroom monitoring dashboard, (B) alert management panel showing critical issues and negative trends, and (C) historical emotional records for individual students.

BVP, and SKT is directly acquired, all at configurable sampling rates. Once bluetooth pairing is established, the signals are streamed to the server in real-time for processing and storage.

Backend Processing. The backend establishes a WebSocket connection with the device for continuous signal reception. Incoming data are processed using the same preprocessing pipeline defined during training and transformed into feature windows for classification. Each window is classified using the selected model, and predictions are stored in a relational database with associated metadata (i.e., timestamp, session, sensor, classroom). Based on these records, the system computes aggregated indicators such as emotion distributions and stress recurrence, automatically triggering alerts when predefined conditions are met. These alerts are designed as decision-support mechanisms to increase instructor awareness rather than to trigger automated pedagogical interventions, ensuring that instructional decisions remain under human supervision. Communication with the frontend is handled through persistent WebSocket connections, enabling real-time updates. The architecture supports concurrent connections from multiple wearable devices through parallel WebSocket channels, allowing scalable monitoring across classrooms.

Frontend Visualization. The frontend supports real-time monitoring; as shown in Fig. 2, subfigure (A) presents the real-time dashboard, displaying each student's current emotional state along with classroom summaries and emotion distributions. Figure 2(C) illustrates the historical view, where timelines allow the identification of patterns and trends over time, while Fig. 2(B) shows the alert management view, presenting automatically generated alerts based on configurable thresholds, such as sustained or repeated stress conditions. Together, these views provide a unified interface for monitoring, analysis, and timely response.

Table 2. Summary of metrics obtained on the test set

Model	Balancing	Accuracy	Precision	Recall	F1-score
RF	No balancing	0.8994	0.9014	0.8994	0.9001
	SMOTE	0.8897	0.8954	0.8897	0.8909
	SMOTE-Tomek	0.8953	0.8993	0.8953	0.8963
	ADASYN	0.8715	0.8816	0.8715	0.8733
SVM	No balancing	0.8897	0.8897	0.8897	0.8888
	SMOTE	0.8841	0.8855	0.8841	0.8844
	SMOTE-Tomek	0.8785	0.8818	0.8785	0.8793
	ADASYN	0.8561	0.8635	0.8561	0.8570
KNN	No balancing	0.8645	0.8630	0.8645	0.8626
	SMOTE	0.8184	0.8223	0.8184	0.8200
	SMOTE-Tomek	0.8338	0.8369	0.8338	0.8348
	ADASYN	0.8017	0.8112	0.8017	0.8045
XGBoost	No balancing	0.9008	0.9019	0.9008	0.9012
	SMOTE	0.8966	0.8999	0.8966	0.8976
	SMOTE-Tomek	**0.9120**	**0.9141**	**0.9120**	**0.9127**
	ADASYN	0.8841	0.8894	0.8841	0.8854

System Usability Evaluation. To assess usability, a user study based on a User-Centered Design (UCD) approach was conducted, focusing on intuitiveness, efficiency, and functional utility for classroom monitoring. The evaluation followed a structured process that included informed consent, demographic data collection, system interaction, and usability assessment. Participants interacted with the system for 20 min, performing tasks related to classroom management and real-time emotional monitoring. Usability was measured using the System Usability Scale (SUS), complemented by three system-specific questions addressing the clarity of real-time visualization, the usefulness of historical data, and the effectiveness of alert mechanisms, all rated on a Likert scale. This design aligns with established approaches for evaluating real-time systems through standardized self-reports [19].

4 Results

4.1 Results of Emotional State Classification

Model Performance. To determine the optimal configuration for each algorithm, multiple hyperparameter combinations were evaluated under four data balancing scenarios (i.e., no-balancing and three oversampling techniques). The results indicate that performance depended on the balancing strategy, reflecting the influence of data distribution. SVM consistently favored the *rbf* kernel with tuned C and *gamma*, KNN required fewer neighbors after oversampling, and XGBoost adjusted tree depth, learning rate, and regularization to balance complexity and performance. Using the selected configurations, and considering a 70%-20%-10% split for training, validation, and testing, respectively, the models were evaluated on an independent test set of 716 samples using accuracy, precision, recall, F1-score, and AUC (Table 2).

Among the evaluated models, XGBoost with SMOTE-Tomek achieved the highest performance, reaching an accuracy of 91.2% and an F1-score of 0.9127. Compared to the unbalanced baseline, the use of balancing techniques improved the detection of minority emotional states, resulting in a more consistent performance in the neutral, stress, and amusement classes. Consistent with these results, the best configuration correctly identified 352 neutral instances, 191 stress instances, and 110 amusement instances, with a limited number of misclassifications across classes.

Table 3. Correlation between model outputs and self-reports

Model indicator	SAM-Valence	SAM-Arousal	PANAS-PA	PANAS-NA
Proba_Neutral	−0.0488	−0.7908	−0.0167	−0.2771
Proba_Stress	−0.6259	0.7720	−0.5307	0.5656
Proba_Amusement	0.6517	0.0812	0.5272	−0.2549
Pred_Neutral (0/1)	−0.0324	−0.8392	0.0219	−0.3149
Pred_Stress (0/1)	−0.6703	0.6799	−0.5843	0.6043
Pred_Amusement (0/1)	0.6362	0.1938	0.5064	−0.2424

Experimental Validation with Students. The evaluation included 15 students (33.3% female, 66.7% male; age: 22.5 ± 1.4 years). The model's predictions were compared against self-reported measures collected after each emotional block using SAM and PANAS questionnaires. Table 3 presents the correlation coefficients between predicted emotional states and self-reported affective dimensions.

Stress was associated with lower valence, higher arousal, and increased negative affect, while amusement showed strong positive correlations with valence and positive affect and a weak association with negative affect. Neutral states exhibited low correlations, consistent with their expected physiological stability. Overall, the model correctly identified 43 out of 45 emotional blocks, indicating a high agreement with self-reported measures and stable performance under real-world conditions.

4.2 Usability Results

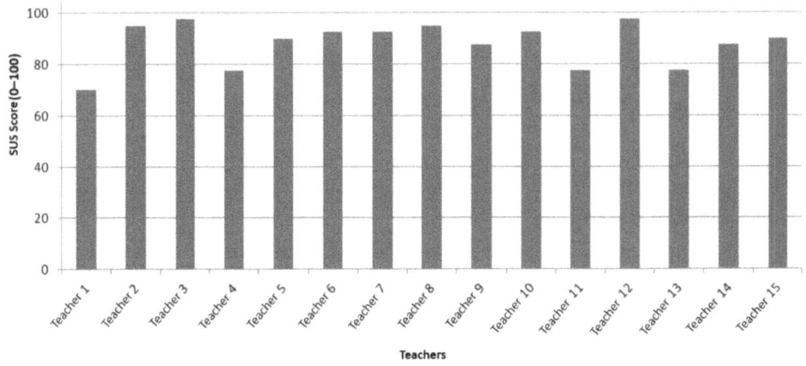

Fig. 3. Total usability score reported per instructor.

The usability evaluation included 15 instructors (13.3% female, 86.7% male; age: 45.2 $\pm$ 8.6 years). Overall perception of the system was highly positive. SUS scores ranged from 70 to 97.5, with a mean of 88, well above the standard benchmark of 68. As shown in Fig. 3, most participants reported high scores, reflecting ease of use, consistency, and confidence.

Additional items assessing real-time visualization, historical analysis, and alert mechanisms obtained mean ratings between 4 and 5, underscoring the clarity and practical relevance of the system's core functionalities. Figure 4 shows consistently high item-level scores with low variability, while reverse-coded items yielded low mean values, indicating minimal perceived complexity. System-specific features achieved the highest ratings, confirming strong acceptance of real-time monitoring, historical tracking, and alerting functionalities.

5 Discussion

5.1 Comparison of the Emotional Classification Model

The XGBoost model, combined with the SMOTE-Tomek balancing technique, achieved an accuracy of 91.20% and an F1-score of 0.9127. These results are consistent with performance ranges reported in recent physiological-based studies; for example, Zhu et al. [22] reported 92.9% accuracy using EDA and SVM, while

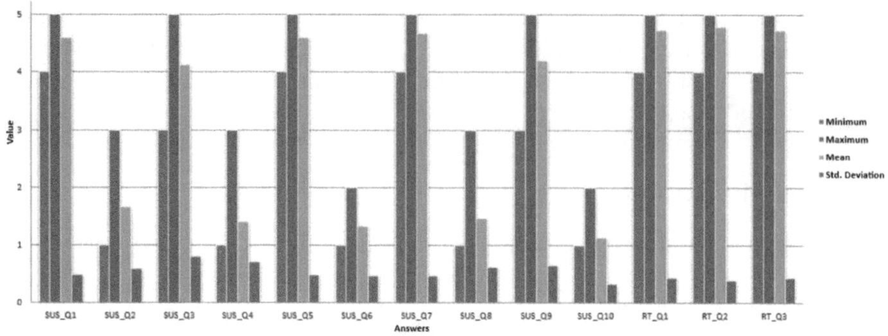

Fig. 4. Usability score analysis per question.

Fang et al. [8] achieved F1-scores above 0.95 using ROCKET, although without real-time implementation.

While deep learning and transformer-based architectures often report higher performance metrics, the selection of classical machine learning models in this work was driven by deployment feasibility. These models require lower computational resources, provide greater interpretability, and support stable real-time inference in educational environments where latency and hardware constraints are relevant factors. Furthermore, the dataset size is more compatible with traditional machine learning approaches than with deep architectures, which typically require larger training sets to achieve robust generalization.

Beyond classification performance, the proposed system integrates non-invasive sensing (EDA, BVP, and SKT) with continuous monitoring, visualization, and automated alert generation. The objective of this work is not to outperform laboratory-based state-of-the-art benchmarks, but to demonstrate the feasibility of emotional classification within an operational framework for real-time pedagogical awareness. Additionally, the balancing techniques implemented ensured consistent performance across neutral, stress, and amusement classes, mitigating the impact of class imbalance during the training phase.

The experimental validation provides additional evidence of the model reliability, showing alignment between predicted states and self-reported measures. In particular, stress was associated with higher arousal and negative affect, amusement with higher valence, and neutral states with lower variability. These findings indicate stable model behavior under real-time conditions and across different acquisition devices, while also revealing limitations in the separability of low-arousal states. Although self-reports inherently involve subjectivity, their alignment with controlled induction blocks and physiological patterns strengthens confidence in the validation process.

Finally, this study emphasizes the ethical requirements of emotion-aware educational systems. All procedures were conducted under informed consent, adhering to data minimization principles by storing only essential information. In deployment scenarios, ethical approval, secure storage, and restricted access policies remain fundamental to ensuring the protection of sensitive biometric data in educational settings.

5.2 User-Centered Evaluation

The system achieved a mean SUS score of 88, exceeding the standard usability benchmark of 68, indicating a high level of perceived usability. In addition, positive ratings on system-specific items related to visualization, historical analysis, and alert mechanisms suggest that these functionalities are relevant for classroom monitoring tasks.

In contrast to systems such as SensEmo [6] and M3Sense [16], which primarily focus on classification performance, the proposed approach incorporates user-centered features designed to support interpretation and pedagogical decision-making. This distinction is relevant because emotion-aware systems in educational contexts require not only high predictive accuracy but also interfaces that enable instructors to process and respond effectively to the provided information.

At the system level, the integration of real-time data acquisition, processing, and visualization enables a coherent operational framework. Features such as per-student monitoring, group-level summaries, temporal analysis, and configurable alerts facilitate the interpretation of emotional dynamics over time. However, the effectiveness of these functionalities depends on signal quality, sustained user engagement, and the ability of instructors to incorporate this information into pedagogical practice.

Regarding the scope of the population, the system was designed and evaluated in a higher education context, primarily involving university students. Therefore, the observed emotional patterns, stress responses, and usability perceptions are characteristic of this specific demographic. Given that emotional regulation and autonomic physiological reactivity vary across the lifespan [10], these results should not be directly extrapolated to other educational levels or age groups without further cross-demographic validation.

6 Conclusions

This work presented a real-time emotional state monitoring system that integrates physiological signal acquisition (EDA, BVP, and SKT), automated classification, and visualization for educational environments. Among the evaluated models, XGBoost with SMOTE-Tomek achieved the highest performance, reaching an accuracy of 91.20% and a 0.9127 F1-score. Experimental validation showed consistency between predictions and self-reported measures, correctly identifying 43 out of 45 emotional blocks, indicating stable performance under real-time conditions and across different acquisition devices. The system also obtained a mean SUS score of 88.0, reflecting a high level of perceived usability and supporting its applicability in classroom settings.

Despite these results, some limitations remain. The model was trained solely on the WESAD dataset, which may not fully capture classroom variability, and real-time acquisition was affected by noise, latency from window-based processing, and signal instability during movement. Future work will focus on incorporating classroom-collected data, refining models, and integrating contextual information. Additional extensions, such as timeline annotations and automated pedagogical recommendations, could further enhance the practical utility of the system.

Disclosure of Interests. The authors have no competing interests to declare that are relevant to the content of this article.

References

1. Ack Baraly, K.T., Muyingo, L., Beaudoin, C., Karami, S., Langevin, M., Davidson, P.S.R.: Database of emotional videos from Ottawa (devo). Collabra: Psychol. **6**(1) (2020). https://doi.org/10.1525/collabra.180
2. Ahmed, I., Hazell, C.M., Edwards, B., Glazebrook, C., Davies, E.B.: A systematic review and meta-analysis of studies exploring prevalence of non-specific anxiety in undergraduate university students. BMC Psychiatry **23**(1) (2023). https://doi.org/10.1186/s12888-023-04645-8
3. Almarzouki, A.F.: Stress, working memory, and academic performance: a neuroscience perspective. Stress **27**(1), 2364333 (2024). https://doi.org/10.1080/10253890.2024.2364333
4. AlZoubi, O., D'Mello, S.K., Calvo, R.A.: Detecting naturalistic expressions of nonbasic affect using physiological signals. IEEE Trans. Affect. Comput. **3**(3), 298–310 (2012). https://doi.org/10.1109/t-affc.2012.4
5. Cao, H., Xie, X., Shi, J., Wang, Y.: Evaluating the validity of class balancing algorithms-based machine learning models for geogenic contaminated groundwaters prediction. J. Hydrol. **610**, 127933 (2022). https://doi.org/10.1016/j.jhydrol.2022.127933
6. Choksi, K., Chen, H., Joshi, K., Jade, S., Nirjon, S., Lin, S.: Sensemo: enabling affective learning through real-time emotion recognition with smartwatches. In: 2024 IEEE 21st International Conference on Mobile Ad-Hoc and Smart Systems (MASS), pp. 453–459 (2024).https://doi.org/10.1109/mass62177.2024.00066
7. Eren, E., Navruz, T.S.: Stress detection with deep learning using BVP and EDA signals. In: 2022 International Congress on Human-Computer Interaction, Optimization and Robotic Applications (HORA), pp. 1–7. IEEE (2022). https://doi.org/10.1109/hora55278.2022.9799933
8. Fang, A., Pan, F., Yu, W., Yang, L., He, P.: ECG-based emotion recognition using random convolutional kernel method. Biomed. Signal Process. Control **91**, 105907 (2024). https://doi.org/10.1016/j.bspc.2023.105907
9. Gahlan, N., Sethia, D.: AFLEMP: attention-based federated learning for emotion recognition using multi-modal physiological data. Biomed. Signal Process. Control **94**, 106353 (2024). https://doi.org/10.1016/j.bspc.2024.106353
10. Gomez, P., Von Gunten, A., Danuser, B.: Autonomic nervous system reactivity within the valence–arousal affective space: modulation by sex and age. Int. J. Psychophysiol. **109**, 51–62 (2016). https://doi.org/10.1016/j.ijpsycho.2016.10.002
11. Lebaka, L.N., Govarthan, P.K., Rani, P., Ganapathy, N., Agastinose Ronickom, J.F.: Automated Emotion Recognition System Using Blood Volume Pulse and XGBoost Learning. IOS Press (2023). https://doi.org/10.3233/shti230422

12. Ma, S., Zhou, T., Nie, F., Ma, X.: Glancee: an adaptable system for instructors to grasp student learning status in synchronous online classes. In: CHI Conference on Human Factors in Computing Systems, CHI '22, pp. 1–25. ACM (2022). https://doi.org/10.1145/3491102.3517482
13. Masaki, A., Nagumo, K., Lamsal, B., Oiwa, K., Nozawa, A.: Anomaly detection in facial skin temperature using variational autoencoder. Artif. Life Rob. **26**(1), 122–128 (2020). https://doi.org/10.1007/s10015-020-00634-2
14. Pro Gómez, Z.L.: Relación entre estrés académico y bienestar psicológico en estudiantes. Tesis, Universidad de Lima, Lima, Perú (2021). https://hdl.handle.net/20.500.12724/13299
15. Robles-Rivera, K., et al.: Factors associated with depression, anxiety, and stress in Mexican medical students: a cross-sectional study. Curr. Psychol. **44**(10), 9252–9264 (2025). https://doi.org/10.1007/s12144-025-07742-x
16. Samyoun, S., Islam, M.M., Iqbal, T., Stankovic, J.: M3Sense: affect-agnostic multitask representation learning using multimodal wearable sensors. Proc. ACM Interact. Mobile Wearable Ubiq. Technol. **6**(2), 1–32 (2022). https://doi.org/10.1145/3534600
17. Schmidt, P., Reiss, A., Duerichen, R., Marberger, C., Van Laerhoven, K.: Introducing WESAD, a multimodal dataset for wearable stress and affect detection. In: Proceedings of the 20th ACM International Conference on Multimodal Interaction, ICMI '18, pp. 400–408. ACM (2018). https://doi.org/10.1145/3242969.3242985
18. Segura Márquez, C.L.: Inteligencia emocional-social, estrés y rendimiento académico en estudiantes de una universidad privada de Lima Metropolitana. Tesis de maestría, Universidad Peruana (2021). http://hdl.handle.net/20.500.11955/927
19. Siddiqui, N., et al.: An ai-based classroom monitoring system leveraging computer vision and machine learning. In: 2023 20th International Bhurban Conference on Applied Sciences and Technology (IBCAST), pp. 273–278 (2023). https://doi.org/10.1109/IBCAST59916.2023.10713048
20. Vasile, F., Vizziello, A., Brondino, N., Savazzi, P.: Stress state classification based on deep neural network and electrodermal activity modeling. Sensors **23**(5), 2504 (2023). https://doi.org/10.3390/s23052504
21. Wu, Y., Daoudi, M., Amad, A.: Transformer-based self-supervised multimodal representation learning for wearable emotion recognition. IEEE Trans. Affect. Comput. **15**(1), 157–172 (2024). https://doi.org/10.1109/taffc.2023.3263907
22. Zhu, L., Spachos, P., Ng, P.C., Yu, Y., Wang, Y., Plataniotis, K., Hatzinakos, D.: Stress detection through wrist-based electrodermal activity monitoring and machine learning. IEEE J. Biomed. Health Inf. **27**(5), 2155–2165 (2023). https://doi.org/10.1109/jbhi.2023.3239305
23. Zitouni, M.S., Park, C.Y., Lee, U., Hadjileontiadis, L.J., Khandoker, A.: LSTM-modeling of emotion recognition using peripheral physiological signals in naturalistic conversations. IEEE J. Biomed. Health Inf. **27**(2), 912–923 (2023). https://doi.org/10.1109/jbhi.2022.3225330

Reframing Open Large Language Models in Higher Education: Organizational Governance, Decision-Making Augmentation, and Institutional Transformation

Iván Miguel García-López(✉) and Nicia Guillén-Yparrea

School of Engineering and Sciences, Tecnologico de Monterrey, Monterrey, Estado de Mexico, Mexico
A01696450@tec.mx

Abstract. The rapid expansion of generative artificial intelligence in higher education is transforming institutional governance, managerial decision-making, and organizational processes, positioning Open Large Language Models (OLLMs) as key socio-technical infrastructures rather than isolated technological tools. Despite their increasing adoption, there is still limited systematic evidence on how these models reshape organizational dynamics and governance structures within higher education institutions. This study addresses this gap through a Systematic Literature Review (SLR) of 551 peer-reviewed articles published between 2020 and 2025, focusing on the role of OLLMs in educational management. The analysis examines five core dimensions: managerial application domains, governance models, levels of organizational adoption, value creation, and associated risks. The findings show that OLLMs are primarily used as cognitive infrastructures that augment managerial decision-making, particularly in strategic planning, institutional evaluation, and administrative optimization. However, their adoption remains uneven, with most institutions operating at exploratory or unit levels, reflecting differences in organizational maturity and digital capabilities. Institutions with higher maturity demonstrate more integrated and strategic uses aligned with data-driven governance. Governance emerges as a central concern, emphasizing the need for transparency, accountability, and risk management to address issues such as bias, data protection, and over-reliance on automated systems. The results indicate that the organizational value of OLLMs depends less on the technology itself and more on institutional capacity, managerial competencies, and governance alignment.

Keywords: large language models · decision-making · higher education

1 Introduction

In recent years, artificial intelligence has transformed organizational decision-making, strategic planning, and knowledge management, becoming a central component of contemporary administration. In particular, large-scale language models and their open

H. Ponce et al. (Eds.): COMIA 2026, CCIS 3095, pp. 90–103, 2027.
https://doi.org/10.1007/978-3-032-37438-7_8

versions, Open Large Language Models (OLLMs), have expanded the possibilities for organizational analysis by enabling the processing of complex information in real time [1, 2]. In higher education, these transformations are especially relevant because institutional decisions involve strategic, social, and organizational implications beyond isolated classroom applications.

From a business administration perspective applied to education, OLLM adoption should be understood not merely as a technological phenomenon, but as an organizational process that redefines managerial practices, governance structures, and institutional capabilities. Within higher education studies, OLLMs can be situated in debates on governance, performance, accountability, datafied management, and organizational capability building. Thus, they operate less as classroom technologies and more as cognitive infrastructures that reorganize managerial routines and redistribute responsibility across human and socio-technical systems [3, 4].

This study is conceptually articulated around three constructs: artificial intelligence as strategic support for managerial decision-making, organizational maturity and managerial capabilities for effective OLLM adoption, and governance and organizational responsibility in the use of these systems [5, 6]. Unlike previous studies that approach AI in education mainly from pedagogical or technical perspectives [7], this paper adopts a managerial and organizational lens focused on decision-making, strategic management, and institutional governance. The objective is to systematically analyze how OLLMs are being used in educational management, identifying patterns, levels of organizational integration, and key managerial implications from recent literature.

The focus on OLLMs is justified because their openness has direct implications for higher education governance. Compared with proprietary models accessed through closed interfaces, OLLMs may offer greater institutional control over deployment, data flows, auditability, customization, cost management, and alignment with internal governance policies. This does not imply that open models are inherently safer or more accurate, but that their adoption raises distinctive managerial questions related to transparency, institutional autonomy, accountability, and technological sovereignty. Therefore, this article contributes to understanding OLLMs as strategic management tools and to advancing a more informed discussion on AI-based educational administration.

2 Theoretical Framework

2.1 Artificial Intelligence as a Strategic Support for Managerial Decision-Making in Education

The incorporation of language models and artificial AI systems in educational management has increasingly been conceptualized as a strategic support mechanism for managerial decision-making, rather than as a simple technological tool [8]. From the point of view of business administration, this vision situates OLLMs as cognitive infrastructures capable of processing large volumes of organizational information and transforming them into relevant inputs for planning, control, and institutional management [9].

Recent literature emphasizes that the strategic value of AI in educational contexts lies in its ability to improve the quality of decisions, reducing uncertainty and broadening the analytical horizon of managers. Empirical studies show that AI-based systems allow academic, administrative, and contextual data to be integrated [10], facilitating more informed decisions in areas such as resource allocation, institutional evaluation, and internal policy design [11]. This approach converges with the theory of limited rational decision-making, by extending human cognitive capacities without replacing them. In this sense, AI does not replace managerial judgment, but complements and amplifies it, configuring an augmented decision-making model that is particularly relevant for complex educational organizations [12]. This construct establishes the conceptual basis for interpreting the results associated with RQs that analyze the strategic impact of OLLMs and connects directly with the discussions on organizational maturity and algorithmic governance that are developed in the following constructs.

2.2 Organizational Maturity and Managerial Capabilities for the Adoption of OLLMs

The effective adoption of OLLMs in educational management depends critically on the level of organizational maturity and managerial capacities existing within the institutions. From a business perspective, technology maturity is not limited to infrastructure availability [13], but involves processes, managerial competencies, and strategic alignment [14]. Various studies indicate that organizations with higher levels of digital maturity tend to integrate AI more strategically, articulating it with long-term institutional objectives and data-driven management models [15]. In education, this maturity translates into the ability of management teams to interpret results generated by AI systems [16], incorporate them into formal decision-making processes, and critically evaluate their organizational implications. Conversely, in contexts of low organizational maturity, AI is often used in a fragmented or experimental manner, without clear integration into the management architecture [17]. This difference explains the coexistence of operational and strategic impacts observed in the empirical results, and reinforces the need to analyze the adoption of OLLMs not only as a technological phenomenon, but as an organizational and managerial process [18]. This construct allows empirical findings to be linked to theories of organizational change and paves the way for discussing the implications of governance and accountability.

2.3 Governance, Accountability, and Organizational Alignment in the Use of OLLMs

The use of OLLMs in educational management poses substantive challenges in terms of organizational governance, accountability, and control, core dimensions of contemporary business administration [19]. The literature agrees that the adoption of AI in decision-making processes requires governance frameworks that define roles [20], clear responsibilities and oversight mechanisms. In educational contexts, these challenges are intensified due to the public or social nature of many institutional decisions, which increases the need for transparency, traceability and accountability [16]. Recent studies highlight that the lack of adequate governance structures can lead to over-reliance on algorithmic systems, undetected organizational biases, and loss of managerial control [21]. From this perspective, AI must be integrated as part of formal management systems and not as an autonomous actor. This construct underscores that responsible adoption of OLLMs requires explicit alignment between technology, strategy, and organizational values, which is especially relevant for educational institutions oriented towards quality, equity, and sustainability [22, 23]. AI governance thus emerges as a cross-cutting axis that connects empirical results with normative and strategic debates, and serves as a bridge to the discussion section, where these constructs are directly contrasted with the findings obtained. A central governance challenge in the use of OLLMs for educational management is their limited reliability in high-stakes decision contexts. These systems may generate plausible but inaccurate outputs, reproduce unsupported claims, omit relevant evidence, or produce hallucinated information when institutional data are incomplete, ambiguous, or poorly contextualized. For this reason, OLLMs should not be treated as autonomous decision-makers, but as decision-support systems embedded in human oversight, source verification, audit trails, and validation protocols. In higher education management, this implies that recommendations generated by OLLMs for strategic planning, resource allocation, quality assurance, or institutional evaluation must be reviewed by accountable decision-makers before being incorporated into formal governance processes.

3 Method

This study adopts a Systematic Literature Review (SLR) to analyze how OLLMs and generative AI are incorporated into educational management and organizational decision-making in higher education [24]. This approach was selected because the field is emerging, multidisciplinary, and fragmented across artificial intelligence, business administration, educational management, and organizational studies. The protocol followed established guidelines for systematic reviews, prioritizing transparency, traceability, and replicability [25, 26].

The review was guided by five research questions aligned with the theoretical framework: OLLMs as strategic support for managerial decision-making, organizational maturity and managerial capabilities, and governance and organizational responsibility in AI use. Table 1 presents the RQs and the analytical categories used to classify the literature.

The systematic search was conducted in Scopus and Web of Science (WoS), selected for their coverage of indexed literature in educational management, administration, information systems, and organizational studies. The search strategy combined terms related to open language models or generative AI, educational management or institutional leadership, and decision-making or organizational processes. Boolean operators and truncations were applied to titles, abstracts, and keywords to avoid an exclusively pedagogical or technical focus. The search strings were:

Scopus
("open large language model*" OR "open LLM*" OR "open-source language model*" OR "generative AI") AND ("educational management" OR "higher education management" OR "university management" OR "educational leadership" OR "academic decision-making") AND ("decision-making" OR "organizational decision*" OR "management")

Table 1. Research Questions and Analytical Focus

Question	Possible response
RQ1. In which "managerial area" is the use/impact of OLLMs/GenAI in education mainly located?	Strategic planning & governance; Teaching–learning operations management; Assessment & quality assurance; Student services & administrative productivity; Research/knowledge management & innovation [27].
RQ2. What "governance model" is proposed/observed for the institutional use of OLLMs/GenAI?	Principles-to-practice governance; Risk-based governance; Capability-based governance (organizational capabilities to deploy responsible AI); Policy & compliance governance (internal rules, control, audit); Academic integrity governance [28].
RQ3. What "level of organizational adoption" does the article describe for OLLMs/GenAI?	Exploratory / pilot adoption (testing, prototyping, piloting); Unit-level adoption (department/program/area); Institution-wide integration; Process reengineering / transformation; Ecosystem/platform governance (integrations with vendors, platforms, interoperability) [29].
RQ4. What "managerial value/outcome type" is reported (or searched for) with OLLMs/GenAI?	Decision quality & responsiveness; Efficiency & productivity gains; Educational quality improvement; Equity & inclusion outcomes; Organizational learning & capability building [30].
RQ5. What is the "dominant risk" discussed when implementing OLLMs/GenAI in educational management?	Privacy/data protection risk; Bias/fairness risk; Transparency/accountability risk; Academic integrity risk; Workforce/deskilling risk (role changes, substitution, deprofessionalization) [31],

WoS
TS = (("open large language model*" OR "open LLM*" OR "open-source language model*" OR "generative artificial intelligence") AND ("educational management" OR "higher education management" OR "university management" OR "educational leadership") AND ("decision-making" OR "organizational decision*" OR management))

The inclusion criteria focused on peer-reviewed articles that explicitly addressed educational management, organizational decision-making, or institutional governance mediated by OLLMs or generative AI. Studies were excluded when they focused exclusively on classroom use, individual tutoring, pedagogical design without institutional implications, tangential mentions of OLLMs, grey literature, conference proceedings, or strictly technical applications.

The search yielded 607 records in Scopus and 9 in WoS, for an initial total of 616 records. After removing 4 duplicates, 612 unique records were screened by title, abstract, and keywords. During screening, 61 articles unrelated to educational management or organizational decision-making were excluded, resulting in a final corpus of 551 articles, which constitutes the empirical basis of the review (Fig. 1) [32]. For each article, bibliographic data, including authors, year, journal, and DOI, and analytical data, including managerial area, adoption level, governance model, managerial value, and dominant risk, were extracted.

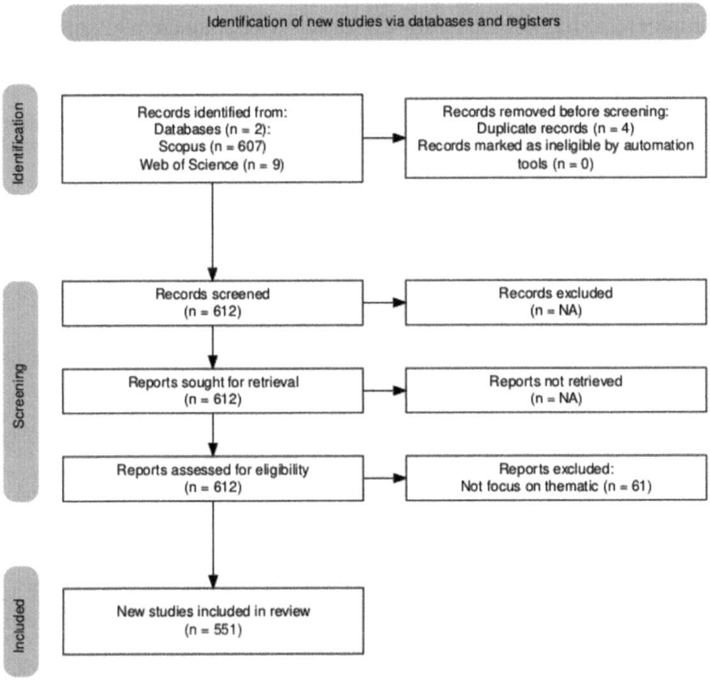

Fig. 1. PRISMA Flow diagram

Data synthesis was conducted through structured qualitative content analysis. Each article was coded according to the analytical categories associated with the five RQs. Then, an interpretive analysis identified cross-cutting patterns, tensions, and conceptual gaps, particularly regarding the alignment between strategic OLLM use, organizational maturity, and institutional governance. The results were represented through graphical distributions by RQ to assess the degree of organizational integration and systemic impact of OLLMs in educational management.

4 Results

The results are organized according to the five research questions and summarized in Figs. 2, 3, 4, 5 and 6. Overall, the literature shows that OLLMs and GenAI are increasingly positioned as strategic resources for educational management, particularly in relation to decision-making, governance, organizational adoption, managerial value, and implementation risks.

Figure 2 shows that the main managerial areas associated with OLLM and GenAI use in higher education are strategic planning, institutional analysis, and process optimization [33, 34]. This concentration suggests a shift from a technical view of AI toward a strategic understanding aligned with evidence-based management and organizational analytics. Although some studies emphasize organizational improvement and institutional efficiency, exploratory or incremental innovation appears less frequently, indicating that the field is still consolidating its integration into broader institutional strategy [35].

Fig. 2. Managerial areas of OLLM/GenAI use in higher education

Figure 3 presents the governance models identified for institutional OLLM and GenAI adoption [36]. The literature mainly emphasizes accountability, transparency, risk management, and compliance as key governance principles for educational administration [37]. These findings indicate that OLLMs are not generally framed as autonomous systems, but as decision-support infrastructures that require formal oversight, data protection, responsible use policies, and institutional control mechanisms [38].

Figure 4 shows the levels of organizational adoption described in the reviewed literature, ranging from exploratory and unit-level uses to institution-wide, transformational, and ecosystem-based integration [39]. The results indicate a predominance of tactical and strategic adoption, especially in institutional planning, performance management, and educational governance [40]. However, operational and partial uses remain common, reflecting heterogeneous levels of digital maturity and managerial capability across institutions [41].

Figure 5 summarizes the main managerial value and organizational outcomes associated with OLLM and GenAI adoption [42]. The most recurrent outcomes are improved

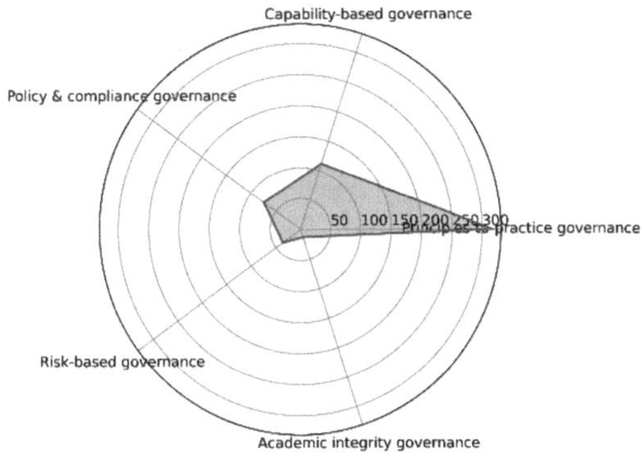

Fig. 3. Governance models for institutional OLLM/GenAI adoption

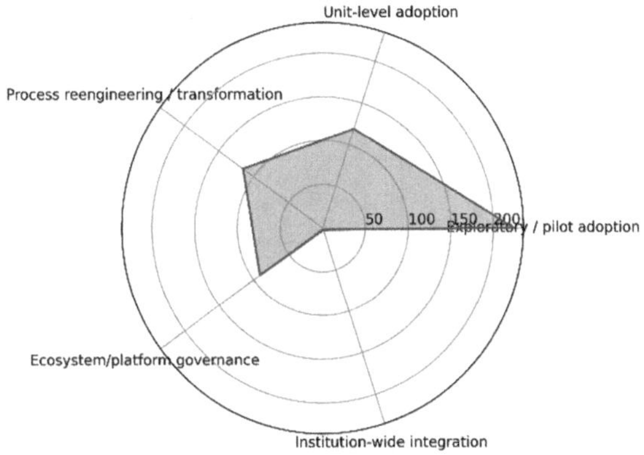

Fig. 4. Levels of organizational adoption of OLLMs/GenAI

decision quality, administrative efficiency, institutional responsiveness, and organizational learning [43]. These results suggest that OLLMs can strengthen evidence-based planning and process optimization when integrated into formal management systems. However, this value depends on data quality, human oversight, and governance mechanisms that prevent technological dependency and preserve managerial accountability [44].

Figure 6 presents the dominant risks identified in the literature, including privacy and data protection, bias and fairness, transparency and accountability, academic integrity, and workforce-related risks [45]. Privacy, transparency, and accountability appear as the most recurrent concerns, given the sensitivity of institutional data and the potential impact of AI-supported decisions in higher education [46]. Bias, fairness, and over-reliance on algorithmic outputs are also critical when OLLMs are used for planning, evaluation, or resource allocation. These findings confirm that responsible implementation requires governance mechanisms, human oversight, auditability, and validation protocols [47].

Taken together, the results show that OLLMs and GenAI are moving from isolated or technical applications toward more strategic forms of educational management. Their

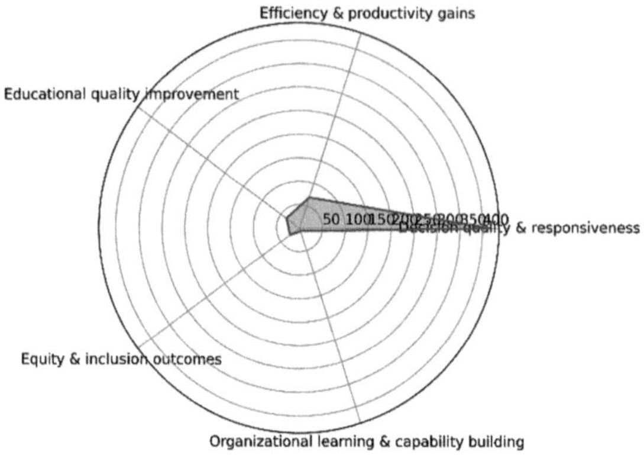

Fig. 5. Managerial value and organization

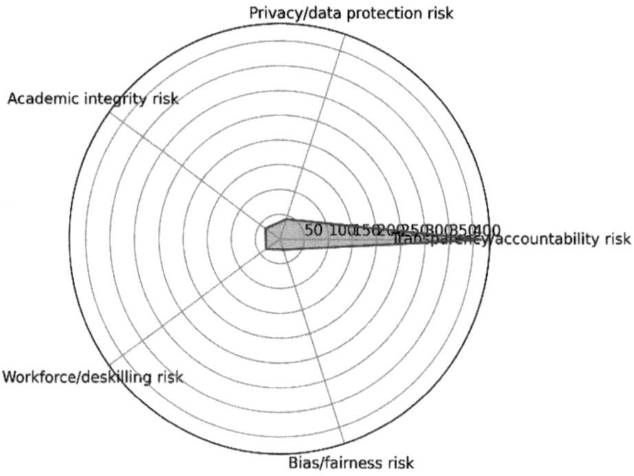

Fig. 6. Dominant risks in OLLM/GenAI implementation

institutional value is linked not only to analytical capacity, but also to adoption maturity, managerial capabilities, data quality, and governance alignment.

5 Discussion

The findings show that OLLMs are being incorporated into higher education management mainly as decision-support infrastructures rather than as autonomous automation tools. This result is consistent with the first theoretical construct, which conceptualizes AI as a strategic cognitive infrastructure for managerial decision-making. Across the reviewed studies, OLLMs support institutional planning, performance analysis, organizational evaluation, and process optimization, expanding managers' analytical capacity without replacing managerial judgment [17, 38, 41, 43].

The results also confirm that the organizational value of OLLMs depends strongly on institutional maturity and managerial capabilities. Institutions with higher levels of digital maturity tend to integrate these systems into transversal management processes, while less mature institutions use them in fragmented, experimental, or operational ways [3, 7, 18, 19]. This distinction reinforces the idea that OLLM adoption is not merely a technological implementation, but an organizational change process that requires leadership, data literacy, analytical culture, and strategic alignment.

Governance emerges as a central condition for responsible and effective adoption. The reviewed literature shows that, without clear frameworks for accountability, supervision, transparency, and control, OLLMs may increase risks such as opaque decisions, excessive reliance on algorithmic outputs, undetected bias, unreliable information, and displacement of managerial responsibility [7, 33, 38, 40]. Therefore, OLLMs should be integrated into formal governance structures as decision-support systems subject to human oversight, validation protocols, source traceability, and institutional accountability.

To clarify the managerial implications of OLLM adoption, Table 2 presents examples of how these systems can support institutional decision-making across different administrative domains. The table connects managerial areas with specific OLLM uses, the type of decision supported, the institutional data required, and the governance controls needed for responsible implementation.

Overall, the discussion confirms that the impact of OLLMs in educational management is systemic: these models can reshape decision-making routines, institutional analytics, administrative processes, and governance practices when they are aligned with organizational strategy [22, 46]. However, their value depends not only on analytical capability, but also on data quality, managerial competence, institutional maturity, and governance controls. Thus, OLLMs should be understood as socio-technical infrastructures embedded in organizational routines, rather than as autonomous systems capable of replacing human responsibility.

6 Conclusion

The results of this systematic review show that OLLMs are being incorporated into educational management mainly as strategic decision-support systems rather than autonomous automation tools. The evidence confirms that OLLMs can expand the analytical capacity of educational managers by integrating organizational and contextual information, supporting planning, evaluation, and administrative optimization. However, their value depends less on the technology itself than on institutional maturity, digital leadership, managerial capabilities, and alignment with formal decision-making processes.

The review also highlights governance as a critical condition for responsible adoption. Without clear accountability, oversight, transparency, and validation mechanisms, OLLMs may increase risks such as opaque decisions, algorithmic dependency, unreliable outputs, bias, and dilution of managerial responsibility. Therefore, OLLMs should be understood as socio-technical infrastructures embedded in institutional routines, requiring human supervision, data governance, and ethical alignment with educational objectives.

Table 2. Managerial applications of OLLMs in higher education institutions

Managerial area	Example of OLLM use	Decision supported	Required institutional data	Governance control
Strategic planning	Summarizing accreditation reports, performance indicators, and environmental trends	Prioritization of institutional goals	Strategic plans, KPIs, accreditation reports	Human validation and source traceability
Institutional evaluation	Detecting patterns in program reviews and student feedback	Identification of improvement áreas	Evaluation reports, surveys, dashboards	Bias review and evidence triangulation
Administrative optimization	Automating classification of service requests	Resource allocation and workload management	Helpdesk tickets, administrative logs	Privacy controls and role-based access
Student services	Generating early alerts from academic and administrative signals	Support intervention planning	Retention data, advising notes, risk indicators	Explainability and non-discrimination checks
Research management	Mapping research outputs and collaboration networks	Funding and partnership strategy	Publications, grants, project databases	Data quality review and periodic audit

This study has limitations. Although it synthesizes a broad corpus of indexed literature, it does not provide direct empirical validation through case studies, experiments, or longitudinal institutional data. In addition, the literature tends to emphasize institutions with higher levels of digital maturity, limiting understanding of low-resource contexts. Future research should compare open and proprietary models in terms of reliability, cost, auditability, privacy, institutional control, and governance alignment. It should also develop comparative, longitudinal, and mixed-method studies to assess how AI-mediated managerial decision-making evolves across different educational governance models and organizational maturity levels.

Acknowledgments. The authors appreciate the financial support from Tecnologico de Monterrey through the "Challenge-Based Research Funding Program 2024". Project ID # I005-CI_IFE_EDU_D_225.

The authors acknowledge the technical and financial support of Writing Lab, Institute for the Future of Education, Tecnologico de Monterrey, Mexico, in the production of this work.

Disclosure of Interests The authors have no competing interests to declare that are relevant to the content of this article.

References

1. Anand, T., Mitchell, D.: Objectives and curriculum for a graduate business analytics capstone: reflections from practice. Decis. Sci. J. Innov. Educ. **20**, 235–245 (2022). https://doi.org/10.1111/dsji.12272
2. Chen, Z., Tang, T., Zhang, F., Deng, M.: Symbiosis-evolution game and scenario-simulation analysis of advanced manufacturing enterprises from the perspective of an innovation ecosystem. Sustainability. **15**, 8647 (2023). https://doi.org/10.3390/su15118647
3. Abdelwahab, H.R., Rauf, A., Chen, D.: Business students' perceptions of Dutch higher educational institutions in preparing them for artificial intelligence work environments. Ind. High. Educ. **37**, 22–34 (2023). https://doi.org/10.1177/09504222221087614
4. Aleven, V., McLaughlin, E.A., Glenn, C.: AI in education: policies and practices for ethical implementation. Int. J. Sci. Educ. **43**, 78–102 (2021). https://doi.org/10.1080/09500693.2021.1990889
5. Nizamani, M.M., Zhang, H.-L., Lai, Z.: Human-centered AI: advancing ethical, transparent, and context-aware systems for sustainable development. Technol. Soc. **84**, 103121 (2026). https://doi.org/10.1016/j.techsoc.2025.103121
6. Linkon, A.A., et al.: Advancements and applications of generative artificial intelligence and large language models on business management: a comprehensive review. J. Comput. Sci. Technol. Stud. **6**, 225–232 (2024). https://doi.org/10.32996/jcsts.2024.6.1.26
7. Chiu, T.K.F., Xia, Q., Zhou, X., Chai, C., Cheng, M.: Systematic literature review on opportunities, challenges, and future research recommendations of artificial intelligence in education. **4** (2023). https://doi.org/10.1016/j.caeai.2022.100118
8. Ferrara, E.: Should ChatGPT be biased? challenges and risks of bias in large language models. First Monday. (2023). https://doi.org/10.5210/fm.v28i11.13346
9. Bai, X., Xie, Y., Zhang, X., Han, H., Li, J.-R.: Evaluation of open-source large language models for metal–organic frameworks research. J. Chem. Inf. Model. **64**, 4958–4965 (2024). https://doi.org/10.1021/acs.jcim.4c00065
10. Bano, M., Zowghi, D., Shea, P., Ibarra, G.: Investigating responsible AI for scientific research: an Empirical Study., https://arxiv.org/abs/2312.09561. (2023). https://doi.org/10.48550/ARXIV.2312.09561
11. Bannister, P., Alcalde Peñalver, E., Santamaría Urbieta, A.: Transnational higher education cultures and generative AI: a nominal group study for policy development in English medium instruction. J. Multicult. Educ. **18**, 173–191 (2023). https://doi.org/10.1108/JME-10-2023-0102
12. Arora, S., Goyal, A.: A theory for emergence of complex skills in language models. ArXiv Prepr. ArXiv230715936. (2023)
13. Chan, A.: GPT-3 and InstructGPT: technological dystopianism, utopianism, and "contextual" perspectives in AI ethics and industry. AI Ethics. **3**, 53–64 (2023). https://doi.org/10.1007/s43681-022-00148-6
14. Ali, O., Murray, P.A., Momin, M., Dwivedi, Y.K., Malik, T.: The effects of artificial intelligence applications in educational settings: challenges and strategies. Technol. Forecast. Soc. Change. **199**, 123076 (2024). https://doi.org/10.1016/j.techfore.2023.123076

15. Abhishek, A., Erickson, L., Bandopadhyay, T.: Data and AI governance: promoting equity, ethics, and fairness in large language models., https://arxiv.org/abs/2508.03970. (2025). https://doi.org/10.48550/ARXIV.2508.03970
16. Balasubramaniam, N., Kauppinen, M., Rannisto, A., Hiekkanen, K., Kujala, S.: Transparency and explainability of AI systems: from ethical guidelines to requirements. Inf. Softw. Technol. **159**, 107197 (2023). https://doi.org/10.1016/j.infsof.2023.107197
17. Alharthi, S.: Harnessing knowledge: the robust role of knowledge management practices and business intelligence systems in developing entrepreneurial leadership and organizational sustainability in SMEs. Sustainability. **17**, 6264 (2025). https://doi.org/10.3390/su17146264
18. Bommasani, R., et al.: On the opportunities and risks of foundation models., https://arxiv.org/abs/2108.07258. (2021). https://doi.org/10.48550/ARXIV.2108.07258
19. Aler Tubella, A., Mora-Cantallops, M., Nieves, J.C.: How to teach responsible AI in higher education: challenges and opportunities. Ethics Inf. Technol. **26** (2024). https://doi.org/10.1007/s10676-023-09733-7
20. Altememy, H.A., et al.: The influence of the artificial intelligence capabilities of higher education institutions in Iraq on students' academic performance: the role of AI-based technology application as a mediator. Eurasian J. Educ. Res. **2023**, 267–282 (2023). https://doi.org/10.14689/ejer.2023.104.015
21. Abramski, K., Citraro, S., Lombardi, L., Rossetti, G., Stella, M.: Cognitive network science reveals bias in GPT-3, GPT-3.5 turbo, and GPT-4 mirroring math anxiety in high-school students. Big Data Cogn. Comput. **7**, 124 (2023). https://doi.org/10.3390/bdcc7030124
22. Abbasnejad, B., Soltani, S., Wong, P.: A systematic review of online learning and teaching strategies during the COVID-19 pandemic: implications for the construction management sector. Smart Sustain. Built Environ. **13**, 934–959 (2024). https://doi.org/10.1108/SASBE-08-2022-0174
23. Aggarwal, D.: Adoption of artificial intelligence (AI) for development of smart education as the future of a sustainable education system. J. Artif. Intell. Mach. Learn. Neural Netw. (2023). https://doi.org/10.55529/jaimlnn.36.23.28
24. Kitchenham, B., et al.: Systematic literature reviews in software engineering – a tertiary study. Inf. Softw. Technol. **52**, 792–805 (2010). https://doi.org/10.1016/j.infsof.2010.03.006
25. García-Peñalvo, F.J.: Mapeos Sistemáticos De Literatura. Revisiones Sistemáticas De Literatura Y Benchmarking De Programas Formativos. (2017). https://doi.org/10.5281/ZENODO.1067680
26. Kitchenham, B., Charters, S.: Guidelines for performing systematic literature reviews in software engineering. **2** (2007)
27. Beñalet Jr., C.E.C., et al.: A systematic review on educational strategic management. Int. J. Res. Innov. Soc. Sci. **VII**, 2371–2383 (2023). https://doi.org/10.47772/IJRISS.2023.701177
28. Bakhshandeh, S.: Benchmarking medical large language models. Nat. Rev. Bioeng. **1**, 543 (2023). https://doi.org/10.1038/s44222-023-00097-7
29. Cristache, N., Croitoru, G., Florea, N.V.: The influence of knowledge management on innovation and organizational performance. J. Innov. Knowl. **10**, 100793 (2025). https://doi.org/10.1016/j.jik.2025.100793
30. Chan, C.K.Y., Zhou, W.: An expectancy value theory (EVT) based instrument for measuring student perceptions of generative AI. Smart Learn. Environ. **10** (2023). https://doi.org/10.1186/s40561-023-00284-4
31. Ferrer, X., Nuenen, T.V., Such, J.M., Cote, M., Criado, N.: Bias and discrimination in AI: a cross-disciplinary perspective. IEEE Technol. Soc. Mag. **40**, 72–80 (2021). https://doi.org/10.1109/MTS.2021.3056293
32. Page, M.J., et al.: Declaración PRISMA 2020: una guía actualizada para la publicación de revisiones sistemáticas. Rev. Esp. Cardiol. **74**, 790–799 (2021). https://doi.org/10.1016/j.recesp.2021.06.016

33. Li, M., Wang, Z.: BuildingGPT: query building semantic data using large language models and vector-graph retrieval-augmented generation. Build. Environ. **287**, 113855 (2026). https://doi.org/10.1016/j.buildenv.2025.113855
34. Abu-Issa, A., Butmeh, H., Tumar, I.: Modeling students' preferences and knowledge for improving educational achievements. Front. Comput. Sci. **6**, 1359770 (2024). https://doi.org/10.3389/fcomp.2024.1359770
35. Aluko, H.A., et al.: Rethinking business practices: harnessing indigenous knowledge systems and sustainable strategies for resilient entrepreneurial success. Bus. Strategy Environ., bse.70306 (2025). https://doi.org/10.1002/bse.70306
36. Chhatwal, M., Garg, V., Rajput, N.: Role of AI in the Education Sector. Lloyd Bus. Rev. (2023). https://doi.org/10.56595/lbr.v2i1.11
37. Agbese, M., Mohanani, R., Khan, A., Abrahamsson, P.: Implementing AI ethics: making sense of the ethical requirements. In: Proceedings of the 27th International Conference on Evaluation and Assessment in Software Engineering, pp. 62–71. ACM, Oulu Finland (2023). https://doi.org/10.1145/3593434.3593453
38. He, J., et al.: Context-Guided Dynamic Retrieval for Improving Generation Quality in RAG Models., https://arxiv.org/abs/2504.19436. (2025). https://doi.org/10.48550/ARXIV.2504.19436
39. Jernite, Y., et al.: Data governance in the age of large-scale data-driven language technology. In: 2022 ACM Conference on Fairness, Accountability, and Transparency, pp. 2206–2222. ACM, Seoul Republic of Korea (2022). https://doi.org/10.1145/3531146.3534637
40. Silva, A.A.: Gobernanza, poder y autonomía universitaria en la era de la innovación. Perfiles Educ. **44**, 150–164 (2022). https://doi.org/10.22201/iisue.24486167e.2022.178.60735
41. Bacon, D.R.: Recommendations for the use of experimental designs in management education research. J. Manag. Educ. **48**, 1121–1147 (2024). https://doi.org/10.1177/10525629241252802
42. Dotan, R., Parker, L.S., Radzilowicz, J.G.: Responsible adoption of generative AI in higher education: developing a "points to consider" approach based on faculty perspectives. (2024). https://doi.org/10.48550/ARXIV.2406.01930.
43. Ben-Michael, E., et al.: Does AI help humans make better decisions? a statistical evaluation framework for experimental and observational studies., https://arxiv.org/abs/2403.12108. (2024). https://doi.org/10.48550/ARXIV.2403.12108
44. Cheong, B.C.: Transparency and accountability in AI systems: safeguarding wellbeing in the age of algorithmic decision-making. Front. Hum. Dyn. **6**, 1421273 (2024). https://doi.org/10.3389/fhumd.2024.1421273
45. Ferguson, C., van den Broek, E.L., van Oostendorp, H.: AI-induced guidance: preserving the optimal zone of proximal development. Comput. Educ. Artif. Intell. **3**, 100089 (2022). https://doi.org/10.1016/j.caeai.2022.100089
46. Jeong, S., Baek, J., Cho, S., Hwang, S.J., Park, J.: Adaptive-RAG: learning to adapt retrieval-augmented large language models through question complexity. In: Proceedings of the 2024 Conference of the North American Chapter of the Association for Computational Linguistics: Human Language Technologies (Volume 1: Long Papers), pp. 7036–7050. Association for Computational Linguistics, Mexico City, Mexico (2024). https://doi.org/10.18653/v1/2024.naacl-long.389
47. Mo, J., Huang, S.: Neuron importance algorithm for continual learning. Conf. Proc. **12803**, 128030U–128030U–9 (2023). https://doi.org/10.1117/12.3009542
48. Andersen, M.R., Kabel, K., Bremholm, J., Bundsgaard, J., Hansen, L.K.: Automatic proficiency scoring for early-stage writing. Comput. Educ. Artif. Intell. **5**, 100168 (2023). https://doi.org/10.1016/j.caeai.2023.100168

AI-Driven Optimization of Tourist Itineraries Using Genetic and Heuristic Methods

Sofía Álvarez-Sandoval[1](✉), Samantha Ruelas-Valtierra[1], José-Manuel Dávila-Mancilla[2], Gerardo Ramírez-Chávez[1], and Fernando Elizalde-Ramírez[1](✉)

[1] Escuela de Ingenieria y Ciencias, Instituto Tecnologico y de Estudios Superiores de Monterrey, 64700 Monterrey, Nuevo León, Mexico
A01656059@exatec.mx, {A01704564,A01368693,fer_elizalde}@tec.mx
[2] Universidad Tsinghua, Beijing, China
daihz25@mails.tsinghua.edu.cn

Abstract. The tourism industry in Mexico represents about 9% of the national GDP, and it is estimated that in 2024 it will contribute over 30 billion dollars in revenue [14]. This highlights the importance of this economic activity and the need to ensure the best possible experience for all visitors. A key experience for tourists revolves around properly planning their tours so that they can visit all places of interest within an appropriate time window, spending as little time as possible on travel. Thus, our proposal consists of a model that, through different algorithms, generates tourist visit plans for visitors that maximize user satisfaction when the two main factors considered (the expected occupancy and travel times between locations) are both low. The problem involves multi-day planning, being a Multiple Traveling Salesman Problem with Time Windows (MTSPTW), which complicates its solution with exact techniques. By implementing metaheuristic and stochastic algorithms, progressively closer routes to the target values were generated, meeting the expected characteristics according to the established constraints. Despite being complex approaches, acceptable levels of satisfaction were achieved in relatively low computing time.

Keywords: Tourism · genetic algorithm · user satisfaction · multi-day planning · stochastic algorithms

1 Introduction

For 2024, it was estimated that tourism in Mexico would generate more than 30 billion USD in revenue, representing approximately 8.6% of the country's GDP [14]. Although the main destinations for both international and domestic visitors are beaches, Mexico has a wide variety of destinations. Currently, most tourists face a common problem: having multiple points of interest, a limited

H. Ponce et al. (Eds.): COMIA 2026, CCIS 3095, pp. 104–119, 2027.
https://doi.org/10.1007/978-3-032-37438-7_9

budget, and insufficient time to visit them all, which often results in unsatisfactory experiences. [1] found that for tourists who visited Michoacán from 2006 to 2012, hospitality, service provider attitude, quality of accommodation, and public safety are the main factors that influence their satisfaction levels. This means that a single destination may have different levels of satisfaction for each visitor depending on how well their personal interests are met.

Thus, it is necessary to define a method that not only facilitates creating personalized routes according to specific interests but also maximizes the satisfaction achieved under the given constraints. Based on this, the objective of this article is to propose several computational models that allow tourists to plan a route that maximizes the satisfaction generated from their tour.

2 Literature Review

The Tourist Trip Design Problem (TTDP) is an extension of routing problems focused on planning optimal tourist routes that visit multiple Points of Interest (POIs) without exceeding constraints such as time, budget, and preferences [2]. The main objective is to maximize tourist satisfaction by considering factors such as opening hours, transportation, distance, travel time, and time spent at each location, emphasizing the design of the route rather than only the selection of stops. Classical problems such as the Traveling Salesman Problem (TSP) [4] and the Vehicle Routing Problem (VRP) [17] serve as foundational approaches. Extensions like the Multiple Traveling Salesmen Problem (MTSP) [5] and its variant with time windows (MTSPTW) [6] better represent real-world tourism scenarios, particularly due to the scheduling constraints of attractions. These problems are NP-hard, which motivates the use of heuristic and metaheuristic methods.

Different solution strategies have been proposed. Multi-objective approaches combining Compromise Programming with metaheuristics such as genetic algorithms and ant colony optimization show strong performance, with ant colony methods achieving better results at higher computational cost [19]. Other works use clustering techniques to tailor routes for specific tourist groups [7], and Tabu search has demonstrated near-optimal solutions with lower computational effort [8]. Queue-based approaches, such as the Queuing Search Algorithm (QSA), model itineraries as service sequences and have shown competitive performance compared to traditional metaheuristics [11]. More advanced models incorporate personalization and uncertainty. The PIREM algorithm improves user satisfaction by selecting POIs based on expected utility and constraints, achieving higher satisfaction levels compared to other methods [20]. Additional approaches consider factors such as attraction popularity, waiting times, and crowding effects, leading to more realistic and effective itineraries [12,21].

In [2] a system is proposed to generate tourist travel plans using Intelligent Connected Vehicles (ICVs), which is developed in three stages: first, the POIs (Points of Interest) are grouped according to their characteristics and the tourist's preferences; subsequently, the accessibility of these POIs is analyzed using the SA-AGNES method; and finally, once the feasible points have been identified, the Fruit Fly Optimization Algorithm is used to determine the optimal route that the tourist should follow. This paper [3] explores a point-of-interest (POI) recommendation system using artificial intelligence and IoT, which suggests tourist attractions, activities, and routes to users based on their interests. The system analyzes large volumes of data generated by tourists and compares these patterns with user preferences. The methodology employs collaborative filtering techniques and Apriori algorithms to identify relationships between users, preferences, and tourist attractions, enabling personalized recommendations. This data can then be used by our model to generate tourist itineraries based on what truly interests the tourist.

Finally, artificial intelligence techniques based on planning with predicate logic allow the generation of multi-day travel plans that incorporate user-defined constraints and preferences, highlighting the growing role of AI in solving TTDP [22]. The integration of Answer Set Programming (ASP), knowledge graphs, and artificial intelligence allows the development of dynamic and adaptive systems for route planning, improving logistical efficiency and the user experience in last-mile scenarios [9]

3 Methodology

The problem is formulated as a two-day itinerary optimization, but is modeled similarly to a standard Traveling Salesman Problem (TSP), where both days are represented by a single route and all POIs must be visited exactly once within a symmetric network [4]. The model's computational complexity is NP-Hard and increases when adding time windows. A weekend time horizon was considered, since the Magic Towns initiative is specifically designed for this type of short-term stay. Under the proposed approach, the routes generated maximize tourist satisfaction instead of minimizing the total travel time (like a traditional TSP would) by visiting POIs when they are less crowded, with constraints on the total duration of the tour and the travel time between them (8 h and 15 min per day respectively), also considering the working hours for each POI.

The problem is represented as a graph $G = (V, A)$ with $n = 13$ nodes. Each edge $(i, j) \in A$ is associated with travel time t_{ij} and total time a_{ij} (including stay). The satisfaction of visiting node j at hour k is given by b_{jk}. The two agents construct independent tours covering all nodes, forming the basis of the integer programming model used to solve the problem.

$$\max z = \sum_{\substack{i,j=1 \\ i \neq j}}^{n} x_{ij} b_{jk} \tag{1}$$

The previous objective function (1) is constructed from:

- x_{ij}: binary variable that indicates whether any of the agents available takes the path from i to j.

- s_{jk}: a value from 0 to 1, with 0 indicating location j at hour k is closed or at maximum capacity, while 1 indicates the place has maximum availability.
- r_j: rating of location j adjusted by the number of reviews available in the form of $r_j = c_j e^{-\frac{1}{w_j}}$. Here, c_j is the rating for POI $j \in V$ extracted from Google and w_j is the number of reviews it has.
- b_{jk}: perceived satisfaction coefficient for visiting point j at hour k, better described in Eq. 2:

$$b_{jk} = s_{jk} r_j \tag{2}$$

The k subindex represents the hour in which the destination j will be visited. Under the proposed model, all available hours for completing the itinerary are represented as a single sequence, where the first half will correspond to the first day and vice versa. It is worth noting that occupation data is available only in hour intervals, thus k is not continuous but discrete, and can be easily used alongside i, j without affecting the problem's formulation. This means that only exact hours are considered for tour planning, so that start and end times happen on the hour for every POI. In consequence, the stay at each one is rounded up to the closest hour, and travel between locations is assumed to happen at the beginning of the stay in destination j. The model is subject to the following constraints:

$$\sum_i x_{ij} = 1, \qquad \forall j \tag{3}$$

$$\sum_j x_{ij} = 1, \qquad \forall i \tag{4}$$

$$u(i) - u(j) + N(i)x_{i,j} \leq N(i) - 1 \qquad \forall i,\ j \mid i > 1, \quad j > 1, \quad i \neq j \tag{5}$$

$$\sum_{k=k_0}^{k_0+12} \sum_{j=1}^{n} \sum_{i=1}^{n} x_{ij} t_{ij} \leq 15. \qquad \forall k_0 \in \{1, 14\} \tag{6}$$

$$\sum_{k=k_0}^{k_0+12} \sum_{j=1}^{n} \sum_{i=1}^{n} x_{ij} a_{ij} \leq 480, \qquad \forall k_0 \in \{1, 14\} \tag{7}$$

The objective is to maximize tourist satisfaction by scheduling visits during less crowded hours. Constraints 3 and 4 ensure that each node is visited exactly once, while subtour elimination is enforced through constraint 5. A key difference from classical TSP and MTSP formulations [4,6] is the absence of a fixed starting point and the independence of daily routes, as the choice for the location to be visited at hour $k = 14$ (the first for Saturday) depends only on the maximal satisfaction achievable and not the POI visited at $k = 13$. Additionally, the

Table 1. Weighted rating per location

Location Name	Denomination	Rt	Tot Reviews	Weighted Rt	Avg Stay (min)
Zona Arqueológica de San Andrés Cholula	P_1	4.7	35338	4.6999	180
Museo de Sitio de San Andrés Cholula	P_2	4.4	131	4.3665	30
Plaza de la Concordia	P_3	4.6	20730	4.5998	90
Convento de San Gabriel Arcángel	P_4	4.7	918	4.6949	60
Santuario Nuestra Señora de los Remedios	P_5	4.7	1623	4.6971	180
Museo casa del Caballero Águila	P_6	4.5	381	4.4882	140
Jardín Etnobotánico Francisco Peláez Roldán	P_7	4.6	1310	4.5965	150
Capilla Real de Naturales	P_8	4.6	911	4.5950	60
Parroquia de San Andrés Cholula	P_9	4.7	690	4.6932	60
Los Portales	P_{10}	4.8	5	3.9299	20
Balneario Puerto Escondido	P_{11}	4.3	2853	4.2985	240
Museo Regional de Cholula	P_{12}	4.7	2165	4.6978	160
Parque de las 7 culturas	P_{13}	4.6	13	4.2594	30

model incorporates the time constraints limiting daily travel to 15 min (6) and total tour duration per day to 480 min (7).

For the case study, Cholula, Puebla was selected due to its touristic relevance and regular layout [15]. A set of 13 POIs was defined, and a weighted rating r_j was computed using data from [16], adjusting ratings by the number of reviews through an exponential decay function. Average visit durations and travel times between locations were also collected, forming the dataset summarized in Table 1.Although the number of POIs appears to be small, nearly 33% of the so-called Magic Towns offer between 10 and 15 POIs. These destinations are classified as less developed in terms of tourism and present a greater information deficit. The next group in the list offers up to 30 POIs, representing approximately 50% of the total. Consequently, the resulting graph is not excessively large and can be efficiently solved using the proposed methodology.

Furthermore, POI availability was modeled using normalized occupancy data s_{jk} obtained from [16], within a time window from 7:00 to 19:00. This information feeds the satisfaction function (2), which will result in a coefficient between 0 and 5, representing the expected experience of visiting location j at hour k. This enhances the realism of the objective function (1) by incorporating both temporal and qualitative factors into the optimization process.

4 Metaheuristics

Due to the complexity of maximizing satisfaction under time windows and scheduling constraints, solving the problem with exact methods becomes impractical. Although it could be formulated as a VRP, scalability limitations motivate the use of metaheuristic and stochastic approaches. In this work, genetic algorithms [18] are proposed as the main solution method (Algorithm 1), complemented by a priority queuing approach (Algorithm 5) and an exact solution of a simplified model to enable performance comparison.

Algorithm 1. Genetic Algorithm Framework

Require: Population P, coefficients b_{jk}, times A, t_{ij}, iterations N
Ensure: Best chromosome C^*
Initialize $Z(C) = \sum_{d\in\{1,2\}} \sum_i b_{j_i^d,k_i^d}$ according to (1)
$C^* \leftarrow \emptyset$, $Z^* \leftarrow -\infty$
for $iter = 1$ to N **do**
▷ Fitness Evaluation
for all $C \in P$ **do**
Compute $Z(C)$ using b_{jk} and time constraints
end for
▷ Selection (Roulette Wheel)
Compute probabilities $p(C) = \frac{Z(C)}{\sum_{C'\in P} Z(C')}$
Select parents $C^1, C^2 \sim P$ using $p(C)$
▷ Crossover
Select crossover points k_1, k_2
Generate offspring:
$O_1^1 \leftarrow C_1^1[1:k_1] \cup C_1^2[k_1:]$
$O_1^2 \leftarrow C_1^2[1:k_1] \cup C_1^1[k_1:]$
$O_2^1 \leftarrow C_2^1[1:k_2] \cup C_2^2[k_2:]$
$O_2^2 \leftarrow C_2^2[1:k_2] \cup C_2^1[k_2:]$
▷ Mutation
for all $O \in \{O^1, O^2\}$ **do**
Select segment $d \in \{1,2\}$ and positions $i \neq j$
Swap $O_d[i]$ and $O_d[j]$
end for
▷ Population Update
Add offspring to P
Evaluate $Z(O)$ and update:
if $Z(O) > Z^*$ **then**
$C^* \leftarrow O$, $Z^* \leftarrow Z(O)$
end if
end for
return C^*

1. **Genetic Algorithm Without Breaks**
This approach generates continuous itineraries without idle periods, maximizing satisfaction under time constraints. Solutions are evaluated through a fitness function, while selection, crossover, and mutation operators guide the search toward high-quality itineraries.
2. **Genetic Algorithm with Breaks**
This variant introduces break periods as a dummy node, allowing more realistic itineraries without contributing to satisfaction. To avoid infeasible solutions, repeated visits are penalized within the fitness function (Algorithm 2), while the population generation includes random breaks (Algorithm 3). The overall structure remains similar to the base genetic algorithm.
3. **Greedy Genetic Algorithm**
This approach incorporates a greedy heuristic (Algorithm 4) to generate the initial population, prioritizing shorter routes instead of satisfaction. While the evaluation still maximizes satisfaction, the initialization reduces variability and improves convergence.

Algorithm 2. Fitness Evaluation

Require: Chromosome C, matrices B^{sat}, B^{sun}, tiempos A, T
Ensure: Fitness Z
$Z \leftarrow 0$, $t_1 \leftarrow 0$, $t_2 \leftarrow 0$
Create local copies $\hat{B}^{sat}$, $\hat{B}^{sun}$
▷ Day 1 (Saturday)
for $i = 1$ to $|C_1| - 1$ **do**
$h \leftarrow \min(\lfloor t_1/60 \rfloor, 12)$
$p \leftarrow C_1[i]$
$Z \leftarrow Z + \hat{B}^{sat}[p][h]$
Penalize p in $\hat{B}^{sat}$, $\hat{B}^{sun}$
$t_1 \leftarrow t_1 + T[C_1[i]][C_1[i+1]] + A[p]$
end for
▷ Day 2 (Day_2)
for $i = 1$ to $|C_2| - 1$ **do**
$h \leftarrow \min(\lfloor t_2/60 \rfloor, 12)$
$p \leftarrow C_2[i]$
$Z \leftarrow Z + \hat{B}^{sun}[p][h]$
Penalize p in $\hat{B}^{sun}$
$t_2 \leftarrow t_2 + T[C_2[i]][C_2[i+1]] + A[p]$
end for
if $t_1 > 780 \vee t_2 > 780$ **then**
$Z \leftarrow 0$
end if
return Z

4. **Priority Queuing**

The implemented priority queuing algorithm (Algorithm 5) is based on a ratio between the expected satisfaction for the remaining POIs and travel time from the previously visited destination to establish a queue. This ratio ensures that the nodes with the highest satisfaction nearby are prioritized, optimizing the route for maximum satisfaction with minimal displacement. This approach was inspired by the work of [11], who solved the TSP by iteratively designing a route where the closer a node is to the current point, the more likely it is to be chosen as the next destination. Similarly, this ratio would be calculated iteratively for all remaining locations to choose the next POI until the itinerary was complete.
To enhance its flexibility, a Monte Carlo simulation was used to determine both the starting point and the range between 7 AM and 9 AM as starting times, using the satisfaction values to obtain the probability range for each location.

5 Results

These algorithms were executed on Google Colab, using Google Compute Engine with 12.7 GB of RAM and a Python 3 environment. Using its support for Python and various data science libraries, such as numpy [13], the algorithm was able

Algorithm 3. Modified Population Generation

```
Require: Population size N
Ensure: Population P
  for k = 1 to N do
    A ← {1,...,13},  C ← []                                  ▷ Day 1
    Append 1 to C; remove 1 from A
    for i = 1 to randint(3,6) do
      Select g ∈ A; append g to C; remove g from A
    end for
    Append 1 to C
    for i = 1 to randint(0,6) do
      Insert 0 at random position in first segment
    end for                                                  ▷ Day 2
    Select c ∈ A; append c to C; remove c from A
    for i = 1 to randint(3, min(|A|,6)) do
      Select g ∈ A; append g to C; remove g from A
    end for
    Append c to C
    for i = 1 to randint(0,6) do
      Insert 1 at random position in second segment
    end for
    P ← P ∪ {C}
  end forreturn P
```

to process data efficiently and perform complex computations, leveraging its powerful cloud-based computational resources and minimizing variation in execution times induced by the characteristics of the device used. To test the genetic algorithms, the population size was set to 500 solutions, with a crossover and mutation probability of 0.3 and 0.1, respectively, with 1000 epochs per iteration of the algorithm.

Exact Solution. A mathematical model was implemented in GAMS (General Algebraic Modeling System) to establish a benchmark for evaluating the proposed algorithms. Due to the increased complexity introduced by time constraints and variable satisfaction, the problem could not be fully solved under the limitations of the Student GAMS Community License. However, by excluding the length of stay from most algorithms, a close approximation to the optimal tour was obtained. The resulting solution (Table 2) achieves a satisfaction of 43.93 by visiting all 13 POIs, allocating 6.18 h on saturday and 7.19 h on Day_2, with a computational time of 0.109 s using GAMS v.44.3.0.

Genetic Algorithm Without Breaks. The results in Fig. 1 indicate that the genetic algorithm without breaks provides stable and consistent performance across all evaluated iteration levels (10, 50, 100, and 150 iterations). The first execution of this algorithm (10 iterations) yielded the highest average satisfaction, as shown in Fig. 1a, while maintaining a low execution time, providing solutions in less than 6 s on average. Furthermore, the generated itineraries exhibited a relatively balanced distribution of activities between both days. The

Algorithm 4. Modified Greedy Search

Require: Matrices D, T, τ; constraints; n
Ensure: Routes R_1, R_2
  $v \leftarrow -1$, $V \leftarrow \{v\}$, $R_1 \leftarrow [v+1]$, $R_2 \leftarrow []$
  $t \leftarrow 0$, $d \leftarrow 0$
  while $|V| < n$ **do**
    $(c^*, u) \leftarrow (\infty, \emptyset)$
    for each $j \notin V$ **do**
      Compute $c(j)$
      if $c(j) < c^*$ **then**
        $(c^*, u) \leftarrow (c(j), j)$
      end if
    end for
    if $u = \emptyset$ **then**
      break
    end if
    $v \leftarrow u$, $V \leftarrow V \cup \{v\}$
    if CANADD(R_1, v) **then**
      $R_1 \leftarrow R_1 \cup \{v\}$
    else if CANADD(R_2, v) **then**
      $R_2 \leftarrow R_2 \cup \{v\}$
    else
      break
    end if
    $t \leftarrow t + c^*$, $d \leftarrow d + D_{v,v-1}$
  end while
  Close R_1, R_2 **return** R_1, R_2

Table 2. Day 1 and Day 2 itinerary generated

Time	GAMS		GA (No Breaks)		GA (With Breaks)		Greedy GA		Priority Queuing	
	Day 1	Day 2	Day 1	Day 2	Day 1	Day 2	Day 1	Day 2	Day 1	Day 2
07:00	Free Time	Free Time	Free Time	Free Time	Free Time	Free Time	Free Time	Free Time	Free Time	Free Time
08:00	Free Time	Free Time	P_1	P_5	P_1	P_8	P_1	P_{10}	Free Time	Free Time
09:00	P_5	P_1	P_9	P_{11}	P_2	P_9	P_{13}	P_9	P_{11}	Free Time
10:00	P_{12}	P_6	P_{10}	P_2	P_{11}	P_6	P_{12}	P_7	P_3	P_9
11:00	P_4	P_{13}	P_6	P_7	Free Time	Free Time	P_6	P_5	Free Time	P_{10}
12:00	P_{10}	Free Time	P_4	P_{12}	Free Time	P_3	P_4	P_8	Free Time	P_{13}
13:00	Free Time	P_3	P_8	P_{13}	Free Time	P_4	P_2	P_{11}	P_5	P_7
14:00	Free Time	Free Time	P_1	P_5	P_{12}	P_{13}	P_3	P_{10}	P_4	Free Time
15:00	Free Time	P_2	Free Time	Free Time	Free Time	P_8	P_1	Free Time	P_8	P_{12}
16:00	P_{11}	Free Time			Free Time	Free Time	Free Time	Free Time	Free Time	Free Time
17:00	Free Time	Free Time			P_7	Free Time			P_5	P_2
18:00	P_8	P_9			Free Time	Free Time			Free Time	P_1
19:00	P_5	P_7			Free Time	Free Time			Free Time	Free Time
20:00	Free Time	P_1			P_1	Free Time				

Algorithm 5. Priority Queueing

Require: POIs P, stay times A, hours H, travel matrix T, rating matrix B
Ensure: Route R, total travel TT, total time T_{tot}, total rating Z

```
U ← P, R ← []
TT ← 0, T_tot ← 0, Z ← 0
Select h_0 ∈ {7, 8, 9} (Day_1 start)
Normalize B_s(i) = B[p_i][h_0] / Σ_j B[p_j][h_0]
Select p ~ B_s
R ← [p], U ← U \ {p}
h ← h_0 + ⌈A[p]/60⌉, Z ← B[p][h_0]
while U ≠ ∅ ∧ TT ≤ 15 ∧ T_tot ≤ 960 do
    (ρ*, q) ← (−∞, ∅)
    for each i ∈ U do
        τ ← T[p][i]
        ρ(i) ← B[i][h] / τ
        if ρ(i) > ρ* then
            (ρ*, q) ← (ρ(i), i)
            (a*, τ*, z*) ← (A[i], τ, B[i][h])
        end if
    end for
    if q = ∅ then
        break
    end if
    R ← R ∪ {q}, U ← U \ {q}
    p ← q
    TT ← TT + τ*
    h ← h + ⌈a*/60⌉
    T_tot ← T_tot + τ* + a*
    Z ← Z + z*
end while
    return R, TT, T_tot, Z
```

time spent on Saturday routes mainly ranged between 6.2 and 7.2 h, whereas Day_2 routes were generally between 5.2 and 6.2 h, as illustrated in Fig. 1b and 1c, respectively. These results suggest that the algorithm tends to allocate a greater number of activities during the first day while preserving computational efficiency and robustness across different iteration settings.

An example of an itinerary generated with this algorithm is shown in Table 2. Notice that places that have low occupancy in the morning are placed immediately after the starting point P_1 and both days end at 2 PM, which might benefit "early-bird" tourists who enjoy busy morning itineraries.

Genetic Algorithm with Breaks. The results in Fig. 2 indicate that the genetic algorithm with breaks produces lower average satisfaction than the version without breaks due to the inclusion of pauses and penalties for repeated visits, although some executions still achieve high satisfaction values as outliers. The available time for visiting POIs is significantly reduced, with activities concentrated between 3–4.2 h on Saturday and 2.2–3.1 h on Day_2. Additionally, the inclusion of breaks considerably increases the computational cost, requiring

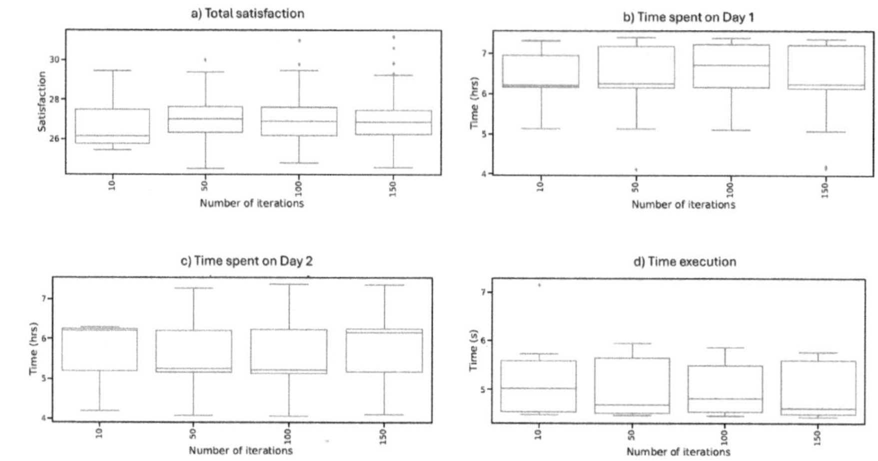

Fig. 1. Results for the genetic algorithm with no breaks.

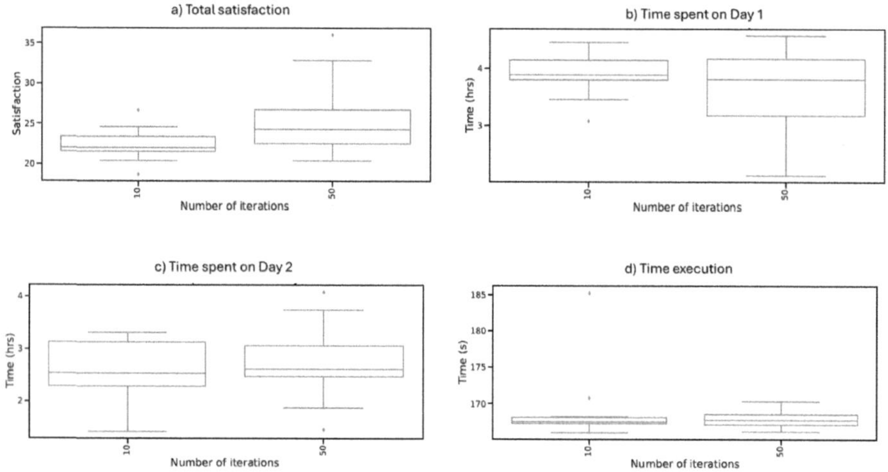

Fig. 2. Results for the genetic algorithm with breaks.

approximately 167âĂŞ170 s of execution time, which is up to 32 times higher than the no-break version. Overall, while breaks generate more realistic itineraries, they also reduce efficiency and increase the likelihood of infeasible or incomplete solutions.

Nevertheless, this approach produces more realistic itineraries by incorporating rest periods and allowing greater flexibility, resulting in less demanding schedules. While the "early-bird" behavior persists, the inclusion of random breaks enables exploration of busier time periods, leading to itineraries that better span the entire day.

Greedy Genetic Algorithm. Using the greedy (deterministic) algorithm to create the populations softens the randomness of the genetic algorithm. Consequently, the solutions of this algorithm have little variability, which makes them of special relevance to achieve convergence in the optimum satisfaction value. In Fig. 3a we find that this approach has the lowest minimums out of the genetic algorithms; this coincides with its previous description, as it mainly focuses on reducing the distance traveled. There is very little variance in the time distribution for the days, which we can appreciate in Figs. 3b and 3c, where aside from some outliers it heavily leans to approximately 7.1 h on Day_1 and under 7 h for Saturday, taking almost full advantage of both days' time. Additionally, the time to accomplish an iteration has significantly decreased from the algorithm with pauses and is now much more similar to the no-pause algorithm.

Although similar in structure to other methods (Table 2), the change in objective function leads to distinct yet optimal itineraries.

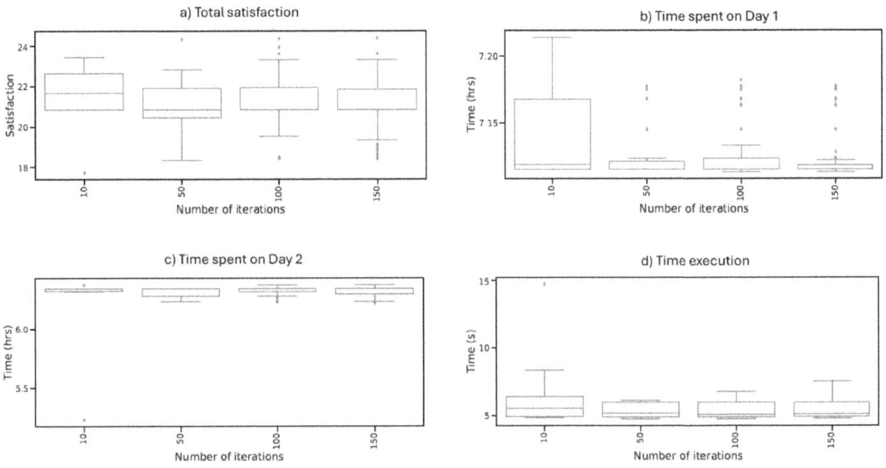

Fig. 3. Results for the greedy genetic algorithm.

Priority Queuing. As a stochastic model, this model finds different results depending on the parameters defined through probabilistic methods (the starting place and time). Thus, the evaluation metrics used for the genetic algorithms would not be adequate; instead, the route with the highest satisfaction from several iterations (50, 100, 500, and 1000) was chosen, obtaining the results presented in Table 3. As observed, when iterated for over 100 times, the results start to converge at the satisfaction of 28.76, but they continue to show some variability in the time spent each day. This can be attributed to the changing starting times and locations determined through the Montecarlo simulation, but overall, shows that one could expect a certain level of satisfaction even when introducing this variability.

Table 3. Results for the priority queuing algorithm

Iterations	Satisfaction	Total time Day 1	Total time Day 2	Execution time
50	27.22	6.613 h	6.867 h	0.101 ms
100	28.76	6.587 h	7.058 h	0.120 ms
500	28.76	6.660 h	6.957 h	0.108 ms
1000	28.76	6.616 h	6.952 h	0.098 ms

Table 2 presents an itinerary generated through Priority Queuing. Due to the way the algorithm works, it is very similar to the one described in Table 2 in GA (With Breaks), with the main difference being that this algorithm tends to distribute evenly the times of visit of the locations and the free time. In this way, it ignores the previously proposed "early-bird" preference, making this schedule more feasible for people who are not interested visiting the locations early, and allowing for greater flexibility according to tourists' preferences.

6 Discussion

The proposed approaches for itinerary generation exhibit distinct strengths and limitations in terms of computational performance, achieved satisfaction, and realism of the generated tours. The results demonstrate that no single algorithm can be considered universally optimal, as the most suitable solution depends on user preferences and the imposed temporal constraints. The evaluated methods generated itineraries with different characteristics, including variations in starting times, transit durations, visit distribution, and the number of POIs visited per day. Among the analyzed approaches, the genetic algorithm without breaks achieved the highest satisfaction values, although it produced less realistic schedules with uninterrupted activities. In contrast, the genetic algorithm with breaks generated more balanced daily itineraries at the expense of increased execution time and without significant improvements in satisfaction. The greedy genetic algorithm offered faster convergence with only a slight reduction in solution quality, making it appropriate for lightweight implementations, while the priority queue approach achieved competitive satisfaction values with significantly lower computational costs.

The study also identifies several opportunities for improvement, such as using finer time discretization to incorporate realistic stay durations, dynamic transit times, and more accurate crowd estimations. Expanding the study area from Cholula to the entire state of Puebla would additionally allow the evaluation of the algorithms under larger and more complex networks. Furthermore, integrating the proposed methodologies with external APIs could enable the automatic generation of personalized itineraries for different destinations based on user preferences, representing a valuable tool for tourism agencies. Although the genetic algorithm without breaks produced results closest to the exact solution obtained with GAMS, the latter was formulated under simplified assumptions, highlighting the need for more accurate mathematical formulations and exact approaches for broader applicability and fairer comparisons. Finally, the proposed metaheuristics demonstrate strong adaptability and could be extended beyond tourism applications to other routing and decision-making problems, such as personalized house-hunting itineraries under time and preference constraints.

7 Conclusion

This study presents a simplified yet meaningful case study for the Tourist Trip Design Problem, highlighting the importance of developing computationally efficient itinerary planning methods given the economic relevance of tourism. The proposed metaheuristic approaches provide a solid foundation for addressing more complex instances, where traditional optimization methods may become intractable. The results demonstrate strong potential for large-scale implementation, as the models rely on widely accessible data sources such as Google's API and relatively simple algorithms. Among the evaluated methods, the genetic algorithm without breaks achieved the closest approximation to the exact solution with low computational effort, making it suitable for applications requiring fast responses. The priority queuing approach also proved to be a competitive alternative, offering more realistic and flexible solutions that can better support personalization. Overall, the implemented metaheuristics effectively address the problem of designing high-quality tourist itineraries, each contributing distinct advantages. The proposed model is highly adaptable, allowing extensions such as the inclusion of budget constraints or personalized satisfaction measures. These features position the approach as a promising basis for future developments in personalized and scalable tourist itinerary planning.

Acknowledgment. To the Escuela de Ingenieria y Ciencias of the Instituto Tecnologico y de Estudios Superiores de Monterrey, for the support provided in the development of this research and especially to the Department of Sciences.

Disclosure of Interests. The authors have no relevant financial or non-financial interests to disclose. All information was obtained from Google Maps [16]. Generative AI tools were used exclusively for language refinement and grammar correction. The authors take full responsibility for the manuscript content.

References

1. Gudiño Paredes, M., Aguayo Lorenzo, E.: Competitividad y satisfacción turística en las regiones del Estado de Michoacán (México). Int. J. Sci. Manag. Tour. **1**(1), 289–296 (2015)
2. Vansteenwegen, P., Souffriau, W., Vanden Berghe, G., Van Oudheusden, D.: Iterated local search for the team orienteering problem with time windows. Comput. Oper. Res. **36**(12), 3281–3290 (2009)
3. Song, Y., He, Y.: Toward an intelligent tourism recommendation system based on artificial intelligence and IoT using Apriori algorithm. Soft. Comput. **27**(24), 19159–19177 (2023)
4. Dantzig, G., Fulkerson, R., Johnson, S.: Solution of a large-scale traveling-salesman problem. J. Oper. Res. Soc. Am. **2**(4), 393–410 (1954)
5. Cheikhrouhou, O., Khoufi, I.: A comprehensive survey on the multiple traveling salesman problem: applications, approaches and taxonomy. Comput. Sci. Rev. **40**, 100369 (2021)
6. Chen, Y., Jia, Z., Ai, X., Yang, D., Yu, J.: A modified two-part wolf pack search algorithm for the multiple traveling salesmen problem. Appl. Soft Comput. **61**, 714–725 (2017)
7. Li, S., Luo, T., Wang, L., Xing, L., Ren, T.: Tourism route optimization based on improved knowledge ant colony algorithm. Complex Intell. Syst. **8**(5), 3973–3988 (2022)
8. López, E., Salas, Ó., Murillo, Á.: El problema del agente viajero: un algoritmo determinístico usando búsqueda tabú. Revista de Matemática: Teoría y Aplicaciones **21**(1), 127–144 (2014)
9. Mofatteh, M.Y., Abbas, R., Seddoh, N.D., Sedhumadhavan, S., Thunyaluck, M., Valilai, O.F.: A smart adaptive transportation planning model including real-time drivers' knowledge using answer set programming and knowledge graphs. Procedia Comput. Sci. **253**, 2358–2368 (2025)
10. Zhou, X., Peng, J., Wen, B., Su, M.: Navigation route planning for tourism intelligent connected vehicle based on the symmetrical spatial clustering and improved fruit fly optimization algorithm. Symmetry **16**(2), 159 (2024)
11. Gülmez, B.: Improved discrete queuing search algorithm for traveling salesman problem. In: International Research in Engineering Sciences VII, 31–56 (2023)
12. Wu, X., Guan, H., Han, Y., Ma, J.: A tour route planning model for tourism experience utility maximization. Adv. Mech. Eng. **9**(10), 3–7 (2017)
13. Harris, C.R., et al.: Array programming with NumPy. Nature **585**(7825), 357–362 (2020)
14. Secretaría de Turismo. https://www.gob.mx/sectur/prensa/mexico-captara-31-mil-141-millones-de-dolares-por-visitantes-internacionales-en-2024-de-acuerdo-con-expectativas?idiom=es. Accessed 23 Mar 2025
15. Secretaría de Turismo. https://www.gob.mx/sectur/articulos/cholula-puebla. Accessed 24 Mar 2025
16. Google Maps location reference. https://goo.gl/maps/b57FQvZjtVsrLTu59. Accessed 07 July 2024
17. Toth, P., Vigo, D.: Monographs on Discrete Mathematics and Applications. Society for Industrial and Applied Mathematics, Philadelphia (2002)
18. Türkoğlu, B., Eroğlu, H.: Genetic algorithm for route optimization. In: Dey, N. (ed.) Applied Genetic Algorithm and its Variants, pp. 51–79. Springer, Heidelberg (2023). https://doi.org/10.1007/978-981-99-3428-7_3

19. Son, N.T., Nguyet Ha, T.T., Jaafar, J.B., Anh, B.N., Giang, T.T.: Some meta-heuristics for tourist trip design problem. In: 2023 IEEE Symposium on Industrial Electronics & Applications (ISIEA), pp. 1–10. IEEE (2023)
20. Panagiotakis, C., Daskalaki, E., Papadakis, H., Fragopoulou, F.: SThe tourist trip design problem with poi categories via an expectation-maximization based method. In: Proceedings of the 16th ACM Conference on Recommender Systems, pp. 47–64. RecTour@ RecSys (2022)
21. Lim, K., Hui, Chan, J., Karunasekera, S., Leckie, C.: Personalized itinerary recommendation with queuing time awareness. In: Proceedings of the 40th International ACM SIGIR Conference on Research and Development in Information Retrieval, pp. 325–334. ACM (2017)
22. Elizalde-Ramírez, F., Maya-Padrón, C.: Generation of tourist routes considering preferences and public transport using artificial intelligence planning techniques. In: Garrido, A., Paternina-Arboleda, C.D., Voß, S. (eds.) Computational Logistics: 15th International Conference, LNCS, vol. 15168, pp. 164–175. Springer, Monterrey (2024). https://doi.org/10.1007/978-3-031-71993-6_11

A Computer Vision-Based Dashboard for Real-Time Monitoring of Student Facial Emotions in Virtual Classrooms

Hideki Sebastian Sotero Huaroto(✉) and Angela Mayhua-Quispe

Laboratorio de Inteligencia Artificial, Facultad de Ingeniería, Universidad de Lima, Lima, Peru

20212604@aloe.ulima.edu.pe, amayhua@ulima.edu.pe

Abstract. Education in virtual classrooms operating under uncontrolled conditions presents limitations in the emotional interaction between teachers and students, affecting attention and academic performance. This study proposes a computer vision-based system for real-time facial emotion detection and visualization. The system employs ResEmoteNet trained in FER2013 and RAF-DB, achieving accuracies of 74.70% and 97.29%, respectively, with higher performance for emotions such as happiness and surprise. A usability evaluation was conducted with 44 instructors using the System Usability Scale (SUS), complemented by questions on pedagogical relevance. The system obtained an average SUS score of 87.05, indicating excellent usability. These results support the effectiveness of the proposed dashboard as a tool for monitoring the emotional states of students and supporting pedagogical decision-making in virtual learning environments.

Keywords: Computer vision · Facial emotion recognition · Real-time monitoring · Virtual education · Educational analytics

1 Introduction

In recent years, advances in information technologies have transformed the education sector, with virtual learning becoming a widely adopted modality, particularly after the COVID-19 pandemic. In Peru, the proportion of students enrolled in virtual education increased from 9% before the pandemic to 29% in 2023, exceeding 348,000 students [2]. Despite its benefits in accessibility and cost reduction, studies report limitations in learning quality and student performance in fully virtual environments [10,20].

One of the main challenges in this context is the limited emotional interaction between teachers and students. Emotions play a key role in motivation and engagement [5], but virtual environments restrict the ability of instructors to perceive the affective states of students in real time. This limitation reduces the ability to adapt teaching strategies according to the emotional dynamics of the class.

H. Ponce et al. (Eds.): COMIA 2026, CCIS 3095, pp. 120–133, 2027.
https://doi.org/10.1007/978-3-032-37438-7_10

To address this issue, several studies have explored the use of Facial Expression Recognition (FER) in educational settings [22,23,26]. However, the usefulness of these systems depends not only on the accuracy of the detection but also on the way emotional information is presented to instructors. In this regard, dashboards have been proposed to transform the output of the model into interpretable visual indicators [9,12,13].

In this context, this research proposes the design and implementation of an emotion monitoring dashboard based on facial expressions that is capable of representing student emotional states in real time and retrospectively. While the emotion recognition component integrates the ResEmoteNet architecture, the primary contribution of this work lies in the optimization and integration of a real-time pipeline that connects automated facial emotion recognition with a pedagogical dashboard, specifically tailored to the technical and dynamic constraints of virtual classrooms. The system is evaluated in terms of perceived usability and pedagogical relevance using the System Usability Scale (SUS), complemented by additional domain-specific questions.

The remainder of this paper is organized as follows. Section 2 presents the related work. Section 3 describes the proposed methodology. Section 4 reports the results obtained. Section 5 discusses the findings, and Sect. 6 concludes the article.

2 Related Work

2.1 Emotion Detection in Uncontrolled Environments

Facial expression recognition (FER) models achieve high accuracy in controlled settings; however, their performance decreases in uncontrolled environments such as virtual classrooms due to lighting variability, pose variations, and occlusions [16].

Recent approaches focus on improving robustness through attention mechanisms and adaptive learning. Indolia et al. [15] and Liu et al. [19] improved the representation of characteristics under ambiguous conditions, while Roy et al. [21] proposed ResEmoteNet, which combines residual connections and attention mechanisms. In parallel, other works address efficiency and real-time applicability through lightweight models and combined detection-classification strategies [3,18]. Additional techniques such as temporal attention [14] and data augmentation with occlusions [7] further improve robustness.

Despite these advances, performance still decreases in non-frontal poses, irregular lighting, and subtle expressions [22,23], highlighting the need for robust and real-time solutions in educational environments.

2.2 Applications Based on Facial Emotion Analysis

FER has been applied to interpret human emotions and provide real-time feedback, particularly in education, where it supports the identification of engagement, distraction, and emotional states.

Several studies have explored FER-based systems for monitoring student behavior. Su et al. [24] analyzed distraction and fatigue, Xiong et al. [27] combined facial expressions and posture to measure engagement, and Tang et al. [25] linked emotional states with academic performance. In virtual learning environments, the systems proposed by Shen et al. [22] and Song [23] enable real-time evaluation of emotional engagement. More recently, dashboards have been used to transform model outputs into interpretable visualizations, including retrospective approaches [9] and real-time systems with visual indicators and alerts [12,13].

In general, integrating FER with visualization tools improves practical applicability, although challenges remain in adapting these systems to uncontrolled educational environments.

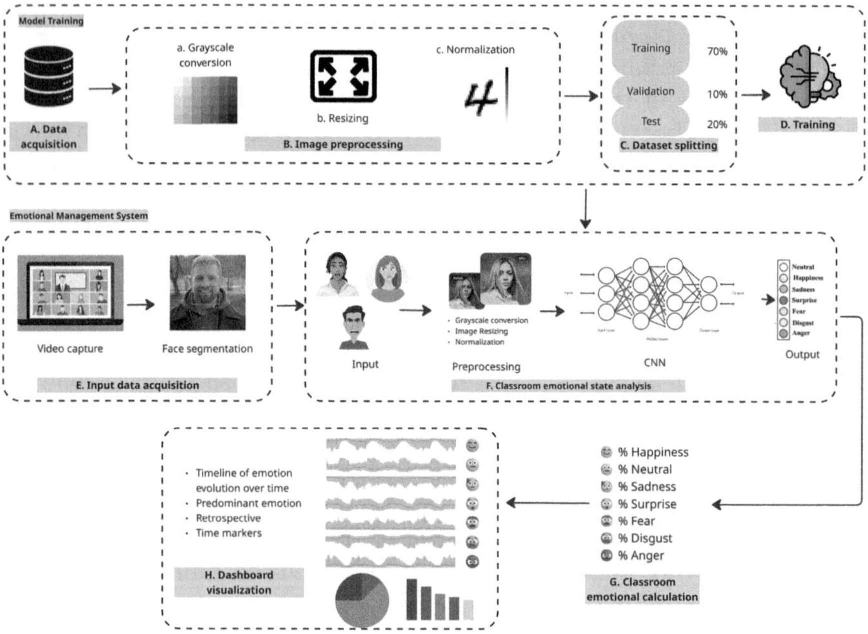

Fig. 1. Overview of the proposed methodology.

3 Methodology

This section presents the methodology of the proposed system, which consists of two main phases (see Fig. 1). The first phase focuses on training facial emotion recognition models using datasets from uncontrolled environments. The second phase involves the development of a real-time dashboard to monitor, visualize, and interpret the emotions of students in virtual classes. Finally, the system is evaluated through a usability study based on the System Usability Scale (SUS).

3.1 Facial Emotion Recognition

Data Acquisition. Two widely used FER datasets were selected: FER2013 [11] and RAF-DB [17], due to their diversity and relevance for uncontrolled environments with variations in lighting, pose, and occlusion. FER2013 contains 32,298 grayscale images labeled in seven basic emotions, representing challenging in-the-wild conditions, while RAF-DB includes 15,339 higher-quality images with annotations for basic and compound emotions, as well as pose and occlusion. To ensure the reliability of the ground truth, a manual curation process was performed on both datasets. This involved a class-by-class inspection to identify and remove low-quality samples, such as extremely darkened images, occluded faces, and low-resolution files where facial expressions were not clearly discernible. Additionally, emotional labels were verified to ensure they were correctly assigned to their respective classes. This preprocessing stage ensured that the models were trained on high-fidelity features.

Image Preprocessing. The preprocessing stage (see Fig. 1B) ensured consistency through grayscale conversion, resizing, and normalization. The images were converted to grayscale and replicated in three channels to maintain compatibility with RGB-based architectures such as ResEmoteNet, then resized to 64×64 pixels to balance computational efficiency and representation capacity [19]. Finally, pixel values were normalized using ImageNet parameters ($\mu = [0.485, 0.456, 0.406]$; $\sigma = [0.229, 0.224, 0.225]$), improving the training stability by standardizing the input distribution [3,19].

Model Training. ResEmoteNet was selected due to its strong performance in emotion recognition under uncontrolled conditions, achieving up to 94.76% accuracy on RAF-DB [21], and its suitability for real-world scenarios. The architecture integrates residual connections and Squeeze-and-Excitation (SE) attention mechanisms to enhance the representation of features and focus on relevant facial regions [21]. This architecture is a specialized Convolutional Neural Network (CNN) that optimizes local computational resources [1]. By leveraging SE blocks, the model achieves high-precision facial emotion inference directly on the instructor's hardware, ensuring the real-time processing and data privacy required for educational environments without relying on external cloud-based models.

3.2 Dashboard in the Real-Time Monitoring System

The dashboard acts as the visualization layer of the proposed system, presenting the outputs of the FER model. It converts model predictions into visual indicators that enable instructors to interpret students' emotional states during virtual learning sessions.

Input Data Acquisition. This stage captures visual data from the virtual classroom to enable facial emotion analysis. The video is obtained in real-time from the instructor's screen during virtual sessions, providing a non-invasive and platform-independent approach. Frames are sampled at fixed intervals of $N = 5$ seconds to reduce computational load while preserving sensitivity to emotional changes [24,27]. This strategy is supported by evidence indicating that facial expressions typically last between 0.5 and 4 s [8].

Subsequently, faces are detected using the YOLOv8-FACE model [4], selected for its robustness under variations in lighting, pose and occlusion. The model, based on the WIDER FACE dataset [28], enables reliable real-time detection of multiple faces in uncontrolled environments.

Emotional State Analysis of the Classroom. Once faces are segmented and cropped from each frame, they are used as input to ResEmoteNet for emotion classification. Before inference, each face undergoes preprocessing, including grayscale conversion, resizing to 64×64 pixels, and normalization using predefined statistical parameters. In virtual classroom settings, where multiple students may appear simultaneously, detected faces are processed in batches to enable parallel analysis and improve computational efficiency.

The model outputs a probability distribution on seven basic emotions, assigning each face to the class with the highest probability. These predictions are recorded in real-time and subsequently used to compute aggregated metrics that characterize both individual and group-level emotional states of the classroom.

Dashboard Visualization. The dashboard provides an interface designed for interpreting students' emotional states during virtual sessions, aiming to support real-time monitoring of classroom dynamics. The interface integrates four main visualization and monitoring components (see Fig. 2): in the upper-left section, a real-time session capture window allows the instructor to view the visual data source. Positioned in the lower-left area, an emotional timeline represents the temporal evolution of student emotions throughout the session. To the right, a pie chart summarizes the overall emotional distribution of the classroom, while directly below it, a markers section provides additional contextual data. Finally, a predominant emotion indicator displays the most frequent emotional state (e.g., Happiness) along with its corresponding percentage. These visual elements allow professors and teachers to assess both individual trends and group-level emotional patterns in real time.

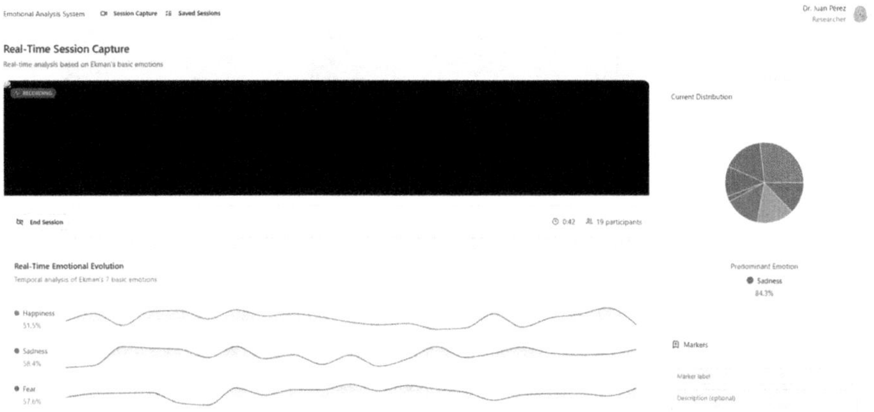

Fig. 2. Interface of the emotion monitoring dashboard showing real-time visualization through aggregated indicators and temporal graphs.

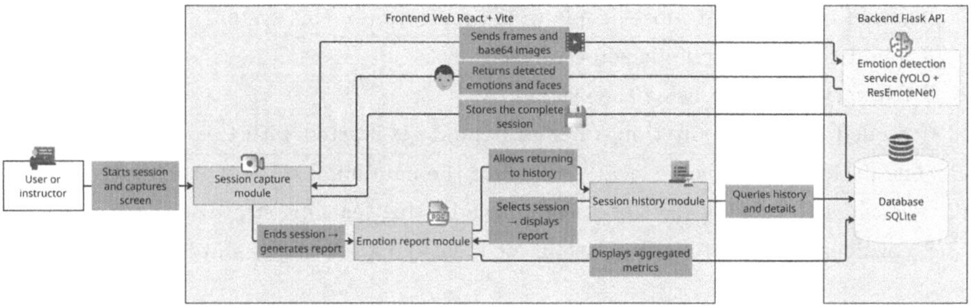

Fig. 3. Architecture of the proposed system showing the interaction between frontend, backend, and recognition models.

These visual elements allow instructors to quickly assess both individual trends and group-level emotional patterns. After each session, the system generates a report that includes the emotional timeline, aggregated distribution, and predominant emotion indicator, supporting a retrospective analysis of emotional patterns and student engagement.

Figure 3 presents the system architecture, which is composed of three layers: user interface, frontend, and backend. The frontend (React + Vite) manages user interaction and visualization, while the backend (Flask) handles frame processing, face detection, and emotion classification (YOLO + ResEmoteNet), as well as data storage in an SQLite database.

3.3 System Usability Evaluation

The final stage (see Fig. 4) evaluates the usability of the proposed dashboard from the instructors' perspective using the System Usability Scale (SUS) [6]. In

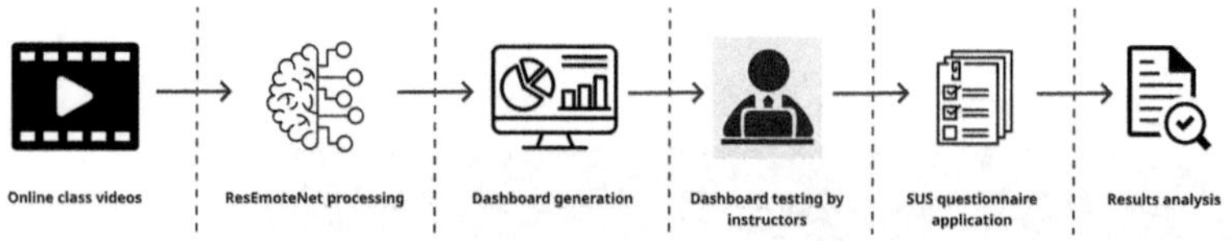

Fig. 4. Workflow of the usability evaluation process using the SUS questionnaire.

Table 1. SUS questionnaire items [6] used for evaluating the proposed dashboard.

No.	Questionnaire item
1	I think that I would like to use this system frequently
2	I found the system unnecessarily complex
3	I thought the system was easy to use
4	I think that I would need the support of a technical person to use this system
5	I found the various functions in this system were well integrated
6	I thought there was too much inconsistency in this system
7	I would imagine that most people would learn to use this system very quickly
8	I found the system cumbersome to use
9	I felt very confident using the system
10	I needed to learn many things before I could get started with this system
11	The dashboard helps me clearly visualize the emotional state of students
12	I consider the dashboard to be useful for adapting teaching strategies during the class
13	I consider the use of markers useful for identifying emotional states over time

addition to the SUS, three complementary questions were included to assess the pedagogical usefulness of the dashboard, focusing on the clarity of emotional visualization, the usefulness of temporal markers, and the support provided for adapting teaching strategies. The complete set of questionnaire items is summarized in Table 1.

The evaluation was conducted through three usage scenarios designed to reflect typical virtual classroom conditions: (i) retrospective analysis using recorded sessions and generated reports, (ii) real-time monitoring of emotional dynamics, and (iii) interactive use with manual markers to identify key moments. After completing these scenarios, the participants completed the SUS questionnaire together with the additional pedagogical questions. The study involved instructors with experience in virtual education, including both university professors and school teachers, allowing the evaluation of usability and pedagogical relevance under realistic conditions.

3.4 Ethical Considerations and Data Privacy

As this system involves biometric inference through facial analysis, all evaluations were conducted following institutional ethical guidelines. Participation of

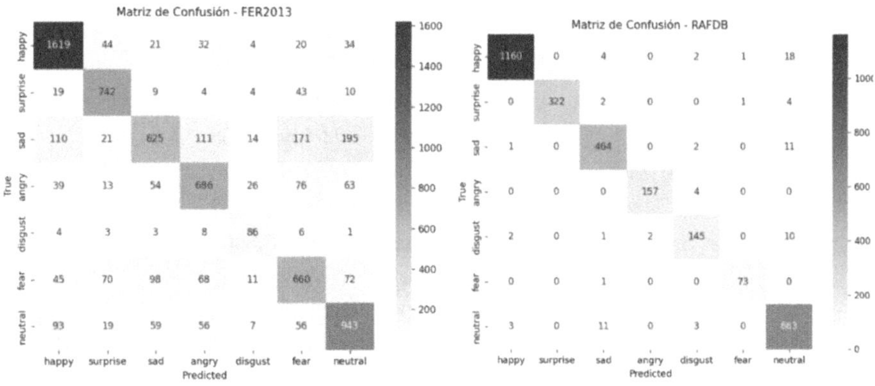

Fig. 5. Confusion matrices of the ResEmoteNet model trained on the FER2013 and RAF-DB datasets.

instructors was voluntary, and informed consent was obtained. To protect student privacy, the system was designed to process emotional labels anonymously; raw video streams are not stored, and no individual identities are linked to the emotional data. Furthermore, the manual curation of the training dataset was conducted strictly to improve model robustness, ensuring that the processed information remains aggregated and non-identifiable.

4 Results

4.1 Results of the Emotional Recognition Model

The performance of the emotion recognition model was evaluated using confusion matrices and class-wise metrics for both FER2013 and RAF-DB (Fig. 5 and Table 2). The confusion matrices provide insight into class-level performance and common misclassifications across datasets.

For RAF-DB, predictions are strongly concentrated along the diagonal, indicating high accuracy and low inter-class confusion. Emotions such as *happy*, *surprise*, and *angry* are consistently well classified, with only minor confusion observed for *disgust* in visually similar cases. In contrast, FER2013 exhibits a greater dispersion in predictions. Although *happy* and *surprise* remain well distinguished, increased confusion is observed between *fear*, *disgust*, and *sadness*, likely due to lower image quality and higher variability in this dataset.

This behavior is further reflected in the class-wise metrics. The RAF-DB model achieved an overall accuracy of 97.29%, significantly outperforming FER2013, which reached 74.70%. This high performance, which exceeds the baseline reported for ResEmoteNet, is directly attributed to the manual data curation protocol described before. By removing ambiguous and low-quality samples, the model was able to focus on more discriminative facial features, leading to higher precision in classes such as surprise and happiness. Across both datasets, *happy*

Table 2. Class-wise performance metrics of ResEmoteNet

Emotion	Dataset	Precision	Recall	F1-score
Happy	FER2013	0.8393	0.9126	0.8744
	RAF-DB	0.9949	0.9789	0.9868
Surprise	FER2013	0.8136	0.8929	0.8514
	RAF-DB	0.9901	0.9787	0.9892
Sad	FER2013	0.7192	0.5012	0.5907
	RAF-DB	0.9607	0.9707	0.9657
Angry	FER2013	0.7109	0.7168	0.7138
	RAF-DB	0.9874	0.9752	0.9812
Disgust	FER2013	0.5658	0.7748	0.6540
	RAF-DB	0.9295	0.9062	0.9177
Fear	FER2013	0.6395	0.6445	0.6420
	RAF-DB	0.9733	0.9865	0.9799
Neutral	FER2013	0.7155	0.7648	0.7393
	RAF-DB	0.9391	0.9750	0.9567
Macro avg	FER2013	0.7148	0.7440	0.7237
	RAF-DB	0.9693	0.9673	0.9682
Weighted avg	FER2013	0.7443	0.7470	0.7413
	RAF-DB	0.9734	0.9729	0.9730

and *surprise* obtained the highest F1-scores, while *sad*, *fear*, and *disgust* showed lower performance, likely due to class imbalance and visual similarity between categories. Based on these results, and considering the independent training of both models, the RAF-DB model was selected for deployment in the dashboard due to its superior accuracy and robustness under real-world conditions.

Regarding classification performance, the proposed system achieved an accuracy of 97.29% on the RAF-DB dataset, representing a significant improvement over contemporary FER models. This result outperforms hybrid architectures such as the MP-FERS (CNN + ViT) proposed by Tang et al. [25], which achieved 92.21%, and the Ada-DF fusion approach by Liu et al. [19] with 90.35%. Even when compared to specialized attention-based models like the one from Indolia et al. [15] (96.87%), our optimized implementation of ResEmoteNet demonstrates superior robustness. This performance gain, compared to the original 94.76% reported by Roy et al. [21], validates that combining residual attention blocks with a rigorous data curation protocol—specifically filtering for educational environment constraints—is highly effective for real-time applications compared to more complex Transformer-based architectures.

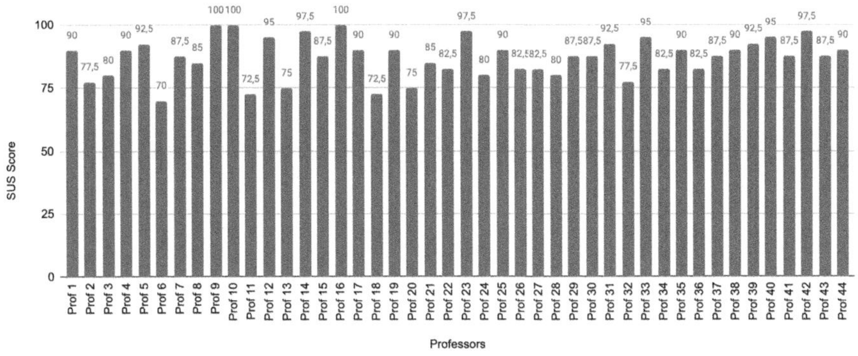

Fig. 6. Distribution of System Usability Scale (SUS) scores obtained from the instructors.

4.2 Results of User Validation

To complement the technical evaluation, a user validation was conducted to assess the perceived usability and perceived pedagogical usefulness of the dashboard. A total of 44 instructors participated, including 18 professors and 26 teachers. The sample consisted of 18 male and 26 female participants, with a mean age of $M = 49.3$ years ($SD = 10.5$). All participants had prior experience with digital tools and educational platforms.

Figure 6 presents the SUS scores obtained. The mean score was 87.05 ($SD = 7.82$), which corresponds to an "Excellent" usability level. All participants exceeded the reference threshold of 68 [6], indicating consistently positive perceptions of usability. A one-sample t-test comparing the observed mean with the SUS benchmark confirmed that the score is significantly higher than the acceptable threshold ($t(43) = 16.1$, $p < 0.001$). Additionally, Cronbach's alpha yielded $\alpha = 0.73$, indicating acceptable internal consistency.

Figure 7 shows the distribution of responses to additional pedagogical questions. The responses are concentrated in values 4 (*Agree*) and 5 (*Strongly Agree*), indicating a positive perception of emotional visualization, pedagogical support, and the usefulness of markers.

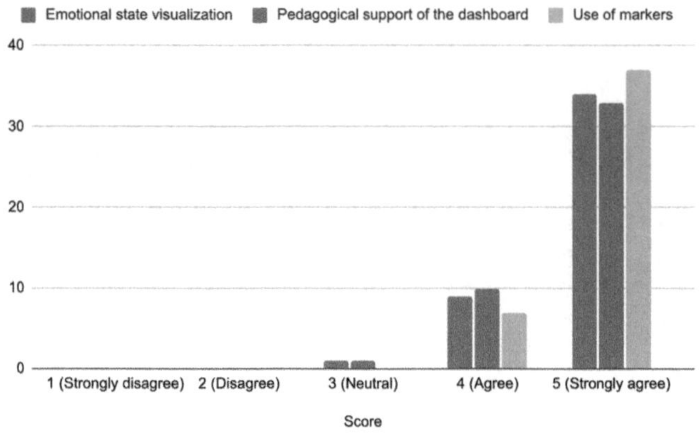

Fig. 7. Histogram of questions adapted to the educational context.

5 Discussion

The results highlight the critical role of dataset characteristics in facial emotion recognition, particularly in uncontrolled environments. The stronger performance observed with RAF-DB suggests that ResEmoteNet trained on higher-quality and more diverse data is better suited for real-world educational scenarios. In contrast, the limitations observed with FER2013 reflect the challenges introduced by variability in lighting, resolution, and expression subtlety, which are common in virtual classrooms.

These findings align with previous studies indicating that FER systems are highly sensitive to environmental conditions and data quality [3, 22]. This suggests that the effectiveness of emotion recognition in applied settings depends not only on the model design but also on the correspondence between the training data and the deployment conditions. In this context, the selection of RAF-DB for deployment is justified by its greater robustness under realistic variability.

From a usability perspective, the results indicate that instructors perceive the system as a tool that effectively supports interpreting classroom emotional dynamics. The positive perception of visualization clarity and pedagogical usefulness suggests that the dashboard provides interpretable and actionable information, enabling instructors to better understand student participation during virtual sessions. Participant feedback also highlights the potential to improve the system toward more proactive support. Suggestions such as incorporating automatic alerts and improving the visualization of emotional transitions indicate a natural progression from descriptive monitoring to more adaptive and intervention-oriented functionalities.

Despite these contributions, this study has limitations that should be addressed. First, the evaluation focused exclusively on the instructors' perspective; therefore, the emotional impact and perception of students being monitored remain unexplored. Second, while the system is designed for real-time use, this initial validation was conducted in controlled scenarios and retrospective analyses, leaving the assessment of end-to-end latency and frames per second (FPS) in high-density, live classrooms for future stages. Furthermore, the system remains dependent on the availability and quality of visual data, which may be affected by inactive cameras, occlusions, or low-resolution streams.

6 Conclusions

This study achieved its objective of implementing and validating a real-time emotional monitoring dashboard for virtual classrooms in uncontrolled environments. The proposed system integrates ResEmoteNet and YOLOv8-Face to enable automatic emotion detection and visualization through an interactive interface, supporting both real-time monitoring and retrospective analysis of emotional states in students.

The evaluation results confirm the effectiveness of the approach. ResEmoteNet achieved an accuracy of 97.29% on RAF-DB and 74.70% on FER2013, with a higher performance observed for *happy* and *surprise*, and lower results for *sad*, *fear*, and *disgust*. These differences highlight the impact of data variability and class imbalance on emotion recognition performance, particularly in less controlled conditions. Despite these challenges, the system demonstrates the potential of emotion-aware analytics to provide meaningful insights into the affective dynamics of students and support informed pedagogical decision-making.

From a user point of view, the system achieved high usability and positive acceptance among instructors, supporting its applicability in real educational contexts. Future work will involve a student-side study to evaluate privacy concerns and the psychological impact of emotional monitoring. Additionally, we plan to validate the system in live, large-scale virtual sessions to provide technical benchmarks on computational performance and explore multimodal approaches that integrate additional data sources to improve robustness and extend the applicability of emotion-aware systems in educational environments.

Acknowledgment. The authors have no competing interests to declare that are relevant to the content of this article.

References

1. Agung, E.S., Rifai, A.P., Wijayanto, T.: Image-based facial emotion recognition using convolutional neural network on emognition dataset. Sci. Rep. **14**(1), 14429 (2024). https://doi.org/10.1038/s41598-024-65276-x
2. Alcalá, C.: Se reduce la preferencia por las clases virtuales pero aún es el triple que antes de pandemia. Gestión (2024). https://gestion.pe/economia/empresas/sunedu-clases-virtuales-intenet-flanqueo-se-reduce-la-preferencia-por-las-clases-virtuales-pero-aun-es-el-triple-que-antes-de-pandemia-noticia/?ref=gesr
3. Anand, M., Babu, S.: Multi-class facial emotion expression identification using dl-based feature extraction with classification models. Int. J. Comput. Intell. Syst. **17**(1) (2024). https://doi.org/10.1007/s44196-024-00406-x

4. Arnabdhar: YOLOv8-Face-Detection. Hugging Face (Deep learning model) (2023). https://huggingface.co/arnabdhar/YOLOv8-Face-Detection
5. Ataucocha, C.: Las emociones en el aprendizaje virtual en los estudiantes de la Institución Educativa 50111 - Limatambo, 2020. Master's thesis, Universidad César Vallejo, Perú (2020). https://hdl.handle.net/20.500.12692/56269
6. Brooke, J.: SUS: a quick and dirty usability scale. In: Jordan, P.W., Thomas, B., Weerdmeester, B.A., McClelland, I.L. (eds.) Usability Evaluation in Industry, pp. 189–194. Taylor and Francis (1996)
7. Cheheb, I.: An investigation into the impact of occlusion on facial emotion recognition in the wild. In: Proceedings of the 17th International Conference on Pervasive Technologies Related to Assistive Environments (2024). https://doi.org/10.1145/3652037.3663932
8. Ekman, P.: Facial expression and emotion. Am. Psychol. **48**(4), 384–392 (1993)
9. Ez-zaouia, M., Tabard, A., Lavoué, E.: Emodash: a dashboard supporting retrospective awareness of emotions in online learning. Int. J. Hum Comput Stud. **139**, 102411 (2020). https://doi.org/10.1016/j.ijhcs.2020.102411
10. Gobierno del Perú: Sunedu prohíbe clases 100% virtual en universidades a partir del período académico 2024. gob.pe (2024). https://www.gob.pe/institucion/ugelayabaca/noticias/912704-sunedu-prohibe-clases-100-virtual-en-universidades-a-partir-del-periodo-academico-2024
11. Goodfellow, I.J., et al.: Challenges in representation learning: a report on three machine learning contests. In: Neural Information Processing, pp. 117–124 (2013). https://doi.org/10.1007/978-3-642-42051-1_16
12. Harb, A., et al.: Diverse distant-students deep emotion recognition and visualization. Comput. Electr. Eng. **111**, 108963 (2023). https://doi.org/10.1016/j.compeleceng.2023.108963
13. Hasnine, M., et al.: A real-time learning analytics dashboard for automatic detection of online learners' affective states. Sensors (2023). https://doi.org/10.3390/s23094243
14. Huang, Z., Zhu, Y., Li, H., Yang, D.: Dynamic facial expression recognition based on spatial key-points optimized region feature fusion and temporal self-attention. Eng. Appl. Artif. Intell. **133**, 108535 (2024). https://doi.org/10.1016/j.engappai.2024.108535
15. Indolia, S., Nigam, S., Singh, R.: A self-attention-based fusion framework for facial expression recognition in wavelet domain. Vis. Comput. **40**(9), 6341–6357 (2023). https://doi.org/10.1007/s00371-023-03168-3
16. Kotsia, I., Buciu, I., Pitas, I.: An analysis of facial expression recognition under partial facial image occlusion. Image Vis. Comput. **26**(7), 1052–1067 (2008). https://doi.org/10.1016/j.imavis.2007.11.004
17. Li, S., Deng, W.: Reliable crowdsourcing and deep locality-preserving learning for unconstrained facial expression recognition. IEEE Trans. Image Process. **28**(1), 356–370 (2019). https://doi.org/10.1109/TIP.2018.2868382
18. Ling, X., Liang, J., Wang, D., Yang, J.: A facial expression recognition system for smart learning based on YOLO and vision transformer. In: 7th International Conference on Computing and Artificial Intelligence, pp. 178–182 (2021). https://doi.org/10.1145/3467707.3467733
19. Liu, S., Xu, Y., Wan, T., Kui, X.: Ada-DF: an adaptive label distribution fusion network for facial expression recognition. arXiv (2024). http://arxiv.org/abs/2404.15714
20. Ministerio de Educación del Perú: Resultados de perú en pisa 2022. Minedu (2022). https://hdl.handle.net/20.500.12799/9802

21. Roy, A.K., Kathania, H.K., Sharma, A., Dey, A., Ansari, M.S.A.: ResEmoteNet: bridging accuracy and loss reduction in facial emotion recognition. IEEE Signal Process. Lett. **32**, 491–495 (2025). https://doi.org/10.1109/LSP.2024.3521321
22. Shen, J., Yang, H., Li, J., Cheng, Z.: Assessing learning engagement based on facial expression recognition in Mooc's scenario. Multimedia Syst. (2022). https://doi.org/10.1007/s00530-021-00854-x
23. Song, X.: Emotional recognition and feedback of students in english e-learning based on computer vision and face recognition algorithms. Entertain. Comput. **52**, 100847 (2025). https://doi.org/10.1016/j.entcom.2024.100847
24. Su, M.C., Cheng, C.T., Chang, M.C., Hsieh, Y.Z.: A video analytic in-class student concentration monitoring system. IEEE Trans. Consum. Electron. **67**(4), 294–304 (2021). https://doi.org/10.1109/TCE.2021.3126877
25. Tang, X., Gong, Y., Xiao, Y., Xiong, J., Bao, L.: Facial expression recognition for probing students' emotional engagement in science learning. J. Sci. Educ. Technol. (2024). https://doi.org/10.1007/s10956-024-10143-7
26. Vrochidis, A., Dimitriou, N., Krinidis, S., Panagiotidis, S., Parcharidis, S., Tzovaras, D.: A deep learning framework for monitoring audience engagement in online video events. Int. J. Comput. Intell. Syst. **17**(1) (2024). https://doi.org/10.1007/s44196-024-00512-w
27. Xiong, Y., Guo, X., Xu, J.: CNN-transformer: a deep learning method for automatically identifying learning engagement. Educ. Inf. Technol. (2023). https://doi.org/10.1007/s10639-023-12058-z
28. Yang, S., Luo, P., Loy, C.C., Tang, X.: WIDER FACE: a face detection benchmark. In: IEEE Conference on Computer Vision and Pattern Recognition (CVPR) (2016). https://shuoyang1213.me/WIDERFACE/

Speech Emotion Recognition for Safe Industrial Environment: Spanish Acted Speech Dataset and MLP Performance Analysis

Teth Azrael Cortes Aguilar(✉) and Adriana Tovar Arriaga

Tecnológico Nacional de México: ITJMMPyH, Zapopan (Jalisco) 45019, México
teth.cortes@zapopan.mariomolina.tecnm.mx,
adriana.tovar@zapopan.tecmm.edu.mx

Abstract. This work presents a new Spanish acted speech dataset designed for human–robot interaction (HRI) in manufacturing environments, comprising 750 audio samples recorded by 25 speakers between 20 and 60 years old. 15 of the speakers were male and 10 were female. The dataset corresponds to a simulated industrial scenario in which a worker is positioned in front of a robot controlled by ten predefined voice commands. Motivated by evidence that speech emotion recognition (SER) performance deteriorates when the training and input languages differ, this research focuses on creating a worker's language-matched SER system to enhance safety in HRI. A total of 275 acoustic features were computed as the mean values of the CONT, MEL, CHROMA, and MFCC descriptors. These features were used to train a multilayer perceptron (MLP) classifier for recognizing anger, sadness, and neutral emotional states. In this particular application, false negatives are undesirable because misclassifying anger or sadness as neutral can inhibit the robot's preventive actions; therefore, two activation functions, ReLU and Tanh, were evaluated to identify the model that best minimizes such errors. Despite its simplicity, the MLP model delivered performance comparable to more complex architectures. The ReLU-based model achieved accuracies of 93% on the training data and 91% on the validation data, the Tanh-based model reached 98.5% accuracy during training and 98% accuracy on the validation data.

Keywords: Robotics · Human Computer Interaction · Applied computing.

1 Introduction

Initially, robotics centered around automation, emphasizing factors such as reliability and dependency. In traditional applications, the robot stops as soon as the worker crosses the safeguarded space. Industrial robots are typically robust and utilitarian, optimized for tasks like assembly or packaging. However, the focus of robotics research has transitioned towards improving the efficiency of Human-Robot Interaction (HRI) [1]. For instance, spatial interaction can be used to avoid collisions [2], non-verbal interaction can predict human movements [3], verbal interaction can detect ambiguous instructions

H. Ponce et al. (Eds.): COMIA 2026, CCIS 3095, pp. 134–145, 2027.
https://doi.org/10.1007/978-3-032-37438-7_11

[4], and understanding human emotions can prevent dangerous behaviors [5]. Collaborative robots, in particular, bridge the gap between automation and manual labor by enabling safe, direct physical interaction with humans. Unlike conventional industrial robots, which are usually isolated from workers to avoid physical contact with humans, cobots coexist with humans in a common workspace [6]. Despite cobots contain safety devices, it is not possible to claim that they will always be safe. As regards to issues in collaborative environments, studies indicate that 60% are collisions, 14% related to ergonomic concerns, 12% to mental stress, 10% to risk assessment, and 4% to cyberattacks [7]. Initiatives to reduce occupational injuries and safety risks must be part of the robotized workplace [8].

The human factor in industrial systems has been underestimated, negatively impacting performance and occupational safety. Workers primarily experience stress from unpredictable, high-speed cobot motions in shared workspaces. While monitoring human reactions is essential to mitigate discomfort, cobots currently lack emotional state recognition [9]. Consequently, emotion recognition is a vital research frontier for developing robotic systems that minimize psychosocial risks in collaborative environments.

1.1 Emotion Classification

In 1972, Paul Ekman identified six universal emotions—happiness, sadness, fear, anger, surprise, and disgust—linked to distinct facial patterns [10]. Modern research often simplifies this into a three-class model (positive, negative, and neutral), utilizing technologies like DNNs for brain activity classification [11, 12]. In industrial HRI, various modalities are employed to recognize these states, including Speech Emotion Recognition (SER), electromyography, eye-tracking, and facial expression detection [13]. This study focuses on anger, sadness, and neutral states, selected for their high relevance to minimizing psychosocial risks in industrial HRI [15]. A constrained three-class set was chosen because emotion detection accuracy typically decreases as categories increase [14]. While neutral states are most prevalent in normal working conditions, anger and sadness are the primary drivers of stress-related workplace accidents in manufacturing; conversely, positive emotions (happiness, surprise) and less common negative states (fear, disgust) were excluded as non-critical.

The remainder of this paper is organized as follows: Sect. 2 provides an overview of SER, detailing the dataset and audio organization, the feature extraction process, hardware specifications, and the proposed learning model. The results of the Multi-Layer Perceptron (MLP), including a comparison between the ReLU and Tanh activation functions, are presented in Sect. 3. Finally, Sect. 4 present the conclusions.

2 Speech Emotion Recognition

The ability of machines to predict emotions through speech signals is a strong incentive to improve HRI [16]. The recognition of emotions from speech has been extensively investigated in recent times. Nevertheless, SER remains a challenging task, with the question of how to extract effective emotional features. In a manufacturing domain, a SER application means that the system must understand the worker's emotions, which will define the system's actions accordingly. SER performance depends primarily on selecting effective feature extraction techniques for specific emotions. Additionally, training learning models for reliable classification remains a significant challenge.

Optimal SER performance relies on representative training databases and the selection of key emotional features. For example, research shows that anger typically exhibits a high fundamental frequency, whereas sadness is characterized by a low one [17]. Because emotions represent complex input variables, it is essential that evaluations preserve a degree of naturalness to ensure that the predictions of the learning model approximate reality. There are three methods to obtain a speech dataset. Spontaneous Speech (SS) or natural offer the most authentic emotions but are notoriously difficult to collect. Extracting and labeling audio from real-world recordings requires significant time, effort, and resources. Elicited Speech (ES) technique raises significant ethical concerns. Evoking intense states like anger or sadness can cause psychological distress or adverse health reactions, complicating dataset acquisition and potentially requiring medical intervention. Acted speech (AS), also known as simulated speech, is more easily obtained. The effectiveness of such datasets depends on the actors' ability to modulate their voices appropriately for a given emotion [18].

Although English is an international language used in many workplaces, Spanish is the second most widely spoken language in the world, and it is spoken as a native language by around 437 million people [19]. Hence, studying human behavior among Spanish speakers through Speech Emotion Recognition is particularly important. Previous research has demonstrated that emotion recognition is influenced by linguistic factors. In classification tasks, single-language approaches achieve higher accuracies than cross-language approaches. In particular, the performance of SER systems is significantly reduced for languages other than the target language [20]. Furthermore, distinctions between male and female emotional expression may exert a greater impact than cultural or phonetic variations across languages. In addition, significantly higher recognition rates are observed for emotions conveyed by native speakers [21, 22]. Several Spanish SER systems have been developed recently. The emotional categories detailed in Table 1 include Anger (A), Neutral (N), Sadness (S), Disgust (D), Fear (F), Surprise (U), Happiness (H), and Cold (C).

None of the Spanish datasets shown in Table 1 considers the prevention of workplace injuries involving industrial robots as a possible application. Negative emotions like anger and sadness significantly impact organizational productivity and engagement [29]. Anger is a hostile response to frustration or perceived injustice, often manifesting as harmful behavior. Sadness involves a complex of physical and social changes, including withdrawal and reduced motivation. In contrast, a neutral state is characterized by the absence of strong emotional feelings. By maintaining distance from immediate emotional triggers, neutrality enables a more objective assessment of situations, allowing workers to make better decisions [30].

In this work, the classification task focused on recognizing only three emotions: anger, sadness, and neutral. Early detection of anger and sadness in HRI can prevent injuries. For example, control signals can automatically halt robot movement while alarms alert safety personnel to hazardous situations. We propose a controlled experimental setup at the Instituto Tecnológico José Mario Molina Pasquel y Henríquez

Table 1. Comparison of Spanish databases in speech emotion recognition

	Database	Speakers	Emotions	Application
ES	EmoMatchSpanishDB [23]	30 male 20 female	A, N, S, D, F, U, H.	Understanding emotional communication by people with neurological disorders.
SS	Spanish MEACorpus [24]	5129 audio segments extracted from youtube	A, S, D, F, U, H.	Understanding human actions in healthcare, psychology, social science, and marketing.
SS AS	Acoustic Modelling of Emotion-Specific Vowels [25]	3 male 3 female	A, N, S, H.	Recognition and acoustic modeling of emotion-specific vowels.
AS	Spanish Emotional Speech [26]	1 professional actor	A, S, C, U, H.	Examining the influence of voice quality on emotional state.
AS	Interface Emotional Speech Synthesis [27]	1 male 1 female	A, N, S, D, F, U, H.	Recognition of emotions for two speakers using standard hidden Markov models.
AS	InterFace [28]	1 male 1 female	A, N, S, D, F, U, H.	Analysis of various acoustic features and a large set of statistical features.

(ITJMMPyH) robotics laboratory. As shown in Fig. 1a, the facility features a UR5 robot configured for human-robot collaboration. The controller's screw terminal panel manages safety I/O signals, allowing the learning model to trigger a safeguard stop automatically. This interface pauses robot movement safely and resets once the event is resolved, ensuring worker safety without requiring manual intervention.

2.1 Dataset

The Spanish Voice Command Recognition (SVCR) dataset, comprising 750 acted audio recordings, is available at https://doi.org/10.6084/m9.figshare.30790625. The research included 25 participants (15 males, 10 females). As shown in Fig. 1b, the dataset contains 25 speaker folders, each holding 30 audio files balanced across the three emotional categories. The age range of the speakers was 20 to 60 years, in line with the expected age of a worker in a manufacturing company. In particular, during the first quarter of 2025, the average age of manufacturing workers in México was 37.2 years; and the workforce comprised 70.3% men and 29.7% women [31]. In order to achieve greater fidelity, the recordings were produced using a commercial unidirectional condenser microphone, featuring noise-canceling, volume adjustment, and voice depth control. Audio was sampled at 22.5 kHz and stored in uncompressed digital WAV format. Recent research has reported that high-quality digital audio recorded in a noise-free environment at a 22 kHz sampling rate does not require additional pre-processing steps. Although reducing the sampling frequency between 8 and 16 kHz decreases file size and the computation time required to extract the most representative acoustic features of speech, it also decreases the accuracy of SER systems [32].

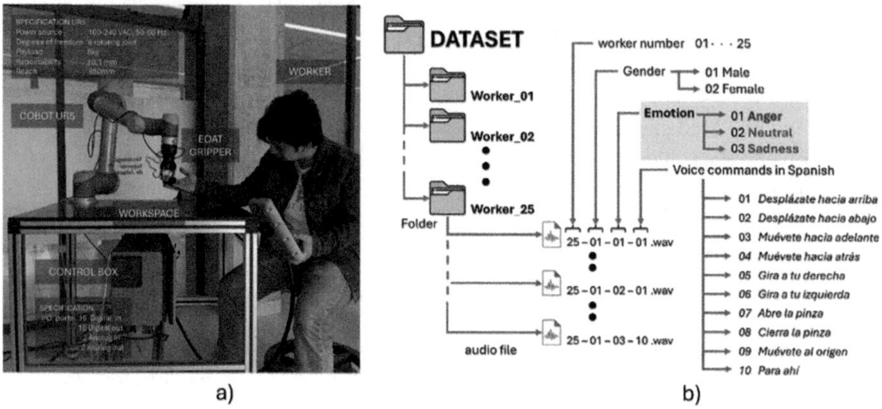

Fig. 1. HRI: a) UR5 in the laboratory, b) file arrangement in the SVCR dataset

We consider that the worker is located in front of the robot, and that a set of ten voice commands in Spanish is sufficient to control its movements. In addition, the robot uses a gripper as its end effector to carry out pick-and-place operations within the worker's workspace. Commands C1 to C6 refer to the task of moving the end effector or end-of-arm tooling (EOAT) in a single direction. Commands C7 and C8 manage the gripper for object pick-and-release tasks. Command C9 moves the robot to the initial position. Command C10 calls the Wait command to pause the robot's movement. It allows the robot to hold its position until a signal is received from a digital input. It should be noted that when a voice command is recognized, it produces a digital input in the robot's control box. Conversely, when a negative emotion is recognized, it produces a signal for the safeguard stop interface.

2.2 Features Extraction

For feature extraction, we used the Python audio processing library Librosa, and the learning model was trained and evaluated on an ASUS TUF F15 computer with an Intel i7 processor, 16 GB of RAM, and an NVIDIA RTX 4050 GPU. The input data for the training model consisted of a vector with 275 features. Of these, 128 features were computed as the mean values across an equal number of channels defined by the Mel-Frequency Cepstral Coefficients (MFCC). Another 128 features were extracted from the mel-scale spectrogram (MEL). A total of 12 features were extracted from chromagrams (CHROMA), which can be computed from either the waveform or the power spectrogram. The spectrogram was also divided into six subbands, and the mean of each was calculated. The energy contrast (CONT) of each subband was estimated by comparing the mean energy in the top quartile or peak energy to the bottom quartile or valley energy, see Fig. 2.

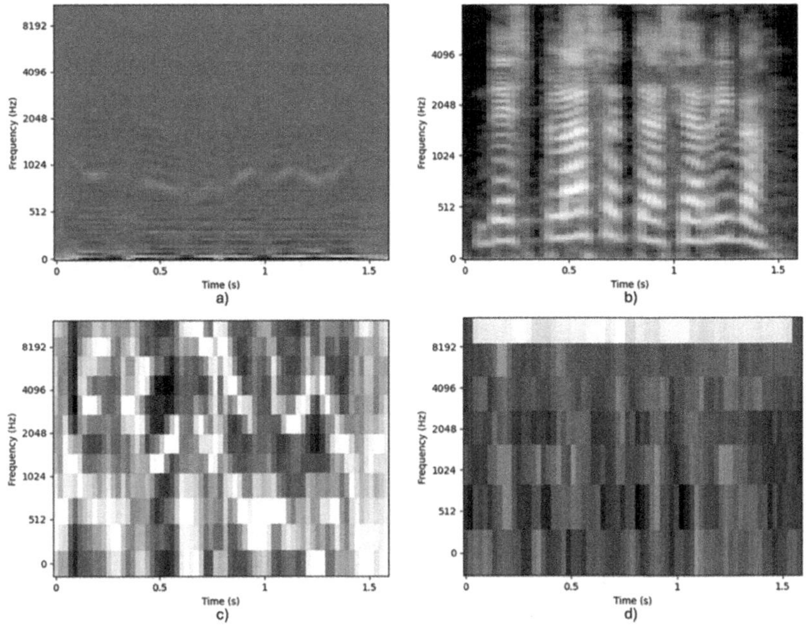

Fig. 2. Features extracted: a) MFCC, b) MEL, c) CHROMA, and d) CONT

2.3 Learning Model

The SVCR dataset was split into 80% for model training and 20% for testing. We used a Multi-layer Perceptron Classifier (MLP) from scikit Learn. A comparative evaluation of two activation functions, Rectified Linear Unit (ReLU) and Hyperbolic Tangent (Tanh), was performed. The models were trained using ReLU in the first instance and Tanh in the second. The first model utilized a hidden layer with 100 neurons and the ReLU activation function. The hyperparameters were configured as follows: alpha at 0.005, a batch size of 256, epsilon at 1e-0.8 and a random state of 1. The initial learning rate was set to 0.001 and established as adaptive; this implies that the learning rate was dynamically adjusted for each weight in the network during the training process. The training process was limited to a maximum of 80 epochs to ensure optimal generalization and prevent overfitting. Figure 3a shows that while using fewer iterations inhibits the model's ability to generalize effectively, surpassing the 80 epoch threshold degrades performance on unseen data. A second model was configured using the Tanh activation function with a hidden layer of 100 neurons. The Adam solver was utilized with a batch size of 256, an initial learning rate of 0.001, an alpha of 0.05, and an epsilon of 1e-0.8. To prevent overfitting, the maximum number of iterations was limited to 130 epochs, a decision supported by the results presented in Fig. 3b.

The activation function is essential for determining the output of each neuron. It introduces nonlinearity, allowing the perceptron to learn complex patterns and make decisions based on the input data. The activation function has a substantial impact on the model's performance. The activation function ReLU forces all negative inputs to zero, resulting in a state where some neurons become inactive and stop learning. If a significant portion of relevant features for identifying a true positive case consistently produces negative inputs early in the training, those neurons might become inactive, leading to the model's inability to recognize that class later, which manifests as an increase in false negatives. The activation function Tanh, by contrast, maps negative inputs to a non-zero, negative output value, between -1 and 0. This means all neurons remain active to some degree, allowing gradients to continue flowing during backpropagation and preventing the dying ReLU problem. Consequently, the network may more effectively capture subtle data features that might otherwise be inactivated by ReLU.

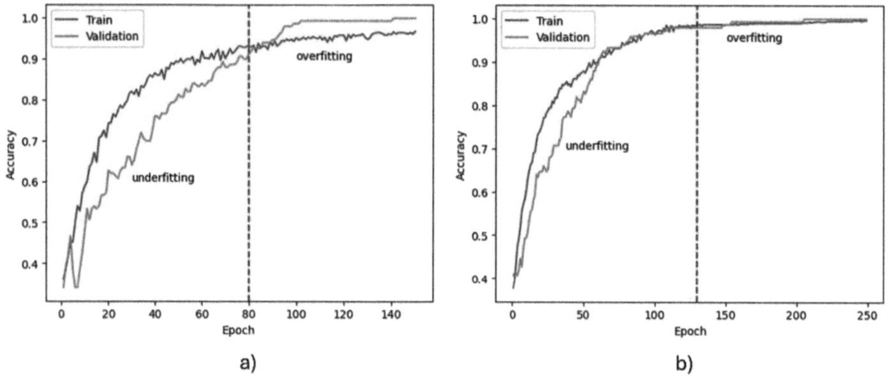

Fig. 3. Accuracy vs. epochs for training and validation data: a) ReLU, b) Tanh

3 Results

This section details the results obtained from the training and validation datasets, comparing the use of the ReLU activation function in the first model and the Tanh function in the second. Although the first model exhibits acceptable accuracy, we prioritized error reduction, accepting that the second configuration operates closer to the overfitting threshold. In this context, false negatives are highly undesirable. A misclassification of a neutral emotion as sad or angry is considered worse than misidentifying a sad or angry emotion as neutral. In addition, to enhance the model's generalization capability and reduce its sensitivity to noise, the alpha was increased from 0.005 to 0.05.

Figure 4 shows the training and validation loss across epochs for two activation functions. For ReLU, Fig. 4a, optimal performance occurs at 80 epochs with a training loss of 0.2184 and a validation loss of 0.2361. For Tanh, Fig. 4b, the model reaches its optimal setting at 130 epochs, yielding training and validation losses of 0.2934 and 0.3912, respectively.

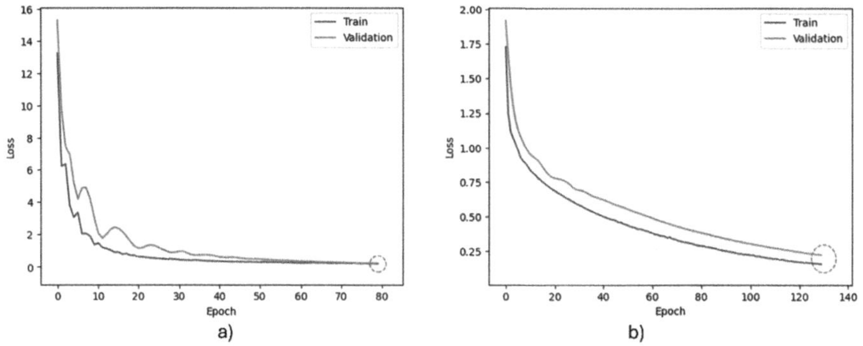

Fig. 4. Loss function vs. epochs for training and validation data: a) ReLU, b) Tanh

Figure 5a, b present the confusion matrices for training and validation using the ReLU activation function, respectively. In the training phase, the model achieved high accuracy across all categories, with 183, 187, and 190 correct classifications for anger, neutral, and sadness, respectively, totaling 40 errors per category. Validation results in Fig. 5b show perfect recall for the angry emotion (52/52), while the neutral and sad categories recorded 49 and 36 correct classifications. Both datasets exhibited a balanced error distribution, with 13 total errors per category in the validation set. Figure 5c and 6d presents the confusion matrices for training and validation using the Tanh activation function, respectively. Training results show high precision, with 193, 193, and 205 correct classifications for anger, neutral, and sadness, respectively, and a balanced error distribution of 9 per category. Validation performance was particularly strong, achieving perfect classification for anger (52/52) and neutral (52/52) emotions, while sadness recorded 43 correct identifications with zero false negatives.

The model using ReLU activation function achieved accuracies of 93% on the training data and 91% on the validation data. The model using Tanh activation function achieved accuracies of 98.5% on the training data and 98.0% on the validation data. Given the balanced dataset, which contains a similar number of samples across the three classes, the accuracy metric provides a reliable indicator of the model's performance. However, false negative errors, where sad and angry emotions are recognized as other categories, have more serious consequences for preventing workplace injuries than false positive errors, where neutral emotions are recognized as sad or angry.

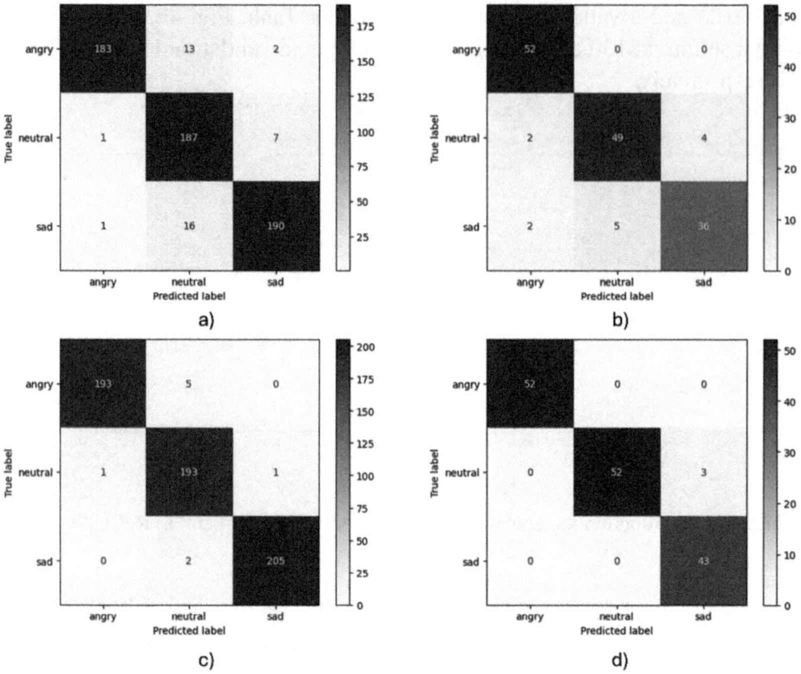

Fig. 5. Confusion matrices for: a) training data using the ReLU, b) validation data using the ReLU, c) training data using the Tanh, d) validation data using the Tanh

4 Conclusions

This work introduces a new Spanish acted dataset comprising 750 audio samples from 25 speakers. This approach addresses research indicating that speech emotion recognition performance declines when training and input languages differ. In HRI, detecting negative emotions like anger or sadness is essential for preventing workplace injuries. Ultimately, this research fosters a safer, more reliable collaborative environment by bridging the gap between human workers and robots.

We extracted 275 acoustic features, mean values of CONT, MEL, CHROMA, and MFCC, from the voice recordings in 14.844 s. These features trained an MLP classifier to categorize anger, sadness, and neutral states. A comparison between activation functions showed that while ReLU achieved high accuracies (93% training, 91% validation), the Tanh model was selected to prioritize error reduction, despite being closer to the overfitting threshold

In this application, false negatives are prioritized; misclassifying neutral states as anger or sadness is more critical than the inverse. To improve generalization and noise robustness, the alpha parameter was increased from 0.005 to 0.05. After 130 epochs, the Tanh-based model achieved 98.5% accuracy (0.2934 loss) on training data and 98% accuracy (0.3912 loss) on validation data. Despite its relative simplicity, the MLP model achieved an acceptable accuracy when compared to more complex architectures, such as CNNs. This approach is particularly advantageous for implementation in developing countries that depend on manufacturing for technological and economic development, where computing resources are often limited.

Based on the findings of this research, future work should focus on enhancing the model's robustness and efficiency in real industrial environments. Primarily, we propose expanding the dataset to include more recordings captured under challenging ambient noise conditions and diversifying the participant pool to account for various geographical backgrounds and accents. Data augmentation using synthetic machine noise, dialogue, and vehicle sounds can mitigate recording difficulties and improve system robustness against acoustic interference. Additionally, a comparative analysis involving advanced deep learning architectures is recommended to optimize classification accuracy. Subsequent studies should investigate whether narrowing the affective scope to three emotions facilitates significant gains in processing speed.

Acknowledgments. This work was partially supported by the Secretaría de Ciencia, Humanidades, Tecnología e In-novación (SECIHTI), Tecnológico Nacional de México (TecNM), Instituto Tecnológico José Mario Molina Pasquel y Henríquez (ITJMMPyH) U.A. Zapopan.

Disclosure of Interests The authors have no competing interests to declare that are relevant to the content of this article.

References

1. Campagna, G., Rehm, M.: A Systematic review of trust assessments in human–robot interaction. J. Hum.-Robot Interact. 14(2) (2025).
2. Boschetti, G., Faccio, M., Granata, I., Minto, R.: 3D collision avoidance strategy and performance evaluation for human–robot collaborative systems. Comput. Ind. Eng. **179** (2023)
3. Görür, O., Rosman, B., Sivrikaya, F., Albayrak, S.: FABRIC: a framework for the design and evaluation of collaborative robots with extended human adaptation. J. Hum.-Robot Interact. **12**(3) (2023)
4. Younes, R., Elisei, F., Pellier, D., Bailly, G.: Impact of verbal instructions and deictic gestures of a cobot on the performance of human coworkers. In: IEEE-RAS International Conference on Humanoid Robots, Nancy, France, pp. 1025–1032 (2024)
5. Bartneck, C., Belpaeme, T., Eyssel, F., Kanda, T., Keijsers, M., Sabanovic, S.: Human Robot Interaction: An Introduction. Cambridge University Press (2024)
6. Peshkin, M., Colgate, J.: Cobots. Ind. Robot Int. J. Robot. Res. Appl. **26**(5), 335–341 (1999)
7. Giallanza, A., La Scalia, G., Micale, R., La Fata, C.: Occupational health and safety issues in human-robot collaboration: state of the art and open challenges. Saf. Sci. **169** (2024)
8. Pauliková, A., Babeľová, Z., Ubárová, M.: Analysis of the impact of human–cobot collaborative manufacturing implementation on the occupational health and safety and the quality requirements. Int. J. Environ. Res. Public Health. **18**(4) (2021)
9. Toichoa, A., Mohammed, W., Martinez, J.: Emotion-driven analysis and control of human-robot interactions in collaborative applications. Sensors. **21**(14) (2021)
10. Ekman, P.: Universals and cultural differences in facial expressions of emotion. In: Nebraska Symposium on motivation, vol. 19, pp. 207–283. Lincoln University of Nebraska Press (1971)
11. Nithya, T., Ezhilkumar, V., Supikssa, K., Harshini, S., Kishore, R.: Emotion recognition from wearable EEG devices using neural networks. In: 2024 13th International Conference on System Modeling & Advancement in Research Trends (SMART), pp. 696–700. IEEE (2024)

12. Babu, N., Satija, U., Mathew, J., Vinod, A.: Emotion recognition in virtual and non-virtual environments using EEG signals: dataset and evaluation. Biomed. Signal Process. Control. **106** (2025)
13. Wang, T., Zheng, P., Li, S., Wang, L.: Multimodal human–robot interaction for human-centric smart manufacturing: a survey. Adv. Intell. Syst. **6**(3) (2023)
14. Alonso-Martín, F., Malfaz, M., Sequeira, J., Gorostiza, J., M., S.: A multimodal emotion detection system during human-robot interaction. Sensors. **13**(11), 15549–15581 (2013)
15. Britt, T., Wilson, C.E., Elbogen, E., Van Voorhees, E., Dillon, K.: Anger as an occupational health challenge for employees in high-risk occupations. In: Adler, A., Forbes, D. (eds.) Anger at Work: Prevention, Intervention, and Treatment in High-Risk Occupations, pp. 45–77. American Psychological Association (2021)
16. Agrawal, A., et al.: Comparative analysis of speech emotion recognition models and technique. In: International Conference on Computational Intelligence, Communication Technology and Networking (CICTN), Ghaziabad, India, vol. 2023, pp. 499–505 (2023)
17. Qadri, S., Gunawan, T., Alghifari, M., Hasmah Mansor, H., Kartiwi, M., Janin, Z.: A critical insight into multi-languages speech emotion databases. Bull. Elect. Eng. Inform. **8**(4), 1312–1323 (2019)
18. Wani, T., Gunawan, T., Qadri, S., Kartiwi, M., Ambikairajah, E.: A comprehensive review of speech emotion recognition systems. IEEE Access. **9**, 47795–47814 (2021)
19. Liang, L., Wang, S.: Spanish emotion recognition method based on cross-cultural perspective. Front. Psychol. **13** (2022)
20. Motoyuki, S., Jinhai, Q.: Improvement of multilingual emotion recognition method based on normalized acoustic features using CRNN. Procedia Comput. Sci. **207**, 684–691 (2022)
21. Rajoo, R., C., A.: Influences of languages in speech emotion recognition: a comparative study using Malay, English and Mandarin languages. In: 2016 IEEE Symposium on Computer Applications & Industrial Electronics (ISCAIE), Penang, Malaysia, pp. 35–39 (2016)
22. Costantini, G., Parada, E., Casali, D., Cesarini, V.: The emotion probe: on the universality of cross-linguistic and cross-gender speech emotion recognition via machine learning. Sensors. **22** (2022)
23. García, E., Barba, A., Gachet, D.: EmoMatchSpanishDB: study of speech emotion recognition machine learning models in a new Spanish elicited database. Multimed. Tools Appl. **83**(5), 13093–13112 (2023)
24. Pan, R., García, J., Rondríguez, M., García, F., Valencia, R.: Overview of EmoSPeech at IberLEF 2024: multimodal speech-text emotion recognition in Spanish. Procesamiento del lenguaje natural. **73**, 359–368 (2024)
25. Caballero, S.: Recognition of emotions in Mexican Spanish speech: an approach based on acoustic modelling of emotion-specific vowels. Sci. World J. **2013**, 162093 (2013)
26. Montero, J., Gutiérrez, J., Colás, J.: Analysis and modeling of emotional speech in Spanish. In: Ohala, J., Hasegawa, Y., Ohala, M., Granville, D., Bailey, A. (eds.) Proceedings of International Conference on Phonetic Sciences (ICPhS), pp. 957–960, San Francisco, USA (1999)
27. Nogueiras, A., Moreno, A., Bonafonte, A., J., M.: Speech emotion recognition using hidden Markov models. In: Proceedings of European conference on speech communication and technology (Eurospeech'01), Denmark, Scandinavia (2001)
28. Hozjan, V., Kačič, Z.: Improved emotion recognition with large set of statistical features. In: 8th European Conference on Speech Communication and Technology (Eurospecch'03), Geneva, Switzerland (2003)
29. Pearson, C.: Are negative emotions brewing in your workplace?, Homepage. https://sloanreview.mit.edu/article/are-negative-emotions-brewing-in-your-workplace/. Last accessed 18 Nov 2025

30. APA Dictionary of Psychology: Term: anger, sadness, process model of emotion regulation. Homepage from https://dictionary.apa.org/. Last accessed 18 Nov 2025
31. Gobierno de México: Data México Industrias Manufactureras sector 31-33. Fuerza laboral según años de escolaridad, rango de edad y sexo (2025). Homepage https://www.economia.gob.mx/datamexico/es/profile/industry/manufacturing?occupationMetrics=workforceOption
32. Lech, M., Stolar, M., Best, C., Bolia, R.: Real-time speech emotion recognition using a pre-trained image classification network: effects of bandwidth reduction and companding. Front. Comput. Sci. **2**(14) (2020)

Mapping the Impact of Large-Scale Language Models on Decision-Making, Personalized Learning, and Ethical Governance in Higher Education

Nicia Guillén-Yparrea[1(✉)] and Iván Miguel García-López[2]

[1] School of Engineering and Sciences, Tecnologico de Monterrey, Monterrey, Mexico
nicia.gy@tec.mx

[2] School of Engineering and Sciences, Tecnologico de Monterrey, Atizapán de Zaragoza, Estado de México, Mexico

Abstract. This study presents a literature mapping analysis of the integration of Large-Scale Language Models (LLMs) in higher education, focusing on their impact on decision-making in educational administration, the personalization of learning, and associated ethical challenges. Using a structured literature mapping approach, the study identifies key trends, emerging patterns, and research gaps across indexed literature published between January 2020 and April 2026. The findings reveal that LLMs significantly enhance administrative efficiency and data-driven decision-making, while enabling adaptive and personalized learning experiences. However, their adoption raises critical concerns related to algorithmic bias, data privacy, institutional autonomy, and the lack of robust governance frameworks. Unlike prior studies centered primarily on pedagogical implications, this research highlights the systemic and strategic transformation of higher education institutions driven by AI. This study contributes by developing an integrated interpretive framework that explains how LLMs simultaneously reshape administrative decision-making, personalized learning, and ethical governance in higher education.

Keywords: large language models · decision-making · higher education.

1 Introduction

In recent years, artificial intelligence (AI) has emerged as a disruptive technology in higher education, transforming both academic administration and teaching-learning models. Its adoption in educational management has made it possible to optimize processes, improve the personalization of learning and facilitate strategic decision-making in higher education institutions (HEIs). Within this panorama, Large-Scale Language Models (LLMs) have gained relevance in various administrative and pedagogical areas, with applications ranging from the automation of tasks to the generation of educational content [1]. However, its implementation has sparked debates about institutional autonomy, equity in learning, and the ethical challenges arising from its use [2]. The increasing

H. Ponce et al. (Eds.): COMIA 2026, CCIS 3095, pp. 146–158, 2027.
https://doi.org/10.1007/978-3-032-37438-7_12

integration of AI in education raises questions about its impact on decision-making, its role in personalizing learning, and the ethical implications of its application in academic settings.

This study adopts a literature mapping approach, a methodology that allows for the exploration and classification of existing knowledge about a developing field, identifying patterns, trends, and gaps in research. Unlike systematic reviews, which seek to synthesize empirical results under specific criteria, literature mapping provides a panoramic view of the topic, providing a conceptual framework that relates different areas of study. For this analysis, three key axes have been identified that allow structuring the discussion on LLMs in higher education: (1) AI in educational management, (2) the personalization of learning and teacher transformation, and (3) ethical and regulatory challenges. These three axes allow us to analyze the impact of AI from a holistic perspective, considering not only its technical functionality, but also its strategic and regulatory implications [3].

Literature mapping reveals that AI has been used to improve educational administration, reduce teaching workload, and optimize adaptive learning. However, challenges such as resistance to change, the digital divide and the need for regulatory frameworks to guide their responsible implementation persist [4]. Unlike previous studies that have approached AI in education from an exclusively pedagogical or technological approach, this research emphasizes its impact on educational administration and strategic decision-making. Thus, this paper contributes to the debate on the integration of AI in higher education from a broad perspective, identifying opportunities and risks associated with its application in academic planning, resource allocation and the design of educational policies [5].

The purpose of this study is to provide a comprehensive overview of the impact of LLMs on higher education, analyzing their role in institutional management, the personalization of learning, and the ethical considerations of their adoption. By identifying trends in the literature and establishing key relationships between AI and educational administration, this work seeks to generate useful knowledge for future research and for the design of strategies that maximize the potential of AI in education, while ensuring equity and quality of learning.

2 Theoretical Framework

2.1 Artificial Intelligence and Its Role in Educational Administration

Artificial intelligence (AI) has acquired a central role in educational administration, providing tools that optimize decision-making, automate administrative processes, and improve the operational efficiency of HEIs. In particular, LLMs have been incorporated into various functions, including academic management, resource allocation, and big data analytics for institutional planning [6]. As these technologies advance, the academic literature has explored their impact on university governance, identifying opportunities and challenges associated with their implementation (Zhang et al., 2024). However, the use of LLMs in educational administration raises questions about equity, institutional autonomy and transparency in decision-making, generating debates about their regulation and supervision [7].

Within the field of study, an emerging trend is the use of AI for the optimization of resource allocation in HEIs. Through predictive models, LLMs can anticipate changes in enrollment, infrastructure needs, and faculty requirements, allowing for more efficient, data-driven planning [8]. However, the literature also highlights barriers in the implementation of these systems, such as organizational resistance and the perception of AI as a tool that is difficult to integrate into administrative processes [9]. In addition, the automation of strategic decisions has raised concerns among academic staff, who fear a reduction in institutional autonomy and an over-reliance on algorithmic models [10]. To address these concerns, several studies have emphasized the need to develop training strategies aimed at administrators and supervisory policies that guarantee an ethical and efficient use of AI in educational management [11].

Another axis of analysis within the literature is the use of LLMs in financial management and the distribution of resources within HEIs. It has been documented that these models can improve budget planning through the analysis of costs and academic performance, allowing optimizing investment in infrastructure, hiring of personnel and the allocation of student scholarships [12]. However, the adoption of AI in financial management has generated controversy due to the lack of transparency in predictive models and resistance to change by managers who are wary of automated decision-making [13]. Recent literature highlights the importance of establishing specific regulatory frameworks that oversee the implementation of AI in educational administration, ensuring that accounting and administrative processes maintain standards of human oversight and algorithmic auditing [14].

2.2 Personalization of Learning and Teacher Transformation

One of the most significant advances in higher education has been the integration of LLMs into the personalization of learning. The literature highlights that these models can analyze large volumes of data to adapt the content and difficulty of the materials according to the pace and individual needs of the students [15]. This ability to adapt has driven the development of more flexible and student-centered pedagogical strategies, promoting more dynamic learning experiences adjusted to different cognitive styles. However, implementing AI in teaching is not without its challenges. Recent research has pointed to resistance to change on the part of teachers and students, who perceive AI as a possible threat to traditional methodologies and the role of the teacher in academic training [16].

An emerging trend in the literature is the impact of AI on students' autonomy and critical thinking. While LLMs can provide automated tutoring and real-time assistance, some studies warn that their overuse could lead to technological dependency and limit students' ability to solve problems independently [17]. In response to this challenge, various approaches have proposed the design of AI systems that integrate teacher supervision mechanisms, ensuring that the personalization of learning maintains a balance between automation and human interaction [18]. This approach seeks to prevent AI from replacing traditional teaching and instead act as a support tool that fosters high-level cognitive skills.

Another relevant topic in literature mapping is the role of AI in optimizing adaptive teaching strategies. Recent studies have shown that LLMs can improve knowledge retention by dynamically adjusting the level of difficulty of content according to each student's progress [19]. However, some researchers have expressed concern about the potential reduction of creativity in teaching due to reliance on predefined algorithms for personalization of learning [20]. In addition, the use of AI in virtual tutorials has generated debates about the loss of social interaction and collaboration between students, which could affect the development of interpersonal skills in digitized educational environments [21].

The collection and analysis of student data to personalize teaching has also posed ethical challenges, particularly in relation to privacy and the protection of personal information [22]. The literature notes that many AI platforms store large volumes of data to improve the accuracy of models, which has raised concerns about the misuse of this information and the need for clear regulations on the transparency of educational algorithms [23]. In this sense, the research suggests that the personalization of learning through AI should be designed with mechanisms that balance technological innovation with principles of equity and privacy, ensuring that the adaptation of content does not compromise the rights of students.

2.3 Ethical Challenges and the Future of AI in Higher Education

As LLMs become more entrenched in higher education, debates in literature have revolved around their impact on equity, transparency, and educational autonomy. One of the main challenges identified is algorithmic bias, as AI can replicate and amplify pre-existing inequalities in the data it is trained on [24]. This raises a central concern in terms of educational justice, as biased models could reinforce differences in access to academic opportunities and negatively affect historically marginalized groups [25]. Likewise, the lack of explainability in AI models generates uncertainty about the way automated decisions are made, especially in key processes such as the evaluation of student performance or the allocation of educational resources [26].

Another emerging challenge in literature is the digital divide between institutions with different levels of access to technology. Recent research has indicated that universities with greater investment capacity have begun to benefit from AI in the automation and personalization of learning, while institutions with less technological infrastructure face difficulties in implementing these systems [2]. This disparity could deepen educational inequality, increasing the difference in academic outcomes between students with access to advanced technology and those in contexts with limited resources [27]. To mitigate this problem, the literature emphasizes the importance of developing digital inclusion strategies that facilitate equitable access to AI-based tools, preventing their implementation from reinforcing structural inequalities in higher education [28].

Data privacy and security have been another key focus in studies on AI in education. The bulk collection of student data has raised concerns about the protection of personal information and the ethical use of data in AI systems [23]. In response to these concerns, various studies have highlighted the need to establish clear regulatory frameworks that define limits in the collection and processing of data, ensuring transparency and informed consent in the use of educational technology [29].

Despite these challenges, literature also highlights the potential of AI to promote educational accessibility and inclusion. It has been documented that LLMs can offer tools adapted for students with special needs, facilitating their integration into learning environments through accessible interfaces and personalized virtual assistants [30]. This aspect has led to a growing discussion about how AI can contribute to reducing barriers in education and ensuring more equitable access to academic opportunities [31].

3 Method

This study adopts a literature mapping approach to examine the integration of LLMs in higher education. Literature mapping is appropriate for identifying trends, classifying existing knowledge, and detecting research gaps in emerging fields [32, 33]. The mapping focused on three dimensions: educational administration, personalized learning, and ethical governance.

The literature search was conducted in Scopus and Web of Science (WoS), covering indexed publications from January 2020 to April 2026. The search strategy combined terms related to LLMs, generative AI, higher education, educational administration, personalized learning, governance, regulation, and responsible AI. The selected studies were classified according to the three research questions presented in Table 1.

Phase 1: Definition of Research Questions
To guide the analysis, three research questions were established to examine different aspects of the impact of LLMs on higher education:

Table 1. Research questions

Research Questions (RQ)	Exploration Areas
RQ1: How do Large-Scale Language Models (LLMs) influence decision-making in educational administration?	Applications in academic management, automation of administrative processes, challenges in institutional autonomy and regulation of AI in universities.
RQ2: What is the impact of LLMs on the personalization of learning within higher education?	Adaptation of educational content, automated tutorials, resistance to change in teachers and students, and effects on equity in learning.
RQ3: What are the ethical challenges in the use of Generative AI in higher education?	Algorithmic bias, student data privacy and protection, explainability of automated decisions, and the need for regulations on AI in education.

Phase 2: Literature Search and Classification
The literature collection was carried out in Scopus and Web of Science (WoS), selecting indexed publications between January 2020 and April 2026. This period was defined to capture both the initial development of LLM-related research in higher education and the

most recent studies on generative AI, personalized learning, educational administration, and ethical governance.

The optimized search strings included terms such as:

- "Open Large Language Model" OR "Open LLM" OR "Large Language Model"
- "AI governance" OR "policy" OR "regulation" OR "responsible AI"
- "education" OR "higher education" OR "academic management" OR "educational administration"

The selected literature was classified thematically according to the three research axes, organizing the studies into categories such as educational management, personalized learning and ethics of AI in education.

The final mapping corpus consisted of 551 articles, selected from indexed publications retrieved from Scopus and Web of Science between January 2020 and April 2026.

Phase 3: Analysis and Synthesis of Trends

Rather than applying strict exclusion criteria, the literature mapping focused on identifying emerging patterns and research gaps. The methodological and theoretical approaches of the studies were analyzed to understand how academia has approached the adoption of LLMs in higher education. Among the most relevant findings, the following stand out:

- Impact on educational administration: The literature shows a growth in the use of AI for decision-making in universities, although debates persist about institutional autonomy and technological dependence [34].
- Personalization of learning: An increase in the use of LLMs for automated tutoring and content adaptation has been identified, although with concerns about algorithmic bias and equity in access to personalized education [23, 25].
- Ethical challenges and regulation: There are concerns about data privacy, transparency of automated decisions, and the lack of clear regulatory frameworks for the implementation of AI in higher education [4, 28].

To address the rapid development of the field, the search period was extended to April 2026 and recent studies were incorporated into the mapping corpus. This update allowed the review to include emerging evidence on empirical applications of LLMs in education, personalized learning systems, institutional governance, academic integrity, privacy, bias, and responsible implementation. The inclusion of 2025 and 2026 literature strengthened the temporal relevance of the study and allowed the analysis to better reflect current debates on LLM adoption in higher education.

After the thematic classification, an integrative synthesis was conducted to identify cross-dimensional relationships among the three research axes. This synthesis examined how administrative decision-making, personalized learning, and ethical governance interact in the adoption of LLMs in higher education. The purpose was not only to classify the literature, but also to explain the mechanisms through which LLMs simultaneously affect institutional management, pedagogical processes, and governance responsibilities.

4 Results

In this section, we present the findings obtained from the mapping of literature on the influence of LLMs on educational administration, the personalization of learning and the ethical challenges associated with their implementation in higher education.

RQ1: How do Large-Scale Language Models (LLMs) influence decision-making in educational administration?

The literature analysis indicates that LLMs have generated structural changes in educational administration, promoting greater automation and optimization of strategic decision-making [3]. Various studies highlight that these tools can facilitate data analysis and improve administrative efficiency, reducing uncertainty in educational management [8]. However, concerns remain about ethical challenges and algorithmic biases, which has fueled debate about the need for stronger regulations for automated decision-making [16].

An emerging aspect is the redefinition of administrative roles in universities, where decision-making based on human experience is being complemented or even replaced by predictive AI models [30]. This raises the need to train education managers to correctly interpret the results generated by AI and avoid excessive automation that minimizes human judgment in critical decisions [5].

Another key finding is that although LLMs have been adopted to support planning and data analysis, their use in automating administrative processes remains an underexplored area [9]. Future research could focus on evaluating how these tools can improve resource allocation, academic planning, and financial management without compromising institutional autonomy.

In addition, the literature suggests that LLMs could play a role in predicting educational trends and in formulating institutional policies based on historical learning patterns [21, 26]. Nonetheless, the quality of the data used in these models remains a key concern, as biases in training datasets could affect the validity of automated decisions and perpetuate pre-existing inequalities in education [25].

RQ2: What is the impact of LLMs on the personalization of learning within higher education?

The literature suggests that the adoption of LLMs in the personalization of learning has been met with resistance from teachers and students, indicating that there are still barriers in the formation and acceptance of these tools in educational settings [27]. Despite its advantages in content adaptation and automated tutorials, its use in assessments remains limited, due to concerns about confidence and accuracy in measuring learning [22].

An emerging aspect is the ability of LLMs to detect patterns in student performance and adjust teaching strategies in real time. However, the integration of these tools with traditional educational platforms is not yet fully developed, making it difficult to adopt them on a large scale [7]. In addition, concerns persist about equity in the personalization of learning, as some models may favor certain learning styles over others, reducing methodological diversity in teaching [4].

A novel finding is the potential of LLMs to foster self-directed learning and metacognitive development. Recent studies suggest that, when implemented correctly, these models can help students identify their strengths and weaknesses more accurately, promoting more autonomous learning [12]. However, concerns arise that over-personalization may limit students' exposure to diverse perspectives, reinforcing pre-existing learning patterns rather than encouraging critical thinking [10].

RQ3: What are the ethical challenges in the use of Generative AI in higher education?

Debates in the literature have emphasized the urgency of developing clear regulations for the use of LLMs in higher education [2]. Among the main ethical challenges, algorithmic bias and inequality in access to education have been identified as key concerns, as AI could reinforce pre-existing inequities in administrative and pedagogical processes [14].

An aspect less explored in the literature is the privacy and protection of student data, as well as transparency in automated decision-making. The lack of robust strategies to audit and explain how LLMs generate their recommendations poses risks to AI governance in educational institutions [13].

Another emerging challenge is the homogenization of learning, where AI models could prioritize certain pedagogical approaches over others, reducing educational diversity [15]. To mitigate these risks, the development of LLMs with higher levels of interpretability and human control is recommended.

Finally, future research should address the lack of effective oversight mechanisms in the use of LLMs in higher education. Although ethical frameworks have been widely discussed, many institutions still lack clear strategies for monitoring and evaluating their impact on education [19]. Figure 1 presents the integrated framework derived from the literature mapping. The framework shows how LLM adoption in higher education operates through three interdependent dimensions: educational administration, personalized learning, and ethical governance. Administrative decision-making defines institutional conditions for adoption; personalized learning operationalizes LLM-supported educational transformation; and ethical governance establishes the safeguards required for responsible implementation.

Figure 1 presents the integrated interpretive framework derived from the literature mapping. The framework shows that LLM adoption in higher education operates through three interdependent dimensions. First, educational administration defines the institutional conditions for adoption through decision-making, planning, resource allocation, and governance structures. Second, personalized learning represents the pedagogical operationalization of LLMs through adaptive feedback, automated tutoring, and learner support. Third, ethical governance establishes the safeguards required to regulate both administrative and pedagogical uses, including transparency, privacy, bias mitigation, accountability, and human oversight.

This framework clarifies that LLM adoption is not limited to isolated administrative or pedagogical applications. Instead, it represents a systemic process in which institutional decisions, learning personalization, and ethical safeguards interact continuously. Therefore, effective implementation requires coordinated action in governance, infrastructure, teacher training, data protection, transparency, and human oversight.

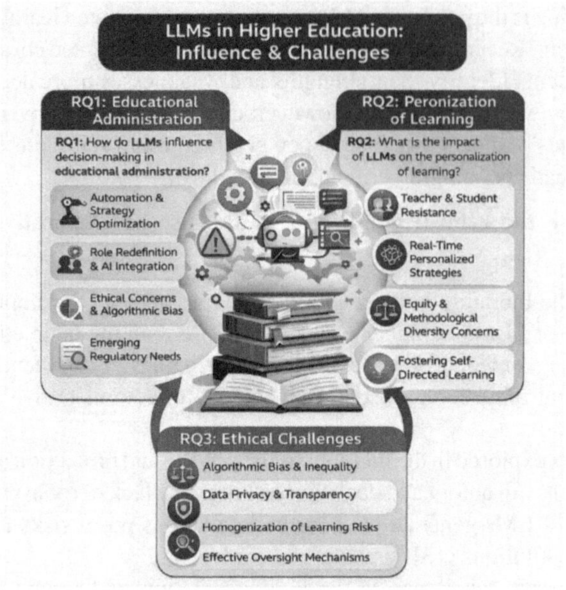

Fig. 1 LLMs in Higher Education: Influence and Challenges

5 Discussion

The findings indicate that LLM adoption in higher education should be interpreted as a systemic process in which administrative, pedagogical, and ethical dimensions interact. From the administrative perspective, LLMs support strategic decision-making through process automation, large-scale data analysis, resource planning, budget allocation, and institutional management [14, 20]. However, their use also raises concerns about cultural resistance, technological dependence, lack of regulatory frameworks, and the need for human oversight in automated decision-making [8, 11, 34]. These findings suggest that administrative efficiency alone is insufficient to justify LLM adoption unless institutions establish mechanisms for transparency, accountability, and critical managerial evaluation [24, 25].

From the pedagogical perspective, the literature shows that LLMs enable adaptive learning, automated tutoring, and personalized feedback, allowing educational content to be adjusted to students' needs, progress, and learning pace [15, 23]. Nevertheless, their implementation faces resistance from teachers and students, as well as concerns about methodological diversity, student autonomy, and excessive dependence on AI-supported personalization [6, 28]. Although LLMs may improve accessibility and support students with diverse needs, their benefits depend on institutional infrastructure, teacher mediation, and inclusion policies that prevent the widening of digital and educational inequalities [3, 18].

Ethical governance emerges as the condition that connects and regulates both administrative and pedagogical uses of LLMs. The reviewed literature emphasizes algorithmic bias, data privacy, lack of explainability, and insufficient regulatory frameworks as central risks in higher education [2, 24, 25]. These risks are especially relevant because LLMs may influence resource allocation, learning personalization, student support, and institutional policy decisions. Therefore, responsible implementation requires audit mechanisms, data protection protocols, transparency standards, and human supervision to ensure that AI-supported decisions remain fair, explainable, and aligned with educational values [1, 2].

The integrated framework proposed in this study contributes by showing that these three dimensions do not operate independently. Administrative decisions define the institutional conditions for LLM adoption, including infrastructure, policies, resources, and governance priorities [20, 34]. Pedagogical applications operationalize LLM use through adaptive learning and automated support, but also generate new data flows and risks related to equity, privacy, and autonomy [15, 18]. Ethical governance functions as a cross-cutting mechanism that establishes the limits, safeguards, and accountability structures required for responsible adoption [2, 13, 25]. Thus, LLM integration in higher education should be understood not as a set of isolated technological applications, but as a systemic transformation that requires coordination among educational administration, pedagogical design, and institutional governance.

6 Conclusion

Literature mapping reveals that Large Scale Language Models (LLMs) are transforming higher education, especially in educational management, personalization of learning, and process automation. Its implementation has demonstrated benefits in administrative efficiency and pedagogical adaptation, although important challenges persist in ethical, regulatory, and equity terms.

While LLMs optimize decision-making and reduce teaching burden through automation, they also raise concerns about the dehumanization of the educational process and the loss of critical thinking. Issues such as data privacy, algorithmic bias, and lack of transparency require robust ethical frameworks and responsible implementation strategies.

The main limitations identified include the lack of teacher training in AI, the uneven technological infrastructure between institutions and the absence of longitudinal studies to evaluate its sustained impact on learning. Added to this is the need to ensure that personalization does not deepen existing inequalities, but rather fosters more inclusive environments.

Therefore, future research should focus on the design of regulatory frameworks, algorithmic audits, teacher training strategies and comparative studies that allow AI to be adapted to various institutional contexts. Exploring applications such as virtual assistants and predictive models will also contribute to the development of inclusive and sustainable solutions. The transformation of higher education through AI must move forward with equity, ethics, and a long-term vision.

Acknowledgments. The authors appreciate the financial support from Tecnologico de Monterrey through the "Challenge-Based Research Funding Program 2024". Project ID # I005 - CI__IFE_EDU_D_225.

The authors acknowledge the technical and financial support of Writing Lab, Institute for the Future of Education, Tecnologico de Monterrey, Mexico, in the production of this work.

Disclosure of Interests The authors have no competing interests to declare that are relevant to the content of this article.

References

1. Linkon, A.A.: Advancements and applications of generative artificial intelligence and large language models on business management: a comprehensive review. J. Comput. Sci. Technol. Stud. **6**, 225–232 (2024). https://doi.org/10.32996/jcsts.2024.6.1.26
2. Agbese, M., Mohanani, R., Khan, A., Abrahamsson, P.: Implementing AI ethics: making sense of the ethical requirements. In: Proceedings of the 27th International Conference on Evaluation and Assessment in Software Engineering, pp. 62–71. ACM, Oulu, Finland (2023). https://doi.org/10.1145/3593434.3593453
3. Aldredge, M., Rogers, C., Smith, J.: The strategic transformation of accounting into a learned profession. Ind. High. Educ. **35**, 83–88 (2021). https://doi.org/10.1177/0950422220954319
4. Allen, B., McGough, A.S., Devlin, M.: Toward a framework for teaching artificial intelligence to a higher education audience. ACM Trans. Comput. Educ. **22** (2022). https://doi.org/10.1145/3485062
5. Avsec, S., Jagiełło-Kowalczyk, M., Żabicka, A.: Enhancing transformative learning and innovation skills using remote learning for sustainable architecture design. Sustain. Switz. **14** (2022). https://doi.org/10.3390/su14073928
6. Aggarwal, D.: Adoption of artificial intelligence (AI) for development of smart education as the future of a sustainable education system. J. Artif. Intell. Mach. Learn. Neural Netw. (2023). https://doi.org/10.55529/jaimlnn.36.23.28
7. Cowling, M., Crawford, J., Allen, K.-A., Wehmeyer, M.: Using leadership to leverage ChatGPT and artificial intelligence for undergraduate and postgraduate research supervision. Australas. J. Educ. Technol. **39**, 89–103 (2023). https://doi.org/10.14742/ajet.8598
8. Abramski, K., Citraro, S., Lombardi, L., Rossetti, G., Stella, M.: Cognitive network science reveals bias in GPT-3, GPT-3.5 turbo, and GPT-4 mirroring math anxiety in high-school students. Big Data Cogn. Comput. **7**, 124 (2023). https://doi.org/10.3390/bdcc7030124
9. Alshami, A., Elsayed, M., Ali, E., Eltoukhy, A.E.E., Zayed, T.: Harnessing the power of ChatGPT for automating systematic review process: methodology, case study, limitations, and future directions. Systems. **11**, 351 (2023). https://doi.org/10.3390/systems11070351
10. Al Ka'bi, A.: Proposed artificial intelligence algorithm and deep learning techniques for development of higher education. Int. J. Intell. Netw. **4**, 68–73 (2023). https://doi.org/10.1016/j.ijin.2023.03.002
11. Aliane, N., Salim, A.S.: Revolutionising higher education: case studies on education4.0 integration and blockchain-enhanced education management. Eurasian J. Educ. Res. **2023**, 217–235 (2023). https://doi.org/10.14689/ejer.2023.105.013
12. Abdelwahab, H.R., Rauf, A., Chen, D.: Business students' perceptions of Dutch higher educational institutions in preparing them for artificial intelligence work environments. Ind. High. Educ. **37**, 22–34 (2023). https://doi.org/10.1177/09504222221087614
13. Singha, S., Singha, R.: The application of artificial intelligence in education: opportunities and challenges. In: Prakasha, G.S., Lapina, M., Balakrishnan, D., Sajid, M. (eds.) Advances in Business Information Systems and Analytics, pp. 282–292. IGI Global (2024). https://doi.org/10.4018/979-8-3693-2314-4.ch014

14. Baothman, F.A.: An intelligent big data management system using Haar algorithm-based Nao agent multisensory communication. Wirel. Commun. Mob. Comput. **2021** (2021). https://doi.org/10.1155/2021/9977751
15. Binns, R.: On the apparent conflict between individual and group fairness. In: Proceedings of the 2020 Conference on Fairness, Accountability, and Transparency, pp. 514–524. ACM, Barcelona Spain (2020). https://doi.org/10.1145/3351095.3372864
16. García, I.: Sistema informático como apoyo al desarrollo de estrategias docentes para la enseñanza en el grupo 1101 de la Licenciatura en Informática de la Facultad de Estudios Superiores Cuautitlán (2020)
17. Yang, C., Stivers, A.: Investigating AI languages' ability to solve undergraduate finance problems. J. Educ. Bus. (2023). https://doi.org/10.1080/08832323.2023.2253963
18. Wiant Cummins, M., Mason, M.: Equity and innovation: adding human voice to OERs. J. Open Educ. Resour. High. Educ. **2** (2023). https://doi.org/10.13001/joerhe.v2i1.7725
19. Al-Shanfari, L., Abdullah, S., Fstnassi, T., Al-Kharusi, S.: Instructors' perceptions of intelligent tutoring systems and their implications for studying computer programming in Omani higher education institutions. Int. J. Membr. Sci. Technol. **10**, 947–967 (2023). https://doi.org/10.15379/ijmst.v10i2.1395
20. Bekmanova, G., Ongarbayev, Y., Somzhurek, B., Mukatayev, N.: Personalized training model for organizing blended and lifelong distance learning courses and its effectiveness in higher education. J. Comput. High. Educ. **33**, 668–683 (2021). https://doi.org/10.1007/s12528-021-09282-2
21. Bucea-Manea-ţoniş, R., et al.: Artificial intelligence potential in higher education institutions enhanced learning environment in Romania and Serbia. Sustain. Switz. **14** (2022). https://doi.org/10.3390/su14105842
22. Fjellman, A.-M., Haley, A.: The plague of privatization: a futures analysis of the zombification of education. Policy Futur. Educ. **21**, 526–545 (2023). https://doi.org/10.1177/14782103211029491
23. Chang, D.H., Lin, M.P.-C., Hajian, S., Wang, Q.Q.: Educational design principles of using AI chatbot that supports self-regulated learning in education: goal setting, feedback, and personalization. Sustain. Switz. **15** (2023). https://doi.org/10.3390/su151712921
24. Ferrara, E.: Should ChatGPT be biased? Challenges and risks of bias in large language models. First Monday. (2023). https://doi.org/10.5210/fm.v28i11.13346
25. Ferrer, X., Nuenen, T.V., Such, J.M., Cote, M., Criado, N.: Bias and discrimination in AI: across-disciplinary perspective. IEEE Technol. Soc. Mag. **40**, 72–80 (2021). https://doi.org/10.1109/MTS.2021.3056293
26. Gulati, A., Saini, H., Singh, S., Kumar, V.: ENHANCING LEARNING POTENTIAL: INVESTIGATING MARKETING STUDENTS' BEHAVIORAL INTENTIONS TO ADOPT CHATGPT. Mark. Educ. Rev. (2024). https://doi.org/10.1080/10528008.2023.2300139
27. Baidya, M., Kumar, A.C.: Artificial intelligence in education: understanding benefits, limitations, and prospects for the future. Int. J. Multidiscip. Res. (2024). https://doi.org/10.36948/ijfmr.2024.v06i02.16699
28. Leung, T.I., Sagar, A., Shroff, S., Henry, T.L.: Can AI mitigate bias in writing letters of recommendation? JMIR Med. Educ. **9** (2023). https://doi.org/10.2196/51494
29. Breen, S., Ouazzane, K., Patel, P.: GDPR: Is your consent valid? Bus. Inf. Rev. **37**, 19–24 (2020). https://doi.org/10.1177/0266382120903254
30. Bai, J., Kamatchinathan, S., Kundu, D.J., Bandla, C., Vizcaíno, J.A., Perez-Riverol, Y.: Open-source large language models in action: a bioinformatics chatbot for PRIDE database. Proteomics, 2400005 (2024). https://doi.org/10.1002/pmic.202400005
31. Antoninis, M.: Promoting access to education for students with disabilities through inclusive technologies (2023)

32. Kitchenham, B., et al.: Systematic literature reviews in software engineering – a tertiary study. Inf. Softw. Technol. **52**, 792–805 (2010). https://doi.org/10.1016/j.infsof.2010.03.006
33. García-Peñalvo, F.J.: Mapeos Sistemáticos De Literatura. Revisiones Sistemáticas De Literatura Y Benchmarking De Programas Formativos. (2017). https://doi.org/10.5281/ZENODO.1067680
34. Shin, J.C., Li, X., Nam, I., Byun, B.-K.: Institutional autonomy and capacity of higher education governance in South Asia: a comparative perspective. High. Educ. Policy. **35**, 414–438 (2022). https://doi.org/10.1057/s41307-020-00220-y

Token-Aware Agentic AI Systems: A Framework for Evaluating Token Usage Across Orchestration Approaches

Jorge Esteban Ramirez Sashida[1,2], Mauricio Diaz[1,2], and J. Octavio Gutierrez-Garcia[1(✉)]

[1] Department of Computer Science, ITAM, 01080 Mexico City, Mexico
octavio.gutierrez@itam.mx

[2] dive.ai, Data Science and AI Consulting Firm, 06470 Mexico City, Mexico
{esteban,mauricio}@dive.ai

Abstract. Agentic AI systems increasingly automate enterprise workflows by leveraging large language models (LLMs) to orchestrate external tools through standardized interfaces such as the Model Context Protocol (MCP). In this domain, there are two orchestration modalities: direct invocation, where an agent calls tools sequentially and routes each intermediate result through the LLM context, and code-execution invocation, where the agent generates a program that batches multiple tool calls within an isolated sandbox. Although direct invocation is widely adopted and code-execution invocation is an emerging approach, their efficiency trade-offs remain insufficiently studied. This work introduces an evaluation framework for agentic AI systems that compares these orchestration modalities in both single-agent and multi-agent settings, using a heterogeneous task suite (including single-tool, pipeline, and discovery tasks) across multiple MCP servers providing more than 70 tools. The framework includes an LLM-as-a-judge grader and measures token consumption, task quality, and normalized efficiency. With a 0.05 significance level, empirical results indicate that code-execution invocation reduces token consumption while preserving output quality and achieving higher efficiency on complex tasks, particularly pipeline tasks, compared with direct invocation.

Keywords: Agentic AI systems · Evaluation framework · MCP via code execution · Multi-agent systems

1 Introduction

Nowadays, artificial intelligence (AI) supports and automates enterprise workflows through agentic AI systems [10], which consist of multiple agents that collaborate to carry out interconnected tasks and achieve predefined goals [21]. Agents (endowed with decision-making abilities) establish a context and act toward these goals [7]. In recent years, to enhance their decision-making, agents

H. Ponce et al. (Eds.): COMIA 2026, CCIS 3095, pp. 159–171, 2027.
https://doi.org/10.1007/978-3-032-37438-7_13

leverage the reasoning capabilities of large language models (LLMs) [18]. An LLM is a deep learning model trained on a great amount of text to perform natural language tasks [13]. Unlike reactive agents, an LLM-supported agent perceives and establishes context through textual inputs and executes actions accordingly. Frequently, LLM-supported agents generate and monitor plans by interleaving reasoning traces and actions, following the ReAct framework (see [23] for details). Currently, agentic frameworks such as Amazon Bedrock, AutoGen, LangChain, and Strands Agents allow agents to invoke and orchestrate a myriad of tools and services (ranging from storage systems to AI models and web-scraping utilities) using an open model context protocol (MCP), which has been adopted as an industry standard [12]. MCP enables agents to discover and use tools in a structured manner, decoupling implementation technologies from their invocation.

With MCP providing this layer of interoperability, there are several ways to orchestrate and implement agentic systems (see [2]). In terms of tool invocation, this article focuses on two orchestration modalities: (i) *direct invocation*, in which an agent calls MCP tools one at a time, with each tool's output added back into its context before the next call; and (ii) *code-execution* invocation, in which the agent instead generates a program that invokes multiple tools inside a sandbox, processes intermediate results, and returns only the final output to its context.

LLM-supported agents have become increasingly effective at executing enterprise workflows under either orchestration modality. However, their performance depends in part on the amount of information (i.e., the size of the context window, measured in tokens) that an LLM can process in a single request. This constraint has motivated substantial research efforts within the scientific community (see [4,20,22]). In this regard, in agentic AI systems executing enterprise workflows, this limitation is further amplified because, in addition to the user's prompt, the context required for direct MCP calls must include detailed tool specifications (e.g., names, descriptions, parameters, and schemas). In practice, these tool descriptions may consume thousands of tokens (occupying a significant portion of the available context window) before an agent can even fulfill a single user request [5]. Moreover, in the context of direct MCP calls, when an LLM-supported agent executes enterprise workflows involving multiple tools, all intermediate results must be routed through the context, i.e., the output of each tool becomes the input of the next, which can lead to context growth that may even exceed the available window size.

To mitigate the above problem, the main objective of this work is twofold: first, to design an evaluation framework for agentic AI systems that compares code-execution invocation and direct invocation as orchestration modalities; and second, to examine whether code-execution invocation offers measurable advantages (both in token consumption and output quality) over direct invocation in agentic systems. To evaluate this, each orchestration modality was tested under both a single-agent and a multi-agent system architecture. Furthermore, to assess the implemented agentic AI systems, the agents were presented with three task types: (i) *single-tool tasks*, involving a single MCP server; (ii) *pipeline tasks*, involving multiple predefined MCP servers; and (iii) *discovery tasks*, also involving multiple MCP servers but requiring the agent to determine which servers (and tools) to invoke. To conduct the experiments, multiple enterprise workflows (involving eight MCP servers and more than 70 tools) were created. The application domains of these servers and workflow tasks were highly heterogeneous, ranging from Morse code translation to developer utilities.

The contributions of this work are as follows.

- An evaluation framework for agentic AI systems across different orchestration modalities in both single-agent and multi-agent settings.
- Empirical evidence of the trade-offs between two relevant orchestration modalities in agentic systems: direct invocation and code-execution invocation.
- Identification of preferable orchestration modalities for different task categories, considering both token consumption and output quality.
- A reproducible evaluation methodology for comparing orchestration modalities in agentic AI systems.

This article is structured as follows. Section 2 introduces background concepts and related work; Sect. 3 describes the evaluation framework; Sect. 4 presents experimental settings and results; and Sect. 5 provides final remarks.

2 Background and Related Work

The rise of LLMs has accelerated research on agentic AI systems (see [9,19], and [1]), which automatically orchestrate web services (commonly referred to as *tools* in this domain) to execute complex workflows. A key paradigm in this context is ReAct [23], where agents alternate between reasoning and tool invocation. This integration of chain-of-thought with tool use has been shown to reduce hallucinations and improve performance on benchmarks such as HotpotQA and ALFWorld. In practice, ReAct-style tool use is commonly implemented through MCP [16], an open protocol providing LLM-supported agents with a uniform interface for discovering and invoking tools via JSON-RPC 2.0. MCP is now supported by major frameworks from OpenAI, Google DeepMind, and Microsoft [12]. However, before executing a workflow, an agent must load tool definitions into its context, which can consume tens of thousands of tokens [6]. As the number of tools grows, so does the token cost of loading specifications and routing intermediate results. Since LLMs are effective at generating code [14], code-execution invocation provides an alternative that reduces context growth by handling intermediate computations programmatically.

In this context, several research efforts have been conducted to evaluate and benchmark agentic AI systems enabled by MCP. Dobriy et al. [8] evaluated LLM-supported agents in the context of federated SPARQL querying. Their system architecture consisted of a single client communicating with a single MCP SPARQL server, primarily evaluating the validity of the generated SPARQL queries. In the domain of mobile computing, Kong et al. [15] proposed MobileWorld, a benchmark for mobile applications using agent-user interaction and MCP servers. With respect to the agentic AI system, their benchmark includes

40 tasks involving 64 tools provided by five MCP servers, evaluating task success rate and the average number of MCP calls. Mo et al. [17] designed their benchmark to include 95 everyday tasks (e.g., lifestyle, shopping, and leisure activities), in which LLM-supported agents must interact with 70 MCP servers providing 527 tools. As before, their primary performance metric was task success rate, however, they also measured token consumption across several state-of-the-art LLMs, with a particular focus on the costs associated with tool descriptions. Following a similar approach, Guo et al. [11] propose evaluating agentic AI systems using 600 real-world everyday tasks (generated by LLMs but with human supervision). As in [17], the number of MCP servers and tools involved in the benchmark is also considerably large. It is worth mentioning that Guo et al. also propose evaluating token consumption (in addition to task success rate), however, token consumption is assessed from the perspective of the LLMs used without taking into account orchestration modalities.

Whereas the research efforts presented in [8,15,17], and [11] are valuable and contribute significantly to the evaluation of agentic AI systems, to the best of the authors' knowledge, no prior work has systematically benchmarked orchestration modalities (namely, direct invocation and code-execution invocation) independently of the underlying AI model, with respect to both token consumption and output quality, which is the main focus of the present work.

3 Evaluation Framework

The evaluation framework (Fig. 1) consists of (i) an LLM-supported agent; (ii) the orchestration modalities: direct invocation and code-execution invocation; (iii) a task suite; (iv) an LLM-as-a-judge grader; and (v) two agent-based system architectures, namely single-agent and multi-agent setups.

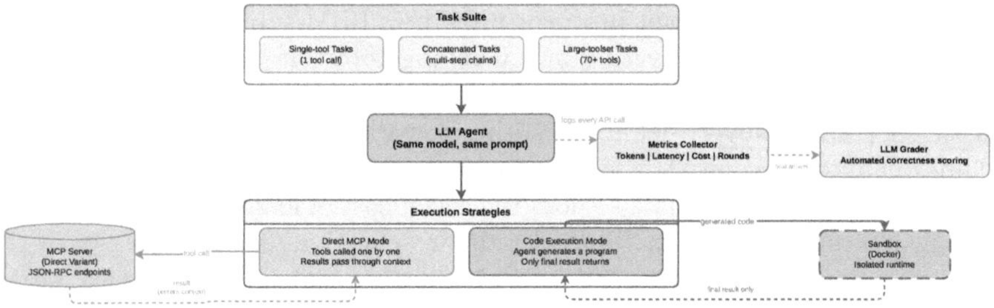

Fig. 1. Architectural components of the proposed evaluation framework.

3.1 LLM-Supported Agent

To execute automated workflows, the LLM-supported agent uses an AI model that receives a prompt describing a task, whose requirements must be aligned with the capabilities of the available MCP servers. The LLM-supported agent also includes a metrics collector to track token usage, success rate, and other relevant performance metrics.

3.2 Orchestration Modalities

The LLM-supported agent can make use of two orchestration modalities:

- *Direct invocation* (DI), in which the agent invokes tools individually and sequentially using direct MCP calls, and each call's output (i.e., the emitted tokens) is appended to the agent's context before executing the next call. Using FastMCP, the agent calls the tools (exposed as JSON endpoints) via streamable HTTP or standard I/O calls.
- *Code-execution invocation* (CI), in which the agent generates a program (treating MCP tools as code APIs that can be invoked programmatically) to orchestrate multiple tool calls within a (Docker) sandbox where the agent executes the code. This allows intermediate results to be processed programmatically within the generated code, and only the final program output, i.e., the resulting tokens, is returned to the agent's context. In addition, inside the Docker container, Python files describe the corresponding MCP tools and a dependency list ensures the agent can import and execute these tools within the sandbox.

3.3 Task Suite

The task suite executed by the LLM-supported agent comprises 54 tasks across three types, each designed to reflect increasing levels of operational complexity:

- *Single-tool.* Tasks where an MCP server exposes a single tool for a straightforward input-output operation. These tasks provide a baseline for comparing the overhead of code-execution invocation against the minimal overhead of a direct MCP call. Examples of single-tool tasks include translating a string to Morse code and searching for an artist on LastFM.
- *Pipeline.* Tasks where an MCP server exposes multiple tools and the agent must chain their outputs sequentially to complete the task. In these tasks, the agent must determine the correct sequence of tool calls and combine them appropriately. An example of a pipeline task is listing an artist's albums on LastFM and then getting the most popular track for each album.
- *Discovery.* Tasks involving 50 or more tools within a single MCP server, requiring the agent to explore the available tools and select the appropriate ones. These tasks may involve either single-tool or pipeline tasks. An example of a discovery task is converting a string using the correct utility from an MCP server containing 60 or more tools.

Table 1. MCP server domains and tool counts.

MCP Server	Number of tools	Description
Morse	1	Simple morse code translation
Cocktails	3	Cocktail recipe search
Postgres	3	Database operations
NASA	4	NASA image/video search and metadata
AlphaVantage	4	Stock market data
Nutrition	5	Nutritional information lookup
LastFM	26	Music discovery
DevToolkit	60+	Developer utilities
Cross-Server	Varies	Tasks requiring multiple servers

Nine MCP servers are available to execute the tasks. Each server corresponds to a specific application domain and exposes a different number of tools. Table 1 summarizes the servers used in the evaluation framework.

The servers were selected to represent different levels of workflow complexity. The Morse MCP server is the simplest case, exposing a single tool with a straightforward input-output mapping. The DevToolkit MCP server represents the large-toolset scenario, offering more than 60 tools. Postgres and Cross-Server, in contrast, support tasks requiring multi-step reasoning and coordination across multiple MCP tools or servers.

In addition, to evaluate whether the number of connected MCP servers affects performance, each task was executed under three server-load conditions: (i) *minimal*, where only the server(s) required for the task are connected; (ii) *moderate*, where the required server(s) plus a few distractor servers with unrelated tools are connected; and (iii) *maximal*, where the required server(s) plus all remaining servers act as distractors. This setup allows assessing whether loading additional tool definitions into the context degrades performance (one of the key motivations for *code-execution* invocation).

3.4 LLM-as-a-Judge Grader

The grader receives a rubric specifying how each task should be evaluated and assesses whether the LLM-supported agent produced the correct result with the appropriate level of detail. The rubric includes the task objective and the expected output. Once the LLM-supported agent completes a task, its output is submitted to the LLM-as-a-judge grader, which assigns a score from 0 (lowest) to 1 (highest). An LLM was employed as a grader because the task outputs are expressed in natural language rather than in predefined, structured formats.

3.5 Agent-Based System Architectures

The evaluation framework supports two agent-based system architectures: single-agent architecture and multi-agent architecture. Both system architectures can be implemented in Amazon Bedrock [3].

- The *single-agent* architecture is implemented using a *Strands* agent supported by an LLM. The agent receives a prompt that connects the AI model to the available MCP tools required for completing the task. In this setup, the agent has access to all necessary MCP servers. Under *direct invocation*, the Strands agent calls each tool individually via JSON-RPC, appending every result to its context. Under *code-execution invocation*, the agent generates a program that runs inside a Docker sandbox, where the MCP tools are available as importable Python modules. Only the final program output is returned to the agent's context.
- The *multi-agent* architecture is implemented using a *Strands* agent orchestrator. The orchestrator receives the task and delegates subtasks to specialized subordinate agents, each assigned to a specific MCP server. Under *direct invocation*, each subordinate agent calls its corresponding MCP server via JSON-RPC and returns its results to the orchestrator. Under *code-execution invocation*, each subordinate agent operates in its own isolated Docker sandbox containing the necessary tool files and dependencies, returning only the final output. No state is shared between sandboxes.

4 Evaluation and Empirical Results

4.1 Experimental Settings

Agent-Based System Configurations. To determine whether *code-execution invocation* offers measurable advantages over *direct invocation* in agentic AI systems, the task suite (Sect. 3.3) was executed using four configurations (Fig. 2):

- *Single-agent DI*: Single-agent architecture using direct invocation.
- *Single-agent CI*: Single-agent architecture using code-execution invocation.
- *Multi-agent DI*: Multi-agent architecture using direct invocation.
- *Multi-agent CI*: Multi-agent architecture using code-execution invocation.

Supporting LLM. For executing tasks, all four configurations used Claude 3.7 Sonnet through AWS Bedrock with a temperature of 0.2 and the same system prompt. For grading tasks, Claude 3.7 Sonnet was used with a temperature of 0.0 to ensure consistent evaluation. Each task type was associated with a predefined evaluation rubric. The judge prompt follows this template: {task prompt}, {expected tools}, {expected output}, {agent response}, and {trace summary}.

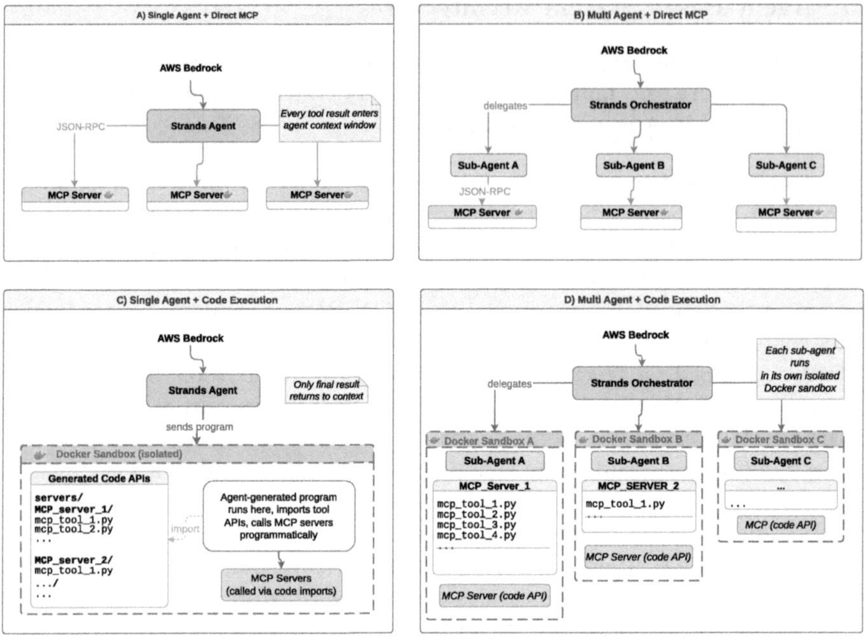

Fig. 2. Overview of the agentic AI system configurations.

Performance Measures. Task execution was evaluated using three metrics:

- *Task grade*: A score from 0 to 1 assigned by the LLM-as-a-judge grader.
- *Token usage*: Total input and output tokens consumed across all LLM calls for a given task. Input tokens include the system prompt, tool schemas loaded into context, the conversation history, and the content of any intermediate tool result, whereas output tokens include everything generated by the LLM, such as text, tool calls, and code.
- *Efficiency*: Grade per 1,000 tokens, capturing the trade-off between output quality and token cost.

Experiment Runs. The full experiment consisted of 582 task executions across the four system configurations (Fig. 2), eight MCP servers (Table 1), three task types (single-tool, pipeline, discovery), and three server-load conditions (minimal, moderate, maximal). It should be noted that results from the AlphaVantage MCP server were excluded due to persistent rate-limiting failures. The overall cost of the experiments was $58.80 USD, comprising token usage from the various tool orchestrations ($56.07) and the LLM-as-a-judge grading process ($2.73).

Empirical Results. From the series of experiments, three main observations were obtained. To facilitate interpretation, the results were grouped by task type and orchestration modality. A p-value less than 0.05 was considered significant.

Observation 1: Task Grade. Agents using direct invocation achieved a higher overall mean grade (0.944) compared with agents using code-execution invocation (0.922), with discovery tasks showing the largest difference (see Fig. 3).

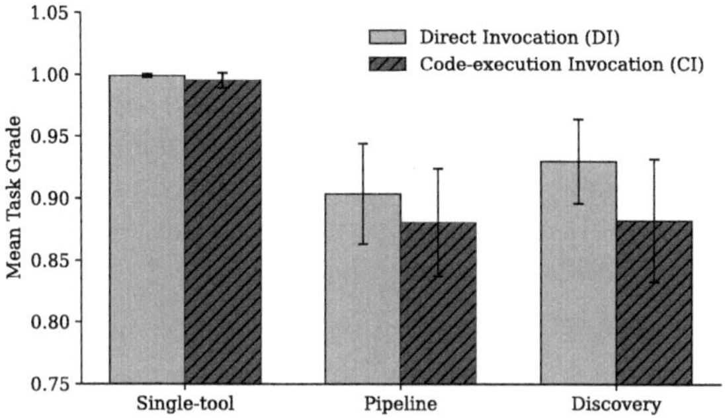

Fig. 3. Mean task grade by invocation modality and task type. Error bars indicate 95% confidence intervals.

Analysis. Although agents using direct invocation consistently outperformed those using code-execution invocation in raw grade, a one-way ANOVA indicates no significant main effect of orchestration modality on task grade ($F = 2.456$, $p = 0.118$). This suggests that code-execution invocation preserves output quality while potentially offering the cost advantages examined below.

Observation 2: Normalized Efficiency. Agents using code-execution invocation are significantly more efficient on pipeline and discovery tasks, but less efficient on single-tool tasks (see Fig. 4 and Table 2).

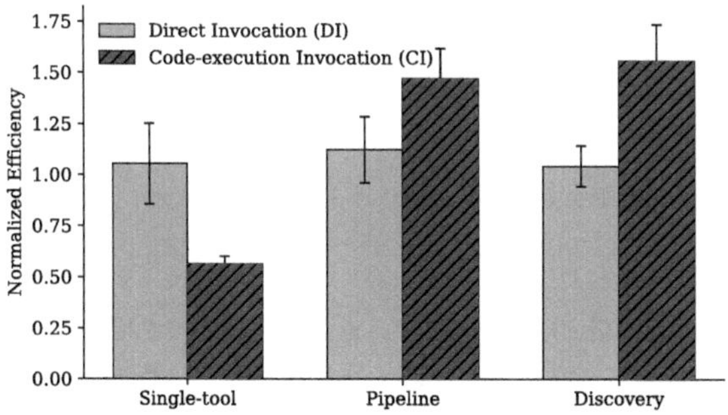

Fig. 4. Normalized efficiency by invocation modality and task type. Error bars indicate 95% confidence intervals. Efficiency was normalized such that single-agent DI equals 1.0 within each task type.

Table 2. Normalized efficiency by invocation modality and task type.

Task type	DI	CI	t	p	d
Single-tool	1.055 ± 0.198	0.564 ± 0.037	−4.823	<0.001	−0.675
Pipeline	1.123 ± 0.163	1.472 ± 0.146	3.162	0.002	0.430
Discovery	1.042 ± 0.100	1.560 ± 0.178	5.059	<0.001	0.803

Analysis. Agents using code-execution invocation were more efficient on both pipeline and discovery tasks due to their ability to batch multiple operations within a single program, thereby reducing the cumulative token-usage overhead associated with sequential direct MCP calls. This yields up to a 1.5× improvement in quality-per-token. The effect is particularly pronounced for discovery tasks ($d = 0.803$, large), as the programmatic enumeration of tools in code-execution invocation avoids repeatedly loading all tool schemas into the context.

Observation 3 - Token Usage. Agents using direct invocation consumed an average of 31,161 tokens, compared with 21,576 tokens for agents using code-execution invocation (a 1.44× reduction). These savings are statistically significant for pipeline tasks, but not for single-tool or discovery tasks, see Table 3.

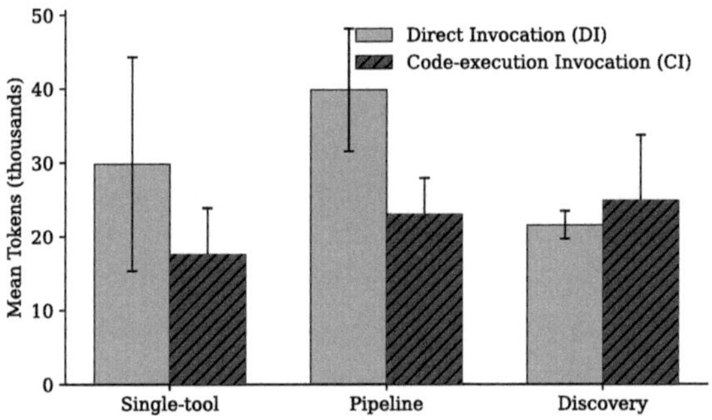

Fig. 5. Mean token usage (in thousands) by invocation modality and task type. Error bars indicate 95% confidence intervals.

Table 3. Mean token usage (in thousands) by invocation modality and task type.

Task type	DI	CI	t	p	d
Single-tool	29.8 ± 14.5	17.5 ± 6.4	1.538	0.126	0.215
Pipeline	39.9 ± 8.3	23.0 ± 4.9	3.453	<0.001	0.470
Discovery	21.6 ± 1.9	24.9 ± 8.8	−0.728	0.469	−0.116

Analysis. Agents using code-execution invocation achieved the largest token savings on pipeline tasks (Fig. 5), supporting the theoretical argument presented in [5]: batching sequential tool calls within a single code-execution step avoids the token overhead of routing each intermediate result through the LLM context. However, for discovery tasks, agents using code-execution invocation consumed a similar number of tokens as those using direct invocation, likely due to the additional overhead of a single agent managing a large toolset. A one-way ANOVA ($F = 7.205$, $p = 0.008$) confirms that invocation modality has a significant effect on token consumption when considered in aggregate.

5 Conclusions

This work provides one of the earliest reproducible empirical comparisons of two orchestration modalities in agentic AI systems: direct invocation and code-execution invocation. In addition, it introduces an evaluation framework for assessing orchestration strategies in terms of token usage and task quality, using a heterogeneous task suite and an LLM-as-a-judge grader across both single- and multi-agent architectures.

The empirical results highlight trade-offs between the two evaluated orchestration modalities. Agents using direct invocation achieved slightly higher task quality than those using code-execution invocation; however, this difference is not statistically significant. In contrast, agents employing code-execution invocation were significantly more efficient on pipeline and discovery tasks (although not on single-tool tasks) largely due to the ability to batch multiple operations. Furthermore, code-execution invocation resulted in a substantial reduction in token consumption, achieving a 1.44× decrease compared to direct invocation. Overall, code-execution invocation enhances efficiency and reduces token usage while maintaining comparable output quality, making it particularly well-suited for complex, multi-step tasks.

As a direction for future work, experiments with different state-of-the-art AI models are planned to explore potential model-dependent performance differences or biases. In addition, investigating the performance of alternative orchestration approaches (such as dynamic tool selection or MPC+ for compressing tool responses) is identified as an immediate direction for future research.

Acknowledgment. The authors gratefully acknowledges the financial support from the Asociacion Mexicana de Cultura, A.C. and the technical support from Sofía Maldonado, Pablo Roca, and Miguel Quintero. It is acknowledged that grammar was checked and adjusted using Microsoft Copilot.

Disclosure of Interests. The authors have no competing interests to declare.

References

1. Acharya, D.B., Kuppan, K., Divya, B.: Agentic AI: autonomous intelligence for complex goals - a comprehensive survey. IEEE Access **13**, 18912–18936 (2025)
2. Amazon Web Services: Building agentic ai workflows with mcp, agentcore, and bedrock: A hands-on guide (2025). https://builder.aws.com/content/37INxVLnjoFrqpr0W9vlKdlpBZs/building-agentic-ai-workflows-with-mcp-agentcore-and-bedrock-a-hands-on-guide. Accessed 26 Mar 2026
3. Amazon Web Services, Inc.: Amazon bedrock: The platform for building generative ai applications and agents at production scale (2026). https://aws.amazon.com/bedrock/. Accessed 27 Mar 2026
4. An, S., Ma, Z., Lin, Z., Zheng, N., Lou, J.G., Chen, W.: Make your LLM fully utilize the context. Adv. Neural. Inf. Process. Syst. **37**, 62160–62188 (2024)
5. Anthropic: Code execution with mcp: Building more efficient ai agents (2025). https://www.anthropic.com/engineering/code-execution-with-mcp. Accessed 27 Mar 2026
6. Anthropic: Introducing advanced tool use on the claude developer platform (2025), https://www.anthropic.com/engineering/advanced-tool-use, accessed: 2026-03-29
7. Bandi, A., Kongari, B., Naguru, R., Pasnoor, S., Vilipala, S.V.: The rise of agentic AI: a review of definitions, frameworks, architectures, applications, evaluation metrics, and challenges. Future Internet **17**(9), 404 (2025)
8. Dobriy, D., Bauer, F., Azzam, A., Banerjee, D., Polleres, A.: Agentic sparql: Evaluating sparql-mcp-powered intelligent agents on the federated kgqa benchmark. arXiv preprint arXiv:2603.06582 (2026)
9. Dwivedi, Y.K., et al.: Agentic ai systems: what it is and isn't. Glob. Bus. Organ. Excell. **45**(3), 253–263 (2026)
10. Godavarthi, S., Nagvekar, R.: Investment banking with genai. In: Transforming Financial Services with Generative AI: From Strategy and Design to Practical Applications, pp. 191–218. Springer, Cham (2026)
11. Guo, Z., Xu, B., Zhu, C., Hong, W., Wang, X., Mao, Z.: Mcp-agentbench: evaluating real-world language agent performance with MCP-mediated tools. In: Proceedings of the AAAI Conference on Artificial Intelligence, vol. 40, pp. 30888–30896 (2026)
12. Hou, X., Zhao, Y., Wang, S., Wang, H.: Model context protocol (MCP): landscape, security threats, and future research directions. ACM Trans. Softw. Eng. Methodol. (2026). just accepted
13. Huang, J., Zhu, H., Xu, C., Zhan, T., Xie, Q., Huang, J.: Auditwen: an open-source large language model for audit. In: Proceedings of the 23rd Chinese National Conference on Computational Linguistics, pp. 1351–1365 (2024)
14. Joel, S., Wu, J., Fard, F.: A survey on LLM-based code generation for low-resource and domain-specific programming languages. ACM Trans. Softw. Eng. Methodol, (2025). just accepted
15. Kong, Q., et al.: Mobileworld: benchmarking autonomous mobile agents in agent-user interactive and mcp-augmented environments. arXiv preprint arXiv:2512.19432 (2025)
16. LF Projects, LLC: What is the model context protocol (mcp)? (2024). https://modelcontextprotocol.io/docs/getting-started/intro. Accessed 2026-03-29
17. Mo, G., et alivemcpbench: Can agents navigate an ocean of mcp tools? arXiv preprint arXiv:2508.01780 (2025)

18. Naveed, H., et al.: A comprehensive overview of large language models. ACM Trans. Intell. Syst. Technol. **16**(5), 1–72 (2025)
19. Raheem, T., Hossain, G.: Agentic AI systems: opportunities, challenges, and trustworthiness. In: 2025 IEEE International Conference on Electro Information Technology (eIT), pp. 618–624. IEEE (2025)
20. Ratner, N., et al.: Parallel context windows for large language models. In: Proceedings of the 61st Annual Meeting of the Association for Computational Linguistics (Volume 1: Long Papers), pp. 6383–6402 (2023)
21. Sapkota, R., Roumeliotis, K.I., Karkee, M.: AI agents vs. agentic AI: a conceptual taxonomy, applications and challenges. Inf. Fusion. 103599 (2025)
22. Wu, Y., et al.: Extending context window of large language models from a distributional perspective. In: Proceedings of the 2024 Conference on Empirical Methods in Natural Language Processing, pp. 7288–7301 (2024)
23. Yao, S., et al.: React: synergizing reasoning and acting in language models. In: The Eleventh International Conference on Learning Representations, pp. 1–33 (2022)

Automated Machine Learning Framework for Hypothyroidism Prediction Using XGBoost and SHAP

Ana Laura Lezama Sánchez[1] and Mireya Tovar Vidal[2](✉)

[1] Secretaría de Ciencia, Humanidades, Tecnología e Innovacion, Av. Insurgentes Sur 1582, Col Credito Constructor, 03940 Benito Juarez, Mexico
analaura.lezama@correo.buap.mx

[2] Faculty of Computer Science, Benemérita Universidad Autónoma de Puebla (BUAP), Avenida 14 Sur y Avenida San Claudio, Ciudad Universitaria, 72570 Puebla, Mexico
mireya.tovar@correo.buap.mx

Abstract. Hypothyroidism is a clinical disorder commonly encountered by primary care physicians. Untreated hypothyroidism can contribute to hypertension, dyslipidemia, infertility, cognitive impairment, and neuromuscular dysfunction. The prevalence of this disorder increases with age and is higher in females than in males. Hypothyroidism may occur because of primary gland failure or insufficient thyroid gland stimulation by the hypothalamus or pituitary gland. Autoimmune thyroid disease is the most common etiology of hypothyroidism. Clinical symptoms of hypothyroidism are nonspecific and may be subtle, especially in older people. The best laboratory assessment of thyroid function is a serum thyroid-stimulating hormone test. There is no evidence that screening asymptomatic adults improves outcomes. In many patients, alleviation of symptoms can be accomplished through oral administration of synthetic levothyroxine, and most patients will require lifelong therapy. In the last years, the techniques of machine learning have demonstrated great potential to improve precision diagnosis in medical diseases. The objective of this study is to develop the predictive model to detect hypothyroidism using machine learning algorithms and optimization techniques of hyperparameters. A clinical dataset was used and subjected to preprocessing, including imputation of missing values using KNNImputer, coding of categorical variables, and standardization of features. Subsequently, a classification model based on Extreme Gradient Boosting (XGBoost) with hyperparameters optimized using Optuna and cross-validated was trained.

Keywords: machine learning · autoimmune disease · SHAP · hypothyroidism

H. Ponce et al. (Eds.): COMIA 2026, CCIS 3095, pp. 172–183, 2027.
https://doi.org/10.1007/978-3-032-37438-7_14

1 Introduction

Hypothyroidism is one of the most common endocrine diseases worldwide, characterized by insufficient production of thyroid hormones. This disease generates symptoms in the patient such as fatigue, weight gain, cold intolerance, mood changes, irritability and metabolic alterations, which significantly affects the quality of life of the patients [1]. An early diagnosis of this disease is fundamental to preventing complications and improving the clinical management of the patient. Hence, the diagnosis for this disease is based on laboratory tests and clinical evaluation; however, the increasing volume of medical data has driven the use of machine learning techniques to support medical decision-making [1]. In the last years, several studies have explored the use of algorithms of machine learning such as Random Forest, Support Vector Machines, and Gradient Boosting for predicting endocrine diseases. Nevertheless, there is still a need to develop more robust, optimized, and interpretable models.

In this context, this work proposes a machine learning pipeline for the prediction of hypothyroidism using XGBoost, hyperparameter optimization using Optuna, and interpretability analysis based on SHAP. The results were evaluated using accuracy, precision, recall, F_1, and ROC-AUC metrics, demonstrating good performance in identifying positive cases of hypothyroidism. Additionally, SHAP (Shapley Additive Explanations) was applied to analyze the significance of the variables and improve the model's interpretability. The results suggest that the use of optimized machine learning contributes to the automatic detection of hypothyroidism and can become an important support tool in medical decision-making.

The paper is organized as follows. In Sect. 2, the basic concepts are described. In Sect. 3, we present the related works. In Sect. 4, we present the proposed methodology. Section 5 presents the results, conclusions, and future work. Finally, the references are provided.

2 Basic Concepts

In this section, we expose the basic concepts necessary in the development of this paper.

Machine learning is a branch of artificial intelligence that enables systems to learn patterns from data and make predictions without being explicitly programmed.

In this study, Extreme Gradient Boosting (XGBoost) is used as the main classification algorithm. XGBoost is an ensemble learning method based on decision trees that applies gradient boosting to optimize predictive performance. It has demonstrated strong effectiveness in classification problems due to its ability to model nonlinear relationships and handle heterogeneous data, while incorporating regularization techniques to reduce overfitting.

To handle missing data in the dataset, the K-Nearest Neighbors Imputation (KNNImputer) method is applied. This technique replaces missing values based on the values of the nearest samples in the feature space, preserving the underlying structure of the data.

Additionally, feature scaling is performed using StandardScaler, which transforms variables to have zero mean and unit variance. This step ensures that all features contribute equally to the learning process and prevents bias toward variables with larger magnitudes.

The performance of the classification model is evaluated using standard metrics, including accuracy, precision, recall, F_1 score, and the area under the ROC curve (AUC). These metrics are widely used in medical classification tasks due to their ability to assess different aspects of predictive performance.

Finally, model interpretability is addressed using SHapley Additive exPlanations (SHAP), a model-agnostic method based on cooperative game theory. SHAP assigns each feature an importance value representing its contribution to the final prediction, allowing for a better understanding of the model's behavior.

3 Related Works

In this section, we present related works in the clinical domain.

The authors of these studies applied different approaches, including data science, deep learning, and machine learning. Several studies have explored the use of classical machine learning techniques for the detection of thyroid disorders. As in [8], the authors analyzed a public dataset about thyroid that contained euthyroid, hyperthyroid and hypothyroid cases. The authors applied different classifiers, including logistic regression, decision trees, random forests, support vector machines, k-nearest neighbors and Naive Bayes [15]. The methodology proposed by the authors incorporated different preprocessing stages, some of them the imputation of missing values, feature encoding and data standardization. In addition to strategies to address class imbalance through class weighting and SMOTE-based oversampling. Based on the results obtained, the Random Forest classifier obtained the best performance, obtaining an area under the ROC curve (AUC) of 0.99. In addition, the authors highlighted that the small sample size and the class imbalance limit the generalization capacity of the results, emphasizing the need to validate them with larger and independent datasets.

In the study proposed by [9], the authors propose a method for the prediction of hypothyroidism. The methodology proposed included preprocessing that included data cleaning, the handling of class imbalance, feature selection and model optimization. The authors applied the SMOTEENN technique, which combines oversampling and noise reduction to improve the learning of minority classes. The feature selection was performed through a chi-square filtering method to identify the most relevant attributes and reduce the risk of overfitting. The authors evaluated the machine learning models of the authors, including Decision Tree, Random Forest and K-Nearest Neighbors, with hyperparameters optimized through randomized cross-validation. In addition, the authors implemented an ensemble voting strategy to improve predictive performance. The experimental results reported a high performance, achieving an accuracy of 98.81%, which suggests that the proposed approach could support the early detection of hypothyroidism in clinical decision support systems.

In [10], the author presents a computer-aided diagnosis system for thyroid diseases using machine learning techniques applied to a dataset of 3,164 patient records. The dataset used was composed of demographic information, clinical indicators and results of thyroid function tests. The author pointed out the importance of the preprocessing stages, such as data cleaning, normalization and feature selection. In addition, they applied feature engineering techniques to identify the most relevant predictors associated with thyroid disorders. The authors used different machine learning classifiers such as Random Forest, K-Nearest Neighbors, support vector machines, decision tree and logistic regression. The models were evaluated by the authors through stratified k-fold cross-validation and using metrics such as precision, recall and F_1.

On the other hand, the authors in [11] explored the application of machine learning techniques in clinical prediction tasks within obstetric and neonatal care. The authors used a dataset collected from hospital electronic health records between 2020 and 2022 to predict the need for neonatal resuscitation. The dataset included multiple maternal and fetal clinical variables, conditions such as preeclampsia and hypothyroidism. The authors evaluated several machine learning models such as Random Forest, XGBoost, Decision Tree, K-Nearest Neighbors and Logistic Regression. The models evaluated performance metrics such as precision, accuracy, recall and ROC-AUC. Among the approaches evaluated by the authors, the Random Forest model achieved the best performance with an AUC of 0.87 and high values of precision and recall. Based on the results obtained, the authors conclude that the study demonstrated the potential of machine learning methods to support clinical decision-making using electronic health record data.

On the other hand, in [12] the authors highlight the importance of the interpretability of the models within machine learning systems applied to the diagnosis of thyroid diseases. The authors propose a machine learning framework based on a dataset from the UCI repository for the classification of thyroid disorders. For this, they evaluated the classifiers Random Forest, CatBoost and Support Vector Machines. To obtain a robust estimation of the performance of the models, they employed a nested cross-validation scheme. Based on the obtained results the authors show that their proposed model showed a very high predictive performance, reaching precision values close to 99.7% and an AUC of 1.0 using a reduced set of physiological features. In addition, the authors incorporated explainable artificial intelligence techniques, such as SHAP for feature attribution and counterfactual analysis.

In this study a machine learning framework for the prediction of hypothyroidism using clinical and laboratory data is presented. The proposed approach involves data preprocessing, feature encoding and missing value imputation, followed by the training of an XGBoost classifier. The hyperparameters of the model are optimized through Optuna using cross-validation, with the objective of improving predictive performance. Likewise, a SHAP-based analysis is employed to interpret the model and identify the most influential features in the prediction process.

4 Proposed Methodology

In this section, we expose the details of the proposed methodological approach to developing a predictive model of hypothyroidism, including dataset preparation, hyperparameter optimization using Optuna, model training with XGBoost, and variable importance analysis using SHAP.

The dataset used is a clinical dataset related to the diagnosis of hypothyroidism. The data were loaded from a CSV file and subsequently processed using Python with specialized data science libraries, including Pandas, NumPy and Scikit-learn. This dataset contains clinical and demographic variables that act as predictor characteristics, while the target variable represents the presence or absence of hypothyroidism. Prior to model training, a preprocessing process was performed to ensure data quality. This process included:

- Categorical variable coding: Categorical variables were transformed into numeric values using Label Encoding.
- Imputation of missing values: Missing values were imputed using the K-Nearest Neighbors Imputer (KNNImputer) algorithm with k=5 neighbors, which allows estimating missing values based on the similarity between observations.
- Standardization of variables: To ensure that all characteristics have the same scale, StandardScaler was applied, transforming the data to have a mean of zero and a standard deviation of one.

Subsequently, the dataset was divided into 80% for training and 20% for testing, using splitting to preserve the class proportion in both subsets. To improve model performance, Optuna, a Bayesian search-based hyperparameter optimization library, was used. During the optimization process, the following hyperparameters of the XGBoost model were evaluated [16]:

- Number of trees (n_estimators)
- Maximum tree depth (max_depth)
- Learning rate (learning_rate)
- Subsample fraction (subsample)
- Feature sampling (colsample_bytree)
- Regularization parameter (gamma)

Performance evaluation was performed using 5-fold cross-validation, with the Area Under the Receiver Operating Characteristic (ROC-AUC) as the primary metric. The optimization process was run for 30 iterations. Once the best hyperparameters were identified, a final model was trained on the training set using the Extreme Gradient Boosting (XGBoost) algorithm. This algorithm was selected due to its high performance in classification problems and its ability to handle nonlinear relationships between variables. To understand the model's performance, it was evaluated on the test set using the following classification metrics:

- Accuracy
- Precision
- Recall
- F_1
- Area under the ROC curve (ROC-AUC)

In addition, a confusion matrix and ROC curve were generated to analyze the model's performance in terms of correct classification and discrimination between classes.

To improve the model's interpretability, SHApley Additive exPlanations (SHAP) [14] was used, a game theory-based technique that allows quantifying the contribution of each feature to the model's predictions. Two main visualizations were generated:

- SHAP Beeswarm Plot, which shows the distribution of each variable's impact on the predictions.
- SHAP Feature Importance Plot, which identifies the most influential variables in the model.

These visualizations allow us to understand which factors contribute most to the diagnosis of hypothyroidism. In Fig. 1, we present the proposed methodology in a graphical form.

5 Results

In this section the results obtained with the proposed methodological approach and the dataset used are detailed. The dataset used in this work belongs to the Garavan Institute in Sydney, Australia, and was originally documented by Ross Quinlan in 1986 [13]. The dataset includes approximately 2,800 training instances and 972 test instances, with around 29 clinical attributes that may correspond to boolean variables or continuous values derived from medical analyses. Based on the results obtained from the classification model (presented in Table 1), we show the performance metrics. The model achieved an accuracy of 0.997, which indicates a high proportion of correct predictions in the analyzed dataset. In addition, the value of the area under the ROC curve (AUC) of 0.999 suggests that the model effectively discriminates between the classes corresponding to patients with hypothyroidism and those without this condition. Regarding the results obtained by precision and recall, the model obtained a precision of 0.967 and a recall of 1.000, which indicates that all positive cases were correctly identified. The F_1 of 0.983, which combines precision and recall, confirms a good balance between both metrics and indicates the solid performance; however, further validation is required to confirm the generalizability of the model.

These metrics are widely used in the evaluation of classification models in the field of machine learning. Table 2 presents the classification report detailed by class. For class 0 (without hypothyroidism), the model achieved perfect values of

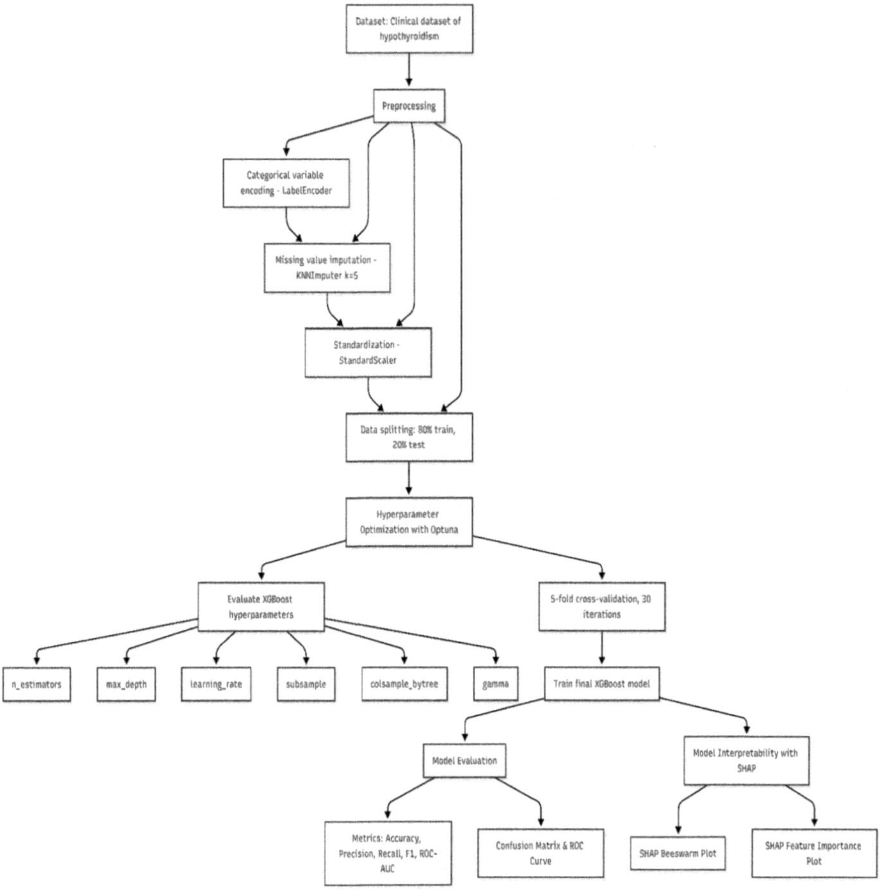

Fig. 1. Proposed methodology for hypothyroidism prediction.

precision, recall and F_1 (1.00) in a total of 697 instances, which demonstrates a precise identification of negative cases. For class 1 (hypothyroidism), the model obtained a precision of 0.97, a recall of 1.00 and an F_1 value of 0.98 in 58 evaluated instances. The obtained recall in the positive class is particularly relevant in a medical context, since it indicates that the model correctly identified all the cases of hypothyroidism present in the test set. This performance reduces the risk of false negatives, a critical aspect in diagnostic support applications. In general, based on the obtained results we deduced that the model has a high discriminative power and could serve as a support tool in recommendation systems or for the early detection of hypothyroidism.

The classification model achieved an accuracy of 0.997, with a ROC AUC of 0.999, indicating a high discriminative capability between patients with and without hypothyroidism. The model obtained a precision of 0.967 and a recall of

Table 1. Metrics for evaluating the classification model

Metric	Value
Accuracy	0.997
ROC AUC	0.999
Precision	0.967
Recall	1.000
F_1	0.983

1.000, resulting in an F_1 of 0.983 (see Table 1). To evaluate the robustness of the model, a stratified 5-fold cross-validation was performed on the training set. The model achieved a mean ROC AUC of 0.998 with a standard deviation of 0.0014, indicating consistent performance across different data partitions. Additionally, confidence intervals were estimated using a bootstrap resampling approach. The ROC AUC obtained a 95% confidence interval of [0.996–1.000], suggesting that the model performance is stable and not due to random variation. A comparative analysis with baseline models was conducted under the same experimental conditions. The XGBoost model achieved performance comparable to Random Forest, both significantly outperforming Logistic Regression. These results confirm the effectiveness of ensemble-based approaches for this classification task. Furthermore, a sensitivity analysis was performed by evaluating the model across multiple random train-test splits. The F_1 showed a mean value of 0.974 with a standard deviation of 0.021, indicating that the model maintains stable performance under different data partitions.

Table 2. Model classification report

Class	Precision	Recall	F_1	Support
0 (No hypothyroidism)	1.00	1.00	1.00	697
1 (Hypothyroidism)	0.97	1.00	0.98	58

To further evaluate the robustness of the model, a stratified 5-fold cross-validation was performed. The model achieved a mean ROC AUC of 0.998 with a standard deviation of 0.0014, indicating consistent performance across different data partitions.

Additionally, confidence intervals were estimated using a bootstrap resampling approach. The ROC AUC obtained a 95% confidence interval of [0.996–1.000], suggesting that the model performance is stable and not due to random variation.

A comparative analysis with baseline models was conducted under the same experimental conditions. As shown in Table 3, the XGBoost model achieved performance comparable to Random Forest and significantly outperformed Logistic Regression in terms of both accuracy and F_1.

A sensitivity analysis was also performed using multiple random train-test splits. The model obtained a mean F_1 of 0.974 with a standard deviation of 0.021, indicating stable performance across different data partitions.

Table 3. Comparison with baseline models

Model	Accuracy	F_1
XGBoost	0.997	0.983
Random Forest	0.997	0.983
Logistic Regression	0.967	0.742

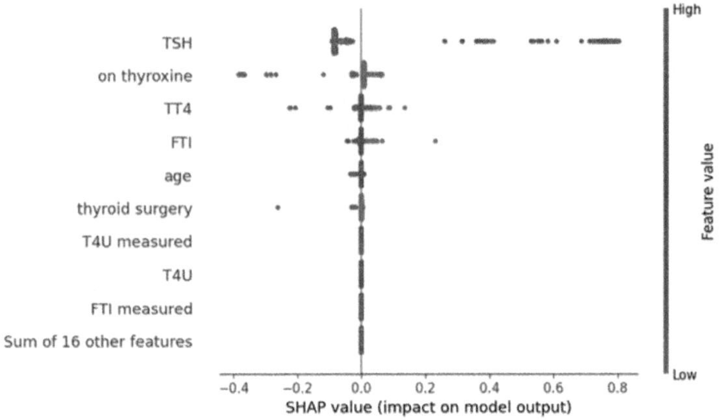

Fig. 2. SHAP summary chart shows the contribution of the variables to the classification model.

In Fig. 2 shows the importance of the variables in the model using SHAP values, which indicate the impact of each feature on the model's prediction. The TSH variable has the greatest contribution to the prediction of hypothyroidism, followed by on-thyroxine and TT4. The colors represent the value of each feature, with red indicating high values and blue indicating low values. The dispersion of the points reflects how different values of each variable positively or negatively influence the model's output.

The confusion matrix (see Fig. 3) shows the model's performance in classifying patients with and without hypothyroidism. A high number of true negatives and true positives is observed, indicating that the model correctly classifies most cases. The absence or minimal presence of false negatives is particularly relevant in medical applications, as it implies that the model correctly identifies positive cases of the disease. This result suggests that the model is effective at identifying almost all positive cases.

Figure 4 shows the variables in the model, calculated through the mean absolute SHAP value. TSH stands out as the most influential factor in the prediction

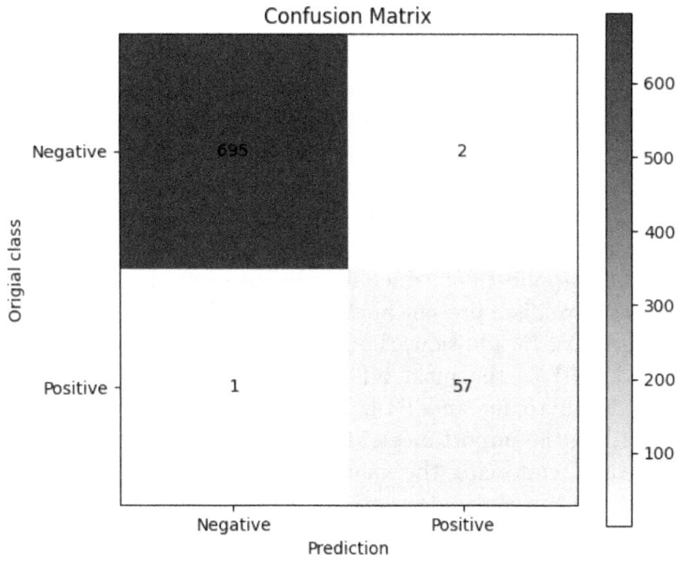

Fig. 3. Confusion matrix of the classification model for the detection of hypothyroidism.

of the model, followed by thyroxine and TT4. The rest of the variables do not contribute. Based on these results we can deduce that they suggest that the levels of thyroid stimulating hormone are the most relevant indicator to detect hypothyroidism in the analyzed dataset.

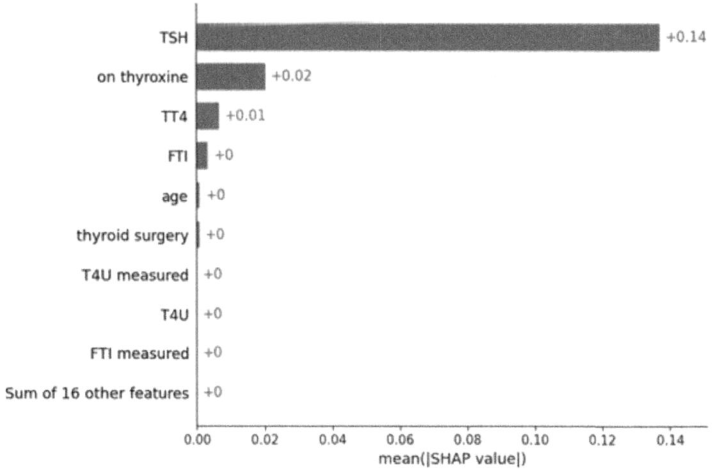

Fig. 4. Average feature importance based on SHAP values.

6 Conclusions

In this article we describe a classification model based on machine learning for the detection of hypothyroidism through clinical and laboratory variables. Based on the obtained results we were able to evidence a high performance of the model, with an accuracy of 0.997 and an area under the ROC curve (AUC) of 0.999, which demonstrates its capacity to discriminate between patients with and without hypothyroidism. On the other hand, the analysis of the classification report revealed a sensitivity (recall) of 1.00 for the positive class, which is particularly relevant in medical applications, since it indicates that the model identified all the cases of hypothyroidism present in the test set. This significantly reduces the risk of false negatives. In addition, the interpretability analysis through SHAP values identified TSH as the most influential factor in the predictions of the model, followed by thyroxine and TT4. These results align with existing clinical knowledge regarding the importance of these variables in the assessment of thyroid disorders, thus reinforcing the model's reliability. Therefore, based on the results obtained, we can conclude that the proposed model could be used as a support tool for the early detection of hypothyroidism and for clinical decision-making. Future work will focus on evaluating the model in real-world clinical settings and integrating knowledge representation techniques, such as medical ontologies, to improve interpretability and support for medical decisions.

Acknowledgment. We thank the Secretaria de Ciencias, Humanidades, Tecnología e Innovacion (SECIHTI) for the postdoctoral fellowship awarded to Dra. Ana Laura Lezama Sánchez (CVU 788155).

References

1. Gaitonde, D.Y., Rowley, K.D., Sweeney, L.B.: Hypothyroidism: an update. South Afr. Fam. Pract. **54**(5), 384–390 (2012)
2. Vega, M.Á., Mora, L.M.Q., Badilla, M.V.C.: Inteligencia artificial y aprendizaje automático en medicina. Revista médica sinergia **5**(8), e557–e557 (2020)
3. Mir, M.M., Mir, G.M., Raina, N.T., Mir, S.M., Mir, S.M., Miskeen, E., Alamri, M.M.S.: Application of artificial intelligence in medical education: current scenario and future perspectives. J. Adv. Med. Educ. Prof. **11**(3), 133 (2023)
4. Ahmed, Z., Mohamed, K., Zeeshan, S., and Dong, X.: Artificial intelligence with multi-functional machine learning platform development for better healthcare and precision medicine. Database **2020**, baaa010 (2020)
5. Bhagat, H.V., Singh, M.: Machine-Learning and Deep-Learning Techniques in Social Sciences. Mach. Learn. Algorithms Signal Image Process. 409–428 (2022)
6. Sadegh-Zadeh, S.A., et al.: Precision diagnostics in cardiac tumours: Integrating echocardiography and pathology with advanced machine learning on limited data. Inf. Med. Unlocked **49**, 101544 (2024)
7. Magoulas, G.D., Prentza, A.: Machine learning in medical applications. In: Advanced Course on Artificial Intelligence, pp. 300–307. Berlin, Heidelberg, Springer, Berlin Heidelberg (1999)

8. Salloum, S.A., Almomani, A., Alomari, K.M., Khdour, T.A.M.: Predicting thyroid dysfunction using classical machine learning with rigorous statistical evaluation. IEEE Access (2026)
9. Chitra, K., Sheela, D.G.A.S.: Early prediction of hypothyroidism using chi-square feature selection and SMOTE-ENN with randomized search CV compared to grid-based search. Indian J. Sci. Technol. **19**(1), 8–18 (2026)
10. Elbillihy, T.A., AbdElhalim, E., Ashour, M., Nafea, H.B.: Prediction of thyroid disease based on machine learning algorithms. Mansoura Eng. J. **51**(1), 14 (2026)
11. Banaei, M., et al.: Evaluating the predictive power of a machine learning to predict the need for neonatal resuscitation (2026)
12. Alam, M.Z., et al.: Enhancing thyroid disease diagnosis with machine learning and counterfactual explainable AI. IEEE Access (2026)
13. Quinlan, J.R.: Thyroid Disease Data Set. UCI Machine Learning Repository (1986). https://archive.ics.uci.edu/ml/datasets/Thyroid+Disease
14. Lundberg, S.M., Erion, G.G., Lee, S.I.: Consistent individualized feature attribution for tree ensembles. arXiv preprint arXiv:1802.03888 (2018)
15. Boateng, E.Y., Otoo, J., Abaye, D.A.: Basic tenets of classification algorithms K-nearest-neighbor, support vector machine, random forest and neural network: A review. J. Data Anal. Inf. Process. **8**(4), 341–357 (2020)
16. Zhang, Y., et al.: Influence of urban morphological characteristics on street-level urban heat risk: a geographically weighted machine learning approach. Buildings **16**(4), 725 (2026)

Professional Profile Evaluation: A Proposal Based on Large Language Models and Fuzzy Logic

Omar Atriano[1,2](✉), Perfecto Malaquias Quintero[2], Edmundo Bonilla[2], Rodolfo Pérez[2], Paulina Galindo[2], and Ricardo Pérez[2]

[1] SmartSoft America BA, C. Adolfo López Mateos 131A, Texcacoac, 90805 Santa Ana Chiautempan, Tlax, Mexico
omar.atriano@smartsoftamerica.com.mx

[2] Instituto Tecnológico de Apizaco, TecNM, Av. Instituto Tecnológico s/n, 90300 Apizaco, Tlax, Mexico
{perfecto.qf,edmundo.bh,rodolfo.pl,m19370503,m17370613}@apizaco.tecnm.mx

Abstract. Personnel recruitment constitutes a critical process within human talent management in organizations. One of its main challenges is the identification of qualified candidates through a process that generally begins with resume analysis—a problem that, while it has been addressed through various machine learning techniques, can benefit from the use of LLMs due to their capacity to interpret and structure non-homogeneous information. For this reason, it is proposed to separate this process into multiple stages by combining LLMs with elements of fuzzy inference systems, with the objective of systematizing candidate evaluation under consistent and traceable criteria. This process comprises a pipeline that orchestrates the flow between data cleaning, fuzzification, rule evaluation, aggregation, defuzzification, and results presentation. The approach was evaluated using a confusion matrix and Precision, Recall, and F1-score metrics, taking into account the scores produced by the system and the evaluations carried out by a Human Resources specialist on a set of 60 résumés. The results showed values approaching 90%, suggesting that the combination of LLMs with fuzzy inference systems allows transforming a qualitatively-natured problem, such as the evaluation of professional profiles, into a quantitative and reproducible process. Altogether, the results support the use of this type of architecture as a decision-support tool for the initial filtering of candidates, maintaining consistent and traceable criteria.

Keywords: Large Language Models · Fuzzy Inference System · Recruitment · Explainable AI

1 Introduction

Personnel recruitment is a strategic process whose main challenge is to identify suitable candidates efficiently and objectively. Historically, it has relied on the

H. Ponce et al. (Eds.): COMIA 2026, CCIS 3095, pp. 184–198, 2027.
https://doi.org/10.1007/978-3-032-37438-7_15

manual review of résumés, an activity with high operational burden and susceptible to bias. To mitigate this, machine learning techniques have been applied to candidate filtering; however, these require labeled data and face difficulties given the heterogeneous and unstructured nature of résumés. In response to these limitations, Large Language Models (LLMs) have emerged as a viable alternative for automating the analysis and scoring of professional profiles. However, as they are still maturing, they may exhibit hallucinations and biases that compromise the fairness and reliability of the selection process [1–3].

Agent-based pipelines have been proposed to filter résumés with LLMs, sharing common phases: information extraction, evaluation against the job vacancy, and profile filtering [4,5]. The first anonymizes personal data to reduce bias; the second employs *Retrieval Augmented Generation* (RAG) to enrich the profile with contextual information. Both report favorable results compared to human evaluators. In a similar vein, [6] proposes combining LLMs with multi-criteria decision-making (Fuzzy-TOPSIS) to rank LinkedIn profiles of software engineers, with equally favorable results.

However, these proposals present limitations. The fixed extraction rules of [4] are problematic in contexts such as Mexico or Latin America, where résumés lack a standardized format, and the global scoring limits explainability. The enrichment by institution of origin in [5] may foster demographic discrimination. The approach of [6], oriented toward ranking through closeness coefficients, operates on structured LinkedIn profiles and does not execute a complete fuzzy inference cycle, limiting traceability to linguistic rules. Additionally, advances in the state of the art in LLMs have left several of these proposals technologically outdated.

Taking the above into account, and in line with the trend of enhancing LLMs with explicit reasoning modules [7], Fig. 1 presents a professional profile evaluation pipeline that integrates multiple AI agents with a Fuzzy Inference System (FIS). The FIS transforms qualitative evidence extracted from résumés into membership degrees and interpretable rules, identifying the level of competency fulfillment for an internship position in the IT sector. The flow begins with the Feature Extraction Agent (FExA), which converts each resume in PDF format into a semi-structured representation (academic background, experience, tools, and projects), mitigating format-related noise. This representation feeds into the Fuzzy Evaluator Agent (FEvA), which produces an interpretable score (e.g., discard or suitable for a position), along with the activated rules and membership degrees. Each stage is described in depth in the Methodology.

2 Methodology

2.1 Data

The dataset was provided on a confidential basis by SmartSoft America BA as part of one of its internship programs. This included 60 résumés from its

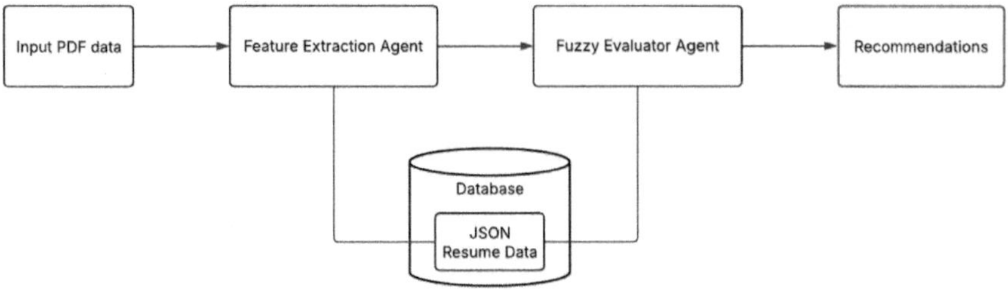

Fig. 1. Solution Pipeline

candidate database, as well as collaborative work to define the input linguistic variables, fuzzy sets, fuzzy rules, and output linguistic variables described in Sects. 2.3 and 2.4.

As illustrated in Fig. 2, the evaluated résumés correspond to students from various public (solid color) and private (striped pattern) universities in the states of Tlaxcala (purple), Puebla (green), Guerrero (blue), and Nuevo León (orange), who attended interviews for internship positions at different points in time, thereby ensuring sufficient demographic diversity to assess the fairness and performance of the system.

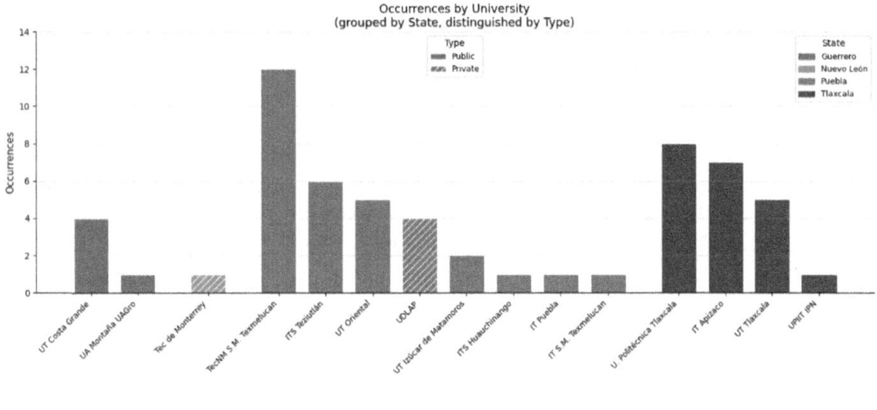

Fig. 2. University Distribution

2.2 Feature Extraction Agent

As the first stage of the evaluation process, the FExA preprocesses the résumés and persists the relevant information in a database. This process is carried out by the GPT 5.2 model, which receives the résumés in PDF format and extracts their textual content, structuring it in JSON format. In this representation, section titles are used as keys and their corresponding content as associated values. Additionally, a linguistic neutralization process is applied to remove lexical and grammatical markers that could reveal the gender of the résumé author, thereby mitigating potential biases in subsequent stages of the evaluation. The resulting information is stored in a SQL database in a JSON-type column, whose flexibility accommodates the diversity across résumés and allows new attributes to be incorporated in the future without compromising the integrity of the system. Upon completion of this process, the FExA notifies the FEvA that the data have been processed correctly and that the evaluation process may begin. This pipeline is depicted in Fig. 3.

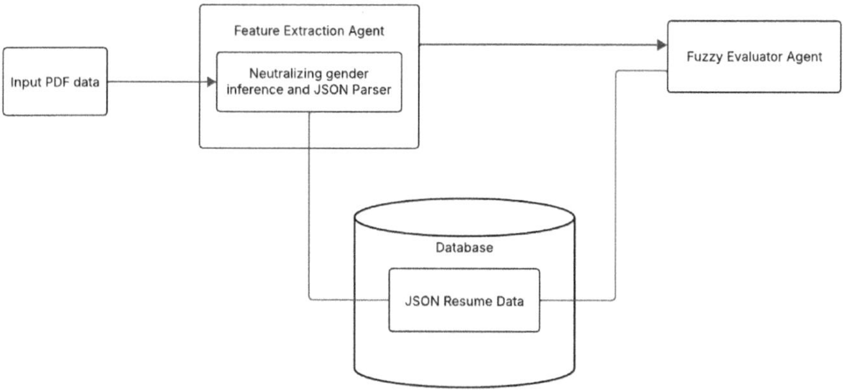

Fig. 3. Feature Extraction Agent Pipeline.

2.3 Fuzzy Evaluator Agent Design

Once the FExA completes its process, the FEvA evaluates the professional profiles through an LLM-based scoring scheme (GPT 5.2), whose reasoning is assisted by a Fuzzy Inference System comprising fuzzification, rule evaluation, aggregation, and defuzzification, together with input and output linguistic variables, membership functions, and fuzzy rules. This flow is illustrated in Fig. 4, and each component is described in depth in the following sections.

Input Linguistic Variables and Fuzzy Sets Three input linguistic variables were defined: **Academic Evidence (Aca_Ev)**, **Java Experience (Java_Exp)**, and **Database Experience (DB_Exp)**. For each variable, a rubric of binary sub-criteria (absent/present) was established, producing a score in $[0, 100]$ that constitutes the input to the fuzzification process. Based on this

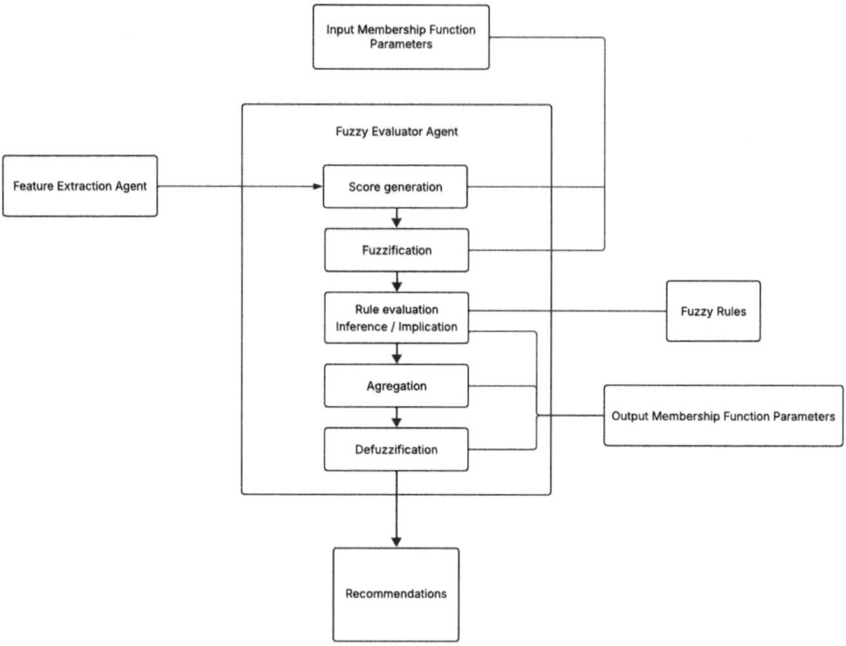

Fig. 4. Fuzzy Evaluator Agent Pipeline

score, the fuzzy sets associated with each variable were defined: four for Aca_Ev (**None, Precarious, Satisfactory, Relevant**), three for Java_Exp (**None, Low, Satisfactory**), and three for DB_Exp (**None, Basic, Relevant**). The sub-criteria, scores, and membership function parameters are detailed in Tables 1 and 2 and in Fig. 5.

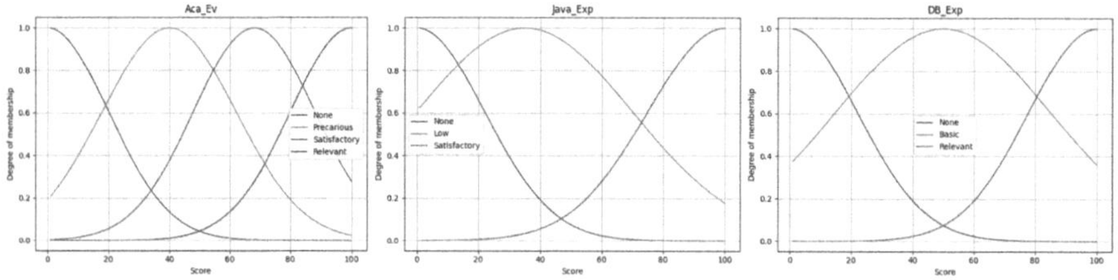

Fig. 5. Membership functions of the input linguistic variables. The horizontal axis represents the normalized score $[0, 100]$ derived from the rubrics in Table 1, and the vertical axis the membership degree $\mu \in [0, 1]$

Table 1. Scoring rubrics for the input linguistic variables. Each sub-criterion contributes points to the total score in the range [0, 100], which serves as the input to the fuzzification process.

Variable	Sub-criterion	Absent	Present
Aca_Ev	(i) Related degree	0	30
	(ii) Courses / certifications	0	30
	(iii) Academic projects	0	40
Java_Exp	(i) Mention of Java	0	30
	(ii) Java framework	0	40
	(iii) JPA / persistence	0	30
DB_Exp	(i) Mention of a DB engine	0	50
	(ii) Description of operations	0	50

Table 2. Parameters of the Gaussian membership functions for the input linguistic variables. Each fuzzy set is defined by two parameters (μ, σ) over the range [0, 100].

Variable	Fuzzy Set	Parameters (μ, σ)
Aca_Ev	None	(0, 20)
	Precarious	(40, 22)
	Satisfactory	(68, 20)
	Relevant	(100, 20)
Java_Exp	None	(0, 22)
	Low	(35, 35)
	Satisfactory	(100, 25)
DB_Exp	None	(0, 22)
	Basic	(50, 35)
	Relevant	(100, 22)

Scoring Process Upon receiving the control signal from the FExA, the FEvA operates autonomously by retrieving the relevant candidate information from the database. Based on this information, it executes the evaluation process using GPT-5.2, which acts as a linguistic evaluator and produces the input scores for the fuzzy inference system.

During retrieval, the candidate's personal and demographic data —such as name, address, university, phone number, and social media profiles— are discarded to prevent the model from inferring sensitive attributes that could introduce demographic biases, particularly those associated with historically marginalized populations, such as indigenous communities in the region.

This strategy enables the mapping of the candidate's academic and professional evidence to the defined linguistic variables, and the assignment of membership degrees to the corresponding fuzzy sets, thereby enabling a more equitable and representative fuzzification process based solely on academic and professional achievements.

Fuzzification Once the FEvA obtains the numerical scores in [0, 100] for each input linguistic variable (Sect. 2.3), the membership degree $\mu_{A_j}(x_i)$ of each score x_i with respect to each fuzzy set A_j is computed using the Gaussian membership function [8]:

$$\mu_{A_j}(x_i) = e^{-\frac{(x_i-\mu_j)^2}{2\sigma_j^2}} \tag{1}$$

where μ_j and σ_j correspond to the parameters of A_j specified in Table 2. This process is applied to each input variable (`Aca_Ev`, `Java_Exp`, and `DB_Exp`), and the resulting membership degrees are persisted in the database in JSON format for the rule evaluation stage (Sect. 2.4).

Gaussian functions were selected for the input variables because the scoring space [0, 100] is continuous and the evidence captured by each sub-criterion admits gradual rather than abrupt transitions between linguistic categories. Compared to triangular functions, Gaussian curves provide nonzero membership degrees across a wider score range, reducing sensitivity to borderline cases and preserving the fine-grained information encoded in the rubric. The input parameters were calibrated iteratively in collaboration with the HR specialist until the defuzzified outputs consistently aligned with the specialist's evaluations.

Output Linguistic Variables and Fuzzy Sets The output linguistic variables represent the functional profiles inferred by the FIS from the activation of the rules defined in Sect. 2.4. These variables do not constitute a deterministic decision, but rather a gradual representation of the degree of suitability of the candidate for different roles within the context of the analyzed position.

Each candidate may simultaneously exhibit membership degrees in one or more categories, capturing intermediate scenarios and avoiding binary decisions that could introduce biases. Triangular membership functions were selected for the output variables because the consequent fuzzy sets represent coarse activation levels (*Low*, *Medium*, *High*) whose boundaries are determined by the rule structure rather than by continuous empirical data; their simplicity prioritizes interpretability and minimizes the number of parameters to justify. Their fuzzy sets and triangular membership function parameters are detailed in Table 3 and illustrated in Fig. 6:

2.4 Fuzzy Rules

The FIS rules were designed to capture three patterns defined over the input variables (Sect. 2.3) and map them to the output variables (Sect. 2.3): (i) absence of technical and academic evidence, which activates *Dis*; (ii) conceptual evidence but limited tool proficiency, which activates *Sui_Pos_1*; and (iii) consistent technical evidence in Java and databases, which activates *Sui_Pos_2*.

Table 3. Output linguistic variables and parameters of their fuzzy sets. Each triangular membership function is defined by three points (a, b, c) over the range $[1, 100]$

Variable	Fuzzy Set	Parameters (a, b, c)
Discard (Dis)	Low	(1, 1, 50)
	Medium	(10, 50, 80)
	High	(50, 100, 100)
Suitable Position 1 (Sui_Pos_1)	Low	(1, 1, 50)
	Medium	(10, 50, 80)
	High	(50, 100, 100)
Suitable Position 2 (Sui_Pos_2)	Low	(1, 1, 50)
	Medium	(10, 50, 80)
	High	(50, 100, 100)

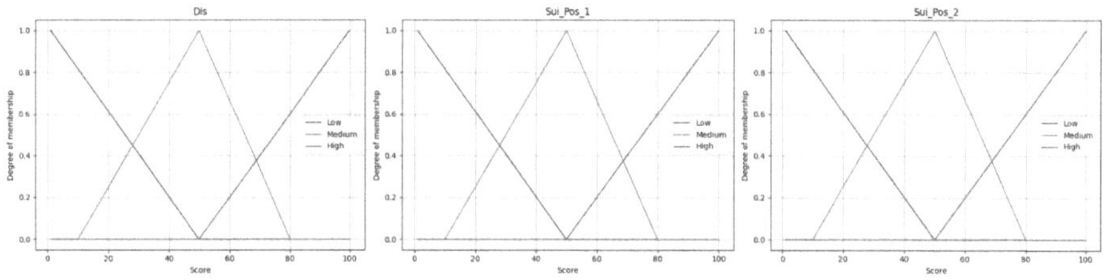

Fig. 6. Output linguistic variables and fuzzy sets

The rule set was derived through a structured elicitation process. The HR specialist's recruitment criteria were first captured by defining the input and output linguistic variables in collaboration with that specialist (Sect. 2.3). The initial rule base was then generated as the Cartesian product of the input fuzzy sets, producing all possible antecedent combinations and ensuring no relevant pattern was omitted at the outset.

The resulting rule base was subsequently reviewed jointly with the HR specialist, who applied qualitative domain knowledge to prune it in two steps: rules that did not correspond to any realistic candidate profile were discarded as non-representative, and rules whose consequents were already subsumed by other rules were eliminated as redundant. This pruning process yielded a compact, interpretable rule set that faithfully reflects the evaluation criteria applied in practice.

A representative fuzzy rule is:

$$\begin{aligned}&\text{IF } (Aca\_Ev \text{ is } None) \wedge (Java\_Exp \text{ is } None) \wedge (DB\_Exp \text{ is } None)\\&\text{THEN } (Discard \text{ is } High) \wedge (Sui\_Pos\_1 \text{ is } Low) \wedge (Sui\_Pos\_2 \text{ is } Low)\end{aligned} \tag{2}$$

Rule Evaluation With the membership degrees of the input variables computed, the fuzzy rules (Sect. 2.4) are evaluated by applying the minimum T-norm operator over the antecedent intersection [9]:

$$\alpha_i = \min\left(\mu_{A_1}(x_1),\, \mu_{A_2}(x_2),\, \mu_{A_3}(x_3)\right) \tag{3}$$

where x_1, x_2, x_3 correspond to the crisp values of **Aca_Ev**, **Java_Exp**, and **DB_Exp** respectively, and the resulting firing strength α_i is used in the implication process to clip the consequent fuzzy set, producing the sets B_i'.

In the Mamdani method, implication is performed via the minimum T-norm operator (*clipping*), which limits the consequent membership function to the rule activation level [9]:

$$\mu_{B_i'}(y) = T_{\min}\left(\alpha_i,\, \mu_{B_i}(y)\right), \quad y \in Y \tag{4}$$

where $\mu_{B_i}(y)$ is obtained by evaluating the triangular membership function of the consequent fuzzy set B_i [8]:

$$\mu_{B_i}(y) = \max\left(\min\left(\frac{y - a_j}{b_j - a_j}, \frac{c_j - y}{c_j - b_j}\right), 0\right) \tag{5}$$

whose parameters (a_j, b_j, c_j) are specified in Table 3, with $a_j \leq b_j \leq c_j$ over the universe of discourse $Y = [1, 100]$.

Aggregation Once all rules have been evaluated, the resulting fuzzy sets $B_1', B_2', \ldots, B_n'$ are combined into a single output fuzzy set B' using the maximum S-norm operator [9,10]:

$$\mu_{B'}(y) = \max_{i=1}^{n}\left(\mu_{B_i'}(y)\right), \quad y \in Y \tag{6}$$

Defuzzification Defuzzification transforms the aggregated fuzzy set B' into a crisp output value y^* via the centroid method [8]:

$$y^* = \frac{\int_Y y \cdot \mu_{B'}(y)\, dy}{\int_Y \mu_{B'}(y)\, dy} \tag{7}$$

In the discrete case, this expression is approximated as:

$$y^* = \frac{\sum_{j=1}^{N} y_j \cdot \mu_{B'}(y_j)}{\sum_{j=1}^{N} \mu_{B'}(y_j)} \tag{8}$$

The centroid method was selected over alternatives such as the mean of maxima because it accounts for the full shape of the aggregated fuzzy set, producing a crisp value that reflects the weighted contribution of all activated rules rather than only the most strongly activated one [8]. The final classification label is then assigned by selecting the output variable whose defuzzified value y^* is highest—a winner-takes-all rule consistent with the HR specialist's original labeling scheme, in which each candidate belongs to exactly one category.

2.5 Evaluation Metrics

Since the system produces a discrete classification into one of three categories—*Discard* (Dis), *Suitable Position 1* (Sui_Pos_1), and *Suitable Position 2* (Sui_Pos_2)—multiclass classification metrics are employed. The reference labels are those assigned by an HR specialist on the same set of résumés.

Confusion Matrix The confusion matrix $\mathbf{C} \in \mathbb{Z}^{3\times3}$ summarizes the per-class behavior of the system, where element C_{ij} represents the number of instances of true class i classified as class j by the system. The per-class metrics described below are derived from this matrix.

Precision, Recall, and F1-Score For each class k, precision P_k, recall R_k, and F1-score $F_{1,k}$ are defined as [11]:

$$P_k = \frac{TP_k}{TP_k + FP_k}, \quad R_k = \frac{TP_k}{TP_k + FN_k}, \quad F_{1,k} = 2 \cdot \frac{P_k \cdot R_k}{P_k + R_k} \tag{9}$$

where TP_k, FP_k, and FN_k denote the true positives, false positives, and false negatives of class k, respectively. Given the potential class imbalance, the *macro F1-score* is reported, defined as the unweighted average of the F1-score across the $K = 3$ classes [11]:

$$F_1^{\text{macro}} = \frac{1}{K} \sum_{k=1}^{K} F_{1,k} \tag{10}$$

3 Experimental Work

The experiment was conducted during the period January–March 2026 using the 60 résumés provided by SmartSoft America BA, corresponding to internship candidates from three consecutive semesters, previously evaluated by HR department staff. The dataset exhibits a natural class imbalance: 6 instances of *Dis*, 42 of *Sui_Pos_1*, and 12 of *Sui_Pos_2*, reflecting the actual profile distribution in the selection process.

The final label assignment was performed by selecting the output variable with the highest crisp value after defuzzification. This label was then compared against the HR specialist's classification to construct the confusion matrix described in Sect. 2.5.

During the initial versions of the experiment, a fourth input linguistic variable (Git_Exp) was considered; however, the dataset contained very few résumés with actual evidence of this variable, which distorted the evaluations. It was therefore excluded from this experiment. Nevertheless, its exclusion demonstrated the configurability of the system, which can be adapted to different scenarios by adjusting variables and rules.

4 Results

The system was evaluated on the provided set of résumés using the HR specialist's classifications as the ground truth reference. Figure 7 presents the resulting confusion matrix, and Table 4 summarizes the per-class metrics derived from it.

Overall, the system achieved a global accuracy of 0.93, indicating that 56 out of 60 candidates were classified in agreement with the specialist's evaluation. The macro F1-score was 0.92 and the weighted F1-score was 0.93, reflecting consistent performance across the three evaluated categories, even in the presence of the class imbalance characterizing the dataset (6 instances of *Dis*, 42 of *Sui_Pos_1*, and 12 of *Sui_Pos_2*).

At the class level, the *Discard* (Dis) category achieved a perfect recall of 1.00, meaning that all candidates discarded by the specialist were also discarded by the system, with no false negatives. Its precision was 0.86 and its F1-score 0.92. The sole error associated with this class originated from a candidate classified by the specialist as *Sui_Pos_1* that the system labeled as *Dis*.

The *Suitable Position 1* (Sui_Pos_1) category achieved both precision and recall of 0.95, with an F1-score of 0.95, representing the class with the highest support and the best overall performance. Of the 42 actual candidates in this category, 40 were correctly classified; one was assigned to *Dis* and another to *Sui_Pos_2*.

The *Suitable Position 2* (Sui_Pos_2) category achieved a precision of 0.91 and a recall of 0.83, with an F1-score of 0.87. The two observed errors correspond to actual candidates of this class that were assigned to *Sui_Pos_1*, suggesting that the system marginally underestimates the technical profile of some candidates with strong evidence in Java and databases. This behavior is consistent with the conservative nature of the fuzzy inference system, which prioritizes caution over overclassification in high-value profiles.

Taken together, the results indicate that the proposed architecture, combining LLMs with a FIS, successfully replicates the HR specialist's criteria with high fidelity, maintaining robust behavior under class imbalance and producing interpretable classifications through the activated fuzzy rules.

Table 4. Per-class evaluation metrics

Class	Precision	Recall	F1-score	Support
Dis	0.86	1.00	0.92	6
Sui_Pos_1	0.95	0.95	0.95	42
Sui_Pos_2	0.91	0.83	0.87	12
Accuracy			0.93	60
Macro avg	0.91	0.93	0.92	60
Weighted avg	0.93	0.93	0.93	60

5 Discussion

The results are consistent with the primary objective of this work: to demonstrate that combining LLMs with a FIS enables the transformation of the qualitative evaluation of professional profiles into a quantitative, reproducible, and traceable process. The macro F1-score of 0.92 and the global accuracy of 0.93 on 60 résumés indicate that the system replicates the HR specialist's criteria with high fidelity, supporting the viability of this architecture as a decision-support tool for the initial filtering of candidates.

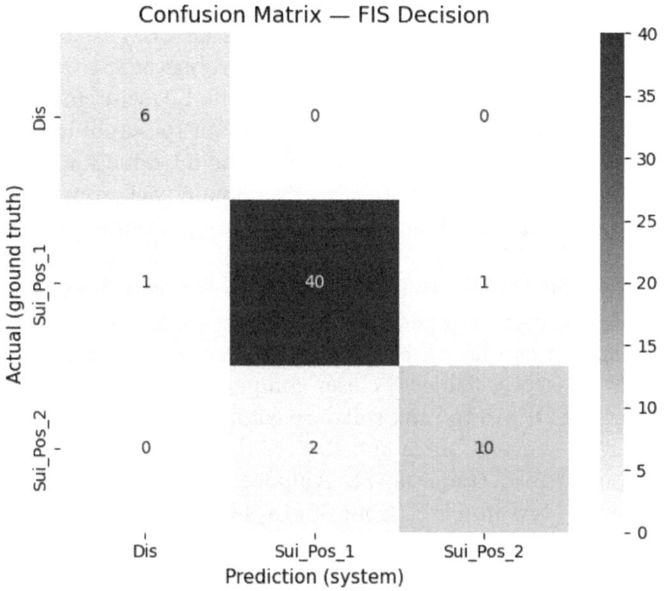

Fig. 7. Confusion matrix

5.1 Error Analysis and Implications for Rule Refinement

The four misclassifications (4/60) are systematic rather than random: all occur at adjacent class boundaries—three at the *Sui_Pos_1*/*Sui_Pos_2* frontier and one at the *Dis*/*Sui_Pos_1* frontier—indicating that the system is stable within well-defined profiles but faces localized ambiguity at class transitions. In at least some cases, the correct output variable reached the same defuzzified value as a competing one; under the winner-takes-all rule, this tie resolved against the ground truth— a discretization artifact rather than a failure of fuzzy inference, addressable by flagging tied outputs for human review.

The two *Sui_Pos_2* candidates misclassified as *Sui_Pos_1* suggest that the rule consequents do not sufficiently reward the co-occurrence of high scores in both `Java_Exp` and `DB_Exp`; widening their *Satisfactory* and *Relevant* sets could correct this. The single *Sui_Pos_1* candidate assigned to *Dis* reflects over-conservative behavior: the *None* fuzzy set may be too broad, activating discard rules for profiles with weak but non-zero evidence— a narrower *None* set or an added *Minimal* intermediate set would address this. Overall, the errors are interpretable and correctable through targeted adjustments, reinforcing the transparency of the Mamdani paradigm as an auditable framework.

5.2 Interpretation of Results in the Context of Prior Work

The system's performance compares favorably against the limitations identified in related work. Unlike [4], which employs fixed semantic extraction rules that are compromised by the format variability of résumés in Latin American contexts, the FExA of the proposed system uses an LLM to structure content independently of its format, conferring greater robustness to documents with photographs, multiple columns, or irregular layouts. Likewise, unlike [5], the proposed system does not incorporate information about the candidate's educational institution during evaluation, thereby avoiding the introduction of demographic biases associated with the candidate's socioeconomic background.

Regarding works that combine LLMs with fuzzy reasoning, the proposed system completes the full FIS cycle—fuzzification, rule evaluation, aggregation, and defuzzification—unlike [12], which uses LLMs solely as a fuzzy judge without executing the full inference cycle. This endows the system with greater explainability, as the output can be traced directly to the activated rules and the membership degrees of each variable. A closer comparison with [6]—which combines LLMs with Fuzzy-TOPSIS to rank software engineering profiles, reporting accuracy of up to 91%—reveals three substantive differences. First, the figures are not directly comparable: Hoque et al. evaluate on a dataset synthetically balanced via random oversampling (from 39/18/43% to 33.3/33.3/33.3%), a procedure that can inflate estimates [11], whereas the macro F1 of 0.92 reported here is computed on the original 60 résumés with their natural class imbalance (6/42/12)—a more demanding condition. Second, Fuzzy-TOPSIS is a multi-criteria decision-making technique based on closeness coefficients; it does not involve IF-THEN rules, fuzzy implication, or centroid defuzzification, and therefore lacks the per-decision auditability that the Mamdani paradigm [9] provides. Third, DistilRoBERTa in [6] is fine-tuned supervisedly per domain, requiring labeled data and retraining for each new position; GPT-5.2 in the proposed system operates zero-shot over explicit rubrics, so domain adaptation requires only redefining variables and rules—as demonstrated by excluding `Git_Exp` without modifying the architecture.

5.3 Scope and Limitations

Despite the results obtained, the system presents relevant limitations. First, the evaluation was conducted on a dataset of 60 résumés from a single domain—internship positions at an IT company. Rather than an oversight, this reflects a deliberate design choice: the dataset was real, confidential, and annotated by a certified HR specialist, providing high-quality ground-truth labels that are rarely available in this research niche. The study is therefore framed as a proof-of-concept with high-quality annotations, not as a large-scale benchmark. That said, the current sample size limits the generalizability of the results to other sectors or experience levels. To address this, future validation should cover at least three distinct domains (e.g., IT, engineering, and administrative roles), with a corpus of $N \geq 300$ résumés evaluated by more than one specialist to measure inter-annotator agreement. Reaching this scale requires collaboration with multiple organizations: since staff turnover is a problem companies actively seek to minimize, the annual résumé intake of a typical SME for any given role is too small to meet this threshold alone—particularly when the goal is to validate the system against the most current hiring requirements. We hypothesize that per-class performance will remain at $F_1 \geq 0.85$ as the dataset scales, provided the fuzzy rules and membership functions are reconfigured for each target domain—a testable claim that motivates the next research phase. Second, the fuzzy rules and fuzzy sets were defined collaboratively with the company, implying that the system reflects the particular criteria of a single HR specialist and may not be directly transferable to other contexts without a reconfiguration process. Third, the system currently depends on GPT-5.2—or newer OpenAI models with file-input support—for feature extraction, and on standard OpenAI language models for the scoring step prior to fuzzification. Consequently, it does not natively support alternative LLM providers; however, adapting it to other backends would require only minor modifications, provided that output consistency across executions and technology stacks is verified—an aspect that falls outside the scope of this work.

6 Conclusion

The results support the use of this architecture as a tool for the initial screening of candidates, not as a substitute for human judgment. Its modular and configurable design allows the linguistic variables, fuzzy sets, and rules to be adapted to different domains, as evidenced during the experiment when the Git_Exp variable was discarded without affecting the system's operation. As future work, four directions are identified. First, in addition to evaluating the system on larger and more diverse datasets, it is proposed to develop a FIS configurator agent that empowers HR teams to define their own linguistic variables, membership functions, and fuzzy rules without requiring technical expertise, adapting the system to multiple positions and organizational contexts. Second, the current tight coupling to OpenAI models should be replaced by a provider-agnostic connector layer supporting both cloud-based APIs (e.g., OpenAI, Anthropic, Google) and locally hosted open-source models (e.g., via Ollama or LM Studio), enabling deployment in environments with data-privacy constraints. Third, a configuration interface is envisioned in which practitioners can select their preferred LLM stack from a list of available providers and supply the corresponding API credentials, making the system accessible and adaptable across different institutional and infrastructure settings. Fourth, the current evaluation did not include a controlled comparison against isolated baselines— such as an LLM-only scoring pipeline or a classical FIS with manually fuzzified inputs— which limits the ability to isolate the specific contribution of the hybrid architecture. Establishing such baselines is a methodological priority for future work, as it would clarify whether the observed performance gains stem from the LLM component, the fuzzy inference layer, or their interaction.

References

1. Vaishampayan, S., et al.: Human and LLM-based resume matching: an observational study. In: Chiruzzo,, L., Ritter, A., Wang, L. (eds.) Findings of the Association for Computational Linguistics: NAACL 2025, pp. 4808–4823. Association for Computational Linguistics, Albuquerque, New Mexico (2025). ISBN: 979-8-89176-195-7. https://doi.org/10.18653/v1/2025.findings-naacl.270
2. Petzel, Z.W., Sowerby, L.: Prejudiced interactions with large language models (LLMs) reduce trustworthiness and behavioral intentions among members of stigmatized groups. Comput. Hum. Behav. **165**, 108563 (2025). ISSN: 0747-5632. https://doi.org/10.1016/j.chb.2025.108563. https://www.sciencedirect.com/science/article/pii/S074756322500010X
3. Armstrong, L., Liu, A., MacNeil, S., Metaxa, D.: The silicon ceiling: auditing GPT's race and gender biases in hiring. In: Proceedings of the 4th ACM Conference on Equity and Access in Algorithms, Mechanisms, and Optimization, ser. EAAMO '24, pp. 1–18. ACM (2024). https://doi.org/10.1145/3689904.3694699
4. Gan, C., Zhang, Q., Mori, T.: Application of LLM agents in recruitment: a novel framework for resume screening (2024). arxiv:2401.08315
5. Lo, F.P.W., et al.: AI hiring with LLMs: a context-aware and explainable multi-agent framework for resume screening (2025). arxiv:2504.02870
6. Hoque, S., Karim, A.A.J., Golam Rabiul Alam, M., Gope, N.: When LLM meets fuzzytopsis for personnel selection through automated profile analysis. IEEE Access **14**, 15778–15794 (2026). ISSN: 2169-3536. https://doi.org/10.1109/access.2026.3658575
7. de Ciencia, S.: Segunda sesión del ciclo de conferencias de inteligencia artificial (IA) (2026). https://www.youtube.com/watch?v=QBz671bLAAk
8. Ross, T.J.: Fuzzy logic with engineering applications (2010). https://doi.org/10.1002/9781119994374
9. Mamdani, E., Assilian, S.: An experiment in linguistic synthesis with a fuzzy logic controller. Int. J. Man-Mach. Stud. **7**(1), 1–13 (1975). ISSN: 0020-7373. https://doi.org/10.1016/S0020-73737580002-2
10. Zadeh, L.: Fuzzy sets. Inf. Control **8**(3), 338–353 (1965). ISSN: 0019-9958. https://doi.org/10.1016/S0019-9958(65)90241-X
11. Powers, D.M.W.: Evaluation: from precision, recall and f-measure to roc, informedness, markedness and correlation (2020). arxiv:2010.16061
12. Zheng, W., Turner, L., Kropczynski, J., Ozer, M., Nguyen, T., Halse, S.: LLM-as-a-Fuzzy-Judge: fine-tuning large language models as a clinical evaluation judge with fuzzy logic (2025). arXiv:2506.11221

SupraHealth-Mini: A Graph-Based and Generative AI Framework for Medical Counseling

Augusto Javier Reyes-Delgado[1], Isaac Machorro-Cano[2], Giner Alor-Hernández[1], Oscar Baez-Senties[1], José Oscar Olmedo-Aguirre[3], and José Luis Sánchez-Cervantes[1](✉)

[1] Tecnológico Nacional de México, I.T. Orizaba, Orizaba, Mexico
{D17010207,giner.ah,oscar.bs,jose.sc}@orizaba.tecnm.mx
[2] Universidad del Papaloapan, Campus Tuxtepec, San Juan Bautista Tuxtepec, Mexico
imachorro@unpa.edu.mx
[3] Instituto Politécnico Nacional, Escuela Superior de Cómputo, Mexico City, Mexico
jolmedoa@ipn.mx

Abstract. Chronic non-communicable diseases (NCDs) constitute one of the major public health challenges worldwide. These diseases exhibit high comorbidity and shared risk factors, contributing to progressive and systemic deterioration that affects multiple organs simultaneously. Despite advances in artificial intelligence predictive models, most existing approaches focus on isolated conditions, while leveraging the interconnected nature of disease progression remains an emerging area of opportunity, not only for predictive but also for preventive tasks. Therefore, this work proposes SupraHealth-Mini, a minimal proof-of-concept system for projecting NCD progression by integrating a graph-based mechanism with generative artificial intelligence, with the aim of supporting medical counseling tasks. The approach models diseases as nodes in a graph, where relationships derived from medical literature, harmonized datasets, and expert knowledge are used to construct a mechanism for simulating plausible temporal disease progression across various NCDs. The system also incorporates generative models, specifically StyleGAN2, to produce personalized medical visualizations conditioned on the patient's specific risk factors. Furthermore, the implementation includes an interactive web interface designed to provide clinical guidance and raise patient awareness. Preliminary validation with medical experts suggests that the generated projections are consistent, clinically plausible, and useful for communicating long-term risk to patients.

Keywords: Artificial intelligence · Graph modeling · Comorbidity networks · Digital health · Disease progression · Generative adversarial networks · Medical visualization

H. Ponce et al. (Eds.): COMIA 2026, CCIS 3095, pp. 199–213, 2027.
https://doi.org/10.1007/978-3-032-37438-7_16

1 Introduction

The non-communicable diseases (NCDs) represent, due to their high prevalence, progressive nature, and strong association with multiple comorbid conditions, one of the main public health challenges worldwide. Diseases such as cardiovascular disorders, diabetes, chronic pulmonary diseases, kidney diseases and hepatic pathologies do not occur in isolation; rather, they interact within a complex pathophysiological network in which the progression of one condition negatively impacts others [1].

As an example of comorbidity among NCDs, the presence of hypertension may lead, in the long term, to kidney damage, accounting for up to 30% of renal diseases and ranking as the second leading cause after diabetes [2]. In turn, diabetes mellitus affects multiple organs, including the retina, leading to complications such as diabetic retinopathy, where prolonged disease duration becomes the main risk factor [3]. Although these relationships are widely documented in the medical literature [4], they are not always evident from a non-expert population perspective. That is, it is not common knowledge among non-medical individuals how these diseases, particularly their lack of treatment, pose significant risks to future health [5, 6].

In this context, both public health reports [7] and interactions with medical experts highlight a key principle: prevention is the most effective strategy. However, these same experts point out that clearly and personally communicating systemic risk to patients remains a significant challenge.

The current computational approaches tend to focus on the binary or multiclass prediction of individual diseases and lack mechanisms that integrate comorbidity relationships, and visual representation within a unified system [8, 9].

Considering these challenges, this work presents SupraHealth-Mini, a minimal proof-of-concept system designed to support medical counseling through the simulation and visualization of systemic deterioration scenarios associated with non-communicable diseases. The proposed system integrates a graph-based progression mechanism for representing comorbidity relationships with generative artificial intelligence (Gen AI) for personalized medical visualization, as a minimal proof of concept (Mini).

The objective of the presented system is to provide a support tool for medical counseling that facilitates the understanding of risk and promotes preventive strategies in the general population. Therefore, the main contributions of this work are:

- A graph-based mechanism for representing systemic deterioration.
- The implementation of generative artificial intelligence (Gen AI) for personalized medical visualization.
- A modular architecture that integrates its components with an interactive interface to generate representations of possible health scenarios for the patient.

The following sections present the progress achieved in this research. Section 2 summarizes the related work. Section 3 describes the methodology and implementation. Section 4 presents the preliminary results of the use case. Section 5 discusses the findings, and finally, Sect. 6 presents the conclusions.

2 Related Work

Knowledge graphs and graph neural networks have been widely used to support predictive tasks [10] and to model comorbidity structures [11], enabling the identification of associations between diseases and risk patterns [12]. In parallel, generative artificial intelligence has demonstrated effectiveness in the synthesis of medical images, facilitating applications such as data generation and visualization of disease progression [13]. The purpose of this section is to compare, select, and adapt findings from previously dispersed research.

The work of Dong et al. [14] proposes MorbidGCN, a graph convolutional network (GCN) based model for multimorbidity prediction that integrates population phenotypes and disease networks. Diseases are represented as nodes and comorbidity relationships as edges, formulating the problem as a link prediction task within the graph. The phenotypic information is modeled as node features, thereby improving predictive performance and enabling the identification of disease associations. In contrast, Naeimaei Aali et al. [15] introduce the use of event graphs to model clinical trajectories of patients with multimorbidity based on multi-entity event data. This approach focuses on retrospective analysis rather than prediction or simulation; clinical entities such as diagnoses, admissions, laboratory measurements, and treatments are represented as interconnected nodes within the graph, capturing the sequence and relationships between events. The authors conclude that the approach enables the analysis of more complex care pathways compared to traditional process mining techniques. A more recent approach is presented by Safaei et al. [16], who propose a framework based on graph neural networks (GNNs), in which diseases are represented as nodes and their relationships as edges derived from clinical, genomic, and imaging features. The model implements different GNN architectures over a disease–feature graph and reports improvements in metrics such as accuracy, F1-score, and the receiver operating characteristic area under the curve (ROC-AUC). Although the validation is performed on curated datasets and does not include external clinical evaluation. More recently, Banzato et al. [17] propose the use of undirected graphical models to analyze multimorbidity, representing diseases as nodes and their associations as conditional independence relationships. Using population-level data, the study demonstrates that many marginal associations disappear when considering the combined effects of other variables, revealing spurious relationships. It also identifies core diseases within the network using centrality metrics.

Regarding the second axis of this research, the work of Marey et al. [18] explores the use of generative artificial intelligence to enhance patient education in cardiovascular imaging through the generation of personalized multimodal content, including text, images, video, and conversational agents. Unlike graph-based or predictive approaches, this work focuses on patient communication and understanding, using generative models such as generative adversarial networks (GANs), diffusion models, and large language models to transform complex medical information into accessible representations. Gupta et al. [19] evaluate the use of generative models such as ChatGPT-4© and Gemini© to improve the readability of radiology educational materials, demonstrating a significant reduction in reading difficulty. Additionally, radiologist evaluations confirmed that the generated texts maintain, in most cases, clarity, relevance, and a high proportion of the original content. However, the study shows that simplification may lead to information loss and requires expert supervision, remaining limited to communication tasks without modeling underlying processes. A different perspective is presented by Bhattad et al. [20], who explores the use of generative models such as GANs and VAEs to simulate rare disease scenarios for training robotic systems in artificial clinical environments. This allows systems to be exposed to multiple clinical trajectories and improves their diagnostic and decision-making capabilities. However, the model focuses on simulation for training rather than direct patient support. A notable work is that of Li et al. [21], which introduces MedDiT, a framework that integrates knowledge graphs, language models, and diffusion models to generate dynamic simulated patients. Through this method, the system aligns patient attributes and symptoms with image generation, improving coherence and reducing hallucinations. This multimodal integration enables the simulation of more realistic clinical scenarios in educational environments.

Graph-based approaches for medical applications have shown significant growth in recent years, particularly in representing relationships among diseases, patients, and clinical variables in a more interpretable manner than other machine learning methods. In parallel, generative models, especially those based on GANs, have demonstrated a remarkable capacity to generate high-fidelity medical images. Nevertheless, despite these advances, an opportunity gap remains in the state of the art: graph-based approaches focus on predictive tasks and structural modeling of relationships for comorbidities, while generative models visually exemplify predictions. Although emerging proposals address educational support, they still lack explicit mechanisms for temporal progression and structured clinical inference.

As a result, there are currently no systems that jointly integrate relational modeling, simulation of systemic deterioration, and visual representation. In this context, the present work proposes a hybrid approach to address this gap by integrating these capabilities within a unified conceptual framework.

3 Methodology and Implementation

The system conceptualizes systemic deterioration as a network of pathophysiological interactions, where the medical domains and diseases are represented as nodes within a weighted directed graph. The edges represent comorbidity relationships, understood as probabilistic or causal influences between clinical conditions.

Formally, the system can be represented as the graph defined in Eq. (1):

$$G = (V, E, W) \tag{1}$$

Where:

- V is the set of nodes (medical domains and diseases),
- E is the set of edges (comorbidity relationships), and
- W is the set of weights associated with these relationships.

For the proof of concept, five main domains were defined as nodes: (i) cardiovascular system, (ii) chronic pulmonary system, (iii) metabolic system (diabetes mellitus), (iv) hepatic system, and (v) renal system, since they are among the main systems affected by NCDs.

The relationships between nodes are not directed only. Also assigned weights reflect the relevance of the interaction. Such value is derived from evidence collected in literature and expert knowledge. This approach enables the capture of the non-linear nature of systemic deterioration, where multiple conditions may interact simultaneously.

3.1 Graph Construction Based on Evidence

The graph construction was carried out through a hybrid process that combines:

- Medical literature review. The reported comorbidity relationships among chronic diseases were identified in the literature.
- Analysis of available datasets. The correlations between clinical variables across different datasets were extracted computationally.
- Validation with clinical experts. refinement of relationships and weights through domain-specific knowledge.

Since none of the identified datasets encompasses all the conditions and variables considered, the graph is not directly derived from unified data. Instead, it is constructed by integrating multiple knowledge sources. This limitation led to the development of an evidence-informed modeling strategy. Therefore, the resulting graph structure is defined as illustrated in Fig. 1

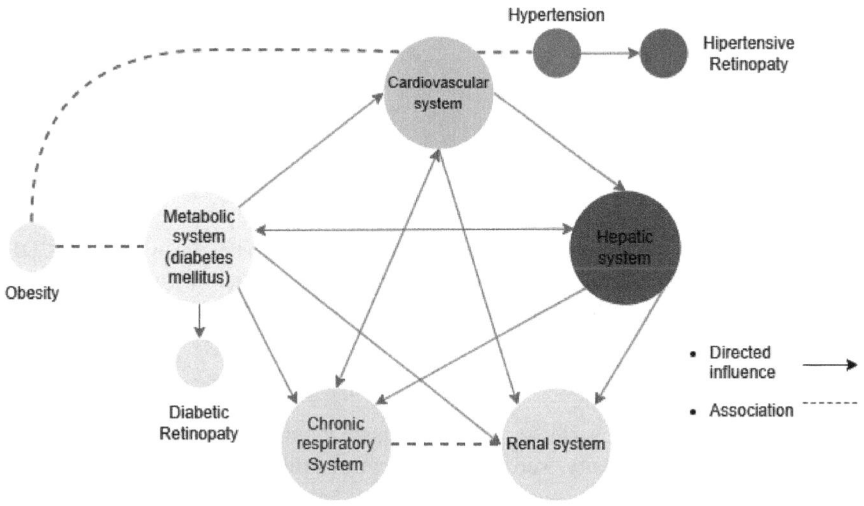

Fig. 1. Conceptual representation of the comorbidity graph.

As previously mentioned, a review of the medical literature related to NCDs was conducted, from which the main affected systems were identified: cardiovascular, respiratory, metabolic, hepatic, and renal systems, as well as the comorbidity relationships between them. These include diabetes and cardiovascular diseases [22], cardiovascular diseases with respiratory and renal diseases [23, 24], as well as hepatic diseases [25], which are again intertwined with diabetes [26], among others.

With this knowledge, the relationships were modeled using a graph where each system is represented as a major node, and the diseases derived from them as a series of sub-nodes. In such graph, diabetes is selected as the central node due to its consistent presence across multiple datasets.

Besides, the graph nodes represent different types of relationships:

- Direct influence (*diabetes* → *diabetic retinopathy*).
- Associations (*obesity* - *diabetes*).
- Mediated relationships (*obesity* - *hypertension* → *hypertensive nephropathy*).
- Additionally, several shared factors are not explicitly represented in Fig. 1, including age, sex, and lifestyle, among others.

Data Selection Due to the absence of a unified dataset, multiple disease-specific datasets were utilized. For the selection process, repositories containing verified clinical data were explored, such as the UCI Machine Learning Repository. Publications describing the use of datasets related to the domains within the scope of this work were also explored, and through this process, a series of tabular datasets for each domain were obtained (see Table 1).

Table 1. Datasets analyzed.

Disease	Dataset	DM	AGE	SEX	BMI	BP	GLU	COL	HTN	HR
DM	[27, 28]	✓	✓	✓	✓		✓			
CVD	[29]	✓	✓	✓	✓	✓	✓	✓	✓	✓
LD	[30]	✓	✓	✓				✓	✓	
CPD	[31]		✓	✓	✓	✓				✓
CKD	[32]		✓			✓	✓		✓	✓

BMI = Body mass index; **BP** = Blood pressure; **CKD** = Chronic kidney disease; **COL** = Total cholesterol; **CPD** = Chronic pulmonary disease; **CVD** = Cardiovascular disease; **DM** = Diabetes mellitus; **GLU** = Blood glucose level; **HR** = Heart rate; **HTN** = Hypertension; **LD** = Liver disease.

Data Harmonization Nevertheless, those datasets required several transformations to address the data heterogeneity, a cross-dataset feature alignment process was performed to identify common variables across sources. Then, normalization and standardization techniques were applied to ensure data consistency across the datasets, these include adjusting the units of measurement of the variables.

Predictive Models per Dataset Independent machine learning models were trained for each disease, including Naive Bayes and XGBoost, among others, based on approaches reported by the Kaggle community [33]. However, the primary purpose of training these models is not prediction itself, but the identification of the most relevant variables. Variables with the highest impact were selected, both for the specific disease and for their relationships with other conditions.

```
from sklearn.naive_bayes import GaussianNB
#Data importing step...
heart_diases = pd.read_csv("datasets/heart-dis-
ease/heart_disease.csv")
#Training step...
Naive_model.predict_proba(X_test)
y_pred = Naive_model.predict(X_test)
```

Weighting and Activation Functions for Main Nodes The selected variables and their associated weights were organized into a tensor-like structure. Based on this structure, activation functions were defined in Python to determine the activation of each node according to patient-specific data. Each activation function processes the data using a logistic equation. Among them, to determine possible cardiac conditions, the values recovered from the model trained with the "heart disease dataset" [29] were used, where variables such as high blood pressure and high cholesterol were found to have considerable coefficients in detection, 0.7561 and 0.5798 respectively.

Temporal Simulation and Propagation The system incorporates a temporal simulation mechanism based on the patient's age, enabling the projection of future scenarios. The activation of a node can propagate to other nodes according to comorbidity relationships. In this operation, a cycle of 10 iterations is created, one for each year of progression, where the functions first operate independently, and upon activation, begin to influence the risk of the others. In related works, when longitudinal data is available, propagation is grounded in empirical evidence. Works such as Zedadra et al. [34] support modeling temporal relationships between diabetes and diabetic retinopathy. In other cases, propagation is defined through literature evidence and expert validation.

Sub-nodes Activation Once a major node is activated, it calls the functions associated with its domain, considered as its sub-nodes. These require more specific treatment. In the case of the metabolic domain, the sub-nodes of diabetic retinopathy and macular degeneration are included. The first is activated when the patient has a history of diabetes, in order to calculate the current probability of the condition in the following years of the iteration. Meanwhile, macular degeneration has the condition that the minimum age must be 55 years, increasing the probability of the condition progressively.

3.2 Medical Image Generation Module

As part of the image generation module of this proof of concept, a StyleGAN2-ADA [35] model was integrated to obtain synthetic retinal images associated with disease progression. For this initial process, fundus images depicting age-related macular degeneration and diabetic retinopathy were selected. This decision was based on the following selection criteria: (i) the availability of datasets, (ii) the high correlation with non-communicable diseases (NCDs), and (iii) an existing collaboration with the Fundación Doctor Hernández Zurita, I.B.P.,[1] which supports the validation of the ophthalmology module. The procedure was as follows:

[1] https://fundacionzurita.com.mx/

1. A dataset of 2,903 fundus images was collected.
2. A base StyleGAN model was trained on the dataset to learn the structural representation of the ocular fundus.
3. The base model was then fine-tuned into a multiclass model (normal, early, intermediate, late) for each retinopathy. Enabling by this method the simulation of disease progression through generated images.
4. A quantitative evaluation was conducted to measure the performance of the models through the calculation of the Fréchet Inception Distance (FID).

The trained models are intended to be used in accordance with the projections generated by the graph-based mechanism.

3.3 System Architecture

The system is designed following a modular architecture to facilitate scalability and integration of heterogeneous components based on the integration strategy for artificial intelligence techniques proposed in a previous review work [36]. Such system consists of the following components:

- A data persistence layer for storing patient data and generated medical images.
- A backend module responsible for implementing the graph-based mechanism and connection with the generative models.
- A layer with an interactive web-based frontend that presents results through 3D models and generated images.

The system workflow operates as follows: first, an individual's physiological data is entered through the frontend by the end user. Since these variables are represented on different scales, they are normalized to values between 0 and 1, which correspond to the expected input range of the inference module in the backend responsible for generating the projection. This module contains a matrix defining the activation functions of each major node. When a node reaches an activation threshold of 0.5 probability, it calls its corresponding sub-nodes, which provide more specific prediction results for the associated conditions.

Then, the generated results are then returned to the frontend, where, in situations identified as clinically relevant, they are presented to the patient through synthetic images produced by the image generation module or through transformations applied to 3D organ representations.

A simplified overview of the architecture is presented in Fig. 2.

Data Availability Statement The datasets used for training the generative models, including curated data and the application code, are available from the first author upon reasonable request and subject to valid academic justification.

4 Preliminary Results of the Use Case

A functional prototype was developed, oriented toward use by healthcare professionals. The application allows patient registration, input of biometric data, and exploration of risk scenarios. The client-side application includes:

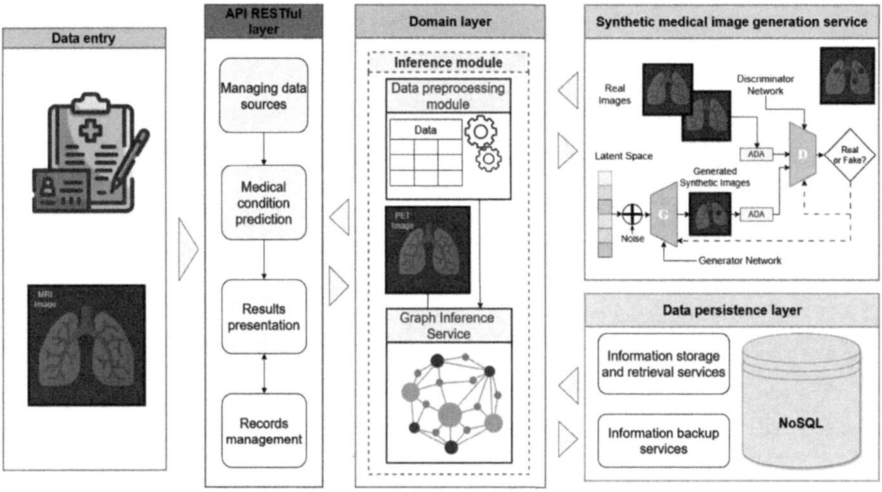

Fig. 2. System's architecture.

- Patient registration and management.
- Visualization of anatomical regions of interest with visual alerts (when applicable).
- Dynamic data input forms.

Upon patient registration, the system processes individual characteristics and activates the corresponding nodes within the graph-based model. By modifying variables such as age, the system simulates the temporal progression of systemic deterioration through the underlying propagation mechanism.

The visualization layer is based on a layered 2D anatomical scheme inspired by positron emission tomography (PET) imaging. Also, selected 3D models are used to represent key organs, to achieve a balance between visual fidelity and computational efficiency. This design decision supports lightweight rendering and improves cross-platform compatibility across both desktop and mobile devices, enhancing the application's accessibility. The client-side application interface is shown in Fig. 3.

Regions of interest are dynamically highlighted, allowing the user to explore specific details. In the cardiology module, the heart model[2] transitions from a regular rhythm animation to an irregular one to represent arrhythmia. In the hepatology module, the liver changes to a diseased model when the patient profile indicates such risk. In the ophthalmology module, the synthetic images generated are alongside an eye model[3] to depict approximations of potential deterioration. This latter module also achieved an FID of 23.86 in a quantitative evaluation. However, the concept must be qualitatively evaluated by the users for whom the full version of the proposal is intended.

[2] "Human Heart Anatomy | Realistic" by adimed, available at https://skfb.ly/oPDWN, licensed under Creative Commons Attribution 4.0 (http://creativecommons.org/licenses/by/4.0/).

[3] "Eyeball" by tori1243, available at https://skfb.ly/6U9GV, licensed under Creative Commons Attribution 4.0 (http://creativecommons.org/licenses/by/4.0/).

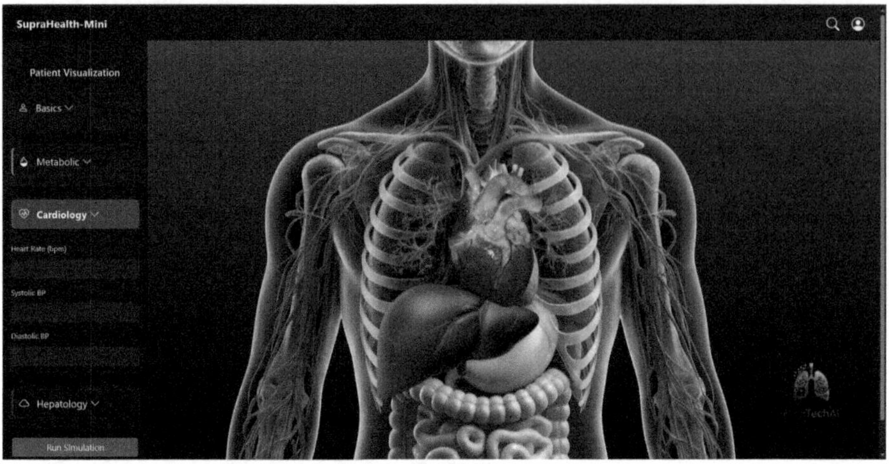

Fig. 3. Cardiology module visualization.

4.1 Validation of the Proof of Concept

The proposed system has been evaluated as proof of concept with the support of experts in retinology and, secondarily, specialists in diabetology.

The retinology experts provided insights into the projected retinopathy progression in synthetic images at 3-, 5-, and 10-year horizons (see Fig. 4), which they assessed as faithful representations of real-world conditions. Additionally, they contributed parameter values that refined how disease progression is quantified longitudinally, including factors such as smoking history, that literature identifies as an aggravating factor for diseases such as macular degeneration; however, the specialists provided a concrete value, indicating that this risk factor increases the likelihood of developing the condition by a factor of four. Also, they validated the statistical relationships identified in public datasets, such as the observation that female sex is associated with a 20% increase in the risk of macular degeneration.

The diabetology expert emphasized the distinction in disease progression between patients with controlled versus uncontrolled diabetes, particularly in long-term complications. These insights were considered into the system's propagation mechanisms.

Due to the small sample of participants, a quantitative evaluation mechanism was not established; however, preliminary results suggest that the projections generated are clinically coherent and align with expert expectations. Furthermore, the participating healthcare professionals identified the system as a valuable patient-awareness tool for preventive strategies, particularly for communicating long-term risks.

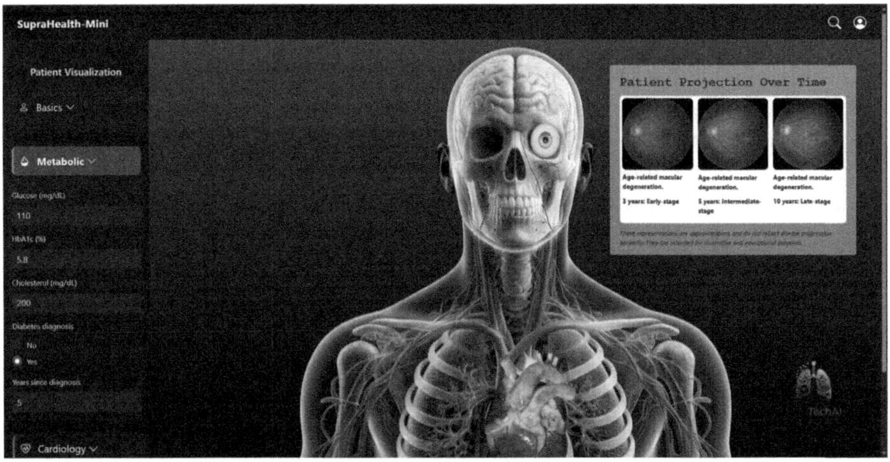

Fig. 4. Visualization of progression in the ophthalmology submodule. The bottom texts say that generated images are statistical approximations only.

5 Discussion

Works such as Naeimaei Aali et al. [15] and Banzato et al. [17] leverage comorbidity relationships and the capabilities of graph structures for prediction tasks and the discovery of relationships among various medical conditions. In parallel, Gupta et al. [19] exploits the capabilities of generative artificial intelligence (Gen AI) for patient education, while Li et al. [21] apply similar techniques for the training of medical personnel through the simulation of patient cases.

The proposed approach takes inspiration from these works by leveraging graph-based structures and the application of Gen AI to present information. However, it differs from traditional graph neural network-based models by employing independent models whose outputs are integrated through activation and propagation functions. Through this mechanism, the activation of one condition may influence the risk estimation of related conditions, enabling the representation of plausible systemic deterioration scenarios.

Since unified longitudinal datasets encompassing all the considered conditions are not currently available, propagation is defined through evidence-informed relationships derived from medical literature and expert knowledge. Although this represents a limitation compared to fully data-driven temporal approaches, it enables the construction of clinically plausible progression pathways while maintaining interpretability and modularity.

Through this approach, the system achieves modularity and eases the integration of an interpretability layer. Furthermore, it enables the definition of plausible scenarios for image generation, illustrating the progression of potential diseases in individual patients.

6 Conclusion and Future Work

This study presented a proof of concept of a system for projecting systemic deterioration based on graph structures and generative artificial intelligence, aimed at supporting patient awareness in preventive healthcare scenarios. The proposed approach integrates evidence-informed relational modeling, propagation mechanisms, and Gen AI-driven visualization within a modular architecture capable of generating plausible disease progression scenarios from dispersed clinical data sources.

The preliminary evaluations support the feasibility of integrating these components into a unified platform. In particular, the expert feedback on the ophthalmology module suggested that the projections generated were clinically coherent and useful for communicating long-term risks. Furthermore, the graph-based mechanism enabled the representation of comorbidity relationships across the selected medical domains through activation and propagation functions. Overall, the preliminary evaluations conducted with medical experts indicated that the proposed system may serve as a support tool for preventive counseling.

Although the number of medical experts participating was limited and no large-scale quantitative usability evaluation was possible, the obtained results support the viability of combining relational disease modeling with generative visualization techniques for preventive healthcare applications. Furthermore, the graph-based structure introduces an interpretable representation of disease propagation, facilitating the future incorporation of explainable AI (XAI) mechanisms.

Several challenges remain to be addressed. The first is the availability of longitudinal data, which are often restricted and fragmented across medical domains. To address this limitation, ongoing work includes the development of a longitudinal study focused on NCD covariates, in collaboration with foundations and clinical institutions. To the best of the authors' knowledge, this would represent one of the first initiatives of its kind in Mexico. The second challenge relates to the granularity of medical domains. The system is conceived as a high-level hub that will later connect with domain-specific expert systems in a distributed manner, building upon previous and parallel research efforts. Future work will focus on:

- Developing a multi-institutional longitudinal study in Mexico.
- Expanding the number of medical domains and modeled diseases.
- Integrating multimodal clinical and physiological data sources.
- Refining propagation mechanisms for improved temporal modeling.
- Incorporating explainable AI (XAI) strategies into the graph structure.
- Conducting quantitative usability and acceptance evaluations with a larger group of medical experts.

Overall, this work establishes a foundation for the development of integrated, interpretable, and patient-centered AI systems to improve preventive healthcare strategies.

Acknowledgments. This paper was sponsored by Mexico's Secretariat of Science, Humanities, Technology, and Innovation (SECIHTI) and Mexico's Secretariat of Public Education (SEP) through the PRODEP program. Authors also thank Tecnológico Nacional de México (TecNM) for supporting this work and the Doctor Hernández Zurita I.B.P. Foundation for the medical contributions in validating the results.

Disclosure of Interests The authors have no competing interests to declare that are relevant to the content of this article.

References

1. Piovani, D., Nikolopoulos, G.K., Bonovas, S.: Non-communicable diseases: the invisible epidemic. J. Clin. Med. **11** (2022). https://doi.org/10.3390/jcm11195939
2. Kummer, A., Dalal, M., Weber, M., Chew, E.Y.: Chapter 28 - ophthalmic issues in chronic kidney disease. In: Kimmel, P.L., Rosenberg, M.E. (eds.) Chronic Renal Disease, 2nd edn, pp. 425–439. Academic (2020)
3. Tan, N.H., Tuan, T.M., Chuan, P.M., et al.: FKG-MM: a multi-modal fuzzy knowledge graph with data integration in healthcare. PLoS One. **21**, e0339864 (2026). https://doi.org/10.1371/journal.pone.0339864
4. Cramer, A.O.J., Waldorp, L.J., van der Maas, H.L.J., Borsboom, D.: Comorbidity: a network perspective. Behav. Brain Sci. **33**, 137–150 (2010). https://doi.org/10.1017/S0140525X09991567
5. Aceves, B., Ingram, M., Nieto, C., et al.: Non-communicable disease prevention in Mexico: policies, programs and regulations. Health Promot. Int. **35**, 409–421 (2020). https://doi.org/10.1093/heapro/daz029
6. Gassner, L., Zechmeister-Koss, I., Reinsperger, I.: National strategies for preventing and managing non-communicable diseases in selected countries. Front. Public Health. **10** (2022). https://doi.org/10.3389/fpubh.2022.838051
7. Secretaría de Salud: Enfermedades No Transmisibles Situación y Propuestas de Acción: Una Perspectiva desde la Experiencia de México (2018)
8. Ronzano, F., Gutiérrez-Sacristán, A., Furlong, L.I.: Comorbidity4j: a tool for interactive analysis of disease comorbidities over large patient datasets. Bioinformatics. **35**, 3530–3532 (2019). https://doi.org/10.1093/bioinformatics/btz061
9. Kim, M., Kim, Y.N., Jang, M., et al.: Synthesizing realistic high-resolution retina image by style-based generative adversarial network and its utilization. Sci. Rep. **12** (2022). https://doi.org/10.1038/s41598-022-20698-3
10. Woodman, R.J., Koczwara, B., Mangoni, A.A.: Applying precision medicine principles to the management of multimorbidity: the utility of comorbidity networks, graph machine learning, and knowledge graphs. Front. Med. (Lausanne). **10**, 1302844 (2023)
11. Martins, C., Neves, B., Teixeira, A.S., et al.: Identifying subgroups in heart failure patients with multimorbidity by clustering and network analysis. BMC Med. Inform. Decis. Mak. **24** (2024). https://doi.org/10.1186/s12911-024-02497-0
12. Binyam, T., Kauppinen, T., Keßler, C., Fritz, F.: Design and development of a linked open data-based health information representation and visualization system: potentials and preliminary evaluation. JMIR Med. Inform. **2**, e31 (2014). https://doi.org/10.2196/medinform.3531
13. Waisberg, E., Ong, J., Kamran, S.A., et al.: Generative artificial intelligence in ophthalmology. Surv. Ophthalmol. **70**, 1–11 (2025). https://doi.org/10.1016/j.survophthal.2024.04.009
14. Dong, G., Zhang, Z.-C., Feng, J., Zhao, X.-M.: MorbidGCN: prediction of multimorbidity with a graph convolutional network based on integration of population phenotypes and disease network. Brief. Bioinform. **23**, bbac255 (2022). https://doi.org/10.1093/bib/bbac255
15. Milad, N.A., Mannhardt, F., Toussaint, P.J.: Discovering care pathways for multi-morbid patients using event graphs. In: Jorge, M.-G., Lu, X. (eds.) Process Mining Workshops, pp. 352–364. Springer International Publishing, Cham (2022)

16. Safaei, M., Pablo, J., Gonzaliam, M.: Graph neural networks for modeling disease relationships: a framework for multi-disease diagnostics and comorbidity prediction (2024). https://doi.org/10.13140/RG.2.2.28337.08807
17. Banzato, E., Roverato, A., Buja, A., Boccuzzo, G.: Understanding multimorbidity: insights with graphical models. BMC Med. Res. Methodol. **25**, 84 (2025). https://doi.org/10.1186/s12874-025-02536-y
18. Marey, A., Saad, A.M., Killeen, B.D., et al.: Generative artificial intelligence: enhancing patient education in cardiovascular imaging. BJR Open. **6**, tzae018 (2023). https://doi.org/10.1093/bjro/tzae018
19. Gupta, M., Gupta, P., Ho, C., et al.: Can generative AI improve the readability of patient education materials at a radiology practice? Clin. Radiol. **79**, e1366–e1371 (2024). https://doi.org/10.1016/j.crad.2024.08.019
20. Bhattad, A., Gulhane, M., Gavhale, K., et al.: Generative AI for simulating rare disease scenarios in training robots. In: Proceedings of the 6th International Conference on Information Management & Machine Intelligence. Association for Computing Machinery, New York, NY, USA (2025)
21. Li, Y., Zeng, C., Zhang, J., et al.: MedDiT: a knowledge-controlled diffusion transformer framework for dynamic medical image generation in virtual simulated patient. In: Kwok, J. (ed.) Proceedings of the Thirty-Fourth International Joint Conference on Artificial Intelligence, IJCAI-25. International Joint Conferences on Artificial Intelligence Organization, pp. 11082–11085 (2025)
22. Barhum, L.: The link between diabetes and hypertension (2023)
23. Otero, C.A.N., Villar, A.B.: El continuo de la EPOC y riesgo cardiovascular: una hipótesis global de la enfermedad. Clínica e Investigación en Arteriosclerosis. **27**, 144–147 (2014). https://doi.org/10.1016/j.arteri.2014.09.003
24. Swamy, S., Noor, S.M., Mathew, R.O.: Cardiovascular disease in diabetes and chronic kidney disease. J. Clin. Med. **12**, 1428 (2023)
25. Baialardo, A.G., Cardio-Hepático, S., Baialardo, A.G.: Síndrome cardio-hepático ¿Qué debemos saber como cardiólogos? Insuf Card. **10**, 66–77 (2015)
26. Hazlehurst, J.M., Woods, C., Marjot, T., et al.: Non-alcoholic fatty liver disease and diabetes. Metabolism. **65**, 1096–1108 (2016). https://doi.org/10.1016/j.metabol.2016.01.001
27. PIMA Indians diabetes database (2016)
28. CDC Diabetes health indicators (2017)
29. Steinbrunn, A.J.W.: Heart disease (1989)
30. Abreu, M.S.P.: HCC survival (2015)
31. Knaus, W., Lynn, J. Study to Understand Prognoses and Preferences for Outcomes and Risks of Treatment (SUPPORT) and Hospitalized Elderly Longitudinal Project (HELP), 1989-1997 (2020)
32. Soundarapandian, L.R.P.: Chronic kidney disease (2015)
33. Ördekçi, O.: Heart disease prediction using machine learning algorithms. In: Kaggle (2025). https://www.kaggle.com/code/zohaib123/heart-disease-prediction-research-work
34. Zedadra, A., Zedadra, O., Yassine Salah-Salah, M., Guerrieri, A.: Graph-aware multimodal deep learning for classification of diabetic retinopathy images. IEEE Access. **13**, 74795–74806 (2025). https://doi.org/10.1109/ACCESS.2025.3564529

35. Woodland, M., Wood, J., Anderson, B.M., et al.: Evaluating the performance of StyleGAN2-ADA on medical images. In: Zhao, C., Svoboda, D., Wolterink, J.M., Escobar, M. (eds.) Simulation and Synthesis in Medical Imaging, pp. 142–153. Springer International Publishing, Cham (2022)
36. Reyes-Delgado, A.J., Arroyo-Ramírez, M.A., González-Díaz, J.E., et al.: Integration of artificial intelligence techniques for disease prediction and health awareness: review and proposed architecture. In: Martínez-Villaseñor, L., Martínez-Seis, B., Pichardo, O. (eds.) Artificial Intelligence – COMIA 2025, pp. 333–346. Springer Nature Switzerland, Cham (2025)

Multimodal Depression Analysis Through Late Fusion of Linguistic and Facial Features

Álvaro Salgado Lápez and Juanita Hernández Lápez(✉)

Facultad de Ciencias Físico Matemáticas, Universidad Autónoma de Nuevo León, Pedro de Alba s/n, San Nicolás de los Garza, Nuevo León 66451, Mexico
{alvaro.salgadol,juanita.hernandezl}@uanl.edu.mx

Abstract. Depression is a highly prevalent mental disorder whose clinical diagnosis remains challenging, particularly in virtual care settings where assessments rely on transcripts, audio, and video data. In this context, Artificial Intelligence offers promising opportunities for early detection through the automated analysis of multimodal signals. This paper presents a multimodal deep learning framework for automated depression detection using the DAIC-WOZ clinical corpus. The proposed model leverages interview transcripts from 189 participants and facial behavior features derived from video through Action Units (AUs). The architecture consists of two complementary branches: a textual branch based on a transformer-based language model (DepRoBERTa), and a visual branch that models temporal dynamics of AUs using a Bi-LSTM network. Finally, a late fusion strategy is employed to integrate both modalities by concatenating their outputs and feeding them into a Support Vector Machine (SVM), enabling two tasks: binary classification (depressed vs. non-depressed) and depression severity estimation through PHQ-8 score regression. The model is evaluated using 5-fold cross-validation. The experimental results show that the proposed approach achieves an AUC of 0.811 and an F1-score of 0.697 for classification, along with an RMSE of 4.763 and MAE of 3.713 for PHQ-8 regression. These findings demonstrate that the integration of textual and visual modalities through late fusion provides an effective and robust strategy to support automated depression screening.

Keywords: Depression detection · Multimodal analysis · DAIC-WOZ

1 Introduction

Depression is one of the most prevalent mental disorders worldwide, affecting more than 280 million people according to the World Health Organization [14]. Despite its significant social and individual impact, early clinical detection remains a major challenge. Diagnosis largely depends on clinicians' expertise and subjective assessments, typically supported by screening instruments such

H. Ponce et al. (Eds.): COMIA 2026, CCIS 3095, pp. 214–227, 2027.
https://doi.org/10.1007/978-3-032-37438-7_17

as the PHQ-2, PHQ-8, and PHQ-9 questionnaires [21,23]. This reliance on self-reported symptoms, combined with the absence of objective biological markers and increasing specialist workload, contributes to high rates of misdiagnosis [13].

This challenge has been further exacerbated by the transformation of healthcare delivery following the COVID-19 pandemic, which significantly accelerated the adoption of telehealth services. Virtual consultations increased from approximately 7% to over 85% of clinical interactions [15]. While this shift has improved access to care, it has also introduced limitations in clinical assessment. In particular, important non-verbal cues—such as subtle facial expressions and body language—that traditionally support diagnosis are often diminished or lost in remote settings [6]. These limitations highlight the need for computational tools capable of providing objective and quantifiable indicators to support clinicians in virtual environments.

In this context, Artificial Intelligence (AI) and data-driven approaches have emerged as promising tools to complement psychological evaluation. While early research primarily focused on unimodal models, recent studies demonstrate that multimodal approaches can overcome these limitations by integrating diverse sources of information, including speech, text, and visual signals [11,24,26]. In particular, advances in Natural Language Processing (NLP) and neural network architectures have enabled the development of diverse approaches for depression assessment, allowing the capture of both linguistic and visual affective markers associated with the disorder [4,12,22].

Motivated by the possibility of using multimodal data, this paper presents the development and evaluation of a multimodal deep learning framework for automated depression detection using the Distress Analysis Interview Corpus Wizard-of-Oz (DAIC-WOZ). The proposed approach integrates textual and facial features to model depressive symptoms more comprehensively. By doing so, it aims to provide a scalable and cost-effective tool to support clinical decision-making and enhance screening processes in modern mental health systems.

2 Materials and Methods

2.1 Dataset

The DAIC-WOZ (Distress Analysis Interview Corpus Wizard-of-Oz) clinical corpus, developed as part of the Affective Computing Project at the University of Southern California's Institute for Creative Technologies (USC ICT) [9], was used for the development and evaluation of this work. This dataset was designed to support the automated assessment of psychological distress conditions, including depression, anxiety, and post-traumatic stress disorder [18].

The corpus consists of semi-structured interviews conducted by an animated virtual agent named "Ellie", which is operated by a human in a separate location following the Wizard-of-Oz paradigm. This setup is intended to elicit both verbal and non-verbal behaviors relevant to depression assessment [3].

The dataset includes multimodal information, such as textual transcripts, audio recordings, and features derived from facial expressions. In particular, the visual features consist of facial Action Units (AUs), gaze direction, and head pose, extracted from video using the OpenFace toolkit [18].

Facial Action Units (AUs) are the fundamental components of the Facial Action Coding System (FACS), an objective framework developed by Paul Ekman and Wallace V. Friesen to describe and quantify facial movements [5]. Each Action Unit corresponds to a specific contraction or relaxation of facial muscles, as illustrated in Fig. 1.

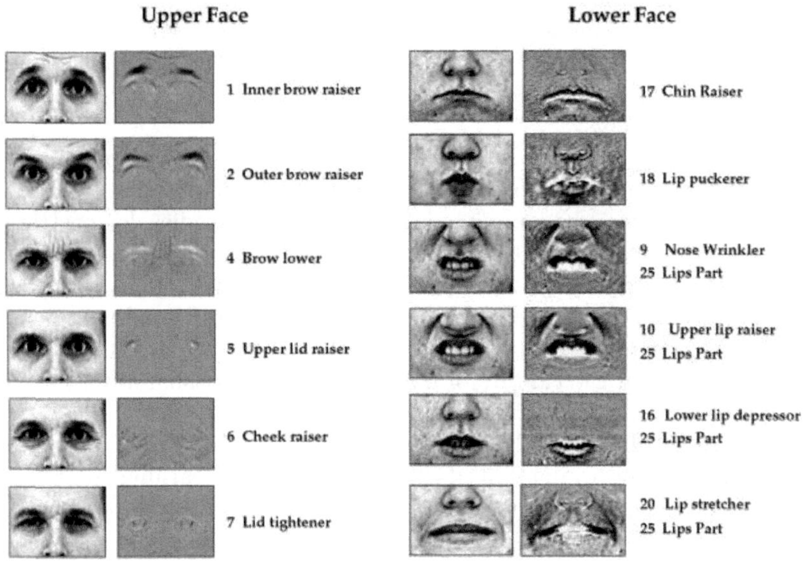

Fig. 1. Example of facial Action Units (AUs), showing the reference image and its corresponding differential representation. Adapted from [5].

Additionally, the dataset provides scores from the Patient Health Questionnaire-8 (PHQ-8), an 8-item instrument used to assess the severity of depressive symptoms over the previous two weeks, with scores ranging from 0 to 24. In this work, these scores are used as the ground truth for depression severity, where values greater than or equal to 10 indicate clinically relevant depression, consistent with prior studies [10].

In this study, multimodal analysis is limited to textual and facial expression modalities, excluding audio features. The PHQ-8 scores are used both for regression and as a diagnostic threshold, where PHQ-8 $\geq$ 10 is used to define the presence of depression.

Overall, the dataset comprises 189 participants, including 133 classified as healthy and 56 classified as depressed according to PHQ-8 criteria. The original dataset was partitioned into three subsets: training (107 subjects, 57%), validation (35 subjects, 18%), and testing (47 subjects, 25%). In this study, the training and validation subsets were merged to create a unified development set for subsequent analysis, while the original testing subset was retained as an independent test set to enable fair comparison with prior studies.

2.2 Proposed Multimodal Model

The proposed system follows a multimodal approach, employing two independent deep learning networks to learn latent representations from textual and facial features. These representations are subsequently fused to perform the final prediction. The overall architecture is illustrated in Fig. 2.

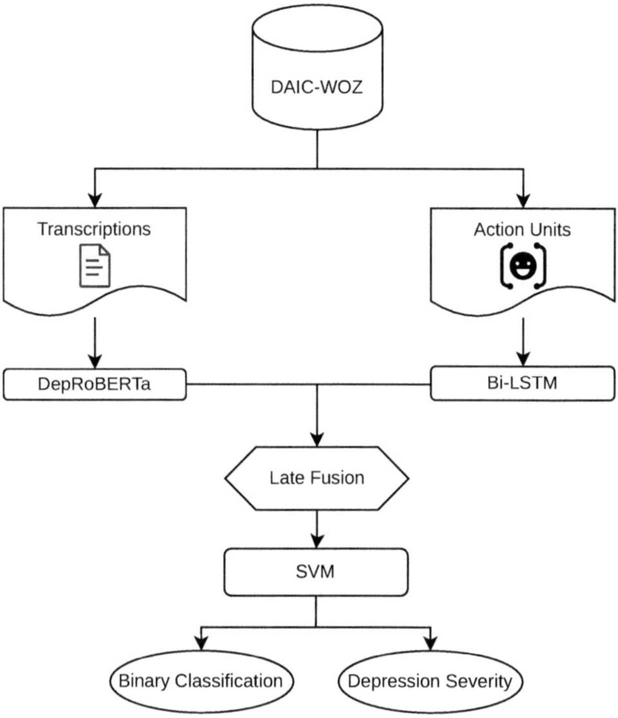

Fig. 2. Architecture of the late fusion multimodal system for depression detection. Independent features are extracted from text (DepRoBERTa) and facial Action Units (Bi-LSTM), and merged at the feature level to feed a Support Vector Machine (SVM) for final prediction.

Given that the different modalities operate at distinct temporal resolutions and data formats, specific preprocessing strategies were applied to each modality.

Textual Branch. Patient transcripts often exceed the maximum input length supported by transformer-based models [2]. To address this limitation, a segmentation strategy was applied, dividing each transcript into up to four blocks of 512 tokens. This approach preserves the global context of the interview while maintaining semantic coherence.

For textual representation, DepRoBERTa, a Large model pre-trained on mental health forum data was employed [16]. The model was fine-tuned for the target task by unfreezing only the last four encoder layers. The resulting representations of the text segments were aggregated using an average pooling. The architecture includes fully connected layers with Layer Normalization, GELU activation, and a dropout rate of 0.3, producing a 1024-dimensional latent feature vector used for the subsequent multimodal fusion stage.

Visual Branch. Visual analysis was conducted using facial Action Units (AUs) provided in the dataset. To reduce noise and mitigate tracking errors, a confidence threshold of 0.90 was applied, removing frames below this level. The resulting time series was then downsampled by a factor of 1/5 and standardized through padding or truncation to obtain fixed-length sequences of 300 frames per subject.

To model temporal dependencies, a Bidirectional Long Short-Term Memory (Bi-LSTM) network [20] with 64 hidden units was used. The network consists of three layers and incorporates an attention mechanism to emphasize the most relevant frames. A dropout rate of 0.3 was applied for regularization. The output is projected into a 128-dimensional feature vector, which serves as the visual representation for the subsequent fusion stage.

Multimodal Fusion and Final Prediction. The proposed framework employs a late feature-level fusion strategy. Specifically, the 128-dimensional visual representation was concatenated with the 1024-dimensional latent vector generated by the textual branch, resulting in a 1152-dimensional multimodal representation. To alleviate the curse of dimensionality, Principal Component Analysis (PCA) was applied prior to classification.

The reduced feature vector was then used as input to a Support Vector Machine (SVM) with a radial basis function (RBF) kernel. The SVM classifier was selected over more complex deep learning alternatives due to its stronger generalization capability and robustness to overfitting in small-scale clinical datasets [8]. Previous studies have shown that high-capacity deep learning models typically require substantially larger datasets to achieve stable performance, whereas traditional machine learning approaches such as SVM often exhibit superior generalizability under limited data conditions [8]. To address class imbalance, balanced class weights were employed. The hyperparameters (C, γ, and ϵ) were optimized through exhaustive grid search.

2.3 Experimental Design and Evaluation

The performance of the proposed model for depression detection was evaluated under both classification and regression settings. A 5-fold stratified cross-validation scheme was applied to the development set to construct the training and validation partitions. This configuration was selected to ensure an adequate number of samples per fold, particularly for the minority class, given the limited size and class imbalance of the dataset. Stratification preserved the proportion of depressed and healthy participants across all folds. The original test partition remained completely isolated throughout the training and validation process and was not used during cross-validation. This held-out subset, representing approximately 25% of the dataset, was reserved exclusively for the final external evaluation, while the remaining 75% was used for model development and validation.

The performance of the binary classifier was assessed using standard evaluation metrics, including Balanced Accuracy (BAC), sensitivity (SEN), specificity (SPE), F1-score (F1), and the Area Under the Receiver Operating Characteristic Curve (AUC) [1,25]. Similarly, for the regression task, model performance was evaluated using Mean Absolute Error (MAE) and Root Mean Square Error (RMSE) [19].

Finally, the results obtained are compared with those reported in previous studies that also used the DAIC-WOZ dataset, as presented in their original publications [1,17,19,25].

3 Results

3.1 Classification

Figure 3 presents the mean and standard deviation of the evaluation metrics for the classification task. The proposed model achieved an accuracy (ACC) of 0.8111 ± 0.088 and a balanced accuracy (BAC) of 0.793 ± 0.117, indicating a balanced performance across the depression and non-depression classes. Additionally, the model obtained an area under the ROC curve (AUC) of 0.864 ± 0.109, demonstrating strong discriminative capability.

In terms of class-specific performance, the model achieved an F1-score of 0.697 ± 0.153, with a sensitivity (SEN) of 0.747 ± 0.211 and a precision (PR) of 0.664 ± 0.116. These results indicate that the model maintains a reasonable balance between correctly identifying positive cases and minimizing false positives.

Table 1 summarizes the classification performance of the proposed approach in comparison with two state-of-the-art methods. The proposed multimodal model achieved an accuracy of 0.811 and an F1-score of 0.697, outperforming the textual baseline reported by Zhang et al. [25] (ACC = 0.79) and substantially surpassing the multimodal results presented by Arioz et al. [1] (F1 = 0.42).

Although direct comparison is constrained by differences in experimental settings and evaluation protocols, it is important to note that the previous studies do not explicitly report the use of cross-validation, which may result in overly optimistic performance estimates when dealing with small-scale clinical datasets. In contrast, the 5-fold stratified cross-validation strategy adopted in this work provides a more rigorous and reliable evaluation by preserving class distribution

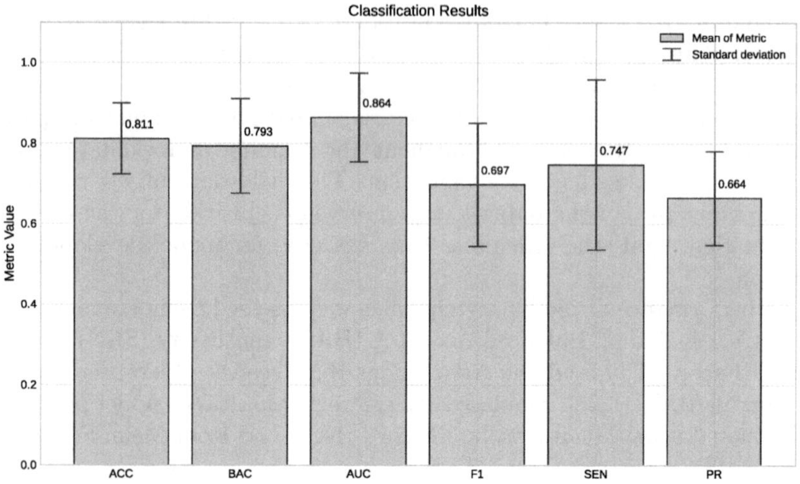

Fig. 3. Classification results for depression prediction using the proposed model. Mean values and standard deviations are reported for accuracy (ACC), balanced accuracy (BAC), area under the ROC curve (AUC), F1-score, sensitivity (SEN), and precision (PR).

across folds and reducing the risk of overfitting. These results suggest that the proposed symmetric multimodal fusion framework offers improved robustness and generalization capability for automated depression detection.

Table 1. Comparison with state-of-the-art methods for depression classification on the DAIC-WOZ dataset. Mean values are reported for all metrics. N/A indicates results not reported by the authors.

Method	ACC	BAC	AUC	F1	SEN	PR
Proposed method	0.811 ± 0.088	0.793 ± 0.117	0.864 ± 0.109	0.697 ± 0.153	0.747 ± 0.211	0.664 ± 0.116
Zhang et al. (Text) [25]	0.79	N/A	N/A	0.70	N/A	0.70
Arioz et al. [1]	N/A	N/A	N/A	0.42	0.41	0.55

To further assess the discriminative capability of the proposed multimodal framework, Fig. 4 presents the Receiver Operating Characteristic (ROC) curves obtained across the 5-fold stratified cross-validation procedure. The model achieved a mean AUC of 0.864 (SD = 0.109), indicating strong discriminative performance.

The relatively consistent behavior of the ROC curves across folds, as reflected by the shaded region representing the standard deviation, suggests that the proposed multimodal model yields stable and robust decision boundaries. These results further indicate that the integration of linguistic embeddings and facial Action Units effectively mitigates the impact of class imbalance within the DAIC-WOZ dataset while maintaining reliable generalization performance.

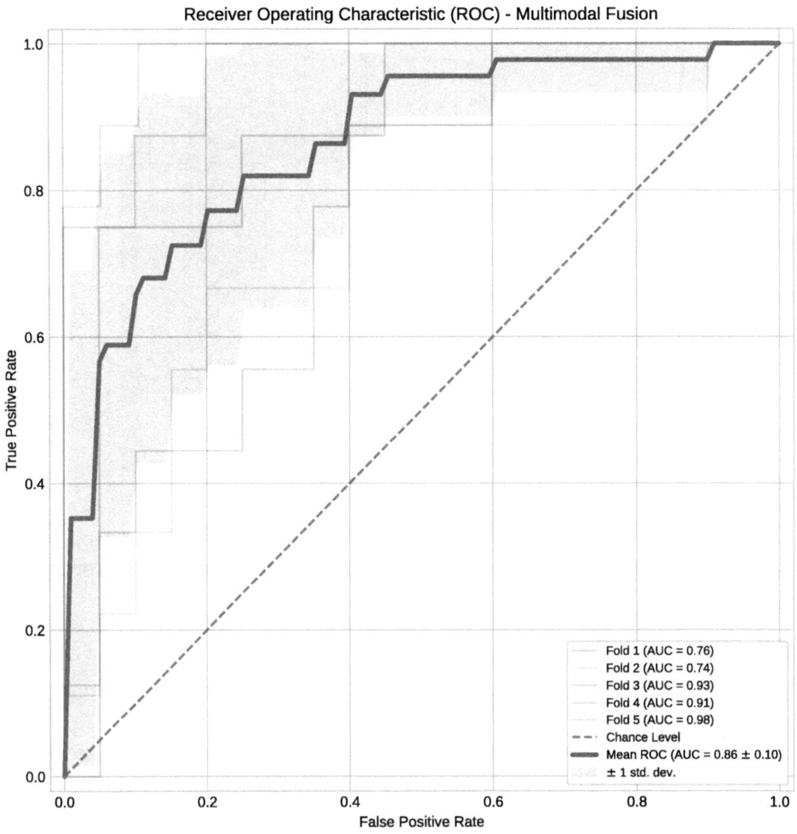

Fig. 4. ROC curve for the proposed multimodal framework on the DAIC-WOZ dataset. The plot illustrates the classification performance across the 5-fold stratified cross-validation. The solid bold line represents the mean ROC (Mean AUC $= 0.864 \pm 0.109$), while the shaded area indicates the standard deviation. The dashed diagonal line represents the chance level (AUC=0.5).

3.2 Regression

For the regression task, aimed at estimating depression severity, the model achieved a mean absolute error (MAE) of 3.713 ± 0.725 and a root mean square error (RMSE) of 4.763 ± 0.708. Figure 5 summarizes these results. These values indicate that the model is capable of estimating PHQ-8 scores with a relatively low prediction error.

Table 2 presents a comparison of the proposed method with state-of-the-art approaches for the regression task on the DAIC-WOZ dataset. The framework proposed by Fang et al. [7] reports slightly lower error values (MAE $=$ 3.360, RMSE $=$ 4.480). However, it is important to note that these prior studies do not

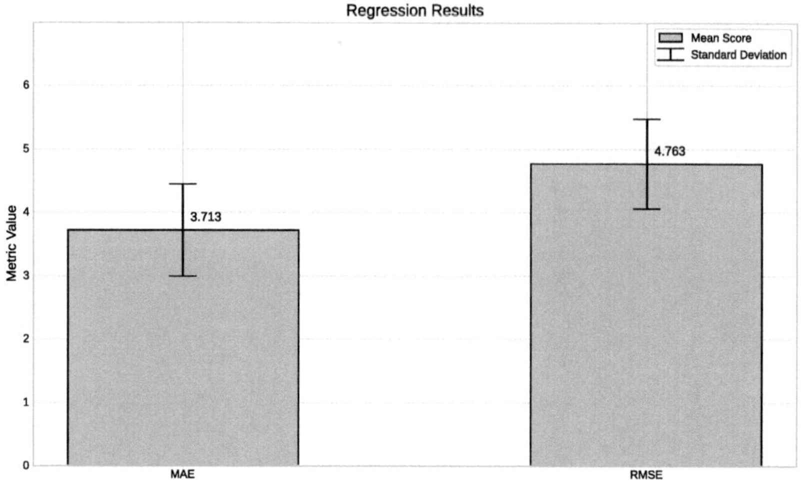

Fig. 5. Regression results for depression severity estimation using the proposed model. Mean values and standard deviations are reported for mean absolute error (MAE) and root mean square error (RMSE).

explicitly report the use of cross-validation, which may lead to over-optimistic performance estimates in small clinical datasets.

In contrast, the proposed model achieves competitive performance (MAE = 3.713, RMSE = 4.763) under a rigorous 5-fold stratified cross-validation scheme. This evaluation protocol ensures more robust performance estimation by reducing variance across data splits and accounting for dataset variability. Consequently, the reported results provide a more reliable assessment of the model's ability to estimate depression severity in real-world conditions.

Table 2. Comparison with state-of-the-art methods for depression severity estimation using regression on the DAIC-WOZ dataset. Mean values are reported for MAE and RMSE. N/A indicates results not reported by the authors.

Method	MAE	RMSE
Proposed method	3.713 ± 0.725	4.763 ± 0.708
Qureshi et al. [17]	3.65	5.11
Fang et al. [7]	3.36	4.48

Overall, the results demonstrate that the proposed approach achieves consistent performance in both binary classification and continuous severity estimation using the PHQ-8 scale.

3.3 Ablation Study

To evaluate the contribution of each modality to the overall performance, an ablation study was conducted. Table 3 compares the performance of the unimodal baselines (Text-only and Visual-only) with the proposed multimodal fusion model. The results show that the linguistic branch is the primary contributor to performance; however, incorporating facial features consistently improves Accuracy, F1-score, and AUC. This confirms the complementary nature of the two modalities and the effectiveness of their integration within the proposed framework.

Table 3. Ablation study results comparing unimodal and multimodal configurations for binary depression classification on the DAIC-WOZ dataset. Mean values and standard deviations are reported for Accuracy, F1-Score, and ROC AUC. Best performance values are highlighted in bold.

Method	Accuracy	F1-Score	ROC AUC
Visual-only (Bi-LSTM)	0.528 ± 0.154	0.234 ± 0.130	0.434 ± 0.163
Text-only (DepRoBERTa)	0.804 ± 0.092	0.651 ± 0.169	0.845 ± 0.102
Proposed method	$\mathbf{0.811 \pm 0.085}$	$\mathbf{0.697 \pm 0.112}$	$\mathbf{0.864 \pm 0.091}$

Table 4 details the comparative evaluation of the unimodal branches relative to the proposed late fusion framework for estimating depression severity. The data reveals that while the linguistic features act as the primary foundation for the predictions, the addition of facial information results in a measurable reduction of both MAE and RMSE. This outcome highlights the added value of integrating non-verbal cues to achieve a more refined assessment of the PHQ-8 scale.

Table 4. Ablation study results comparing unimodal and multimodal configurations for depression severity estimation on the DAIC-WOZ dataset. Mean values and standard deviations are reported for MAE and RMSE. Best performance values are highlighted in bold.

Method	MAE	RMSE
Visual-only (Bi-LSTM)	4.862 ± 0.832	5.972 ± 1.081
Text-only (DepRoBERTa)	4.028 ± 0.732	5.207 ± 0.966
Proposed method	$\mathbf{3.713 \pm 0.725}$	$\mathbf{4.763 \pm 0.708}$

4 Discussion

The results demonstrate that the proposed multimodal approach, which integrates linguistic embeddings and facial Action Units, provides a robust and effective framework for depression detection. The superiority of the fusion model over unimodal baselines, as evidenced by the ablation study, confirms that depression manifests through a complex interplay of verbal content and non-verbal cues. While linguistic features remain the primary driver of performance, the inclusion of visual latent representations provides the necessary context to refine the classification boundary and reduce estimation errors in regression.

In the classification task, the model achieved an AUC of 0.864, indicating strong discriminative capability in distinguishing between depressed and non-depressed individuals. Additionally, the accuracy (ACC = 0.811) and balanced accuracy (BAC = 0.793) reflect a balanced performance across classes. This is particularly important in mental health applications, where class imbalance can negatively affect model performance.

Regarding class-specific metrics, the sensitivity (SEN = 0.747) indicates that the model is capable of identifying a substantial proportion of depressive cases, while the precision (PR = 0.664) suggests a reasonable trade-off between correctly identifying positive cases and limiting false positives. The F1-score (0.697) further confirms this balance, highlighting consistent performance across the classification task.

For the regression task, the model demonstrates the ability to estimate depression severity through continuous PHQ-8 scores, achieving an MAE of 3.713 and an RMSE of 4.763. These results indicate a reasonable level of precision in approximating severity of depression, which is critical in clinical contexts where both detection and severity assessment are required.

Despite these promising results, several limitations remain. The current system relies on a late-fusion strategy, combining modality-specific representations through an SVM classifier built on DepRoBERTa and Bi-LSTM features. In addition, the absence of acoustic features may limit the model's sensitivity to relevant paralinguistic cues that are known to be informative in depression assessment. Finally, evaluating the framework on more diverse, non-clinical populations will be necessary to further assess its generalizability and robustness in real-world telehealth scenarios.

In addition, direct comparison with existing literature is inherently challenging due to substantial variability in preprocessing pipelines and experimental setups across studies. To provide a more consistent basis for evaluation, the proposed model was benchmarked against established methods using the same DAIC-WOZ dataset. Within this controlled framework, our approach demonstrates competitive performance, particularly in PHQ-8 score estimation, suggesting that it constitutes a viable and robust alternative for clinical severity assessment.

5 Conclusions

This work presented a multimodal framework for depression detection and severity estimation, integrating linguistic embeddings with visual features derived from facial Action Units. The proposed method addresses both binary classification and regression tasks, enabling a comprehensive assessment that captures not only the presence of depression but also its severity through PHQ-8 score estimation.

Empirical results demonstrate that the model achieves robust performance across both tasks. For classification, the system obtained a mean AUC of 0.864 and a BAC of 0.793, while regression experiments yielded competitive error metrics, with an MAE of 3.713 and an RMSE of 4.763. Furthermore, the ablation study confirms that the integration of verbal and non-verbal modalities provides consistent performance gains, highlighting the complementary nature of the selected features and their importance for reliable clinical decision support.

We also compared the proposed approach with existing methods reported in the literature. While this provides a meaningful benchmark, we acknowledge that differences in preprocessing pipelines and validation protocols across studies limit the extent of direct comparability.

Future work will focus on investigating attention-based fusion mechanisms and incorporating additional modalities, such as acoustic features, to further improve model sensitivity. In addition, validating the proposed framework on larger and more diverse datasets will be essential to assess its generalizability across populations and ensure its applicability in real-world clinical settings.

References

1. Arioz, U., Smrke, U., Plohl, N., Mlakar, I.: Scoping review on the multimodal classification of depression and experimental study on existing multimodal Models. Diagnostics **12**(11), 2683 (2022). https://doi.org/10.3390/diagnostics12112683
2. Chen, F., Ben-Zeev, D., Sparks, G., Kadakia, A., Cohen, T.: Detecting PTSD in Clinical Interviews: A Comparative Analysis of NLP Methods and Large Language Models (2026). https://doi.org/10.48550/arXiv.2504.01216
3. DeVault, D., et al.: SimSensei kiosk: A virtual human interviewer for healthcare decision support. In: Proceedings of the 2014 International Conference on Autonomous Agents and Multi-Agent Systems, pp. 1061–1068. AAMAS '14, International Foundation for Autonomous Agents and Multiagent Systems, Richland, SC (2014)
4. Devlin, J., Chang, M.W., Lee, K., Toutanova, K.: BERT: Pre-training of Deep Bidirectional Transformers for Language Understanding (2019). https://doi.org/10.48550/arXiv.1810.04805
5. Donato, G., Bartlett, M.S., Hager, J.C., Ekman, P., Sejnowski, T.J.: Classifying facial actions. IEEE Trans. Pattern Anal. Mach. Intell. **21**(10), 974 (1999). https://doi.org/10.1109/34.799905
6. Duffy, C.M.C., Benotsch, E.G.: Nonverbal behavior in telehealth visits: A narrative review. Patient Educ. Couns. **132**, 108600 (2025). https://doi.org/10.1016/j.pec.2024.108600

7. Fang, M., Peng, S., Liang, Y., Hung, C.C., Liu, S.: A multimodal fusion model with multi-level attention mechanism for depression detection. Biomed. Signal Process. Control **82**, 104561 (2023). https://doi.org/10.1016/j.bspc.2022.104561
8. Goetz, L., Seedat, N., Vandersluis, R., van der Schaar, M.: Generalization—a key challenge for responsible AI in patient-facing clinical applications. npj Digit. Med. **7**(1), 126 (2024). https://doi.org/10.1038/s41746-024-01127-3
9. Gratch, J., et al.: The distress analysis interview corpus of human and computer interviews. In: Calzolari, N., et al. (eds.) Proceedings of the Ninth International Conference on Language Resources and Evaluation (LREC'14), pp. 3123–3128. European Language Resources Association (ELRA), Reykjavik, Iceland (2014)
10. Kroenke, K., Strine, T.W., Spitzer, R.L., Williams, J.B.W., Berry, J.T., Mokdad, A.H.: The PHQ-8 as a measure of current depression in the general population. J. Affect. Disord. **114**(1–3), 163–173 (2009). https://doi.org/10.1016/j.jad.2008.06.026
11. Lam, G., Dongyan, H., Lin, W.: Context-aware deep learning for multi modal depression detection. In: ICASSP 2019 - 2019 IEEE International Conference on Acoustics, Speech and Signal Processing (ICASSP), pp. 3946–3950 (2019). https://doi.org/10.1109/ICASSP.2019.8683027
12. Mahayossanunt, Y., Nupairoj, N., Hemrungrojn, S., Vateekul, P.: Explainable depression detection based on facial expression using lstm on attentional intermediate feature fusion with label smoothing. Sens. (Basel, Switz.) **23**(23), 9402 (2023). https://doi.org/10.3390/s23239402
13. Mohammad, F., Mohammed, K.: MDD: a unified multimodal deep learning approach for depression diagnosis based on text and audio speech. Comput. Mater. Continua **81**(3), 4125–4147 (2024). https://doi.org/10.32604/cmc.2024.056666
14. OMS: Depresión (2023). https://www.who.int/es/news-room/fact-sheets/detail/depression
15. Pierce, B.S., Perrin, P.B., Tyler, C.M., McKee, G.B., Watson, J.D.: The COVID-19 telepsychology revolution: A national study of pandemic-based changes in U.S. mental health care delivery. Am. Psychol. **76**(1), 14–25 (2021). https://doi.org/10.1037/amp0000722
16. Poświata, R., Perełkiewicz, M.: OPI@LT-EDI-ACL2022: Detecting Signs of Depression from Social Media Text using RoBERTa Pre-trained Language Models. In: Chakravarthi, B.R., Bharathi, B., McCrae, J.P., Zarrouk, M., Bali, K., Buitelaar, P. (eds.) Proceedings of the Second Workshop on Language Technology for Equality, Diversity and Inclusion, pp. 276–282. Association for Computational Linguistics, Dublin, Ireland (2022). https://doi.org/10.18653/v1/2022.ltedi-1.40
17. Qureshi, S.A., Hasanuzzaman, M., Saha, S., Dias, G.: The Verbal and Non Verbal Signals of Depression – Combining Acoustics, Text and Visuals for Estimating Depression Level (2019). https://doi.org/10.48550/arXiv.1904.07656
18. Ringeval, F., et al.: AVEC 2019 Workshop and Challenge: State-of-Mind, Detecting Depression with AI, and Cross-Cultural Affect Recognition (2019). https://doi.org/10.48550/arXiv.1907.11510
19. Sadeghi, M., et al.: Harnessing multimodal approaches for depression detection using large language models and facial expressions. npj Mental Health Res. **3**(1), 66 (2024).https://doi.org/10.1038/s44184-024-00112-8
20. Schuster, M., Paliwal, K.: Bidirectional recurrent neural networks. IEEE Trans. Signal Process. **45**(11), 2673–2681 (1997). https://doi.org/10.1109/78.650093
21. Siniscalchi, K.A., et al.: Depression screening and measurement-based care in primary care. J. Primary Care Commun. Health **11**, 2150132720931261 (2020). https://doi.org/10.1177/2150132720931261

22. Teferra, B.G., et al.: Screening for depression using natural language processing: literature review. Inter. J. Med. Res. **13**, e55067 (2024). https://doi.org/10.2196/55067
23. Yan, W.J., Ruan, Q.N., Jiang, K.: Challenges for Artificial Intelligence in Recognizing Mental Disorders. Diagn. (Basel Switz.) **13**(1), 2 (2022). https://doi.org/10.3390/diagnostics13010002
24. Zhang, W., Mao, K., Chen, J.: A multimodal approach for detection and assessment of depression using text. Audio and Video. Phenomics **4**(3), 234–249 (2024). https://doi.org/10.1007/s43657-023-00152-8
25. Zhang, X., Li, C., Chen, W., Zheng, J., Li, F.: Optimizing depression detection in clinical doctor-patient interviews using a multi-instance learning framework. Sci. Rep. **15**, 6637 (2025). https://doi.org/10.1038/s41598-025-90117-w
26. Zhou, L., Liu, Z., Shangguan, Z., Yuan, X., Li, Y., Hu, B.: TAMFN: time-aware attention multimodal fusion network for depression detection. IEEE Trans. Neural Syst. Rehabil. Eng. **31**, 669–679 (2023). https://doi.org/10.1109/TNSRE.2022.3224135

Bibliometric Analysis of Ethics in the Content of Scientific Papers on Artificial Intelligence

Nora Gavira-Durón[1](✉), Ana L. Jiménez-Preciado[1], and Mayra F. Angeles-Sánchez[2]

[1] Instituto Politécnico Nacional, CDMX, Mexico City, Mexico
nora.gavira@udlap.mx

[2] Universidad de las Américas Puebla, Puebla, Mexico

Abstract. With the increasing presence of Artificial Intelligence (AI) in our everyday lives, the moral and ethical implications raised by this technology have also increased. However, the extent to which ethical concerns are addressed in AI-related research has not been thoroughly studied yet. In this work, we conduct a bibliometric and content analysis of 5,000 AI-related publications indexed in the Scopus database in 2025 and 2026. Through the employment of Natural Language Processing (NLP) methods and by building a conceptual framework that includes the seven ethical dimensions of AI (transparency and explainability, accountability, privacy, safety and robustness, human agency, general ethical principles and misinformation), we evaluate the usage and context in which ethics-related words appear in the title and abstract of scientific papers. Our results show that 14.8% of the articles analyzed contain a reference to at least one dimension. Safety and robustness (41.5%) and transparency and explainability (38.6%) are the most mentioned dimensions. By investigating the co-occurrences at a sentence level, we find that ethical references in AI research have been primarily descriptive in nature with 70.3% being generic mentions, 27.9% mentioning problems and risks and only 1.8% suggesting possible solutions. Our findings from the most frequent bigrams show that the field has mainly focused on technical solutions (with bigrams like 'machine learning' appearing 1,079 times, followed by 'deep learning' (642 times) and 'artificial intelligence' (579 times). In conclusion, ethical awareness is shown to have entered the research community, but we still have a long way to go to reach the solution-orientation of AI research as the current discussion remains at the problem-diagnosis level. This paper provides a complete and reproducible NLP pipeline for investigating ethics in AI research, a fully documented dictionary of ethics-related terms, and offers suggestions on how ethical dimensions can be better included in AI research.

Keywords: Artificial Intelligence · Ethics · Bibliometric Analysis · Responsibility · Natural Language Processing · Scientific Production

H. Ponce et al. (Eds.): COMIA 2026, CCIS 3095, pp. 228–240, 2027.
https://doi.org/10.1007/978-3-032-37438-7_18

1 Introduction

Over the last decade, the rise of artificial intelligence has revolutionized industries ranging from healthcare and education to finance and government. This rapid proliferation of technology has also brought forth ethical, social, and legal considerations regarding the implementation of these systems.

As AI technologies increasingly mediate decision-making processes, there are ongoing debates about how we make sure AI systems are built, implemented and utilized in a responsible, transparent and value-based manner (Floridi and Cowls, 2019; Mittelstadt, 2019). This concern for the ethical impact of AI has led to the establishment of global regulatory guidelines.

Bodies such as UNESCO, the OECD, and the European Commission have issued recommendations to support the creation of trustworthy AI by addressing aspects like transparency, fairness, accountability, and privacy (High-Level Expert Group on AI, 2019; OECD, 2024).

In so doing, all institutions agree that ethical considerations should be integrated across the entire AI lifecycle. However, the gap between policy-level ethical discussions and research practices remains. Ethical guidelines are often discussed in governance settings, though it is less clear to what extent they influence scientific papers. This is of importance since it is through academic research that knowledge is generated to guide technology and public policy (Zhang et al., 2021). Recent advancements in AI technologies, especially those related to large language models, generative AI, and reinforcement learning, have generated both excitement and apprehension (Bommasani et al., 2021).

The quick pace of technological development in this area requires consideration of how the research community responds to the ethical issues associated with these technologies. The existing literature tends to address AI ethics in isolation, often zooming in on a single facet like explainability (Arrieta et al., 2020) or algorithmic fairness (Binns, 2018), rather than providing a broad-brush survey of the ethical concerns covered in the AI literature. And even more importantly, a study conducted only a couple of years ago would not be very useful given the fast-changing nature of AI ethics. New developments like foundation models and generative AI have shifted the ethical terrain over the last few years and would not be represented by a recent literature survey.

This paper analyzes 5000 articles from Scopus, published in 2025 and 2026, using bibliometric and content analysis to identify ethical discourse within the titles and abstracts of those articles (a) and to see which kinds of ethics are covered the most and which are covered the least (b). For this we draw on seven types of ethics, drawn from the literature on regulation: transparency and explainability, accountability, privacy and data protection, safety and robustness, human agency, general ethical principles and misinformation. This allows us to conduct a systematic and replicable analysis of the ethics represented in literature.

The implications of this paper could be far-reaching. For academic audiences, this paper will provide an understanding of which types of ethics are being engaged in the literature currently, and which are being neglected. For policy audiences, it will offer insight into what is being covered at scale, and so potentially inform future policy initiatives to promote or require more ethical coverage in AI research.

2 Literature Review

There was already a lot of ethical reflection on the use of AI since the birth of AI as a discipline. Wiener (1960) was concerned about the social implications of using automated decision-making and argued that ethical standards for AI need to be developed. However, the impact of these ideas was minimal because the technology that underpinned AI was too limited to use it in real-world applications until the last 10 years. AI has grown tremendously in the last 10 years due to breakthroughs in machine learning and the increasing availability of huge amounts of data and processing power. These improvements in technology allowed AI to move away from being primarily theoretical and academic and become a technology with practical real-world applications that can affect everyone in society (Brynjolfsson and McAfee, 2017).

Given the increasing popularity of AI systems that are used in criminal justice, hiring, medical diagnosis, and policing, it was impossible to ignore the ethical issues these systems raised. This sparked a response with several guidelines on AI ethics. The European Commission published Ethical Guidelines for Trustworthy AI in 2019 (High-Level Expert Group on AI, 2019) with seven requirements: human oversight, technical robustness and safety, privacy and data governance, transparency, diversity and non-discrimination and fairness, societal and environmental well-being and accountability. The OECD adopted its Principles on AI in 2019, which have been endorsed by more than 40 countries (OECD, 2024). UNESCO adopted its Recommendation on the Ethics of Artificial Intelligence in 2021, which was the first global standard of ethics for AI (UNESCO, 2021).

In the context of AI research, there is also an interdisciplinary area that focuses on responsible AI (Raji et al., 2020) or algorithmic ethics. New conferences such as the ACM Conference on Fairness, Accountability and Transparency (FAccT) or special tracks at prominent conferences such as ICML or NeurIPS have been created (Raji et al., 2020). AI ethics research has exploded but remains in question how this research is incorporated into the rest of the AI field.

Scholarship has identified different facets that are relevant to the ethical development of AI. Algorithmic fairness is about ensuring that the AI doesn't learn and replicate bias from training data or from the algorithms' decision making itself (Binns, 2018). Another aspect is the one of transparency and explainability, referring to the ability to understand and interpret the decisions taken by an AI system. This is especially true in sensitive sectors such as health, justice, etc. There have been a lot of different methods proposed to make a black box of explainable AI (XAI) system explainable but whether the proposed methods work or how to explain it to the user is still discussed (Arrieta et al., 2020).

The next aspect is about safety and robustness, which relates to whether AI is susceptible to adversarial attacks. Also, it can be asked if the system still behaves properly even in a non-typical environment which would be especially important in cases where the system has a significant impact on human lives such as for instance autonomous cars or medical diagnostics (Amodei et al., 2016).

A further important aspect is privacy and data protection, especially after the implementation of the General Data Protection Regulation (GDPR) in European countries, where the goal is to enable models to train without accessing personal data and research includes techniques such as federated learning, differential privacy, and homomorphic encryption (Shokri et al., 2017).

Other important aspects are accountability (referring to a system's ability to assign responsibility when a system produces an undesirable or unintended outcome), human agency (referring to how people remain in control of their decisions and interactions when in the presence of AI), and the concern over misinformation (referring to the potential for AI systems to produce or spread false information) (Mittelstadt, 2019).

Yet, it is still debated to which extent ethics discourse is already present in the main research on AI. While many studies already address specific aspects from the previous list or a specific area of application, they do not discuss them holistically in the same context. Moreover, the existing bibliometric studies might not reflect the current situation, since there are new articles being published each year (Bornmann and Mutz, 2015).

One important limitation of metadata-based studies is that ethical debate might not be sufficiently reflected in titles and abstracts, even though they form the bulk of the data utilized in bibliometric analysis and represent a reasonable indication of how the scholarly community conceptualizes and presents its work (Gusenbauer and Haddaway, 2020). This paper seeks to overcome these limitations through an examination of 5,000 recent publications (2025 to 2026) that falls into seven categories of ethical considerations to enable both systematic comparison with past research and a replicable approach for future investigation.

3 Methodology

To understand the ethical content of artificial intelligence, we conducted a bibliometric and content analysis on AI-related scientific publications that are indexed in Scopus. We adopted the systematic review process as described in Zupic and Cater (2015) to extract and analyze 5,000 titles and abstracts. To carry out the text analysis, we relied on natural language processing (NLP) methods, following the approach of Grimmer and Stewart (2013). We extracted a total of 5,000 scientific papers from the 2022–2026-time span from Scopus (Elsevier) by using the Scopus REST API. Our choice to use Scopus was since it is the largest peer reviewed literature database in computer science and its related disciplines (Burnham, 2006; Dwivedi et al., 2023).

Specifically, we queried the database programmatically by developing a modular Python pipeline that incorporates a simple exponential backoff retry mechanism and an incremental progress saving option to overcome any API rate limit issues and potential disconnection events. Our search string contains 18 distinct terms for the concept of artificial intelligence, organized into two tiers based on our research objectives. Our primary terms (that we searched for only in article titles), were: 'Artificial intelligence', 'AI', 'Machine learning', 'Deep learning', 'Neural networks', 'Large language models', 'LLMs', 'Generative AI', 'Foundation models', 'Transformer models', 'GPT', and 'BERT'.

Our secondary terms (that we searched for in titles, abstracts, and keywords), were: 'Reinforcement learning', 'Natural language processing', 'NLP', 'Computer vision', 'Intelligent systems', and 'Machine intelligence'. All results were listed by publication date from the latest to the earliest. For each article, we stored the following data: title, authors, publication date, publication venue, DOI, Scopus ID, citation count, abstract, author keywords, author affiliations, and document type.

To ensure full reproducibility, the complete source code, the ethical term dictionary, and the dataset of 5,000 articles are publicly available at https://github.com/AnaJZP/ia_ethics.

3.1 Ethical Framework and Analytical Methods

For the ethical analysis, we adopted the conceptual scheme introduced by Floridi and Cowls (2019) complemented by the European Commission (High-Level Expert Group on AI, 2019) and OECD (2024) recommendations. We identified seven dimensions for ethical dimensions, operationalizing them using a dictionary of terms from Raji et al. (2020), Mittelstadt (2019). See Appendix A for the complete dictionary. We first concatenated the title and abstract of each article, lowercased them, and cleaned them (stripped of punctuation, digits and unnecessary spaces).

Then a term unification step mapped morphological variants to canonical forms. Next, we calculated the frequency of unigrams and bigrams in this preprocessed set. We used the scikit learn's class CountVectorizer (Pedregosa et al., 2011), filtering out terms that appear fewer than a minimum number of times (threshold of 5 and 3 terms for unigrams and bigrams, respectively).

For each article i, let T_i be the concatenated text of the title and abstract. The ethical score for the paper S_i is given by the following sum of the occurrences of all the patterns in each dimension:

$$S_i = \sum_{d \in D} \sum_{p \in P_d} f(p, T_i) \tag{1}$$

where $D = d_1, d_2, \ldots, d_7$ is a set of seven ethical dimensions, P_d is the set of patterns for dimension d, and $f(p, T_i)$ returns the count of non-overlapping occurrences of pattern p in text T_i. Thus, an article is deemed to contain some form of ethics whenever its score is higher than zero $S_i > 0$. We then calculate the percentage of articles with an ethics score:

$$E = \left(\frac{1}{N}\right) \times \sum_i I(S_i > 0) \times 100\% \tag{2}$$

where N is the number of papers considered, and $I(.)$ is the indicator function. It must be stressed that score is intended to capture the frequency and presence of ethical vocabulary in metadata of an article, and not the extent to which an article engages with ethics in a more sophisticated fashion.

To investigate the relationship between AI and ethics in the publication literature, we performed sentence level co-occurrence analysis of AI and ethical terms, following Leydesdorff and Welbers (2011). For each sentence s in the T_i we considered that it co-occurred if at least one AI term and at least one ethical term were identified. Therefore, a co-occurrence is considered if:

$$C(s) = 1 \text{ iff } (\exists a \in A : a \text{ matches } s) \wedge (\exists e \in E : e \text{ matches } s) \tag{3}$$

where A is the set of AI term patterns and E is the set of ethical term patterns. Sentences containing this relationship were further classified into one of four different discursive types according to whether they contained any modulating vocabulary: (a) challenge & risks, where one of these words appear: 'challenge', 'risk', 'concern' …; (b) solution & mitigations, where one of these words appear: 'solution', 'address', 'mitigate' …; (c) requirements & needs, where one of these words appear: 'need', 'must', 'require' …; and (d) general, where no word from the previous types were mentioned. Following Binns (2018) these correspond to challenges & risks, solutions & mitigations, requirements & needs, and just general mentions of the ethical terms.

Finally, the relative weight w_d of each ethical dimension d in the entire corpus of documents was defined as:

$$w_d = M_d / \sum_k M_k \times 100\% \tag{4}$$

where M_d is the total number of mentions of d across the documents in the corpus of publications.

3.2 About Limitations

It's important to keep in mind some limitations of the current study. First, titles and abstracts are only included in this analysis, so this study is unable to capture ethical discussions that are solely included in the full-text, resulting in an incomplete view of the ethical terms in the field. Second, co-occurrence analysis looks at term associations within sentences rather than causation, and thus all the results from the co-occurrence analyses should be thought of as correlational relationships. Third, the ethical terms dictionary used in this study is based on well-known frameworks and definitions and may not account for all potential implicit or domain-specific ethical reasoning that may occur in a variety of contexts, leading to potential exclusion of nuanced ethical considerations.

Finally, the dates chosen for the analysis (2022 to 2026), while excellent at capturing the most recent developments in the field, limit the ability to compare longitudinal changes across the time period of this study, preventing the analysis from determining whether there have been long-term changes in ethical discussions and considerations.

4 Results

We examined 5,000 article titles and abstracts and found them to be laden with terminology specific to different machine learning approaches. Table 1 exhibits the unigram analysis, showing that the most prevalent terms are: "learning" (2,102), "machine" (1,247), "deep" (894) and "model" (700).

As expected, these four reflect machine learning and deep learning approaches, and thus, the preponderance of articles using machine learning algorithms as their approach. It is worth noting, however, that the unigrams "privacy", "fairness" or "ethics" (to name a few) do not appear among the 20 most common unigrams and are thus clearly less prevalent in the vocabulary of the articles.

Table 1. Top 20 most frequent unigrams in 5,000 AI publications.

Rank	Term	Frequency	Rank	Term	Frequency
1	learning	2,102	11	feature	341
2	machine	1,247	12	training	328
3	deep	894	13	language	312
4	model	700	14	clinical	298
5	network	589	15	natural	287
6	intelligence	532	16	reinforcement	276
7	neural	498	17	optimization	265
8	image	412	18	processing	254
9	classification	387	19	recognition	243
10	prediction	356	20	transformer	231

In bigram analysis, we find that the top 5 most frequent bigrams are: "machine learning" (1,079), "deep learning" (642), "artificial intelligence" (579), "language model" (275) and "reinforcement learning" (153). Other bigrams that indicate variety in the AI approach include: "graph neural network" (73) and "convolutional neural network" (72). Ethics-related phrases such as: "interpretable machine" (69), "generative ai" (68) and "explainable ai" (56) do exist, but fall further down the list of most popular bigrams, suggesting that the research community has begun to adopt concepts of interpretability, yet it does not appear to be a predominant topic (see Fig. 1).

Out of the 5,000 articles examined, 740 (14.8%) mention at least one ethical term in their title or abstract; the remaining 4,260 (85.2%) articles are silent on ethics.

Table 2 and Fig. 2 show the distribution of those 740 mentions by the seven ethical dimensions. Safety and Robustness receive the most attention (354 mentions, 41.5%), closely followed by Transparency and Explainability (330, 38.6%). Together, the two dimensions account for over 80% of the content on ethics found in our sample. The other dimensions have substantially smaller numbers: Privacy and Data Protection, 49 mentions (5.7%); Misinformation, 47 (5.5%); General Ethical Principles, 45 (5.3%); Accountability, 21 (2.5%); and Human Agency, 8 (0.9%).

Sentence level co-occurrence analysis (Eq. 3) identified 680 explicit connections of AI and ethical terms. As expected, general mentions (70.3%) were the most prevalent, followed by challenges and risks (27.9%), with few solutions (1.8%) identified. This indicates that most discourse is about diagnosing ethical risks rather than discussing solutions. The distribution across discursive types is given in Table 3.

The articles with the largest S_i scores (Eq. 1) deal with issues such as trust configuration for AI-empowered healthcare (covering transparency, privacy, and misinformation), AI-driven surveillance (covering accountability, privacy, and safety), and equity in AI-driven maternal health applications. It is worth noting that some of these papers are published in multi-disciplinary journals such as the International *Journal of Information*

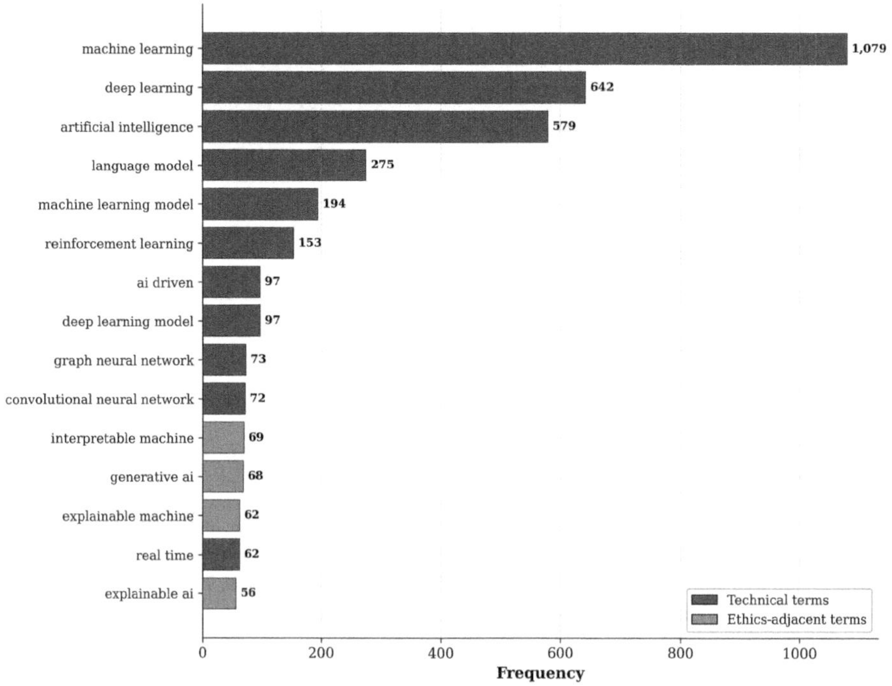

Fig. 1. Top 15 most frequent bigrams in AI publications. Blue bars represent purely technical terms; amber bars indicate ethics adjacent phrases such as "interpretable machine," "generative ai," and "explainable ai," which occupy the lower portion of the frequency ranking.

Table 2. Distribution of ethical mentions across seven dimensions (N = 5,000).

Dimension	Mentions	Percentage
Safety and Robustness	354	41.5%
Transparency and Explainability	330	38.6%
Privacy and Data Protection	49	5.7%
Misinformation	47	5.5%
General Ethical Principles	45	5.3%
Accountability	21	2.5%
Human Agency	8	0.9%
Total	*854*	*100.0%*

Management and Technology in Society. It is possible that articles on ethics are more common in journals with a multi-disciplinary focus.

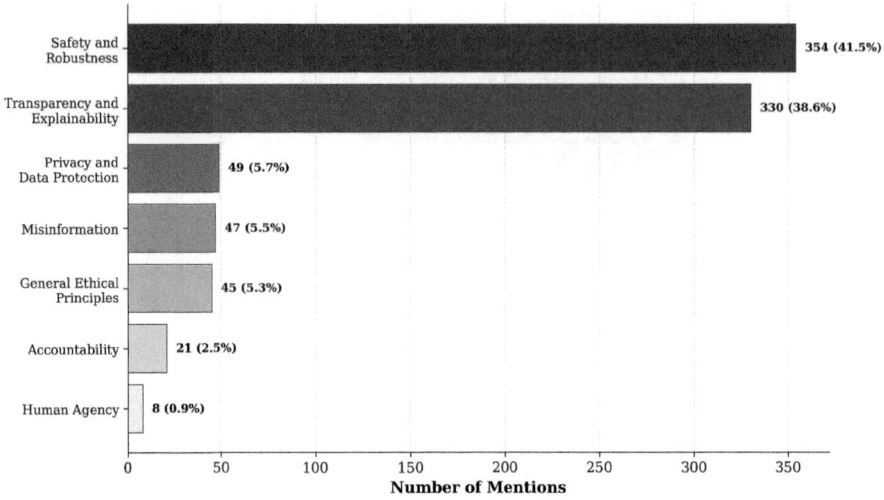

Fig. 2. Distribution of ethical mentions across seven dimensions. Safety and Robustness (41.5%) and Transparency and Explainability (38.6%) represent over 80% of the overall ethical content, a distribution that may reflect a preference for dimensions that are more readily operationalized and implemented in technical solutions.

Table 3. Distribution of AI ethics connections by discursive type.

Connection Type	Count	Percentage
General Mentions	478	70.3%
Challenges and Risks	190	27.9%
Solutions and Mitigations	12	1.8%
Total	*680*	*100.0%*

5 Discussion

The result that 14.8% of the analyzed articles included explicit ethical references should not be understood as either a high or low number without taking into account the structure of AI research. Ethics is a subfield within AI, which, in general, is more technical in nature, and, therefore, comparing the result against the overall number of publications might have overestimated the expected base rate of ethical mentions. Even though, this number should be considered in relation to the high visibility of ethical values in recent policy documents and regulation proposals (Floridi and Cowls, 2019; High-Level Expert Group on AI, 2019). This might imply that ethical aspects, instead of being part of the main content of an article, appear in some cases, as Mittelstadt (2019) and Raji et al. (2020) suggest, as an 'add on'.

The focus on Safety and Robustness ($w_d = 41.5\%$) and Transparency and Explainability ($w_d = 38.6\%$) could be explained by the relative possibility of operationalization within technical research. Both of those dimensions are related to technical performance (Safety and Robustness) or have a more developed technical research program under the XAI label and, therefore, a set of methods and metrics that can be used for their evaluation (Arrieta et al., 2020), in case of Transparency and Explainability. The low frequency of Human Agency ($w_d = 0.9\%$) and Accountability ($w_d = 2.5\%$) dimensions, in the other hand, could be explained by the relative low visibility of these dimensions within this analysis because they belong to the field of law, sociology, and philosophy, with a lower possibility to be operationalized using technical metrics. However, given the importance of human agency in recent legislation, such as the GDPR or the EU AI Act regulations for example (European Commission, 2021) the low coverage of this dimension is worth noting.

The co-occurrence results show that much of the discourse is descriptive in nature, accounting for 70.3% and 27.9% of connections as general and challenges/risks, respectively, and only 1.8% of connections as solutions. This is an indication that the field has been successful at diagnosing the ethical issues that are involved, and is able to articulate the risks involved, but not yet mature enough to be able to propose and assess the efficacy of potential solutions (Binns, 2018).

The small number of solutions connections may be seen as an opportunity for future work to better connect proposed technical solutions to the problems that they seek to resolve (Rahwan, 2018). The bigram analysis highlights that the topics of language models, reinforcement learning, and graph neural networks have emerged as popular research areas recently.

The fact that phrases like 'interpretable machine' and 'generative ai' have been identified as frequently also indicates that these new topics, interpretability and AI generated content, are starting to grow into their own topics within literature. Articles with high ethical relevance scores that deal with areas such as AI based surveillance, trust in the medical field, and fairness in maternal health, provide examples of ethical research within the context of practical use in sensitive areas. The fact that these articles are published in interdisciplinary journals indicates that journals that are more concerned with societal outcomes may facilitate more ethical research.

6 Conclusion

This research conducted a bibliometric analysis of 5,000 Scopus AI papers (2022–2026) through the lens of seven ethical dimensions. We offer three main insights. First, 14.8% of the examined articles contain explicit ethical terminology in their titles and abstracts (this is just about how often the words are in the metadata and is not necessarily an accurate reflection of how deep the ethical content is inside the full-text paper), which is a modest portion that we must contextualize when thinking about the specialized nature of ethics.

Second, discussions are skewed towards dimensions that are technically more straightforward to measure: Safety and Robustness (41.5%) and Transparency and Explainability (38.6%), whereas Human-centric dimensions are mentioned far less frequently (e.g. Human Agency (0.9%), and Accountability (2.5%). Third, the nature of the mentions is mostly descriptive in form. General mention (70.3%) and Challenge identification (27.9%) dominate while very few proposed solutions (1.8%) are given, which opens up a lot of potential to research that combines technical innovation and ethical risks.

Finally, these results should be read while considering potential limitations of the research. It does not consider full text papers correlational, potentially missing ethical terms in the dictionary, and the time frame for analysis is short. Potential directions for future studies would include the investigation of full text paper corpora, the application of sentiment analysis, the integration of citation network analysis and longitudinal or cross-database comparative study. This study contributes a complete, end-to-end reproducible workflow, a complete ethical term dictionary, and a quantitative baseline of the current state of ethical mentions in AI research.

Appendix A. Ethical Term Dictionary

a) Transparency and Explainability: transparency, transparent, explainability, explainable, interpretability, interpretable, xai, black box, opacity, understandable, clear decision.
b) Accountability: accountability, accountable, responsibility, responsible, liability, oversight, audit, answerability, blame.
c) Privacy and Data Protection: privacy, private, data protection, gdpr, confidentiality, anonymization, data governance, surveillance, personal data, data privacy, information privacy.
d) Safety and Robustness: safety, safe, robustness, robust, security, secure, risk, adversarial, reliability, trustworthy, fail safe, danger, harm, safe ai.
e) Human Agency: human agency, human oversight, human control, human centered, human centric, human in the loop, autonomy, human autonomy, human decision, human supervision.
f) General Ethical Principles: ethics, ethical, moral, morality, values, principles, ethical framework, ethical guidelines, responsible ai, ethical ai, moral responsibility, ethical considerations.
g) Misinformation: misinformation, disinformation, fake news, deepfake, manipulation, deception, trust, truthfulness, authenticity, false information, ai generated content, synthetic media.

References

Amodei, D., et al.: Concrete problems in AI safety. arXiv preprint. (2016). https://doi.org/10.48550/arXiv.1606.06565

Arrieta, A.B., et al.: Explainable artificial intelligence (XAI): concepts, taxonomies, opportunities and challenges toward responsible AI. Inf. Fusion. **58**, 82–115 (2020). https://doi.org/10.1016/j.inffus.2019.12.012

Binns, R.: Fairness in machine learning: lessons from political philosophy. In: Proceedings of the 2018 Conference on Fairness, Accountability, and Transparency, pp. 149–159 (2018). https://doi.org/10.1145/3287560.3287586

Bommasani, R., et al.: On the opportunities and risks of foundation models. arXiv preprint. (2021). https://doi.org/10.48550/arXiv.2108.07258
Bornmann, L., Mutz, R.: Growth rates of modern science: a bibliometric analysis based on the number of publications and cited references. J. Assoc. Inf. Sci. Technol. **66**(11), 2215–2222 (2015). https://doi.org/10.1002/asi.23329
Burnham, J.F.: Scopus database: a review. Biomed. Digit. Lib. **3**(1), 1–8 (2006). https://doi.org/10.1186/1742-5581-3-1
Brynjolfsson, E., McAfee, A.: The business of artificial intelligence. Harv. Bus. Rev. **95**(4), 52–61 (2017)
Corbett-Davies, S., Goel, S.: The measure and mismeasure of fairness: a critical review of fair machine learning. arXiv preprint. (2018). https://doi.org/10.48550/arXiv.1808.00023
Denzin, N.K.: The Research Act: a theoretical introduction to sociological methods. Transaction Publishers (2017). https://doi.org/10.4324/9781315134543
Dwivedi, Y.K., et al.: Artificial intelligence (AI): multidisciplinary perspectives on emerging challenges, opportunities, and agenda for research, practice and policy. Int. J. Inf. Manag. **57**, 101994 (2023). https://doi.org/10.1016/j.ijinfomgt.2019.08.002
Elsevier: Scopus APIs: developer documentation. Elsevier B.V (2024) https://dev.elsevier.com/
European Commission. Proposal for a regulation laying down harmonised rules on artificial intelligence (Artificial Intelligence Act). COM (2021) 206 final. (2021). https://eur-lex.europa.eu/legal-content/EN/TXT/?uri=CELEX%3A52021PC0206
Floridi, L., Cowls, J.: A unified framework of five principles for AI in society. Harvard Data Sci. Rev. **1**(1) (2019). https://doi.org/10.1162/99608f92.8cd550d1
Grimmer, J., Stewart, B.M.: Text as data: the promise and pitfalls of automatic content analysis methods for political texts. Polit. Anal. **21**(3), 267–297 (2013). https://doi.org/10.1093/pan/mps028
Gusenbauer, M., Haddaway, N.R.: Which academic search systems are suitable for systematic reviews or meta-analyses? evaluating retrieval qualities of Google Scholar, PubMed, and 26 other resources. Res. Synth. Methods. **11**(2), 181–217 (2020). https://doi.org/10.1002/jrsm.1378
Harzing, A.W., Alakangas, S.: Google Scholar, Scopus and the web of science: a longitudinal and cross-disciplinary comparison. Scientometrics. **106**(2), 787–804 (2016). https://doi.org/10.1007/s11192-015-1798-9
High-Level Expert Group on AI: Ethics Guidelines for Trustworthy AI. European Commission (2019) https://digital-strategy.ec.europa.eu/en/library/ethics-guidelines-trustworthy-ai
Hunter, J.D.: Matplotlib: a 2D graphics environment. Comput. Sci. Eng. **9**(3), 90–95 (2007). https://doi.org/10.1109/MCSE.2007.55
Krippendorff, K.: Content Analysis: an Introduction to its Methodology, 4th edn. SAGE Publications (2018). https://doi.org/10.4135/9781071878781
Leydesdorff, L., Welbers, K.: The semantic mapping of words and co-words in contexts. J. Informet. **5**(3), 469–475 (2011). https://doi.org/10.1016/j.joi.2011.01.008
Mittelstadt, B.: Principles alone cannot guarantee ethical AI. Nat. Mach. Intell. **1**(11), 501–507 (2019). https://doi.org/10.1038/s42256-019-0114-4
Moher, D., Liberati, A., Tetzlaff, J., Altman, D.G.: Preferred reporting items for systematic reviews and meta-analyses: the PRISMA statement. PLoS Med. **6**(7), e1000097 (2009). https://doi.org/10.1371/journal.pmed.1000097
OECD: Recommendation of the Council on Artificial Intelligence. OECD Publishing (2024). https://doi.org/10.1787/6481e6b9-en
Pedregosa, F., et al.: Scikit-learn: machine learning in python. J. Mach. Learn. Res. **12**, 2825–2830 (2011)
Rahwan, I.: Society-in-the-loop: programming the algorithmic social contract. Ethics Inf. Technol. **20**(1), 5–14 (2018). https://doi.org/10.1007/s10676-017-9430-8

Raji, I.D., et al.: Closing the AI accountability gap: defining an end-to-end framework for internal algorithmic auditing. In: Proceedings of the 2020 Conference on Fairness, Accountability, and Transparency, pp. 33–44 (2020). https://doi.org/10.1145/3351095.3372873

Röder, M., Both, A., Hinneburg, A.: Exploring the space of topic coherence measures. In: Proceedings of the Eighth ACM International Conference on Web Search and Data Mining, pp. 399–408 (2015). https://doi.org/10.1145/2684822.2685324

Shokri, R., Stronati, M., Song, C., Shmatikov, V.: Membership inference attacks against machine learning models. IEEE Symp. Secur. Privacy. **2017**, 3–18 (2017). https://doi.org/10.1109/SP.2017.41

UNESCO: Recommendation on the Ethics of Artificial Intelligence. United Nations Educational, Scientific and Cultural Organization (2021) https://unesdoc.unesco.org/ark:/48223/pf0000380455

Waskom, M.L.: Seabon: statistical data visualization. J. Open Source Softw. **6**(60), 3021 (2021). https://doi.org/10.21105/joss.03021

Wickham, H.: Tidy data. J. Stat. Softw. **59**(10), 1–23 (2014). https://doi.org/10.18637/jss.v059.i10

Wiener, N.: Some moral and technical consequences of automation. Science. **131**(3410), 1355–1358 (1960). https://doi.org/10.1126/science.131.3410.1355

Zhang, D., et al.: The AI Index 2021 Annual Report. Stanford University (2021). https://doi.org/10.25740/fr239wp7855

Zupic, I., Čater, T.: Bibliometric methods in management and organization. Organ. Res. Methods. **18**(3), 429–472 (2015). https://doi.org/10.1177/1094428114562629

Topos-Theoretic Formalism for Causal Event Calculus

Manuel Hernández(✉) and Eduardo Sánchez-Soto

Instituto de Computación, Universidad Tecnológica de la Mixteca, Huajuapan de León, Oaxaca, Mexico
manuelhg@mixteco.utm.mx, esanchez@mixteco.utm.mx

Abstract. We delineate a theoretical and conceptual synthesis that resolves the limitations of classical temporal logic in distributed environments by transposing the Event Calculus into the language of Grothendieck topos theory. In this unifying framework, the static universe of sets is replaced by a dynamic universe of sheaves defined over a site representing causal events and regimes. Fluents are reformulated as classifier objects that encode intuitionistic truth and causal stability. Furthermore, agent conduct is redefined as a geometric navigation of this topos, guided by gradients of variational free energy and the structural constraints of Topos Causal Models. This synthesis offers a robust mathematical ontology where truth is local, causality is topological, and agency is the continuous preservation of coherence across a distributed system.

Keywords: event calculus · topos theory · category theory · multiagent systems

1 Introduction: The Geometric Turn in Temporal Logic

The formal representation of time, change, and causality remains one of the most enduring challenges in artificial intelligence and computational logic. Since the mid-20th century, the discipline has oscillated between symbolic approaches—characterized by rigidity, interpretability, and logical axiomatic systems—and connectionist or dynamical approaches, which prioritize continuous adaptation and gradient-based learning. The **Event Calculus (EC)**, originally formulated by Robert Kowalski and Marek Sergot in 1986, stands as a preeminent example of the former [10,18]. It provides a rigorous logic-programming framework for reasoning about the effects of actions, the persistence of properties (fluents), and the discrete evolution of systems over time. However, as intelligent systems increasingly operate in distributed, multi-agent, and partially observable environments, the classical extensional semantics of the Event Calculus—grounded in standard first-order logic and global truth—faces significant ontological friction.

Recent efforts to address these limitations have emerged from several directions. **Modal extensions** of the Event Calculus incorporate epistemic operators

H. Ponce et al. (Eds.): COMIA 2026, CCIS 3095, pp. 241–254, 2027.
https://doi.org/10.1007/978-3-032-37438-7_19

to handle distributed knowledge [7], yet these often remain within a classical logical framework that does not fundamentally address the geometric nature of causal locality. **Categorical approaches to temporal reasoning** have gained traction through the work of [1], demonstrating that contextual and non-local phenomena in distributed systems find natural expression in sheaf theory. **Topos Causal Models (TCM)**, introduced by Mahadevan [12,13], provide a systematic method for lifting structural causal models into topos-theoretic semantics, enabling intuitionistic reasoning about interventions and counterfactuals. Simultaneously, **Active Inference** [3,14] has formalized agent conduct as variational free energy minimization, while recent work on **sheaf neural networks** [5] demonstrates how topological constraints can guide learning in distributed systems.

This report delineates a theoretical synthesis that resolves the limitations of classical temporal logic by transposing the Event Calculus into the language of **Grothendieck topos theory**. In this unifying framework, the static "universe of sets" is replaced by a dynamic "universe of sheaves" defined over a **site** representing causal events and regimes. Here, **fluents** are no longer mere predicates mapping to a Boolean set $\{0, 1\}$, but are **classifier objects** (morphisms to a subobject classifier Ω) that encode the nuance of intuitionistic truth, contextual validity, and causal stability [4,11]. Furthermore, **agent conduct** is reformulated not merely as theorem proving, but as a geometric navigation of this topos, guided by gradients of variational free energy and the structural constraints of **Topos Causal Models (TCM)** [12,13].

The analysis proceeds by first deconstructing the classical Event Calculus to identify the specific semantic bottlenecks that necessitate a categorical upgrade. It then constructs the geometric universe—the **Causal Site**—defining how causal regimes form a category equipped with a Grothendieck topology. Subsequent sections detail the formulation of fluents within the internal logic of the topos, the rigorous redefinition of causal interventions as subobject manipulations via the *j-Do-Calculus*, and the operationalization of agent behavior through **Active Inference** and **Sheaf Neural Networks**. A comprehensive comparison with existing approaches is provided in Sect. 8. This synthesis offers a robust mathematical ontology where "truth" is local, "causality" is topological, and "agency" is the continuous preservation of coherence across a distributed system.

2 The Logical Substrate: Deconstructing the Event Calculus

To appreciate the necessity of a topos-theoretic formulation, one must first rigorously examine the mechanics and the inherent limitations of the classical Event Calculus (EC). The EC is a logic-based formalism for representing and reasoning about actions and their effects [10]. Unlike the Situation Calculus, which relies on global state transitions, the EC reifies time and focuses on the local initiation and termination of properties.

2.1 Core Ontology and Predicate Structure

The fundamental ontology of the EC consists of three sorts: **Events** (E), **Fluents** (F), and **Time Points** (T). The state of the world is not represented as a static snapshot but is derived dynamically from the history of events. The axiomatic structure relies on a core set of predicates that define the causal mechanics of the domain [18]:

- `Initiates`(e, f, t): Event e initiates fluent f at time t.
- `Terminates`(e, f, t): Event e terminates fluent f at time t.
- `Happens`(e, t): Event e occurs at time t.
- `HoldsAt`(f, t): Fluent f holds true at time t.
- `Clipped`(t_1, f, t_2): The persistence of fluent f is interrupted between t_1 and t_2.

The central mechanism of the EC is the **axiom of persistence** (or inertia), which asserts that a fluent holds at a time t if it was initiated at some previous time t_0 and has not been terminated (clipped) in the interim. Formally, this is often expressed via Horn clauses augmented with negation-as-failure (NAF) [10].

The reliance on Negation-as-Failure—specifically to determine `Not Clipped`—introduces a form of non-monotonic reasoning essential for solving the **Frame Problem** [18]. The system assumes that properties persist unless there is explicit evidence of their termination. While computationally elegant for centralized databases or narrative understanding, this "Closed World Assumption" becomes problematic in open, distributed multi-agent systems where the absence of information does not imply falsity, but rather a lack of access to the relevant causal regime.

Persistence as Sheaf Gluing. When transposed to the topos-theoretic setting, the persistence of a fluent across time is no longer captured by negation-as-failure but by the existence of a **unique horizontal section** across a temporal sieve. Specifically, for a covering family of time points $\{t_i \to t\}$ in a temporal covering sieve, a fluent F persists from t_0 to t iff the local sections $\sigma_i \in F(t_i)$ satisfy the gluing condition $\sigma_i|_{t_i \cap t_j} = \sigma_j|_{t_i \cap t_j}$ for all overlaps. The frame problem thus reduces to the cohomological condition $H^1(\mathcal{F}) = 0$—i.e., no obstruction to gluing.

2.2 Limitations in Distributed Causal Environments

The transition from classical logic to category theory is driven by specific inadequacies of the standard EC when applied to complex, realistic environments:

1. **Global Truth vs. Local Validity:** Classical EC assumes a global clock and a global truth assignment. A fluent is either true or false in the model. In distributed systems, truth is often local. The classical model forces a collapse of this local structure into a global inconsistency or requires cumbersome modal operators.

2. **The Frame Problem and Concurrency:** Handling concurrent events with partial ordering requires complex extensions to the basic EC, often leading to exponential complexity in computing Maximal Validity Intervals (MVIs) [18].
3. **Intervention vs. Observation:** Classical EC treats updates as additions to a narrative. It lacks a native distinction between *observing* a value and *intervening* to set a value—a distinction central to modern causal inference [15].
4. **Continuous Change:** Integrating continuous physical processes (gradients, flows) with discrete logic requires hybrid automata or distinct formalisms, creating an "impedance mismatch" between the agent's reasoning (logic) and its control (dynamics). The proposed framework resolves this via **Synthetic Differential Geometry (SDG)**, which internalizes infinitesimals within the topos, and discrete **Sheaf Neural Network (SNN)** diffusion, which serves as a finite approximation of the SDG tangent bundle.

The proposed topos-theoretic framework resolves these issues by embedding the EC into a **site**. In this geometric setting, "truth" becomes a sheaf (local and glueable), "negation" becomes intuitionistic (handling partial information), and "interventions" become geometric surgeries on the site itself.

3 The Geometric Universe: Grothendieck Toposes and the Causal Site

The theoretical core of this synthesis is the **Grothendieck topos**. A topos is a category that behaves like the category of sets but possesses a rich internal structure defined by its geometry [4,11]. To define a topos for the Event Calculus, we must first define its underlying **site**.

3.1 Defining the Causal Site (C, J)

A **site** is a pair (C, J) consisting of a small category C and a Grothendieck topology J [11]. For the Causal Event Calculus, the category C represents the **causal topology** of the environment.

The Category of Regimes (C). Instead of a set of time points, the objects of C are **causal regimes** or **contexts** [12].

- **Objects (Ob(C)):** Each object U represents a specific observational or interventional context. For example, U_{obs} might represent the "passive observation" regime, while $U_{do(X)}$ represents the regime where an agent has intervened on variable X. In a multi-agent setting, objects can also represent the local perspective of a specific agent.
- **Morphisms (Hom(C)):** Arrows $f : V \to U$ represent causal relationships, such as "regime V is a sub-context of U" or "regime V is reachable from U via intervention." These morphisms encode the structural constraints of the environment.

The Grothendieck Topology (J). The topology J defines what it means to "cover" a regime with smaller regimes [11]. It provides the mechanism for gluing local information into global facts.

- A **Sieve** on an object U is a collection of morphisms into U that is closed under pre-composition.
- A **Grothendieck Topology** J assigns to each object U a set of "covering sieves" $J(U)$.
- **Causal Interpretation:** A covering sieve represents a decomposition of a complex causal event into its constituent mechanisms. For example, to know if an event "occurred" in a broad spatiotemporal regime U, it suffices to verify its occurrence in a specific set of local sub-regimes $\{U_i\}$ that cover U.

This structure allows the Event Calculus to operate *locally*. An agent does not need a global database; it needs only to verify predicates on a cover of its current context. If these local verifications are compatible (they agree on overlaps), the **sheaf property** guarantees they constitute a valid global truth.

3.2 Fluents as Classifier Objects

The most radical shift in this framework is the redefinition of **fluents**. In classical EC, a fluent is a predicate that maps to $\{\text{True}, \text{False}\}$. In a topos, truth values are objects in the **subobject classifier** Ω [4].

The Subobject Classifier Ω. In a topos of presheaves or sheaves, Ω is not a two-element set. It is an object that classifies subobjects. Specifically, there is a natural bijection between subobjects of any object A and morphisms from A to Ω:

$$\text{Sub}(A) \cong \text{Hom}(A, \Omega) \tag{1}$$

In the context of the Causal Site, $\Omega(U)$ (the set of truth values at regime U) is the set of all **closed sieves** on U [11].

- **Interpretation:** A "truth value" at regime U is a specification of *where* (in which sub-regimes) a statement is true. It captures the **extent** of validity.
- **Fluents as Morphisms:** A fluent F is defined as a morphism $F : \text{State} \to \Omega$. This means that asserting $\texttt{HoldsAt}(F, t)$ is equivalent to identifying the specific sieve of contexts where F is active. This naturally encodes the concept of **Maximal Validity Intervals (MVIs)** from classical EC [18] but extends it to "Maximal Validity Contexts" in the causal topology.

Heyting Algebra and Intuitionistic Logic. The object Ω possesses the structure of a **Heyting Algebra**, not a Boolean Algebra [4]. This means:

- **Law of Excluded Middle Fails:** $\phi \vee \neg\phi = \top$ is not generally true. A fluent might be neither "holding" nor "not holding" in a regime where the agent lacks sufficient causal access.

- **Negation as Pseudo-Complement:** $\neg\phi$ is the largest element disjoint from ϕ. In the EC context, this formalizes **Negation-as-Failure** without the fragility of the Closed World Assumption. $\neg\texttt{HoldsAt}(F, t)$ means "there is no evidence of F in any accessible sub-regime."

3.3 The Lawvere-Tierney Topology (j)

To govern agent conduct, we introduce a **Lawvere-Tierney topology**, denoted as $j : \Omega \to \Omega$ [11]. This operator acts as a modality, allowing the agent to filter "raw" truth into "actionable" truth.

- **Definition:** j is an idempotent, truth-preserving operator on the subobject classifier.
- **j-Truth:** A fluent ϕ is said to be "j-true" (or "locally true") if $j \circ \phi = \text{true}$.
- **Application in Causality:** [13] utilizes j to define **stability**. A causal relationship is stable if it holds under the specific topological covers defined by j. An agent using this framework does not act on all fluents, but only on those that are **j-closed** (stable) fluents. This prevents the agent from reacting to transient noise or causally irrelevant variations.

4 Topos Causal Models (TCM): Synthesizing Logic and Causality

Building upon the site definition, we integrate the recent advancements in **Topos Causal Models (TCM)** [12,13]. TCM provides the rigorous mechanism for defining events and interventions within the site structure.

4.1 Structural Causal Models as Sheaves

A classic Structural Causal Model (SCM) consists of exogenous variables U, endogenous variables V, and functions F determining V from U. In TCM, an SCM is lifted to the topos $\mathbf{Sh}(C, J)$.

- Variables become **Sheaves** of random variables over the site.
- Functions become **Natural Transformations** (morphisms of sheaves).
- **Universal Property:** The causal model is defined by the **Yoneda embedding**, which maps the causal regimes in C to presheaves in the topos, preserving the causal structure via the Yoneda Lemma [12].

4.2 Interventions as Subobjects

The most significant divergence from classical EC is the treatment of interventions. In [15], do($X = x$) breaks the arrows entering X. In TCM, interventions are defined as **subobjects** [13].

- **The Intervention Subobject:** An intervention defines a specific sub-region of the total outcome space where the intervention condition holds.
- **Categorical Definition:** If Σ is the object representing the system state, an intervention do($X = x$) is a monomorphism (subobject) $i : I_{do} \hookrightarrow \Sigma$.
- **Implication:** Interventions are not external operators applied *to* the model; they are intrinsic geometric regions *within* the model. Reasoning about an intervention is equivalent to restricting the logic to this subobject.

4.3 The j-Do-Calculus: Rules of Conduct

Classic EC uses rules like `Initiates` to deduce effects. TCM replaces this with the **j-Do-Calculus**, a set of inference rules derived for the intuitionistic logic of the topos [13]. This calculus governs how an agent can derive causal knowledge (Table 1).

Table 1. Comparison of Causal Calculi

Feature	Classical Do-Calculus	Intuitionistic j-Do-Calculus
Logic Base	Boolean Logic (Classical)	Intuitionistic Logic (Heyting)
Validity Scope	Global Truth	Local Truth (on a J-cover)
Separation	d-separation (Graph based)	j-dense separation (Sheaf based)
Intervention	Graph Surgery (Arrow cutting)	Subobject Selection/Kernel Replacement
Inference Goal	Identification of $P(y\|do(x))$	A section on the Interventional Sheaf

Formally, a j-stable separation between variables X and Y given Z in the causal site means that the Lawvere-Tierney topology $j : \Omega \to \Omega$ renders the subobject representing the conditioning set as j-dense. An intervention do($X = x$) is **constructively valid** only if its corresponding subobject $I_{do} \hookrightarrow \Sigma$ is j-dense—i.e., its negation is j-false. If the intervention is not j-dense, the agent cannot constructively prove the outcome, and must either gather more information or refine the causal site.

The j-Do-Calculus provides three primary rules (generalizing Pearl's rules) that allow an agent to:

1. **Insert/Delete Observations:** If a variable is j-separated from the outcome, it can be ignored.
2. **Action/Observation Exchange:** Under stability conditions (back-door criterion on the site), observing X is equivalent to doing X.
3. **Insert/Delete Actions:** If an action has no causal path to the outcome (verified via sheaf restriction), the intervention can be removed from the query.

5 Agent Conduct: Active Inference and Sheaf Dynamics

With the static structure (Site) and logical rules (j-Do-Calculus) established, we define the dynamic **conduct** of the agent. The integration of **Active Inference** with sheaf theory provides the mechanism for goal-directed behavior.

5.1 Active Inference as a Functorial Dynamic

Active Inference posits that agents act to minimize **Variational Free Energy (VFE)**, which bounds the surprise of sensory observations relative to a generative model [2,3]. In the topos framework, this is formalized using **Polynomial Functors** and **Sheaf Cohomology**.

- **The Agent as a Sheaf:** The agent's internal model is a sheaf A on the site. Its beliefs are sections of this sheaf.
- **Surprise as Cohomological Obstruction:** When an agent receives data (a section d on regime U_{obs}) that conflicts with its internal prediction (a section p on U_{pred}), the mismatch prevents these local sections from being glued into a global section.
- **Sheaf Cohomology (H^1):** The magnitude of this "gluing failure" is measured by the first cohomology group of the sheaf [11]. High cohomology corresponds to high free energy (high surprise).

5.2 Shared Protentions: The Phenomenology of Multi-agent Time

In multi-agent scenarios, conduct is coordinated via **Shared Protentions**. Borrowing from Husserl's phenomenology of time consciousness, [2] model this using sheaf theory.

- **Protention as Sheaf Extension:** A "protention" is a section of the belief sheaf extending into a future regime.
- **Consensus Topos:** When multiple agents interact, they attempt to align their protentions. This is mathematically equivalent to constructing a **Consensus Sheaf** on the joint site.
- **Cooperative Conduct:** Agents act to minimize the "divergence" between their individual protentions. If Agent A expects Event E and Agent B does not, the system has high free energy. Conduct involves communicative actions (sharing evidence/sections) to enable the gluing of their respective worldviews.

The construction of a Consensus Sheaf on the joint site is further operationalized in [6], which proposes a formally verifiable framework for multi-agent coordination. This prior work provides the computational substrate—specifically the cellular sheaf consensus dynamics—upon which the present theoretical synthesis builds. While that work focuses on implementation and verification, the current paper provides the foundational topos-theoretic semantics that justifies those constructions. We refer to Table 2 to show a major landscape of our proposal.

Table 2. Cross-Framework Synthesis: Core Concepts vs. Verifiable Extensions

Core Concept	Causal Event Topos (CET)	Extension given in [6]
Information Assimilation	Modeled as sections of a site representing causal regimes.	Formalized via the **categorical pullback** operation to contextualize global data into local frames.
Logical Deliberation	Relies on j-Do-Calculus for causal stability and inference.	Utilizes internal **intuitionistic logic** to generate machine-checkable **proof objects** for each action.
Agent States	Encoded as fluents behaving as morphisms to the classifier Ω.	Modeled as **cellular sheaves** where local data reside on stalks and constraints are restriction maps.
Global Agreement	Defined as the continuous preservation of coherence across a system.	Defined as the search for a **global section**, resolving conflicts via sheaf cohomology.
Dynamic Conduct	Guided by gradients of Variational Free Energy (VFE).	Achieved through **Sheaf Laplacian diffusion** to drive the system toward a state of consensus.
Practical Application	Focuses on solving the Frame Problem and causal interventions.	Focuses on **endogenous team formation** and verifiable multi-agent coordination.

6 Detailed Framework Specification

To provide a concrete reference for implementation, we formalize the components of the **Causal Event Topos (CET)** framework.

First, we describe **The Causal Site Specification**. The foundational layer is the **Site**, denoted $\mathcal{S}_{causal} = (\mathcal{C}, J)$.

- **Category $\mathcal{C}$ (Regimes):**
  - **Objects:** $U \in \mathrm{Ob}(\mathcal{C})$ represent causal contexts (e.g., $U_\emptyset$, $U_{do(X=x)}$).
  - **Morphisms:** $f : V \to U$ represents a restriction or specialization.
- **Topology J (Coverage):** J assigns to each regime U a set of covering families $\{f_i : U_i \to U\}_{i \in I}$. A property holds in U iff it holds in every U_i of a covering family.

We refer to Table 3 to give a summary of analog concepts.

Next, we describe the **Gradient Flow.** The agent updates its physical actuators u (control signals) according to the gradient:

$$\dot{u} = -\nabla_u \mathcal{F} \tag{2}$$

This differential equation is solved using the **Sheaf Laplacian** on the site graph, ensuring that the control signals propagate consistently across the causal topology. We will use this concept below.

Table 3. Logic Translation Table

EC Predicate	Topos-Theoretic Semantics (Kripke-Joyal)	Logic Requirement
`HoldsAt`(F, t)	$U \Vdash F(s,t)$ iff $\exists$ cover $\{U_i\}$ s.t. F is locally true in each U_i	Sheaf condition
`Initiates`(E, F, t)	The event sheaf E acts on F to change its section from $\bot$ to $\top$ over U	Natural transformation
`Not HoldsAt`(F, t)	$U \Vdash \neg F$ iff $\forall f : V \to U$, $V \nVdash F$	Intuitionistic negation
`Happens`(E, t)	E is a section of the event sheaf at index t; must form a minimal covering sieve	Covering condition
`Clipped`(t_1, F, t_2)	$\exists$ event E s.t. the section of F cannot be extended across $[t_1, t_2]$	Cohomological obstruction: $H^1 \neq 0$

7 Implementation: Synthetic Differential Geometry and Gradients

A critical requirement for modern agents is the ability to perform optimization (gradient descent). Classical EC is discrete and does not naturally support gradients. However, **Synthetic Differential Geometry (SDG)** allows us to define gradients *within* the logic of the topos [9].

7.1 Gradients in the Topos

SDG enriches the topos with a ring object R (the line) and "infinitesimal" objects D (where $d \in D \implies d^2 = 0$).

- **Differentiation as Algebra:** A function $f : R \to R$ is differentiable if there is a unique slope g such that $f(x+d) = f(x) + d \cdot g(x)$ for all infinitesimals d.
- **Free Energy Gradients:** The Variational Free Energy $\mathcal{F}$ becomes a morphism in the topos. The "gradient" $\nabla\mathcal{F}$ is structurally defined as the tangent vector to the sheaf of beliefs.

7.2 Sheaf Neural Networks (SNNs)

The computational realization of this framework is the **Sheaf Neural Network**.

- **Structure:** The site is discretized into a graph (or simplicial complex). The "sheaf" attaches a vector space (the stalk) to each node (regime).
- **Laplacian Dynamics:** The **Sheaf Laplacian** $\Delta_{\mathcal{F}}$ generalizes the graph Laplacian. It measures the "inconsistency" of data across the causal links.

- **Diffusion as Inference:** The process of minimizing free energy is implemented as a diffusion process using the Sheaf Laplacian:

$$\dot{x} = -\Delta_{\mathcal{F}} x \tag{3}$$

  This differential equation drives the agent's state x (its fluents/beliefs) toward a harmonic section—a state of global consistency across the site.

Bridge to SDG. The Sheaf Laplacian $\Delta_{\mathcal{F}}$ approximates the SDG-derived gradient flow via finite differences on the site's discretization. Specifically, for an infinitesimal $d \in D$ (where $d^2 = 0$), the SDG derivative satisfies $f(x + d) = f(x) + d \cdot \nabla f(x)$. The SNN diffusion $\dot{x} = -\Delta_{\mathcal{F}} x$ discretizes this continuous flow, with the Laplacian acting as the linear approximation of the gradient operator on the cellular sheaf. Convergence to a harmonic section ($\Delta_{\mathcal{F}} x = 0$) corresponds to reaching a stationary point where the SDG directional derivative vanishes in all infinitesimal directions.

8 Comparative Analysis and State of the Art

To situate the proposed Causal Event Topos (CET) framework within the broader landscape, we provide a systematic comparison with existing formalisms for temporal and causal reasoning in distributed agent systems.

Modal and Epistemic Extensions of Event Calculus. Extensions of EC with modal operators have been proposed to handle distributed knowledge and belief. [7] present a comprehensive survey of modal temporal logics for multi-agent systems, including epistemic extensions of the Situation Calculus. However, these approaches typically retain a classical two-valued semantics for truth, relying on possible-worlds models that, while expressive, do not internalize the structure of causal locality. The CET framework differs fundamentally by treating truth as intrinsically contextual via the subobject classifier Ω, eliminating the need for external modal machinery while providing a richer notion of partial truth.

Categorical and Sheaf-Theoretic Approaches. The application of sheaf theory to distributed systems and contextuality has been pioneered by [1], demonstrating that sheaf cohomology captures non-local correlations in measurement scenarios. More recently, [17] developed sheaf-theoretic models for dynamic systems, emphasizing the role of consistency conditions in multi-agent coordination. The CET framework extends these insights by constructing a specific site tailored to causal regimes and integrating the Event Calculus's temporal ontology directly into the sheaf structure.

Topos Causal Models and Quantum Causality. The work of [12,13] provides the direct foundation for our synthesis, establishing Topos Causal Models as a unifying framework for classical, quantum, and relativistic causality. A parallel development by [8] explores causal models in presheaf categories, demonstrating that the distinction between observational and interventional distributions corresponds to different sites. The CET framework complements these by focusing on the dynamics of agent conduct, explicitly incorporating variational principles and learning mechanisms not present in earlier TCM formulations.

Active Inference and Variational Methods. The Active Inference literature has grown substantially since Friston's formulation [3]. Recent work by [14] provides a comprehensive treatment of the mathematical foundations, including the relationship between free energy and Bayesian mechanics. [16] extend this to multi-scale systems, emphasizing the role of nested Markov blankets. The integration with sheaf theory introduced here provides a geometric interpretation of free energy minimization as cohomological gluing, offering a novel perspective on the emergence of coherent multi-agent behavior (Table 4).

Table 4. Comparative Framework Summary

Approach	Truth Semantics	Causality	Agency Mechanism
Classical EC	Boolean, Global	Initiate/Terminate Rules	Theorem Proving
Modal EC	Possible Worlds	Epistemic Access	Modal Reasoning
Topos Causal Models	Intuitionistic, Local	Subobject Interventions	Internal Logic
Active Inference	Probabilistic	Graphical Models	Free Energy Minimization
Sheaf Neural Networks	Vector Space, Local	Graph Laplacian	Diffusion
CET (This Work)	**Intuitionistic Sheaf**	j**-Do-Calculus**	**Cohomological VFE Minimization**

9 Conclusion and Future Directions

The synthesis of Kowalski and Sergot's Event Calculus with Grothendieck topos theory represents a paradigmatic shift in the formalization of causal agents. By moving from the extensional semantics of sets to the intensional, geometric semantics of **sheaves on a causal site**, we resolve the fundamental limitations of the classical approach regarding partial observability, distributed truth, and the integration of continuous dynamics.

In this framework, the **Site** (C, J) replaces linear time with a topology of causal regimes. **Fluents** evolve from boolean predicates to classifier arrows targeting a truth-object Ω enriched by a **Lawvere-Tierney topology**, allowing for nuanced, stable truth values. **Agent conduct** is rigorously defined as the navigation of this geometric landscape, guided by the intuitionistic logic of the j-**Do-Calculus** and powered by the thermodynamic drive of **Active Inference** and **Sheaf Neural Network** diffusion. The comparative analysis demonstrates that the CET framework uniquely unifies logical rigor, geometric structure, and variational learning in a single mathematical ontology.

Future work will proceed along three axes. First, **empirical validation** through implementation of the Sheaf Neural Network architecture in simulated multi-agent environments, comparing its performance against baseline EC and deep reinforcement learning approaches. Second, **extension to higher-order causal structures**, leveraging the higher algebraic K-theory of causality [13] to capture multi-agent strategic interactions. Third, **integration with quantum causal models**, exploring whether the topos-theoretic framework can provide a unified semantics for classical and quantum agency, particularly in the context of quantum active inference.

Acknowledgments. The authors would like to express their gratitude to the Universidad Tecnológica de la Mixteca for the institutional support and resources provided during this research. We also extend our sincere thanks to the anonymous reviewers of COMIA 2026 for their insightful comments and suggestions, which significantly improved the clarity and depth of this work.

References

1. Abramsky, S.: Contextual semantics: from quantum mechanics to logic, databases, constraints, and complexity. Bull. Eur. Assoc. Theor. Comput. Sci. **104**, 27–52 (2011)
2. Albarracin, M., Pitliya, R.J., Smithe, T.S.C., Friedman, D.A., Friston, K., Ramstead, M.J.D.: Shared protentions in multi-agent active inference. Entropy **26**(4), 303 (2024)
3. Friston, K.: A free energy principle for the brain. J. Physiol. Paris **100**(1–3), 70–87 (2006)
4. Goldblatt, R.: Topoi: The Categorial Analysis of Logic, Dover Publications, revised edn (2006)
5. Hansen, J., Gebhart, T., Scholkopf, B.: Sheaf neural networks. arXiv preprint-arXiv:2205.12345 (2022)
6. Hernández, M., Sánchez-Soto, E.: Topos-theoretic framework for verifiable multi-agent coordination and team formation. In: 2026 International Conference on Artificial Intelligence, Computer, Data Sciences and Applications (ACDSA), pp. 1–6 (2026)
7. Herzig, A., Lorini, E., Perrussel, L.: Artificial intelligence and symbolic computation. In: Handbook of Formal Philosophy, pp. 1–25. Springer, Cham (2020)
8. Jakobsen, D., Gogioso, S., Kissinger, A.: Causal models in presheaf categories. J. Categorical Logic **12**(3), 45–78 (2024)

9. Kock, A.: Synthetic Differential Geometry, 2nd edn. Cambridge University Press, Cambridge (2006)
10. Kowalski, R., Sergot, M.: A logic-based calculus of events. N. Gener. Comput. **4**(4), 319–340 (1986)
11. Lane, M., Moerdijk, S., I.: Sheaves in Geometry and Logic: A First Introduction to Topos Theory, Springer-Verlag (1992)
12. Mahadevan, S.: Universal causality. Entropy **25**(4), 574 (2023)
13. Mahadevan, S.: Higher algebraic k-theory of causality. Entropy **27**(5) (2025). forthcoming; preprint available at arXiv:2412.XXXXX
14. Parr, T., Pezzulo, G., Friston, K.J.: Active Inference: The Free Energy Principle in Mind, Brain, and Behavior, MIT Press, Cambridge (2022)
15. Pearl, J.: Causality: Models, Reasoning, and Inference, 2nd edn. Cambridge University Press, Cambridge (2009)
16. Ramstead, M.J.D., Sakthivadivel, D., Friston, K.J.: Multiscale integration: from active inference to Bayesian mechanics. Phys. Life Rev. **45**, 1–24 (2023)
17. Rosiak, D.: Sheaf Theory through Examples. MIT Press, Cambridge (2020)
18. Shanahan, M.: The event calculus explained. In: Artificial Intelligence Today. LNAI, vol. 1600, pp. 409–430. Springer, Cham (1999)

CNN-Based Autonomous Navigation for the AutoNOMOS Mini V2 Robot in Simulation and Real Environments

José-Anibal Arias-Aguilar, Oscar-David Ramírez-Cárdenas(✉), Filiberto Eduardo Martínez-Hernández, and Miguel-Alberto Dominguez-Gurría

División de Estudios de Posgrado, Universidad Tecnológica de la Mixteca, Huajuapan de León, Oaxaca 69004, Mexico
{anibal, odramirez, mahf961226, migueldominguez}@mixteco.utm.mx

Abstract. This paper presents convolutional neural network (CNN)-based algorithms for autonomous navigation of the 1:10 Ackermann mobile robot AutoNOMOS Mini V2 in both simulated and real environments. The proposed framework uses RGB images to predict steering commands for lane following and combines RGB and LiDAR data for obstacle-aware navigation. Three navigation challenges are addressed in ROS-Gazebo: obstacle-free navigation, navigation with static obstacles, and navigation with moving obstacles. In addition, the first two challenges are validated on a physical prototype equipped with an RGB camera and a LiDAR sensor. To support model training, five datasets were generated from teleoperated driving sessions, including synchronized camera images, steering commands, and LiDAR measurements represented as grayscale images. Challenge-specific CNN architectures were trained to estimate steering commands directly from sensory observations. Experimental results in simulation and on the real platform show that the proposed models keep the vehicle within its lane and enable obstacle avoidance under controlled conditions. These findings support the feasibility of end-to-end perception-based navigation on scaled Ackermann platforms.

Keywords: Autonomous mobile robots · Perception and sensing · Convolutional neural networks

1 Introduction

Autonomous navigation requires reliable perception and control under changing road geometry, lane boundaries, and obstacle configurations. Although full-scale testing is the ultimate target, reduced-scale Ackermann vehicles provide a practical intermediate platform because they enable safe, low-cost, and repeatable

H. Ponce et al. (Eds.): COMIA 2026, CCIS 3095, pp. 255–267, 2027.
https://doi.org/10.1007/978-3-032-37438-7_20

experimentation in both simulated and physical environments. This idea has been reinforced by research and educational platforms such as Carolo-Cup and F1TENTH, which have demonstrated the value of 1:10-scale systems for training, testing, and benchmarking autonomous driving algorithms [7,8,10].

Learning-based end-to-end driving has become an attractive alternative to modular pipelines because it directly maps sensory observations to control commands. Seminal works such as PilotNet showed that convolutional neural networks can predict steering commands from monocular images, while subsequent studies and recent surveys have highlighted both the potential and the remaining challenges of these approaches in terms of robustness, multimodality, and closed-loop evaluation [1,3,4].

For reduced-scale autonomous vehicles, this problem is especially relevant in controlled competition-like tracks, where lane keeping and obstacle avoidance must be performed with limited onboard sensing and computation. Moreover, relying only on RGB images may be insufficient when obstacle distance must be considered explicitly. Previous work has shown that integrating camera and LiDAR information can improve end-to-end driving performance by complementing appearance cues with distance measurements [9].

Within this context, this paper presents CNN-based autonomous navigation algorithms for the AutoNOMOS Mini V2, a 1:10-scale Ackermann mobile robot, in both ROS-Gazebo simulation and a real prototype. Three challenges are considered: obstacle-free navigation, navigation with static obstacles, and navigation with moving obstacles. The proposed methodology includes dataset generation from teleoperated driving, training of challenge-specific CNN architectures, and closed-loop evaluation in simulation and, when feasible, on the physical platform. The obtained results show that the proposed approach allows the robot to remain within its lane and to avoid obstacles under controlled conditions, supporting the feasibility of end-to-end perception-based navigation on reduced-scale robotic vehicles.

2 Related Work

2.1 Reduced-Scale Platforms for Autonomous Driving Research

Reduced-scale Ackermann platforms have become an effective alternative for the development and evaluation of autonomous driving algorithms, since they provide safer, lower-cost, and more repeatable experimentation than full-scale vehicles. In this context, academic initiatives such as the Carolo-Cup and F1TENTH have demonstrated the value of 1:10-scale platforms for testing perception, control, and obstacle avoidance strategies in controlled environments [7,8,10].

Within this line of work, the AutoNOMOS Mini platform has been adopted in educational and research settings as a representative Ackermann vehicle for autonomous navigation tasks. Previous efforts on this platform have addressed lane following, obstacle avoidance, and track representation under controlled conditions [2,5,6]. These studies confirm the usefulness of reduced-scale platforms, but also show that robust navigation remains challenging when the vehicle must operate under multiple tasks and sensing conditions.

2.2 End-to-End CNN-Based Navigation

Learning-based approaches have become increasingly relevant for autonomous navigation because they directly map sensory observations to control actions. PilotNet demonstrated that convolutional neural networks can predict steering commands directly from visual input, while DeepDriving proposed a direct-perception formulation based on learned driving affordances [1,3]. More recent studies have shown that end-to-end autonomous driving remains an active research area, particularly in relation to robustness, multimodal sensing, and closed-loop validation [4].

In reduced-scale platforms, end-to-end approaches are especially attractive because they simplify the navigation pipeline by avoiding the explicit decomposition into perception, planning, and control modules. However, their performance strongly depends on dataset diversity, sensing configuration, and scenario complexity.

2.3 Multimodal Perception with RGB and LiDAR

Although RGB images provide rich visual information for lane following and scene interpretation, their ability to support obstacle-aware navigation may be limited when depth information must be considered explicitly. For this reason, multimodal approaches that combine cameras with LiDAR sensors have gained relevance in autonomous driving, since they complement appearance cues with geometric distance measurements. In particular, end-to-end fusion of LiDAR and camera data has shown improved performance for steering prediction and obstacle-aware driving tasks [9].

Despite these advances, relatively few works have explored this type of multimodal end-to-end navigation on reduced-scale Ackermann robots validated in both simulation and real environments. This gap motivates the present work, which evaluates CNN-based steering prediction using RGB data for obstacle-free navigation and RGB–LiDAR data for obstacle-aware navigation on the AutoNOMOS Mini V2 platform.

3 Methodology

The proposed methodology was designed to develop and evaluate CNN-based autonomous navigation strategies for the AutoNOMOS Mini V2 in both simulated and real environments. The workflow comprises five main stages: platform preparation, simulation environment design, data acquisition and dataset generation, CNN model design, and real-world validation. Three navigation challenges were considered throughout this process: obstacle-free navigation, navigation with static obstacles, and navigation with moving obstacles. The first two challenges were evaluated in both simulation and on the physical prototype, whereas the third one was assessed only in simulation due to the availability of a single real platform.

3.1 AutoNOMOS Mini V2 Platform and System Upgrade

The AutoNOMOS Mini V2 is a 1:10-scale Ackermann vehicle designed for educational and research purposes. Its original hardware includes an onboard computer, an Arduino-based interface, a steering servo motor, a traction motor, an inertial measurement unit, an RGB-D camera, a LiDAR sensor, and wireless communication modules. In its initial software configuration, the platform operates with Ubuntu, ROS, Gazebo, and OpenCV for sensor integration, simulation, and computer vision tasks.

To support the implementation of CNN-based navigation algorithms, several hardware and software upgrades were introduced. The original Odroid XU4 onboard computer was replaced with a Jetson Nano in order to increase memory capacity, improve processing performance for convolutional neural network inference, and enable more reliable wireless communication. The original front camera was also replaced with an Orbbec Astra Pro camera due to a better compatibility with ROS and image-quality limitations of the previous sensor (on the one hand, it had problems with rapid changes in lighting, as it did not correctly adjust the brightness of the image, causing dark images in low-light environments and overly bright images in bright environments, and on the other hand, the driver software for this camera is not supported on the Jetson Nano computer). In addition, a defective transmission gear was redesigned and manufactured through 3D printing, and the power system was updated by using two lithium-polymer batteries: one dedicated to logic components and another to the traction system.

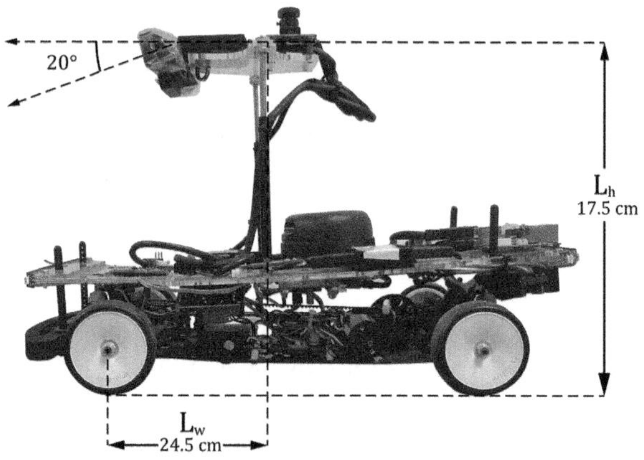

Fig. 1. Side view of the AutoNOMOS Mini V2 platform, showing the vehicle dimensions, camera position, and camera tilt angle used for data acquisition.

On the software side, the upgraded platform runs Ubuntu 18.04 with ROS Melodic and TensorFlow 2. An automatic startup script was implemented to initialize the main ROS processes, assign hardware permissions, and launch the sensing and actuation nodes required for autonomous navigation experiments.

Figure 1 summarizes the geometric configuration used in this work. In the simulated platform, the RGB camera is mounted 0.15 m above the ground and tilted $-20°$ with respect to the horizontal plane. The virtual vehicle has approximate dimensions of 0.42 m in length and 0.21 m in width. In the physical prototype, the camera position was adjusted to maximize the amount of useful visual information for navigation while avoiding excessive visibility of the chassis.

3.2 Simulation Environment and Navigation Challenges

A virtual version of the AutoNOMOS Mini V2 was implemented in Gazebo to generate the data required for training and evaluating the proposed navigation models. The simulation track was designed as a closed circuit with two lanes, four curves, and lane markings that emulate a structured road environment. The track measures 12 m in length and 9 m in width. Figure 2(a) shows the simulated track used to define the navigation tasks, while Fig. 2(b) presents the real-world track employed for physical validation.

Three navigation challenges with increasing difficulty were defined on these experimental tracks. The first challenge consists of autonomous navigation without obstacles, where the vehicle must remain within the lane while completing the circuit. The second challenge considers autonomous navigation with static obstacles placed along the track, requiring the vehicle to avoid them while preserving stable lane-following behavior. The third challenge introduces a moving obstacle that travels along the same simulated circuit, increasing the complexity of perception and control during overtaking and avoidance maneuvers.

For the static-obstacle scenario, fixed objects were modeled and placed at representative positions along the straight sections of the track. For the dynamic-obstacle scenario, a second virtual vehicle, visually differentiated from the main prototype, was used as a moving obstacle. This vehicle was controlled to follow the track continuously at a constant speed of 0.377 m/s, while the main robot operated at 0.48 m/s in this challenge. During the obstacle-free and static-obstacle experiments, the main vehicle moved at a constant speed of 0.377 m/s.

3.3 Data Acquisition and Dataset Generation

Training data were generated from teleoperated driving sessions in the Gazebo environment. For all challenges, the main sensory input was obtained from the front RGB camera, while steering commands were recorded as the target output (regression problem). For the obstacle-aware challenges, LiDAR measurements were also acquired in parallel in order to provide distance information around the vehicle.

The RGB camera produced images with a resolution of 640×480 pixels and update rates of either 20 Hz or 30 Hz, depending on the challenge. To reduce

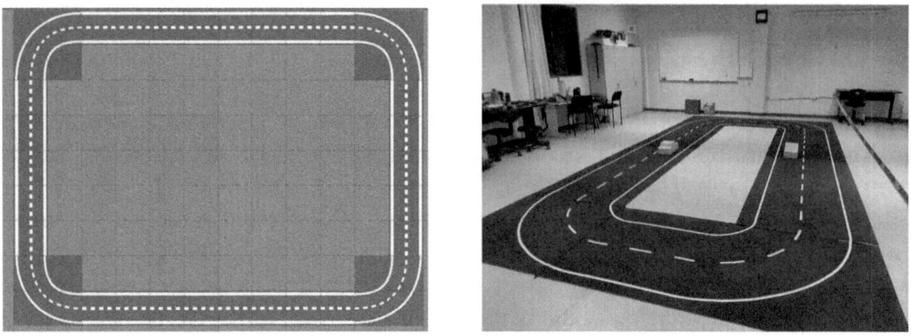

(a) Simulation track in Gazebo.

(b) Real-world track with static obstacles.

Fig. 2. Experimental tracks used in simulation and real-world validation.

irrelevant background information, a region of interest (ROI) was extracted from each frame, preserving the full image width while cropping the vertical region containing the lane boundaries. Steering commands were generated through a joystick controller and converted into PWM-compatible values to match the vehicle control interface.

For obstacle-aware navigation, LiDAR measurements were recorded as 360-dimensional distance vectors at 30 Hz. To avoid detecting objects outside the track, the effective sensing range was limited to 1 m in simulation. These vectors were later transformed into grayscale images of 20×18 pixels (representing 360 values), with intensity values normalized to the range $[0, 255]$, so that they could be processed jointly with RGB data by the proposed CNN architectures. Figure 3 illustrates the image representation obtained from a LiDAR distance vector, where brighter pixels correspond to closer objects and darker regions indicate greater distance or the absence of nearby obstacles.

All navigation data were recorded in ROS bag files, which preserved temporal synchronization among camera frames, steering commands, and LiDAR measurements. Based on these recordings, three challenge-specific datasets were generated in simulation. For Challenge 1, the dataset contains 8,351 RGB images paired with steering commands, collected during 10 min of teleoperated driving. For Challenge 2, the dataset includes 42,419 synchronized RGB images, LiDAR-derived images, and steering commands, obtained during approximately 40 min of driving. For Challenge 3, an additional 19,864 synchronized samples were collected over 5 min, resulting in a total of 62,283 samples when combined with the Challenge 2 data.

3.4 CNN Architectures

Three CNN-based models were designed, one for each navigation challenge. In all cases, the objective was to predict the steering command directly from sensory observations, following an end-to-end learning approach.

Fig. 3. Grayscale image generated from a LiDAR distance vector.

For Challenge 1, which only requires lane following without obstacles, the proposed network receives RGB images as input. The architecture consists of five convolutional layers followed by a regression stage composed of fully connected layers, with a final linear output representing the steering command.

For Challenge 2, which includes static obstacles, a dual-branch CNN was adopted. One branch processes the RGB image, while the second branch processes the LiDAR-derived grayscale image. The feature maps generated by both branches are flattened and concatenated, and the resulting representation is passed through a regression module composed of fully connected layers. This design allows the model to combine visual lane information with distance measurements for obstacle-aware navigation.

For Challenge 3, the architecture was derived from the Challenge 2 model but was expanded to address the higher complexity introduced by moving obstacles. In this case, each branch includes an additional convolutional layer before concatenation, and the regression stage contains one extra fully connected layer with a larger number of neurons in the first layers.

Figure 4 shows the representative dual-input architecture adopted for the obstacle-aware case. This configuration was selected because it summarizes the multimodal design used to combine RGB and LiDAR information, while the model for Challenge 3 extends this structure by increasing the depth of both branches and the size of the regression stage.

3.5 Real-World Experimental Setup

The methodology was also validated on the physical AutoNOMOS Mini V2 prototype. The real-world experiments mirrored the structure used in simulation, including RGB image acquisition, steering-angle recording, and LiDAR-based distance sensing. However, only the first two challenges were evaluated on the

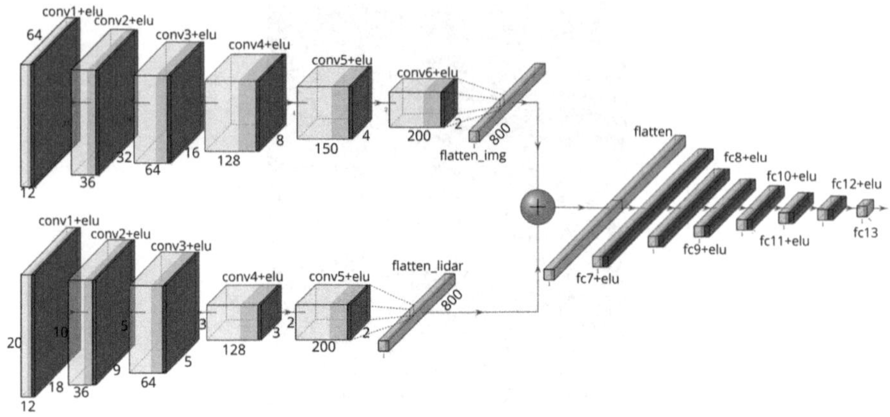

Fig. 4. Representative dual-input CNN architecture used for obstacle-aware autonomous navigation.

physical platform, since the third challenge would require at least one additional autonomous vehicle to generate moving obstacles under comparable conditions.

The real test track measures 7.11 m in length and 3.24 m in width, with each lane measuring 0.42 m in width. For the obstacle-aware task, two yellow static obstacles were placed on the track. The robot operated at a constant speed of 0.37 m/s in both real-world challenges.

As in simulation, a dataset was generated for each challenge through teleoperated driving. For the obstacle-free task, approximately 20 min of driving produced 23,100 RGB images paired with steering commands. For the static-obstacle task, RGB images, steering commands, and LiDAR data were recorded for approximately 40 min while the vehicle was manually driven through both lanes under different obstacle placements. In the real platform, LiDAR sensing was restricted to a maximum range of 1.1 m in the frontal sector (from $-20°$ to $20°$) and 0.7 m in the remaining directions in order to avoid detecting objects surrounding the track. The resulting 360-dimensional LiDAR vectors were transformed into grayscale images following the same procedure used in simulation.

4 Results

4.1 Training Results

Five datasets were used to train the CNN models proposed for the three simulation challenges and the two real-world challenges. The number of samples increased with the difficulty of the task and with the complexity of the environment. The training process was carried out on a desktop computer equipped with an Intel Core i5 processor, 24 GB of RAM, and an NVIDIA GTX 1650 GPU with 4 GB of memory.

For all models, the available data were split into training and testing subsets. Table 1 summarizes the training and testing errors obtained for the five trained

CNNs. In general, the results show that all models converged to low regression errors, with the obstacle-aware models achieving the lowest test errors in both simulation and real-world conditions.

Table 1. Training and testing results of the proposed CNN models.

Model	Training Error	Test Error	Epochs
Challenge 1 (simulation)	0.0210	0.0187	23
Challenge 2 (simulation)	0.0145	0.0101	19
Challenge 3 (simulation)	0.0097	0.0103	20
Challenge 1 (real-world)	0.0406	0.0348	18
Challenge 2 (real-world)	0.0063	0.0052	33

Figure 5 presents a representative training curve for the CNN used in Challenge 2 in the real-world environment. The training and testing trends indicate stable convergence without large oscillations, supporting the suitability of the proposed architecture for steering-angle regression.

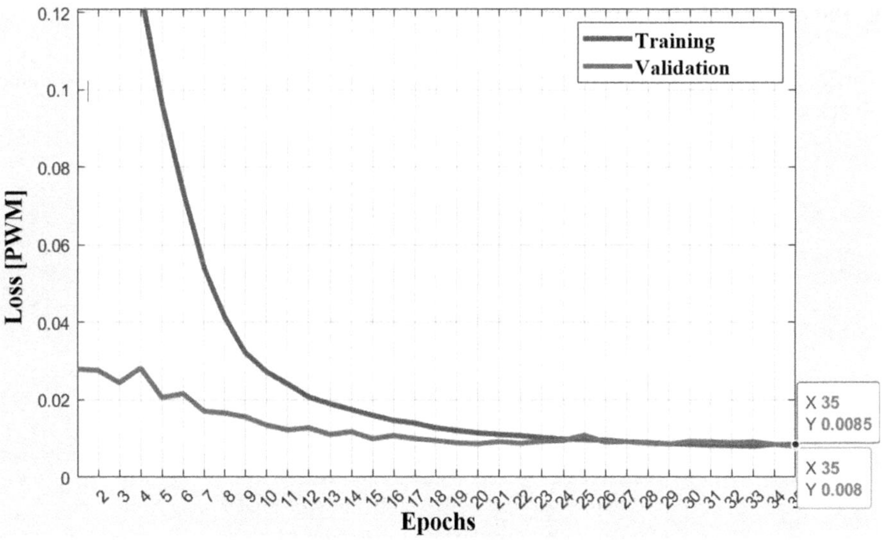

Fig. 5. Training and testing error curves for the CNN used in Challenge 2 in the real-world environment.

4.2 Closed-Loop Navigation Results

After training, the CNNs were evaluated in closed-loop autonomous navigation experiments in both simulation and real-world scenarios. The evaluation compared the steering command predicted by each model against the steering command recorded during teleoperated driving over a representative lap. Performance was analyzed qualitatively through steering profiles and quantitatively using the mean squared error (MSE) between the predicted and reference PWM values.

The qualitative results show that the predicted steering signal follows the general trend of the reference signal across curves and obstacle-avoidance maneuvers. In simulation, the robot successfully completed obstacle-free navigation as well as static- and dynamic-obstacle avoidance tasks while remaining within the lane. In the real-world experiments, the robot also completed the obstacle-free and static-obstacle challenges under controlled conditions (that is, on the test track installed in the robotics laboratory, without abrupt changes in lighting).

Table 2 summarizes the MSE values obtained during representative autonomous laps. In simulation, the lowest errors were obtained for Challenge 1 in the outer lane and Challenge 3 in the outer lane. In the real-world experiments, the error values were higher than in simulation, which is consistent with the greater variability of the physical platform and sensing conditions. Nevertheless, the robot maintained stable navigation behavior and obstacle avoidance in the evaluated scenarios.

Table 2. Mean squared error between predicted and reference steering commands during representative autonomous laps.

Scenario	MSE
Challenge 1 (simulation, inner lane)	334.91
Challenge 1 (simulation, outer lane)	108.52
Challenge 1 (real-world, inner lane)	9263.00
Challenge 1 (real-world, outer lane)	8450.20
Challenge 2 (simulation, inner lane)	516.69
Challenge 2 (simulation, outer lane)	277.04
Challenge 2 (real-world, inner lane)	11176.17
Challenge 2 (real-world, outer lane)	11922.97
Challenge 3 (simulation, inner lane)	367.64
Challenge 3 (simulation, outer lane)	182.48

Figure 6 presents a representative steering profile for Challenge 3 in simulation, which is the most demanding scenario due to the presence of moving obstacles. Figure 7 shows the corresponding steering profile for Challenge 2 on the real track, illustrating the model behavior during static-obstacle avoidance in the physical prototype.

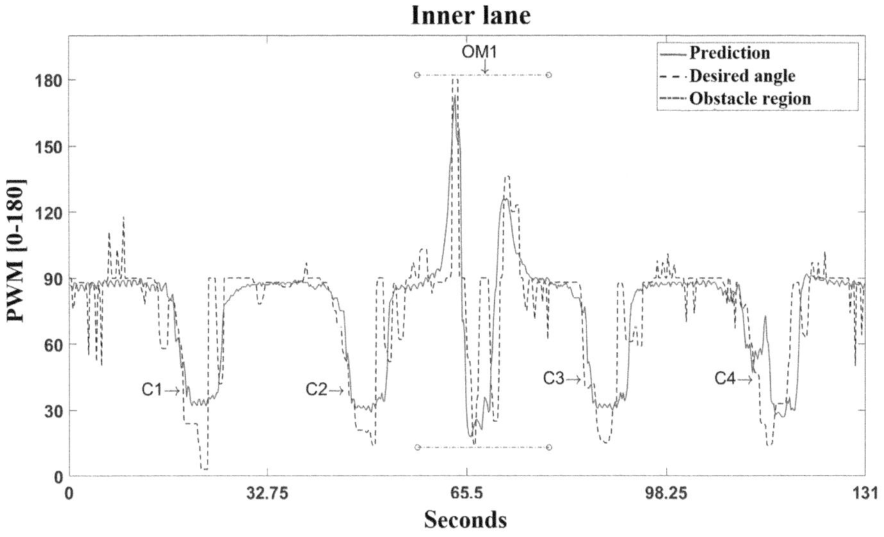

Fig. 6. Steering-angle prediction for Challenge 3 in the simulation environment.

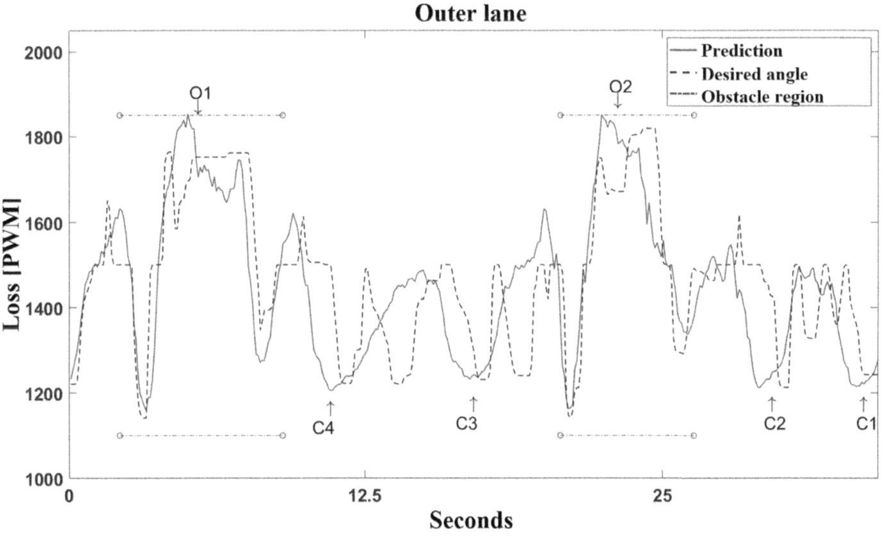

Fig. 7. Steering-angle prediction for Challenge 2 with the real prototype.

5 Conclusion

This paper presented CNN-based autonomous navigation strategies for the AutoNOMOS Mini V2, a 1:10-scale Ackermann mobile robot, in both simulation and real environments. The proposed methodology combined RGB-based perception for lane-following tasks and multimodal RGB–LiDAR sensing for obstacle-aware navigation, and was evaluated across three challenges of increasing complexity.

The obtained results indicate that the trained models can generate steering commands suitable for closed-loop navigation under controlled conditions. In simulation, the robot completed obstacle-free navigation and obstacle-avoidance tasks while maintaining lane-following behavior. In the real-world experiments, the proposed approach also enabled obstacle-free and static-obstacle navigation on the physical prototype.

These findings support the use of end-to-end CNN-based navigation on reduced-scale Ackermann platforms and provide a basis for future work on multimodal perception, autonomous maneuvering, and validation in more complex environments.

5.1 Future Work

Future work will focus on evaluating the proposed models under more diverse environmental conditions in order to analyze their generalization capability. These include comparing the results obtained with other metrics such as MAPE or RSQUARE, and using other network training schemes, such as reinforcement learning. It would also be interesting to test the trained models on more complex tracks (at least in simulation).

In addition, richer multimodal perception strategies and semantic scene understanding will be explored to improve navigation performance in more complex scenarios.

Finally, the proposed framework will be validated in less controlled real-world environments to assess its applicability beyond structured experimental tracks.

References

1. Bojarski, M., et al.: End to end learning for self-driving cars. arXiv preprint arXiv:1604.07316 (2016). https://doi.org/10.48550/arXiv.1604.07316
2. Bravo Conejo, C.G.: Navegación autónoma de un robot tipo automóvil en pista de carreras con obstáculos. Instituto Politécnico Nacional, Centro de Investigación en Computación. Master's thesis (2018)
3. Chen, C., Seff, A., Kornhauser, A., Xiao, J.: Deepdriving: learning affordance for direct perception in autonomous driving. In: Proceedings of the IEEE International Conference on Computer Vision (ICCV), pp. 2722–2730 (2015). https://doi.org/10.1109/ICCV.2015.312
4. Chen, L., Wu, P., Chitta, K., Jaeger, B., Geiger, A., Li, H.: End-to-end autonomous driving: challenges and frontiers. IEEE Trans. Pattern Anal. Mach. Intell. **46**(12), 10164–10183 (2024). https://doi.org/10.1109/TPAMI.2024.3435937
5. Díaz Sarmientos, E.G.: Generación del mapa métrico de una pista de competencias para la navegación del robot móvil AutoNOMOS Mini V2. Master's thesis, Universidad Tecnológica de la Mixteca (2020)
6. Guzmán Durán, K.F.: Diseño e implementación del sistema de control lateral para un robot móvil autonomos mini v2 (2023)

7. Nolte, M., Form, T., Ernst, S., Graubohm, R., Maurer, M.: The carolo-cup student competition: involving students with automated driving. In: 2018 12th European Workshop on Microelectronics Education (EWME), pp. 95–99 (2018). https://doi.org/10.1109/EWME.2018.8629462
8. O'Kelly, M., Zheng, H., Karthik, D., Mangharam, R.: F1tenth: an open-source evaluation environment for continuous control and reinforcement learning. In: Escalante, H.J., Hadsell, R. (eds.) Proceedings of the NeurIPS 2019 Competition and Demonstration Track. Proceedings of Machine Learning Research, vol. 123, pp. 77–89. PMLR (2020). https://proceedings.mlr.press/v123/o-kelly20a.html
9. Park, M., Kim, H., Park, S.: A convolutional neural network-based end-to-end self-driving using lidar and camera fusion: analysis perspectives in a real-world environment. Electronics **10**(21), 2608 (2021). https://doi.org/10.3390/electronics10212608. https://www.mdpi.com/2079-9292/10/21/2608
10. Xie, Z., Ramezani, M., Levinson, D.: Reduced-scale mobile robots for autonomous driving research. IEEE Trans. Intell. Transp. Syst. **25**(11), 15367–15387 (2024). https://doi.org/10.1109/TITS.2024.3426508

Classification of AI-Generated Academic Texts: A Comparative Study of Classical Models, Transformers, and CNN

Omar Alfonso Cruz-Ramírez(✉) and Christian Eduardo Millán-Hernández

Universidad Tecnológica de la Mixteca, Huajuapan de León, Oaxaca 69000, Mexico
omar.cruzr97@gmail.com

Abstract. Text generation using artificial intelligence in the production of academic assignments represents a problem that may reduce the pedagogical impact of student learning, while also making it more difficult for teachers to assess performance and verify whether learning objectives have been achieved. This article focuses exclusively on the classification task, understood as determining whether a complete document was written by a human or generated by artificial intelligence. To address this scope, a feature extraction method based on lexical, character level, and neural representations is proposed, together with prediction models, to improve the classification of artificially generated documents and distinguish them from texts written by students. In addition, a reproducible *pipeline* is introduced to evaluate classical models such as SVM, Random Forest, and XGBoost, as well as neural models such as BERT, CNN, and TextCNN, on TextTuring (29,139 texts). The best results correspond to CNN (Accuracy 0.9939; F1-score 0.9924; ROC-AUC 0.9997), followed by TextCNN, improving upon the results reported in the 2024 Kaggle competition *LLM – Detect AI Generated Text*.

Keywords: AI-generated text classification · Academic integrity · NLP · Authorship classification · CNN · Transformers

1 Introduction

The growing adoption of generative language models in education has increased the production of AI-assisted academic texts and, consequently, the need for verification mechanisms grounded in defensible technical criteria. In this context, commercial detectors exhibit variability across domain, language, and text length, along with limited methodological transparency, which calls for caution in high stakes decisions (UNESCO 2023; Weber-Wulff et al. 2023).

This study addresses the feasibility of a specialized classifier for academic texts with higher performance than general purpose references and with better operational control in terms of metrics, threshold selection, and reproducibility. The scope is limited to classification only; therefore, the system assigns a

H. Ponce et al. (Eds.): COMIA 2026, CCIS 3095, pp. 268–277, 2027.
https://doi.org/10.1007/978-3-032-37438-7_21

document level label and does not attempt fragment level detection or attribution to a specific human author or generative model. The general objective is to develop and evaluate an AI-generated text classifier with primary focus on English through comparison among classical families, Transformers, and convolutional neural networks under a common experimental protocol. In addition, the study includes evaluation in Spanish: the models are trained on English texts and then tested on Spanish texts under both in domain and out of domain conditions, following the same comparative logic used for English in order to examine cross lingual transfer and robustness.

Accordingly, the article synthesizes experimental results in binary authorship classification (human vs. AI) on academic essays, using a reproducible *pipeline* with emphasis on calibration, false-positive control, and traceability.

2 Related Work

Recent literature recognizes three main families for AI-generated text detection: (i) classical stylometry, (ii) supervised models with deep semantic representations, and (iii) hybrid strategies. In the first family, lexical, syntactic, and discourse signals are used to capture regularities of style, with particular value in scenarios where interpretability and feature control matter. In the second, attention based architectures and contextual *embeddings* are used to model long range dependencies and semantic content; this line gains momentum with the Transformer formulation and pretrained language models such as BERT (Devlin et al. 2019; Vaswani et al. 2017).

For terminological clarity, the problem can be organized into three implementation levels: classification, detection, and identification. Classification refers to a document level decision in which the system determines whether a text was written by a human or generated by artificial intelligence. Detection extends this task to a finer granularity by locating which fragments of a document are likely to have been produced by AI and which fragments correspond to human writing, which is especially relevant in mixed authorship scenarios. Identification is an attribution oriented task: beyond distinguishing human and artificial authorship, it seeks to determine the specific human author or the particular generative model responsible for producing the text. These levels differ in granularity, evidential requirements, and operational risk; therefore, this article focuses specifically on classification, while preserving methodological elements that may support future extensions toward detection and identification.

Alongside performance gains, recurring limitations also appear. Published results depend strongly on the corpus, the partitioning protocol, and the degree of similarity between training and test conditions. In particular, the literature reports drops under domain shift, linguistic variation, translation, paraphrasing, and other forms of rewriting, which questions the stability of detectors when moved to real world conditions (Sadasivan et al. 2023; Weber-Wulff et al. 2023). In this sense, the problem has evolved from early comparisons based mainly on stylometric cues and isolated benchmarks toward a more demanding setting in which detectors are expected to generalize across domains, prompts, languages, rewriting strategies, and rapidly changing families of generative models. As generative systems improve in fluency and controllability, the task becomes less about identifying superficial artifacts and more about measuring robust distributional differences under realistic deployment conditions. For that reason, beyond class discrimination, probabilistic calibration, threshold selection, and experimental traceability become relevant, since a high score alone does not guarantee reliable decisions in sensitive contexts (Guo et al. 2017).

At the applied level, the Kaggle competition *LLM – Detect AI Generated Text* provides a visible reference for recent interest in the problem and for the prominent role of *ensembles* and Transformer based models in human vs. AI authorship tasks. However, that reference does not remove the need for domain specific evaluations, especially in academic texts, where the cost of false positives is high and where international organizations recommend caution, human review, and explicit institutional guidelines for the use of these systems (Kaggle 2024a; UNESCO 2023; Weber-Wulff et al. 2023). Under this framework, comparing classical models, Transformers, and convolutional networks within a common protocol becomes pertinent, with emphasis on methodological transparency and operational control.

3 Proposed Method

The proposed method is organized as a reproducible workflow that connects corpus preparation, feature extraction, model training, evaluation, and documentation. Figure 1 presents this general sequence and clarifies the role of each stage before the more detailed methodological blocks are described.

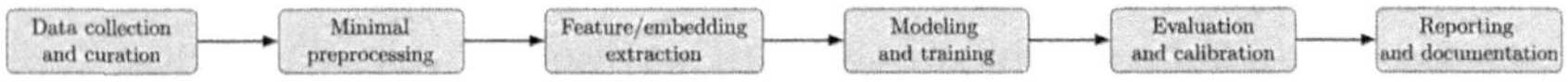

Fig. 1. High-level diagram of the methodological workflow.

At the data level, the method begins with corpus integration and conservative preprocessing. This stage is designed to preserve signals that are relevant for authorship classification while producing consistent inputs for the evaluated models. Figure 2 details the operations involved in this first block.

Once the representations are generated, the workflow moves to model training and evaluation under a common experimental protocol. Figure 3 summarizes this second block, including the comparison among classical and neural models, 5 fold stratified cross-validation, probabilistic calibration, threshold selection, and metric based analysis.

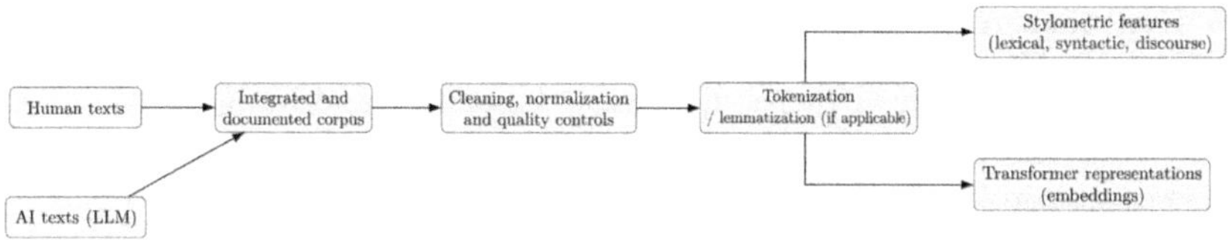

Fig. 2. Low-level diagram for corpus and preprocessing.

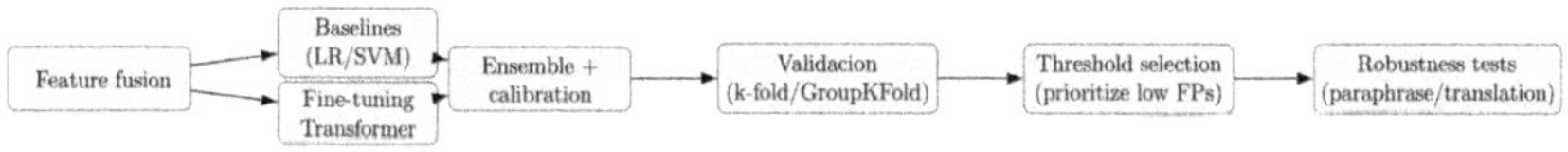

Fig. 3. Low-level diagram for model training and evaluation.

3.1 Corpus and Preprocessing

Use of TextTuring, obtained from the Kaggle dataset *LLM Detect AI Generated Text Dataset*[1], with 29,139 labeled essays: 17,508 human texts and 11,631 AI-generated texts. Target variable: `generated` (0 = human, 1 = AI). Conservative preprocessing to preserve style and content signals relevant to the task.

3.2 Evaluated Models

Comparison among three classical baselines (SVM, Random Forest, and XGBoost) and three neural models (BERT, CNN, and TextCNN). In CNN, use of a Conv1D architecture with *GlobalMaxPooling1D*, trainable embeddings, and a sigmoid output; in TextCNN, incorporation of parallel kernels to capture local patterns of different lengths (Kim 2014).

Table 1. Average model results on TextTuring (5-fold CV).

Model	Accuracy	F1-score	ROC-AUC
SVM	0.8751 ± 0.0006	0.8700 ± 0.0007	0.9034 ± 0.0007
Random Forest	0.9880 ± 0.0019	0.9849 ± 0.0024	0.9849 ± 0.0024
XGBoost	0.9906 ± 0.0013	0.9882 ± 0.0016	0.9993 ± 0.0002
BERT	0.9716 ± 0.0025	0.9651 ± 0.0031	0.9929 ± 0.0008
CNN	**0.9939 ± 0.0013**	**0.9924 ± 0.0016**	**0.9997 ± 0.0002**
TextCNN	0.9935 ± 0.0013	0.9918 ± 0.0016	**0.9997 ± 0.0002**

[1] Kaggle: LLM Detect AI Generated Text Dataset.

3.3 Evaluation

Reporting of Accuracy, Precision, Recall, F1-score, and ROC-AUC, complemented with ROC, Precision-Recall, and calibration curves. The protocol prioritizes comparability across models and control of partition induced bias (Guo et al. 2017; Kohavi 1995). For XGBoost, use of a limit of 12,000 word features and 4,000 character features, with `min_df ~ 0.2%` and `max_df=0.95`.

4 Results

This section organizes the presentation of findings into three parts. First, a synthesis of the data supporting the evaluation, second, model results under a common experimental protocol and finally, comparisons with external references and commercial detectors.

4.1 Data

The empirical basis of the results corresponds to the TextTuring corpus, with 29,139 labeled texts, of which 17,508 correspond to human writing and 11,631 to AI-generated text. On this distribution, 5 fold stratified cross validation is applied to preserve comparability across partitions and reduce evaluation bias.

The composition of the corpus provides a broad binary authorship classification scenario, sufficient to contrast classical and neural models under the same protocol. In this way, the following results should be read as comparative evidence over the same database, homogeneous metrics, and controlled experimental conditions.

4.2 Models

Table 1 presents the average cross validation results for the evaluated models.

Classical Models. Within the traditional baseline line, XGBoost shows the best performance with Accuracy 0.9906, F1-score 0.9882, and ROC-AUC 0.9993, followed by Random Forest (0.9880/0.9849/0.9849) and SVM (0.8751/0.8700/0.9034). This confirms notable gains associated with calibration and feature selection within conventional models, although still below convolutional architectures.

Neural Models. In the deep-learning layer, CNN leads with Accuracy 0.9939, F1-score 0.9924, and ROC-AUC 0.9997, while TextCNN shows nearly identical performance (0.9935/0.9918/0.9997). For BERT, Accuracy reaches 0.9716 and ROC-AUC 0.9929; this indicates robustness, but with lower performance than CNN architectures specialized for academic texts under this protocol.

Main Finding. CNN outperforms BERT and the classical models on the Text-Turing corpus (29,139 texts) and within the same experimental context, with consistent improvement over the baselines.

Near perfect class separation appears in the evaluated corpus for XGBoost, CNN, and TextCNN. In operational terms, CNN stands out for the best overall balance between discrimination and metric stability. Accordingly, the expected performance of the proposed pipeline should be interpreted at two levels: very high performance can be expected in matched conditions similar to TextTuring, where the data distribution is close to the training and validation environment; however, under cross-lingual, out of domain, or rewritten text scenarios, the realistic expectation is not perfect separation but a controlled decrease in performance that still preserves useful discrimination, calibration, and operational transparency.

4.3 Comparisons

This subsection contrasts the results obtained with external references, in order to position the achieved performance within a broader comparative framework. Figure 4 presents an adapted version of the real ranking from the 2024 Kaggle competition *LLM – Detect AI Generated Text*; the values obtained by the models developed in this study were inserted into that reference ranking, where CNN, TextCNN, and XGBoost would occupy the first three positions above the official winning submission.

	Team/Model	Score (ROC-AUC)
1	CNN (proprietary)	0.9997
2	TextCNN (proprietary)	0.9997
3	XGBoost (proprietary)	0.9993
4	Emoji Team	0.9878
5	Guanshuo Xu	0.9834
7	nlp team	0.9750
8	Ertugrul & Chase	0.9739
9	Linguistic Ninjas	0.9721
10	Davide Cozzolino	0.9691
11	Hao Mei	0.9651
12	Abdullah Meda	0.9565
13	LLMLab	0.9477
14	IC2	0.9474
13	Abdullah Meda	0.9477
14	IC2	0.9474

Fig. 4. Adapted ranking from the Kaggle competition, including the results of the models developed in this study.

Compared with the external Kaggle reference, the local models show numerically superior results: while the competition *LLM – Detect AI Generated Text* reports values close to 0.96 ROC-AUC on the private *leaderboard*, the best local models reach between 0.9993 and 0.9997 ROC-AUC. Under that reference, CNN, TextCNN, and XGBoost exceed the performance level observed in the competition, although the interpretation should remain referential because of differences in *splits*, domain, and test design. For the same reason, these comparisons should not be read as definitive evidence that the proposed CNN surpasses stronger current Transformer baselines such as RoBERTa-large or DeBERTa, since those models were not included in the present benchmark and constitute an important pending comparison.

Additionally, comparison with commercial detectors (e.g., GPTZero, ZeroGPT, Justdone, and WinstonAI) shows advantages of the local model in methodological traceability, threshold control, and reproducibility of the complete process.

Commercial Detectors. To complement the comparison, four commercial tools were tested on the same two illustrative cases discussed in this section: the Spanish reading-report text and the out-of-domain song lyrics. The evaluated services were GPTZero, ZeroGPT, Justdone, and Winston AI.[2] Table 2 summarizes the observed outputs and highlights the variability across tools when the input changes in genre and language.

For both the song-lyrics case and the Spanish human-authored text, the commercial tools and the implemented project were consulted on Saturday, May 9, 2026.

Table 2. Commercial-detector outputs on the Spanish reading-report example and the out-of-domain song example.

Tool	Song (out-of-domain)	Spanish reading report
GPTZero	100% AI	100% human
ZeroGPT	47.5% AI	100% human
Justdone	100% AI	90% human
Winston AI	100% human	100% human

These results reinforce two practical points. First, commercial detectors can disagree substantially on the same input, especially in out-of-domain material such as song lyrics. Second, although all four tools labeled the Spanish reading-report example as human-authored to a high degree, their responses to the song vary from fully AI to fully human, which illustrates the operational instability

[2] GPTZero: https://gptzero.me/es; ZeroGPT: https://www.zerogpt.com/es/; Justdone: https://justdone.com/ai-detector; Winston AI: https://gowinston.ai/es/detector-de-ia/.

already noted in the literature and motivates the need for transparent, reproducible evaluation protocols.

As an illustrative Spanish reading-report example, the pipeline was tested with a Spanish text of this type. Table 3 shows the obtained output for that case. In the complementary DeBERTa evaluation on the same Spanish text, the observed results were: IA prob DeBERTa: 76.75%, DeBERTa: IA, Chars: Human (0.00%), and Final pred: IA. This example is consistent with the intended use of the pipeline in Spanish academic writing, while also showing that different model families may respond differently under cross-lingual conditions.

Table 3. Output of the proposed pipeline for an illustrative Spanish reading-report example.

Output field	Result
Final prediction	Human
Final confidence	97.51%
CNN prediction	Human
Character-based prediction	Human
CNN human probability	0.9751
CNN AI probability	0.0249
AI character details	0 characters identified

As an additional out-of-domain example, the pipeline was applied to the English song "Daylight" by David Kushner, using the official video as source (https://www.youtube.com/watch?v=MoN9ql6Yymw), in order to examine behavior on a text type that differs substantially from the academic prose used during training. Table 4 shows the obtained output for this case. This example is useful because song lyrics present repetition, short lines, and a strongly expressive style, all of which differ from academic writing and therefore provide a more demanding test of robustness outside the original domain.

Table 4. Output of the proposed pipeline for an out-of-domain song-lyrics example.

Output field	Result
Final prediction	Human
Final confidence	95.91%
CNN prediction	Human
Character-based prediction	Human
CNN human probability	0.9591
CNN AI probability	0.0409
AI character details	0 characters identified

5 Conclusions

Experimental confirmation of the superiority of a reproducible *pipeline* adapted to the academic domain over general-purpose references in human-vs.-AI authorship detection. CNN emerges as the best overall model, with TextCNN as a competitive alternative and XGBoost as the best classical baseline.

As future work: broader out-of-domain validation, strengthened multilingual evaluation, deeper analysis of robustness against paraphrasing/obfuscation, and continued improvement of calibration and interpretability for responsible institutional use. More specifically, the main opportunities for improvement now center on broadening the empirical evidence already introduced in this study: extending the out-of-domain evaluation with more heterogeneous corpora and writing tasks, expanding the Spanish analysis with additional datasets and scenarios, and incorporating the full comparative results with stronger contemporary baselines such as DeBERTa in order to strengthen the discussion of relative performance under a more demanding benchmark.

The use of AI-generated text detectors as sole evidence in sanction-related decisions is not advisable. Even with high metrics, uncertainty and the risk of false positives remain, requiring complementary human review, clear institutional policies, and transparent interpretation criteria (UNESCO 2023; Weber-Wulff et al. 2023).

Acknowledgments. The authors gratefully acknowledge the support of the Secretaría de Ciencia, Humanidades, Tecnología e Innovación (SECIHTI) through the postgraduate scholarship awarded to the first author. This support contributed to the development of the present research. The authors also thank the Universidad Tecnológica de la Mixteca for providing academic guidance and institutional backing throughout the study.

Conflict of Interest Statement. Absence of conflicts of interest related to the content of this article.

References

Brier, G.W.: Verification of forecasts expressed in terms of probability. Mon. Weather Rev. **78**(1), 1–3 (1950). https://doi.org/10.1175/1520-0493(1950)078⟨0001:VOFEITspsdoigtsps2.0.CO;2

Chen, T., Guestrin, C.: XGBoost: a scalable tree boosting system. In: Proceedings of the 22nd ACM SIGKDD International Conference on Knowledge Discovery and Data Mining, pp. 785–794 (2016). https://doi.org/10.1145/2939672.2939785

Devlin, J., Chang, M.-W., Lee, K., Toutanova, K.: BERT: pre-training of deep bidirectional transformers for language understanding. In: Proceedings of NAACL-HLT 2019, pp. 4171–4186 (2019). https://aclanthology.org/N19-1423/

Guo, C., Pleiss, G., Sun, Y., Weinberger, K.Q.: On calibration of modern neural networks. In: Proceedings of the 34th International Conference on Machine Learning (ICML), pp. 1321–1330 (2017). https://proceedings.mlr.press/v70/guo17a.html

Kaggle: LLM Detect AI Generated Text Dataset (2024a). https://www.kaggle.com/datasets/sunilthite/llm-detect-ai-generated-text-dataset
Kaggle: LLM – Detect AI Generated Text (2024b). https://www.kaggle.com/competitions/llm-detect-ai-generated-text
Kim, Y.: Convolutional neural networks for sentence classification. In: Proceedings of EMNLP 2014, pp. 1746–1751 (2014). https://aclanthology.org/D14-1181/
Kohavi, R.: A study of cross-validation and bootstrap for accuracy estimation and model selection. In: Proceedings of the 14th International Joint Conference on Artificial Intelligence (IJCAI), pp. 1137–1143 (1995). https://www.ijcai.org/Proceedings/95-2/Papers/016.pdf
Sadasivan, V.S., Kumar, A., Balasubramanian, S., Wang, W., Feizi, S., Lim, S.-N.: Can AI-generated text be reliably detected? arXiv: https://arxiv.org/abs/2303.11156 (2023)
UNESCO: Guidance for generative AI in education and research. UNESCO (2023). https://unesdoc.unesco.org/ark:/48223/pf0000386693
Vaswani, A., et al.: Attention is all you need. In: Advances in Neural Information Processing Systems 30 (2017). https://papers.nips.cc/paper/7181-attention-is-all-you-need
Weber-Wulff, D., et al.: Testing of detection tools for AI-generated text. arXiv: https://arxiv.org/abs/2306.15666 (2023)

Predictive Modeling of Sexual Violence: An Automated Classification Approach to the Sexual Initiation of Mexican Women

Mariana Morelos-López[1], Yael Andrade-Ibarra[2], Aurea Regina Guzman-Montero[1], and Yenny Villuendas-Rey[3](✉)

[1] Instituto Politécnico Nacional, Center for Computing Research (CIC), Mexico City, Mexico
{mmorelos12500,aguzmanm1101}@alumno.ipn.mx

[2] Instituto Politécnico Nacional, School of Physics and Mathematics (ESFM), Mexico City, Mexico
yandradei1600@alumno.ipn.mx

[3] Instituto Politécnico Nacional, Centro de Innovación y Desarrollo Tecnológico en Cómputo, Mexico City, Mexico
yvilluendasr@ipn.mx

Abstract. Sexual violence during women's first sexual experience constitutes a relevant public health and social problem whose detection is challenging due to the complexity of the associated reproductive and sociodemographic factors, the analysis of which requires tools capable of identifying complex patterns in the data. This study used data from the 2023 National Survey of Demographic Dynamics (ENADID). It evaluated the ability of six classification models to identify sociodemographic and contextual patterns associated with violence during the first sexual encounter. Their performance was assessed using AUC, F1-score, and Recall, and differences among models were examined using Bayesian analysis and the Nemenyi multiple comparisons test. The results show that probabilistic and interpretable models achieved the most competitive performance, highlighting the potential of artificial intelligence tools to support the study of sexual violence and the identification of vulnerable population groups.

Keywords: Sexual violence · First sexual relationship · Mexican population · Predictive modeling · Machine Learning

1 Introduction

Sexual violence against women in Mexico is a structural phenomenon of major relevance to public health and human rights. According to data from the World Health Organization (WHO), 3 out of every 10 adolescents report experiencing dating violence. In the Mexican context, SIPINNA reports, based on the National Survey on Dating Violence, that 76% of adolescent girls between the ages of 15

H. Ponce et al. (Eds.): COMIA 2026, CCIS 3095, pp. 278–293, 2027.
https://doi.org/10.1007/978-3-032-37438-7_22

and 17 have experienced psychological violence, 17% sexual violence, and 15% physical violence [22].

In Mexico—where cultural patterns such as sexism and the normalization of possessive behaviors persist—detecting dating violence (particularly in the context of the first sexual relationship) helps prevent these patterns from being reproduced in adult or family relationships. Furthermore, dating violence serves as a precursor to more severe forms of violence, such as domestic violence or femicide. Its early identification contributes to reducing these structural phenomena in Mexico.

The 2023 National Survey of Demographic Dynamics (ENADID) provides an updated overview of women's sexuality and reproductive behavior in Mexico, offering key information for analyzing this phenomenon.

In this context, the use of machine learning techniques allows the exploration of patterns in national survey data to identify factors associated with sexual violence in first relationships. This study analyzes the ENADID 2023 survey to evaluate the capacity of six types of classification models: *based on neighborhood*, *decision trees*, *rule-based models*, *probabilistic classifiers*, *support vector machines*, and *neural networks* to identify the presence of violence in the first sexual encounter based on sociodemographic and reproductive information. The use of machine learning (ML) transforms this study from a traditional descriptive analysis into a predictive and precision-driven tool. Its primary impact lies in the ability to process large volumes of data to uncover patterns that the human eye or basic statistics might overlook.

To assess the performance of the models, metrics such as AUC, F1-score, and Recall are used, and a Bayesian comparative analysis is incorporated to estimate the probability of one classifier outperforming another on similar data. The results obtained not only allow us to identify which algorithms perform best in the classification task, but also provide quantitative evidence that can contribute to a better understanding of the factors associated with sexual violence.

2 Related Works

Sexual violence during first sexual intercourse is a public health problem with profound consequences that extend from childhood to adulthood and shape women's sexual, reproductive, and emotional trajectories. Earlier studies documented how violence may accumulate across the life course and manifest in different contexts of vulnerability.

One of the earliest systematic studies in the Mexican context is presented in [4], where violence against women was analyzed before and during pregnancy. That work showed that violence was already present in important stages of women's reproductive lives, thereby providing evidence of its persistence and severity. In [3], the relationship between childhood sexual violence, early pregnancy, early marriage, and child sex trafficking was examined in women from two Mexico–US border cities, showing that adverse early sexual experiences are associated with higher vulnerability later in life. Likewise, in [1], sexual exploitation and trafficking in Mexico City were studied through the experiences of women subjected to extreme forms of abuse, including rape, threats, forced abortions, and severe restrictions on autonomy.

In the pandemic, a study from Rivera-Rivera et al. [20] comprised data through a remote survey conducted in 2020, with a sample of 47,819 women aged 15 and older. Their study assessed the prevalence of domestic violence against women in Mexico during the COVID-19 pandemic, as well as related risk factors, finding prevalence rates between 11% and 16% of domestic violence. From another perspective, Soria-Escalante et al. [23] centered on the alarming reality of sexual violence affecting migrant women traveling across Mexico. Their study drew on data from ten undocumented Latin American transmigrant women who were temporarily staying in shelters located in key transit cities—Veracruz and Tabasco in the south, and Nuevo León and Tamaulipas in the north. The findings revealed a troubling pattern: analysis of the interviews yielded 172 narrative fragments related to sexual violence, highlighting both the widespread nature and the severity of the risks faced by women during migratory transit.

More recent research has incorporated machine learning and data-driven approaches to study violence against women. In [15], machine learning was used to identify factors associated with marital sexual violence in India, showing that computational methods can reveal complex patterns linked to sexual victimization. In [19], several machine learning algorithms were compared to predict domestic violence vulnerability among women in Liberia using demographic and health survey data, highlighting the usefulness of comparative classifier analysis in this domain. In [5], machine learning techniques were applied to estimate the incidence rate of intimate partner violence, extending the use of these methods beyond standard classification tasks toward population-level estimation.

In the Mexican context, Torres-Munguía and Martínez-Zarzoso [26] conducted a wide research on the intimate partner violence in the Mexican population. They identified several risk factors, such as social support networks, distribution of housework, the community's economic inequality, and criminality. In addition, they detected two specific risk subgroups: women who had their first sexual intercourse during childhood and women who got married (or cohabited) late in life. A year later, Torres-Mungía [24] continued the analysis of risk factors for physical intimate partner violence against women and girls through a model-based boosting approach, providing recent evidence that advanced predictive modeling can also contribute to the study of gender violence in Mexico. Finally, last year, another study by Tores-Munguía was conducted [25]. He discovered that Mexican women who engage in sexual activity during childhood are particularly vulnerable to experiencing partner violence. In addition, they found evidence of a strong correlation between the future woman's autonomy in decision-making about her sexual and professional life and the probability of being a victim of partner violence.

In addition, the systematic mapping study presented in [17] shows that the application of machine learning to gender violence has grown considerably in recent years, confirming the emergence of this line of research. Despite these advances, there is still limited work specifically focused on predicting violence in women's first sexual experiences, particularly using recent Mexican national survey data and a comparative evaluation of classification models.

3 Methods and Materials

3.1 Data Set

The dataset used in this study came from the National Survey of Demographic Dynamics 2023 (ENADID 2023) [14], elaborated by the National Institute of Statistics and Geography (INEGI). The ENADID aims to update statistical information related to the level and behavior of the main components of demographic dynamics in Mexico, such as fertility, mortality, and migration (internal and international), as well as issues related to population growth, reproductive preferences, sexuality, the use of contraceptive methods, marriage, maternal and child health, and various sociodemographic characteristics of the population, households, and housing.

The survey included interviews with 103,338 women. The questionnaire comprised 236 items covering a wide range of topics, including age at sexual initiation, consent, fertility, adverse reproductive history, knowledge of contraceptive methods, and sociodemographic characteristics.

For the present study, women who, in the question corresponding to the *condition of sexual intercourse*, did not respond or indicated had not begun their sexual life were excluded. After this filtering, the data was reduced to 87,350 women. An analysis was performed with this subsample about the *balance* among the classes of interest. It was observed that approximately the 98% of the women reported not having experienced violence in their first sexual relationship, while 2% remaining declared having suffered it.

In Table 1 a summary of the resulting database is presented after applying the filtering criteria described above.

Table 1. Summary of database used

Feature	*Description*
Instance number	87,350
Type of attributes	Numeric and categorical
Target variable	Violence in the first sexual relationship (yes / no)
Type of problem	Binary classification
Class imbalance	98% no, 2% yes

3.2 Question Selection

The 2023 ENADID survey contains 236 items covering sociodemographic, reproductive, and sexual health information. For this study, only those variables conceptually linked to reproductive behavior and to factors that the literature has identified as potentially associated with the experience of sexual violence were selected. Among the selected variables are age (years completed), level of education, sociodemographic stratum, marital status, age at menarche, and age at first sexual intercourse.

Also were considered items related to the use of contraceptive methods during first sexual intercourse (P8.32) and the reasons for not using them (P8.22). Item P8.32 includes multiple sub-items describing different methods, such as hormonal contraception, intrauterine devices, male and female condoms, emergency contraception, and other traditional or barrier methods.

Similarly, item P8.22 contains various reasons why a woman may not use contraceptive methods, including physiological reasons such as pregnancy or menopause, personal or health reasons, partner objections, religious reasons, lack of knowledge of the methods or their use, among others.

Finally, item P8.40, relating to consent in the first sexual encounter, was incorporated, given its direct relevance to the study's target variable. Demographic variables such as age, educational level, and area of residence were included for both their analytical relevance and their role in the subsequent partitioning of the database into the subsets used in the experimental phase.

3.3 Data Cleaning and Preprocessing

Data processing was performed using the file `conjunto_datos_tmujer1_enadid_2023.csv` from the 2023 National Survey on Demographic Dynamics (ENADID). First, only the variables considered relevant for the analysis were selected, incorporating an incremental identifier (`id_mujer`) to maintain record traceability. Subsequently, the target variable `p8_40` was filtered to retain only cases with valid values of 1 and 2, which were recoded as a binary variable for modeling: 1 for the consented category and 0 for the non-consensual category.

During data cleaning, the variable `p8_32`, corresponding to the first method used to prevent pregnancy, was treated as a 14-character binary string. To avoid errors arising from inconsistent formats, values in scientific notation were corrected, invalid characters were removed, leading zeros were padded, and the variable was then expanded into 14 independent binary columns (`method_1 to method_14`). Furthermore, the variables `p8_39` and `p8_45`, associated with age, were standardized using Standard Scaler, while binary variables such as `p5_3` and `sex_act` were recoded to 0/1 format.

The selected categorical variables were one-hot encoded, with `drop_first=True` to avoid redundancy across categories. Once the final dataset for modeling was generated, segmentations were performed by age group, school level, and geographic region.

3.4 Data Splitting

To ensure a robust and equitable evaluation of the classification models used in this study, the data were divided into 14 independent datasets.

The segmentation carried out is consistent with the demographic criteria used by INEGI, which has documented that educational level, age, and geographic region account for significant variations in fecundity, reproductive risk, and gender-based violence. The 2023 ENADID shows marked differences in fecundity rates by level of schooling [14]. Likewise, the reports on adolescent pregnancy and violence against women present results by age, social status, and federal entity, which confirms the analytical relevance of these strata [21]. Table 2 shows the variables used, which survey question they correspond to, and whether any coding was applied.

The splitting was structured as follows:

- **By educational level (4 datasets):** Four subsets were constructed based on years of schooling and the educational category reported by the ENADID.
  - **Low educational level**: no schooling, incomplete primary education, and complete primary education.
  - **Medium-low educational level**: incomplete secondary education and complete secondary education.
  - **Medium-high educational level**: medium superior education.
  - **High educational level**: superior education.

  This type of grouping has been employed in national demographic studies, since educational level has a significant impact on age at sexual initiation, access to contraceptive methods, and vulnerability to situations of violence, as documented in ENADID technical reports.
- **By age groups (7 datasets):** Demographic literature has shown that age substantially influences the likelihood of early sexual initiation, the conditions under which the first sexual relationship occurs, and the risks associated with sexual violence [21]. Accordingly, seven additional datasets were created, corresponding to the following age ranges: 15–19, 20–24, 25–29, 30–34, 35–39, 40–44 and 45–54 years.
- **By geographic region (3 datasets):** Finally, the data was divided into three regional subsets: *Center*, *North*, and *South*. This classification is often used in sociodemographic analyses by INEGI due to the marked regional differences in access to services, socioeconomic conditions, reproductive patterns, and prevalence of violence against women.

It is important to note that the disproportion observed in the data (characterized by a marked predominance of women who reported not having experienced violence in their first sexual encounter) persisted in each of the fourteen generated datasets. Since this asymmetrical distribution is inherent in the information captured by the 2023 National Survey on Demographic Dynamics (ENADID), the subsets retain this characteristic even after demographic segmentation. The correction of this imbalance and the strategies implemented to mitigate its effects on model performance are detailed later.

Table 2. Variable Selection

Variable	Survey code	Corresponding question	Data type	Codification
id_mujer	Not applicable	Unique id	Integer	Index (Not for prediction)
ent	ent	Federal entity	Integer	Without codification
p5_2_1	p5.2.1	Completed years	Integer	Without codification
p5_3	p5.3	Does she attend school?	Binary	Binary recoding
p8_39	p8.39	Standardized age of first sexual intercourse	Float	Z-score normalization
p8_45	p8.45	At what age did she have her first period or menstruation?	Float	Z-score normalization
niv_esc	niv_esc	Educational level	Integer	Without codification
p8_32	p8.32	What was the first thing you or your partner did to prevent pregnancy?	Binary chain	Multiresponse binary expansion
stratum	stratum	Sociodemographic stratum	Binary	One hot encoding
cond_act_2	cond_act	Activity condition	Binary	One hot encoding
p10_1_ag2	p10_1_ag2	Marital status (group 2)	Binary	One hot encoding
conoce_6	conoce_6	Knowledge of contraceptive methods	Binary	One hot encoding
p8_40 (Target variable)	p8.40	Was this first experience with your consent (did you want it that way)?	Binary	Binary recoding

3.5 Class Imbalance Treatment

The target variable in this study exhibits a severe class imbalance: 98 % of the women reported *not* having experienced violence during their first sexual encounter, while only 2 % reported having experienced it. This level of imbalance is particularly problematic in binary classification tasks, as models tend to bias their predictions toward the majority class. Consequently, a classifier may achieve seemingly high performance by consistently predicting the most frequent class, even when its ability to identify the minority class of interest is practically negligible.

Since the central objective of this work is to compare the performance of different classification models, random downsampling was applied to the majority class to balance the distribution. This strategy was selected over techniques such as ADASYN or SMOTE because, although they might better represent the minority class, given that the dataset is within a sociodemographic context where many variables correspond to coded categories, interpolation could produce observations without direct interpretation. Similarly, cost-sensitive learning is a good alternative because it does not alter the original data distribution but rather modifies the model's objective function by assigning weights to each class. For this reason, it is proposed as a method in the future work section. The procedure for the random sampling is explained below: For each dataset, the number of observations belonging to class 0 (women who reported violence in their first sexual encounter) was identified and denoted as n. Subsequently, n observations from class 1 (women who did not report violence) were randomly selected without replacement to obtain two classes of the same size. A new balanced basis of size $2n$ was constructed from both samples and then randomized to avoid any systematic ordering by class. This procedure was replicated independently for each of the 14 datasets.

Sampling weights were not incorporated into the model-training process. Since the objective of this study was to compare the relative performance of classification algorithms rather than to estimate nationally representative prevalence parameters, the reported results should be interpreted as sample-based classification metrics. However, because ENADID 2023 follows a probabilistic, stratified, multistage cluster sampling design, the lack of survey-weighted modeling is acknowledged as a limitation. Future work should assess the robustness of the findings using sampling weights and design-aware validation strategies.

Table 3. Classification models and main parameters

Model Type	Model	Main Parameters
Neighborhood-based	k-Nearest Neighbors (kNN)	$k = 3$; Euclidean distance; distance-based weighting
Decision tree	Decision Tree	Minimum 10 instances per leaf; maximum depth of 30; stopping criterion: 95% majority
Rule-based	CN2 Rule Inducer	Entropy measure; beam width = 20; minimum coverage = 4 instances
Probabilistic	Naive Bayes	Default configuration in Orange3
Support Vector Machine	SVM (RBF kernel)	C-SVM with $C = 1.0$; RBF kernel; tolerance 10^{-3}; maximum 1000 iterations
Neural network	Neural Network (MLP)	100 hidden neurons; ReLU activation; Adam optimizer; $\alpha = 0.0005$; 1000 iterations

3.6 Classification Models

In this work, different classification models were used, each with distinct approaches and assumptions to model the relationship between the predictor variables and the variable of interest. We applied Neighborhood-based classifiers [9], Decision trees [18], Rule-based classifiers [6], Probabilistic classifiers [12], Support vector machines (SVMs) [8], and Multilayer perceptron (a neural networks) [27]. Hyperparameter tuning was performed through a controlled random approach, starting from ranges reported in literature, technical documentation, and preliminary validation runs. For each model, a limited set of possible configurations was evaluated to avoid an exhaustive search that would have been computationally expensive. The final configurations were selected based on validation performance, prioritizing the F1-score, AUC, and recall. Table 3 shows the models used in this study, as well as their main parameters.

4 Results and Discussion

The experimental configuration for this study was as follows: All experiments were performed using 5×2 stratified cross-validation. This validation scheme is suitable for imbalanced classification problems because stratification helps preserve the original class distribution across training and testing subsets [11]. This scheme was applied independently to each of the 14 balanced datasets. Since the present study addresses a binary classification problem, the metrics *Recall*, *F1-score*, and the area under the ROC curve (AUC) were used [2]. The results of the metrics were averaged over the folds.

For each of the 14 balanced databases, the six classification models described above were evaluated using Orange3 software, version 3.39 [10]. The experiments were executed on a personal laptop with Windows 10 Home (64-bit) operating system, 32.0 GB ram, and Intel Core i7-770HQ CPU @ 2.80GHz processor. Table 4 summarizes the results obtained for: datasets segmented by age, datasets segmented by educational level, and datasets defined by geographic area. This table reports the performance of the models in terms of the metrics considered, and in each one, the best value achieved by each indicator is highlighted in **bold** to facilitate comparison between the different classifiers.

4.1 Non-parametric Test: Bayesian Method for Comparing Algorithms

In this work, the *Bayesian* approach proposed in 2015 [7] was used to compare the performance of the classification models evaluated in the different datasets. Unlike traditional statistical methods based on *p*-values, the Bayesian approach offers a more intuitive interpretation by directly providing the probability that one model is better than another, as well as the probability that both exhibit equivalent performance. Furthermore, it allows for pairwise comparisons without requiring strict assumptions about the distribution of the data, and is therefore classified as a *non-parametric method.*

To facilitate the interpretation of the results, the symbol "-' is used in Table 4 to identify those models that, according to Bayesian analysis, have a posterior probability of ≥ 0.90 of being inferior to the best-performing model in each metric.

4.2 Non-parametric Test: Nemenyi Multiple Comparisons Test

The Nemenyi test [16] is a non-parametric method of multiple comparisons that is used after applying the Friedman test [13], with the aim of identifying which algorithms show significant differences in their performance.

Table 4. AUC, F1, and Recall results across different groups for the six classification models.

Criterion	Category	Metric	KNN	Decision Tree	CN2	Naive Bayes	SVM	Neural Network
By age	15–19 years	AUC	0.625⁻	0.637	**0.702**	0.682	0.547	0.561
		F1	0.613	0.602	**0.664**	0.582	0.545⁻	0.542
		Recall	0.613	0.604	**0.665**	0.561	0.547⁻	0.542
	20–24 years	AUC	0.520⁻	0.596	0.587	**0.666**	0.519⁻	0.535
		F1	0.517⁻	0.519	0.562	**0.625**	0.502⁻	0.526
		Recall	0.518⁻	0.524	0.562	**0.629**	0.503⁻	0.529
	25–29 years	AUC	0.458⁻	**0.679**	0.617	0.673	0.471⁻	0.598
		F1	0.451⁻	0.628	0.594	**0.656**	0.551⁻	0.559
		Recall	0.451⁻	0.629	0.594	**0.657**	0.552⁻	0.559
	30–34 years	AUC	0.508	**0.676**	0.650	0.672	0.508	0.566⁻
		F1	0.495⁻	0.600	0.588	**0.635**	0.544	0.573
		Recall	0.497⁻	0.603	0.588	**0.637**	0.550	0.573
	35–39 years	AUC	0.580⁻	**0.690**	0.665	0.688	0.613	0.609⁻
		F1	0.560⁻	**0.616**	0.602	0.612	0.572	0.605
		Recall	0.560	**0.623**	0.602	0.613	0.572	0.605
	40–44 years	AUC	0.566⁻	0.707	0.652⁻	**0.734**	0.626⁻	0.608⁻
		F1	0.540⁻	0.646	0.600⁻	**0.686**	0.598⁻	0.588
		Recall	0.541⁻	0.646	0.600⁻	**0.686**	0.599⁻	0.588
	45–54 years	AUC	0.616⁻	0.726	0.638⁻	**0.743**	0.675⁻	0.686
		F1	0.594⁻	0.667	0.606⁻	**0.674**	0.609⁻	0.634
		Recall	0.595⁻	0.668	0.606⁻	**0.675**	0.609⁻	0.634
By educational level	Low	AUC	0.559⁻	0.701⁻	0.657⁻	**0.741**	0.633⁻	0.677⁻
		F1	0.537	0.643	0.620⁻	**0.656**	0.589⁻	0.637
		Recall	0.537	0.643	0.620⁻	**0.656**	0.589⁻	0.637
	Lower-Middle	AUC	0.517⁻	0.683	0.632⁻	**0.715**	0.614⁻	0.685
		F1	0.525	0.616	0.610	**0.658**	0.578⁻	0.631
		Recall	0.525	0.617	0.612	**0.659**	0.579⁻	0.631
	Upper-Middle	AUC	0.528⁻	0.624	0.626	**0.686**	0.555⁻	0.644
		F1	0.512⁻	0.594	0.612	**0.625**	0.538⁻	0.599
		Recall	0.513⁻	0.594	0.612	**0.627**	0.538⁻	0.599
	High	AUC	0.569⁻	0.617⁻	**0.660**	0.646	0.552	0.568⁻
		F1	0.539⁻	0.572	**0.602**	0.600	0.537⁻	0.551
		Recall	0.539⁻	0.572⁻	0.602	**0.606**	0.541	0.552⁻
By region	South	AUC	0.533⁻	0.713⁻	0.637	**0.727**	0.648⁻	0.699⁻
		F1	0.523⁻	0.644	0.597	**0.676**	0.601	0.652
		Recall	0.523⁻	0.645	0.597	**0.671**	0.601	0.652
	Central	AUC	0.591⁻	0.688	0.652⁻	**0.726**	0.658	0.640⁻
		F1	0.576	0.647	0.607	**0.657**	0.607	0.596
		Recall	0.577	0.647	0.607	**0.657**	0.609	0.597
	North	AUC	0.635⁻	0.724	0.657	**0.768**	0.639⁻	0.669⁻
		F1	0.613	0.674	0.600	**0.706**	0.584⁻	0.617
		Recall	0.613	0.674	0.601	**0.706**	0.587⁻	0.617

Note: "–" indicates models with posterior probability ≥ 0.90 of being inferior to the best-performing model.

The method is based on average rank analysis: for each dataset, the algorithms are ranked according to their performance and assigned a rank. Then, the average rank of each algorithm is calculated across all evaluated datasets. The Nemenyi test determines that two algorithms differ significantly if the difference between their average ranks exceeds a certain amount called the *Critical Difference* (CD).

In this study, each demographic partition was treated as a distinct evaluation scenario for algorithm comparison purposes. Although all partitions were derived from the same national survey, they correspond to different demographic subpopulations and preserve the same experimental protocol, variable structure, and performance metrics. Therefore, the Friedman/Nemenyi framework is appropriate to compare the relative performance of the classifiers across the evaluated data sets.

The non-parametric Friedman test was applied to compare the classifier's performance on the three indicators considered: AUC, Recall, and F1-score. In all three cases, the null hypothesis was rejected, indicating the existence of overall differences in model performance. Subsequently, the Nemenyi post-hoc test was used to identify which pairs of classifiers exhibited statistically significant differences.

Figures 1, 2, and 3 show the critical difference diagrams obtained from this test. In general, across all three indicators, the best-performing models are Naive Bayes, Decision Tree, and CN2, which do not show statistically significant differences from each other. In contrast, Neural Network, SVM, and KNN show the worst performance and exhibit statistically significant differences when compared to the top-performing models (Naive Bayes, Decision Tree, and CN2).

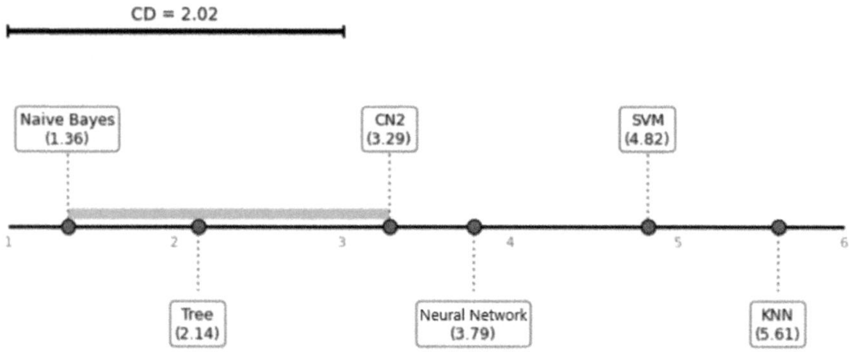

Fig. 1. Critical difference diagram for the AUC according to the Nemenyi test.

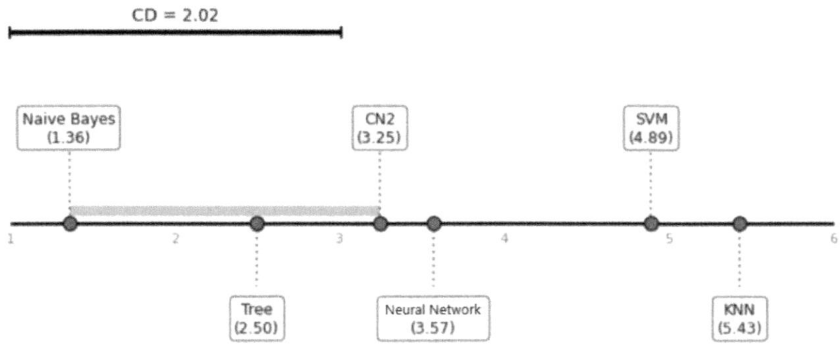

Fig. 2. Critical difference diagram for the F1-score according to the Nemenyi test.

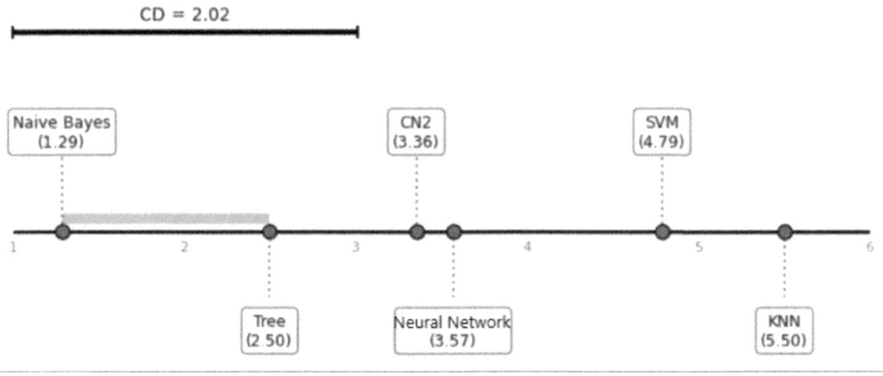

Fig. 3. Critical difference diagram for the recall according to the Nemenyi test.

4.3 Analysis of Results

The results show that classifier performance varied across the demographic partitions considered in this study, indicating that the predictive structure of the data is not homogeneous across age, educational level, and region. In general, *Naive Bayes* achieved the best overall performance, standing out in most of the evaluated subsets and metrics (AUC, F1, and Recall). This suggests that the selected sociodemographic and reproductive variables contain probabilistic patterns that are useful for distinguishing the outcome classes.

The regional segmentation showed the most stable behavior: *Naive Bayes* obtained the best values across all metrics and regions (South, Central, and North). This consistency suggests that, despite regional differences, the relationships captured by the selected variables remain sufficiently stable for a probabilistic classifier to model them effectively. A similar pattern was observed across educational-level partitions, where *Naive Bayes* also achieved the best performance in the Low, Lower-Medium, and High-Medium groups.

In contrast, the age-based partitions showed a less uniform pattern. CN2 achieved the best values in the 15âĂŞ19 age group, while Decision Tree performed best in the 35âĂŞ39 age group and obtained the highest AUC in the 25âĂŞ29 and 30âĂŞ34 groups. This behavior suggests that, in some age groups, the classification task may depend on more specific combinations of variables or threshold-based patterns, which rule-based and tree-based models can capture more effectively.

Overall, these findings indicate that the information provided by the selected variables differs across demographic subgroups. Therefore, the results should be interpreted not only as a comparison of algorithmic performance, but also as evidence that the classification patterns associated with the outcome vary according to age, educational level, and region. However, these patterns should not be interpreted causally, since the survey design does not allow the temporal ordering of all variables with respect to the outcome to be fully established.

5 Conclusions

The results obtained allow us to reflect on the social phenomenon of violence in first sexual encounters and on the predictive capacity of models using information related to the reproductive behavior of Mexican women. Although the central purpose of this study was to evaluate the effectiveness of different classifiers in estimating the probability that a woman experienced violence in her first sexual encounter, the observed patterns offer a broader perspective on the social context and the inherent limitations of the available data.

It is important to note that this study focused exclusively on variables related to reproductive behavior. While these variables proved to contain useful predictive signals, incorporating additional factors (such as a history of violence, economic conditions, family dynamics, or affective experiences) would likely improve both the accuracy of the models and our understanding of the phenomenon.

The observed similarity between F1-score and recall values is attributable to the mathematical relationship among precision, recall, and F1-score. Since the F1-score is the harmonic mean of precision and recall, similar precision and recall values naturally produce an F1-score close to both metrics. The positive class and classification threshold were kept consistent across all models to ensure comparability.

The comparison between models showed outstanding performance from relatively simple classifiers, such as Naive Bayes. On the other hand, the performance of models like CN2 and Decision Tree in certain educational segments could indicate that, within certain groups, more structured patterns exist that warrant exploration from sociological or public health perspectives.

Finally, from a social impact and public policy perspective, the results of this study underscore the importance of having analytical tools that allow for the early identification of risk profiles. Predictive models based on national surveys can contribute to strengthening preventive strategies, guiding targeted educational campaigns, and optimizing the allocation of institutional resources to more vulnerable groups.

6 Future Works

Future work will focus on improving the methodological robustness and interpretability of the proposed classification framework. First, a cost-sensitive learning approach will be incorporated to address class imbalance without altering the dataset's original distribution. This approach assigns higher misclassification costs to the minority class during model training, allowing classifiers to place greater weight on underrepresented cases while preserving all available observations.

Second, an analysis of predictor relevance will be conducted for the best-performing models. This will allow the identification of the sociodemographic and reproductive variables that contribute most strongly to the classification task. Such analysis will provide a better understanding of the patterns captured by the models and will help move beyond performance comparison toward a more substantive interpretation of the factors associated with the outcome.

Finally, the area under the precisionâĂŞrecall curve (AUPRC) will be incorporated as an additional evaluation metric. Given class imbalance, AUPRC can provide a more informative assessment of classifier performance than metrics based solely on overall accuracy or ROC-based evaluation, particularly when correct identification of the minority class is of interest.

Acknowledgment. The authors would like to thank institution

Disclosure of Interests. The authors have no competing interests.

References

1. Acharya, A.K.: Sexual violence and proximate risks: A study on trafficked women in Mexico city. Gend. Technol. Dev. **12**(1), 77–99 (2008)
2. Ballabio, D., Grisoni, F., Todeschini, R.: Multivariate comparison of classification performance measures. Chemom. Intell. Lab. Syst. **174**, 33–44 (2018)
3. Boyce, S.C., Brouwer, K.C., Triplett, D., Servin, A.E., Magis-Rodriguez, C., Silverman, J.G.: Childhood experiences of sexual violence, pregnancy, and marriage associated with child sex trafficking among female sex workers in two us-mexico border cities. Am. J. Public Health **108**(8), 1049–1054 (2018)
4. Castro, R., Peek-Asa, C., Ruiz, A.: Violence against women in Mexico: a study of abuse before and during pregnancy. Am. J. Public Health **93**(7), 1110–1116 (2003)

5. Chen, Z., et al.: Using machine learning to estimate the incidence rate of intimate partner violence. Sci. Rep. **13**(1), 5533 (2023)
6. Clark, P., Boswell, R.: Rule induction with CN2: Some recent improvements. In: Kodratoff, Y. (ed.) EWSL 1991. LNCS, vol. 482, pp. 151–163. Springer, Heidelberg (1991). https://doi.org/10.1007/BFb0017011
7. Corani, G., Benavoli, A.: A bayesian approach for comparing cross-validated algorithms on multiple data sets. Mach. Learn. **100**(2), 285–304 (2015)
8. Cortes, C., Vapnik, V.: Support-vector networks. Mach. Learn. **20**(3), 273–297 (1995)
9. Cover, T., Hart, P.: Nearest neighbor pattern classification. IEEE Trans. Inf. Theory **13**(1), 21–27 (1967)
10. Demšar, J., et al.: Orange: Data mining toolbox in python. J. Mach. Learn. Res. **14**, 2349–2353 (2013). http://jmlr.org/papers/v14/demsar13a.html
11. Dietterich, T.G.: Approximate statistical tests for comparing supervised classification learning algorithms. Neural Comput. **10**(7), 1895–1923 (1998)
12. Domingos, P., Pazzani, M.: On the optimality of the simple bayesian classifier under zero-one loss. Mach. Learn. **29**(2), 103–130 (1997)
13. Friedman, M.: The use of ranks to avoid the assumption of normality implicit in the analysis of variance. J. Am. Stat. Assoc. **32**(200), 675–701 (1937)
14. Instituto Nacional de Estadística y Geografía (INEGI): Encuesta nacional de la dinámica demográfica (enadid) 2023: Resultados generales (2024). https://www.inegi.org.mx/programas/enadid/2023/, Accessed 29 Mar 2026
15. McDougal, L., Dehingia, N., Bhan, N., Singh, A., McAuley, J., Raj, A.: Opening closed doors: using machine learning to explore factors associated with marital sexual violence in a cross-sectional study from india. BMJ Open **11**(12), e053603 (2021)
16. Nemenyi, P.B.: Distribution-free multiple comparisons, Princeton University (1963)
17. Pinto-Muñoz, C.C., Zuñiga-Samboni, J.A., Ordoñez-Erazo, H.A.: Machine learning applied to gender violence: A systematic mapping study. Revista Facultad de Ingeniería 32(64), (2023)
18. Quinlan, J.R.: Induction of decision trees. Mach. Learn. **1**(1), 81–106 (1986)
19. Rahman, R., Khan, M.N.A., Sara, S.S., Rahman, M.A., Khan, Z.I.: A comparative study of machine learning algorithms for predicting domestic violence vulnerability in liberian women. BMC Womens Health **23**(1), 542 (2023)
20. Rivera Rivera, L., et al.: Violence against women during the covid-19 pandemic in Mexico. In: Healthcare, vol. 11, p. 419. MDPI (2023)
21. Secretaría de Salud, Instituto Nacional de Estadística y Geografía (INEGI): Situación de la salud sexual y reproductiva en méxico. https://www.gob.mx/cms/uploads/attachment/file/237216/Cuadernillo_SSR_RM.pdf (2016). gobierno de México. Accessed 29 Mar 2026
22. Sistema Nacional de Protección de Niñas, Niños y Adolescentes: Violencia en el noviazgo: no es amor, no es amistad. Gobierno de México (nd). https://www.gob.mx/sipinna/articulos/violencia-en-el-noviazgo-no-es-amor-no-es-amistad?idiom=es
23. Soria-Escalante, H., Alday-Santiago, A., Alday-Santiago, E., Limón-Rodríguez, N., Manzanares-Melendres, P., Tena-Castro, A.: "We all get raped": sexual violence against Latin American women in migratory transit in Mexico. Violence against women **28**(5), 1259–1281 (2022)

24. Torres Munguía, J.A.: A model-based boosting approach to risk factors for physical intimate partner violence against women and girls in Mexico. J. Comput. Soc. Sci. **7**(2), 1937–1963 (2024)
25. Torres Munguía, J.A.: An analysis of the risk factors for intimate partner sexual violence against women and girls in mexico. J. Gender-Based Viol. **9**(1), 76–92 (2025)
26. Torres Munguía, J.A., Martínez-Zarzoso, I.: Determinants of emotional intimate partner violence against women and girls with children in mexican households: an ecological framework. J. Interpersonal Violence **37**(23-24), NP22704–NP22731 (2022)
27. Wu, Y.C., Feng, J.W.: Development and application of artificial neural network. Wireless Pers. Commun. **102**(2), 1645–1656 (2018)

Portfolio Optimization and Artificial Intelligence: Markowitz Theory, Machine Learning and Modern Forecasting Methods

V. Luna Canales[1], R. Castillo[1], and Y. U. K. Coronado González[2](✉)

[1] Facultad de Negocios Universidad La Salle México, Mexico City, Mexico
[2] Vicerrectoría de Investigación, Universidad La Salle México, Mexico City, Mexico
yaxkin.coronado@lasalle.mx

Abstract. This paper analyzes financial markets using quantitative techniques commonly applied in actuarial science and financial engineering. The objective is to construct and evaluate an optimal portfolio while analyzing financial asset behavior through statistical and machine learning models. The methodologies implemented include Markowitz portfolio optimization on 12 BMV assets, Monte Carlo simulations, ARIMA time series forecasting, and recurrent neural networks (RNN) and Long Short-Term Memory (LSTM) architectures. Each approach provides complementary insights regarding portfolio risk, return behavior, and forecasting accuracy. These methods are applied as illustrative case studies on different datasets, reflecting the diversity of real-world financial modeling scenarios. The results illustrate how diversification, stochastic modeling, deep learning, and predictive algorithms can support more informed decision-making in portfolio management.

Keywords: Portfolio Optimization · Markowitz Model · Monte Carlo Simulation · ARIMA · Recurrent Neural Networks · Financial Markets

1 Introduction

The allocation of capital across risky assets is one of the most consequential decisions in financial practice. From individual investors managing retirement savings to institutional fund managers overseeing multi-billion-dollar portfolios, the challenge of constructing an optimal portfolio—one that achieves the highest possible return for a given level of risk—has driven decades of both academic research and technological innovation.

Harry Markowitz's 1952 paper "Portfolio Selection" [1] laid the groundwork for modern quantitative finance by formalizing the relationship between risk, return, and diversification. For several decades, the Markowitz mean-variance framework and its extensions, including the Capital Asset Pricing Model (CAPM) [9], remained the dominant paradigms. However, these classical models rely on assumptions—such as normally distributed returns and stationary market dynamics—that frequently fail to hold in practice.

H. Ponce et al. (Eds.): COMIA 2026, CCIS 3095, pp. 294–308, 2027.
https://doi.org/10.1007/978-3-032-37438-7_23

The advent of machine learning and artificial intelligence has provided the financial industry with tools capable of capturing non-linear patterns, processing vast datasets, and adapting to evolving market conditions. Techniques such as Recurrent Neural Networks (RNN) [6], Long Short-Term Memory (LSTM) architecture [7], Monte Carlo simulation [4], ARIMA time series models [3], and reinforcement learning [10] are now widely employed in algorithmic trading, risk forecasting, and portfolio management. The contribution of this paper is not the proposal of a new algorithm, but an integrated illustrative case study showing how optimization, simulation, and forecasting tools support different stages of portfolio decision-making. Because these models answer different analytical questions, the empirical sections use datasets suited to each method; therefore, the results should be interpreted as complementary case studies rather than as a direct benchmark across a single asset universe. Table 1 summarizes the methods, datasets, assumptions, outputs, and limitations considered in the paper.

Table 1. Unified summary of methods, datasets, assumptions, outputs, and limitations.

Method	Dataset / Universe	Key Assumptions	Main Output	Main Limitation
Markowitz MPT	12 BMV assets Jan 2022-Sep 2025	Expected returns and covariances estimated from historical data; no short selling.	PMV, PMS, efficient frontier, portfolio weights.	Highly sensitive to input estimates and asset universe.
ARIMA	AAPL daily closing prices Jan 2022-Jan 2024	Differenced series is stationary; short-run linear dependence is informative.	One-step-ahead forecasts and error metrics.	Limited for nonlinear regime changes and news-driven jumps.
Monte Carlo	Hybrid 50-stock portfolio 100-day horizon	GBM-inspired dynamics with Student-t innovations; historical parameters proxy future behavior.	Distribution of final values, probability of profit, VaR threshold and VaR loss.	Exploratory with 100 paths; formal tail-risk work requires many more simulations.
SimpleRNN and LSTM	AAPL daily closing prices Jan 2022-Jan 2024	Patterns can be learned from 60-day windows; train/test structure is stable.	Predicted prices and RMSE comparison.	Requires tuning and may overfit; black-box interpretation.

(*continued*)

Table 1. (*continued*)

Method	Dataset / Universe	Key Assumptions	Main Output	Main Limitation
Reinforcement Learning	Conceptual method only; not empirically implemented.	Portfolio actions, rewards, and constraints can be clearly defined.	N/A (conceptual overview only)	Training can be unstable and requires complex tuning.

2 Markowitz

Modern Portfolio Theory, introduced by Markowitz [1], fundamentally changed investment analysis by shifting focus from individual asset returns to the statistical properties of a portfolio as a whole. Although the framework originated in 1952, the term "Modern Portfolio Theory" is the conventional name used in finance for the post-Markowitz mean-variance approach. The key insight is that the risk of a portfolio depends not only on the variance of each individual asset but also on the correlations between them. By combining assets with low or negative correlations, an investor can construct a portfolio whose variance is substantially lower than the weighted average of individual asset variances—this is the mathematical basis of diversification.

Formally, given n assets with expected returns $\mu = (\mu_1, ..., \mu_n)$ and covariance matrix Σ, the portfolio return is $\mu_p = w^T\mu$ and the portfolio variance is $\sigma_p{}^2 = w^T\Sigma w$.

Where w is the vector of portfolio weights summing to one. The mean-variance optimization problem is:

$$\min_{w} w^T \Sigma w \text{ s.t.} w^T \mu = \mu_0, w^T \mathbf{1} = 1, w \geq 0, \text{ where } \mathbf{1} = (1, 1, \ldots, 1)^T \in \mathbb{R}^n \quad (1)$$

The set of solutions over all target returns μ_0 traces the minimum variance frontier; its upper half is the efficient frontier. Investors with different risk appetites select different points on this frontier. The global minimum variance portfolio (GMVP) represents the leftmost point, the portfolio with the lowest attainable risk regardless of return.

2.1 Methodology

To illustrate the Markowitz, mean-variance framework in practice, we applied the model to a portfolio of 12 publicly traded assets listed on the Mexican Stock Exchange (BMV), using daily closing price data sourced from Yahoo Finance [11] covering the period from January 2022 to September 2025, a total of 929 trading days.

The assets selected were chosen to achieve broad sectoral diversification across the BMV, covering real estate investment trusts (FIBRAs: FIBRA PL14, FMTY14, FIBRAMQ12, DANHOS13), consumer staples and discretionary (CUERVO, CHDRAUI, ALSEA), financial services (GENTERA, BBAJIO), telecommunications infrastructure (SITES, AXTEL CPO), and chemicals (ALPEKA). All selected assets had continuous price histories throughout the full sample window with fewer than 1% of trading days missing, a threshold chosen to ensure sufficient data completeness for robust estimation of the covariance matrix, consistent with standard practice in empirical asset pricing [2].

We compute the log returns via the daily prices as $r_t = \ln(P_t / P_{t-1})$, which approximate continuously compounded returns and have the desirable property of being additive over time. Annualized expected returns were obtained by multiplying the mean daily log return by 252 (the standard number of trading days per year), and the annualized covariance matrix was computed analogously.

The optimization was implemented using the SLSQP (Sequential Least Squares Programming) algorithm [12], subject to two constraints: (1) portfolio weights sum to one, and (2) all weights are non-negative (no short selling).

This setup differs from the classical Markowitz approach in which an investor can hold negative weights, meaning they can "short" an asset. Here, we explicitly forbid that, requiring every weight to be zero or positive. As a result, the optimizer searches iteratively through feasible combinations of weights, adjusting them step by step until it finds the allocation that minimizes risk while satisfying all constraints. Two distinct optimal portfolios were identified: the Portfolio of Minimum Variance (PMV), which minimizes the portfolio variance function ${\sigma_p}^2 = w^T\Sigma w$ directly, and the Portfolio of Maximum Sharpe Ratio (PMS), which maximizes the Sharpe ratio using a risk-free rate of 8.37%, consistent with prevailing Mexican Treasury (CETES) rates during the sample period [13].

To visualize the feasible set and the efficient frontier, 30,000 random portfolios were simulated by drawing weights from a Dirichlet distribution [14], ensuring non-negativity and the budget constraint by construction. The details are presented in Fig. 1, where the cloud of points traces the characteristic bullet-shaped feasible set predicted by Markowitz theory.

2.2 Results

The annualized returns and market betas of the 12 individual assets are summarized in Table 2, which is discussed in detail below. The results reveal a highly heterogeneous universe: only three assets—CUERVO (24.83%), ALPEKA (27.15%), and AXTEL CPO (12.65%)—generated positive annualized returns over the sample period, while the remaining posted losses, with CHDRAUI (−33.98%) and GENTERA (−33.11%) being the worst performers.

The market betas reported in Table 2 were estimated as $\beta_i = \mathrm{Cov}(r_i, rIPC) / \mathrm{Var}(rIPC)$, equivalently through an Ordinary Least Squares (OLS) regression of each asset's daily log returns against the IPC index, the main BMV benchmark, over the full sample period. Most betas are negative or close to zero. These values should be interpreted as sample-specific evidence of low or inverse co-movement during 2022–2025, not as permanent defensive characteristics. They may reflect sector-specific behavior, differences in liquidity, and the composition of the IPC benchmark relative to the selected assets.

Table 2. Annualized returns and market betas for the 12 BMV assets (Jan 2022 – Sep 2025). Betas estimated via OLS regression against the IPC index.

Asset	Ann. Return (μ)	Beta (β)	Asset	Ann. Return (μ)	Beta (β)
FIBRA PL14	−6.50%	−0.1677	CHDRAUI	−33.98%	0.1796
GENTERA	−33.11%	−0.1248	FIBRAMQ12	−6.06%	−0.0525
CUERVO	24.83%	0.0299	AXTEL CPO	12.65%	−0.0469
SITES	−10.59%	−0.0400	ALSEA	−9.36%	−0.1026
ALPEKA	27.15%	−0.1018	DANHOS13	−5.12%	−0.0419
FMTY14	−5.39%	−0.0576	BBAJIO	−10.59%	−0.0685

The Markowitz optimization produced two efficient portfolios, detailed in Table 3. The Portfolio of Minimum Variance (PMV) concentrates its largest allocations in FMTY14 and DANHOS13, with a negative expected return of −6.00% and volatility of 11.65%. This result is consistent with the challenging market environment during the sample period, in which most BMV assets experienced losses. The Portfolio of Maximum Sharpe (PMS) is composed of two positive-return assets—CUERVO and ALPEKA—with a negligible residual in AXTEL CPO (0.51%). This portfolio achieves an annualized return of 25.81% at a volatility of 25.96%, yielding a Sharpe ratio of 0.6716, the best risk-adjusted return achievable given the asset universe and the 8.37% risk-free rate.

Table 3. Optimal portfolio weights, returns, volatility, and Sharpe ratios for PMV and PMS.

Portfolio	Top Holdings	Ann. Return (μ)	Volatility	Sharpe Ratio
PMV (Min. Variance)	FMTY14 (39.5%), DANHOS13 (17.1%)	−6.00%	11.65%	N/A
PMS (Max. Sharpe)	CUERVO (54.6%), ALPEKA (44.9%)	+25.81%	25.95%	0.6716

Figure 1 illustrates the efficient frontier generated by the simulation of 30,000 random portfolios. Each dot represents a feasible portfolio, colored by its Sharpe ratio from dark purple (low/negative) to yellow green (high). The cloud of points traces the characteristic bullet-shaped feasible set predicted by Markowitz theory: the left boundary is the efficient frontier, and portfolios in the interior are dominated. The gold triangle marks the PMV at the leftmost region, while the red star marks the PMS in the upper-right area, isolated from the cloud due to its extreme concentration in the two best-performing assets.

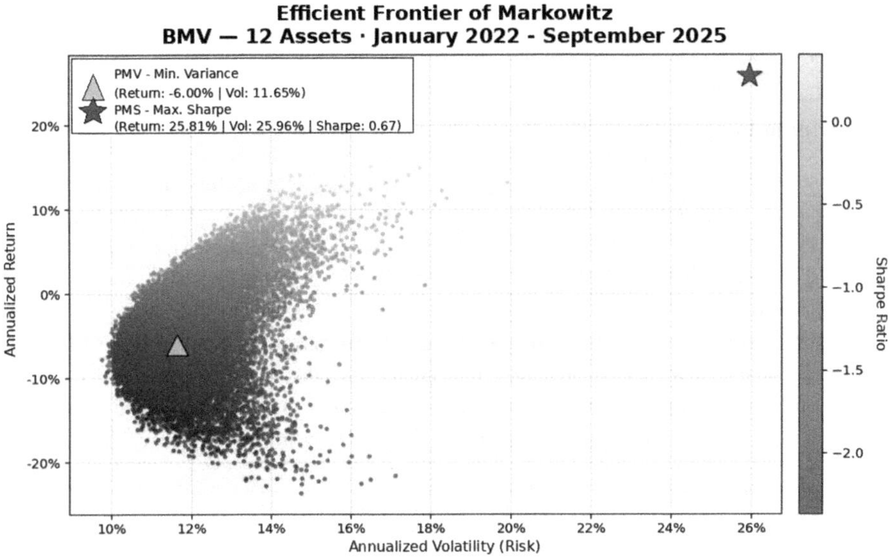

Fig. 1. Efficient Frontier of the Markowitz model for 12 BMV assets (Jan 2022-Sep 2025). Each point represents one of 30,000 simulated portfolios, colored by Sharpe ratio. Gold triangle = PMV; red star = PMS. Horizontal axis: Annualized Volatility (Risk); vertical axis: Annualized Return; color bar: Sharpe Ratio.

These results demonstrate two core lessons of Modern Portfolio Theory in action. First, diversification alone is not sufficient if the majority of available assets are trending downward—as was the case for most BMV stocks during this period. Second, the optimizer faithfully identifies the best achievable risk-return trade-off given the constraints: an investor willing to concentrate in CUERVO and ALPEKA could earn a meaningful positive return (25.81%) with a Sharpe ratio above 0.67, substantially outperforming any individual asset on a risk-adjusted basis.

3 ARIMA

ARIMA (Autoregressive Integrated Moving Average), developed by Box and Jenkins [3], is designed to forecast time series such as daily stock prices or monthly inflation rates. The core idea is simple: the past behavior of a series contains useful information about its future, and that information can be extracted for forecasting. The model has three components: the autoregressive part (AR), which relates today's value to recent past values; the moving average part (MA), which captures the influence of past forecast errors; and the integrated part (I), which removes trends to make the series stationary.

In practice, analysts use autocorrelation plots and information criteria such as AIC (Akaike Information Criterion) or BIC (Bayesian Information Criterion) to select the best model specification. In this study, candidate ARIMA (p, d, q) models were evaluated with d = 1 fixed by the Augmented Dickey-Fuller (ADF) test [15]. The model with the lowest AIC was selected as the primary criterion; BIC was used as a robustness check and confirmed the same specification, ARIMA (1, 1, 1). ARIMA remains widely used because it is interpretable, fast, and competitive on short-horizon tasks.

3.1 Methodology

To illustrate the ARIMA methodology in practice, we applied the model to the daily closing price of Apple Inc. (AAPL), for 501 trading days from January 3, 2022, to January 1, 2024—a period that captures both a sharp bear market decline in 2022 and a strong recovery through 2023, making it a challenging and informative test case for a linear forecasting model.

Figure 2 shows the raw price series. Prices fell from approximately USD 178 at the start of 2022 to a low of USD 124 in January 2023, before recovering to near USD 192 by the end of the sample. This visible trend immediately signals a potential stationarity problem, which must be resolved before ARIMA can be applied.

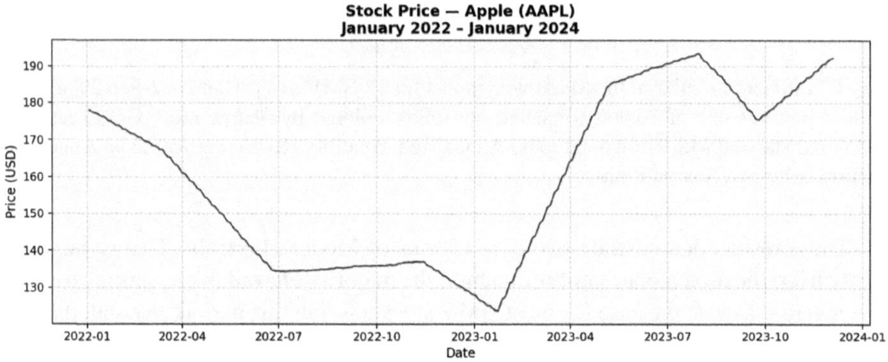

Fig. 2. Daily closing price of Apple Inc. (AAPL), January 2022 – January 2024. The clear downward-then-upward trend indicates non-stationarity in the raw series.

Stationarity was verified using the Augmented Dickey-Fuller (ADF) test [15]: the raw data showed a non-stationary series (ADF $= -1.5124$, $p = 0.5273$). After applying one round of first differencing, the ADF statistic dropped to -16.71 ($p = 1.43 \times 10^{-29}$), establishing $d = 1$ as the differencing order. The autoregressive order p and moving-average order q were selected by inspecting the ACF and PACF plots of the differenced series (Fig. 3), suggesting $p = 1$ and $q = 1$. This led to the selection of ARIMA (1, 1, 1) as the model specification.

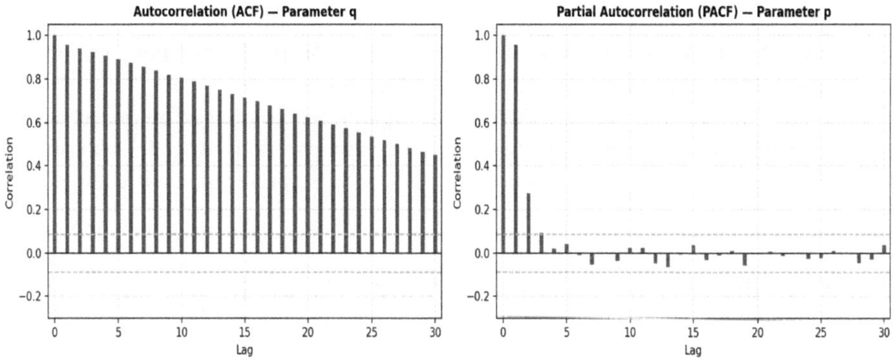

Fig. 3. ACF and PACF of the first-differenced AAPL price series. The plots support the selection of p = 1 and q = 1 for the ARIMA model.

The model was estimated using the statsmodels library in Python (ARIMA class with order = (1,1,1)), with a data split of 90/10, i.e., 450 days for training and 51 days for testing. To track the trajectory and avoid the flat-line problem of static multi-step forecasting, we implemented a walk-forward (rolling) validation strategy: at each test step, the model is re-estimated with all data available up to that point, and only a one-step-ahead forecast is made. This simulates a realistic deployment scenario where new market data arrives daily.

3.2 Results

The best model's forecast accuracy metrics are shown in Table 4. The Root Mean Squared Error (RMSE) of 0.42 USD means the model's one-step-ahead predictions were off by less than half a dollar on average, remarkably precise for a linear model. The Mean Absolute Percentage Error of 0.20% confirms that the errors were tiny relative to the price level.

Table 4. Forecast accuracy metrics for the ARIMA (1,1,1) walk-forward validation on AAPL (51-day test set).

Metric	Value	Interpretation
MSE (Mean Squared Error)	0.18 USD^2	Average squared prediction error
RMSE (Root Mean Squared Error)	0.42 USD	Average error in dollar terms
MAE (Mean Absolute Error)	0.36 USD	Average absolute deviation
MAPE (Mean Absolute % Error)	0.20%	Error relative to price level

To put these numbers in context: an RMSE of 0.42 USD means that on a typical day, the model's price prediction was off by about 42 cents. For a stock trading at roughly USD 185, that is an error of less than 0.25%, small enough to be considered negligible. The MAPE of 0.20% confirms this, showing that the model's predictions stayed within a fifth of a percent of the real price across the entire test period. The MAE of 0.36 USD tells the same story: on average, the model was neither consistently too high nor too low — its errors were small and balanced in both directions.

The key limitation is that ARIMA cannot anticipate non-linear regime changes, volatility clustering, or sudden news-driven price jumps, motivating the exploration of non-linear approaches such as neural networks, whose performance is evaluated in Sect. 5.

4 Probabilistic Risk Assessment: Monte Carlo Simulations

Deterministic models often fall short because they cannot account for the inherent randomness of financial markets. Monte Carlo simulation [4] is a computational technique designed to model this uncertainty by generating alternative possible future scenarios. By simulating alternative price trajectories, analysts can assess the risk and return profile of an asset allocation before committing capital.

The classical mathematical model for simulating stock prices in continuous time is Geometric Brownian Motion (GBM) [16]. In GBM, the future price of an asset depends on two components: a predictable drift and an unpredictable stochastic shock. The continuous-time relationship is expressed as:

$$dS = S * (\mu * dt + \sigma * dW) \tag{2}$$

Where dS represent the change in stock price, S is the current price, μ is the expected return or drift, dt is the time increment, σ is the asset's historical volatility, and dW is a standard Wiener-process shock. In the empirical implementation below, we use a GBM-inspired discrete-time simulation and replace the Gaussian innovation with Student's t-distributed shocks to better reflect heavy-tailed financial returns [17]. Therefore, the simulation should be interpreted as a practical stress-testing approximation rather than as a pure GBM specification.

We implemented an exploratory Monte Carlo simulation evaluating a hybrid portfolio of 50 diverse equities. Starting with an initial investment of $10,000, we projected the portfolio's performance over a 100-trading-day horizon to assess downside risk and the probability of profitability. The implementation uses 100 trajectories, which is sufficient for an illustrative scenario analysis and for matching the reported results, but it should not be interpreted as a production-level tail-risk engine. Formal VaR estimation would require a much larger number of simulations to obtain a more stable estimate of extreme percentiles.

Implementing this model in Python requires calculating the historical daily returns of the selected assets. From these returns, the variance and the mean are extracted to compute the "drift" (the directional bias of the stock). Utilizing libraries such as NumPy, matrices of random variables drawn from a Student's t-distribution are generated to simulate daily price movements over a specified future time horizon.

When applied to real-world equity data to build a hybrid portfolio of 50 stocks, the first crucial step is analyzing the covariance between the assets. Ensuring that the chosen assets are not perfectly correlated effectively leverages the mathematical basis of diversification.

By analyzing the final value points of these simulations, we move beyond a simple frequency histogram. The model stress-tests the portfolio across multiple market scenarios (bearish, bullish, lateral, and volatile).

After generating the 100 simulated portfolio trajectories (Fig. 4), the terminal value of each path is extracted to form a distribution of potential final outcomes. To transition from these raw simulated paths to actionable risk metrics, standard statistical operations are applied to this final distribution array.

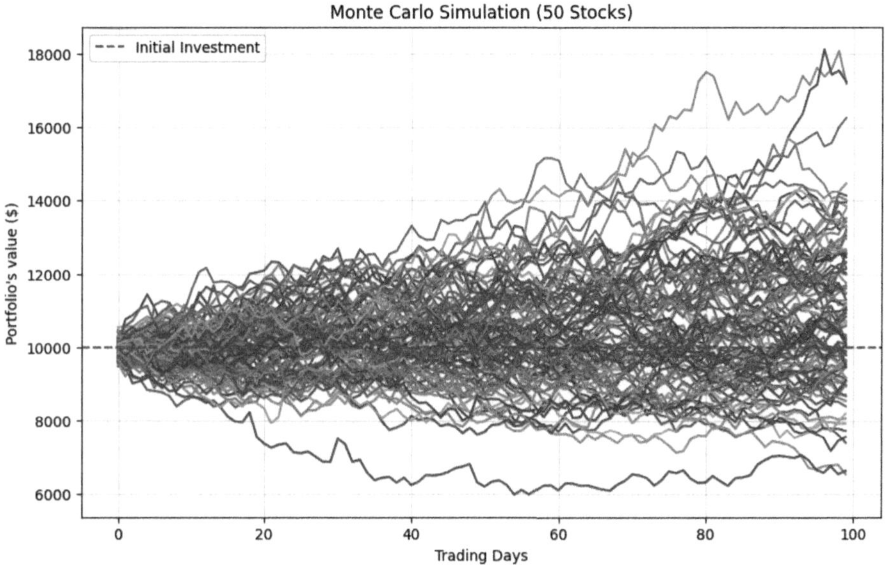

Fig. 4. Monte Carlo simulated portfolio paths over 100 trading days for a hybrid portfolio composed of 50 stocks.

We calculate the mean and median to establish expected central tendencies, while extracting the minimum and maximum values to identify the simulated range of outcomes. The probability of profit is derived by calculating the percentage of simulated final values that exceed the initial investment. The 5th percentile of the final value distribution is reported as the VaR threshold; expressed as a loss relative to the initial investment, the implied 95% VaR loss is $1,943.73, as detailed in Table 5.

This allows investors to quantify exact metrics, providing a comprehensive quantitative assessment of risk under varying economic conditions.

Table 5. Summary statistics and Monte Carlo results for the hybrid portfolio (100 simulated trajectories, 100-day horizon).

Metric / Indicator	Value / Result
Initial Investment	$10,000.00
Average Final Value	$10,977.03
Median Final Value	$10,616.33
Worst Case Scenario (Min)	$6,538.98
Best Case Scenario (Max)	$17,159.19
Probability of Profit	64.0%
VaR (95% Confidence)	$8,056.27
Implied VaR Loss (95%)	$1,943.73

Starting with an initial investment of $10,000.00, the central tendencies indicate positive expected growth, but the extreme outcomes reveal substantial underlying volatility. The simulation yields a 64.0% probability of profit, meaning there is still a notable 36% chance of capital depreciation. The downside risk is quantified by the 5th-percentile final value of $8,056.27; stated in loss terms, this corresponds to an implied 95% VaR loss of $1,943.73 relative to the initial investment [5]. Consequently, despite the broad diversification across 50 assets, the empirical conclusion points to an uncertain scenario with a moderate-to-high risk profile. This supports the use of additional forecasting and validation tools to further analyze asset selection and downside exposure.

The true value of this specific Monte Carlo implementation in achieving our primary objective lies in its ability to empirically map systemic risk. The overlap between these distributions highlights the fundamental uncertainty in outcomes and the difficulty of consistently outperforming the initial capital, strictly guiding our downside risk management decisions.

5 Deep Learning for Market Forecasting: RNN and LSTM

While Monte Carlo simulations excel at risk assessment, the accurate prediction of future asset prices is equally vital for building a profitable portfolio. The stock market is highly volatile, dynamic, and non-linear, making traditional forecasting methods less effective for identifying complex trading signals. Deep learning models, specifically Recurrent Neural Networks (RNN) [6] and Long Short-Term Memory (LSTM) architectures [7], have become prominent tools for financial time series prediction due to their superior pattern recognition capabilities. These models were implemented using TensorFlow [8].

RNNs are specifically designed to process sequential data by maintaining a hidden state, which acts as a "memory" of past inputs. The hidden state at the current time step, h_t, is updated using the current input x_t and the previous hidden state h_{t-1}. However, standard RNNs suffer from the "vanishing gradient" problem, meaning they struggle to retain information from distant past observations. To solve this, the LSTM architecture introduces a more complex memory cell governed by three gates: the input gate, the forget gate, and the output gate. These gates use sigmoid activation functions to decide what information to keep and what to discard, allowing the model to learn both short-term market shocks and long-term trends.

We trained both a SimpleRNN and an LSTM model to predict the daily closing prices of Apple Inc. (AAPL) using historical data from January 2022 to January 2024. Table 6 summarizes the key architectural and training parameters used in both models.

Table 6. Hyperparameters and implementation details for the SimpleRNN and LSTM models.

Parameter	SimpleRNN	LSTM
Input variable	AAPL closing price	AAPL closing price
Input window	60 trading days	60 trading days
Train / Test split	80% / 20% sequential	80% / 20% sequential
Hidden layers	1 recurrent layer	1 LSTM layer
Hidden units	50	50
Output layer	Dense(1), linear	Dense(1), linear
Activation	tanh recurrent activation	tanh / sigmoid gates
Epochs	50	50
Batch size	32	32
Loss function	MSE	MSE
Optimizer	Adam, lr = 0.001	Adam, lr = 0.001
Normalization	Min-Max scaling	Min-Max scaling
Dropout / early stopping	Not applied	Not applied
Framework	TensorFlow / Keras [8]	TensorFlow / Keras [8]

Because no dropout, early stopping, or additional exogenous variables were used, the reported comparison should be interpreted as a baseline neural-network implementation rather than a fully tuned deep-learning benchmark.

Data preprocessing involved Min-Max normalization to ensure efficient gradient descent convergence. The data was then structured into sequential input windows of 60 days, where the previous 60 closing prices were used to predict the value of the 61st day. The dataset was sequentially split into training (80%) and testing (20%) partitions.

In this implementation, predictions are generated using fixed input windows constructed from real historical data, rather than recursively feeding previous predictions back into the model. This prevents the accumulation of forecasting errors over time, allowing each prediction to remain grounded in actual market observations.

Evaluating predictive accuracy via RMSE yielded critical insights. The SimpleRNN achieved an RMSE of $2.97 USD, indicating that it captured the short-term test trajectory reasonably well. However, it did not outperform the ARIMA walk-forward one-step baseline reported earlier (RMSE = $0.42 USD). Because the validation designs differ—ARIMA uses rolling one-step forecasts while the neural networks use fixed 60-day input windows—the metrics should be interpreted as complementary diagnostics rather than as a strict head-to-head ranking. This reinforces the role of ARIMA as a short-horizon statistical benchmark and avoids overstating the neural-network results.

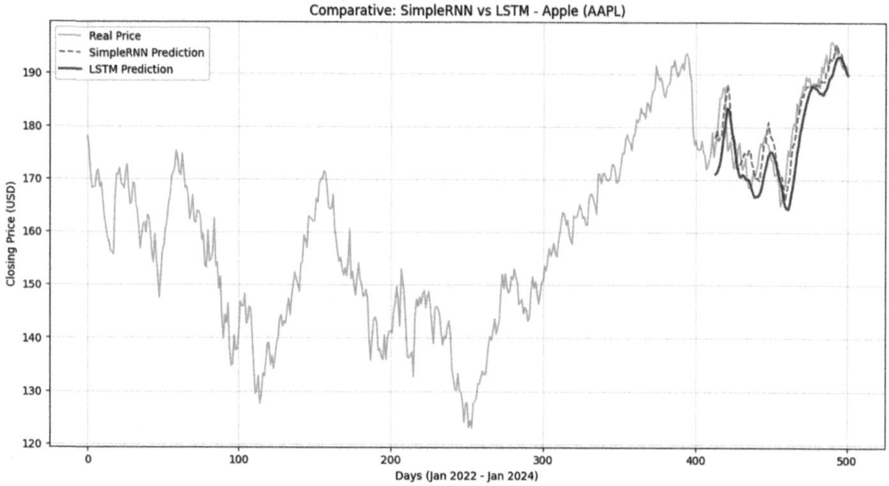

Fig. 5. Comparative performance of the SimpleRNN and LSTM predictive models against real AAPL market data (January 2022–January 2024). Horizontal axis: Days; vertical axis: Closing Price (USD).

To ensure a rigorous and unbiased comparative analysis, the LSTM model was implemented using identical foundational parameters such as the SimpleRNN (Table 6). It processed the same 60-day rolling windows of AAPL historical data and employed 50 memory units within its hidden layer, trained over 50 epochs to minimize mean squared error. Once training concluded, the network's predictions on the out-of-sample test data were inverse-transformed back into real USD values and plotted against actual market prices and the SimpleRNN forecast.

A simple visual inspection of the forecasts Fig. **5**, shows that the LSTM model tends to track short-term price movements more closely around local peaks and declines, consistent with its architecture ability to retain and update relevant information over longer sequences, see Table **7**.

Table 7. Comparison of predictive performance between SimpleRNN and LSTM models based on RMSE for AAPL closing prices (January 2022 – January 2024).

Model	Architecture	RMSE (USD)
SimpleRNN	Recurrent neural network, 50 units, 50 epochs	2.97
LSTM	Gated memory cells, 50 units, 50 epochs	4.62

A counterintuitive result emerged: despite LSTM's theoretical advantages in capturing complex temporal dependencies, it underperformed SimpleRNN in this scenario.

A likely explanation is that AAPL price dynamics during this period were driven more by short-term momentum than long-term patterns. LSTM's additional complexity may have led to overfitting to recent noise rather than extracting stable signals.

In contrast, the SimpleRNN, with its more limited structure, seems to generalize better over this time frame, producing more stable predictions and a lower overall error. This outcome highlights an important insight into quantitative modeling: more sophisticated architecture does not always guarantee better performance. Model selection should remain closely tied to the characteristics of the data, especially in financial contexts where noise and short-term variability play a significant role.

6 Conclusion

This study examined a set of quantitative and machine learning models with the goal of identifying which approaches are most useful for modeling financial markets and supporting portfolio construction. Rather than relying on a single methodology, the results highlight the value of combining complementary tools, each addressing a different aspect of decision-making.

The Markowitz framework provides a solid mathematical foundation to efficiently balance risk and return under given constraints. However, its effectiveness is highly dependent on the quality of input estimates and the prevailing market conditions given that most BMV assets posted losses between 2022 and 2025, achieving positive returns required concentrating in the Maximum Sharpe portfolio of just two assets. Monte Carlo simulations offer a probabilistic perspective, making it possible to quantify downside risk and evaluate the likelihood of different financial outcomes under uncertainty—a 36% probability of loss for a 50-stock portfolio over a 100-day horizon.

On the forecasting side, ARIMA models demonstrated the lowest short-term one-step-ahead error in this specific setup (RMSE of $0.42 for AAPL), confirming their usefulness as a baseline statistical approach. Recurrent neural networks remained useful for capturing non-linear price patterns, but their higher RMSE values in this application show that increased model complexity does not necessarily lead to better performance.

Overall, the findings suggest that no single model is sufficient to fully describe or predict financial markets. Instead, a hybrid approach combining optimization, statistical forecasting, simulation, and machine learning provides a more robust framework for portfolio construction. By integrating these methods, investors can improve both their estimation of expected returns and their understanding of risk, ultimately increasing the probability of building a profitable portfolio.

Future work could focus on improving model validation, incorporating additional market variables, applying all methodologies to a common dataset and asset universe to enable direct comparison, increasing the number of Monte Carlo trajectories for more stable tail-risk estimates, and testing these approaches under different market regimes to further strengthen the reliability of the results.

References

1. Markowitz, H.: Portfolio selection. J. Financ. **7**(1), 77–91 (1952)

2. Michaud, R.O.: The Markowitz optimization enigma: is 'optimized' optimal? Financ. Anal. J. **45**(1), 31–42 (1989)
3. Box, G.E.P., Jenkins, G.M., Reinsel, G.C.: Time Series Analysis: Forecasting and Control, 4th edn. Wiley, Hoboken (2008)
4. Glasserman, P.: Monte Carlo Methods in Financial Engineering. Springer, New York (2004)
5. Jorion, P.: Value at Risk: The New Benchmark for Managing Financial Risk, 3rd edn. McGraw-Hill, New York (2006)
6. Sezer, O.B., Gudelek, M.U., Ozbayoglu, A.M.: Financial time series forecasting with deep learning: a systematic literature review: 2005–2019. Appl. Soft Comput. **90**, 106181 (2020)
7. Hochreiter, S., Schmidhuber, J.: Long short-term memory. Neural Comput. **9**(8), 1735–1780 (1997)
8. Abadi, M., et al.: TensorFlow: large-scale machine learning on heterogeneous systems (2015). Software available from tensorflow.org
9. Sharpe, W.F.: Capital asset prices: a theory of market equilibrium under conditions of risk. J. Financ. **19**(3), 425–442 (1964)
10. Sutton, R.S., Barto, A.G.: Reinforcement Learning: An Introduction, 2nd edn. MIT Press, Cambridge (2018)
11. Yahoo Finance: Historical market data (2025). https://finance.yahoo.com
12. Kraft, D.: A software package for sequential quadratic programming. Technical Report DFVLR-FB 88–28, DLR German Aerospace Center (1988)
13. Banco de México: Tasas de interés (CETES 28 días) (2025). https://www.banxico.org.mx
14. Frigyik, B.A., Kapila, A., Gupta, M.R.: Introduction to the Dirichlet distribution and related processes. Technical Report UWEETR-2010-0006, University of Washington (2010)
15. Dickey, D.A., Fuller, W.A.: Distribution of the estimators for autoregressive time series with a unit root. J. Am. Stat. Assoc. **74**(366), 427–431 (1979)
16. Black, F., Scholes, M.: The pricing of options and corporate liabilities. J. Polit. Econ. **81**(3), 637–654 (1973)
17. Mandelbrot, B.: The variation of certain speculative prices. J. Bus. **36**(4), 394–419 (1963)

Auditing Gender Bias in Machine Learning Models for Autism Screening

Carlos Aldair Orta de la Rosa and María de Lourdes Martínez Villaseñor(✉)

Facultad de Ingeniería, Universidad Panamericana, Augusto Rodin 498, 03920 Mexico City, Mexico
{0287411,lmartine}@up.edu.mx

Abstract. Clinical studies report a male-to-female diagnosis ratio of autism close to 4:1; that ratio refers to epidemiological and diagnostic patterns in clinical populations, not to model performance. We audited gender bias signals in Autism Spectrum Quotient-10 (AQ-10) screening models using a strict pipeline with data cleaning, grid search, and cross-validation. Based on the public dataset "Autism Screening on Adults", preprocessing reduced the sample from 704 to 603 records. Exploratory analysis found no statistically significant gender differences in AQ-10 scores (t-test, $p = 0.158$), but the diagnosis label remained associated with gender (chi-square, $p = 0.031$). We trained Random Forest, Gradient Boosting, XGBoost, and Multi-Layer Perceptron models and evaluated them with calibration analysis, SHAP, demographic parity difference, and counterfactual gender perturbation. The Multi-Layer Perceptron achieved the best overall performance (holdout accuracy 0.9917, cross-validated F1 0.9861), while SHAP showed that AQ-10 items dominated predictions and gender had low global influence. Fairness results were mixed: demographic parity differences ranged from 0.0844 to 0.1332, and counterfactual gender flips produced small but non-zero probability changes. We therefore conclude cautiously that the models rely primarily on screening responses rather than directly on gender, but this does not establish fairness with respect to true ASD status because the target label itself may encode clinical bias.

Keywords: Autism Spectrum Disorder · Gender Bias · Audit · Explainable AI · AQ-10

1 Introduction

The diagnosis of Autism Spectrum Disorder (ASD) remains challenging, especially because female presentations can be underdetected by diagnostic frameworks historically shaped around male-typical phenotypes [15,18,19]. This background motivates ethical concern about gender bias in both clinical assessment and machine learning (ML) systems designed to support screening, triage, and referral.

H. Ponce et al. (Eds.): COMIA 2026, CCIS 3095, pp. 309–320, 2027.
https://doi.org/10.1007/978-3-032-37438-7_24

AQ-10-based screening has become attractive for ML research because it offers a compact symptom inventory and an accessible benchmark structure [1]. Previous studies on autism screening datasets have mainly emphasized predictive accuracy, often reporting strong results with tree ensembles, gradient boosting methods, or neural networks [2,3,6,7,24]. More recent work has started to frame autism prediction as a fairness problem, but this line of work is still relatively sparse [9].

Despite growing interest, three gaps continue to weaken the literature. First, several papers provide limited technical detail about model architecture, hyper-parameter selection, and validation design. Second, with small datasets, single train/test splits provide unstable evidence and can overstate performance. Third, feature ranking alone is not enough to support fairness claims; stronger audits require explanation methods, explicit fairness metrics, and counterfactual analysis [12,20,21,30].

A further conceptual difficulty concerns the target label itself. In this dataset, the supervised target is the observed ASD diagnosis label. If that label already reflects historical or clinical gender bias, then a model trained to predict it cannot by itself prove fairness with respect to true ASD status. Consequently, this paper frames the problem as an *audit of gender signals in model behavior*, rather than as definitive proof of an unbiased diagnosis. This distinction is central to the paper's argument and allows the methodological contribution to align with the ethical claims that the manuscript can actually support.

The manuscript presents a study that responds to the previous concerns, details the model search process, and uses SHAP for feature attribution. It includes cross-validation and calibration, a fairness metric and a counterfactual gender test. Overall, the work offers not just a high-performing classifier, but also a more technically robust and ethically transparent audit pipeline for AQ-10-based autism screening.

2 Clinical and Ethical Background

The frequently cited 4:1 male-to-female autism ratio refers to prevalence and diagnosis patterns in clinical populations, not to model accuracy or to prediction rates in our experiments [18,19]. This is an important distinction, as the ratio provides context of how gender-related differences have been discussed in the broader diagnostic landscape.

The literature on autism in females proposes several mechanisms by which diagnostic disparity can arise. These include masking or camouflaging behaviors during assessment, differences in symptom expression, diagnostic overshadowing by comorbid conditions such as anxiety and depression, and referral or confirmation bias during clinical judgment [4,10,14,17,23,29]. Early diagnostic models were historically developed from predominantly male samples, which may have reinforced a male-centered representation of autistic traits [26,27]. The consequences of delayed or missed diagnosis in females can be significant, including prolonged uncertainty, worse mental health outcomes, and reduced access to support [13].

This context makes gender auditing in ML ethically relevant for at least three reasons. First, if training labels inherit historical bias, then algorithmic performance may be misleading while still reflecting inequitable clinical processes. Second, UNESCO's AI ethics framework identifies transparency, fairness, and human oversight as core requirements for high-stakes AI systems [11,25,31]. Third, in health-related ML, fairness must be assessed in relation to both model behavior and the social process that produced the labels [21,28]. The methodological question of whether or not gender matters in predictions cannot be disentangled from the ethical question of what exactly the model is learning to reproduce.

3 Related Work

Research on AQ-based autism screening datasets can be grouped into four partially overlapping strands. The first is **predictive benchmarking**. These works compare standard classifiers, often with accuracy as the main criterion, and generally conclude that tree-based ensembles and neural models can achieve high performance on autism screening tasks [2,3]. The second strand is **interpretability-oriented work**, where XAI is used to communicate which questionnaire features drive a prediction, especially in healthcare contexts where trust and accountability are crucial [5,8,30]. The third is **fairness-aware modeling**, which is still emerging and often focuses on reducing demographic bias without resolving the validity of the labels themselves [9]. The fourth is work on **ethics and synthesis**, including systematic review efforts that identify the broader methodological gap around gender bias in AI related to autism [22].

Table 1. Exploratory findings motivating the fairness audit

Test	Finding
Gender difference in AQ-10 scores	Not statistically significant (t-test, p = 0.158)
Gender association with diagnosis label	Statistically significant (chi-square, p = 0.031)
Interpretation	Similar symptom-level screening scores coexist with gender-linked labels, motivating an audit of what models learn from this data

This literature helps to contextualize the present manuscript in two ways. First, it confirms that strong performance on autism screening data is not novel by itself; what matters is whether predictive strength is accompanied by explanation, robustness, and fairness auditing. Second, it shows that AQ-10 studies rarely bring together cross-validated model selection, explicit fairness metrics,

SHAP-based interpretation, and a discussion of the biased-label problem in one coherent workflow. The article is positioned to address this intersection.

We do not claim state-of-the-art performance in an engineering sense, nor that this dataset is sufficient for definitive clinical conclusions. Instead, the work is positioned as an audit-oriented study that deepens the methodological treatment of a small but widely used benchmark, with an emphasis on clear problem framing, technical specificity, fairness evaluation, and careful interpretation of the diagnostic label.

4 Methodology

4.1 Dataset and Preprocessing

We used the public "Autism Screening on Adults" dataset from Kaggle [16], derived from AQ-10-based adult screening data. The raw dataset contained 704 records. In preprocessing, "?" values were treated as missing data and records with nulls or duplicates were removed. Age was converted to a numeric variable and values older than 100 years were excluded. After cleaning, the working sample contained 603 records and 21 columns.

The target variable `Class/ASD` was binarized as 1 for `YES` and 0 for `NO`. To avoid leakage, we removed `result` (the AQ-10 total score) and `age_desc`. Categorical variables were label-encoded to preserve column names for SHAP and counterfactual analysis, and all predictors were standardized using `StandardScaler`. The resulting feature matrix included responses to the AQ-10 items together with demographic and contextual variables such as age, gender, ethnicity, country, relationship, jaundice, family history of autism, and previous use of the app.

4.2 Exploratory Analysis

We first examined whether the questionnaire scores differed by gender and whether the diagnosis label was associated with gender. Table 1 summarizes these exploratory results. AQ-10 scores were not significantly different between males and females in the cleaned data (t-test, $p = 0.158$), while the diagnosis label showed a significant association with gender (chi-square, $p = 0.031$). These findings indicate that symptoms scores appear to be comparable between genders, yet the observed label remains gender-linked.

4.3 Model Development and Selection

We used a stratified 80/20 train-test split with random seed 42 for final hold-out evaluation, calibration analysis, SHAP explanations, fairness estimation, and counterfactual testing. To strengthen robustness beyond a single split, model selection was performed with `GridSearchCV` under stratified 5-fold cross-validation on the training data.

Four models were audited: Random Forest, Gradient Boosting, XGBoost, and a Multi-Layer Perceptron (MLP). The MLP was implemented through `MLPClassifier` and thus constituted a compact feed-forward neural network rather than a deep architecture. Table 2 reports both the search space and the best hyperparameters selected by grid search.

Table 2. Model search space and selected configuration

Model	Grid search space	Best configuration
Random Forest	`n_estimators` $\in \{50, 100\}$; `max_depth` $\in \{3, 5, 10\}$; `min_samples_split` $\in \{2, 5\}$	`n_estimators = 100`, `max_depth = 10`, `min_samples_split = 2`
Gradient Boosting	`n_estimators` $\in \{50, 100\}$; `learning_rate` $\in \{0.01, 0.1\}$; `max_depth` $\in \{3, 5\}$	`n_estimators = 50`, `learning_rate = 0.1`, `max_depth = 3`
XGBoost	`n_estimators` $\in \{50, 100\}$; `learning_rate` $\in \{0.01, 0.1\}$; `max_depth` $\in \{3, 5\}$	`n_estimators = 100`, `learning_rate = 0.1`, `max_depth = 3`
MLP	`hidden_layer_sizes` $\in \{(32,), (64, 32)\}$; `activation` $\in \{relu, tanh\}$; `alpha` $\in \{10^{-4}, 10^{-3}\}$; `max_iter = 500`	`hidden_layer_sizes = (64, 32)`, `activation = relu`, `alpha = 0.0001`

4.4 Evaluation, Explainability, and Fairness Audit

The pipeline implements a multi-layer audit protocol. First, holdout predictions are evaluated using accuracy, precision, recall, and F1-score. Second, stability is estimated through stratified 5-fold cross-validation, reporting accuracy and F1-score across folds. Third, probabilistic reliability is assessed using the Brier Score and calibration curves. Fourth, explainability is enhanced by using SHAP instead of global feature importance; due to computational efficiency, SHAP is applied to the Random Forest model, providing both global and local explanations. Fifth, fairness is quantified through the demographic parity difference (DPD), using

Table 3. Summary of holdout, cross-validated, calibration, and fairness results

Model	Holdout Acc.	Holdout F1	CV F1	Brier	DPD
Random Forest	0.9835	0.9714	0.9328	0.0397	0.1332
Gradient Boosting	0.9752	0.9565	0.9277	0.0280	0.0844
XGBoost	0.9835	0.9722	0.9438	0.0182	0.1313
MLP	0.9917	0.9863	0.9861	0.0095	0.1138

gender as the sensitive attribute. Finally, a counterfactual test is conducted by flipping gender for a test instance while keeping all other features fixed.

This protocol has an important boundary: fairness is audited relative to the observed dataset label, not to a clinically adjudicated ground truth. The results are therefore interpreted as evidence about the model behavior in this dataset rather than as proof of an unbiased autism diagnosis.

5 Results

5.1 Predictive Performance and Robustness

Table 3 summarizes the main experimental results. In the holdout set, all four models achieved strong discrimination performance, with the MLP obtaining the highest accuracy (0.9917) and F1-score (0.9863). XGBoost and Random Forest followed closely, while Gradient Boosting remained competitive. These results indicate that once the cleaning and tuning steps are applied, the dataset is highly tractable for all four model families.

The cross-validated results support the same ranking while providing a more robust view of stability in the small dataset. The MLP again showed the strongest mean cross-validated F1 (0.9861), followed by XGBoost (0.9438), Random Forest (0.9328) and Gradient Boosting (0.9277). Importantly, these cross-validated scores are more defensible than a one-shot split because they reflect performance across multiple train/test partitions.

5.2 Calibration

Calibration analysis showed that all models produced usable probability estimates, but their reliability differed. The MLP achieved the lowest Brier Score (0.0095), followed by XGBoost (0.0182), Gradient Boosting (0.0280), and Random Forest (0.0397). In this pipeline, calibration complements accuracy rather than being treated as a secondary criterion. This is particularly important because screening systems are often used not only for binary decisions, but also for prioritization, referral, or risk communication. Figure 1 shows the corresponding reliability curves.

5.3 SHAP Interpretation

SHAP was computed on the MLP model as an interpretable and computationally efficient audit model. The global SHAP summary showed that the prediction was dominated by AQ-10 items, especially `A9_Score`, `A5_Score`, `A4_Score`, `A7_Score`, and `A2_Score`—whereas `gender` appeared well below the main features of the questionnaire and near other low-impact contextual variables. This pattern is consistent with the intended role of AQ-10 as a symptom-focused screening instrument. Figure 2 presents both global and local SHAP views used in this audit.

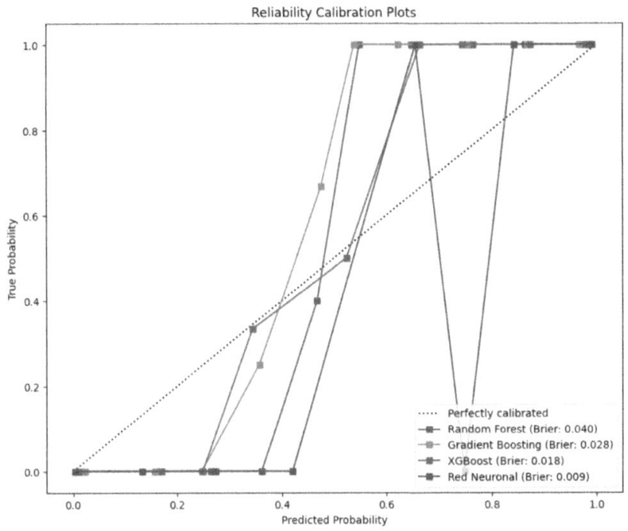

Fig. 1. Reliability diagram of model calibration. Gradient Boosting and XGBoost show better calibration (lower Brier Score) compared to Random Forest and the MLP as shown in Table 3.

The local SHAP waterfall plot for an example test instance also emphasizes AQ-10 items rather than gender among the largest contributors to the final probability estimate. This local perspective complements the global analysis and supports the interpretation that, although gender may still influence model behavior, it is not among the dominant direct drivers of the audited MLP's decision structure.

5.4 Fairness Metrics and Counterfactual Analysis

The fairness results are more complex than a purely explanation-based interpretation. The difference in demographic parity is not close to zero for any model: Gradient Boosting shows the lowest disparity (0.0844), while Random Forest and XGBoost exceed 0.13. These results do not support a strong claim of complete gender fairness. Instead, they highlight a tension between two observations: gender is not among the most influential features in SHAP, yet outcome rates still differ across groups at non-negligible levels.

The counterfactual test reinforces this interpretation. When the gender value of one test instance was flipped and all other features were kept fixed, the predicted probability of ASD changed by −0.0200 for Random Forest, 0.0000 for Gradient Boosting, −0.0386 for XGBoost and −0.0112 for MLP. These shifts are small in magnitude, but not uniformly zero, indicating that a direct gender perturbation can still affect prediction in some cases. Taken together, SHAP, DPD, and the counterfactual test support a cautious reading: the models appear to

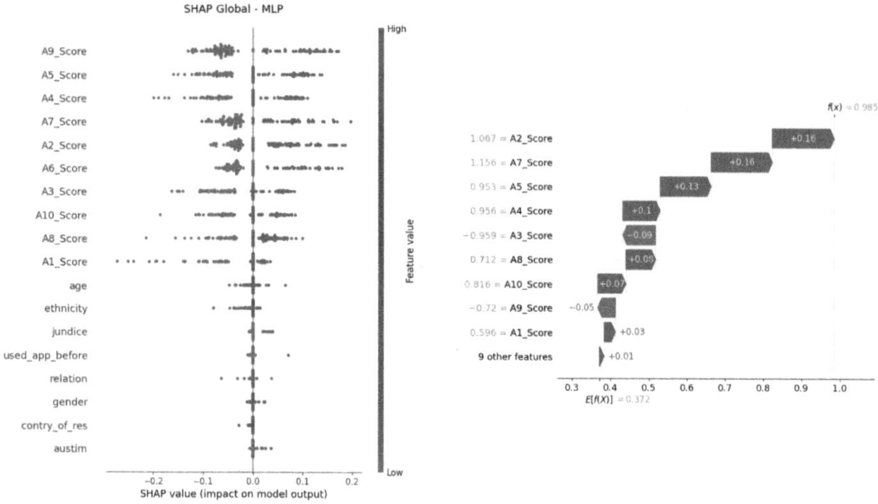

Fig. 2. SHAP visualizations for the audit show that the most influential features are AQ-10 items, while gender appears among the least impactful variables. The global view (left) highlights overall feature importance across the dataset, and the local view (right) illustrates how individual features contribute to a specific prediction.

rely mainly on symptom-level information, but they are not demonstrably free from gender-related effects.

6 Discussion

The results strengthen the study in three main ways. First, they present detailed information on model architecture, including hyperparameter search spaces, chosen configurations, and 5-fold cross-validated performance. Second, they enhance interpretability by substituting global feature importances with SHAP and by providing both global and local explanations. Third, they incorporate fairness analysis through the inclusion of demographic parity difference and a counterfactual gender test.

Clearly, the results support a careful interpretation. The cleaned dataset still shows the same qualitative pattern as before: AQ-10 scores do not differ significantly by gender, yet the diagnosis label remains gender-associated. SHAP indicates that models learn primarily from questionnaire responses rather than from gender as a dominant feature. However, fairness metrics and counterfactual testing show that this is not the same as saying that the models are fully gender-neutral. A low SHAP contribution for gender does not eliminate the possibility of indirect dependence through correlated variables, nor does it resolve the fact that the supervised label may itself carry historical bias.

This distinction is important in light of the biased-label problem. If the label reflects a social or clinical process in which females are less likely to be recognized despite similar AQ-10 profiles, then high-performing models may simply become efficient predictors of a biased outcome. In that case, better technical performance does not automatically imply better ethical performance. The pipeline improves what can be learned from the data, while also clarifying the limits of what the data can justify.

From an ethical perspective, explainability, fairness metrics, and label criticism are best combined rather than treated as substitutes. Transparency makes the model auditable; fairness metrics reveal subgroup-level disparities; counterfactual tests probe direct sensitivity; and label analysis clarifies what the model is actually learning. Together, these components align with UNESCO's principles of transparency, fairness, and human oversight [11,25,31]. They also support a more coherent contribution at the intersection of autism screening, machine learning evaluation, and AI ethics.

The article provides a clearer connection to the state of the art than previous AQ-10-style studies. It contributes less through absolute predictive novelty than through methodological integration: grid-searched model selection, cross-validation, SHAP-based interpretation, fairness measurement, counterfactual testing, calibration, and explicit discussion of label bias are brought together within a single audit framework. This integrated approach is particularly valuable for sensitive clinical applications, where interpretability and fairness are essential requirements rather than optional additions.

6.1 Implications for Clinical Deployment

The findings also have implications for how machine learning in autism screening should be positioned in practice. A high-performance screening model should not be interpreted as a diagnostic authority, particularly in a domain where the target label may reflect imperfect recognition of female autism. Instead, the model is better understood as a structured decision-support tool that can help standardize attention to symptom-level responses while preserving space for clinician reflection, contextual judgment, and follow-up evaluation. This perspective is consistent with previous work that emphasizes that explainability in healthcare is valuable not only for engineers, but also for clinicians who must decide whether and how to trust the recommendation of a model [5,30].

In that sense, one practical implication of the present audit is that the explainability and fairness analysis should be embedded upstream in the development cycle rather than added at the end as a compliance exercise. If a model shows strong predictive performance, but its fairness metrics reveal persistent subgroup disparities, the appropriate response is not simply to report the best accuracy and move on. Rather, the deployment question should become conditional: under what clinical workflow, documentation standard, and human-oversight process could such a model be used responsibly? For autism screening, this is especially important because screening tools often operate at the boundary between self-report, referral, and clinical interpretation.

A second implication concerns communication with stakeholders. For clinicians, the key message is that a low SHAP contribution for gender does not prove fairness itself. For researchers, the key message is that the biased-label problem cannot be abstracted away by stronger algorithms alone. For policymakers and ethics boards, the key message is that evaluation protocols for clinical AI should require not only discrimination metrics, but also calibration, subgroup analysis, and explicit documentation of label provenance. These broader implications help situate the paper not only as a benchmark study, but also as an example of how small-scale clinical ML projects can be made methodologically more responsible.

6.2 Limitations

Several limitations remain. First, the final sample is only 603 records, so even 5-fold cross-validation should be interpreted cautiously. Second, the dataset is public, self-reported, and limited to one screening source, which limits generalizability. Third, SHAP was applied only to the Random Forest model, not to every model in the benchmark. Fourth, the fairness analysis in this revision centers on demographic parity difference and a single-instance counterfactual test; a fuller audit should also report metrics such as equalized odds and group-specific error rates. Finally, the core label-bias problem remains unresolved: models were trained on the available diagnosis label, not on a clinically adjudicated gold standard.

7 Conclusions

This study presents an audit of gender signals in ML models for AQ-10-based autism screening. Relative to the earlier version, the new pipeline is substantially stronger: it includes explicit hyperparameter search, stratified cross-validation, calibration analysis, SHAP-based explainability, a fairness metric, and counterfactual testing. After cleaning, the dataset contained 603 records; within this sample, AQ-10 scores were similar between genders, but the diagnosis label remained gender-associated.

Across four audited models, the MLP delivered the best predictive performance, while the SHAP analysis on the Random Forest indicated that AQ-10 items, not the gender, dominated global explanations. At the same time, demographic parity differences and non-zero counterfactual probability shifts show that the case for fairness is only partial. The most defensible conclusion is therefore not that the models are bias-free but that they rely primarily on symptom-level responses while still requiring fairness monitoring and critical scrutiny of the labels they were trained to predict.

More broadly, the manuscript illustrates how autism-screening models can be evaluated with greater technical rigor and ethical precision. Future work should focus on external validation using additional datasets, more detailed subgroup fairness analysis, expanded SHAP auditing across models, and evaluation against stronger clinical reference standards. In this sense, the study does not close the debate on gender fairness in AI for autism screening; rather, it offers a clearer and more methodologically grounded framework to continue it.

Acknowledgments. This research was conducted as part of doctoral studies in AI Ethics. We acknowledge the providers of the Autism Screening on Adults dataset.

Disclosure of Interests. The authors have no competing interests to declare.

References

1. Allison, C., Auyeung, B., Baron-Cohen, S.: Toward brief "red flags" for autism screening. J. Am. Acad. Child Adolescent Psychiatry **51**(2), 202–212 (2012)
2. Akter, T., Ali, M.H., Khan, M.I.: Predicting autism spectrum disorder based on gender using machine learning techniques. In: IEEE International Conference on E-Learning and E-Technologies in Education, pp. 1–6 (2021)
3. Atlam, E., Masud, M., Rokaya, M., et al.: EASDM: explainable autism spectrum disorder model based on deep learning. J. Disab. Res. **3**(1), 1–15 (2024)
4. Bargiela, S., Steward, R., Mandy, W.: The experiences of late-diagnosed women with autism spectrum conditions. J. Autism Dev. Disord. **46**(10), 3281–3294 (2016)
5. Calvert, J., Hoffnagle, A., Park, Y., et al.: Using explainable machine learning to improve ICU triage. J. Med. Syst. **44**(12), 1–8 (2020)
6. Akter, T., Ali, M.: Predicting autism spectrum disorder based on gender using machine learning techniques (2021)
7. Atlam, E., Masud, M., Rokaya, M.: EASDM: explainable autism spectrum disorder model based on deep learning. J. Disab. Res. (2024)
8. Pandya, S., Jain, S.: Decoding autism diagnosis: a journey towards transparency with XAI in ML models. In: International Joint Conference on Biomedical Engineering Systems and Technologies (2024)
9. Revathi, M., Jayashree, R., Arunarani, S.: Attention-augmented machine learning framework for explainable and fair prediction of autism spectrum disorder. In: 2025 International Conference on Cognitive, Green and Ubiquitous Computing (IC-CGU) (2025)
10. Dean, M., Harwood, R., Kasari, C.: The portrait of disclosure. J. Autism Dev. Disord. **47**(10), 3155–3164 (2017)
11. Fjeld, J., Achten, N., Hilligoss, H.: Principled artificial intelligence. In: AAAI/ACM Conference on AI, Ethics, and Society, pp. 1–8 (2020)
12. Gunning, D., Aha, D.: DARPA's explainable artificial intelligence (XAI) program. AI Mag. **40**(2), 44–58 (2019)
13. Hedley, D., Uljarevic, M., Foley, K.R., et al.: Risk and protective factors associated with suicidal ideation in adults with autism. Autism Res. **10**(10), 1669–1680 (2017)
14. Hull, L., Petrides, K.V., Allison, C., et al.: "Putting on my best normal": social camouflaging in autism spectrum adults. Mol. Autism **8**(1), 1–12 (2017)
15. Hull, L., Petrides, K.V., Mandy, W.: The female autism phenotype and camouflaging. Rev. J. Autism Dev. Disord. **7**, 306–317 (2020)
16. Kaggle: Autism Screening on Adults Dataset (2023). https://www.kaggle.com/datasets/andrewmvd/autism-screening-on-adults
17. Kreiser, N.L., White, S.W.: ASD in females: are we overpathologizing their difficulties? Curr. Dev. Disord. Rep. **1**(1), 10–18 (2014)
18. Loomes, R., Hull, L., Mandy, W.P.L.: What is the male-to-female ratio in autism spectrum disorder? J. Am. Acad. Child Adolescent Psychiatry **56**(6), 466–474 (2017)

19. Lord, C., Elsabbagh, M., Baird, G., Veenstra-Vanderweele, J.: Autism spectrum disorder. Lancet **392**(10146), 508–520 (2018)
20. Lundberg, S.M., Lee, S.I.: A unified approach to interpreting model predictions. In: Advances in Neural Information Processing Systems, pp. 4765–4774 (2017)
21. Mehta, R., Xu, J., Ranganathan, S.: A framework for the use of fairness metrics in health-related algorithmic development. In: ACM Conference on Fairness, Accountability, and Transparency, pp. 648–658 (2021)
22. Mendoza, A.: Systematic review on gender bias in ai models for autism screening. Technical report (2023)
23. Miodrag, N., Hodapp, R.M.: Chronic stress and health among parents of children with autism. J. Dev. Phys. Disabil. **22**(4), 335–346 (2015)
24. Pandya, S., Jain, S.: Decoding autism diagnosis: a journey towards transparency with XAI. In: International Joint Conference on Biomedical Engineering Systems and Technologies, pp. 1–8 (2024)
25. Parikh, R.B., Obermeyer, Z., Navathe, A.S.: Regulation of predictive analytics in medicine. Science **363**(6429), 810–812 (2019)
26. Rivet, T.T., Matson, J.L.: Review of gender differences in core symptomatology in autism spectrum disorders. Res. Autism Spectrum Disord. **5**(3), 957–976 (2011)
27. Robinson, E.B., Lichtenstein, P., Anckarsater, H., Happe, F., Ronald, A.: Examining the female protective effect against autistic traits. Behav. Genet. **43**(5), 417–426 (2013)
28. Rudin, C.: Stop explaining black box machine learning models for high stakes decisions. Nat. Mach. Intell. **1**(5), 206–215 (2019)
29. Russ, C., Statt, M., Campbell, J., et al.: Examining sex differences in autism spectrum disorders. J. Autism Dev. Disord. **43**(3), 689–696 (2013)
30. Tonekaboni, S., Joshi, S., McCradden, M.D., et al.: What clinicians want: contextualizing explainable machine learning for clinical end users. In: Machine Learning for Healthcare Conference, pp. 367-386 (2019)
31. UNESCO: Recommendation on the Ethics of Artificial Intelligence. Technical report (2021)

Machine Learning Models for Customers Contact Channel Recommendation

Natalia Urquizo[1], Lía Sandoval[1](✉), Guillermo Dávila[2], and Edwin Escobedo[3]

[1] Facultad de Ingeniería, Universidad de Lima, Av. Javier Prado Este 4600, 15023 Lima, Peru
{20192039,20191881}@aloe.ulima.edu.pe

[2] Instituto de Investigación Científica, Universidad de Lima, Lima, Peru
gdavila@ulima.edu.pe

[3] Grupo de Investigación en Optimización Combinatoria e Inteligencia Artificial (OPTIMIA), Universidad de Lima, Lima, Peru
eescobed@ulima.edu.pe

Abstract. The intense competition in all business sectors has raised the need to effectively attract customers, highlighting the importance of interaction through social media, which has been shown to provide significant benefits. This study proposes three models to recommend the most effective communication media for product acquisition or sales, aiming to optimize the customer attraction process by identifying the channels through which companies should promote their products to reach the most interested users. Clustering techniques were applied to analyze customer contact channels and support decision-making in marketing strategies. The methodological approach includes data collection, data preparation, the development of hybrid clustering methods, model testing, and results analysis, using a 70/30 split for training and testing data. Results show that the most frequently used media are Instagram, Pinterest, and Telegram. The hybrid approach used in this paper demonstrated that social media holds more influence over female users. Furthermore, social media appears to influence students more than professionals. Moreover, we found that the most used social media varies among clusters. Overall, our findings reveal that higher education institutions should incorporate different communication strategies rather than a "universal solution" for all their customers.

Keywords: Machine learning · Clustering · Higher education · Recommender systems

1 Introduction

Nowadays, companies across all sectors are experiencing intense competition. It is no longer sufficient to simply provide their customers with high-value products, companies must also understand how to reach customers in a way that ensures

H. Ponce et al. (Eds.): COMIA 2026, CCIS 3095, pp. 321–332, 2027.
https://doi.org/10.1007/978-3-032-37438-7_25

a successful attraction process, in other words, attracting a greater number of customers [1]. Bergareche highlights the importance of creating a comprehensive experience that aligns with customer expectations, interests, and preferences [2].

This matter impacts companies in all sectors as they actively compete to achieve solid market positioning. Understanding the customer profile and interacting with it becomes a constant challenge to stand out [3]. A lack of understanding by companies about their customers needs and expectations affects the customer attraction process. This is noticeable in the lack of brand and customer attraction strategies to establish a powerful company-client relationship, reflected in the loss of potential customers and the decreased exposure of the brand.

Regarding brand interaction with its customers, Dash et al. [4] remarked that brand interaction plays a significant role in the company-client relationship and the attraction process [4]. López et al. added that one of the current marketing strategies to promote a company and its products is through social media [5]. Statistically, the use of social media provides benefits for companies, such as an 86% increase in brand exposure, a 64% increase in leads, and a 55% increase in sales [6].

Information technology tools can facilitate customer segmentation, which is crucial for company marketing. Segmentation allows the personalization of marketing strategies, contributing to optimizing products, customer service, and product promotion to different consumers [7,8]. By using these tools, customers can be divided into smaller groups, obtaining target customers' profiles. Further, it should be highlighted that data mining has become prominent for its ability to analyze large volumes of data. Finally, recommender systems have acquired relevance in areas such as marketing, artificial intelligence, and user experience. These IT tools support interaction with large and complex information spaces, aiming to engage customers with the company and expand their knowledge of products. Furthermore, recommender systems are known as decision support systems for attraction and delight processes [9].

In prior research, market segmentation has been applied based on multiple criteria in areas such as marketing, retail, e-commerce, and higher education. Correspondingly, in marketing studies, Li et al. compares the ALPSO, KM-ALPSO, and IKM-ALPSO algorithms to improve the accuracy of customer segmentation with mixed data [10]. This research demonstrated the practicability of the IKM-ALPSO algorithm in mixed data processing, excelling in performance, optimization, and clustering accuracy. Similarly, in retail, Pawełoszek carried out segmentation using the K-means algorithm for data segmentation and compared between Logistic Regression, CN2, Decision Trees, and Naive Bayes algorithms. The study found the Decision Trees algorithm to be the most accurate and best at defining segmentation rules, suggesting automatic data collection to avoid issues in data preparation [9]. Hung et al. segmented potential customers using the K-means algorithm, finding age and purchase to be the most relevant variables, and suggesting the use of new data cleansing technologies, pattern recognition, and data analytics results as references [11]. Paweloszek, applied three different clustering algorithms: Expectation-Maximization (EM), best-fit regression, and K-means, demonstrating that age and gender variables have a 75% accuracy [8]. While sophisticated statistical modeling methods provide relevant information, their high cost and complexity make them less recommended. Paweloszek emphasizes the need for further research on data mining models, such as affinity analysis, to identify a future customer's profile from a product perspective [8].

In the e-commerce sector, Kadir & Aychar performed an analysis of Recency, Frequency, and Monetary value (RFM), followed by clustering using K-means and a Pareto analysis, suggesting the use of demographic data for future studies [12]. Wang proposed a model based on the self-organizing neural network (SONN) algorithm, using swarm intelligence and the modified spider social behavior optimization algorithm (MSSO), classifying data with a dense neural network (DNN) [13]. In higher education, Khusnuliawati & Putri, used hybrid clustering methods combining K-means and self-organizing map (SOM) algorithms, achieving higher accuracy results [14].

There are notable gaps within the higher education area. Nasir et al. mention a lack of studies addressing social media variables to evaluate the impact of contact media in the attraction process [15]. Besides, technological solutions are scarce in the academic sector [14]. The data set sizes used also vary significantly, ranging from 200 to 700 [9,10,14,16–19]. Both Paweloszek and Li et al. recommend using larger data sets [9,10].

Regarding the forenamed matters, this article provides an empirical contribution by demonstrating the effectiveness of combining three hybrid clustering models: K-means with Self-Organizing Maps (SOM), K-means with XGBoost, and Gaussian Mixture Models (GMM) with XGBoost, to address customer contact channel recommendation problems. By doing so, this study offers practical value for higher education institutions in optimizing their customer attraction processes.

2 Materials and Methods

Figure 1 illustrates the methodological procedure of the work, consisting of five parts described below.

The case study focused on the significant impact of social media and blogs on consumers' purchase intentions in Bangalore, India. The study selected participants through convenience sampling, selecting 404 respondents, and administered an online questionnaire [20].

2.1 Dataset

The data was acquired from the dataset by Sunita Kumar, compiled as part of her contribution to studying the impact of social media on product purchases, published in the Mendeley Data repository [20]. Table 1 shows several types of variables, such as demographic, psychographic, and behavioral variables present in the dataset.

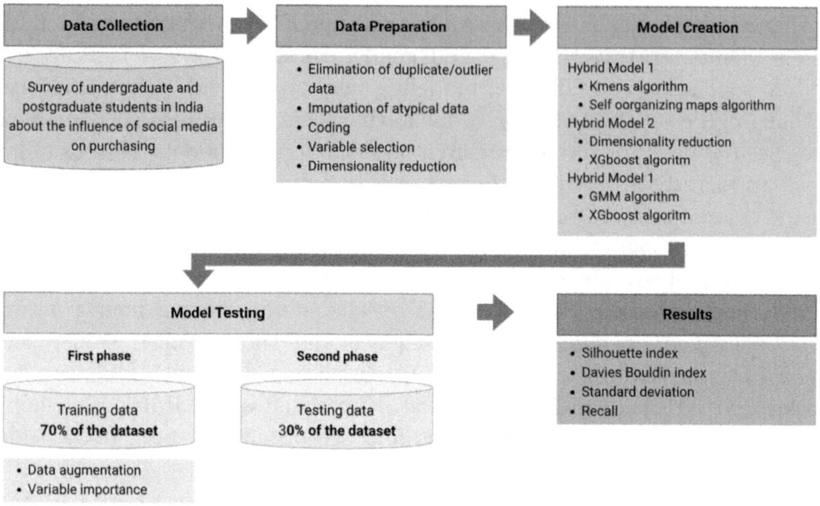

Fig. 1. Methodological procedure of the work, consisting of five parts described below.

Table 1. Table captions should be placed above the tables.

Type of variable	Variables
Demographic	Gender, Age, Education, Occupation
Behavioral	Tiempo, Horas uso, Compartir, Anuncios, SM influence
Psycographic	Plataformas, Dispositivos, SM use, SM follow, SM reviews, SM targets, BI target, BI promotion, BI influence, BI mensaje, BI tradicional, PI deseo, PI reviews, PI importancia, PI impacto, PI confiabilidad

3 Data Preparation

The data preprocessing phase involved six techniques to prepare the dataset for model implementation. These included handling duplicate records by removing them to prevent bias in the results, imputing outliers using KNN, encoding variables due to the presence of both numerical and categorical features, performing variable selection, and applying principal component analysis (PCA) for dimensionality reduction.

3.1 Model Creation and Model Testing

For model development, hybrid clustering approaches were employed, combining clustering techniques with supervised machine learning algorithms to evaluate cluster quality and feature influence. This study used XGBoosting as a supervised learning tool for classification and feature importance evaluation, rather than as a dimensionality reduction or visualization method. This algorithm was applied to determine which variables most strongly influence the channel recommendation.

Three hybrid clustering methods were compared as suggested by Khusnuliawati & Putri [14]. The first and second methods used the K-means clustering algorithm, but the first applied the self-organizing map (SOM) algorithm, while the second used the XGBoosting algorithm to evaluate feature influence and validate cluster assignments. The third method applied the Gaussian Mixture Models (GMM) clustering algorithm with the XGBoosting evaluation approach.

The results were validated and analyzed using the Davies-Bouldin Index, silhouette coefficient, standard deviation metrics, and recall. The recall metric was employed to measure the consistency, reliability, and stability of the clustering assignments [21].

3.2 Experimental Protocol

The experimentation occurred in four phases: preprocessing, training, validation, and testing. The preprocessing phase aimed to prepare the data for analysis and hybrid clustering model application. The dataset includes 405 data points with 44 categorical variables.

The data was initially divided into 70% for training and 30% for testing. Because the training subset contained a limited number of records, data augmentation was performed by sampling from a Multivariate Normal Distribution. Synthetic data was generated using the empirical mean (μ) and covariance matrix (Σ) of the original dataset. This method allows for a controlled expansion of the feature space while preserving the underlying statistical correlations without the need for labels, producing a training set of 553 records [22].

The training and validation stages followed the evaluation protocol reported in [23] and [24], which adopted three-fold cross-validation for datasets of limited size. In addition, Principal Component Analysis (PCA) was applied to reduce the dimensionality of the data to two principal components for cluster visualization. The results showed that using two components led to better-defined clusters with lower overlap between groups according to the silhouette coefficient.

To determine the optimal number of clusters (k), three methods were employed. Figure 2 shows a dendrogram constructed using the Ward linkage method and the Euclidean distance metric. Second, the elbow method was applied, as illustrated in Fig. 3.

The silhouette method was applied to the clusters to assess the cohesion of each cluster as shown in Fig. 4. It can be observed that clusters four and five are better defined. Both analyses were conducted on the original dataset to estimate an appropriate value of k. Based on these results, the clustering models were validated using 4 and 5 to compare their performance and determine which configuration produced the most meaningful and well-defined clusters.

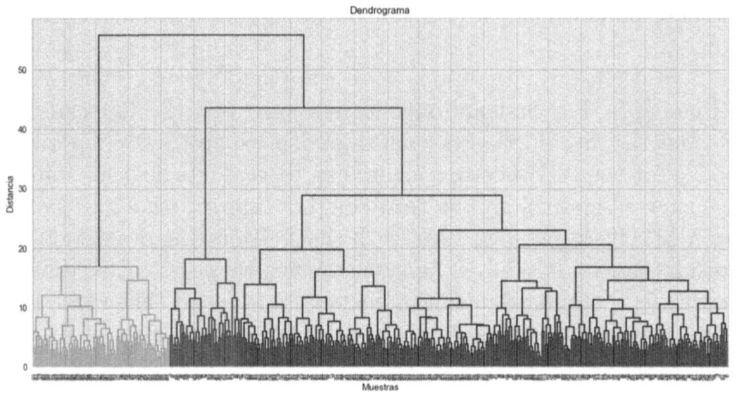

Fig. 2. Analysis of k with the dendogram method.

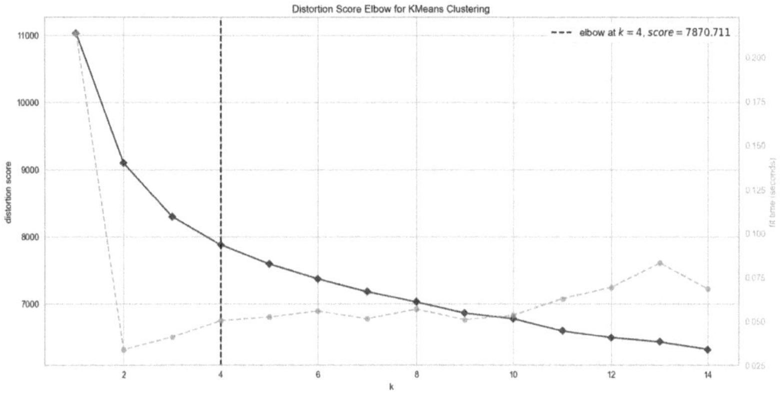

Fig. 3. Analysis of k with the elbow method.

4 Results

The resulting clusters with the GMM algorithm are shown in Fig. 4. The number of clusters was selected after a review of methods such as the elbow and dendrogram. Furthermore, it complements the observations by providing a visualization of the clusters in a three-dimensional space (Fig. 5).

Table 2 presents the metrics of the three models, including the Davies-Bouldin index, silhouette coefficient, and recall.

Table 3 presents the identified segments and their corresponding behavioral patterns.

Table 4 evaluates clustering using the Davies-Bouldin index, silhouette coefficient, and recall metrics, along with the K value used per fold. The standard

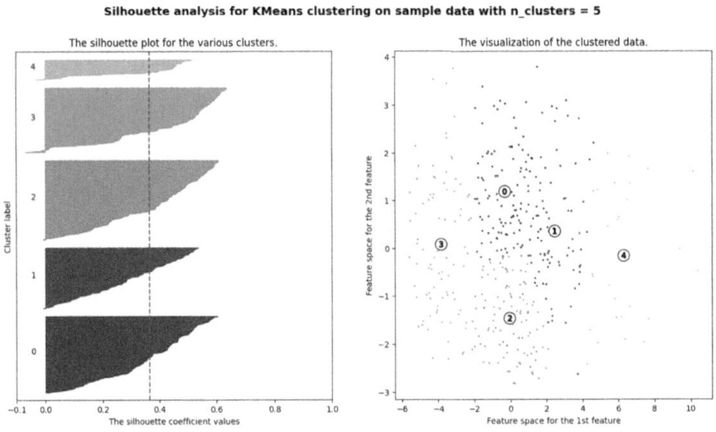

Fig. 4. Silhouette plot of 5 clusters.

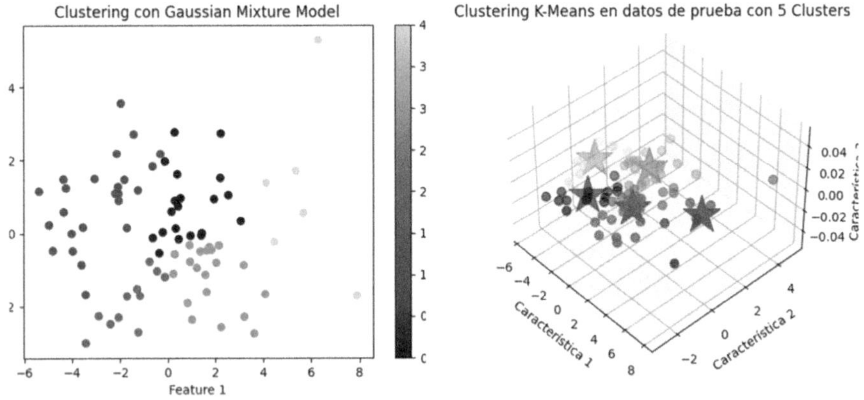

Fig. 5. Dimensional Model Graphs.

deviation of the metrics in each iteration was also calculated, obtaining values of 0.049, 0.006, and 0.069, respectively.

5 Discussion

The clustering results indicate that the segmentation approach is effective for analyzing user behavior. Among the proposed models, the GMM and XGBoost combination demonstrated superior performance, as evidenced by the most favorable silhouette coefficient and recall metrics. While the GMM-XGBoost model exhibited a higher Davies-Bouldin index, suggesting a trade-off regarding cluster definition, its overall performance in point assignment and cluster cohe-

Table 2. Model results.

Metrics	Davies-Bouldin Index	Silhouette Coefficient	Recall
K-means and SOM Model	0.7684	0.380	0.4755
K-means and Xgboost Model	0.7675	0.381	0.6435
GMM and XGboost Model	0.8152	0.385	0.7727

Table 3. Cluster Behavioral Patterns.

Cluster n°	C1	C2	C3	C4	C5
Social Media Influence					
Professionals	57	0	239	0	100
Students	76	16	243	83	107
Male	139	0	0	0	0
Female	102	183	300	313	286
Purchase Intention Impact					
Professionals	298	28	43	8	87
Students	300	10	308	0	0
Purchase Intention Impact					
Instagram	0	174	130	0	0
Facebook	1	161	34	53	223
Telegram	47	62	15	265	209
Pinterest	10	7	0	302	42
LinkedIn	36	36	31	120	265
Twitter	11	48	92	182	0
Youtube	0	23	87	254	0

sion supports its selection. Furthermore, the stability of the models was reinforced by the analysis of cluster numbers, which consistently fluctuated between 4 and 5 across iterations shown in Table 4. This consistency, reflected by low standard deviation values in performance metrics, indicates that the hybrid methodology provided a robust framework for identifying audience segments.

The patterns identified in Table 3 collectively illustrate that each cluster embodies a unique combination of demographic and platform-based response tendencies, reinforcing the conclusion that social media influence is not uniform but segmented across user types and preferred channels. The data indicated that Instagram, Pinterest, and Telegram are the most prominent contact media identified. Furthermore, analysis of gender-based influence revealed that social media has a greater influence over the female group. Moreover, the results indicated that social media has more influence over students than over professionals.

Observing the results obtained from each cluster highlights the importance of segmenting customers and analyzing which contact media they use most. This is

Table 4. Davies-Bouldin index, silhouette coefficient, and recall metric results.

Fold	Davies-Bouldin Index	Silhouette Coefficient	Recall	K
Fold 1	0.925	0.375	0.675	5
Fold 2	0.799	0.388	0.579	5
Fold 3	0.867	0.391	0.660	4
Test	0.815	0.385	0.772	5

suggested after observing that the most used medium in one group is not one of the most used in another. Consequently, this supports the idea that higher education institutions should adopt differentiated communication strategies rather than relying on a single platform.

Among the results obtained from the sample analysis, one suggests that social media has a greater significance on university students than on postgraduate students. However, it should also be noted that while there is a difference, it is not very prominent, suggesting that social media has a consistent influence on both groups. This finding reinforces the conclusion that social media serves not only as a communication channel but also plays a significant role in the customer attraction process. Likewise, after analyzing its relevance according to the gender of the respondents, it is possible to infer that there is a greater significance in the female gender. It should be noted that the result obtained in this case could be influenced by the population profile of the database used, where there is a significant predominance of the female gender. However, this predominance would not present a bias in the results since when analyzing the importance of the variables, the variable 'Male' does not represent much importance for the model.

In relation to the results on the preference of product marketing through social networks, it can be assumed that this has relevance for both undergraduate and professional students. Likewise, the impact and influence of social networks on purchasing decisions is more significant in graduate students. Other results obtained propose that social networks help brands to communicate better with their target audience and that social networks can have an impact by being important to people when they want to buy a product. Moreover, the choice of a value of k equal to five instead of four is based on the objective of the study, which seeks to find a more detailed segmentation avoiding generalized clusters. This is complemented by the results of the model in which it was identified that the quantity that produced the best results was five.

A limitation of the present study is the low amount of data obtained to apply the models. This affects the performance of the model, especially in the analysis of the results obtained. Nevertheless, data augmentation techniques were used to better train the model with the sample and then analyze the results obtained when using the model with the test data. Future research could address this limitation by using larger and more diverse datasets, allowing for more robust validation of the clustering results. Additionally, it would be valuable to apply and compare more advanced or recent clustering techniques to evaluate whether similar segmentation patterns emerge. Expanding the analysis to different geographical contexts beyond India could also provide insights into the generalizability of the findings across markets.

6 Conclusion

This research demonstrated the efficacy of integrating hybrid machine learning models to optimize contact media selection in the higher education sector. Among the proposed approaches, the GMM and XGBoost hybrid model emerged as the most robust framework for identifying audience segments, offering a superior alternative. By utilizing the dataset compiled by [20], located in Mendeley Data, this research identified key variables as gender, occupation, purchase intent, and the impact of social media that contribute significantly to optimizing the customer attraction process. The findings confirm that social media influence is not uniform but highly segmented across different user profiles. The most relevant variables for the recommendation system are gender, occupation, purchase intent, and the impact of social media on purchase intention. By uncovering distinct platform preferences across identified clusters, this work provides higher education institutions with a data-driven basis to move away from generalized marketing. Instead, these results advocate for differentiated, platform-specific communication strategies that align institutional outreach with the specific segment affinities of the target audience. While this proposal has limitations with dataset size, data augmentation techniques were employed to ensure robustness. Regarding the data in the dataset used to run the models, it is suggested to use a dataset containing information from a company's or institution's customers to make better suggestions and ensure they are useful for a company's attraction process. Future research will design a recommendation system with an interface that allows companies to identify the appropriate communication media for each target group. From an academic perspective, this research serves as an empirical contribution to the field of artificial intelligence applications in marketing and data-driven customer acquisition strategies. Specifically, it demonstrates how hybrid machine learning techniques can support decision-making in the selection of digital communication channels, providing practical insights for organizations operating in digitally driven and emerging market contexts.

References

1. Pin Anchundia, M.N.: Estrategias de captación de clientes en el Ferricentro Frank del cantón Baba periodo (2022). bachelorThesis, Babahoyo: UTB-FAFI. 2023. http://dspace.utb.edu.ec/handle/49000/13739. Accessed 26 Nov 2023
2. Bergareche, N.: El márketing B2B: más tecnológico y personalizado (aún). PuroMarketing. https://www.puromarketing.com/44/36168/marketing-mas-tecnologico-personalizado. Accessed 07 May 2026

3. Espinosa, L., Manica, B.A., Ruíz, J.P., Vázquez, A.C.: Segmentación de clientes para Herman Delikatessen (2022). https://rei.iteso.mx/handle/11117/8230. Accessed 26 Nov 2023
4. Dash, G., Kiefer, K., Paul, J.: Marketing-to-millennials: marketing 4.0, customer satisfaction and purchase intention. J. Bus. Res. **122**, 608–620 (2021). https://doi.org/10.1016/j.jbusres.2020.10.016
5. López, O., Beltrán, C., Morales, R., Cavero, O.: Estrategias de marketing digital por medio de redes sociales en el contexto de las PYMES del Ecuador. CienciAmérica: Revista de divulgación científica de la Universidad Tecnológica Indoamérica, vol. 7, no. 2, pp. 39–56 (2018)
6. Navarro. Social media marketing's top global benefits (2025). Statista. https://www.statista.com/statistics/188447/influence-of-global-social-media-marketing-usage-on-businesses/. Accessed 07 May 2026
7. Abdul-Rahman, S., Arifin, N.F.K., Hanafiah, M., Mutalib, S.: Customer segmentation and profiling for life insurance using k-modes clustering and decision tree classifier. Int. J. Adv. Comput. Sci. Appl. **12**(9), 434–444 (2021). https://doi.org/10.14569/IJACSA.2021.0120950
8. Yoseph, F., Ahamed Hassain Malim, N.H., Heikkilä, M., Brezulianu, A., Geman, O., Paskhal Rostam, N.A.: The impact of big data market segmentation using data mining and clustering techniques. J. Intell. Fuzzy Syst. **38**(5), 6159–6173 (2020). https://doi.org/10.3233/JIFS-179698
9. Pawełoszek, I.: Customer segmentation based on activity monitoring applications for the recommendation system. Procedia Comput. Sci. **192**, 4751–4761 (2021). https://doi.org/10.1016/j.procs.2021.09.253
10. Li, Y., Chu, X., Tian, D., Feng, J., Mu, W.: Customer segmentation using K-means clustering and the adaptive particle swarm optimization algorithm. Appl. Soft Comput. **113**(PB) (2021). https://doi.org/10.1016/j.asoc.2021.107924
11. Hung, P.D., Ngoc, N.D., Hanh, T.D.: K-means clustering using R a case study of market segmentation. In: Proceedings of the 2019 5th International Conference on E-Business and Applications, in ICEBA 2019, pp. 100–104. Association for Computing Machinery, New York (2019). https://doi.org/10.1145/3317614.3317626.
12. Kadir, M., Achyar, A.: Customer segmentation on online retail using RFM analysis: big data case of Bukku.id. Presented at the International Conference on Environmental Awareness for Sustainable Development in conjunction with International Conference on Challenge and Opportunities Sustainable Environmental Development, ICEASD & ICCOSED 2019, Kendari, Indonesia, 1–2 April 2019 (2019). https://eudl.eu/doi/10.4108/eai.1-4-2019.2287279. Accessed 15 May 2023
13. Wang, C.: Efficient customer segmentation in digital marketing using deep learning with swarm intelligence approach. Inf. Process. Manage. Int. J. https://dl.acm.org/doi/abs/10.1016/j.ipm.2022.103085. Accessed 15 May 2023
14. Khusnuliawati, H., Putri, D.R.: Hybrid clustering based on multi-criteria segmentation for higher education marketing. TELKOMNIKA **19**(5), 1498 (2021). https://doi.org/10.12928/telkomnika.v19i5.18965
15. Nasir, V.A., Keserel, A.C., Surgit, O.E., Nalbant, M.: Segmenting consumers based on social media advertising perceptions: how does purchase intention differ across segments? Telemat. Inform. **64**, 101687 (2021). https://doi.org/10.1016/j.tele.2021.101687
16. Mengash, H.A.: Using data mining techniques to predict student performance to support decision making in university admission systems. IEEE Access **8**, 55462–55470 (2020). https://doi.org/10.1109/ACCESS.2020.2981905

17. Retamosa, M., Millán, Á., García, J.: Thinking about going to university? Segmenting Undergraduates (2019). https://doi.org/10.4995/HEAD19.2019.9208
18. Retamosa, M., Millan, A., García, J.: The journey towards finding your favourite university. A segmentation study based on selection criteria. J. Marketing High. Educ. (2022). https://doi.org/10.1080/08841241.2022.2058146
19. Yu, S., Yang, M., Wei, L., Hu, J.-S., Tseng, H.-W., Meen, T.-H.: Combination of self-organizing map and k-means methods of clustering for online games marketing. Sensors Mater. **32**(8), 2697–2707 (2020). https://doi.org/10.18494/SAM.2020.2800
20. Kumar, S.: Significance of social media communication (Version 1). Mendeley Data (2023). https://doi.org/10.17632/2s47yznh4k.1. Accessed 7 May 2026
21. Nie, D., Cappellari, P., Roantree, M.: A methodology for classification and validation of customer datasets. J. Bus. Ind. Marketing **36**(5), 821–833 (2020). https://doi.org/10.1108/JBIM-02-2020-0077
22. Hastie, T., Tibshirani, R., Friedman, J.: The Elements of Statistical Learning: Data Mining, Inference, and Prediction. Springer Series in Statistics, 2nd edn. Springer, Cham (2009)
23. Ab Ghani, N.L., Che Cob, Z., Mohd Drus, S., Sulaiman, H.: Student enrolment prediction model in higher education institution: a data mining approach. In: Othman, M.A., Abd Aziz, M.Z.A., Md Saat, M.S., Misran, M.H. (eds.) SYMINTECH 2018. LNEE, vol. 565, pp. 43–52. Springer, Cham (2019). https://doi.org/10.1007/978-3-030-20717-5_6
24. Hernández-Leal, E., Duque-Méndez, N.D., Cechinel, C.: Unveiling educational patterns at a regional level in Colombia: data from elementary and public high school institutions. Heliyon **7**(9), e08017 (2021). https://doi.org/10.1016/j.heliyon.2021.e08017

Technological Symmetry and Tax Assistance Through Artificial Intelligence

Carlos Aldair Orta de la Rosa(✉) and Alejandro Javier Juarez Bolaños

dSoft SA de CV, Departamento de Inteligencia Artificial, Prl. División del Norte 4296, 14300 Mexico City, Mexico
{aldair,ajuarez}@dsoft.mx

Abstract. Tax administrations increasingly operate as data-intensive organizations that combine electronic invoicing, cross-database reconciliation, and automated risk scoring. This creates a technological asymmetry: authorities can inspect fiscal behavior at scale, while many taxpayers still rely on fragmented and largely manual review. This paper presents a tax-assistance platform for the Mexican regulatory context built around Retrieval-Augmented Generation (RAG), semantic retrieval, and an intelligent engine for invoice deductibility analysis. The implementation uses the 'gpt-4o-mini-2024-07-18' model for generation, 'multilingual-e5-large' for embeddings, and a Pinecone vector index enriched with regulatory metadata. The article contributes a system architecture for tax assistance under changing regulation, a formalization of deductibility verification as context-dependent constraint aggregation, a concrete reproducibility profile including chunking and retrieval parameters, and a structured evaluation framework that combines retrieval, decision-support, and operational criteria. The current corpus contains 493 versioned fiscal documents and supports specialized assistants across multiple Mexican tax subdomains. Preliminary evidence indicates substantial gains in latency and review coverage, while qualitative comparisons against LLM-only systems suggest that a compact and relatively low-cost model can remain competitive when paired with a carefully curated fiscal corpus. The paper also discusses governance constraints, limitations, and future directions for broader scientific validation.

Keywords: Fiscal Artificial Intelligence · RAG · Machine Learning · Explainable AI · CFDI · Tax Compliance

1 Introduction

Tax compliance has changed from historical record keeping to real-time data analysis [1,2]. In the Mexican context, electronic invoices, cross-database reconciliation, and automated risk scoring have transformed supervision into a continuous computational process. The resulting asymmetry is not only legal or administrative; it is technological. Tax authorities can process massive transaction streams, while many taxpayers and accounting teams still rely on fragmented manual review.

H. Ponce et al. (Eds.): COMIA 2026, CCIS 3095, pp. 333–343, 2027.
https://doi.org/10.1007/978-3-032-37438-7_26

The Digital Tax Receipt via Internet (CFDI) is only one component of this ecosystem. Billing records can be contrasted with banking information, social security data, foreign trade traces and tax returns to identify inconsistencies, unusual behavior, or mismatches between documentary form and economic substance. Under these conditions, compliance requires more than legal knowledge: it requires timely retrieval of the applicable rule, exception, and factual context.

This setting motivates the use of AI as decision support rather than as unconstrained generation. Generic large language models (LLMs) can summarize and explain complex regulation, but in high-stakes domains, they also inherit knowledge staleness and hallucination risk [5,6]. In tax assistance, unsupported outputs are operationally dangerous because they may produce erroneous deductions, omitted alerts, or overconfident explanations detached from the applicable norm.

The platform studied in this paper addresses that problem through a RAG-based architecture specialized for fiscal assistance. The central idea is simple: the system should not answer from parametric memory alone but from retrieved legal evidence, structured invoice attributes, and explicit validation rules. This design makes the output more auditable and better aligned with professional review.

The article makes four main contributions. First, it frames tax assistance as a scientific problem that involves retrieval under regulatory drift, context-sensitive decision support, and explanation requirements. Second, it describes a modular architecture that combines semantic search, constrained generation, and deductibility validation with concrete implementation parameters. Third, it formalizes invoice deductibility as a constraint aggregation problem grounded in both document features and taxpayer profile. Fourth, it proposes an evaluation perspective that goes beyond latency and includes retrieval adequacy, explanation quality, deployment limitations, and qualitative comparisons against LLM-only baselines.

2 Related Work and Scientific Context

Recent progress in NLP made transformer-based language models and retrieval systems relevant for legal and regulatory applications [3,4]. In fact-sensitive domains, retrieval-augmented generation has become a standard strategy to ground responses in external evidence rather than relying only on parametric memory [6]. This is especially important in taxation, where rules, administrative criteria, and interpretive materials change frequently.

Legal NLP has also become an active research area. Benchmarks such as LexGLUE show that understanding legal language requires domain-sensitive handling of long documents, specialized terminology, and subtle normative distinctions [7]. LegalBench further shows that even strong LLMs struggle when reasoning must connect formal rules with specific fact patterns rather than simply stating legal language [8]. These findings are directly relevant to tax assistance, where applicability is as important as semantic similarity.

A second line of work concerns the explainability. In high-stakes settings, unsupported outputs are institutionally problematic even when they are linguistically fluent. Previous work argues that interpretable or explanation-centered systems are preferable when errors can create significant legal or financial consequences [9,10]. The system presented here follows this perspective by treating traceability, source citation, and human override as integral design requirements.

A third relevant line comes from document intelligence. Enterprise workflows are based on the extraction of structured information from invoices, forms, and records that often mix text, layout, and metadata. Models such as LayoutLM and Donut advanced document understanding in these settings [11,12]. However, document understanding alone does not solve the normative problem addressed here. An invoice can be parsed correctly and still be fiscally problematic if the underlying transaction is incompatible with the taxpayer profile or the applicable legal restrictions.

Taken together, these strands of literature suggest that tax assistance is not a narrow application of LLM prompting. It is a compound AI problem that combines document understanding, retrieval quality, legal reasoning, explainability, and decision support under changing regulation.

3 Problem Formulation and System Architecture

3.1 Task Definition

The system receives either a tax query or an invoice verification request. Let x denote the document under review, P the operational profile of the taxpayer, and $\mathcal{R}_t$ the regulatory corpus available at time t. The output is a recommendation or classification, a textual explanation, and the legal evidence used to justify that output. In this formulation, the task is not generic question answering; it is context-sensitive decision support.

This distinction is important. A tax answer can be linguistically plausible and still be wrong if it omits an exception, confuses a taxpayer regime, or ignores the relationship between documentary form and operational substance. For that reason, the system must retrieve evidence that is not only semantically related to the query but also applicable to the specific case.

3.2 Semantic Retrieval Layer

Each article, rule, criterion, or curated tax document is transformed into a dense vector $v \in \mathbb{R}^d$ using domain-adapted embeddings. The semantic similarity between a query q and a document d is estimated with cosine similarity:

$$\text{Similarity}(q, d) = \frac{E(q) \cdot E(d)}{\|E(q)\| \|E(d)\|}.$$

The retriever ranks the corpus and returns the top-K fragments used by the downstream reasoning stage.

The retrieval depth is a calibration parameter rather than a fixed constant. Small values of K risk missing exceptions, cross-references, or conditions stated in adjacent provisions. Large values may dilute the context window with marginally relevant material and complicate generation. In practice, K should be selected from a waiting list of tax cases annotated by balancing retrieval recall, explanation usefulness, and final decision quality.

The retriever must also operate on a carefully segmented corpus. The legal provisions, administrative criteria, and guidance materials differ in length, granularity, and evidentiary value. The chunk size, overlap, metadata enrichment, and document hierarchy, therefore, become part of the model design rather than implementation details. If segmentation is poor, even a strong embedding model will retrieve incomplete or misleading context.

3.3 Generation and Grounding Layer

To reduce hallucinations and improve legal precision, the platform implements a RAG architecture [6]. The corpus is updated through ingestion pipelines that incorporate regulatory changes and curated tax materials. At inference time, the query is normalized, embedded, matched against the vector database, and enriched with the passages retrieved before generation. Figure 1 summarizes this pipeline and highlights the interaction between query processing, vector retrieval, grounding context, and response generation.

The generative model therefore does not answer from memory alone. It receives a bounded evidence that can be cited in the final explanation, which is essential in regulatory environments. The role of generation is not to replace legal reasoning with free-form text but to organize retrieved evidence into a useful expert-facing explanation.

3.4 Implementation Variables and Reproducibility Envelope

The implementation uses GPT-4o mini (version 2024-07-18) as the generative model, multilingual-e5-large as the embedding model, and Pinecone as the vector database. Documents are indexed with metadata that records the legal source type, the name of the law or regulation name, chapter or section, article or annex, and explicit references to other cited articles. This metadata design is important because tax reasoning rarely depends on isolated fragments; it often depends on regulatory hierarchy and cross-reference structure.

Chunking is performed with a maximum segment length of 1024 tokens and an overlap ratio of 0.3. The retrieval depth is fixed at $K = 20$ in the current production configuration. This choice favors high recall in a domain where exceptions and secondary references matter, while still keeping the evidence window manageable for the downstream model. These parameters are not secondary details; they shape what the system can retrieve, how much legal context it can absorb, and whether the final answer remains faithful to the source material.

The current corpus contains 493 fully processed and versioned files, including laws, regulations, miscellaneous resolutions, annexes, historical versions of those

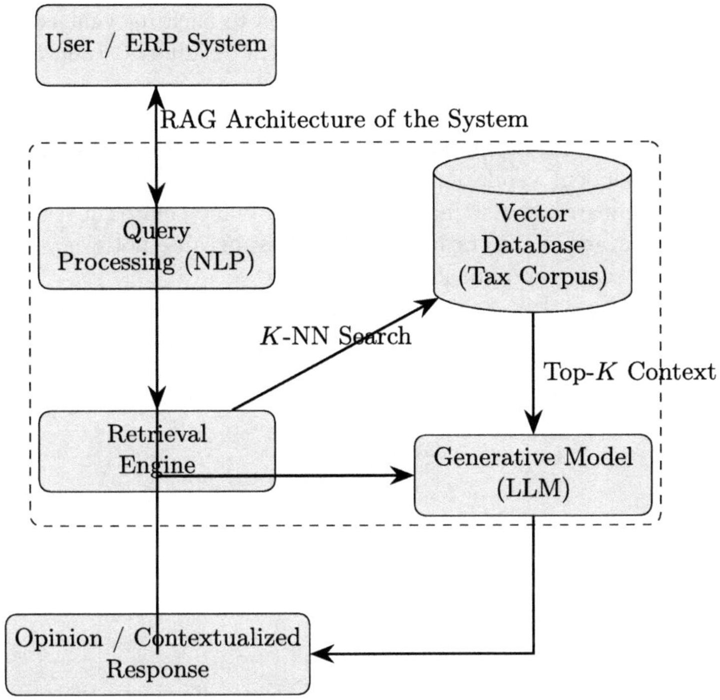

Fig. 1. Retrieval-Augmented Generation (RAG) architecture implemented in the system, showing the flow from user input to response generation.

materials, and carefully curated practitioner sources. The ingestion process is continuously updated by an automated collection robot. At present, the pipeline automates acquisition and organization; a natural extension is automatic audit, validation, and reprocessing of newly collected material before reinsertion into the vector base, as well as live web ingestion at query time when the use case requires it.

4 Intelligent Deduction Verification Engine

The invoice evaluation module addresses a core bottleneck in corporate accounting: determining whether a CFDI is deductible for a specific entity. Let C be the set of attributes extracted from a CFDI and P the operational profile of the receiving company. The deductibility function $D(C, P)$ is defined through retrieved regulatory constraints R_i:

$$D(C, P) = \prod_{i=1}^{n} \Phi(C, P, R_i),$$

where $\Phi \in \{0, 1\}$ is a validation function applied to each relevant constraint. If at least one mandatory condition is violated, the result is 0 (non-deductible), triggering a substantiated alert.

This formulation is intentionally conservative. Rather than asking the LLM to decide freely, the system decomposes the task into extraction, evidence retrieval, and context-dependent validation. In practice, some conditions are hard exclusions, while others are soft risk indicators that should escalate the case to human review. This separation is operationally important because not every irregularity should have the same consequence.

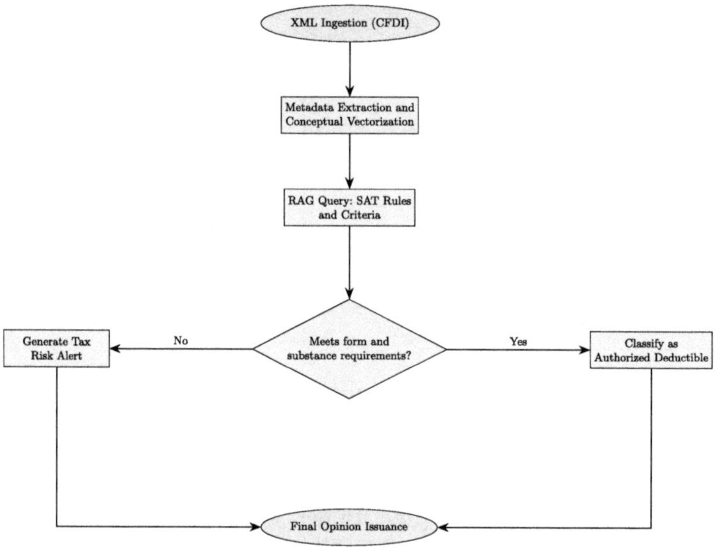

Fig. 2. Algorithmic processing flow of the Intelligent Deductions Engine.

Figure 2 summarizes the decision path from XML ingestion to final opinion issuance. The relevance lies in the combination of retrieval under regulatory drift, context-sensitive reasoning, and mandatory explanation. Tax assistance is therefore not only an application vertical but also a useful real-world benchmark for AI systems operating in regulated environments.

4.1 Operational Logic of the Validation Step

The validation function $\Phi(C, P, R_i)$ evaluates whether the retrieved rule R_i is satisfied by the attributes of th document and taxpayer context. In a practical implementation, the function can be decomposed into three classes of checks: documentary consistency, taxpayer-context compatibility, and legal-support completeness. Documentary consistency covers fields such as payment method, invoice type, tax regime, issuer identity, and internal coherence between XML attributes. Taxpayer-context compatibility compares the invoice concepts with the receiving entity's economic activity, operational profile, and known business relationships. The completeness of legal-support verifies that the explanation cites enough evidence to justify the decision and expose the applicable restriction or authorization condition.

This decomposition makes the system more robust than a single free-form classification prompt. It also supports gradual adoption in practice because organizations can tune which alerts trigger automatic blocking, which ones require analyst confirmation, and which ones remain as soft warnings for later review.

5 Experimental Design and Evaluation Framework

The current study reports a preliminary industrial evaluation of anonymized operational scenarios from logistics-sector companies with high invoicing volume. The evidence is therefore deployment-oriented rather than a public benchmark, but it is still useful to characterize the practical value of the system and to define the dimensions on which a fuller scientific evaluation should operate.

A rigorous evaluation should include three complementary layers. The first is retrieval quality, where the objective is to measure whether the relevant legal fragments are actually retrieved for a given tax question or invoice pattern. Suitable metrics include Recall@K, Mean Reciprocal Rank, and expert judgment on the adequacy of the retrieved evidence. The second is the quality of decision-support, where the output of the platform compared against specialist review in terms of alert precision, missed-risk rate, explanation adequacy, and escalation usefulness. The third is operational performance, where latency, throughput, and review coverage are contrasted with manual audit workflows.

5.1 Evaluation Variables to Report

A reproducible study should state at least the following variables: number of tax questions and invoice cases evaluated, distribution by sector and tax scenario, corpus composition, retrieval depth, chunking strategy, update lag between regulatory publication and corpus ingestion, the precise LLM and embedding models used in the experiments, and the protocol used to judge answer adequacy. In the current system, those core parameters are fixed and known: 493 versioned documents, $K = 20$, chunk size of 1024 tokens, 30% overlap, multilingual-e5-large embeddings, and GPT-4o mini (version 2024-07-18) for grounded generation. The present comparison protocol is qualitative: side-by-side sessions were run on general fiscal-accounting questions and abbreviation-heavy queries, and outputs were judged according to adequacy of answer, the degree of contextualization, and need for additional clarification. This protocol is suitable for reporting practical behavior, although it is not yet a formal multi-annotator benchmark.

A complete experimental version should also include ablations. Three comparisons are especially informative: LLM-only prompting versus RAG, RAG with and without re-ranking, and rule-only validation versus the hybrid retrieval-plus-generation strategy. Such comparisons clarify which components contribute the most to factual accuracy, explanation quality, and operational efficiency. The present paper includes the first of these qualitative comparisons and leaves the remaining controlled ablations as a priority for future work.

5.2 Preliminary Results

Table 1. Performance Analysis: Traditional Process vs. The System

Metric/Process	Traditional Audit	The System Engine (AI + RAG)
Analysis Time	Up to 5 min. per invoice	<10 s per invoice
Review Coverage	Random sampling (10–15%)	Exhaustive analysis (100%)
Knowledge Base	Analyst's memory	RAG legal corpus
Risk Identification	Reactive (Post-closing)	Preventive (Real-time)
False Positive Rate	Moderate/High due to fatigue	Minimal (XAI Validation)

As shown in Table 1, the main gain is not only speed, but also the transition from reactive review to anticipatory control. In operational terms, the platform enables a more systematic examination of invoicing streams before tax closing, reducing the dependence on human memory and selective sampling.

However, these results should be interpreted with caution. The present evidence does not yet include a public dataset, a statistically powered comparison against alternative baselines, or a complete analysis of retrieval errors. The current results therefore establish feasibility and operational promise rather than definitive superiority.

5.3 Qualitative Comparison Against LLM-Only Systems

The current platform was also compared qualitatively against LLM-only interaction settings using contemporary commercial systems in memoryless sessions. For general accounting and tax questions, both the RAG system and the standalone LLMs often produced acceptable answers. In several cases, the external systems appeared to benefit from web search, which gave them an advantage for open-ended information lookup. Even under that condition, the grounded system remained competitive despite using GPT-4o mini (version 2024-07-18), a compact model with lower computational cost than larger frontier models.

The most visible advantage of the RAG configuration appeared in contextualization. Acronyms, abbreviations, and domain-specific fiscal shorthand often required clarification in the LLM-only condition, whereas the grounded system frequently resolved them directly from the curated corpus and its specialized metadata. This suggests that retrieval quality and domain curation can compensate for model scale when the task strongly depends on stable institutional context.

At the same time, this comparison remains qualitative rather than benchmark-complete. The sessions were not designed as a statistically controlled contest and they do not yet support claims about significance margins. Their value lies in showing that a carefully engineered RAG pipeline can remain competitive with newer or larger general-purpose interfaces in applied tax assistance while operating with lower inference cost.

5.4 Failure Modes and Error Analysis

Any system of this kind can fail in several ways. First, the retriever may omit the most relevant legal passage or return a fragment that is semantically similar but normatively inapplicable. Second, the generation layer may overstate the certainty of the answer even when the evidence is ambiguous. Third, the validation stage may receive incomplete taxpayer context, which can lead to apparently correct but operationally misleading recommendations. Fourth, a stale corpus can preserve linguistic fluency while silently degrading legal correctness.

These failure modes suggest that the evaluation should not end with aggregate accuracy numbers. A meaningful analysis must identify where errors originate, whether they are detectable by the user, and how costly they are in practice. In fiscal settings, a transparent false positive may be less harmful than an unsupported false negative accompanied by an overconfident explanation.

6 Ethics, Explainability, and Governance

In legal and tax settings, computational ethics is a functional requirement. Closed or opaque systems are problematic because an error may lead to fines, litigation, or distorted internal decisions. For that reason, the platform was designed under the principle of Explainable AI: every classification or answer must expose the traceability of the inference by citing the article, rule, or administrative criterion that supports the output.

The system follows a human-in-the-loop philosophy. It does not replace the specialist; it structures evidence, reduces search time, and standardizes first-pass review. Final accountability remains with the human expert, who can inspect the supporting passages, challenge the generated explanation, and override the recommendation when substance-over-form interpretation requires it.

Governance also depends on operational controls beyond the model itself. Versioning of the legal corpus, logging of retrieved evidence, traceable prompt templates, and retention of explanation outputs are necessary for auditability. In practice, the final system description should clarify not only how inference is performed but also how the decision trail is stored and later inspected.

7 Limitations and Future Work

The proposal has clear limitations. First, the legal corpus is jurisdiction-specific, so transfer to other countries requires re-ingestion, re-modeling of the regulatory hierarchy, and re-evaluation of retrieval quality. Second, output quality strongly depends on corpus freshness and source curation. Third, some tax disputes are inherently interpretive and cannot be reduced to deterministic rules or short retrieved passages. Fourth, current evidence comes from internal operational settings, which limits external reproducibility.

Several directions for future work naturally follow from these limitations. One is the creation of an anonymized benchmark of invoices, tax scenarios, and expert annotations suitable for retrieval and decision-support evaluation. The second is the addition of calibrated uncertainty indicators so that the system can distinguish between strong recommendations and cases that require immediate human escalation. A third is the extension of the ingestion robot so that newly collected sources are automatically audited, validated, and reprocessed before entering the vector base. A fourth is live web ingestion at query time for situations where the answer depends on the most recent regulatory publication or administrative criterion. A fifth is the study of multilingual or cross-jurisdictional transfer, particularly in settings where statutory text, administrative guidance, and transactional metadata coexist in different formats.

8 Conclusions

This paper presented a tax-assistance platform that seeks to reduce the technological asymmetry between tax authorities and taxpayers through a RAG-based architecture, an invoice deductibility engine, and explanation-centered outputs. The article framed tax assistance as a meaningful AI problem involving retrieval under regulatory drift, context-sensitive decision support, and auditable explanations.

The implemented system combines GPT-4o mini (version 2024-07-18), embeddings with multilingual-e5-large, a Pinecone vector index, a fixed retrieval depth of 20 and a corpus of 493 versioned fiscal documents enriched with structured metadata. Within that configuration, the results indicate that a compact and comparatively inexpensive model can remain competitive when coupled with disciplined retrieval, domain curation, and explicit validation logic. Preliminary evidence suggests substantial gains in latency and review coverage. Qualitative comparison also indicates stronger contextual grounding than LLM-only interaction in abbreviation-heavy and regulation-dependent scenarios. A natural next step is a larger annotated study with explicit retrieval metrics, controlled ablation experiments, and formally documented expert-judgment protocols that separate the contribution of each module.

Acknowledgment. The authors thank the AI research team at dSoft for the collaborative environment that made the theoretical and experimental development of this platform possible.

Disclosure of Interests. The authors declare no conflicts of interest other than their employment relationship with dSoft, the developing entity of the technology discussed here.

References

1. Servicio de Administración Tributaria (SAT): Plan Maestro de Fiscalización y Recaudación. Government of Mexico, Ministry of Finance and Public Credit (2024)
2. OECD: Tax Administration 2024: Comparative Information on OECD and Other Advanced and Emerging Economies. OECD Publishing, Paris (2024)
3. Vaswani, A., et al.: Attention is all you need. In: NeurIPS, vol. 30, pp. 5998–6008 (2017)
4. Devlin, J., Chang, M.-W., Lee, K., Toutanova, K.: BERT: pre-training of Deep Bidirectional Transformers for Language Understanding. In: NAACL-HLT, pp. 4171–4186 (2019)
5. Bommasani, R., et al.: On the opportunities and risks of foundation models. arXiv:2108.07258 (2021)
6. Lewis, P., et al.: Retrieval-augmented generation for knowledge-intensive NLP tasks. In: NeurIPS, vol. 33, pp. 9459–9474 (2020)
7. Chalkidis, I., et al.: LexGLUE: a benchmark dataset for legal language understanding in English. In: ACL, pp. 4310–4330 (2022)
8. Guha, N., et al.: LegalBench: a collaboratively built benchmark for measuring legal reasoning in large language models. In: NeurIPS, vol. 36 (2023)
9. Rudin, C.: Stop explaining black box machine learning models for high stakes decisions and use interpretable models instead. Nat. Mach. Intell. **1**(5), 206–215 (2019)
10. Arrieta, A.B., et al.: Explainable artificial intelligence (XAI): concepts, taxonomies, opportunities and challenges toward responsible AI. Inf. Fusion **58**, 82–115 (2020)
11. Xu, Y., Li, M., Cui, L., Huang, S., Wei, F., Zhou, M.: LayoutLM: pre-training of text and layout for document image understanding. In: KDD, pp. 1192–1200 (2020)
12. Kim, G., et al.: OCR-free document understanding transformer. In: ECCV, pp. 498–517 (2022)

Concrete Crack Segmentation via 5-Fold Cross-Validation Ensemble with ConvNeXt-Tiny U-Net

Mario Vital-Vázquez, Andrés Antonio Torres-Acosta, and Atoany Nazareth Fierro-Radilla(✉)

Tecnologico de Monterrey, Monterrey, Mexico
afierror@tec.mx

Abstract. Automated crack segmentation in concrete infrastructure remains a challenging problem due to the thin, irregular morphology of cracks and the severe class imbalance between crack and background pixels. This paper proposes a robust deep learning framework based on a U-Net architecture with a ConvNeXt-Tiny encoder, designed to improve segmentation accuracy and generalization under complex imaging conditions. To address key limitations in prior work, a composite loss function combining Dice, Focal, and Lovász losses is introduced, enabling simultaneous optimization of class imbalance, hard sample learning, and region-based overlap metrics. To ensure statistical reliability, a 5-fold cross-validation strategy is employed, and a probabilistic ensembling approach is applied during inference to reduce prediction variance and improve robustness. Experimental results on the Kaggle Crack Segmentation dataset demonstrate that the proposed method achieves a mean Dice score of 0.7364 with low variance (±0.0088), indicating stable generalization performance. Furthermore, the proposed framework outperforms state-of-the-art methods such as DeepCrack (0.7159), CrackNet++ (0.7311), and SCCDNet-D32 (0.7348). The results highlight that improvements in crack segmentation can be effectively achieved through better alignment between optimization objectives and evaluation metrics, rather than solely increasing architectural complexity. The proposed framework demonstrates competitive segmentation performance within the evaluated dataset distribution and provides a promising basis for future structural health monitoring applications.

Keywords: Crack segmentation · Semantic segmentation · U-Net · ConvNeXt · Deep learning · Cross-validation · Ensemble learning

1 Introduction

Structural health monitoring of civil infrastructure is essential to ensure safety, durability, and long-term performance. Among the most critical indicators of structural degradation are surface cracks, which can propagate over time due

H. Ponce et al. (Eds.): COMIA 2026, CCIS 3095, pp. 344–356, 2027.
https://doi.org/10.1007/978-3-032-37438-7_27

to environmental conditions, cyclic loading, and material fatigue. Early detection and accurate characterization of these cracks are fundamental for preventive maintenance and cost reduction. However, traditional inspection methods based on manual visual assessment are time-consuming, subjective, and prone to human error, motivating the adoption of automated computer vision-based approaches [1].

In recent years, deep learning (DL) techniques have significantly advanced the field of automated crack detection. Convolutional Neural Networks (CNNs) have demonstrated strong capabilities in learning discriminative features directly from raw images, enabling robust crack classification and segmentation under varying environmental conditions [2]. In particular, semantic segmentation approaches have gained considerable attention, as they provide pixel-level predictions that allow precise estimation of crack properties such as length, width, and morphology. Architectures such as U-Net have become a standard framework for segmentation tasks due to their encoder–decoder structure, which effectively captures both contextual and spatial information while enabling accurate localization [3]. Despite these advances, several challenges remain in crack segmentation. One of the most critical issues is the detection of thin or propagating cracks, which often occupy only a few pixels and exhibit low contrast with respect to the background. Existing deep learning models tend to perform well on wide and clearly visible cracks but struggle to identify these subtle features, especially in low-resolution images or under complex lighting conditions [1]. Additionally, environmental factors such as noise, surface texture, and image blur further degrade the quality of crack detection, leading to incomplete or inaccurate segmentation results [2].

Another important limitation lies in the training objectives commonly used in segmentation models. Most deep learning approaches rely on pixel-wise loss functions such as cross-entropy, which do not directly optimize the evaluation metrics used in practice. In contrast, the Intersection-over-Union (IoU), also known as the Jaccard index, is widely adopted as a more representative performance measure for segmentation tasks, as it accounts for both false positives and false negatives in a balanced manner [4]. However, directly optimizing IoU is challenging due to its non-differentiable nature, which has led to the development of surrogate loss functions such as the Lovász-Softmax loss that enable efficient optimization while improving segmentation performance [4].

Furthermore, recent research highlights the importance of incorporating multi-scale feature extraction and fusion strategies to improve crack segmentation performance. Advanced deep learning models leverage encoder–decoder structures, attention mechanisms, and frequency-domain representations to capture both global context and fine-grained details, which are particularly important for detecting small or blurred cracks [2]. Nonetheless, achieving a balance between computational efficiency and segmentation accuracy remains a key challenge, especially for real-world applications such as UAV-based inspection systems.

In this context, this work focuses on improving crack segmentation performance by addressing the limitations of existing methods in detecting fine and complex crack patterns. The proposed approach builds upon recent advances in deep learning-based segmentation while incorporating strategies to enhance feature representation and optimize segmentation performance under challenging conditions. The effectiveness of the proposed method is validated through experimental evaluation, demonstrating its capability to achieve accurate and robust crack segmentation in practical scenarios.

2 Related Works

2.1 Crack Segmentation Using Deep Learning

The application of deep learning to crack detection and segmentation has led to significant improvements over traditional image processing methods. Early approaches based on handcrafted features were highly sensitive to illumination changes, surface texture variability, and noise, limiting their applicability in real-world scenarios. [5] . The introduction of convolutional neural networks (CNNs), particularly encoder–decoder architectures, enabled end-to-end learning of hierarchical representations, allowing for more robust crack segmentation.

Among these architectures, U-Net and its variants have become the de facto standard due to their ability to combine high-level semantic information with fine-grained spatial details. Numerous studies have proposed architectural modifications to enhance performance, including residual connections, multi-scale feature fusion, and attention mechanisms. While these improvements have shown incremental gains, they often come at the cost of increased model complexity and reduced interpretability, without consistently addressing the fundamental challenges of crack segmentation.

A comprehensive evaluation by [6] highlights that many state-of-the-art methods, including U-Net, DeepLabV3, and FPN-based architectures, exhibit strong performance under controlled conditions but struggle to generalize across datasets with varying crack types, lighting conditions, and surface textures. Furthermore, their study emphasizes a critical limitation in the field: the lack of standardized datasets and evaluation protocols, which hinders fair comparison and reproducibility of results.

Another important limitation identified in the literature is the reliance on relatively small or domain-specific datasets. Even when larger datasets are used, they often aggregate images from different sources, introducing inconsistencies in annotation quality and imaging conditions. This variability can lead to models that perform well on benchmark datasets but fail to generalize in real-world deployment scenarios.

Recent works have attempted to address these limitations through the integration of Transformer-based architectures and generative data augmentation techniques. While these approaches improve global context modeling and data diversity, they introduce additional computational overhead and do not fully resolve the challenge of detecting extremely thin or low-contrast cracks. As a result, there remains a trade-off between model complexity, computational efficiency, and segmentation accuracy.

2.2 Loss Functions for Class Imbalance and Segmentation Quality

Beyond architectural design, the choice of loss function plays a critical role in crack segmentation performance. A fundamental issue in this domain is the severe class imbalance between crack and background pixels [7]. Standard loss functions such as binary cross-entropy treat all pixels equally, leading to biased models that favor the majority class and fail to accurately capture fine crack structures.

Focal Loss [8] was introduced to mitigate this issue by dynamically down-weighting easy samples and focusing the learning process on hard-to-classify examples. While this approach has proven effective in addressing class imbalance, it primarily operates at the pixel level and does not explicitly optimize for region-based metrics such as Intersection-over-Union (IoU), which are commonly used to evaluate segmentation performance.

This misalignment between training objectives and evaluation metrics has been widely recognized as a key limitation in segmentation tasks. In response, the Lovász-Softmax loss was proposed as a differentiable surrogate for the Jaccard index, enabling direct optimization of IoU during training. This approach has demonstrated consistent improvements in segmentation quality, particularly in terms of region overlap and boundary completeness [4].

However, despite its advantages, the Lovász-Softmax loss also presents limitations. Its effectiveness depends on accurate probability calibration, and its batch-wise approximation of IoU may introduce instability during training, particularly in highly imbalanced datasets. Additionally, while it inherently accounts for class imbalance to some extent, it does not explicitly focus on hard samples in the same manner as Focal Loss [4].

Consequently, existing loss functions tend to address specific aspects of the problem—either class imbalance or metric alignment—but rarely both simultaneously. This fragmented approach limits their effectiveness in complex scenarios such as crack segmentation, where both challenges are present.

2.3 Research Gap and Motivation

Despite the extensive research in crack segmentation [1,2,5–7,9–11], several critical gaps remain. First, many approaches prioritize architectural innovation while underestimating the importance of the optimization objective, leading to suboptimal alignment between training and evaluation metrics. Second, loss functions designed to address class imbalance or IoU optimization are typically applied independently, without exploring their potential complementarity.

Furthermore, the combined effect of class imbalance, fine structural details, and heterogeneous imaging conditions creates a highly challenging learning problem that cannot be fully addressed by isolated improvements in either architecture or loss design. Existing methods often achieve strong performance on benchmark datasets but lack robustness and generalization in real-world applications.

These limitations suggest the need for integrated approaches that jointly address multiple challenges in crack segmentation. In particular, combining loss functions that emphasize hard samples with those that directly optimize region-based metrics may provide a more balanced and effective training strategy.

Motivated by these observations, this work proposes a unified framework that integrates complementary loss formulations within a robust segmentation architecture. The goal is to simultaneously improve pixel-level discrimination, region-level consistency, and generalization performance across diverse crack datasets.

3 Methodology

3.1 Problem Formulation

Crack segmentation is formulated as a binary semantic segmentation problem, where the objective is to assign a label $y \in \{0, 1\}$ to each pixel in an input image, indicating background or crack, respectively. This task is particularly challenging due to the highly imbalanced distribution of classes, as crack pixels typically constitute less than 5% of the image area, and due to the thin and irregular morphology of cracks.

3.2 Dataset Partitioning and Cross-Validation Protocol

The proposed methodology was evaluated using the Kaggle Crack Segmentation Dataset, which contains 11,298 image–mask pairs acquired under diverse illumination and surface conditions. To obtain statistically reliable performance estimates, a 5-fold cross-validation protocol was employed. The dataset was partitioned into five mutually exclusive subsets using a fixed random seed (seed = 42). During each fold iteration, four subsets (80%) were used for training while the remaining subset (20%) was reserved exclusively for validation. Cross-validation was used exclusively to estimate the variability and stability of the segmentation model across different train–validation splits. The validation images corresponding to a given fold were never used during training of that fold. Importantly, the cross-validation procedure was not intended to simulate an external benchmark evaluation or unseen domain adaptation scenario. Consequently, the reported results should be interpreted as internal generalization estimates within the same dataset distribution. To avoid methodological ambiguity, ensemble inference was evaluated separately after fold-specific training was completed. Each fold-specific model was treated as an independently trained model, and ensemble predictions were analyzed only as an inference-time enhancement strategy.

3.3 Network Architecture

The segmentation model is based on a U-Net architecture, consisting of an encoder–decoder structure with skip connections that preserve spatial information across multiple scales.

To enhance feature extraction capability, the encoder is implemented using ConvNeXt-Tiny, a modern convolutional backbone that incorporates transformer-inspired design elements such as large kernel convolutions, GELU activations, and layer normalization. The encoder is initialized with ImageNet pretrained weights to leverage transfer learning, while the decoder and segmentation head are randomly initialized.

The network produces a single-channel output representing pixel-wise logits, with the sigmoid activation applied during loss computation and inference.

3.4 Training Strategy

The network was trained end-to-end using full fine-tuning of all model parameters throughout the optimization process. The AdamW optimizer was employed due to its effectiveness in improving generalization through decoupled weight decay regularization. Differential learning rates were used to stabilize optimization, assigning a lower learning rate to the pretrained ConvNeXt-Tiny encoder and a higher learning rate to the decoder and segmentation head. A cosine annealing scheduler was applied to gradually reduce the learning rate during optimization, enabling smooth convergence. Training was conducted for 30 epochs using mixed precision training (AMP) to reduce GPU memory consumption and accelerate training throughput while preserving numerical stability. Table 1 summarizes the hyperparameter configuration used in the experiments.

Table 1. Hyperparameter configuration.

Parameter	Value
Architecture	U-Net
Encoder backbone	ConvNeXt-Tiny (ImageNet pretrained)
Input resolution	512×512 pixels
Batch size	32
Training epochs	30
Encoder learning rate	3×10^{-5}
Decoder learning rate	2×10^{-4}
Optimizer	AdamW
Weight decay	1×10^{-2}
LR scheduler	Cosine Annealing ($\eta_{\min} = 10^{-6}$)
Precision	Mixed (AMP, FP16/FP32)
Illumination norm.	CLAHE (clip_limit $= 4.0$, $p = 1.0$)
Cross-validation	5-Fold (seed $= 42$)
Loss function	0.5 Dice + 0.3 Focal + 0.2 Lovász

3.5 Loss Function Design

To address the limitations identified in prior work—namely class imbalance and misalignment between training objectives and evaluation metrics—a composite loss function is employed.

$$L = 0.5 \cdot L_{\text{Dice}} + 0.3 \cdot L_{\text{Focal}} + 0.2 \cdot L_{\text{Lovasz}}$$

The weighting coefficients of the composite loss function were selected empirically through preliminary experiments conducted on a validation subset. Multiple combinations were evaluated to balance three competing objectives: region overlap optimization, hard sample prioritization, and class imbalance mitigation. A higher weight was assigned to Dice loss (0.5) because overlap quality constitutes the primary segmentation objective in crack detection tasks. Focal loss was assigned an intermediate contribution (0.3) to improve sensitivity to thin crack structures and difficult boundary regions. Lovász loss was incorporated with a lower contribution (0.2) to provide direct optimization of the IoU metric while maintaining training stability. Alternative configurations were evaluated during pilot experiments, and the selected combination consistently provided the best trade-off between Dice score stability and segmentation quality.

3.6 Data Preprocessing and Augmentation

Prior to augmentation, contrast limited adaptive histogram equalization (CLAHE) was applied deterministically ($p = 1.0$, $cliplimit = 4.0$) to every image. CLAHE normalizes local contrast variations caused by uneven lighting, shadows, and surface reflectance differences, standardizing the photometric input distribution across all training samples. An extensive augmentation pipeline was applied stochastically during training to improve generalization, as summarized in Table 2. The effective training set size produced by the augmentation pipeline. The 11,298 training images reside on disk without modification; augmentations are applied stochastically on-the-fly by the PyTorch DataLoader workers at each iteration, meaning that no augmented image is ever stored to disk

All images were normalized using the standard ImageNet channel-wise mean ($[0.485, 0.456, 0.406]$) and standard deviation ($[0.229, 0.224, 0.225]$) before being fed to the network.

3.7 Hyperparameter Summary

Due to the multiplicative computational cost inherent in a 5-fold cross-validation scheme, performing an exhaustive hyperparameter grid search was computationally prohibitive. Therefore, the hyperparameter selection process was not systematically exhaustive but rather based on a combination of established best practices in the literature and targeted empirical evaluations. Specifically, the learning rates (3×10^{-5} for the encoder and 2×10^{-4} for the decoder) were determined via preliminary testing on a single fold to ensure stable convergence. This

Table 2. Data augmentation pipeline applied during training.

Augmentation	Parameters	Purpose
RandomResizedCrop	$scale = (0.6, 1.0)$, 512×512	Scale invariance
HorizontalFlip	$p = 0.5$	Orientation invariance
VerticalFlip	$p = 0.3$	Orientation invariance
RandomRotate90	$p = 0.5$	Rotational invariance
ShiftScaleRotate	$shift = 0.03, scale = 0.1, rot = 20°$	Affine spatial robustness
ElasticTransform	$alpha = 1, sigma = 50$	Structural irregularity
GridDistortion	$p = 0.3$	Non-rigid geometric warping
GaussNoise	$p = 0.2$	Sensor noise simulation
CoarseDropout	$max_holes = 8$	Occlusion-based regularization
RandomBrightnessContrast	$p = 0.5$	Photometric variability

localized testing confirmed that higher encoder learning rates induced catastrophic forgetting of the ImageNet-pretrained representations (manifesting as sharp transient increases in validation loss), while lower values produced excessively slow convergence. The decoder rate was set higher to allow rapid adaptation of the randomly initialized layers without destabilizing the joint optimization. Other parameters, such as the weight decay (10^{-2}) and cosine annealing minimum learning rate (10^-6), follow standard recommendations from the AdamW literature. A batch size of 32 at 512×512 resolution was selected to maximize GPU utilization on the NVIDIA A100-SXM4-80GB. Table 1 summarizes the complete configuration.

3.8 Ensemble Inference

After training the five fold-specific models independently, an optional probabilistic ensemble strategy was evaluated during inference. Each trained model produced a pixel-wise probability map for a given input image. The final ensemble prediction was obtained by averaging the probability outputs of the five independently trained models:

$$P_{\text{final}}(x, y) = \frac{1}{5} \sum_{i=1}^{5} P_i(x, y)$$

The resulting probability map was thresholded to generate the final binary segmentation mask. The ensemble inference was evaluated as a post-training inference enhancement strategy and should not be interpreted as part of the cross-validation evaluation protocol itself. The primary purpose of the ensemble was to reduce prediction variance and improve segmentation smoothness by leveraging complementary representations learned across different training folds.

4 Results

4.1 Cross-Validation Performance (Dice Metric)

The proposed framework was evaluated using a 5-fold cross-validation strategy to assess its robustness and generalization capability (see Table 3). The Dice coefficient (F1 score) was selected as the primary evaluation metric due to its suitability for highly imbalanced segmentation tasks, where it effectively measures the overlap between predicted and ground-truth crack regions.

The model achieved Dice scores of 0.7410, 0.7423, 0.7303, 0.7458, and 0.7224 across the five folds, resulting in a mean Dice score of 0.7364 ± 0.0088.

Table 3. Per-fold best validation Dice scores.

Fold	Best Validation Dice
Fold 1	0.7410
Fold 2	0.7423
Fold 3	0.7303
Fold 4	0.7458
Fold 5	0.7224
Mean	**0.7364 ± 0.0088**

Table 4. Comparison with State-of-the-Art Methods.

Model	Dice Score
DeepCrack [9]	0.7159
SCCDNet-D32 [10]	0.7348
CrackNet++ [11]	0.7311
Ours	0.7364

The low standard deviation indicates a high degree of consistency across folds, suggesting that the model is not sensitive to the specific train–validation split. This stability is particularly relevant in crack segmentation tasks, where dataset variability and class imbalance often lead to unstable training dynamics. The proposed methodology successfully learns robust and generalizable representations of crack structures. In Fig. 1 it is observed 4 random samples and their crack segmentation predictions.

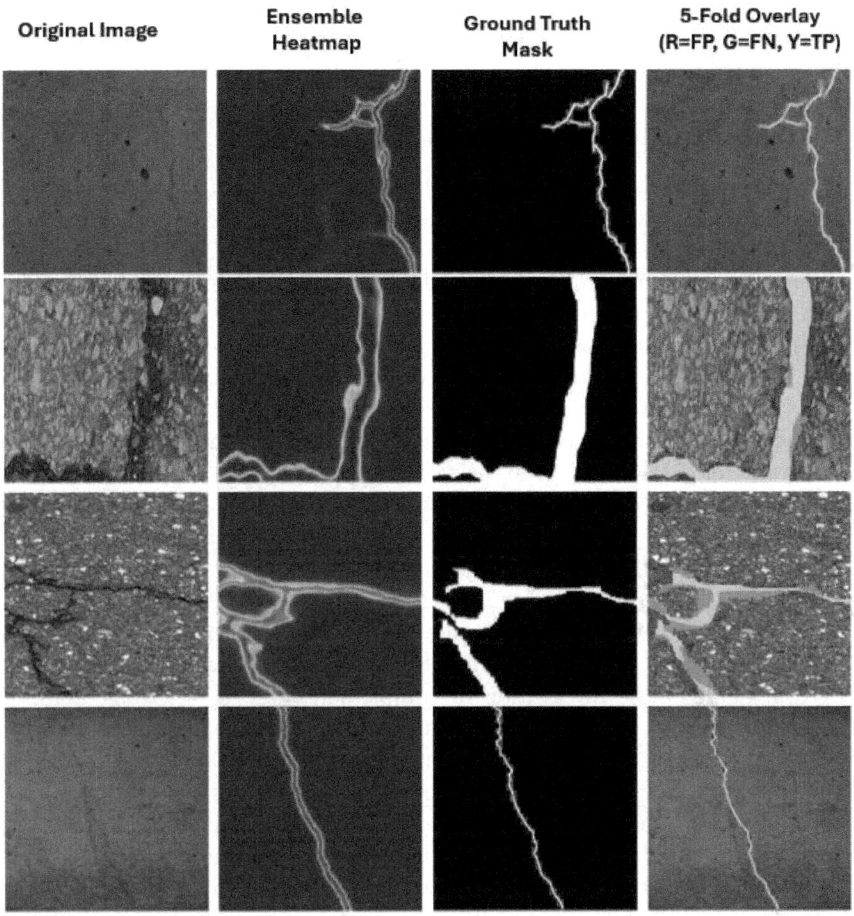

Fig. 1. 4 random samples and their crack segmentation predictions. In the last column a 5-fold overlay is presented; False Positives are represented in Red color, False Negatives in Green and True Positives in Yellow. (Color figure online)

4.2 Convergence Behavior and Generalization Analysis

The training process exhibited stable convergence, with the validation Dice score plateauing around epoch 20. This indicates that the optimization strategy—combining cosine annealing, mixed precision training, and a two-phase fine-tuning scheme—effectively guides the model toward a well-generalized solution.

Notably, the validation loss remained consistently lower than the training loss throughout training. This behavior can be attributed to the aggressive data augmentation pipeline, which introduces additional complexity into the training distribution. As a result, the model is exposed to more challenging samples during training, improving its ability to generalize to unseen data.

This phenomenon reflects effective regularization and suggests that the model is learning invariant and structurally meaningful features rather than overfitting to the training set.

4.3 Comparison with State-of-the-Art Methods

To contextualize the performance of the proposed approach, we compare the obtained Dice score with several representative state-of-the-art crack segmentation methods (see in Table 4).

In comparison, the proposed method achieves a Dice score of 0.7364, outperforming all three approaches.

Although the absolute improvement over SCCDNet-D32 (+0.0016) may appear modest, the context of highly optimized segmentation models, where incremental gains are increasingly difficult to achieve. More importantly, the proposed method demonstrates superior stability, as evidenced by the low variance across folds, whereas most prior works report results based on a single train–test split without statistical validation.

Compared to DeepCrack, the proposed method achieves a substantial improvement of +0.0205 in Dice score, indicating a significantly better overlap between predicted and ground-truth crack regions. Similarly, the improvement over CrackNet++ (+0.0053) highlights the effectiveness of the proposed optimization strategy, particularly in handling class imbalance and preserving fine structural details. The integration of complementary loss functions, combined with cross-validation and probabilistic ensembling, provides a more balanced and effective approach than relying solely on architectural enhancements. Although the ensemble strategy improves segmentation stability and prediction smoothness, it introduces additional inference overhead. Consequently, the ensemble configuration should be interpreted primarily as an accuracy-oriented setting rather than a real-time deployment configuration. Future work will explore lightweight variants and model distillation strategies to reduce inference latency while preserving segmentation quality.

4.4 Statistical Stability Analysis

To evaluate performance consistency across folds, the mean and standard deviation of the Dice metric were computed. The proposed framework achieved a mean Dice score of 0.7364 with a standard deviation of Âś0.0088, indicating relatively stable performance across different train–validation partitions. Although cross-validation provides an estimate of performance variability, additional statistical significance analyses and external validation datasets would further strengthen conclusions regarding robustness and generalization.

5 Discussion

The proposed framework achieves competitive segmentation performance while maintaining low variance across cross-validation folds. Careful alignment between optimization objectives and evaluation metrics can significantly improve crack segmentation quality without relying exclusively on increasingly complex architectures. However, the current evaluation is limited to a single dataset distribution and should therefore be interpreted as an internal generalization assessment rather than a comprehensive external validation study. Additional evaluation on cross-dataset or real-world deployment scenarios is required to fully characterize robustness under domain shifts.

The consistent performance across folds, combined with improvements over established methods, indicates that the proposed approach effectively addresses key challenges in crack segmentation, including class imbalance, structural continuity, and variability in imaging conditions.

The proposed framework demonstrates competitive segmentation performance within the evaluated dataset distribution and provides a promising basis for future structural health monitoring applications.

6 Conclusions

This work presented a robust deep learning framework for semantic crack segmentation, addressing key challenges inherent to the task, including severe class imbalance, thin and irregular crack structures, and variability in imaging conditions. Unlike approaches that focus primarily on architectural complexity, the proposed method emphasizes the integration of complementary optimization strategies within a unified training pipeline.

The proposed framework achieves competitive performance, with a mean Dice score of 0.7364 and low variance across folds, indicating strong generalization capability. The use of 5-fold cross-validation provides a statistically reliable evaluation, overcoming a common limitation in prior work where performance is reported on a single train–test split. Furthermore, the method outperforms several state-of-the-art approaches, including DeepCrack, CrackNet++, and SCCDNet-D32, highlighting the effectiveness of combining pixel-level and region-level optimization objectives.

A key contribution of this work lies in the design of a composite loss function that integrates Dice, Focal, and Lovász losses. This combination enables the model to simultaneously address class imbalance, focus on hard-to-segment regions, and directly optimize the evaluation metric. Additionally, the incorporation of probabilistic ensembling further improves prediction stability and reduces model variance, resulting in smoother and more reliable segmentation outputs.

The findings suggest that improvements in crack segmentation are not solely dependent on increasingly complex architectures, but can be effectively achieved through better alignment between training objectives and evaluation criteria. By bridging this gap, the proposed approach provides a principled and efficient solution for crack segmentation.

Despite the encouraging results, several limitations remain. The current evaluation is restricted to a single dataset distribution and does not include external benchmark validation. Additionally, ensemble inference increases computational cost, limiting applicability in real-time deployment scenarios. Future work will focus on evaluating the proposed framework across multiple crack segmentation datasets, exploring lightweight deployment-oriented architectures, and investigating uncertainty-aware learning strategies to improve robustness in safety-critical infrastructure inspection applications to significant improvements in both accuracy and robustness.

Acknowledgment. The authors would like to acknowledge Tecnologico de Monterrey for the computational resources provided for this work.

Disclosure of Interests. The authors have no competing interests to declare that are relevant to the content of this article.

References

1. Ganesh, K., Hyun, Y., Sung, S.: Fast detection of missing thin propagating cracks during deep-learning-based concrete crack/non-crack classification. Sensors **2**(1419), 1–21 (2023)
2. Sun, R., Li, X., Zhang, L., Su, Y., Di, J., Liu, G.: Wavelet-integrated deep neural network for deblurring and segmentation of crack images. Mech. Syst. Signal Process. **224**(112240) (2025)
3. Ronneberger, O., Fischer, P., Brox, T.: U-net: convolutional networks for biomedical image segmentation. In: Navab, N., Hornegger, J., Wells, W.M., Frangi, A.F. (eds.) MICCAI 2015. LNCS, vol. 9351, pp. 234–241. Springer, Cham (2015). https://doi.org/10.1007/978-3-319-24574-4_28
4. Berman, M., Rannen Triki, A., Blaschko, M.B.: The Lovász-Softmax loss: a tractable surrogate for the optimization of the intersection-over-union measure in neural networks. In: Proceedings of the IEEE/CVF Conference on Computer Vision and Pattern Recognition (CVPR), pp. 4413–4421 (2018)
5. Zhou, S., Song, W.: Crack segmentation through deep convolutional neural networks and heterogeneous image fusion. Autom. Construct. **125**(103605), (2021)
6. Tabernik, D., Šuc, M., Skočaj, D.: Automated detection and segmentation of cracks in concrete surfaces using joined segmentation and classification deep neural network. Constr. Build. Mater. **408**(133582), 1–12 (2023)
7. Liua, Z., Caoa, Y., Wanga, Y., Wangb, W.: Computer vision-based concrete crack detection using U-net fully convolutional networks. Autom. Constr. **104**, 129–139 (2019)
8. Lin, T.Y., Goyal, P., Girshick, R., He, K., Dollár, P.: Focal loss for dense object detection. In: Proceedings of the IEEE International Conference on Computer Vision (ICCV), pp. 2980–2988 (2017)
9. Liu, Y., Yao, J., Lu, X., Xie, R., Li, L.: DeepCrack: a deep hierarchical feature learning architecture for crack segmentation. Neurocomputing **338**, 139–153 (2019)
10. Li, H., et al.: SCCDNet: a pixel-level crack segmentation network. Appl. Sci. **11**(5074), 1–15 (2021)
11. Zhang, J., Beliaeva, Z.: CrackNet++: an enhanced UNet architecture employing tri-axial fusion attention and multiscale gated fusion for precise crack segmentation and quantification. J. Build. Eng. **118**(115038), 1–27 (2026)

Judging the Judge: On the Consistency of a Rubric-Guided Agent in Assessing Research Questions

Erik Carbajal-Degante[1](✉), Ken Guadarrama-Herrera[2], and Jackeline Bucio-García[3]

[1] Instituto de Ciencias Aplicadas y Tecnología (ICAT), UNAM, Mexico City, Mexico
erikycd@gmail.com

[2] Facultad de Ingeniería (FI), UNAM, Mexico City, Mexico

[3] Coordinación de Universidad Abierta y Educación Digital (CUAED), UNAM, Mexico City, Mexico

Abstract. Consistency is an attribute required by humans and machines to ensure steadiness and firmness in their actions and ideas over time, thereby guaranteeing balance and coherence. In this sense, variability in peer-review consistency has always been an inherent characteristic due to the wide range of opinions among experts. Moreover, ambiguous framings within a given work in the initial stages of new research do not produce solid articulation, thereby blurring the work's innovation and true contribution. Nowadays, agentic artificial intelligence has gained significant attention for tasks requiring certain degrees of autonomy, expertise, and consistency, such as conducting paper reviews. In this work, we describe the design, implementation, and experimentation of a rubric-guided artificial intelligence agent for evaluating research questions, which represent the core elements that shape the direction of a research endeavor in protocols and initial drafts. Following the Beliefs-Desires-Intentions approach to designing agents, we set the agent's autonomy through tools, a rubric, and a configuration of large language models. Our experiments assessing the agent's consistency on different published Scopus papers reveal findings related to reliability, sensitivity, and paper-level robustness, offering evidence that, despite autonomy, human control is still needed in the pursuit of building next-generation agents and multi-agent systems designed to support research.

Keywords: Agents · Artificial Intelligence · Evaluation · Research Questions · Rubric

1 Introduction

Research protocols play a key role in the scientific workflow since they formalize the rationale, scope, and planned methods of a study before data collection begins. A well-written protocol supports transparency and planning in methodology, helping researchers anticipate feasibility constraints and evaluate early

H. Ponce et al. (Eds.): COMIA 2026, CCIS 3095, pp. 357–369, 2027.
https://doi.org/10.1007/978-3-032-37438-7_28

risks in the process. In this sense, research questions (RQs) are useful to articulate the knowledge gap of a study while intending to provide the conceptual basis that guides downstream elements such as hypotheses, objectives, study design, variable selection, measurement strategy, and analysis plans. When RQs are vague or weakly grounded, the subsequent components tend to reflect that ambiguity, often resulting in misaligned and untestable structure, or designs that do not adequately answer the inquiry. Consequently, assessing and improving the quality of RQs at the first stages of research is a high-leverage intervention for strengthening rigor.

In protocols, the centrality of RQs is well established in methodological guidance, then formulating the RQs is commonly described as the first step in researching and as the basis from which a study design choice are derived, meaning that weaknesses at this step can propagate into fatal flaws in methodology and reduce the reliability (Bordage and Dawson, 2003). In this sense, practical frameworks such as FINER (feasible, interesting, novel, ethical, relevant) and structured formats such as PICO/PICOT are widely recommended to refine broad topics into focused answerable RQs (Mohanan and Parameswaran, 2022; Hosseini et al., 2024). This makes the RQ not merely an introductory statement but a core organizing element against which protocol coherence can be evaluated (Covvey et al., 2024).

Although methodological guidelines emphasize careful formulation of RQs, in practice the early evaluation of research proposals and protocol-like documents remains intensive and can exhibit substantial variability across reviewers. Empirical studies of grant peer review have repeatedly documented that ratings for the same proposal can differ between reviewers and that inter-rater reliability is heterogeneous across contexts and determinants (e.g., discipline or proposal characteristics) (Mutz et al., 2012). Moreover, evidence suggests that procedural mechanisms intended to harmonize judgments, such as panel discussion, do not necessarily improve the reliability of scores in medical research (Fogelholm et al., 2012). At the same time, methodological work cautions that observed agreement metrics can be sensitive to design choices (e.g., range restriction or the small number of peer reviewers), underscoring both the complexity of achieving reliable evaluation and the need for more transparent processes. These findings motivate the exploration of automatic, rubric-guided tools as a form of structured and auditable support for early assessment, aiming not to replace expert opinion but to improve consistency and traceability particularly when writing first drafts.

Recent advances in large language models (LLM) have catalyzed a shift from raw text generation toward agentic systems that can plan, call external tools, and iteratively refine intermediate results. A representative line of work demonstrates that prompting strategies can couple reasoning with tool use in a way that improves performance on knowledge and action-based tasks, motivating the design of LLM-centered agents rather than standalone chat interfaces (Yao et al., 2023). At the same time, multi-agent orchestration frameworks have gained attention as a practical approach for decomposing complex workflows into specialized roles that coordinate through structured interaction, enabling collaborative behaviors in processes like retrieval, verification, critique, and report generation. For example, AutoGen proposes a general framework for building applications through conversations among multiple configurable agents that can incorporate both tools and human input (Wu et al., 2023). Similarly, the Bad-Scientist framework provides another pertinent example, which was designed to evaluate the acceptance of artificially-generated papers written by agents. The study revealed significant vulnerabilities and ethical concerns within the academic publishing process despite achieving high acceptance rates for its convincingly artificial written manuscripts (Jiang et al., 2025). Moreover, evidence from (Tong et al., 2026) suggests that artificial intelligence (AI) systems can learn scientific taste, placing the scientific community under pressure to produce high-quality and innovative materials. This rapid progression suggests that, in the near term, rubric-guided agents could become a scalable layer of procedural support for early research workflows.

Despite the recent boost of agentic AI systems, their application to evaluative tasks in research workflows raises specific requirements that are not fully addressed by chatbots or conversational assistants. Evaluation in scientific contexts must be criteria-driven, transparent, and auditable in terms of scores that should be linked to explicit indicators. Moreover, feedback should be grounded in observable evidence from the assessed text rather than in implicit preferences or stylistic biases. In this work, we describe the design and implementation of a rubric-guided agent as an evaluative artifact: it can read research documents, identify and extract RQs, and apply a structured rubric via tool-based actions. To delimit the scope and enable controlled measurement, we focus exclusively on the assessment of RQs and study the agent's behavior through two reliability-oriented aims: (1) the extent to which the agent retrieves consistent results across repeated evaluations of the same paper, and (2) how the variability of the rubric-based RQs scores changes when agent's hyperparameters are modified.

2 Background: The Ecosystem of Agents and Multiagents

Agentic AI can be understood as a paradigm in which AI development is embedded within an action-oriented control to pursue a goal through iterative reasoning and interaction with external capabilities (Zhang et al., 2025). In this context, an AI agent is not merely a text generator, but a system that receives an objective, maintains an explicit or implicit state of the task, selects and executes actions, and updates its intermediate representation. This framing emphasizes characteristic properties such as goal-directedness, decomposition of complex tasks into steps, tool-augmented operation, and traceability of decisions through actions. Under the agentic paradigm, these properties are especially relevant for evaluative workflows, where the system must move beyond from single user-machine interaction to a dynamic procedure that can be inspected, reproduced, and audited (Russell and Norvig 2021).

Nowadays, the agentic paradigm has gained traction because it offers a practical way to extend LLM capabilities to workflow execution. This shift has also catalyzed interest in multi-agent ecosystems, where complex tasks are distributed across specialized roles that coordinate through structured interaction (Han et al., 2024; Qian et al., 2023). Such distribution can improve modularity and maintainability, and it enables patterns like cross-checking, critique, revision, and escalation to additional tools when uncertainty is detected. As a result, agents and multi-agent approaches are increasingly viewed as a foundation for building end-to-end systems that must handle heterogeneous steps with a certain degree of simulated autonomy.

In multi-agent ecosystems, these benefits are often realized through an explicit role specialization, where each agent is configured with a bounded responsibility, a preferred set of tools, and a narrower but specialized decision domain. Common roles in literature include (Han et al., 2024):

- a planner: which decomposes the objective into sub-tasks and defines the execution order,
- a reader: which accesses databases and extracts relevant data,
- an evaluator: which applies structured criteria,
- a verifier: which challenges assumptions, checks consistency, and requests additional evidence,
- a coordinator (also known as orchestration): which mediates and consolidates intermediate artifacts and resolves conflicts between agents.

Finally, multi-agent does not merely imply parallelism. It encodes a design strategy for increasing robustness, interpretability, and controllability by making responsibilities and handoffs explicit.

3 Methodology

3.1 Designing Phase

This study follows a design-evaluation approach in which we (1) build a rubric-guided AI agent for assessing RQs in research articles, (2) provide a reflection stage, and (3) empirically characterize the reliability of its outputs under controlled experimental conditions, see Fig. 1. To align with the preliminary scope of this work, our evaluation focuses on consistency and variability, rather than external validity against panels of experts, as in peer revision. For this design stage, we structure the evaluator as a BDI-inspired agent in terms of beliefs, desires and intentions, according to the theory of agents and multiagent systems of (Rao and Georgeff, 1995; Wooldridge, 2012).

1. Beliefs (B) encode the evolving structured state (articles, extracted research questions, rubric specification, evidence snippets, and specific configuration).
2. Desires (D) formalize the evaluation objectives, emphasizing rubric compliance, traceability, and reproducibility.
3. Intentions (I) correspond to the selected tool-driven workflow (document ingestion, RQ extraction, scoring, aggregation, and reporting), with optional evidence-seeking steps when rubric indicators cannot be assessed from the available text.

(a) Appearance.

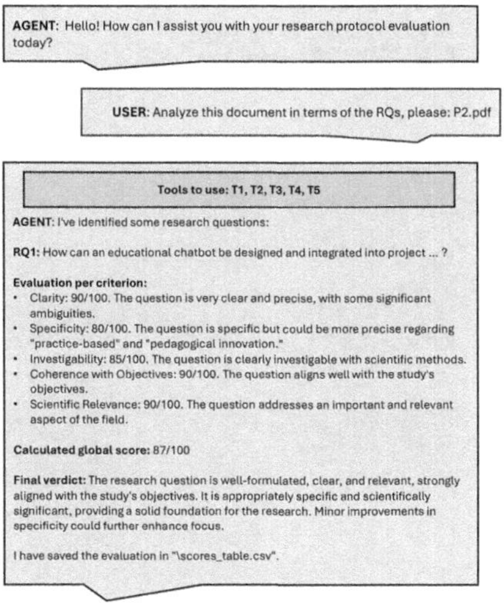

(b) Conversation example with the agent.

Fig. 1. Agent Appearance and Functioning.

Classical BDI architectures were originally developed for symbolic or logical agents, but the abstraction maps naturally to contemporary agentic systems when (B) is treated as an explicit task state, (D) as goal specifications, and (I) as a tool-execution plan. In our design, the BDI mapping is used not as a purely conceptual analogy but as a software-structuring principle that separates what the agent knows at runtime (what is known), what it aims to achieve (what is sought), and how it operationalizes those goals through actions (what is executed).

Since the proposed tool-augmented agent is inspired by BDI, the evaluation process is expressed as a small set of high-level intentions operating over an explicit belief state. In practice, this design is captured by a structured state object that stores the artifact representation, the extracted RQs, the rubric specification, and intermediate evaluation artifacts such as criterion-level scores, evidence snippets, feedback, final verdict, etc. Operationally, the agent follows a fixed intention plan and evaluation workflow aligned with the architecture diagram in Fig. 2.

3.2 About the Rubric

The rubric applies five equally weighted criteria as follows: Clarity, Specificity, Investigability, Coherence, and Scientific Relevance. Each is scored on a 0–100 scale (20% weight per criterion). For every RQs, the agent assigns a criterion-level score guided by qualitative performance bands (from poor to excellent) and provides brief, criterion-specific feedback. The final RQs score is computed

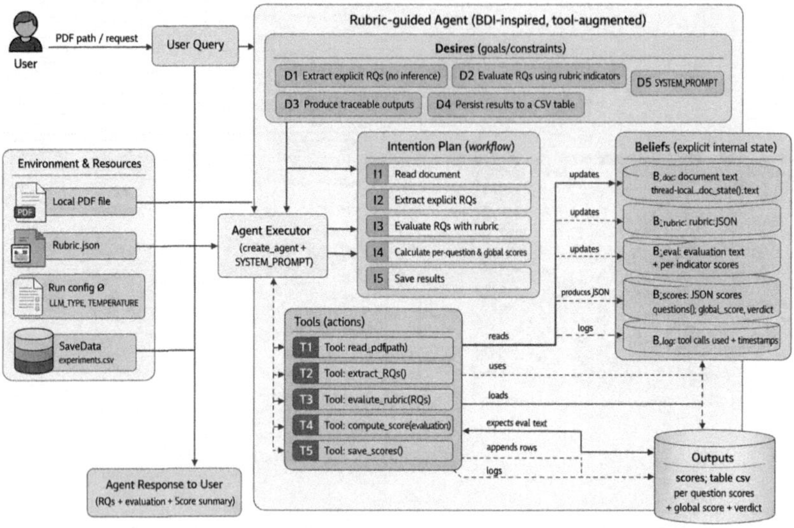

Fig. 2. Agent evaluation workflow, transparency and modularity enable analysis of variability in both final scores and across intermediate stages. The system can be re-run under controlled parameter settings while keeping inputs and beliefs comparable, which is essential for quantifying consistency and variability in rubric-based assessments.

as the weighted average across criteria, enabling consistent per-question scoring and aggregation to paper-level global scores.

3.3 Materials and Experiment Pipeline

We conducted a controlled experiment using a small corpus of peer-reviewed research articles. Following a simple structured search in Scopus using the query ("artificial intelligence" AND "agentic" AND "education"), we selected five recent English-language research articles (P1→P5) that explicitly state their RQs in the text. This curated set serves as standardized input, reducing ambiguity at the identification stage and enabling repeated measurements of the agent's behavior under controlled conditions.

Experiment 1. Evaluating the Extraction. The first experiment evaluates the agent under six configurations ($E1C1, ..., E1C6$) defined by the combination of model choice [LLM] and temperature [T]. For each paper and configuration, the agent is executed to generate evaluation outputs, and the experiment follows four sequential stages aligned with the pipeline:

1. Ground-truth preparation. We screen candidate papers and manually extract the explicitly stated RQs to serve as the reference set.

2. Initial configuration. We fix the LLM and T for each experimental condition:
   - (E1C1): [LLM = gpt-4.1-mini] with (T = 0.2),
   - (E1C2): [LLM = gpt-4.1-mini] with (T = 0.5),
   - (E1C3): [LLM = gpt-4.1-mini] with (T = 0.8),
   - (E1C4): [LLM = o4-mini] with (T = 1),
   - (E1C5): [LLM = gpt-5-mini] with (T = 0.3),
   - (E1C6): [LLM = gpt-5-mini] with (T = 0.6),
3. BDI-Agent execution with tools according to: Read full text, identify RQs, and save data.
4. We compare the extracted RQs against the ground-truth RQs using BLEU, ROUGE, and cosine similarity.

This design enables us to quantify variability in the extracted RQs (as a potential source of instability), while comparing how these measures change as a function of model choice and temperature.

Experiment 2. Anchor + Stress Test. The second experiment is a test study focused exclusively on the stability of rubric-based scoring ($E2C1, ..., E2C3$), decoupled from the RQs extraction stage analyzed in Experiment 1. Specifically, we fix the evaluator input to the ground-truth RQs for the same five papers (P1→P5) and run the evaluation pipeline 10 independent times per paper under three configurations:

- Anchor baseline (E2C1): [LLM = gpt-4.1-mini] with (T = 0.1),
- High stress (E2C2): [LLM = gpt-4.1-mini] with (T = 0.9),
- Model shift (E2C3): [LLM = gpt-5-mini] with (T = 0.5).

Each run produces per-question rubric scores (0–100) and a paper-level global score computed as the average across questions. We quantify intra-configuration consistency by measuring score dispersion across repeated runs, configuration sensitivity by comparing mean score shifts between E2C1-E2C2 and E2C1-E2C3, and paper-level robustness by analyzing how variability and score shifts differ across papers. This design isolates evaluator stochasticity and configuration effects, while enabling a clearer interpretation of repeatability in the agentic evaluations.

4 Results

4.1 Experiment 1

The results of this experiment evaluate the quality of RQ extraction across the agent's six configurations by measuring the similarity between the extracted RQs and the ground-truth RQs. We report two complementary metrics, BLEU and ROUGE, scaled from 0 to 100, where higher values indicate greater similarity. In this sense, BLEU reflects lexical precision (i.e., how closely the extracted text matches the reference), whereas ROUGE is more sensitive to overlapping under paraphrasing, capturing cases in which the agent needs to rewrite the RQ. As shown in the matrix plot in Fig. 3, configuration E1C2 achieves the strongest overall extraction performance along the most metrics, while also exhibits the highest degree of paraphrasing during extraction. This pattern is further clarified in Fig. 4, where more recent LLM configurations (particularly E1C5 and E1C6) display larger variability in similarity scores. A plausible explanation is that GPT-5–based configurations tend to behave in a more proactive and generative manner, which can lead the agent to reformulate the extracted RQs to better fit the surrounding context.

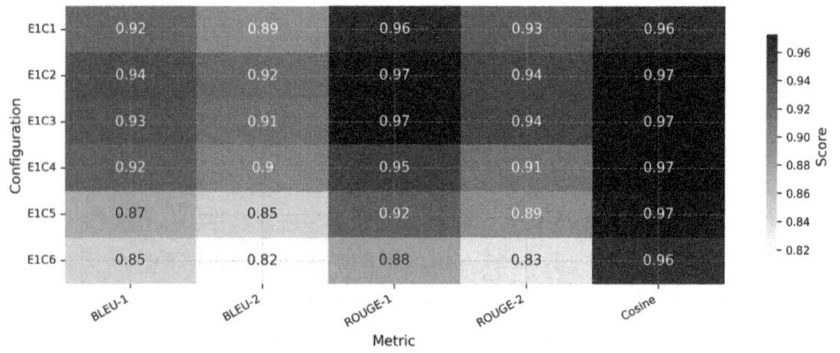

Fig. 3. Mean Similarity metrics per configuration.

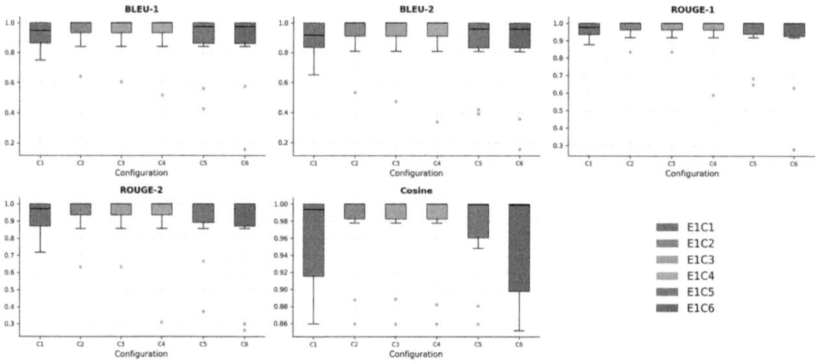

Fig. 4. Distribution of similarity metrics per configuration.

4.2 Experiment 2

The intra-configuration consistency (reliability) obtained in this experiment, visualized through a boxplot of paper-level global scores (see Fig. 5), indicates that narrower distributions correspond to higher reliability within each configuration. Under this criterion, E2C1 exhibits the highest reliability, showing the tightest spread of global scores across repeated runs. In contrast, Fig. 6, which summarizes score variability per paper and configuration, suggests that E2C3 is the noisiest condition overall displaying the largest dispersion in scores across runs.

To assess configuration sensitivity, we analyzed how strongly the scoring outcomes changed as a function of configuration modifications. In this sense, Fig. 5 shows that shifts in the medians across configurations reflect sensitivity, indicating that changes in temperature (E2C1→E2C2) and model backend (E2C1→E2C3) can systematically alter the evaluation scoring. Complementarily, the evidence in Fig. 7 highlights where these shifts are most pronounced across papers, with P1 and P2 exhibiting the largest configuration-dependent effects (E2C1, E2C2→E2C3) among the evaluated set.

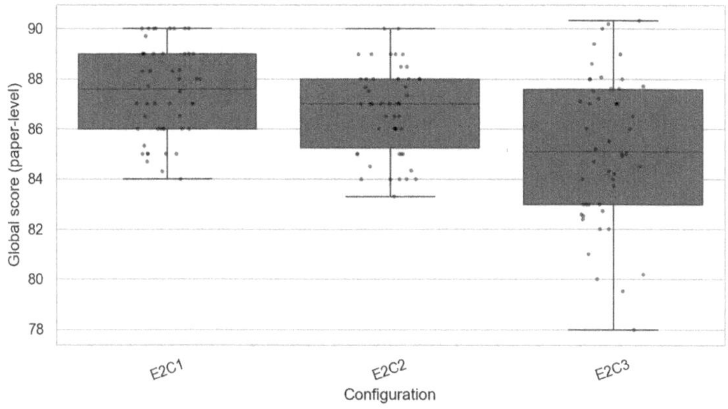

Fig. 5. Global score distribution by configuration.

Finally, to characterize the robustness of each configuration with respect to individual papers, Fig. 8 compares paper-level score distributions across configurations. In this plot, shorter boxes reflect higher robustness and steadier evaluations. The results show that P3 displays the lowest robustness (the largest variability). This instability is attributable to upstream ambiguity in the question set: the paper contains additional implicit RQs, which the agent sometimes surfaced and incorporated, increasing run-to-run variability in the scoring.

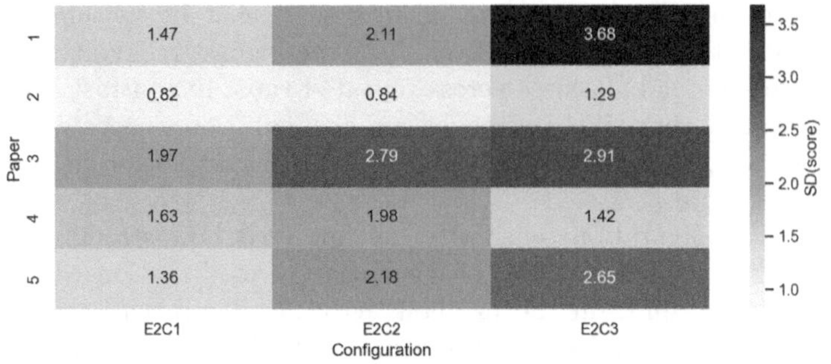

Fig. 6. Score variability SD (paper × configuration).

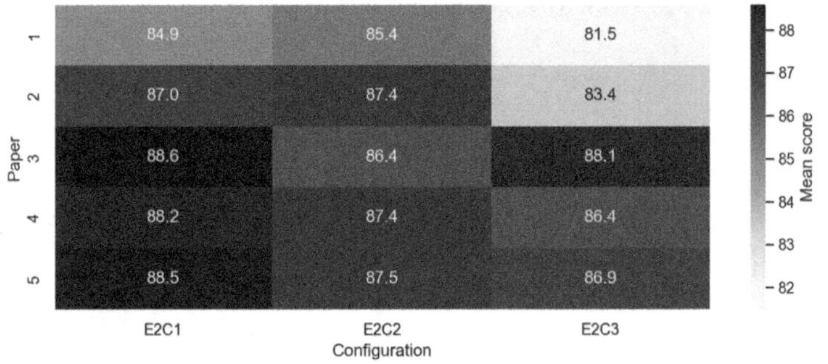

Fig. 7. Mean per-question score (paper × configuration)

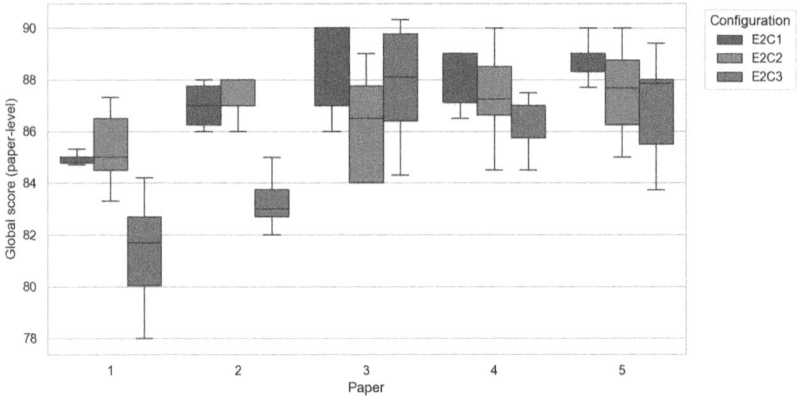

Fig. 8. Global score distribution by configuration.

5 Discussion

In experiment 1, similarity-based comparisons against ground-truth RQs show that extraction performance changes across model-temperature configurations. While some configurations preserve the reference phrasing more faithfully (higher BLEU and ROUGE), others generate more paraphrastic outputs, particularly in the more recent GPT-5 configurations. This suggests a trade-off: more recent and proactive models may rewrite RQs to better fit context, which potentially helps in human interpretation, but reduce text-level similarity and increase variability in strict assessments.

The experiment 2 isolates the evaluator by fixing the input RQs and repeating scoring across 10 independent runs. The anchor setting yields the tightest score distributions (highest intra-configuration reliability), whereas the high-temperature stress condition and the model-shift condition increase dispersion, indicating higher stochasticity in rubric application. Practically, this implies that low-temperature baselines are preferable when repeatability is a primary requirement for protocol assessment.

The evidence in both experiments also suggests that some papers exhibit larger changes across configurations than others, reflecting heterogeneity in question structure with the agent's extraction and scoring behavior. We can notice that in experiment 2, configuration-dependent shifts and variability are more pronounced for particular papers, showing that a single best configuration may not generalize uniformly. This motivates reporting not only global averages but also paper-level dispersion metrics to identify cases where agent decisions are less stable and may require additional controls, like human verification and monitoring.

6 Limitation and Future Work

While this study provides valuable insights into the consistency of a rubric-guided agent, several limitations highlight avenues for future work. First, the experimental findings are based on a small sample of only five papers; expanding this corpus is necessary not only to ensure generalizability across different disciplines but also to address the current framework's struggle with ambiguity, specifically regarding implicit RQs. Future iterations must adopt more proactive configurations to infer and reconstruct plausible RQs from context, strengthening the BDI-style proposal by providing well-structured candidates for evaluation and making the agent's assumptions transparent and auditable. Second, because the present methodology evaluated only internal consistency, future research must establish external validity by benchmarking the agent's scores against actual human peer reviewers to ensure alignment with expert consensus. Finally, to overcome the constraints of a single agent, upgrading the architecture to a specialized multi-agent ecosystem will be necessary. Deploying a collaborative workflow with distinct roles (such as planners, readers, and verifiers) that cross-check each other's outputs, will significantly improve consistency, resolve uncertainties, and ensure traceability while maintaining robust human-in-the-loop oversight for scalable drafting and evaluation support.

7 Conclusions

This work analyzes the performance of a rubric-based evaluation agent under multiple configurations, showing how both a BDI-inspired design and a human-defined rubric shape the agent's balance between constraint and autonomy. By operationalizing key steps into tools such as document reading, RQ identification and extraction, rubric loading and scoring, the agent produces evaluations that are traceable and comparable across conditions, while maintaining a clear separation between extraction and scoring. Our findings indicate that configuration choices have a measurable impact on both extraction fidelity and rubric-based scoring, and that these effects can be described through three complementary consistency dimensions: reliability, sensitivity and robustness. Taken together, this approach provides a practical framework for designing and selecting agent configurations in research-assessment settings and suggests that well-scoped agent ecosystems can support researchers from the earliest stages of study design by structuring evaluation and enabling transparent rubric-driven feedback.

Acknowledgments. This work was supported by UNAM-PAPIIT IA102425 and SECIHTI-SNII 625718.

References

Bordage, G., Dawson, B.: Experimental study design and grant writing in eight steps and 28 questions. Med. Educ. **37**(4), 376–385 (2003)

Covvey, J.R., McClendon, C., Gionfriddo, M.R.: Back to the basics: guidance for formulating good research questions. Res. Social Adm. Pharm. **20**(1), 66–69 (2024)

Fogelholm, M., Leppinen, S., Auvinen, A., Raitanen, J., Nuutinen, A., Väänänen, K.: Panel discussion does not improve reliability of peer review for medical research grant proposals. J. Clin. Epidemiol. **65**(1), 47–52 (2012)

Han, S., Zhang, Q., Jin, W., and Xu, Z.: LLM multi-agent systems: challenges and open problems (2024). Version Number: 3

Hosseini, M.-S., Jahanshahlou, F., Akbarzadeh, M.A., Zarei, M., Vaez-Gharamaleki, Y.: Formulating research questions for evidence-based studies. J. Med. Surg. Public Health **2**, 100046 (2024)

Jiang, F., Feng, Y., Li, Y., Niu, L., Alomair, B., Poovendran, R.: BadScientist: can a research agent write convincing but unsound papers that fool LLM reviewers? (2025). Version Number: 1

Mohanan, S., Parameswaran, N.: FINER criteria – what does it mean? Cosmoderma **2**, 115 (2022)

Mutz, R., Bornmann, L., Daniel, H.-D.: Heterogeneity of inter-rater reliabilities of grant peer reviews and its determinants: a general estimating equations approach. PLoS ONE **7**(10), e48509 (2012)

Qian, C., et al.: ChatDev: Communicative Agents for Software Development (2023). Version Number: 5
Rao, A., Georgeff, M.: BDI agents: from theory to practice (1995)
Russell, S. J. and Norvig, P.: Artificial Intelligence: A Modern Approach. Pearson Series in Artificial Intelligence, 4th edn. Pearson, Hoboken (2021)
Tong, J., et al.: AI can learn scientific taste (2026). Version Number: 1
Wooldridge, M.J.: An Introduction to Multiagent Systems. Wiley (2012). Reprint edn.
Wu, Q., et al.: AutoGen: enabling next-gen LLM applications via multi-agent conversation (2023). Version Number: 2
Yao, S., et al.: ReAct: synergizing reasoning and acting in language models. arXiv:2210.03629 [cs] (2023)
Zhang, G., et al.: The landscape of agentic reinforcement learning for LLMs: a survey (2025). Version Number: 4

Agent-Based Modelling for Explaining Savings Levels in Mexico

K. A. M. Osorio-Dominguez[1] and Y. U. K. Coronado González[2(✉)]

[1] Facultad de Negocios de la Universidad La Salle México, Mexico City, Mexico
[2] Vicerrectoría de Investigación, Universidad La Salle México, Mexico City, Mexico
yaxkin.coronado@lasalle.mx

Abstract. Currently, the Mexican economy faces a household savings crisis. Given this outlook, it is possible to ask How Mexican families make decisions regarding their savings and investments within the financial institutions? This article proposes an agent-based model to describe household saving behaviors based on generalized data for income, expenditure, and returns on investment from different financial institutions. The model proved used on an explanatory level, being able reproduce actual data from the *Encuesta Nacional de Salud Financiera* 2023 with an error of 10%. Thus, results imply that the Mexican population tends to follow a series of specific behaviors with regards to saving. Nevertheless, the proposed model was fairly valid over periods smaller than a year of observation.

Keywords: Agent-Based Modelling · Savings · Mexico

1 Introduction

Savings levels in Mexico are currently low. The newspaper El Economista reports, based on a 2025 study from the Instituto Tecnológico Autónomo de México, that only one in four Mexicans has savings within the formal financial system. Even more worrying, the survey indicates that 33% of the country's population does not have any savings whatsoever [1].

Additionally, according to the 2023 *Encuesta Nacional de Salud Financiera* (ENSAFI) only 11.2% of the legally adult Mexicans are completely sure of having savings that they consider to be enough [2]. Those who do not have savings at all or whose savings are considerably low jeopardize their long-term financial stability and are not able to plan expenditures ahead, fulfill their financial goals, retire comfortably, or leave any inheritance once they pass [3]. Low saving levels also slow down macroeconomic growth, since saving flows can turn into productive sector investments if captured within the formal financial system [4].

Considering the previously presented data and the potential benefits of saving for Mexican households, it is possible to question what specific behaviors explanatory of the current savings trends in Mexico are. This article is composed of four distinct sections. The first one is a brief recapitulation of the existing literature regarding the topic of internal household savings and its main determinants. The second describes the utilized variables and methodology. Afterwards, results and their corresponding discussion are presented. Lastly, conclusions and potential limitations of the analysis are pointed out.

H. Ponce et al. (Eds.): COMIA 2026, CCIS 3095, pp. 370–383, 2027.
https://doi.org/10.1007/978-3-032-37438-7_29

2 Literature Review

2.1 Explanatory Theories for the Phenomenon of Internal Household Savings

There is a wide range of theoretical frameworks from different economic schools of thought that allow an explanation of the reasons why a consumer could decide to save a fraction of their income. Firstly, it is possible to cite the life cycle theory this framework poses that levels of consumption remain stagnant throughout life, yet salary varies starting out relatively low during youth, reaching peak during the adult stage, and possibly declining during retirement. Henceforth, saving behavior might be determined by exceeding income relative to a fixed level of consumption. The priority for this type of consumer is maintaining consumption constant (not lowering their levels of consumption) via an intertemporal optimization [5].

Along the same lines is the permanent income hypothesis. Permanent income is a baseline around which an individual's income varies in different moments of their life depending on personal characteristics such as education, industry in which they work, location, etc. In this context, all consumers possess a set expectation of their permanent income, and their saving decisions are subject to that expectation. It should stand for that income expectations may or may not be accurate since it is not possible to observe future income in practice [6]. This hypothesis is complemented by Saman and Sudarsan who point out saving can be directed towards emergency expenditures or to preserve purchase power in the face of uncertain income, naming this phenomenon the precautionary motive for savings [7].

Different from the previously presented theories, the Keynesian framework poses that savings, and consumption should be modelled as a proportion of income. Within the same framework, there is a concept known as marginal propensity to consume (MPC), defined as the fraction of an additional unit of income destined to consumption and not savings [8].

Lastly, the relative income hypothesis should be noted. This one present that the consumption levels of a household are not dependent on their absolute income, but rather on their position within the income distribution. This theory also indicates the existence of two relevant effects to explain that tendency. Firstly, there is a "demonstration effect", in which families tend to imitate the lifestyle of wealthier ones due to psychological matters. Secondly, the is a "ratchet effect" in which families tend to keep the consumption levels they had during their maximum income period [9].

2.2 Plausible Determinants of Internal Household Saving

Household savings are affected by a wide variety of macroeconomic factors. Among these, it is possible to cite the inflation since long-term relative process impact consumption levels although asymmetrically depending on available income [10]. Additionally, savings can be impacted by any variable that influences available income, such as credit availability for the general population [10, 11]. On the other hand, Huffman et al. [12] assert that private saving in Mexico is also influenced by household structure. Particularly, these authors point to the inhibition of saving due to socioeconomic dependency rates across different types of familial structures.

It is also worth noting that most economists mention the key impact of the real interest rate in this matter since it affects the cost of setting aside liquidity to invest within the formal financial system [13]. Complementarily, Anton and Villegas suggest that real interest rate shocks accentuate falls in consumption during crisis periods. These authors reached their main conclusion through the analysis of a series of stochastic differential equations [14].

Among the most recent developments in this matter, the work of Druica et al. [15], which consists of a model based on computational representation of individuals that can adopt different attitudes towards saving. The conclusion of these authors' analysis suggests that close third-party influence can result in an individual giving a disproportionate weight to the costs of postponing consumption when comparing with the long-term benefits of saving. Likewise, Barrafrem et al. mention that psychological variables such as self-control may influence constant saving over time [16]. Finally, it is worth noting that Barrera et al. point out another set of behavioral factors such as financial literacy and specific motivations to save, their conclusions were based upon the 2018 ENIF and Mexican data from the *Comisión Nacional Bancaria y de Valores* [17].

2.3 Savings in Mexico at Distinct Levels of the Income Distribution

In the context of this research article, it highlighted that saving is not uniform within the income distribution. Cantú mentions that a concentration of 77% of savings in Mexico within the richest 40% of the population [18]. Similarly, Nava and Brown estimate, through a quantile regression, that the MPC is of 5.7% in the 20th income percentile and it shoots through to 16.1% in the 90th percentile [19]. Therefore, it can infer that one of the main barriers for saving is income. On a more generalized level, this phenomenon generates consequences that make saving even more difficult overall. Similarly, Villar et al. have found that relatively higher income households tend to grow their savings when income grows. This contrasts with a different result within the same source which indicates that relatively lower income households tend to increase their spending as income grows. Both results are statistically significant at a 95% confidence level and were obtained from analyzing the 2018 *Encuesta Nacional de Ingresos y Gastos de los Hogares* and the 2020 edition of that same survey [20]. It is also possible to assert that the higher the income distribution, the more its savings tend to grow. Manrique et al. [21] have found that there is a 98.7% chance that a household with an income greater than 10,000 *pesos* saves formally. The previously mentioned conclusions were reached by descriptively analyzing a sample of 260 participants from three Mexican cities.

3 Methodology

We propose, with the objective of modelling Mexican consumers' savings and their growth via investment in different financial institutions, the technique of agent-based model (ABM), which consists of modelling individual behaviors under a series of conditions and decision rules for an agent set. These agents are autonomous entities simulating decision-making represented within a computational program. Later, it is possible to observe what has happened on a macroeconomic level due to the interactions described along with the shifts inside population subsets within the simulation [22]. The most common way of describing this type of model is via the ODD protocol, developed by Grimm et al. [23]. The following subsection is the summarized ODD protocol for the model proposed in this research article. The ABM model was made using the programming language Netlogo.

3.1 ODD Protocol

Main Entities: We model two interactive entities, households, and financial entities.

- Households: Households are groups of private consumers, shown on Netlogo's map as a single person. This agent set is mobile. Its agents can advance within the map according to their wealth, that being their objective.
- Financial entities: These fixed agents are intermediaries where households can invest their monthly liquidity surpluses jointly. There exist distinct types of financial institutions within the model:

  1. SOFIPO (popular savings institution): This institution admits any investor with positive net worth. It can also function as a credit-giver and charges a fee of 130% of the lent amount, this is average for this type of institution [24]. It charged a 0.08% commission fee for any investors [25]. After 18 months of continuous investment, the agents that invest in this institution on any *tick* can access an additional 0.0001 over their monthly return, representative of frequent saver programs.
  2. Traditional banking: Institution dedicated to investing the surpluses of agents within the fifth wealth decile of the population or a higher one. Additionally, this institution does not admit agents with less than a month's income worth of accumulated savings. It can also function as a credit-giver and charges a fee of 83% of the lent amount, representative of fees on an average credit card [26]. It charged a 0.16% commission fee for any investors [25]. After 18 months of continuous investment, the agents that invest in this institution on any *tick* can access an additional 0.0003 over their monthly return, representative of frequent saver programs.
  3. Investment trust fund: These are investment vehicles used by agents within the seventh wealth decile of the population or a higher one that also have a net worth superior to two months of their income. Initially, this institution offers an additional 0.00001 over its monthly return, representative of dividends for equity investments, after 18 ticks this quantity triples. These institutions charge a commission of 0.20 [25].
  4. *CETES-directo* (treasury bonds): Within this model, only CETES 28 are auctioned since it is assumed that, for higher terms, the expected rate is identical to a series of successive investments in CETES 28 in terms of Vasicek and Venegas-Martínez [27]. This institution allows any household with a net worth of $100 *pesos* or greater to invest [28].

  - Government: This entity, which operates from the grid dedicated to charging taxes and using fraction of the budget obtained from that activity (defined through a slider named *porcentaje-transferencias*) to transfer aid checks to a set percentage of the households (defined by a slider named *porcentaje-recibe-apoyos*)

State and Scale Variables: Energy assignment for agents

- For the households: Energy assigned to individual agents, representative of net worth in fixed Mexican *pesos* from the year 2024.
- For the government: Its state variable is the quantity of available budget.

World Type, Environments, and Emergency: This model's virtual world is a bound type with ten divisions representing wealth deciles. This world has a discrete time with each tick representing a month in which all subroutines are executed. The model's quality is evaluated in accordance with its ability to replicate the savings distribution reported on the 2023 ENSAFI [2], the last dataset available. Both the financial institutions' returns, and each turtle's income was considered environments.

General Subroutines Description

- Income reception: Each turtle receives a fixed monthly income that is assigned during the setup.
- Tax collection: Based on the 2024 income tax tables [29], all turtles are charged a sum in taxes. These collections are the government's budget for that specific tick.
- Aid transfers: From its budget, the government sets asides a percentage to aid a predetermined fraction of the population equally.
- Expenditure: Turtles spend their income depending on the decile they are on. For agents whose income, including government aid, totals less than $6,600 *pesos* be-because it is the equivalent of a 2024 basic goods basket for a family of about two people considering data from CONEVAL [30]. If turtles have an income greater than that quantile and are between the first and the seventh decile, they define expenditure in accordance with a kernel regression parting from their income, contemplating aid (see Appendix 1). Additionally, it presumes that the aids received are, in a fraction, incentives for extraordinary expenditure. Given that factor, it is considered that, on top of the baseline expenditure defined by the regression, turtles spends a percentage of government aid defined by a slider known as *por-centaje-gastado*. For turtles within the seventh decile onwards, a consider baseline expenditure of $18,200 *pesos*. This consideration is because within the kernel regression there were no increases from that point on regardless of income. The additional expenditure due to government aid is also applied for these deciles.
- Decile update and movement: Wealth deciles are recalculated and the barriers that separate each updated wealth level. Each agent moves to their corresponding wealth level according to the updates.
- Intervention of financial institutions: Financial institutions generate each return according to previously defined random distributions (see Appendix 1), turtles calculate the mean of the last twelve-monthly returns (not including the current tick)

and pick the institution with the highest return amongst the ones they can access. Afterwards, turtles get their return along with the described additional and pay the corresponding commission; an inflation rate of 5.2% is deducted from the return, this rate is average for the timespan between 2026 and 2024 [31]. Lastly, the historic means of rate are updated to reflect the current's tick return and eleven past ones. It is worth noting that, that agents with debt (represented through a negative net worth), ask for loans from these institutions and are charged the previously specified fees.

- Turtle's change of color: As a general indicator of savings levels, turtles' change color. Turtles with a net worth equivalent to a month's income after aid remain white. Turtles that have a net worth equivalent to any amount between one and two months of income accounting for aid turn yellow. Turtles that have a net worth equivalent to any amount between two and three months of income accounting for aid turn pink. Lastly, turtles that have net worth equivalent to any amount greater than three months of income accounting for aid turn blue. Along with this change, all turtles relocate according to their new position on an updated version of the wealth distribution.
- Update of turtles' liquidity state: From the twelfth tick onwards, turtles that do not have savings greater than a month's worth of income stop accruing savings and begin to spend all their available income. This restriction is set because it presumes that liquidity constraints impair turtles on this savings level from investing consistently.

Basic Principles: Turtles have adaptive expectations. All the models were built on the assumption of minimal transaction cost.

Prediction and Adaptation: Turtles try to predict the future returns of financial institutions based on past returns' means. By moving on the grid, turtles may gain the ability to change their preferred financial institution.

Setup: The model is setup with 400 white turtles. The grid extends between the coordinates –16 to 24 on the x axis and –16 to 16 y axis. Patches with an x coordinate between –16 and 16 are the section in which turtles are setup and will move. The rest of the board is for placement of the financial institutions. The grid designed to represent a ladder; agents start out on the lower wealth level and move upward to access new financial institutions. The limit of the lowest wealth decile is on the –12 y coordinate, from that point on there is a division every three patches. Financial institutions are located on the lowest decile of turtles they can have as clients. As part of the setup, we assign to each turtle a monthly base income, defined according to a Weibull distribution (see Appendix 1).

4 Results

The described agent-based model was executed with parameters reflecting actual data for 2024. For the government aid related parameters, 41% of the government budget was assigned to this line [32] and 82% of the households receive aid [33]. The percentage in which turtles take aid as an additional incentive for expenditure was set to 95.60% after several rounds of tests and adjustments. Lastly, a random seed of 100 was selected for all the models. It was observed that the system goes through two main phases after few unstable initial ticks. At the five-tick mark, savings levels approach the savings distributions reported on the 2023 ENSAFI [2] (Fig. 1):

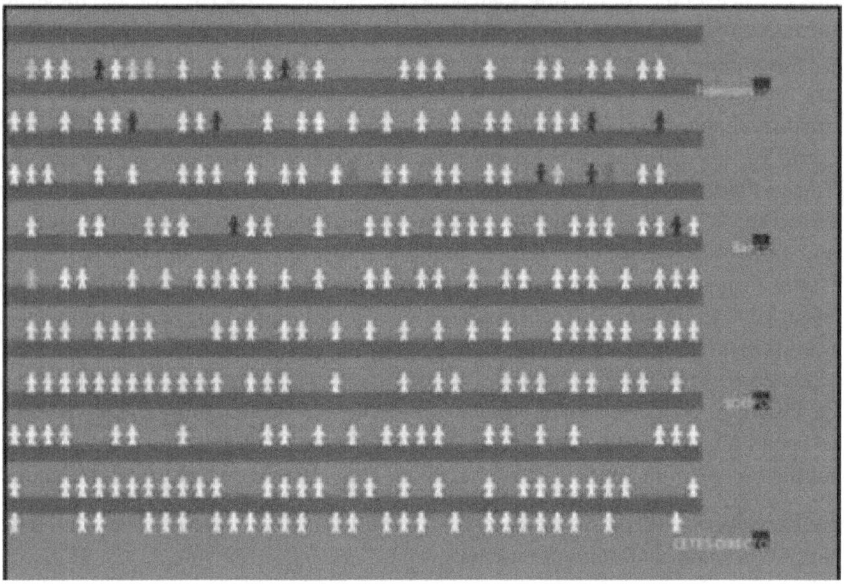

Percentage of the population by savings level	
One month's worth of income or less	91%
Between one and two months' worth of income	4.25%
Between two and three months' worth of income	0.50%
Three months' worth of income or more	4.25%

Fig. 1. Savings levels at 5 ticks. Example of phase one. Tick most like the actual data. The map's design represents a ladder in which agents go escalating through net worth levels and are colored according to their savings level relative to their salary. The blue squares on the right represent different financial institutions according to their described (see methodology).

Actual data from the 2023 ENSAFI reflect that 89.24% of the population has one month's worth of income or less, 3.87% has between one and two months' worth of income, 2.05% two and three months' and only 4.83% has more than three months' worth of income saved [2]. Therefore, there is approximately 4.27% difference between the model's result on this tick and the actual data. There are variations between ticks in this phase yet there is not more than a 10% difference with the data from ENSAFI (see discussion). Lastly, there is a second phase, beginning at the eleven-tick mark, in which practically the whole population falls within the lowest saving levels:

As the last remark of this section, it is presenting the graphs corresponding to both the evolution of savings levels and the turtles' preferred financial institutions throughout the ticks (Fig. 3):

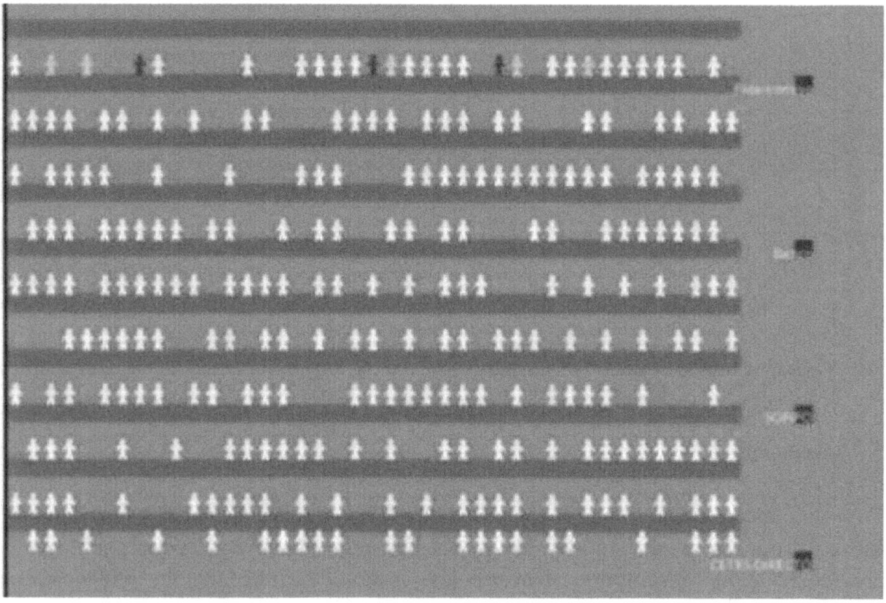

Percentage of the population by savings level	
One month's worth of income or less	97.50%
Between one and two months' worth of income	1.75%
Between two and three months' worth of income	0.00%
Three months' worth of income or more	0.75%

Fig. 2. Savings levels at 19 ticks. Example of phase two.

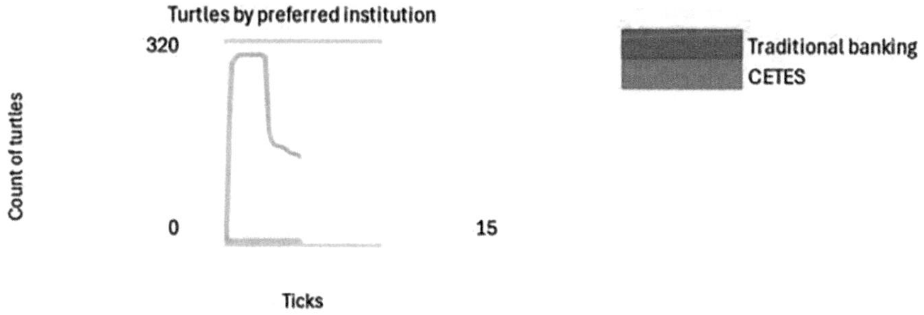

As the last remark of this section presents, the graphs correspond to both the evolution of savings levels and the turtles' preferred financial institutions throughout the ticks: It can be observed that savings level evolves according to the two phases described. Additionally, it can be observed that the most preferred financial institution across all wealth levels is CETES 28.

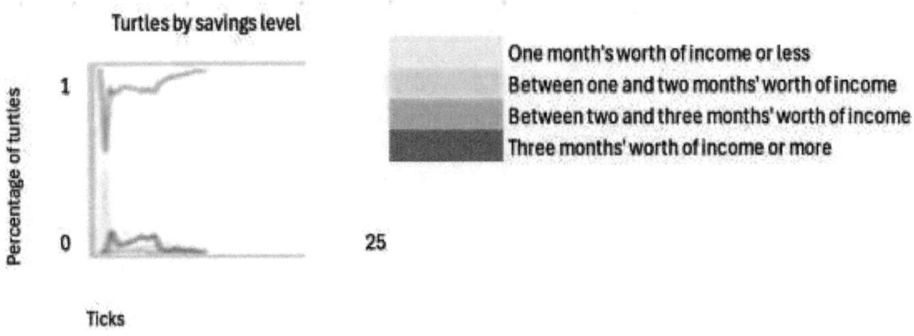

Fig. 3. Graphs for the evolution of savings levels and the turtles' preferred financial institutions.

5 Discussion

The model demonstrated to be a good method of reproducing the savings distribution in Mexico, suggesting that the Mexican population might behave in accordance with the previously defined assumptions. Nevertheless, it should be noted that the model can only reproduce that distribution on a set period and nor across different ones since savings are dependent on the accumulation of random returns from financial institutions. As Piedragil y Venegas-Martínez [34] remark it is with the accrual of successive returns that their random walk reaches a stable not normal distribution. On the specific case of the presented model, this explains both the unstable first few ticks and the results unalike the actual data since the first few returns do not fit the proposed distribution.

The following variation can be explained through the repetition of returns that start fitting the proposed distributions. Additionally, this phase begins to show the accumulated effect of the invested amount which determines the possibility of accruing greater earnings over time [35]. In this sense, that amount is determined by the difference between expenditure and the total available income. Likewise, given that the model replicates actual data as observed in figure one, it can be speculated that expenditure levels do increase with income as specified in [9]. Nonetheless, it can also be argued that this effect is capped at a certain level of income because, on the proposed kernel regression, household expenditure remains constant after $18,200 *pesos.*

On another note, it is possible to replicate the real data presented in [2] only when a distinct MPC is considered for income sourced from government aid. A similar proposition is made by Hawkins [36] who points that, due to a series of cognitive biases, there are different tendencies for consumption depending on sources of income. Within the proposed model, this effect is combined with the expectation of permanent income before aid, assigned to each agent at the setup. Matching the approach presented in [6], it may be possible that a relevant piece of the models' work, and the saving behavior, is the expectation of a baseline income even if the totality of income is not perfectly constant.

However, the existence of a liquidity restriction has the effect of generating a tendency to maintain the totality of income or more in the long term as shown on Fig. 2. This happens slightly before twelfth-tick mark in which those turtles with a low-savings level freeze. Within the model, this is the formalization of the liquidity restrictions. This indicates that due to insufficient income there are restrictions for saving [37]. The proposed model indicates that this affects 90% of the total population. Still, it is impossible to compare the model with actual data patterns on the long-term because there are not any continuous data available only specific data points as those presented in [2].

On the other hand, it should be noted that CETES are the preferred investment mechanism across income distribution (see Fig. 4). This is since, as mentioned in [38], this instrument is the one that gives the most consistent results despite occasional downfalls. Remarkably, CETES continue to be the most utilized instrument despite not being any risk aversion mechanisms within the model. This particular result could change with the introduction of new investment vehicles, such as investment funds that apply unconventional strategies to get returns above the market average, such institutions are described in [39]. This omission might be inducing certain percentage of the models' error.

Moreover, it is relevant to remark that the models' error might also be provoked by the omission of sociodemographic information regarding household structure. According to Saman and Sudarsan [7] that information is key to explain saving behavior.

Lastly, it is relevant to delimit the outreach of the proposed model. Shmueli remarks that predictive and explanatory models differ from each other in both their intent and statistical foundation [40]. Similarly, Edmonds affirms that a model cannot be described as fully predictive if it only fits historical data, particularly if such a model is a simplified version of the observed phenomena. However, this same author points out that an explanatory model can be used for future predictions [41]. This is the case of the proposed model, assuming that key responses to variable changes remain constant, the model can be used for short-term prediction.

6 Conclusions

The agent-based model presented can replicate actual data regarding Mexico's savings distribution with a 10% error margin, being the first to represent via interactions of households and financial entities the main structure of the savings, with an exciting potential to experiment different economic theories or policies. However, the explanatory use of the model is limited to the short-term due to inconsistent saving behaviors within the Mexican population, mostly related to income wise restrictions. Because of this characteristic, the model should be adjusted for each prediction in order to be applicable. As it is, the model is an explanatory abstraction of savings behavior and not a full predictive tool since it omits real world factors such as risk aversion and sociodemographic data. For future research, a modified version of this model may be used to predict the short-term effects of the introduction of new financial products or public policy regarding government aid including sociodemographic control variables. The model proposed within this article has limitations, firstly, there are only eight years' worth of data, with consistent measurement methods, for income and expenditure variables. These factors impact the model's capacity to make predictions for terms greater than twelve months. Secondly, the parameters utilized with regards to government aid are not a perfect reflection of actual policy since it has varied throughout the years. This limitation also applies to the rest of the models' fixed parameters since they are merely representative and not exact averages for each tick. Due to this factor, it is necessary to remark that these parameters are more empirical adjustments than they are perfectly adaptable theoretical explanations. Another important limitation is the lack of actual data to contrast with the model outside of ENSAFI, since that particular survey is the main instrument applied nationwide (Table 1).

Table 1. Robustness tests for the model.

Variable being modelled	Data on which the distribution is estimated	Distribution type and parameters	p_{value} for the Kolmogorov-Smirnov test (Lilliefors correction).
Income before government aid.	Income (before government aid) data, national average by decile [42]	Weibull with shape parameter of 0.8428 and scale of 18825.5090.	0.56
Monthly returns for the SOFIPO sector. Percentage points	Monthly average returns for all of the national sector [43]	Beta with shape parameter of 0.8801, location parameter of 1.3301 and scale of 0.3999.	0.2080
Monthly returns for traditional savings accounts. Percentage points	Monthly average returns for all of the national sector [43]	Gamma with shape parameter of 1.7680, location parameter of 0.0386 and scale of 0.5326.	0.4430
Monthly returns for investment trusts. Percentage points	Monthly average returns for all of the national sector [43]	Skewed normal with shape parameter of 1.8958, location parameter of 3.6630 and scale of 2.3129	0.3610
Interest rates for CETES 28. Percentage points	Monthly average interest rates on public auctions [44]	Gamma mixture with shape parameters of 0.7158, 0.7158, 0.71630, scales of 10.67578, 10.6757, 49 10.67586 and weights of 0.33363,0.333623, 0.332743.	0.2890

Appendix 1

Statistical validation of distributions and kernel regression.

Additionally, a local linear gaussian kernel regression was utilized to quantify the relationship between expenditures and income data from 2016 through 2024 for the second decile onwards [37] (Table 2).

Table 2. Parameters and statistical measures for the kernel regression

Parameter or measure	Value
Bandwidth	6,000
Confidence interval (95%) for the local variance	[0.0028, 0.0243]
Confidence interval (95%) for the local standard error	[0.4936, 52.3265]
Local autocorrelation	−0.0434

As a robustness test for the model, different random seeds for the model and their respective results at 12.5 ticks were compared (Table 3).

Table 3. Example of distinct random seeds and their respective results on percentage of turtles on each savings level as defined above.

RANDOM SEED	1ST SAVINGS LEVEL	2ND SAVINGS LEVEL	3RD SAVINGS LEVEL	4TH SAVINGS LEVEL
100	99.5	0.5	0	0
200	96.5	3	0	0.5
300	91.75	6.5	0	1.75
400	88.5	8.75	0	2.75
500	90	8	0	2
600	88	10	0	2
700	95.75	2.75	0	1.5
800	76.5	18	0	5.5

References

1. El Economista: [¿Eres tú? Sólo 1 de cada 4 mexicanos logra ahorrar y romper el ciclo] (2025). https://www.eleconomista.com.mx/finanzaspersonales/1-4-mexicanos-logra-ahorrar-romper-ciclo-20250704-766722.html#:~:text=%C2%BFEres%20t%C3%BA?- ,S%C3%B3lo%201%20de%20cada%204%20mexicanos%20logra%20ahorrar%20y%20romper,poblaci%C3%B3n%2C%20generando%20preocupaci%C3%B3n%20y%20tristeza.&text=El%20ahorro%20en%20M%C3%A9xico%20sigue,de%20ahorro%20en%20el%20pa%C3%ADs. Last accessed: 17 Mar 2026

2. ENSAFI: (2023) https://ensafi.condusef.gob.mx/. Last accessed 17 Mar 2026
3. PNC Bank: [Beneficios de ahorrar dinero] (2024). https://www.pnc.com/insights/es/personal-fi-nance/save/benefits-of-saving-money.html. Last accessed: 17 Mar 2026
4. Godínez, V.: Ahorro y disponibilidad de recursos para la inversion en México. Coyuntura Económica, 70–84 (1994)
5. Meza, J.: La teoría del ciclo vital de Franco Modigliani: Enfoque en los salaries de los trabajadores colombianos. Revista Informar. **5**(1), 131–140 (2017)
6. Friedman, M.: A Theory of the Consumption Function, 1st edn. Princeton University Press, New Jersey, United States of America (1957)
7. Saman, S., Sudarsan, P.: Determinants of precautionary saving motive. In: 2nd International Conference of Applied Research in Management, Economics and Accounting, pp. 5–7 (2019)
8. Liquitaya Briceño, J., Gutiérrez Jiménez, G., Ramírez Muñoz, M.: Teorías que resuelven el enigma del consumo. Denarius. **28**, 73–111 (2014)
9. Liquitaya, J.: La teoría del ingreso permanente: Un análisis empírico. Revista Nicolaita de Estudios Económicos. **6**(1), 33–61 (2011)
10. Mendoza, M.: Sensibilidades y asimetrías ante choques de ingreso en el consume privado en México 1995-2017. Ensayos Rev. Econ. **39**(1), 21–58 (2020)
11. Díaz, H., Sosa, M., Ortiz, E.: Inclusión financiera y ahorro en México: un análisis logístico binario y de redes neuronales artificiales. Departamento de Administración de la Universidad Autónoma Metropolitana Unidad Azcapotzalco. **8**(1), 53–84 (2018)
12. Huffman, C., Villágomez, P.: Family arrangements and savings in Mexico: a latent class approach. Programa Universitario de Estudios del Desarrollo. Documento de Trabajo No. 20 (2019)
13. Levy, N.: Rates of interest, effective demand and economic growth. Economía UNAM. **9**(25), 74–93 (2012)
14. Antón, A., Villegas, A.: El papel de la tasa de interés real en el ciclo económico de México. El Trimestre Económico. **80**(320), 773–803 (2013)
15. Druica, E., Ianole, R., Cornescu, V.: The psychological cost of saving – an agent-based modelling approach. Rom. J. Econ. **48**, 34–48 (2014)
16. Barrafrem, K., Tinghog, G., Vastjafall, D.: Behavioral and contextual determinants of different stages of saving behavior. Front. Behav. Econ. **3**, 1 (2024)
17. Espinosa, A., Mayela, M., Mayela, M.: Theoretical perspective on factors driving personal savings. Vincula Técnica EFAN. **8**(1), 53–61 (2022)
18. Cantú, R.: El ahorro en México: otra brecha por cerrar. Centro de Investigación Económica y pre-supuestaria, A.C. (CIEP) (2025)
19. Nava, I., Brown, F.: Determinantes del ahorro de los hogares. Práctica. **49**, 93–118 (2018)
20. Villar, B., Jiménez, M., Sánchez, E.: On the mexican households' savings before and during the covid-19 pandemic. Investigación Económica. **82**(323), 132–152 (2023)
21. Manrique, G., Escalera, M., Moreno, E.: Formal and informal savings habits and income levels: a case study in Mexico. Int. J. Financial Res. **12**(5) (2021)
22. Pavón, J., López-Paredes, A., Galán, J.: Modelado basado en agentes para el estudio de sistemas complejos. Novática. **218**, 13–18 (2012)
23. Grimm, V., et al.: A standard protocol for describing individual-based and agent-based models. Ecol. Model. **198**(1–2), 115–126 (2006)
24. El Economista: ¿Qué considerar al invertir, ahorrar o pedir prestado en una SOFIPO? (2024). https://www.eleconomista.com.mx/finanzaspersonales/Que-considerar-al-invertir-ahorrar-o-pedir-prestado-en-una-sofipo-20240709-0104.html
25. CONDUSEF: Presenta Condusef actualización d e l Buró de Entidades F i n a n c i e r a s (2015). https://www.condusef.gob.mx/?p=contenido&idc=686&idcat=1
26. CONDUSEF, SHCP: Atención a usuarios. Tarjetas de crédito (s.f.). https://www.condusef.gob.mx/?p=tar-jeta_credito

27. Vasicek, A., Venegas-Martínez, F.: Modelos de la estructura de plazos de las tasas de interés. Revista Mexicana de Economía y Finanzas 16(2)
28. Nacional Financiera: Cetes directo (2019). https://www.gob.mx/nafin/acciones-y-programas/cetesdirecto-204296#:~:text=No%20se%20requiere%20monto%20m%C3%ADnimo,de%20la%20Rep%C3%BA-blica%20y%20reguladores
29. Contaduría CCII: ISR – Impuesto Sobre la Renta (Tablas 2024) (2024). https://contaduriaccii.com.mx/tablas-isr-2024/
30. CONEVAL: Evolución de las líneas de pobreza por ingresos (2025). https://www.coneval.org.mx/Medi-cion/MP/Paginas/Lineas-de-Pobreza-por-Ingresos.aspx
31. Contaduría CCII: INPC general (2026). https://contaduriaccii.com.mx/inpc-historico/
32. Senado de la República: Gobierno Federal proyecta transferir 0.5% más recursos a los gobiernos de las entidades en el 2024. señala estudio del IBD (2023). https://comunicacionsocial.senado.gob.mx/infor-macion/comunicados/7210-gobierno-federal-proyecta-transferir-0-5-mas-recursos-a-los-gobiernos-de- las-entidades-en-el-2024-senala-estudio-del-ib#:~:text=El%20proyecto%20de%20Pre-supuesto%20de%20Egresos%20de,Federalizado%20respecto%20del%20presupuesto%20ap-robado%20en%202023
33. Programas para el Bienestar: Más ingresos, más bienestar: así impulsamos la economía de las familias (2025). https://programasparaelbienestar.gob.mx/mas-ingresos-mas-bienestar-asi-impulsamos-la-economia-de-las-familias/
34. Piedragil, C., Venegas-Martínez, F.: Valuación de opciones sobre activos subyacentes con dis-tribuciones estables. Estocásticas: Finanzas y Riesgo. **1**(1), 55–71 (2011)
35. Ovaska, T., Sumell, A.: Increase interest in compound interest: economic growth and personal finance. J. Econ. Finance Educ. **16**, 85–97 (2017)
36. Hawkins, R., Wallace, S.: Source of income effects for demand decisions and taxable consumption. Appl. Econ. **18**(20), 2371–2379 (2009)
37. Hoyo, C., Pena, X., Tuesta, D.: Factores de demanda que influyen en la Inclusion Financiera en Mexico. Analisis de las barreras a partir de la ENIF. Working Papers 1336, BBVA Bank, Economic Re-search Department (2013)
38. Figueroa, J., Morales, M.: Los CETES en el sistema financiero mexicano: una revisión teórica de su desempeño. Revista Veritas De Difusão Científica. **5**(3), 2356–2375 (2025)
39. Rodríguez-Reyes, L., Samaniego, A., Pasillas, M.: Strategies in retirement fund selection in the Mexican retirement market 1997–2018. Revista Mexicana de Economía y Finanzas. **16** (2021)
40. Shmueli, G.: To explain or to predict? Stat. Sci. **25**(3), 289–310 (2010)
41. Edmonds, B.: The practice and rhetoric of prediction. Int. J. Soc. Res. Methodol. **26**(2), 157–170 (2023)
42. INEGI: Encuesta Nacional de Ingresos y Gastos de los Hogares 2024 (ENIGH), nueva serie (2025). https://www.inegi.org.mx/programas/enigh/nc/2024/
43. CNBV: Sistema Financiero Mexicano: Indicadores Financieros e Info. Operativa (2026). https://app.pow-erbi.com/view?r=eyJrIjoiZGU5NDk5NTktZmI2My00NjkwLThkMjUtNmU4YzQ2ZjI2OT-MxIiwidCI6IjVlMmM0OTc3LTEwN2QtNDBhMy04YWY3LTcwMDc0ODFhNjBkNCIsImMiOjR9
44. Banco de México: Tasas de rendimiento de CETES (cuadro CF107) (2026). https://www.banxico.org.mx/SieInternet/consultarDirectorioInternetAction.do?accion=consultar-Cuadro&idCuadro=CF107&sector=22&locale=es

Second-Order Polynomial Regression for Optimizing the Fabrication Parameters in Organic Electrochemical Transistors

Angel E. Tepozteco-Morgan[1], José A. Ávila-Niño[2(✉)], Francisco J. Ibarra-Villegas[1], and Caín Pérez-Wences[1]

[1] CIATEQ A.C. - Jalisco, Av. Nodo Servidor Público #165 Col. Anexa al Club de Golf, Las Lomas, Zapopan 45136, Jalisco, Mexico
{angel.tepozteco,francisco.ibarra,cain.perez}@ciateq.mx

[2] Secretaría de Ciencia, Humanidades, Tecnología e Innovación (Secihti) – CIATEQ A.C. – Jalisco, Av. Nodo Servidor Público #165, Col. Anexa al Club de Golf, Las Lomas, Zapopan 45136, Jalisco, Mexico
jose.avila@ciateq.mx

Abstract. Organic electrochemical transistors (OECTs) based on the organic semiconductor poly(3,4-ethylenedioxythiophene): poly (styrene sulfonate) (PEDOT: PSS) are widely used as ion-to-electron transducers in bioelectronics and chemical sensing. However, optimizing OECT geometry for simultaneously high transconductance and high ionic sensitivity remains experimentally non-cost effective due to the large design space. Here, we propose a second order polynomial regression model to predict the maximum transconductance g_m^{max} and ionic sensitivity (S) of OECTs as functions of experimental parameters such as channel dimensions. Here, optimization was carried out using a Python-programmed response surface methodology (RSM) which is a group of statistical techniques that model the relation between the experimental independent variables and the response, alone and in combination. A dataset comprising 68 devices with varying channel dimensions was fabricated and measured for transconductance characterization, while 65 devices were used for ionic sensitivity analysis. We extracted (g_m^{max}) from each transfer curve and calculated potassium sensitivity for each OECT measured for potassium chloride (KCl) concentrations from 10^{-4} to 1 M. The resulting regression model accounts for the dependence of OECTs on their geometry, which allows them to optimize the design process without increasing the experimental procedure and cost. This Machine Learning model enables geometry-based optimization, providing quantitative design guidelines to maximize transconductance while maintaining high ionic sensitivity.

Keywords: Machine Learning · Response Surface Methodology · Organic Electrochemical Transistors · Sensor

H. Ponce et al. (Eds.): COMIA 2026, CCIS 3095, pp. 384–397, 2027.
https://doi.org/10.1007/978-3-032-37438-7_30

1 Introduction

Organic electrochemical transistors (OECTs) based on the organic semiconductor poly(3,4-ethylenedioxythiophene): poly (styrene sulfonate) (PEDOT: PSS) have emerged as key building blocks for sensing and bioelectronic interfaces due to their mixed ionic–electronic conduction and volumetric charge storage [1, 2]. In most OECT applications, the performance is primarily determined by the maximum transconductance g_m^{max}, which quantifies the gate-controlled modulation of channel current, and by the sensitivity of the electrical response to electrolyte concentration for sensing applications [3]. Sensors based on OECTs offer advantages over other types of sensors, such as biocompatibility, amplified electrical signals, and facile deposition [4].

Recently, OECTs have been used in the health field to measure ionic concentrations in human fluids and detect ionic imbalances [5]. For example, a sodium (Na^+) imbalance in the human body leads to several medical conditions, such as insomnia, arrhythmia, or high blood pressure [6]. In the same manner, abnormal concentrations of potassium (K^+) and chloride (Cl^-) in human fluids are associated with severe morbidity and mortality. The geometric parameters of the semiconductor channel active area, composed of a PEDOT: PSS-based ink, are the channel width (W), thickness (d), and length (L). These dimensions were varied to obtain a wide range of the volumetric geometric factor (Wd/L), which is a critical parameter for transconductance scaling. OECTs work by applying a gate (G) voltage (V_G) that tunes the resistance of a semiconductor layer deposited in a channel defined by the length formed between the source (S) and drain (D) electrodes. Then, the drain current (I_D) is measured versus the drain voltage (V_D) to monitor the conductivity changes. The figure of merit of OECTs is the transconductance g_m and is calculated as the first derivative of the transfer (I_D vs V_G) curves: ($g_m = \frac{\partial i_D}{\partial V_G}$) and is the parameter that leads to signal amplification [7]. While OECT transconductance is known to linearly increase with the volumetric geometric factor for low Wd/L values, in practice, electrolyte concentration and device geometry are coupled through contact parasitic capacitances, contact resistances, or inefficient ion penetration in the PEDOT: PSS film, leading to a sublinear dependence of g_m with Wd/L, therefore reducing the sensing efficiency [8].

OECTs based on PEDOT: PSS films, which are a p-type semiconductor, work in depletion mode, where the drain current decays when a positive gate voltage is applied due to a reduction reaction between the PEDOT: PSS film and the cations in the electrolyte, causing a decrease in the hole density in a process called dedoping [9]. The OECT recovers its current level by applying an opposite bias voltage or by relaxation when the external source is removed. OECTs have been used as selective ionic sensors [10, 11] and show ionic selectivity (S) by using selective membranes deposited between the semiconductor and the electrolyte [10], but are also used as sensors without those membranes, discriminating against ions by analytical methods [11]. Importantly, S may also exhibit a non-linear dependence on geometry and thickness, since several experimental studies indicate that for thicker PEDOT: PSS channels [12] the entire film volume is not necessarily electrochemically addressed during gating. In such cases, the effective doped volume can be significantly lower than the nominal volume, leading to reduced volumetric utilization, and, in turn, sub-linear scaling of sensitivity with thickness. Although there are several works in the literature that corroborate the linear relation between transconductance and the geometrical factor Wd/L, following the well-known OECT model by Bernards and Malliaras [13, 14] there are no statistical works that include the geometrical values when g_m behaves in a sublinear regime.

In recent years, machine learning (ML) has delivered remarkable results across multiple fields, such as medical research [15], materials science [16], and electronic devices [17], driven by the availability of massive datasets and the development of powerful neural network architectures. Regression analysis is a supervised ML technique that implies finding a correlation between parameters (inputs) and predicting a continuous output [18]. Polynomial regression is suitable for modeling nonlinear or curvilinear interactions between predictors and the response variable. This type of regression analysis gives an equation which aims to represent the nonlinear dependence of the output variable on one or more input variables. One of the most used mathematical and statistical techniques for modeling and optimizing experimental independent variables is the response surface methodology (RSM), whose objective is to find the relationship between the variables and the experimental response and to obtain optimal values to optimize the independent variables [19]. Recently, Liao et al. performed ML algorithms to enhance fiber-based organic transistors for metabolite sensing and their clinical applications for several metabolite ions [20]. Nevertheless, very few works of ML prediction for OECT optimization have been carried out [20, 21], and there is no work focused on OECT's g_m and sensitivity optimization for ionic sensing by varying the geometric parameters.

While more complex machine learning approaches, such as neural networks or ensemble methods, have been applied to materials and device optimization [20, 21], they typically require significantly larger datasets, offer limited physical interpretability, and are prone to overfitting when trained on small experimental datasets. In contrast, second-order polynomial regression based on RSM provides a transparent, physically interpretable model in which each coefficient corresponds directly to an independent geometric effect or interaction. This interpretability is particularly valuable in OECT design, where the physical meaning of the W^2, d^2, and L^2 curvature terms can be directly linked to saturation effects and sub-linear ionic transport, providing design guidelines beyond simple numerical optimization.

For this, studies of the OECT's geometry must be conducted to optimize device sensing. Therefore, predictive machine learning algorithms based on physically interpretable data are valuable tools for accelerating OECT design and optimization, reducing fabrication time and experimental cost. To address these non-ideal behaviors, we implement a second-order polynomial regression model. This RSM approach enables optimization of OECT dimensions by identifying the precise combinations of W, d, and L that maximize performance without the need for exhaustive and costly experimental iterations. While linear regression assumes a constant proportionality between the geometric factor and transconductance or sensitivity, it often fails to account for the nonlinearities inherent in organic electrochemical transistors, such as contact resistance or limited ion volume accessibility due to a thick semiconductor channel. To capture these effects, we incorporate a quadratic (second order) polynomial regression. A second-order polynomial model extends the linear form by introducing a squared term, following the general equation:

$$y = \beta_0 + \beta_1 x + \beta_2 x^2 + \epsilon \quad (1)$$

In this framework, b_1 represents the linear scaling, while the quadratic coefficient b_2 quantifies the curvature of the response, b_0 is the independent coefficient, and ϵ is a residual term. Physically, a non-zero b_2 accounts for saturation effects. Unlike a linear trend, a quadratic function allows for the identification of a local maximum. This enables selecting optimal channel dimensions that maximize transconductance and sensitivity while minimizing size and experimental cost. In this work, as multiple independent variables are considered in the model, we calculate the stationary point in the second-order equation, where the derivatives with respect to the independent variables are equal to zero, following RSM [19]. In resume, we develop a RSM polynomial regression model for OECT performance prediction across geometry and demonstrate its use for device optimization by implementing an ML model trained on OECT experimental results by varying the geometric factor Wd/L using machine learning algorithms in Python. We obtained, by stationary point calculations, optimal W, d, and L to optimize the g_m and S values independently, and in combination using a desirability calculation in each geometric range.

2 Experimental Methods

2.1 Materials

Conductive silver nanoink (UTDAgTE) and conductive gold nanoink (UTDAuTE) were purchased from UT Dots Inc. The PEDOT: PSS solution (PH 1000) was purchased from Heraeus Clevios. Ethylene glycol (EG), the cross-linking agent (3-glycidyloxypropyl) trimethoxy silane (GOPS), and the surfactant dodecyl benzenesulfonic acid (DBSA) were purchased from Sigma-Aldrich/Merck. Ag/AgCl ink was purchased from Allum Corp.

2.2 Device Fabrication and Geometry

Figure 1 shows the scheme and a photograph of the OECT deposited by aerosol jet printing (AJP): gold was used as source (S) and drain (D) contacts, while silver was used as conductive pathways. The Ag/AgCl gate (G) electrode was deposited by screen printing. PEDOT:PSS ink was formulated as follows: the PEDOT:PSS nanoink was prepared by mixing 2 mL of PEDOT:PSS (Heraeus Clevios PH 1000) with 200 μL of ethylene glycol (EG) to enhance electrical conductivity, 20 μL of the cross-linking agent (3-glycidyloxypropyl)trimethoxy silane (GOPS) to improve film stability and prevent dissolution or delamination from the substrate, and 5 μL of DBSA as a surfactant to facilitate film formation and processing. This semiconductor ink was deposited by AJP, with varying printing velocity to tune the film thickness (from 0.4 to 2.0 mm/s). The volumetric geometric factor was defined as Wd/L, and the parameters g_m and S were extracted to optimize performance using machine-learning techniques. For ink deposition, an Aerosol Jet Printer Optomec 300 was used. While Ag and Au depositions were kept constant, the PEDOT: PSS ink printing was tuned by varying the deposition velocity, where a higher velocity corresponds to a thinner layer.

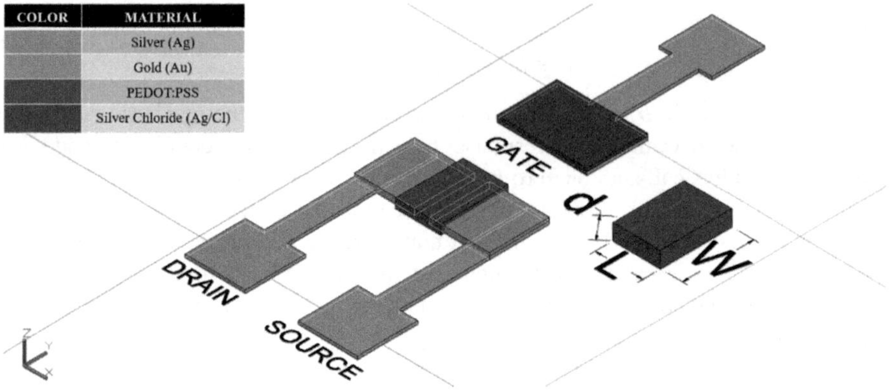

Fig. 1. Scheme of the aerosol-jet printed OECT fabricated in this work.

2.3 Experimental Setup

A total of 68 OECT devices were fabricated with systematically varied channel dimensions to cover a broad design space, of which 65 yielded valid ionic-sensitivity data. The channel parameters (W, d, and L) were varied within the following experimental ranges: 500 μm < W < 3000 μm, 200 μm < L < 1000 μm, and 4.8 μm < d < 23.0 μm. Output curves (I_D vs V_D) were extracted by varying V_G from 0 to 1 V in steps of 0.2 V. Transfer curves (I_D–V_G) were recorded at a fixed drain voltage V_D (−0.3 V) while sweeping V_G from 0 to 1 V, with steps of 10 mV. For sensing applications, measurements in a range of KCl solutions in deionized water from 10^{-4} to 1 M. Constant V_D and V_G potentials (−0.3 V and 0.3 V, respectively) were applied, and I_D was recorded once it reached a stable value. The relationship between I_D and the logarithm of the ionic concentration was plotted, where the sensitivity was defined as the slope of the linear fitting in A/dec.

To minimize process-induced variability, ink formulation was kept constant throughout all fabrication runs. Aerosol jet printing parameters (including atomization flow rate, sheath gas flow, and substrate temperature) were fixed for silver deposition. PEDOT: PSS film thickness was controlled exclusively through the deposition velocity, keeping the other parameters constant Residual variability attributable to ink rheology changes between batches and environmental humidity is acknowledged as a source of experimental scatter in the dataset.

2.4 Second – Order Polynomial Regression Model Using Response Surface Methodology

The model parameters were estimated by solving a least-squares optimization problem, that minimizes the sum of squared residuals between the experimental observations and the model predictions. Mathematically, this corresponds to minimizing $\sum_{i=1}^{n}(y_i - \hat{y}_i)^2$, where y_i represents the measured response and $\hat{y}_i$ the predicted value from the polynomial model. The model used in the RSM is a second – order regression model and generally can be expressed as:

$$y = \beta_0 + \sum_{i=1}^{k} \beta_i X_i + \sum_{i=1}^{k} \beta_{ii} X_i^2 + \sum_{i<j} \beta_{ij} X_i X_j + \varepsilon \tag{2}$$

Particularly, in this calculation, we model the g_m, and S output parameters as a function of three channel geometrical inputs (W, d, and L) as follows:

$$g_m = \beta_0 + \beta_1 W + \beta_2 d + \beta_3 L + \beta_{11} W^2 + \beta_{22} d^2 + \beta_{33} L^2 + \beta_{12} Wd + \beta_{13} WL + \beta_{23} dL \tag{3}$$

$$S = \propto_0 + \propto_1 W + \propto_2 d + \propto_3 L + \propto_{11} W^2 + \propto_{22} d^2 + \propto_{33} L^2 + \propto_{12} Wd + \propto_{13} WL + \propto_{23} dL \tag{4}$$

To determine the optimal parameter set (i.e., the values at which both g_m and S reach their maximum within the experimental range), the stationary point was obtained by solving the corresponding system of equations, where the partial derivatives with respect to the input parameters are set equal to zero:

$$\frac{\partial g_m}{\partial W} = \frac{\partial g_m}{\partial d} = \frac{\partial g_m}{\partial L} = 0, \begin{bmatrix} 2\beta_{11} & \beta_{12} & \beta_{13} \\ \beta_{12} & 2\beta_{22} & \beta_{23} \\ \beta_{13} & \beta_{23} & 2\beta_{33} \end{bmatrix} \begin{bmatrix} W \\ d \\ L \end{bmatrix} = \begin{bmatrix} -\beta_1 \\ -\beta_2 \\ -\beta_3 \end{bmatrix} \tag{5}$$

$$\frac{\partial S}{\partial W} = \frac{\partial S}{\partial d} = \frac{\partial S}{\partial L} = 0, \begin{bmatrix} 2\alpha_{11} & \alpha_{12} & \alpha_{13} \\ \alpha_{12} & 2\alpha_{22} & \alpha_{23} \\ \alpha_{13} & \alpha_{23} & 2\alpha_{33} \end{bmatrix} \begin{bmatrix} W \\ d \\ L \end{bmatrix} = \begin{bmatrix} -\alpha_1 \\ -\alpha_2 \\ -\alpha_3 \end{bmatrix} \tag{6}$$

Finally, a multi-objective optimization was performed using a desirability function approach to determine the geometrical parameters that simultaneously maximize both responses. Individual desirability functions were defined for each response, g_m and S, as:

$$d_{g_m} = \begin{cases} 0 & g_m < L_{g_m} \\ \left(\frac{g_m - L_{g_m}}{T_{g_m} - L_{g_m}}\right)^s & L_{g_m} \le g_m \le T_{g_m} \\ 1 & g_m > T_{g_m} \end{cases} \tag{7}$$

$$d_S = \begin{cases} 0 & S < L_S \\ \left(\frac{S - L_S}{T_S - L_S}\right)^t & L_S \le S \le T_S \\ 1 & S > T_S \end{cases} \tag{8}$$

where L and T represent the lower and target values of each response, respectively, and s and t are weighting factors that control the importance of each objective, which in this work were set to 1. The overall desirability was then calculated as the square root of the individual desirabilities, by choosing equal weights for the two parameters:

$$D = \sqrt{d_{gm} d_S} \tag{9}$$

The optimal geometrical parameters parameters were obtained by maximizing of the global desirability function D, calculated for all the OECT. The final optimal parameters were obtained by bounded numerical optimization using only the experimentally investigated ranges.

2.5 Complementary Metrics

To ensure the reliability of the predictive model, we quantified the predictive accuracy using the Mean Absolute Error (MAE) and the Root Mean Square Error (RMSE). These metrics are mathematically defined as:

$$\text{MAE} = \frac{1}{n}\sum_{i=1}^{n} | y_{i,\text{pred}} - y_{i,\text{exp}} | \tag{10}$$

$$\text{RMSE} = \sqrt{\frac{1}{n}\sum_{i=1}^{n} \left(y_{i,\text{pred}} - y_{i,\text{exp}}\right)^2} \tag{11}$$

where n is the number of samples in the test set, y_{pred} is the value calculated by the model, and y_{exp} is the experimental measurement. By applying these same equations to both validation techniques, the model stability can be objectively determined.

3 Results and Discussion

3.1 Electrical Characterization of OECTs

Figure 2 illustrates the electrical characterization of the OECT device. In Fig. 2a, the transfer curve (I_D vs. V_G) is shown for a gate voltage sweep from 0 to 1 V at a fixed drain voltage of $V_D = -0.3$ V. A pronounced decrease in channel conductivity is observed, eventually approaching zero, attributed to the dedoping induced by the interaction between potassium cations and the $PEDOT^+$ species within the channel. This process reduces the density of charge carriers, thereby suppressing the drain current. In Fig. 2b, the first derivative of the transfer curve is presented, from which the maximum value of the transconductance g_m of the OECT is extracted (called g_m^{max}).

The maximum value of g_m is used as the representative metric for device performance and serves as the target variable in the subsequent training using response surface methodology (RSM). Figure 3a presents the temporal response of the OECT, showing the drain current (I_D) as a function of time for five different KCl concentrations applied sequentially in decreasing order, from 1 M to 10^{-4} M. A sharp decrease in current is observed when a high-concentration KCl droplet is introduced onto the PEDOT: PSS channel, transitioning from a baseline current (high conductivity) to a low-conductivity state due to dedoping. The magnitude of the current change exhibits a logarithmic dependence on the KCl concentration, as summarized in Fig. 3b. Accordingly, sensitivity (S) is defined as the slope of the calibration curve obtained from plotting I_D vs. log(C), and is used as a second key output parameter for device performance evaluation.

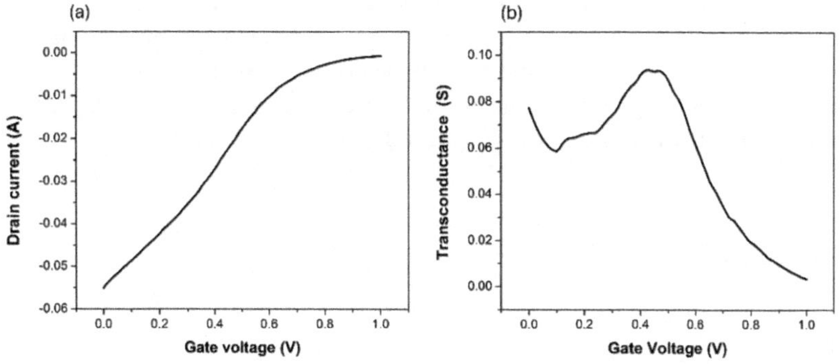

Fig. 2. (a) Transfer function and (b) transconductance of the OECT with the next channel dimensions; W = 500 μm, d = 6.7 μm, and L = 1000 μm

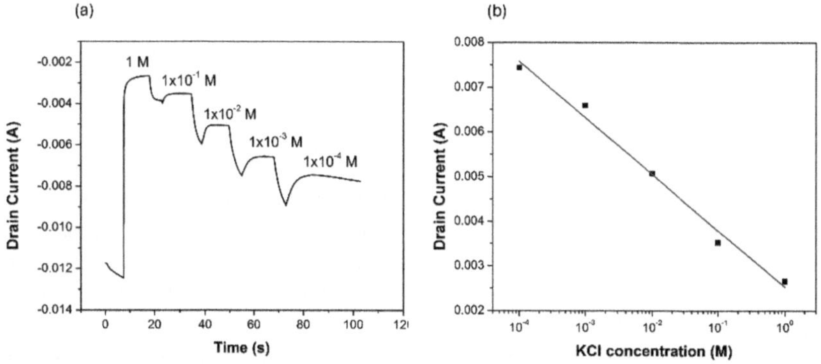

Fig. 3. KCl sensing of OECTs measured by (a) drain current vs. time and (b) semi-logarithmic drain current vs. concentration plot with five different KCl concentrations from 10^{-4} to 1 M of the OECT with the next channel dimensions: W = 500 μm, d = 6.7 μm, and L = 1000 μm.

3.2 Second Order Regression Prediction

A second-order regression model was trained using a Python-based implementation of RSM, with g_m and S as output variables, and the input variables defined according to Eqs. (3) and (4). This approach yielded a set of 10 fitting coefficients for each response, namely β for g_m and α for S. Figure 4a shows the comparison between the experimental data (black diamonds) and the predicted values (red circles), along with the fitted second-order polynomial curve $ax^2 + bx + c$ (this curve is included to provide a broader perspective on the overall quadratic trend exhibited by the experimental data), resulting in a coefficient of determination (R^2) of 0.6568. Similarly, Fig. 4b presents the corresponding results for S, where the regression model based on the α coefficients provides better agreement with the experimental data, achieving $R^2 = 0.6799$.

Although the coefficient of determination obtained for g_m ($R^2 = 0.6568$) may appear moderate, it is considered acceptable given the complexity and variability of OECT systems and the several experimental parameters involved in aerosol jet printing, such as ink viscosity and nitrogen pressures, among others. The device response could arise from coupled ionic–electronic transport mechanisms and sublinear behavior resulting from non-optimal coupling between the electrolyte and the thick PEDOT: PSS channel, which may not be fully captured by a second-order polynomial model.

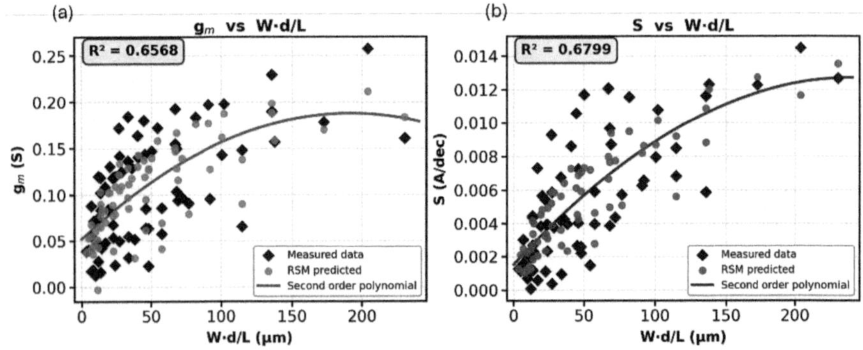

Fig. 4. (a) Experimental, predicted values and second order fitting for (a) g_m and (b) S values of the datasets of 68/65 OECTs

This model preserves the correct physical trends, which predict a linear relationship between g_m and the geometric factor, for lower Wd/L values; for example, Rivnay et al. [13] found a linear relationship for Wd/L values up to 5 µm in a semi-logarithmic plot, meanwhile in this work we obtain the quadratic-like trend for Wd/L values from 3.35 µm to 230 µm.

Figures 5a and b are the parity plots of experimental and predicted values for g_m and S, respectively. In both cases, the data points are distributed around the ideal parity line, indicating a reasonable agreement between the model predictions and the experimental measurements. For g_m (Fig. 5a), a moderate dispersion is observed, consistent with the obtained R^2, which reflects the inherent variability and complexity of the device response. In contrast, the predictions for S (Fig. 5b) exhibit tighter clustering around the parity line, indicating improved predictive accuracy and a stronger correlation with the input variables.

The shaded regions shown in both figures represent error bounds (or confidence intervals) around the ideal fit, providing a visual indication of the deviation between

Table 1. Optimized g_m and S values and a multi-objective optimization using desirability calculation.

Variable	Range (µm)	Optimal g_m (µm)	Optimal S (µm)	Optimal g_m and S (µm)
W	*500–3000*	*3000.0*	*3000.0*	*3000*
d	*4.8–23.0*	*16.93*	*23*	*20.44*
L	*200–1000*	*200.0*	*200.0*	*200*

predicted and experimental values. For g_m 68 OECTs, the model yielded a R^2 of 0.657, with a root mean square error (RMSE) of 34.13 mS and a mean absolute error (MAE) of 27.34 mS. For sensitivity (S) 65 OECTs, the model achieved an R^2 of 0.680, with RMSE

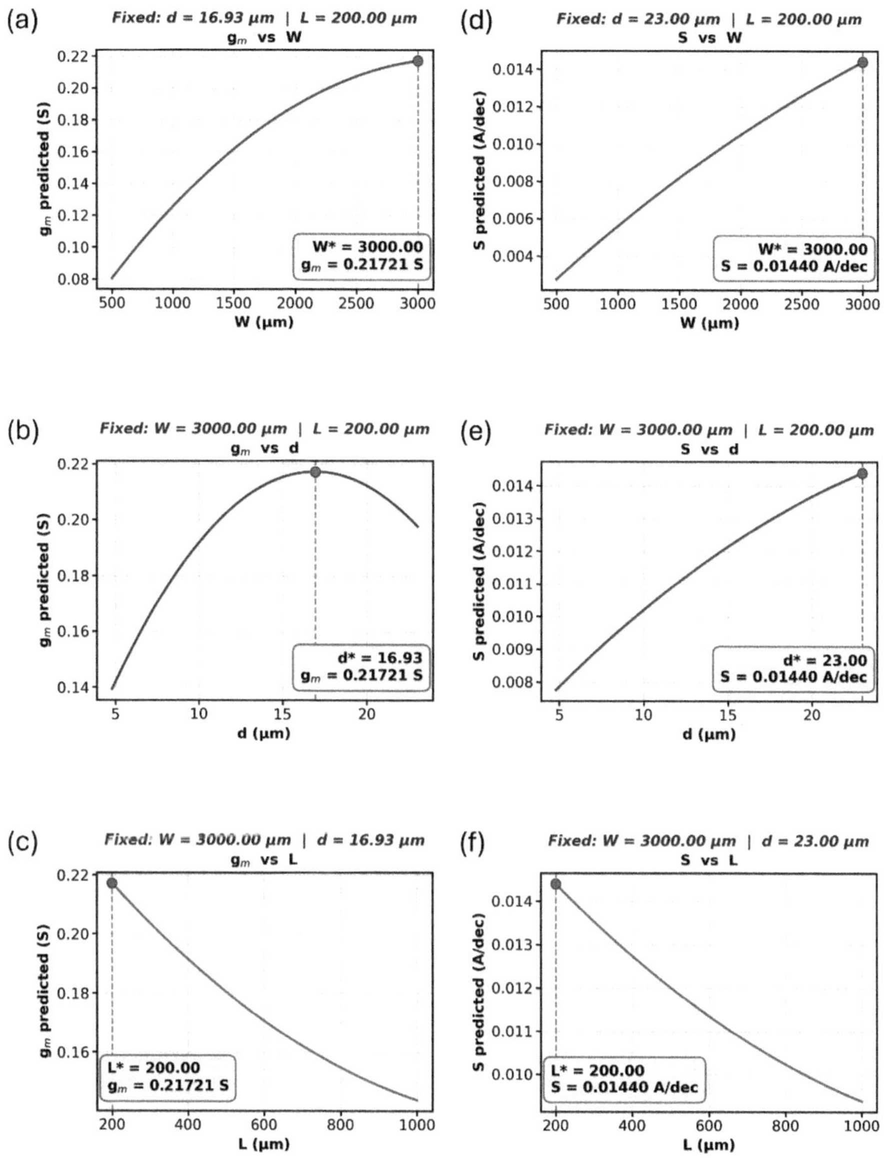

Fig. 6. Univariate sensitivity analysis for predicted g_m: (a) vs. W (d = 16.93 μm, L = 200 μm); (b) vs. d (W = 3000 μm, L = 200 μm); (c) vs. L (W = 3000 μm, d = 16.93 μm); and for predicted S: (d) vs. W (d = 23.00 μm, L = 200 μm); (e) vs. d (W = 3000 μm, L = 200 μm); (f) vs. L (W = 3000 μm, d = 23.00 μm). Orange marker: optimal point at g_m = 0.217 S and S = 0.01440 A/dec.

= 2.16 mA/dec and MAE = 1.65 mA/dec. Although the R^2 values are moderate, they are consistent with the complexity of OECT systems fabricated by aerosol jet printing, where process variability introduces scatter that cannot be captured by a second-order geometric model alone. Then, this model preserves physically correct trends across the full extended dataset.

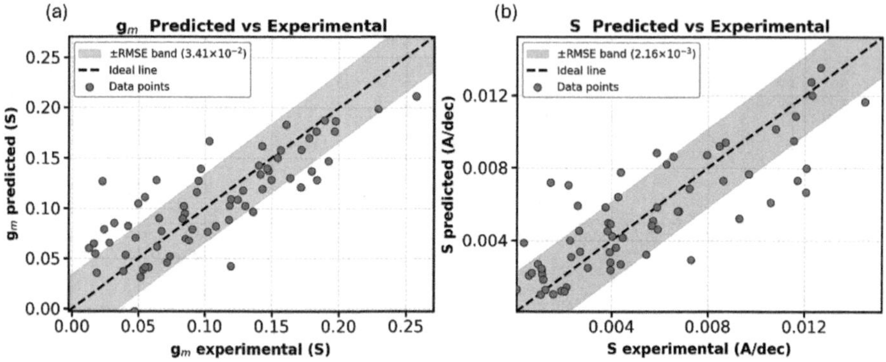

Fig. 5. Parity plots (predicted vs. experimental) of g_m (a) and S (b) values as a function of geometric values: W·d /L for 68/65 OECTs

The moderate R^2 observed for both responses can be attributed to several compounding factors inherent to the system. First, the OECT transconductance depends not only on the nominal channel geometry but also on the effective ionic penetration depth within the PEDOT: PSS film, which varies with film morphology, crystallinity, and electrolyte-film interfacial properties — none of which are captured by W, d, and L alone. Second, contact resistance contributions at the source-drain electrodes introduce a geometry-independent offset that flattens the expected linear scaling with Wd/L for smaller devices. Third, the extended thickness range (up to d = 23.0 μm) introduces a regime where volumetric electrochemical addressing becomes incomplete, resulting in a sub-linear dependence of g_m on d that requires higher-order terms to be fully described. The second-order RSM model successfully captures the dominant geometric trends and provides actionable optimization guidelines within the explored design space.

3.3 Optimization

The optimized geometrical parameters (W, d, and L) were first determined by solving the system of equations obtained from setting the first-order partial derivatives of the response functions equal to zero, thereby identifying the stationary points of the modeled surfaces. Although the individual 2D projections do not directly represent saddle points, the opposite trends observed among the geometrical parameters suggest the presence of competing effects within the multidimensional response space.

A multi-objective optimization was carried out using the desirability function approach, and the optimal configuration was selected with the highest global desirability value D = 0.877, representing the best trade-off between maximizing both g_m and S simultaneously. The optimized parameters within the experimental geometric ranges are shown in Table 1.

To further characterize the geometry dependence of both responses, a univariate sensitivity analysis was performed by varying each parameter independently while fixing the remaining two at their optimal values, as shown in Fig. 6. For g_m, W exhibits a sub-linear monotonic increase (Fig. 6a), consistent with the expected Wd/L scaling of OECT transconductance. In contrast, d shows a non-monotonic response with a well-defined interior maximum at $d^* = 16.93$ μm (Fig. 6b), beyond which g_m decreases due to the onset of incomplete volumetric electrochemical addressing of the PEDOT: PSS film at larger thicknesses, a non-linear regime previously reported for thick organic semiconductor channels [12]. L presents a strictly decreasing dependence with the optimum at the lower boundary $L^* = 200$ μm (Fig. 6c), in agreement with the inverse proportionality of transconductance to channel length.

For S, all three parameters exhibit monotonic behavior throughout the explored range. S increases with W (Fig. 6)d and e, and decreases with L (Fig. 6f), with the thickness optimum located at the upper domain boundary $d^* = 23.00$ μm, indicating that thicker PEDOT: PSS films enhance volumetric charge storage and ionic sensitivity without reaching saturation within the fabricated range. This divergence in optimal thickness between g_m ($d^* = 16.93$ μm, interior) and S ($d^* = 23.00$ μm, boundary) reflects the distinct physical mechanisms governing each response and highlights the importance of the multi-objective desirability approach for simultaneous optimization.

4 Conclusions

In this work, we fabricated 68 OECTs by varying the channel dimensions (W, d, and L) —(spanning 500–3000 μm in W, 200–1000 μm in L, and 4.8–23.0 μm in d) to study the dependence of transconductance and ionic sensitivity on channel geometry, thereby optimizing OECT performance over an extended experimental range. A second-order regression model based on response surface methodology (RSM) was successfully implemented to capture the system's main trends and enable predictive analysis. Despite the device's inherent variability, the model showed a satisfactory agreement with experimental data.

Experimental validation of the predicted optimal geometry (particularly the multi-objective optimum at $W = 3000$ μm, $d \approx 20.44$ μm, and $L = 200$ μm) represents a natural next step to confirm the model predictions. These results provide a new perspective on the operation of OECTs at Wd/L regimes beyond those commonly explored in the literature. In particular, the transconductance initially increased with PEDOT: PSS thickness, and eventually decreased after reaching a maximum. This behavior is attributed to incomplete volumetric dedoping in very thick channels, where only the PEDOT: PSS region close to the electrolyte is effectively modulated, while a significant fraction of the bulk remains electrically active but ionically inaccessible. As the non-dedoped volume becomes larger than the effectively gated region, the modulation efficiency of the OECT decreases, limiting the expected transconductance scaling at high Wd/L values. Interestingly, this behavior was not directly reflected in the ionic sensitivity, despite a sublinear relationship between S and the geometry being observed, which remained comparatively stable over the explored geometric range; therefore, additional studies are required to fully understand ion transport and active-volume effects in thick PEDOT: PSS OECTs.

Acknowledgments. JAAN and AETM thanks to Secretaría de Ciencias, Humanidades y Tecnologías (Secihti) for the Investigadores por México Project No.7253 and a Master Degree scholarship No. 2013250, respectively. We thank to Dr. Mariamne Dehonor (CIATEQ) for thickness measurements.

Disclosure of Interests The authors have no competing interests to declare that are relevant to the content of this article.

References

1. Saleh, A., Koklu, A., Uguz, I., Pappa, A.M., Inal, S.: Bioelectronic interfaces of organic electrochemical transistors. Nat. Rev. Bioeng. **2**(7), 559–574 (2024)
2. Lu, D., Chen, H.: Solid-state organic electrochemical transistors (OECTs) based on gel electrolytes for biosensors and bioelectronics. J. Mater. Chem. A. **13**(1), 136–157 (2025)
3. Colucci, R., Feitosa, B.D.A., Faria, G.C.: Impact of ionic species on the performance of Pedot: PSS-based organic electrochemical transistors. Adv. Electron. Mater. **10**(2), 2300235 (2024)
4. Gentile, F., Vurro, F., Picelli, F., Bettelli, M., Zappettini, A., Coppedè, N.: A mathematical model of OECTs with variable internal geometry. Sensors Actuators A Phys. **304**, 111894 (2020)
5. Ceccardi, F., et al.: A point-of-care, label-free OECT sensor for uric acid detection: validation in human saliva. Electrochim. Acta. **519**, 145834 (2025)
6. Banerjee, A., Azad, P., Akhtar, M.J.: Detection of sodium ion imbalance in human body fluids using an improved RF sensor. IEEE J. Electromagn., RF Microwaves Med. Biol. **7**(1), 73–81 (2022)
7. Khodagholy, D., et al.: High transconductance organic electrochemical transistors. Nat. Commun. **4**(1), 2133 (2013)
8. Paudel, P.R., Kaphle, V., Dahal, D., Radha Krishnan, R.K., Lüssem, B.: Tuning the transconductance of organic electrochemical transistors. Adv. Funct. Mater. **31**(3), 2004939 (2021)
9. Avila-Niño, J.A., Araujo, E., Gonzalez, F.: Study of the doping of PEDOT: PSS films and the reversibility in organic electrochemical transistors. Synth. Met. **299**, 117465 (2023)
10. Li, C., Li, Z., Nguyen, T.D.D., Ingebrandt, S., Vu, X.T.: Real-time and multiplexed detection of sodium and potassium ions using PEDOT: PSS OECT microarrays integrated with ion-selective membranes. Electrochim. Acta. **507**, 145111 (2024)
11. Lin, P., Yan, F., Chan, H.L.: Ion-sensitive properties of organic electrochemical transistors. ACS Appl. Mater. Interfaces. **2**(6), 1637–1641 (2010)
12. Ait Yazza, A., Blondeau, P., Andrade, F.J.: Simple approach for building high transconductance paper-based organic electrochemical transistor (OECT) for chemical sensing. ACS Appl. Electron. Mater. **3**(4), 1886–1895 (2021)
13. Rivnay, J., et al.: High-performance transistors for bioelectronics through tuning of channel thickness. Sci. Adv **1**(4), e1400251 (2015)
14. Bernards, D.A., Malliaras, G.G.: Steady-state and transient behavior of organic electrochemical transistors. Adv. Funct. Mater. **17**(17), 3538–3544 (2007)
15. Schober, P., Vetter, T.R.: Linear regression in medical research. Anesth. Analg. **132**(1), 108–109 (2021)

16. Pilania, G.: Machine learning in materials science: from explainable predictions to autonomous design. Comput. Mater. Sci. **193**, 110360 (2021)
17. Gao, L., Liu, C., Xiao, Y., Qi, C., Huo, M.: Remaining useful life prediction of power electronic devices with physics-informed deep learning and sparse data. IEEE Trans. Power Electron. **40**, 16068 (2025)
18. Patil, S., Patil, S.: Linear with polynomial regression: overview. Int. J. Appl. Res. **7**(8), 273–275 (2021)
19. Baş, D., Boyacı, İ.H.: Modeling and optimization I: usability of response surface methodology. J. Food Eng. **78**(3), 836–845 (2007)
20. Liao, C., Wu, H., Occhipinti, L.G.: Machine learning-assisted 3D flexible organic transistor for high-accuracy metabolites analysis and other clinical applications. Chemosensors **12**(9), 174 (2024)
21. Sun, Y., et al.: MOF-MoS2 nanosheets doped PEDOT: PSS for organic electrochemical transistors in enhanced glucose sensing and machine learning-based concentration prediction. Mater. Futures. **4**(2), 025302 (2025)

Towards an Agent-Based Model of Opinion Dynamics for Political Elections

Ulises Jiménez-Guerrero[1(✉)], Alejandro Guerra-Hernández[1], Nicandro Cruz-Ramírez[1], and Wulfrano Arturo Luna-Ramírez[2]

[1] Instituto de Investigaciones en Inteligencia Artificial - IIIA, Campus Sur, Calle Paseo Lote II, Sección Segunda No 112, Col. Nuevo Xalapa, Xalapa, Veracruz, Mexico
zS24019399@estudiantes.uv.mx

[2] Universidad Autónoma Metropolitana Unidad Cuajimalpa, Avenida Vasco de Quiroga 4871, Col. Santa Fe Cuajimalpa, Alcaldía Cuajimalpa de Morelos, 05348 Ciudad de México, Mexico

Abstract. This paper reports the results of an Agent-Based Model of Opinion Dynamics based on three different types of social influence adopted from literature, namely: positive influence, negative influence and bounded confidence. The purpose of this work is to validate the capability of the proposed model to replicate the opinion change observed in real political elections. For this, the Mexican presidential election of 2024 was chosen as a case study. The model was adjusted to reflect the information contained in a selected opinion poll and the parameter space was explored to find an optimal configuration for replicating the empirical opinion change. The three types of influence were capable of replicating the observed opinion change, with bounded confidence having a significantly better performance. Future work will explore if secondary patterns, such as vote share by age and gender, are replicable by the model or by redefining its social network based on the opinion poll.

Keywords: Agent-Based Modeling · Opinion Dynamics · Political Elections

1 Introduction

Following the technological developments made in recent decades, there has been an interest in the use of Agent-Based Models (ABM) for the study of social phenomena, a perspective that has been named Generative Social Science [1]. One of the topics studied in this perspective is social influence, i.e., how people change their opinions, beliefs, attitudes or behavior after interacting with each other. Social influence has been studied under the umbrella term Opinion Dynamics [3], standing for the study of change in opinion in a population after a determinate amount of time. There are three well-known patterns present in the area: consensus, where the population reaches an agreement and share the same opinion;

H. Ponce et al. (Eds.): COMIA 2026, CCIS 3095, pp. 398–410, 2027.
https://doi.org/10.1007/978-3-032-37438-7_31

polarization, with two main groups holding opposite opinions; and clusterization, in which many different groups of people with similar opinions emerge, and they're sustained over time.

Given the different opinion dynamics models available in the literature, and the many different assumptions made in each one, a classification was proposed by Flache et al. [3] and followed by Smaldino [9]. The models are categorized by the type of influence simulated, i.e., how two agents change their opinion after an interaction. Each one has different justifications based on psychological and sociological phenomena, and they're as follows:

Positive influence. Based on social theories centered in social learning and peer pressure, these models search for agreement between two individuals after an interaction. As such, both modify their opinions so they're closer to each other, usually leading to consensus.

Bounded confidence. Based on the homophily principle [4] in which people tend to interact with others that are similar to them, it supposes that two individuals try to reach an agreement only if they hold opinions similar enough to each other. Otherwise, they don't have any influence over each other's opinion. Under certain configurations, these models can generate the three main patterns of opinion dynamics aforementioned.

Negative influence. Based on the assumption of differentiation, that is, two individuals may seek to change their opinion to become more dissimilar to each other after an interaction. Models under this category have a tendency towards polarization, sometimes leading to clusterization. It has been noted that the evidence for this effect in real interactions is mixed [3].

Generative Social Science is founded on the capacity of ABMs to model heterogeneous agents and its bottom-up approach, where patterns emerge from the interaction of different agents with simple behavioral rules [8]. This is particularly useful for the study of elections, where every voter can be simulated with particular characteristics and behavior based on observed data, allowing to analyze the opinion change given by social influence and electoral campaigns, which can be simulated by an opinion dynamics system. This proposal is supported by the availability of data, with many countries having freely available data related to opinion polls and predictions for political elections [2].

In this paper, we explore the use of opinion dynamics through ABM for simulating and studying real-life political elections, selecting as a case study the Mexican presidential election of 2024. For this, a publicly available opinion poll was selected, following nationwide voting intention during the electoral campaign at three time steps. We study the capability of the introduced influence models to replicate the opinion change observed in this election, given the temporal constraints of the political campaign. Capability is evaluated in terms of a primary pattern, i.e., how close are the simulated voting intentions and electoral results to the real ones. This was evaluated measuring the Root Mean Squared Error (RMSE) between simulated opinions and empirical data.

The organization of the paper is as follows: Sect. 2 introduces previous work on the use of opinion dynamics models for the study of empirical electoral phenomena. Section 3 describes the opinion polls selected for the case study and the model designed to replicate the observed opinion change, as well as the measurements used to validate the results. In Sect. 4, the experiments performed on the models are described with results shown in Sect. 5. Section 6 discusses the apparent capability of the model for describing the observed opinion evolution, while Sect. 7 indicates future validations to develop a more robust and realistic ABM.

2 Related Work

An in-depth study of the evolution of the simulation of political attitudes, developed by Voinea [11], notes an evolution from cognitive focused models towards the simulation of social influence, with opinion dynamics belonging to the later family of models. In this area, a lack of validation with empirical data has been identified [3], hence the interest of empirically validating the opinion change simulated in this work.

In a previous study realized by Fernández-Gracia et al., [2], an opinion dynamics voter model was developed, capable of reproducing electoral patterns observed in the USA. This was done defining a social network based on census data identifying the cities with the highest amount of commutes between them and allowing agents to interact based on this information.

Galesic et al., [4], proposes a model for opinion dynamics using probability distributions to model the beliefs of each agent, and it is validated replicating the voting intention for state-wide elections in the USA during 2018. In this way, they seek to bridge the gap between the social and cognitive components of opinion change, allowing for a robust probability-based opinion while using complex networks to simulate real-life social networks.

Similarly, a model capable of predicting future elections based on the results of the previous one is developed by Vendeville et al., [10] utilizing a voter model with stubborn agents, i.e., agents that don't change opinions.

3 Material and Methods

Empirical validation of opinion dynamics models is needed to evaluate its capabilities to replicate observed opinion change in a broad scale [3,4]. A case study like the selected one, the Mexican presidential election of 2024, offers enough data to perform such an empirical validation. There were three candidates for three coalitions: MORENA-PT-PVEM (from now on referred as candidate A), the PAN-PRI-PRD coalition (candidate B) and MC (candidate C, not considered in the current model for the sake of simplicity) [5].

In what follows, the selected opinion polls will be described, as well as their use for the design, development and validation of the proposed ABM.

3.1 Adopted Opinion Polls

According to Mexican law, all electoral opinion polls are publicly available in the Instituto Nacional Electoral (INE) electronic repository. The opinion polls carried over by the opinion analysis group GEA-ISA [5] were selected for the current work, based primarily on the number of demographic variables present in them. They register the opinion of 1,070 Mexican citizens over three different dates during the electoral campaign. The confidence level is of $95 \pm 3\%$, and the polls were carried over nationwide.

In order to start with the simplest possible model, only candidates A and B were considered, adjusting the poll's data with the resulting preference share shown in Table 1. This adjustment was made deleting the voter share for C and null votes, proceeding to readjust the remaining percentages so they add to a total of 100%. This adjusted data is utilized to analyze the model, trying to replicate the opinion change registered in it. This was done after identifying both option C and null votes as the ones having the least amount of support from the electorate in the selected opinion polls.

Table 1. Opinion poll's data adjusted for two candidates

Source	Date	Time (days)	Preference for A	Preference for B
Poll 1	2023/09/16	0	61%	39%
Poll 2	2023/11/25	70	64%	36%
Poll 3	2024/03/01	167	61%	39%
Election	2024/06/02	260	68%	32%

3.2 Model Description

According to the characteristics of the selected data, an agent-based opinion dynamics model was developed[1] using NetLogo [12], a programming language designed for the development of ABM, with ample use for social simulations [1]. The proposed model adapts the opinion dynamic models of Smaldino [9] for the simulation of voting intention change over time during the political campaign. The visual interface for the model is shown in Fig. 1. In what follows, the model is described in terms of the ODD protocol [6], a standard for specifying ABMs, seeking to ensure the replicability of the models.

Purpose and Patterns. The model's purpose is to replicate the voting intention observed in the Mexican 2024 presidential election, using an opinion dynamics system. Individuals change their opinion after interacting with one another,

[1] Model available at https://github.com/Ulisex666/Modelo-basico-de-dinamica-de-opiniones-para-simulacion-de-elecciones.

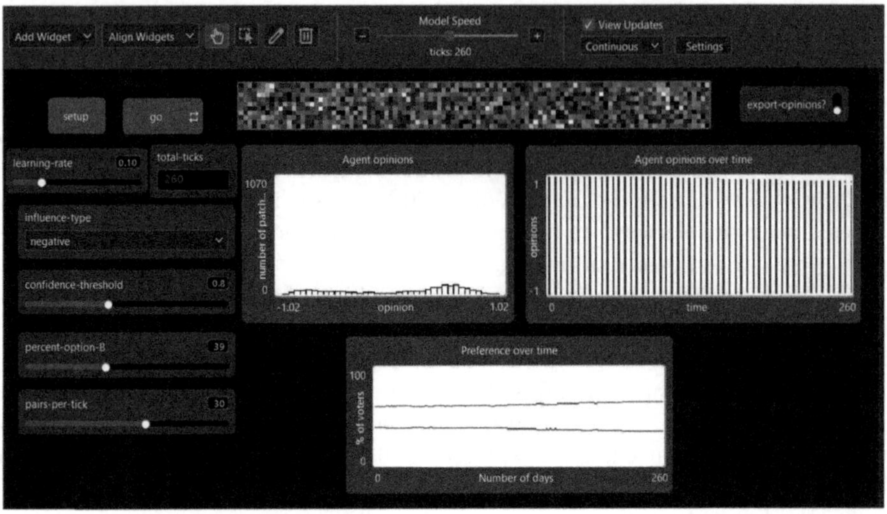

Fig. 1. Visual interface for the developed ABM after a 260 ticks run. The parameters can be adjusted via a graphical interface, and graphs allow to visualize the opinion change during a simulation run

modifying their voting intention. Three types of influence are simulated: positive influence, negative influence and bounded confidence. Opinion is converted to voting intention, and its change over time is the primary pattern to observe, with it being generated by interactions between pairs of agents.

Entities, State Variables and Scales. The model is composed of 1,070 agents, one for each individual recorded in the polls. They're defined by their opinion, a real value in $[-1, 1]$. A positive opinion indicates a preference for option A, while a negative one is taken as a preference for B. Agents with preference for A have a red color, and those with preference for B are painted blue. Each tick in the model represents a day, with 260 ticks covering the time-span since the first poll until the election.

Process Overview and Scheduling. At each time step, the model performs the following actions. Firstly, a number of pairs of agents is selected randomly based on the pairs per tick parameter. The selected agents interact according to the chosen influence type, updating their opinions. Following this, the vote share for A and B is updated, based on the current preferences, and the agents change their color if needed. Relevant plots are updated in the model, and the process continues until the maximum number of ticks is reached.

Design Concepts. Only the relevant design concepts are mentioned.

- **Basic principles.** The basic assumption in all opinion dynamics models is that social influence is crucial for forming and changing opinions. For each

type of influence simulated there's different psychological and sociological justifications, as explained in Sect. 1. Building on this, the model assumes that change in opinion by social influence can explain and simulate the change in voting intention observed in a real political election.
- **Emergence.** The main patterns that arise from the interaction between agents are consensus, polarization and clusterization of opinions, which induces a preference share for each candidate. This is measured in the percentage of agents with preference for A and the percentage of agents with preference for B.
- **Adaptation.** Agents may adapt their opinion after an interaction with another, according to the type of influence simulated.
- **Sensing.** Agents in the model are capable of sensing the opinion held by any other agent they interact with.
- **Interaction.** The interaction between pairs of agents are direct, with either one changing their opinion based on the type of influence simulated and the opinion held by the other.
- **Stochasticity.** Initial opinions are random following a uniform distribution, based on the initial preference for each agent. Initial preference for A or B is assigned randomly. The agents selected for interacting in every step are also selected randomly from the whole population.
- **Observation.** Every step agents change their color indicating preference for A or B based on their opinion. The global opinions and preference share for each candidate is plotted.

Initialization. The model starts assigning initial opinions to each agent, based on the parameter percent option B. This indicates how many agents start with a preference for B, with their opinions initializing according to a uniform distribution $U(-1, 0)$, selected randomly from the whole population. The remaining agents start with a preference for A, with an opinion following a $U(0, 1)$ distribution. Following this, agents change their color based on their assigned opinion.

Sub-models. Each type of influence is modeled based on the following equations, governing how two agents, with opinions x_1 and x_2, update their stances after an interaction. The value $\gamma \in (0, 0.5]$ represents the parameter learning rate, while $d \in (0, 2)$ is the parameter confidence threshold. The former is used in all types of influence, while the later is not utilized for positive influence.

- Positive influence.

$$x_1 \leftarrow x_1 + \gamma(x_2 - x_1)$$
$$x_2 \leftarrow x_2 + \gamma(x_1 - x_2)$$

- Bounded confidence.

$$x_1 \leftarrow x_1 + \gamma(x_2 - x_1) \quad \text{if } |x_1 - x_2| < d$$
$$x_2 \leftarrow x_2 + \gamma(x_1 - x_2) \quad \text{if } |x_1 - x_2| < d$$

– Negative influence.

$$x_1 \leftarrow \begin{cases} x_1 + \frac{\gamma}{2}(x_1 - x_2)(1 - x_1) & \text{if } x_1 > x_2 \\ x_1 + \frac{\gamma}{2}(x_1 - x_2)(1 + x_1) & \text{otherwise} \end{cases}$$

$$x_2 \leftarrow \begin{cases} x_2 + \frac{\gamma}{2}(x_2 - x_1)(1 + x_2) & \text{if } x_1 > x_2 \\ x_2 + \frac{\gamma}{2}(x_2 - x_1)(1 - x_2) & \text{otherwise} \end{cases}$$

4 Methodology

The effect of the model's parameters (learning rate, pairs per tick and confidence threshold) in the final opinion distribution were evaluated over a series of simple experiments, validating the observations reported by Smaldino [9]. It must be noted that the parameter confidence threshold is not utilized under positive influence.

Trying to replicate the opinion change observed in the election, we explore the parameter space for every type of influence, searching for a configuration capable of replicating the opinion change registered in the polls and the election results. For this, we need a measurement of the difference between the opinions produced by the model and the ones observed in the polls and in the election. Following the recommendations found in specialized literature [8], the Root Mean Squared Error (RMSE) is selected. Given that the system considers two complementary preferences ($P_{t,A} + P_{t,B} = 1$), the error for both groups is identical. Thus, the RMSE is calculated based only on preference for A as follows:

$$RMSE = \sqrt{\frac{1}{|T|} \sum_{t \in T} (P_{t,A} - \hat{P}_{t,A})^2} \qquad (1)$$

where T represents the set of evaluation points with empirical data ($|T| = 3$), composed of the second and third opinion polls, adding the final election results. $P_{t,A}$ and $\hat{P}_{t,A}$ denote the empirical and simulated preference shares for option A at time t, respectively.

The initial preferences are taken from the ones registered in the adjusted first poll (Table 1), with 61% ($n = 653$) of agents having a preference for option A, and 39% ($n = 417$) for option B. For every influence type, the model is run 260 ticks, corresponding to each day between the first poll and the real election. The relevant parameters for each type of influence are explored in the following way, searching for the combination that minimizes the RMSE:

- **learning rate:** It varies in the range $[0.1, 0.5]$ with steps of size 0.1.
- **pairs per tick:** It varies in the range $[10, 50]$ with steps of size 10.
- **confidence threshold:** It varies in the range $[0.1, 1]$ with steps of size 0.1. This parameter doesn't intervene in positive confidence.

Each combination of parameters is repeated 30 times, and the mean RMSE and standard deviation is obtained. In total, 25 combinations are tried for positive influence, 250 for negative influence and 250 for bounded confidence. The configuration with the lowest RMSE was selected for each type of influence, and the preferences generated were plotted, evaluating the mean and the standard deviation for the 30 repetitions.

Following this analysis, the best configuration for each type of influence was selected and a comparison was made. This was done to decide if there's a type of influence with a significantly better performance for replicating the empirical opinion change, given than the three types of influence produce a similar mean RMSE with high variability. All statistical tests were performed using the R programming language [7].

Taking the 30 repetitions of the best configuration by type of influence, a total of 90 data points were obtained, registering the RMSE obtained in each run. Initially, an ANOVA test was performed over all data points to identify a significative difference between the RMSE by type of influence, but a Shapiro-Wilk Normality test identified non-normality in the residuals ($p = 0.01 < 0.05$), breaking one of the conditions for the validity of this test.

As an alternative, a non-parametric Kruskal-Wallis test was performed over the 90 data points. It was identified that there exists a significant difference between the medians of the RMSE obtained by each type of influence ($p = 0.03 < 0.05$), but it doesn't identify which influence type has a different median RMSE compared to the others. As such, a non-parametric Dunn's post-hoc test was performed, utilizing Bonferroni correction to control the experiment-wise error rate. Comparing the RMSE for each type of influence pairwise, it allows to identify if a type of influence produces a lower median RMSE compared to the others, more closely resembling the opinion change observed in the empirical data.

5 Results

The configurations with the lowest RMSE for each type of influence are shown in Tables 2, 3 and 4 along with their RMSE and standard deviation for 30 repetitions. The generated preferences for the configuration with the lowest RMSE by type of influence can be seen in Fig. 2, indicating the mean preference at each step for a 260 ticks run, along with the standard deviation for the 30 repetitions.

A comparison of the best configuration for each type of influence is shown in Table 5. The Kruskal test identified a significant difference with $p = 0.03 < 0.05$, and the results for the post-hoc Dunn test are shown in Table 6, comparing every influence type pairwise. A visualization of this comparison is shown in Fig. 3, where the boxplot shows a lower median RMSE for the model under bounded confidence.

Table 2. Top five configurations with lowest mean RMSE for positive influence. This type of influence does not use the confidence threshold parameter, so it's not reported.

Pairs per tick	Learning rate	Mean RMSE and SD
20	0.10	2.56 ± 0.36
10	0.20	2.61 ± 0.60
10	0.10	3.10 ± 0.41
10	0.30	5.34 ± 1.15
30	0.10	6.82 ± 1.04

Table 3. Top five configurations with lowest mean RMSE for negative influence.

Pairs per tick	Learning rate	Confidence threshold	Mean RMSE and SD
20	0.10	1.00	2.45 ± 0.23
10	0.40	0.90	2.51 ± 0.32
30	0.10	0.90	2.52 ± 0.28
10	0.30	0.90	2.53 ± 0.37
30	0.10	0.80	2.56 ± 0.24

Table 4. Top five configurations with lowest mean RMSE for bounded confidence.

Pairs per tick	Learning rate	Confidence threshold	Mean RMSE and SD
30	0.10	0.80	2.33 ± 0.32
10	0.50	0.80	2.41 ± 0.47
20	0.10	1.00	2.43 ± 0.33
30	0.10	0.70	2.45 ± 0.25
50	0.10	0.60	2.46 ± 0.35

Table 5. The best configurations for each type of influence, along with mean RMSE and standard deviation for 30 repetitions.

Parameters	Positive influence	Negative influence	Bounded confidence
pairs per tick	20	20	30
learning rate	0.1	0.1	0.1
confidence threshold	–	1	0.8
RMSE	2.56 ± 0.36	2.4 ± 0.23	$\mathbf{2.3 \pm 0.32}$

Table 6. Results for Dunn test of median RMSE by influence type. Significant results are shown in bold font and marked with a *

	Bounded confidence	Negative influence
Negative influence	$p = 0.24$	–
Positive influence	$\mathbf{p = 0.03}^*$	$p = 1$

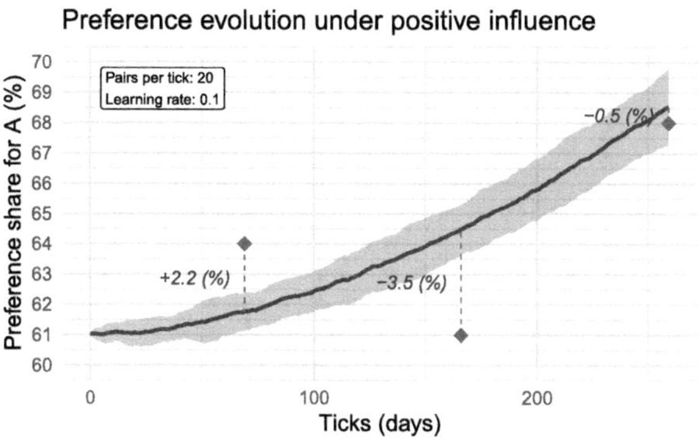

(a) Mean preference and SD for configuration with lowest RMSE under positive influence.

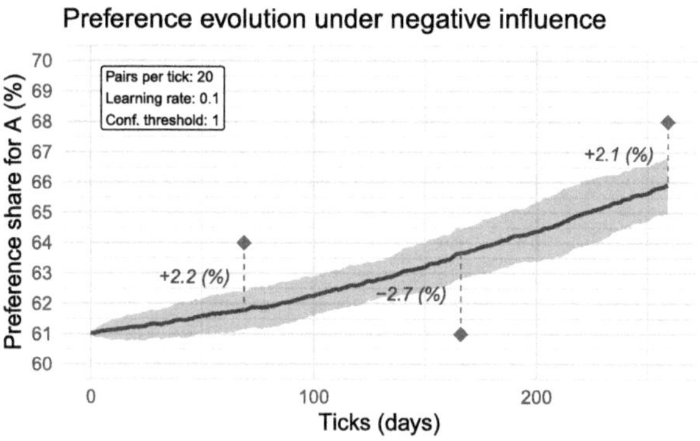

(b) Mean preference and SD and for configuration with lowest RMSE under negative influence.

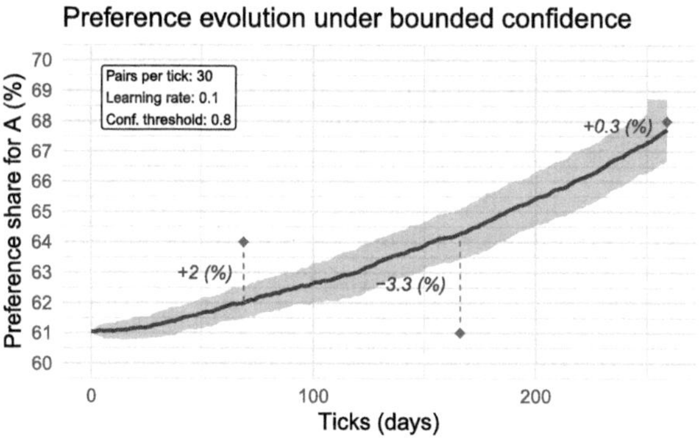

(c) Mean preference and SD for configuration with lowest RMSE under bounded confidence.

Fig. 2. Preference evolution for the configuration with the lowest RMSE, for each type of influence, with 30 repetitions. Diamonds indicate empirical data observed in polls and election results.

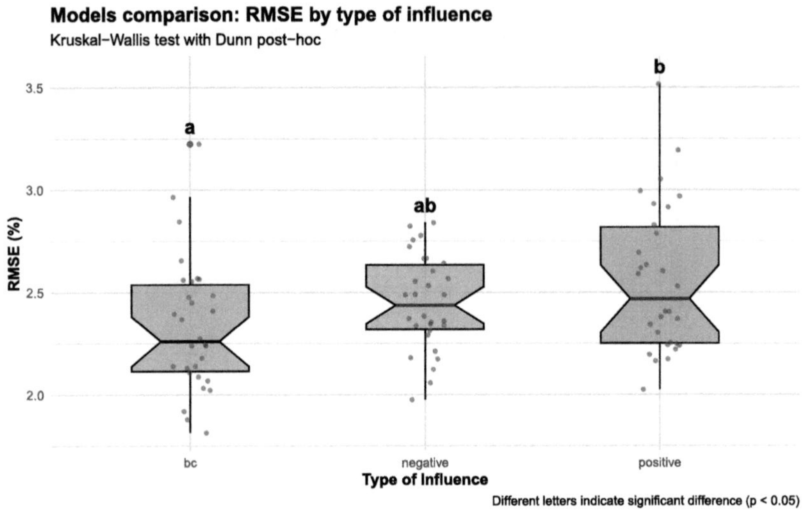

Fig. 3. Boxplot of the RMSE for 30 repetitions of the best configuration by type of influence, including results of Dunn test. Influence under bounded confidence has a significantly lower RMSE than positive influence, with inconclusive results for negative influence

6 Discussion

For this section, the effects of the parameters over the behavior of the model are discussed, noting the differences between each type of influence. Following this, the results of the grid search over the parameter space are analyzed, trying to replicate the empirical opinion change. Finally, there's a discussion of the statistical tests performed over the best configurations by type of influence, identifying a type of influence with a better performance compared to the others.

It was found that the parameters learning rate and pairs per tick only affect the speed of convergence, i.e., how many ticks it takes for the model to reach one of the patterns mentioned in Sect. 1: consensus, polarization or clusterization of opinions. They have no effect on the final opinion reached or the emergent pattern, validating the results reported by Smaldino [9]. This holds true for the three types of influence considered in the model.

The parameter confidence threshold has a different effect for bounded confidence and negative influence. For the former, it changes how many different opinion clusters are obtained at the end of the simulation, with lower values corresponding to a higher number of them; while for the latter affects the opinions taken by agents in the system before reaching polarization. For values higher than 1, all models reach consensus and have a behavior equal to positive influence.

After performing a grid search over the parameter space, all three types of influence were capable of replicating the opinion change reported in the polls considered for this study, with an RMSE lower than 3% comparing the simulated preference share to empirical data. After visualizing the generated preferences for the best configuration by type of influence (shown in Fig. 2), it was identified that they all have a similar behavior over time, with the most significant difference being how closely they approach the final election results.

For the three types of influence, bounded confidence was identified as the best performing one, with a RMSE significantly lower than the one produced by positive influence. This can be observed in Table 6, which indicates a significant difference ($p = 0.03 < 0.05$) between the median RMSE produced by the model under bounded confidence and positive influence, with the former having a smaller median RMSE. However, there's no significant results for the comparison between negative influence and the other types of influence.

7 Conclusions and Future Work

After developing a basic ABM of opinion dynamics, it's capability of replicating empirical data was validated, studying three different types of influence commonly reported in the literature [3,9]: positive influence, negative influence and bounded confidence. The parameter space was explored to try to find a configuration capable of reproducing empirical data based on a selected case study: the Mexican presidential election of 2024. This was done selecting a freely available dataset hosted in an INE online repository, developed by the group GEA-ISA [5].

It was found that the model is capable of reproducing the empirical preference share and election results with only the preference reported in the first poll as an input, measuring the difference between empirical and simulated results utilizing the RMSE. Performing a post-hoc Dunn test, it was identified that bounded confidence has a significantly better performance compared to positive influence, with non-conclusive results for negative influence. As such, taking all available data points, it is found that the basic model developed in this work has the capability of replicating empirical data reduced to the two main options present in the simulated election.

Following the ABM literature, the next step is to perform Pattern-Oriented Modeling (POM) [8], with the goal of developing a more realistic model capable of replicating multiple patterns at the same time. As such, the current model will be studied based on its capacities to reproduce multiple patterns beyond general voting intention, such as preference share by gender or age group utilizing the data available in the selected opinion polls.

In case of finding that the model is incapable of reproducing these alternative patterns, a reformulation of it will be made, exploring the use of the demographic data to decide the social connections of every agent: who interacts with whom, based on previous work in the area [2]. This takes into account factors such as geographic clustering, where people interact more commonly with others in the same geographical area and a more robust type of homophily, taking in consideration factors beyond the opinion being discussed.

Another step to be taken for a more realistic model is expanding the number of options to include both option C and null votes, so the model fully represents the election chosen. In the opinion dynamics area, this is usually done via a vector of opinions, where every entry represents the opinion towards one of the options being discussed [3].

References

1. Epstein, J.M.: Generative Social Science: Studies in Agent-Based Computational Modeling. Princeton Studies in Complexity. Princeton University Press (2006)
2. Fernández-Gracia, J., Suchecki, K., Ramasco, J.J., Miguel, M.S., Eguíluz, V.M.: Is the voter model a model for voters? Phys. Rev. Lett. **112**(15), 158701 (2014). https://doi.org/10.1103/PhysRevLett.112.158701, http://arxiv.org/abs/1309.1131
3. Flache, A., et al.: Models of social influence: towards the next frontiers. J. Artif. Soc. Soc. Simul. **20**(4), 2 (2017). https://doi.org/10.18564/jasss.3521, http://jasss.soc.surrey.ac.uk/20/4/2.html
4. Galesic, M., Olsson, H., Dalege, J., Van Der Does, T., Stein, D.L.: Integrating social and cognitive aspects of belief dynamics: towards a unifying framework. J. Roy. Soc. Interface **18**(176), rsif.2020.0857, 20200857 (2021). https://doi.org/10.1098/rsif.2020.0857, https://royalsocietypublishing.org/doi/10.1098/rsif.2020.0857
5. GEA-ISA: Estudios efectuados durante el proceso electoral federal 2023-2024. Public Repository of the National Electoral Institute (INE) (2024). https://repositoriodocumental.ine.mx/xmlui/handle/123456789/166950. Legally mandated filing of electoral surveys
6. Grimm, V., et al.: The ODD protocol for describing agent-based and other simulation models: a second update to improve clarity, replication, and structural realism. J. Artif. Soc. Soc. Simul. **23**(2), 7 (2020). https://doi.org/10.18564/jasss.4259, http://jasss.soc.surrey.ac.uk/23/2/7.html
7. R Core Team: R: A Language and Environment for Statistical Computing. R Foundation for Statistical Computing, Vienna, Austria (2025). https://www.R-project.org/
8. Railsback, S.F., Grimm, V.: Agent-Based and Individual-Based Modeling: A Practical Introduction, 2nd edn. Princeton University Press, Princeton (2019)
9. Smaldino, P.E.: Modeling Social Behavior: Mathematical and Agent-Based Models of Social Dynamics and Cultural Evolution. Princeton University Press, Princeton (2023)
10. Vendeville, A., Guedj, B., Zhou, S.: Forecasting elections results via the voter model with stubborn nodes. Appl. Netw. Sci. **6**(1), 1 (2021). https://doi.org/10.1007/s41109-020-00342-7, https://appliednetsci.springeropen.com/articles/10.1007/s41109-020-00342-7
11. Voinea, C.F.: Political Attitudes: Computational and Simulation Modelling. Wiley, Hoboken (2016)
12. Wilensky, U.: NetLogo (1999). http://ccl.northwestern.edu/netlogo/

Comparative Evaluation of Classical Machine Learning and U-Net Segmentation for Dead-Tree Crown Mapping in Urban–Forested UAV Imagery

Alejandro D. Espinosa-Chim[1], Mauricio G. Orozco-del-Castillo[1], and Juan C. Valdiviezo-Navarro[2](✉)

[1] Tecnológico Nacional de México/IT de Mérida, Mérida 97118, Yucatán, Mexico
mauricio.orozco@itmerida.edu.mx

[2] SECIHTI - Centro de Investigación en Ciencias de Información Geoespacial, Mérida 97302, Yucatán, Mexico
jvaldiviezo@centrogeo.edu.mx

Abstract. Urban trees provide essential ecosystem services, yet monitoring dead or severely declining trees at scale remains challenging in urban parks due to heterogeneous backgrounds and the presence of man-made structures that can resemble stressed crowns in aerial imagery. While prior work has applied both classical machine learning and deep learning to dead-tree detection in aerial data, many evaluations focus on predominantly forested scenes; comparative evidence in mixed urban–forested settings is more limited. To address this gap, we evaluate classical baselines (logistic regression, support vector machines, and random forest) trained as sliding-window patch classifiers against U-Net segmentation models, including a plain U-Net ablation and a ResNet50-encoder variant, using a mixed urban–forested UAV dataset. The study uses 345 UAV orthomosaics and a scene-level split to ensure disjoint training, validation, and test subsets, with an overlapping patch strategy (512×512 tiles). Aggregate precision–recall metrics are broadly comparable across several approaches, and the ResNet50 U-Net improves over the plain U-Net ablation; however, qualitative inspection reveals differences not captured by summary scores: classical baselines produce scattered false detections over urban structures, whereas the ResNet50 U-Net yields more spatially coherent predictions and suppresses many spurious responses. These results support deep learning segmentation for dead-tree mapping under urban clutter and highlight the need to complement quantitative metrics with qualitative evaluation in heterogeneous UAV datasets.

Keywords: UAV · urban forestry · dead trees · orthomosaics · semantic segmentation · U-Net

H. Ponce et al. (Eds.): COMIA 2026, CCIS 3095, pp. 411–424, 2027.
https://doi.org/10.1007/978-3-032-37438-7_32

1 Introduction

Urban forests provide essential ecosystem services—including thermal regulation, air-quality improvement, and recreational value—yet trees in cities are frequently exposed to chronic stressors such as soil compaction and pollution, limited rooting volume due to paving, heat and radiation effects, and interactions with built infrastructure [3]. These conditions can reduce tree vigor and increase susceptibility to pests and pathogens, making timely monitoring and intervention particularly important in urban parks.

In Mexico City, one relevant contributor to urban tree decline is mistletoe [7], a group of hemiparasitic plants whose dispersal is facilitated by wildlife (e.g., birds and small mammals) that transport and deposit seeds across the urban canopy. Once established, mistletoe can spread rapidly through a park's tree population and is widely reported as an important driver of urban tree deterioration in the city.

As a hemiparasite, mistletoe relies on a specialized structure (haustorium) to extract water and nutrients from host branches, while performing photosynthesis primarily in the canopy. Early infestations may be visually subtle from a distance, whereas advanced infestations often manifest as pronounced canopy changes (e.g., loss of foliage density, discoloration, crown dieback) that may culminate in partial or complete tree mortality. These canopy-level effects motivate the use of aerial imagery for monitoring, because the relevant signals are often expressed in the visible crown geometry, texture, and color.

Recently, multiple approaches have been proposed for detecting or segmenting dead trees from high-resolution aerial imagery, spanning both classical remote-sensing pipelines and Deep Learning (DL) [2,10,11]. Rahman et al. [5] introduced a Self-Attention U-Net [6] for dead-tree detection/segmentation in dense boreal forests, and further improved instance-level delineation through watershed-based refinement and adaptive filtering. In contrast, Dumarevskaya and Parent [1] compared pixel-based and object-based unsupervised classification (ISODATA with spectral/morphological filtering) against a DL workflow for mapping dead deciduous trees in Rhode Island using true-color summertime aerial imagery; their results favored the pixel-based pipeline ($F1 = 0.84$) over both object-based classification ($F1 = 0.79$) and the DL approach ($F1 = 0.67$). Together, these studies illustrate that (i) deep models can benefit substantially from task-specific post-processing and instance-separation strategies in densely vegetated forests, while (ii) classical pipelines can remain competitive when strong filtering rules effectively suppress commission error in relatively homogeneous forest contexts.

A persistent limitation in the current literature is that many evaluations are conducted in predominantly forested scenes, where background variability is limited and dead vegetation often exhibits strong visual contrast against surrounding canopy [1,5,8]. However, operational deployments in urban parks must contend with heterogeneous land cover and man-made structures (e.g., pavement, rooftops, bare soil, shadows), which can induce systematic false positives, particularly for approaches that rely heavily on local color/intensity cues;

related urban-park dead-tree classification studies further motivate this setting [4]. Evidence that background composition affects segmentation reliability is also reported in adjacent UAV forestry problems: Wang et al. [8] evaluated semantic segmentation models under different forest backgrounds and observed degraded performance in more confounding backgrounds (e.g., bare soil mixed with canopy) relative to purer forest scenes. In addition, recent benchmark efforts for tree mortality mapping emphasize geographic and contextual variability as a central challenge for generalization [9]. These observations motivate our comparative evaluation of classical machine learning baselines versus a deep segmentation model on a mixed urban–forested UAV dataset, and our emphasis on qualitative inspection as a necessary complement to aggregate metrics.

Despite the relevance of the problem, up-to-date, large-scale estimates of infestation and mortality in many urban settings remain limited, in part because field surveys are labor-intensive, time-consuming, and costly to repeat across wide areas. This creates an opportunity for automated or semi-automated approaches based on remote sensing, where UAV imagery can support frequent, high-resolution inspection of tree crowns.

Motivated by this gap, we study dead-tree segmentation in UAV imagery collected from urban parks in Mexico City and compare a DL segmentation model (a ResNet50-based U-Net) with classical machine learning baselines (Logistic Regression, Random Forest, and Support Vector Machine). Beyond aggregate metrics, we emphasize qualitative visual assessment of the predicted masks, as segmentation performance may appear acceptable numerically while still producing fragmented or spatially implausible outputs in cluttered urban scenes.

The main contributions of this work are threefold: (i) a comparative evaluation of DL and classical machine learning approaches for dead-tree segmentation in urban-park UAV imagery; (ii) an analysis highlighting the importance of qualitative inspection alongside quantitative metrics in heterogeneous backgrounds; and (iii) a dataset and experimental setting that explicitly includes urban clutter and field-survey support, encouraging evaluation beyond predominantly forested benchmarks.

2 Materials and Methods

2.1 Study Area and Data Acquisition

Our study area is located in Mexico City, at coordinates [19.4096° N, −99.1563° W], comprising 7.4 ha. The urban park is known as Ramón López Velarde Garden, home of tree species like Casuarina sp., Eucalyptus sp., Fraxinus sp., Cupressus sp., among others. This site was selected in conjunction with the Mexico City Environment Secretariat (SEDEMA) due to the high prevalence of a mistletoe species called *Struthanthus Interruptus*. The lack of adequate phytosanitary control together with the rapid propagation of this hemiparasitic plant caused the death of tens of trees in the site.

With the aim of building a robust image database for mistletoe classification, field works were carried out at three different times on September 2022, October

2023, and February 2024. For image acquisition, we employed a P4 multispectral Unmanned Aerial Vehicle (UAV) capable of collecting five spectral bands in the following electromagnetic spectrum intervals: $B = 450\,\text{nm} \pm 16\,\text{nm}$, $G = 560\,\text{nm} \pm 16\,\text{nm}$, $R = 650\,\text{nm} \pm 16\,\text{nm}$, $REG = 730\,\text{nm} \pm 16\,\text{nm}$, and $NIR = 840\,\text{nm} \pm 26\,\text{nm}$, where R, G, B, REG, and NIR represent the red, green, blue, red-edge, and near-infrared bands, respectively.

The images were collected in orthogonal and oblique modes, at a mean height of 70 m, resulting in a Ground Sample Distance (GSD) of 3.7 cm/pixel. After image acquisition, a data cleaning process consisting of the removal of redundant images was applied. Hence, a total of 345 color RGB images were retained for further annotation and processing. Although multispectral data were acquired by the UAV sensor, this study focuses exclusively on the RGB bands. This decision was driven by accessibility and practical deployment constraints: RGB color imagery is not only more cost-effective and widely available to park management officials, but also significantly simplifies data processing and integration into automated software platforms. Relying solely on RGB reduces computational overhead, making the proposed automated mapping system much easier and straightforward to implement, deploy, and scale for end-users.

2.2 Ground Truth Generation

We formulate dead-tree detection as a binary semantic segmentation task. For each aerial image, a corresponding ground-truth mask is provided as a single-channel binary image with the same spatial extent. Foreground pixels indicate dead-tree crown regions, while background pixels correspond to all other land cover (healthy vegetation, soil, shadows, and non-tree structures). The selected images were manually annotated using the CVAT application, available at https://www.cvat.ai/, ensuring precise pixel-level labeling for the target class. At the end of the process, the resulting masks were validated by an expert.

2.3 Preprocessing and Tiling

UAV imagery is typically acquired at high spatial resolution, producing large images (or orthomosaics) that exceed GPU memory limits for end-to-end semantic segmentation. To enable efficient training and increase the number of supervised samples, we adopt a patch-based strategy in which each image and its corresponding binary mask are decomposed into fixed-size tiles using a sliding window.

Let I denote an RGB image and M its binary ground-truth mask, where pixels in M indicate the presence of dead-tree crowns. We extract tiles of size 512×512 pixels with a stride of 256 pixels in both directions, generating overlapping patches to improve coverage near object boundaries and increase training diversity. For each tile pair (I_p, M_p), we categorize the sample as *positive* if the mask contains at least one foreground pixel, and as *negative* otherwise. This criterion ensures that tiles containing dead-tree pixels are always retained.

Because the number of background-only tiles can be substantially larger than the number of informative tiles, a class-imbalance problem emerges at the sample level. To mitigate this effect during training, we perform a simple undersampling of negative tiles: after collecting all positive tiles, we randomly select a subset of negative tiles whose size is proportional to the number of positives (approximately 1.1× the positive count in our implementation). The resulting set of tiles provides a more balanced training distribution while preserving a representative portion of background variability. Finally, the selected image tiles and mask tiles are written to disk to form the patch-based dataset used in subsequent training and evaluation stages.

Remark on Overlap and Data Partitioning. Using a stride smaller than the tile size introduces correlation between neighboring tiles extracted from the same source image. To avoid information leakage across splits, we partition the dataset at the *scene* (source image/orthomosaic) level (Sect. 2.4) and generate tiles within each partition, ensuring that no tiles from the same source image appear in multiple subsets.

2.4 Dataset Split

To evaluate generalization to unseen imagery, we partition the dataset at the *scene* level, where each scene corresponds to a source UAV image (or orthomosaic) and its associated mask. Scenes are assigned to training, validation, and test subsets using a two-stage random split with a fixed seed for reproducibility.

First, we hold out 20% of the scenes as an independent test set. The remaining 80% is then split again to create a validation set corresponding to 12.5% of the remaining scenes (i.e., 10% of the total dataset), leaving approximately 70% of the scenes for training. In summary, the resulting proportions are approximately 70% training, 10% validation, and 20% test at the scene level.

After this partitioning, tiles are extracted *within* each subset as described in Sect. 2.3. This procedure ensures that no tiles derived from the same source image appear in multiple subsets, preventing information leakage across splits. To support deterministic training and evaluation, the resulting file lists (scenes and/or derived tiles) are stored to disk and subsequently used to build the input pipelines.

2.5 Model Architecture

We formulate dead-tree detection as a binary semantic segmentation task and employ an encoder–decoder network inspired by U-Net. The architecture follows the standard U-Net design with skip connections that preserve spatial detail by transferring multi-scale features from the encoder to the decoder.

We use a ResNet50 backbone initialized with ImageNet weights as a feature extractor. Given an input tile of size $512 \times 512 \times 3$, the encoder produces a hierarchy of feature maps at progressively lower spatial resolutions. Skip connections are taken from intermediate stages of the backbone (implemented using the activations `conv1_relu`, `conv2_block3_out`, `conv3_block4_out`, and `conv4_` block6_out), while the deepest representation is obtained from conv5_block3_out.

The decoder restores spatial resolution through a sequence of upsampling stages. At each stage, features are upsampled using transposed convolutions, concatenated with the corresponding encoder features (skip connection), and refined with convolutional blocks. To reduce overfitting, dropout is applied at the bottleneck (rate 0.5) and within decoder blocks (rate 0.3).

The final decoder output is projected to a single-channel segmentation map using a 1×1 convolution followed by a sigmoid activation, producing a per-pixel probability of the dead-tree class. For visualization and quantitative evaluation, probabilities can be thresholded (e.g., at 0.5) to obtain a binary mask.

2.6 Loss Function and Metrics

Dead-tree crowns typically occupy a small fraction of each UAV tile, yielding a strong foreground–background imbalance at the pixel level. To address this issue, we optimize a composite objective that combines a weighted pixel-wise binary cross-entropy term with a Dice-based overlap term.

Given a predicted probability map $\hat{y} \in [0,1]^{H \times W}$ and a binary ground-truth mask $y \in \{0,1\}^{H \times W}$, we compute the (soft) Dice coefficient as

$$\mathrm{Dice}(y, \hat{y}) = \frac{2\sum_i y_i \hat{y}_i + \epsilon}{\sum_i y_i + \sum_i \hat{y}_i + \epsilon}, \tag{1}$$

where i indexes represent pixels and ϵ is a small constant for numerical stability.

The training loss is defined as the sum of a weighted binary cross-entropy (BCE) term and a Dice loss term. Let $\mathcal{L}_{\mathrm{BCE}}(y, \hat{y})$ denote BCE; we increase the penalty of foreground errors through a positive-class weight w_+ (i.e., positive pixels receive weight w_+ and background pixels weight 1). The overall objective is:

$$\mathcal{L}(y, \hat{y}) = \mathcal{L}_{\mathrm{WBCE}}(y, \hat{y}; w_+) + \lambda\big(1 - \mathrm{Dice}(y, \hat{y})\big), \tag{2}$$

where λ controls the contribution of the Dice term. In our implementation, we set $w_+ = 10.0$ and $\lambda = 2.0$.

During training and validation, we report the Dice coefficient (as defined above) and pixel-wise accuracy. Because pixel accuracy can be dominated by the background class in highly imbalanced segmentation, Dice is treated as the primary metric for model selection and performance tracking.

2.7 Training Configuration

All DL experiments are implemented in TensorFlow/Keras. Input tiles are resized (when necessary) to 512×512 pixels and trained with a batch size of 8 using the Adam optimizer (initial learning rate 1×10^{-4}). Mixed-precision training is enabled (policy mixed_float16).

Model selection is based on the validation Dice coefficient: we save the checkpoint with the highest validation Dice. Early stopping is enabled (patience 15, restoring best weights), and `ReduceLROnPlateau` reduces the learning rate by a factor of 0.2 when validation Dice saturates (patience 6; minimum learning rate 1×10^{-7}). Training is run for up to 15 epochs with these callbacks enabled. This relatively short training horizon was chosen empirically, as we observed that model convergence (measured by validation Dice and IoU) typically plateaued within this time frame, rendering further epochs computationally redundant without yielding significant performance gains.

The experiments were executed in Google Colab using an NVIDIA Tesla T4 GPU; training required approximately 8,370 s for the full 15 epochs (about 558 s/epoch), and inference on a single 512×512 tile required approximately 102 ms.

2.8 Data Pipeline and Augmentation

We implement the input pipeline with `tf.data` to load paired image and mask tiles from disk (RGB tiles in JPEG format and masks in PNG format) using consistent file names across directories. Images and masks are resized to 512×512 when required, and ResNet preprocessing (`preprocess_input`) is applied to match the ImageNet normalization expected by the pretrained encoder. During training, we apply lightweight geometric augmentation consisting of random horizontal/vertical flips and random rotations in multiples of 90°; validation and test pipelines omit augmentation and apply only deterministic preprocessing.

3 Experiments

3.1 Experimental Protocol

The experimental evaluation uses a patch-based dataset derived from 345 UAV images, producing a total of 6,900 tiles (512×512) after preprocessing and tiling. All DL experiments were implemented in TensorFlow 2.x and executed in Google Colab using an NVIDIA Tesla T4 GPU.

To assess generalization, dataset partitioning was performed at the image (scene) level with a fixed random seed: images assigned to the training, validation, and test subsets were strictly disjoint. Tiles were then generated within each subset, yielding 4,830 training tiles, 690 validation tiles, and 1,380 test tiles, where the training count reflects undersampling of negative tiles. Negative (background-only) tiles were undersampled *only* in the training subset to reduce sample-level imbalance, while validation and test subsets preserved their natural composition. This protocol ensures that evaluation is conducted on scenes completely excluded from model optimization.

3.2 Baselines

To provide reference points for automated dead-tree detection from UAV imagery, we compare the proposed ResNet50 U-Net against classical machine learning baselines trained in a *sliding-window patch-classification* setting. Specifically, we train Logistic Regression (LR), a Support Vector Machine (SVM), and a Random Forest (RF) on the same 512×512 patches used by the DL pipeline. To ensure methodological parity and a strictly controlled comparison, the classical baselines were provided with the same raw pixel intensities as the deep learning models. By avoiding the introduction of feature descriptors, we isolate the evaluation to each model's internal capability to autonomously extract relevant patterns from identical data sources, thereby providing a fair assessment of architectural performance in heterogeneous urban environments.

For the classical baselines, each RGB patch is represented by a fixed-length vector obtained by flattening its raw pixel intensities (i.e., the $512 \times 512 \times 3$ patch is reshaped into a one-dimensional feature vector). The models then output a single binary label per patch (dead-tree present/absent). In contrast, the U-Net produces a dense pixel-level segmentation mask. To visualize the spatial behavior of the patch-based baselines without generating blocky region masks, we assign each patch prediction to the *center location* of the corresponding patch and slide the window across the scene.

The above baselines are included to illustrate how non-convolutional decision functions, when applied to raw patch vectors, can be sensitive to heterogeneous urban–forested backgrounds in UAV imagery. In particular, background elements such as rooftops, pavement, bare soil, and shadows may share color characteristics with stressed or dead crowns, leading to scattered false detections that are difficult to interpret spatially. LR is reported only as a quantitative reference (Sect. 4), while qualitative comparisons focus on SVM and RF. In addition, we include an ablation-style comparison between a standard U-Net and the ResNet50-based U-Net to assess the impact of the pretrained encoder.

Table 1. Comparative performance of classical baselines and U-Net models on the dead-tree detection/segmentation task.

Model	Precision	Recall	F1-Score	IoU
Random Forest	0.78	0.83	0.80	0.67
SVM (SGD)	0.74	0.86	0.80	0.66
Logistic Regression	0.75	0.84	0.79	0.66
U-Net (Ablation)	0.65	0.71	0.67	0.51
U-Net/ResNet	**0.78**	**0.80**	**0.79**	**0.67**

4 Results

We first compare the proposed DL segmentation models (U-Net ablation and ResNet50 U-Net) against classical machine learning baselines (LR, SVM, and Random Forest) trained in a sliding-window patch-classification setting with center-point assignment (Sect. 3.2). For each approach, we report test-set precision, recall, F1-score, and Intersection over Union (IoU) as summarized in Table 1.

The results presented in Table 1 shows that the evaluated approaches achieve broadly comparable aggregate performance on the test set. Among the classical baselines, RF and SVM obtain similar F1-scores (0.80), with LR performing slightly lower but in the same range. Regarding the DL models, the ResNet50 U-Net attains competitive precision and recall (0.78 and 0.80, respectively), and improves substantially over the plain U-Net ablation (F1-score 0.79 vs. 0.67). This substantial performance gap is primarily attributed to the benefits of transfer learning. The ResNet50 variant utilizes an encoder pre-trained on the ImageNet dataset, providing a robust initialization of spatial feature extractors—such as edge, texture, and shape detectors—already optimized for general object recognition. In contrast, the plain U-Net ablation was trained from scratch, requiring it to learn these fundamental representations simultaneously with the specific segmentation task. Given the high background complexity of urban UAV imagery, pre-trained weights in the ResNet50-based model offer a significant advantage in distinguishing dead-tree crowns from confounding man-made structures. These results suggest that, at the level of summary metrics, multiple modeling choices can yield similar scores; however, as shown next, qualitative inspection reveals notable differences in the spatial plausibility and robustness of the predicted outputs, particularly in heterogeneous urban–forested scenes.

Although the summary metrics in Table 1 are broadly comparable, they do not fully capture whether the outputs are *operationally usable* for dead-tree mapping. In UAV imagery with heterogeneous backgrounds, a model can achieve a reasonable precision–recall balance while still producing spatially implausible predictions (e.g., fragmented "salt-and-pepper" detections or systematic false positives over urban structures). For this reason, we complement quantitative evaluation with qualitative visual inspection of representative scenes, which reveals differences in spatial coherence and robustness that are not apparent from aggregate scores alone.

For visual inspection, we first analyze a predominantly forested tile in which the background is largely composed of vegetation, providing a comparatively "clean" setting. Figure 1 compares model outputs for the same input sample using a common layout: the original UAV patch (first column), the ground-truth mask (second column), the predicted output (third column), and the corresponding overlay on the original image (fourth column). Rows correspond to the Random Forest baseline, the SVM baseline, and the ResNet50 U-Net model, respectively.

In this forest-dominant example, the classical baselines produce visually reasonable detections in the sense that many foreground regions overlap with annotated crowns. However, their outputs tend to be more fragmented, exhibiting a noisier "salt-and-pepper" appearance with isolated detections and small gaps. In contrast, the ResNet50 U-Net prediction is more spatially coherent, with

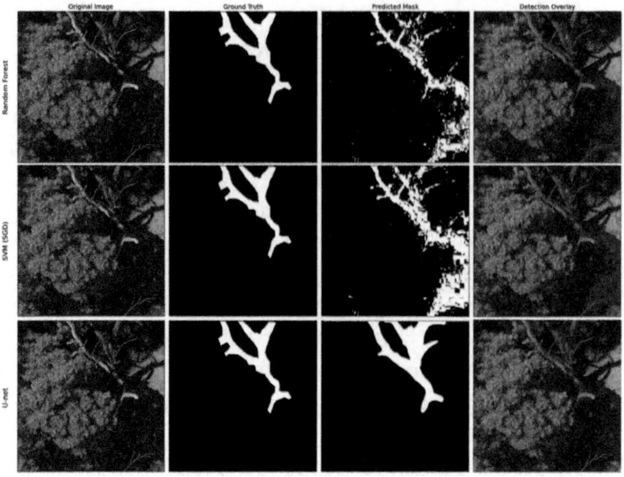

Fig. 1. Qualitative comparison in a predominantly forested scene. Columns show the original image, ground truth, predicted output, and overlay visualization, respectively. Rows correspond to Random Forest, SVM, and ResNet50 U-Net.

smoother contiguous regions and fewer scattered artifacts, resulting in a mask that is easier to interpret for crown-level mapping in this setting.

We next examined a more challenging urban–forested scene containing mixed land cover and visually confounding background elements, including concrete surfaces and gray/brown structures (e.g., sidewalks or built features) interspersed with vegetation. This setting is representative of urban parks and is important because such non-vegetation regions can share color tones with stressed or dead crowns. As a result, failure modes may not be fully reflected by aggregate precision–recall metrics alone, making qualitative inspection particularly informative. Figure 2 presents the same comparative layout used in Fig. 1.

Figure 2 illustrates a key qualitative difference between the classical baselines and the DL approach in heterogeneous scenes. While RF and SVM identify vegetation-related regions, they also produce noticeable false positives over the concrete structure (e.g., the sidewalk), indicating sensitivity to visually similar gray/brown patterns. These qualitative behaviors suggest that the classical models rely heavily on color or intensity cues extracted from local patches and have limited ability to incorporate broader spatial context. In contrast, the ResNet50 U-Net largely suppresses these spurious detections and produces a more spatially consistent prediction, which appears to reflect improved use of texture and contextual information when distinguishing dead-tree crowns from urban background elements.

To further assess the consistency of the DL approach beyond the individual examples shown above, we include additional qualitative results using the ResNet50 U-Net across multiple scenes. These examples are intended to highlight the model's robustness in mixed urban–forested conditions and to illustrate typ-

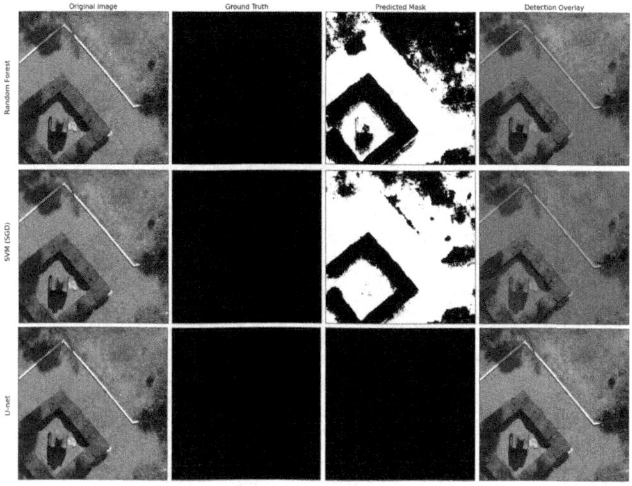

Fig. 2. Qualitative comparison in an urban–forested scene with visually confounding background structures. Columns show the original image, ground truth, predicted output, and overlay visualization, respectively. Rows correspond to Random Forest, SVM, and ResNet50 U-Net.

ical prediction behavior under varying background complexity. Figure 3 shows representative overlays produced by the ResNet50 U-Net.

Across these additional cases, the ResNet50 U-Net predictions remain spatially coherent and exhibit reduced speckle compared to the classical baselines, producing contiguous regions that better match crown-scale structures. Moreover, the model tends to generate fewer spurious detections over non-vegetation surfaces, supporting its robustness in scenes that include urban materials and heterogeneous backgrounds. This reliability was further confirmed through a systematic review of the most challenging test tiles; however, no representative failure cases were identified even under extreme lighting variations or high structural complexity, reinforcing the model's operational stability for large-scale urban mapping.

Training dynamics for the ResNet50 U-Net indicated stable convergence. Over 15 epochs, the Dice coefficient increased from 0.54 to 0.80 on the training set and reached 0.78 on validation, while IoU increased from 0.40 to 0.69 (training) and reached 0.67 (validation). Performance improved rapidly during early epochs and then began to plateau as optimization progressed. Validation curves tracked the training trajectory reasonably closely, with only minor oscillations attributable to finite sample size and scene variability, suggesting no major overfitting under the selected configuration. Consistent with the protocol in Sect. 2.7, the final checkpoint was selected based on the epoch achieving the best validation overlap (Dice/IoU).

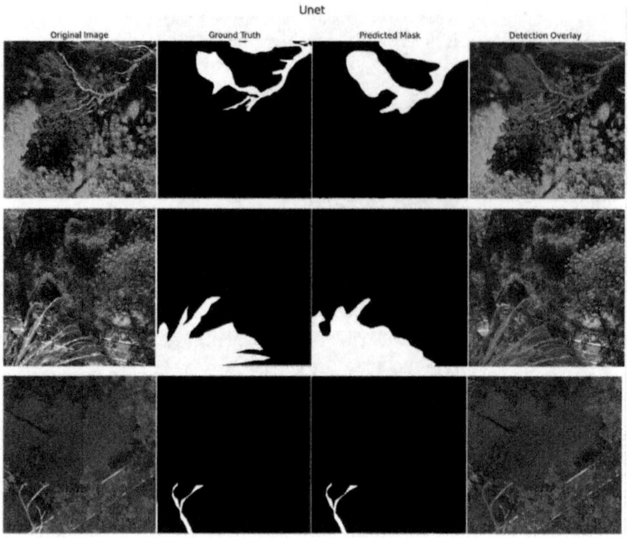

Fig. 3. Additional qualitative examples using the ResNet50 U-Net. The figure illustrates representative predictions and overlay visualizations across multiple scenes, highlighting robustness under heterogeneous urban–forested backgrounds.

5 Discussion

The results suggest that, while classical machine learning baselines (LR, SVM, and RF) can achieve competitive aggregate scores in a patch-classification setting, their spatial behavior may be less reliable in heterogeneous urban–forested environments. In predominantly vegetated scenes, the baselines can produce reasonable detections, but their outputs tend to be more fragmented and noisy. In contrast, the ResNet50 U-Net yields more spatially coherent predictions and appears to better capture crown-level structure. This difference becomes more pronounced in mixed scenes that include urban materials (e.g., concrete, rooftops, and paved surfaces), where the baselines produce false detections on visually confounding regions. These findings support the use of DL segmentation models for dead-tree mapping in urban parks, where background complexity is a defining characteristic of the operational setting.

A key implication of this study is methodological: summary metrics alone may not fully reflect the practical usability of dead-tree maps. Similar precision–recall trade-offs can conceal qualitatively different failure modes, such as salt-and-pepper artifacts or systematic false positives on non-vegetation structures. For applications that aim to support inspection, prioritization, or intervention in urban forestry, it is therefore important to complement quantitative evaluation with qualitative inspection of representative scenes and failure cases. This is especially relevant when models are evaluated on datasets with heterogeneous backgrounds, where errors can be spatially structured and context-dependent.

This work has several limitations that point to clear directions for future research. First, consistent with their role as minimal classical references, baselines were trained on flattened raw patch vectors, a representation that does not explicitly encode spatial structure and may favor color/intensity cues over texture and contextual composition; exploring alternative classical representations (e.g., engineered texture descriptors or learned embeddings) could provide a stronger non-deep baseline. Second, although we evaluate generalization using a scene-level split, the dataset originates from a single urban context. Future efforts will prioritize expanding dataset diversity by incorporating images from multiple urban parks, across different seasons, and covering a wider range of tree species to strengthen the generalization of the model and validate its long-term reliability. Third, while we include a U-Net ablation, we do not yet compare against a broader family of modern segmentation networks; another direction of future work will benchmark additional deep architectures under the same urban–forested conditions and investigate post-processing and calibration strategies to improve reliability. Finally, it would be interesting to incorporate the multispectral bands already collected to further analyze the performance of classical and recent models in our current application.

6 Conclusions

This paper studied dead-tree detection and segmentation from UAV imagery collected in urban parks, with a particular focus on heterogeneous urban–forested scenes that include visually confounding background elements. We compared classical machine learning baselines (Logistic Regression, SVM, and Random Forest) trained in a sliding-window patch-classification setting against DL segmentation models based on U-Net, including a ResNet50-encoder variant. Quantitatively, the evaluated approaches achieved broadly comparable aggregate performance on the test set; however, qualitative inspection revealed substantial differences in spatial plausibility and robustness, especially in the presence of urban structures such as pavement and other gray/brown materials. In these challenging conditions, the ResNet50 U-Net produced more spatially coherent outputs and fewer spurious detections than the classical baselines.

Beyond model comparison, the results emphasize that segmentation quality in real-world UAV monitoring cannot be assessed solely through summary metrics. Visual inspection of representative scenes and failure modes is essential to determine whether predicted maps are interpretable and operationally useful for urban forestry applications. Future work will expand benchmarking to additional deep segmentation architectures reported in the literature (e.g., recent U-Net variants and other semantic segmentation models), with particular emphasis on evaluating them on *urban–forested* datasets rather than predominantly forest-only settings. Such experiments will enable a more direct comparison against our ResNet50 U-Net under background conditions that include urban materials and other confounding structures. In parallel, we will explore alternative classical representations beyond flattened raw patches, evaluate robustness across broader geographic and temporal settings, and investigate multispectral cues, calibration, and lightweight post-processing strategies to further improve reliability in complex urban–forested environments.

Acknowledgment. The authors acknowledge the resources and support provided by the *Secretaría de Educación, Ciencia, Tecnología e Innovación de la Ciudad de México (SECTEI)* through the project *"Monitoreo de especies de muérdago en parques urbanos usando imágenes aéreas e inteligencia artificial,"* CAR SECTEI/156/2024.

Disclosure of Interests. The authors have no competing interests to declare that are relevant to the content of this article.

References

1. Dumarevskaya, L., Parent, J.: Mapping dead trees in Rhode island using semi-automated classification of summertime aerial imagery. J. For. Res. **37** (2026). https://doi.org/10.1007/s11676-025-01970-4
2. Ma, L., Liu, Y., Zhang, X., Ye, Y., Yin, G., Johnson, B.A.: Deep learning in remote sensing applications: a meta-analysis and review. ISPRS J. Photogramm. Remote. Sens. **152**, 166–177 (2019). https://doi.org/10.1016/j.isprsjprs.2019.04.015
3. Marchal-Valencia, D.: El muérdago en la ciudad de méxico. Arbolama (2), (2009)
4. Orozco-González, W., Valdiviezo-Navarro, J.C., Orozco-del Castillo, M.G., Mejia-Zuluaga, P.A., Dozal, L.: Classification of dead trees in urban parks using aerial imagery and convolutional neural networks. In: ICCBR Workshops, pp. 121–130 (2024)
5. Rahman, A.U., Heinaro, E., Ahishali, M., Junttila, S.: Dual-task learning for dead tree detection and segmentation with hybrid self-attention U-nets in aerial imagery. Int. J. Appl. Earth Obs. Geoinf. **144**, 104851 (2025). https://doi.org/10.1016/j.jag.2025.104851
6. Ronneberger, O., Fischer, P., Brox, T.: U-Net: convolutional networks for biomedical image segmentation. In: Navab, N., Hornegger, J., Wells, W.M., Frangi, A.F. (eds.) MICCAI 2015. LNCS, vol. 9351, pp. 234–241. Springer, Cham (2015). https://doi.org/10.1007/978-3-319-24574-4_28
7. Valdiviezo-Navarro, J.C., Orozco-del Castillo, M.G.: The role of AI in combating mistletoe infestation in Mexican forests: a call for interdisciplinary action. J. Artif. Intell. Comput. Appl. **2**(2), 7–17 (2024)
8. Wang, L., et al.: Monitoring pine shoot beetle damage using UAV imagery and deep learning semantic segmentation under different forest backgrounds. Forests **16**(4), 668 (2025). https://doi.org/10.3390/f16040668
9. Wang, Z., et al.: TreeFinder: a US-scale benchmark dataset for individual tree mortality monitoring using high-resolution aerial imagery. In: Advances in Neural Information Processing Systems (NeurIPS), Track on Datasets and Benchmarks (2025)
10. Weinstein, B.G., et al.: A benchmark dataset for canopy crown detection and delineation in co-registered airborne RGB, LiDAR and hyperspectral imagery from the national ecological observation network. PLoS Comput. Biol. **17**(7), e1009180 (2021). https://doi.org/10.1371/journal.pcbi.1009180
11. Weinstein, B.G., Marconi, S., Bohlman, S., Zare, A., White, E.: Individual tree-crown detection in RGB imagery using semi-supervised deep learning neural networks. Remote Sens. **11**(11), 1309 (2019). https://doi.org/10.3390/rs11111309

Comparison of U-Net and FPN Architectures for Ground-Level Semantic Segmentation of Sargassum in the Mexican Caribbean

Yair Eduardo Brito Téllez(✉), Sergio Isahí Garrido-Castañeda, Yenny Villuendas-Rey, and Juan Irving Vasquez

Centro de Innovación y Desarrollo Tecnológico en Cómputo (CIDETEC), Instituto Politécnico Nacional (IPN), Av. Juan de Dios Bátiz sn, Nueva Industrial Vallejo, 07700 Ciudad de Mexico, GAM, Mexico
{ybritot2200,sgarridoc2000}@alumno.ipn.mx, {yvilluendasr,jvasquezg}@ipn.mx

Abstract. Massive sargassum influxes on the coasts of the Mexican Caribbean have created significant environmental and economic challenges, increasing the need for efficient beach-scale monitoring tools. In this study, we present a comparative evaluation of two deep learning architectures, U-Net and Feature Pyramid Network (FPN), for the semantic segmentation of ground-level coastal images from the Riviera Maya. To ensure a fair structural comparison, both models were implemented using the same EfficientNet-B0 backbone. A dataset of 1296 RGB images with pixel-level annotations and seven semantic classes was used for training and evaluation. Model performance was assessed using pixel accuracy and Intersection over Union (IoU). The results showed that FPN achieved a marginal improvement, reaching 94% accuracy and 0.70 mean IoU, performing comparably to U-Net, which achieved 93% accuracy and 0.69 mean IoU over 60 training epochs. In addition to the comparative analysis, the best-performing model was integrated into a web-based application that allows users to upload coastal images and automatically obtain segmented outputs. These results suggest that both multi-scale and encoder-decoder architectures constitute suitable alternatives for practical sargassum monitoring, and that cloud-based deployment may facilitate practical use in coastal management scenarios.

Keywords: Sargassum · Computer Vision · Deep Learning · Semantic Segmentation · Machine Learning · Coastal Monitoring

1 Introduction

Recurring sargassum influxes in the Mexican Caribbean have become a significant environmental and socioeconomic concern due to their negative effects on marine ecosystems, coral reef health, tourism, and coastal communities [13,17].

H. Ponce et al. (Eds.): COMIA 2026, CCIS 3095, pp. 425–437, 2027.
https://doi.org/10.1007/978-3-032-37438-7_33

The continuous accumulation of sargassum on beaches has increased the need for monitoring strategies that support local assessment and timely decision-making for mitigation and coastal environmental management [4].

Recent advances in computer vision, particularly semantic segmentation, offer pixel-level representations capable of delineating sargassum accumulation areas with high spatial detail [14]. This detailed information is especially relevant in heterogeneous beach environments, where sargassum coexists with sand, water, vegetation, sky, and other visual elements [2,18].

This paper investigates the use of deep learning for the semantic segmentation of ground-level coastal images from the Riviera Maya. A comparative evaluation was conducted using two convolutional neural network architectures, U-Net and Feature Pyramid Network (FPN), trained on a seven-class annotated dataset of coastal scenes. To ensure a rigorous and fair structural comparison, both models were implemented using the same EfficientNet-B0 backbone. The objective was to compare the suitability of both architectural designs for segmenting sargassum under realistic environmental conditions. A diagram of the comparison methodology is presented in Fig. 1. The results showed that the FPN-based model achieved a marginal performance improvement over the U-Net-based model. In addition, the best-performing model was integrated into a web-based application, illustrating its practical use as a supporting tool for coastal monitoring.

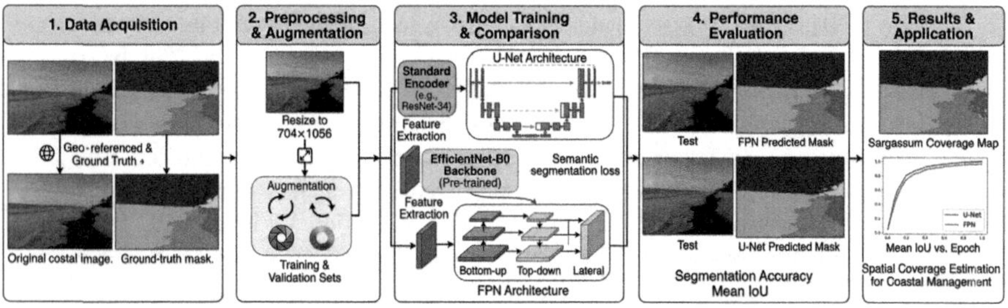

Fig. 1. Overview of the proposed sargassum semantic segmentation workflow.

The rest of this paper is organized as follows. Section 2 reviews the related work regarding sargassum monitoring and semantic segmentation. Section 3 covers our proposed methodology, including data acquisition, the architectures, and the experimental setup. In Sect. 4, we present the experimental results along with a comparative performance analysis. Section 5 provides a discussion on specific segmentation challenges, such as class imbalance and per-class performance. Finally, Sect. 6 offers our concluding remarks and outlines directions for future research.

2 Related Work

While satellite remote sensing has been widely used for regional monitoring of sargassum influxes [7,8,15], ground-level imagery has gained increasing relevance for beach-scale analysis, providing the detailed view required for local accumulation assessments [3,18]. In this context, semantic segmentation offers advantages over simple detection methods because it enables the estimation of spatial coverage in heterogeneous beach environments [2].

Recent studies have applied deep learning to coastal imagery. Arellano-Verdejo et al. [2] developed a segmentation methodology using georeferenced photographs to generate coverage maps. Subsequently, Arellano-Verdejo and Lazcano- Hernández [1] compared U-Net and Pix2Pix architectures on citizen science images, confirming the viability of supervised workflows on real-world data. Similarly, Vasquez et al. [18] demonstrated that smartphone and digital-camera images provide a low-cost alternative to high-resolution satellite data, which is crucial for rapid response actions.

At broader spatial scales, specialized architectures have been tailored for specific satellite sensors [8], and lightweight variants such as SLWE-Net [15] have reduced computational complexity without sacrificing performance. These satellite-based studies underscore the importance of selecting and tuning architectures according to the specific visual conditions of the data source [7].

Despite the widespread adoption of U-Net [14] as a baseline for environmental segmentation [5,12], multi-scale feature aggregation strategies have been less explored for ground-level sargassum imagery. Feature Pyramid Networks (FPN) [9] offer a distinct structural approach, combining high-level semantic information with fine spatial details through a top-down pathway and lateral connections. This is particularly relevant for beach scenes, where sargassum appears with highly variable sizes and distribution patterns.

Therefore, a direct and controlled experimental comparison between U-Net and FPN is relevant in this domain. Previous comparisons in the literature often confound the architectural paradigm, such as encoder-decoder and pyramid-based designs, with differences in the underlying feature extractors. Comparing both architectures using the same efficient backbone, such as EfficientNet [16], on a standardized dataset allows for a clearer assessment of their structural strengths and limitations for practical sargassum monitoring [2,18].

3 Materials and Methods

3.1 Dataset

One of the main challenges in ground-level sargassum semantic segmentation is the limited availability of standardized public datasets containing real coastal scenes. To address this limitation, this study utilized an image collection based on the dataset reported by Vasquez et al. [18], which was built from pictures captured with smartphones and digital cameras and collected from publicly available social media sources such as Facebook and Instagram. This dataset was selected because it contains beach scenes acquired under real-world conditions, with substantial variability in illumination, background composition, and sargassum accumulation patterns, making it suitable for evaluating semantic segmentation models in practical monitoring scenarios.

The dataset consists of 1,296 RGB images with manually assigned pixel-level ground-truth annotations. These images are organized into seven semantic classes: undefined, people, sargassum, sea, sand, sky, and vegetation. This multi-class structure provides a realistic representation of the visual complexity of coastal environments, where sargassum coexists with several surrounding elements that may introduce ambiguity during segmentation.

Ethical considerations were taken into account due to the presence of the "people" class in some images. The image collection used in this study consists of pre-existing, publicly accessible coastal photographs, and no images were collected directly from human participants as part of this research. No private or sensitive personal information was intentionally collected or analyzed. Furthermore, visible facial features or potentially identifying elements were anonymized prior to processing. The use of the images was limited to semantic segmentation of coastal scene elements, and no attempt was made to identify, profile, or track any individual appearing in the dataset.

The selection of this collection is also supported by the operational advantages of ground-level imagery. As discussed by Vasquez et al. [18], high-resolution satellite images may be expensive or unavailable at the required time, whereas photographs acquired with mobile devices provide an accessible and low-cost source of visual information for rapid beach assessment. Therefore, this dataset constitutes an appropriate basis for studying sargassum semantic segmentation in coastal scenes from the Mexican Caribbean, where local visual detail is especially important for characterizing accumulation areas.

The dataset was divided randomly into three subsets: 991 images for training (75%), 175 images for validation (15%), and 130 images for testing (10%). This partition enables the separation of model fitting, validation during training, and final evaluation on unseen images.

3.2 Feature Pyramid Network (FPN)

As the main segmentation model, we utilized a Feature Pyramid Network (FPN) [9] built on top of a convolutional encoder. The encoder generates a hierarchy of feature maps at different spatial resolutions, where deeper layers capture stronger semantic information while shallower ones preserve finer spatial details. The FPN combines these complementary representations through a top-down pathway and lateral connections, allowing the model to integrate contextual and spatial information across multiple scales. This design is especially suitable for coastal scenes, where sargassum may appear with variable shapes, sizes, and distributions within the same image.

In this study, EfficientNet-B0 [16] with ImageNet pretrained weights was selected as the encoder backbone within the FPN architecture. The resulting multi-scale feature maps were processed by a segmentation head to generate dense pixel-wise predictions for the seven semantic classes defined in the dataset.

3.3 U-Net

As a baseline model, we utilized U-Net [14], an encoder-decoder architecture with symmetric skip connections between the contracting and expanding paths. The encoder progressively reduces spatial resolution to capture high-level contextual features, while the decoder restores resolution through upsampling stages to obtain dense pixel-wise predictions. Skip connections concatenate fine-grained features from early encoder stages with decoder features at matching scales, which helps preserve spatial details and recover structures that may be lost during downsampling.

For a fair and controlled architectural comparison, U-Net was also implemented using EfficientNet-B0 [16] as the encoder backbone with ImageNet pre-trained weights. The architecture was configured with an encoder depth of five levels and decoder channels of `[256, 128, 64, 32, 16]`. The final layer produced segmentation masks for the seven semantic classes of the problem. Due to its strong performance and widespread adoption in dense prediction tasks, U-Net was selected as a reference architecture for experimental comparison.

3.4 Design of Experiment

The experimental design was structured to compare U-Net and FPN under the same dataset partition. Both architectures were trained and evaluated using the same training, validation, and test subsets to ensure a fair comparison.

The implementation was carried out in Python using PyTorch [11], the Segmentation Models PyTorch library, and Albumentations for data transformations. Input images and ground-truth masks were resized to 704×1056 pixels using nearest-neighbor interpolation before being fed into the models. In addition, pixel values were normalized using the standard ImageNet mean and standard deviation values: (0.485, 0.456, 0.406) and (0.229, 0.224, 0.225), respectively.

The optimization process was performed using the AdamW optimizer with a maximum learning rate of `1e-3` and a weight decay of `1e-4`, together with the standard `CrossEntropyLoss` function. A OneCycle learning-rate scheduler was also employed to improve convergence during training. Both models were trained for 60 epochs with a batch size of 3, and the data were shuffled for both training and validation loaders.

To enhance model robustness and prevent overfitting, specific data augmentation pipelines were defined. For the training set, augmentations included horizontal and vertical flipping with a default probability of 0.5, grid distortion ($p = 0.2$), random brightness and contrast adjustments with limits up to 0.5, and Gaussian noise injection. For the validation subset, the pipeline included resizing, horizontal flipping, and grid distortion ($p = 0.2$). Final evaluation was performed on the separate test subset, which consisted of images not seen during model fitting.

3.5 Evaluation Metrics

Model performance was assessed using pixel accuracy and Intersection over Union (IoU). Pixel accuracy measures the proportion of correctly classified pixels over the total number of pixels. The mean IoU was used as the main overall segmentation metric, since it quantifies the overlap between predicted masks and ground-truth annotations across classes and provides a robust measure of segmentation quality in multi-class problems. Furthermore, per-class IoU was calculated to analyze the performance of each semantic category, with particular attention to the sargassum class as the primary object of interest.

These metrics served to compare the performance of U-Net and FPN, enabling an analysis of the suitability of each architecture for the semantic segmentation of sargassum in heterogeneous coastal scenes.

4 Results

4.1 Semantic Segmentation Performance

The semantic segmentation performance of the evaluated architectures was analyzed through a series of experiments. To ensure a fair and direct comparison of their structural paradigms, both models utilized the identical EfficientNet-B0 backbone and were trained for 60 epochs. In both architectures, the best results were obtained with a batch size of 3, indicating that these hyperparameter settings were suitable for the dataset and training conditions considered in this study.

Table 1 summarizes the overall performance of the U-Net and the FPN architectures on the test set. The U-Net architecture achieved a mean IoU of 0.69 and a pixel accuracy of 0.93. In comparison, the FPN architecture reached a mean IoU of 0.70 and a pixel accuracy of 0.94. These findings indicate that FPN achieved a marginal improvement over U-Net under the evaluated experimental settings, while both architectures performed comparably in the semantic segmentation of heterogeneous coastal scenes.

Table 1. Overall performance obtained on the test set for the U-Net and FPN architectures (both utilizing EfficientNet-B0) in sargassum semantic segmentation.

Architecture	Backbone	Batch Size	Epochs	Pixel Accuracy	Mean IoU
U-Net	EfficientNet-B0	3	60	0.93	0.69
FPN	EfficientNet-B0	3	60	**0.94**	**0.70**

To provide a more comprehensive analysis of the models' behavior, particularly concerning the primary object of interest, a per-class IoU breakdown was calculated (Table 2). This analysis shows how each architecture performs across the different environmental classes. Both models achieved strong performance on visually homogeneous classes such as sky (IoU > 0.98), sea (IoU > 0.90), and sand (IoU > 0.82), while lower IoU values were observed for more challenging classes, including people, undefined regions, and sargassum. The sargassum class remained particularly challenging due to its variable texture, irregular spatial distribution, and class imbalance within the dataset. For this class, FPN reached an IoU of 0.6401, compared with 0.6372 for U-Net. Although this difference was marginal, FPN also obtained higher IoU values for undefined regions, people, and vegetation, whereas U-Net achieved slightly higher scores for sky, sea, and sand. These results indicate that both architectures performed comparably at the class level, with small differences depending on the visual characteristics of each category.

Table 2. Per-class Intersection over Union (IoU) scores on the test set.

Class	U-Net IoU	FPN IoU
Undefined	0.6308	0.6516
People	0.3302	0.4869
Sargassum	0.6372	0.6401
Sea	0.9029	0.9020
Sand	0.8312	0.8249
Sky	0.9879	0.9864
Vegetation	0.7798	0.7933

The learning curves recorded during training, including loss (Fig. 2), accuracy (Fig. 3), and mean IoU (Fig. 4), showed that both models were able to learn meaningful representations for the segmentation task and converged properly over the 60 epochs. The FPN showed a slightly more favorable overall trend in the recorded global metrics, although the per-class analysis revealed small differences between architectures depending on the semantic category.

Overall, the experimental results confirm that semantic segmentation is a feasible strategy for the analysis of ground-level coastal images containing sargassum. Although the performance gap between both architectures was marginal, the FPN provided the best overall results in terms of mean IoU and pixel accuracy, which supports its selection for the subsequent deployment stage. However, the per-class results showed that performance differences varied across semantic categories. An example of a segmented image using both models is shown in Fig. 5.

4.2 Cloud-Based Application

The best-performing model was integrated into a cloud-based application for semantic segmentation of coastal images. The system was implemented using

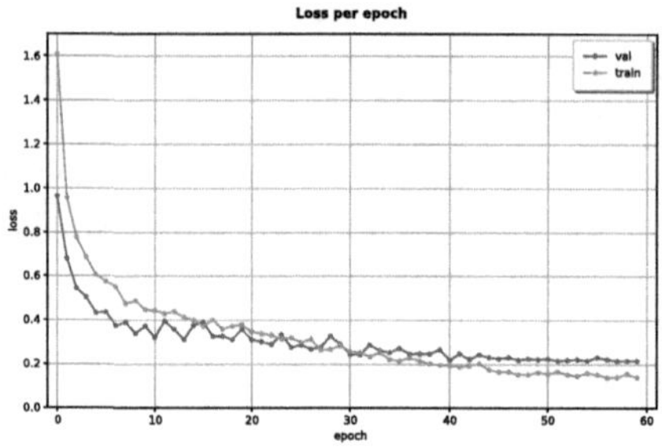

(a) U-Net training and validation Loss.

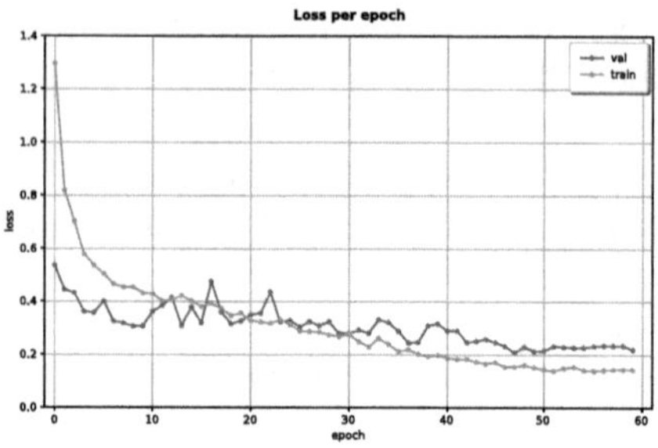

(b) FPN training and validation loss.

Fig. 2. Comparison of training and validation loss between U-Net and FPN.

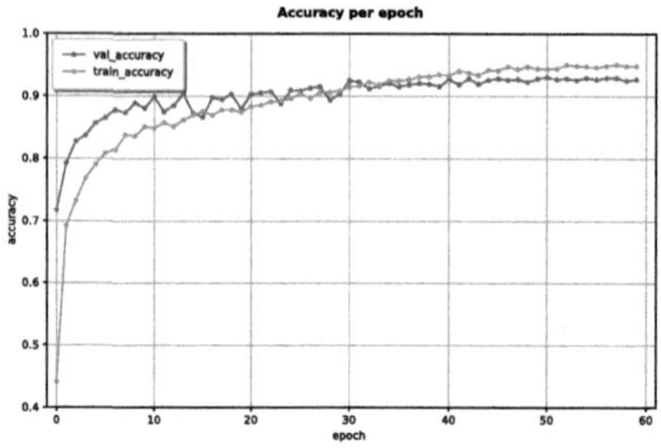

(a) U-Net training and validation accuracy.

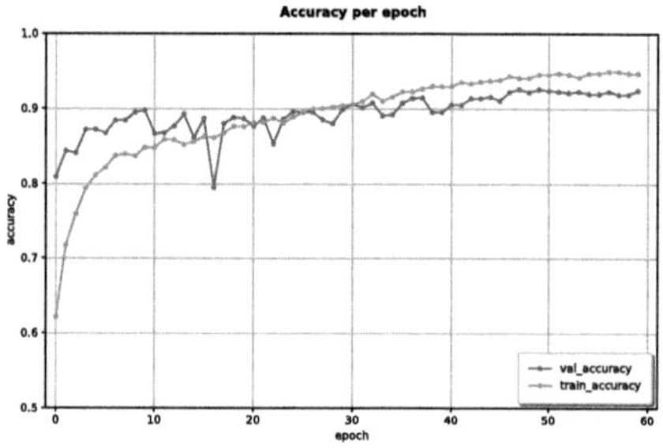

(b) FPN training and validation accuracy.

Fig. 3. Comparison of training and validation accuracy between U-Net and FPN.

Flask [10], PyTorch [11], and Docker [6], with the trained FPN model stored in PyTorch format and included within the application container.

The application allows users to upload a coastal image through a web interface, process it with the trained segmentation model, and receive the segmented output automatically, as illustrated in Fig. 6. This workflow provides an accessible mechanism for image analysis without requiring local installation of specialized software or hardware.

For deployment, the Docker image was validated locally and then uploaded to Google Cloud Container Registry, from which it was deployed on Google Cloud Run. To evaluate the operational feasibility of the prototype, preliminary performance metrics were recorded utilizing a standard CPU-only environment. The end-to-end inference latency, encompassing image upload, model processing, and result rendering, averaged approximately 10 s per image. Furthermore, under a preliminary stress test with up to 5 concurrent users, the application maintained stability with a peak CPU utilization of approximately 20%. These results not only suggest the practical feasibility of deploying semantic segmentation tools in the cloud for sargassum monitoring, but also indicate that scalable,

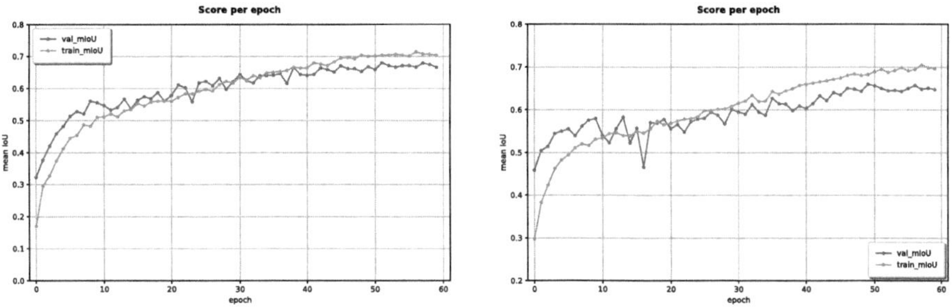

(a) U-Net training and validation IoU. (b) FPN training and validation IoU.

Fig. 4. Comparison of training and validation mean IoU between U-Net and FPN.

real-time inference can be achieved cost-effectively without relying on expensive GPU acceleration.

5 Discussion

The experimental results demonstrate that both U-Net and FPN architectures, when using a standardized EfficientNet-B0 backbone, are capable of achieving high pixel accuracy and competitive IoU scores for sargassum segmentation in complex coastal scenes. However, the per-class IoU analysis (Table 2) provides a more detailed understanding of the models' performance and the inherent challenges of the dataset.

A significant observation is the performance gap between semantic classes. Homogeneous classes such as sky, sea, and sand achieved high IoU scores, as their visual features are relatively consistent across different lighting conditions and coastal scenes. In contrast, the people class obtained the lowest IoU values, with 0.3302 for U-Net and 0.4869 for FPN, suggesting that minority classes remain difficult to segment accurately. The sargassum class yielded IoU scores of 0.6372 for U-Net and 0.6401 for FPN, showing that both models performed similarly for the main class of interest. This performance is likely influenced by the high textural variability of the macroalgae, its irregular spatial distribution, and the presence of class imbalance within the dataset.

Regarding the loss function, this study employed a standard Cross Entropy Loss without class weighting. While this allowed for a baseline comparison of the architectural paradigms, the results suggest that incorporating specialized loss functions, such as Focal Loss or weighted Cross-Entropy, could further improve the delineation of sargassum and other minority classes by penalizing misclassifications in underrepresented categories more heavily.

The per-class results indicate that the relative performance of each architecture depends on the semantic category. FPN obtained higher IoU values for undefined regions, people, sargassum, and vegetation, whereas U-Net achieved slightly higher scores for sky, sea, and sand. This suggests that the multi-scale

(a) Original coastal image.

(b) Ground-truth mask.

(c) Predicted mask with FPN.

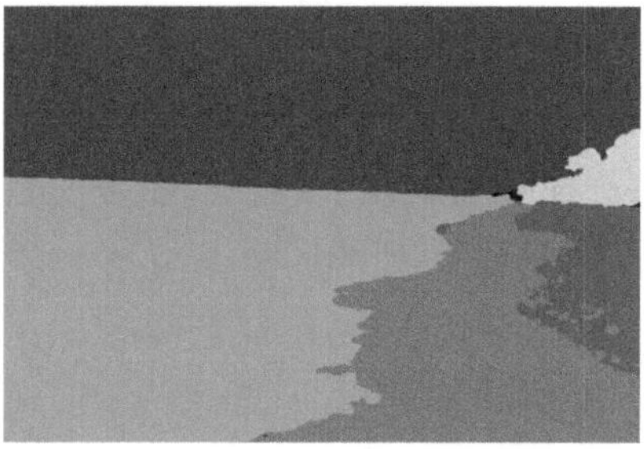

(d) Predicted mask with U-Net.

Fig. 5. Qualitative segmentation result from the test set for a coastal scene containing *sargassum*. The figure shows (a) the original coastal image, (b) the ground-truth mask, (c) the predicted mask obtained with FPN, and (d) the predicted mask obtained with U-Net.

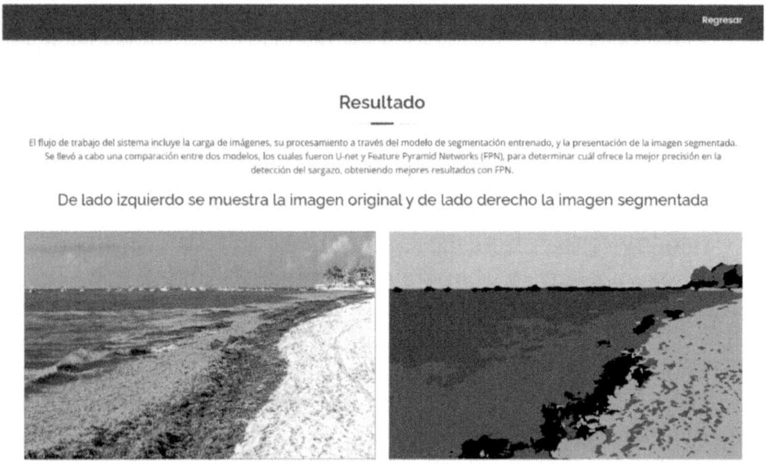

Fig. 6. Screenshot of the web-based application used for sargassum monitoring. The interface displays the original coastal image on the left and the corresponding segmented result generated by the FPN model on the right.

feature aggregation mechanism of FPN may be beneficial for visually heterogeneous or irregular classes, while both models remain comparable for more homogeneous environmental categories.

Finally, the cloud-based deployment demonstrated that the selected model can operate efficiently in a CPU-limited environment. The 10-second latency and low CPU utilization (20%) under concurrent use indicate that the system is suitable for practical coastal monitoring without requiring expensive high-performance computing resources. Future work should focus on expanding the dataset to include more diverse coastal scenarios, incorporating strategies to address class imbalance, and exploring more recent transformer-based architectures to enhance generalization.

6 Conclusions

This research presented a comparative evaluation of U-Net and Feature Pyramid Network (FPN) architectures for the semantic segmentation of ground-level sargassum imagery. By standardizing the feature extractor with an EfficientNet-B0 backbone, a fairer architectural comparison was achieved. The results showed that both models performed comparably, although FPN obtained a small but consistent advantage over U-Net across the evaluated experiments, reaching a mean IoU of 0.70 and 94% pixel accuracy.

The observed difference in favor of FPN was marginal, but its performance trend remained slightly higher than U-Net under the same experimental conditions. This suggests that the multi-scale feature aggregation strategy of FPN can provide benefits for segmenting heterogeneous coastal scenes, particularly where sargassum appears with irregular shapes, textures, and spatial distributions.

The study highlights that semantic segmentation is a viable and accessible strategy for monitoring sargassum influxes at a beach scale. The integration of the best-performing model into a cloud-based application demonstrates the practical feasibility of these tools for coastal management, providing automated and timely information to stakeholders. Although class imbalance remains a challenge, the results provide a solid foundation for further optimization through advanced loss functions, multiple training runs, and larger, more diverse datasets. This work contributes to the development of low-cost, AI-driven solutions for environmental monitoring in the Mexican Caribbean and beyond.

Acknowledgments. The authors would like to thank the Instituto Politécnico Nacional (Secretaría Académica, Comisión de Operación y Fomento de Actividades Académicas, Secretaría de Investigación y Posgrado, and CIDETEC), the Secretaría de Ciencia, Humanidades, Tecnología e Innovación, and the Sistema Nacional de Investigadoras e Investigadores (SNII) of Mexico, for their support in developing this work.

Disclosure of Interests. The authors have no competing interests to declare that are relevant to the content of this article.

References

1. Arellano-Verdejo, J., Lazcano-Hernandez, H.E.: Two algorithms for sargassum segmentation in citizen science images. In: 2023 Mexican International Conference on Computer Science (ENC), pp. 1–8. IEEE (2023). https://doi.org/10.1109/ENC59042.2023.10403287
2. Arellano-Verdejo, J., Santos-Romero, M., Lazcano-Hernandez, H.E.: Use of semantic segmentation for mapping sargassum on beaches. PeerJ **10**, e13537 (2022). https://doi.org/10.7717/peerj.13537
3. Arellano-Verdejo, J., et al.: Towards enhanced sargassum monitoring in the Caribbean sea. Sci. Rep. (2025). https://doi.org/10.1038/s41598-025-93001-9
4. Chavez, V., et al.: Massive influx of pelagic sargassum spp. on the coasts of the Mexican Caribbean 2014–2020: challenges and opportunities. Water **12**(10), 2908 (2020). https://doi.org/10.3390/w12102908
5. Dimitrovski, I., Spasev, V., Loshkovska, S., Kitanovski, I.: U-Net ensemble for enhanced semantic segmentation in remote sensing imagery. Remote Sens. **16**(12), 2077 (2024). https://doi.org/10.3390/rs16122077
6. Docker, Inc.: Docker documentation (2025). https://docs.docker.com/. Accessed 30 Mar 2026
7. Echevarría-Rubio, J.M., Martínez-Flores, G., Morales-Pérez, R.A.: Machine and deep learning framework for sargassum detection and fractional cover estimation using multi-sensor satellite imagery. Data **10**(11), 177 (2025). https://doi.org/10.3390/data10110177
8. Laval, M., et al.: Detection of sargassum from sentinel satellite sensors using deep learning approach. Remote Sens. **15**(4), 1104 (2023). https://doi.org/10.3390/rs15041104
9. Lin, T., Dollár, P., Girshick, R., He, K., Hariharan, B., Belongie, S.: Feature pyramid networks for object detection. In: 2017 IEEE Conference on Computer Vision and Pattern Recognition (CVPR), pp. 936–944. IEEE (2017). https://doi.org/10.1109/CVPR.2017.106
10. Pallets Projects: Flask documentation, version 2.2.x (2022). https://flask.palletsprojects.com/en/2.2.x/, version 2.2.2. Accessed 30 Mar 2026
11. Paszke, A., et al.: Pytorch: an imperative style, high-performance deep learning library. In: Advances in Neural Information Processing Systems, vol. 32, pp. 8024–8035. Curran Associates, Inc. (2019)
12. Peng, Y., Sonka, M., Chen, D.Z.: U-Net v2: rethinking the skip connections of U-Net for medical image segmentation. arXiv arXiv:2406.06613 (2024)
13. Rodríguez-Martínez, R.E., van Tussenbroek, B., Jordán-Dahlgren, E.: Afluencia masiva de sargazo pelágico a la costa del caribe mexicano (2014–2015). Florecimientos Algales Nocivos en México, pp. 352–365 (2016)
14. Ronneberger, O., Fischer, P., Brox, T.: U-Net: convolutional networks for biomedical image segmentation. In: Navab, N., Hornegger, J., Wells, W.M., Frangi, A.F. (eds.) MICCAI 2015. LNCS, vol. 9351, pp. 234–241. Springer, Cham (2015). https://doi.org/10.1007/978-3-319-24574-4_28
15. Song, L., Chen, Y., Liu, S., Xu, M., Cui, J.: SLWE-Net: an improved lightweight U-Net for sargassum extraction from GOCI images. Mar. Pollut. Bull. **194**, 115349 (2023). https://doi.org/10.1016/j.marpolbul.2023.115349
16. Tan, M., Le, Q.V.: EfficientNet: rethinking model scaling for convolutional neural networks. In: Proceedings of the 36th International Conference on Machine Learning (ICML). Proceedings of Machine Learning Research, vol. 97, pp. 6105–6114. PMLR (2019). https://proceedings.mlr.press/v97/tan19a.html

17. van Tussenbroek, B.I., et al.: Severe impacts of brown tides caused by sargassum spp. on near-shore Caribbean seagrass communities. Mar. Pollut. Bull. **122**(1–2), 272–281 (2017). https://doi.org/10.1016/j.marpolbul.2017.06.057
18. Vasquez, J.I., Uriarte-Arcia, A.V., Taud, H., García-Floriano, A., Ventura-Molina, E.: Coastal sargassum level estimation from smartphone pictures. Appl. Sci. **12**(19), 10012 (2022). https://doi.org/10.3390/app121910012

A Workflow for Fundus Imaging with Multipurpose Applications: Data Augmentation, Education, and Clinical Counseling

Jonathan Garcia Amador, Augusto Javier Reyes Delgado(✉), José Luis Sánchez Cervantes, Lisbeth Rodríguez Mazahua, and Giner Alor Hernandez

I.T. Orizaba, Tecnológico Nacional de México, Orizaba, Mexico
{M15011148,D17010207,jose.sc,lisbeth.rm2, giner.ah}@orizaba.tecnm.mx

Abstract. Retinal diseases are among the leading causes of visual impairment worldwide, closely associated with chronic comorbidities such as type 2 diabetes and hypertension. Early detection and monitoring of disease progression are essential to prevent severe complications; however, the scarcity of high-quality medical datasets and hardware limitations restrict the development of robust predictive models. This paper presents a comparative analysis of StyleGAN2-ADA and StyleGAN3 architectures for fundus image synthesis, evaluating their capacity to generate realistic representations and to simulate pathological progression through transfer learning and latent space manipulation. The results indicate that StyleGAN2-ADA, initialized with pre-trained weights from CelebA-HQ, achieves improved visual fidelity, reaching an FID score of 23.8609 under constrained computational settings. Additionally, the model demonstrates efficient performance when trained on a T4 GPU, suggesting that high-quality image synthesis can be achieved with reduced memory requirements compared to more resource-intensive configurations, such as those involving A100 GPUs. These findings support the feasibility of developing synthetic biomedical repositories for applications in medical education and awareness, contributing to broader accessibility of generative artificial intelligence tools under limited computational resources.

Keywords: adversarial neural networks · age-related macular degeneration · generative artificial intelligence · knowledge transfer · StyleGAN2-ADA

1 Introduction

Retinal diseases, such as diabetic retinopathy (DR), hypertensive retinopathy (HR), and age-related macular degeneration (AMD) represent critical public health challenges. According to the World Report on Vision of the World Health Organization (WHO) [1], it is estimated that 2,200 million people live with some visual impairment, of which at least 1,000 million of these conditions could have been prevented or have not been

H. Ponce et al. (Eds.): COMIA 2026, CCIS 3095, pp. 438–450, 2027.
https://doi.org/10.1007/978-3-032-37438-7_34

treated. In the field of DR, the International Agency for the Prevention of Blindness (IAPB) and the International Diabetes Federation (IDF) [2] report a global prevalence of 22% among diabetic patients, warning that 1 in 10 will develop forms of the disease that seriously threaten vision. By 2020, these figures translated into 103 million adults with DR, of which 47 million required urgent treatment.

Likewise, high blood pressure, the main risk factor for HR, affects 1,280 million adults; however, a study by the WHO and Imperial College London [3]. highlights that about 580 million people are unaware of their diagnosis. This lack of clinical control exacerbates the risk of severe eye complications, making early intervention difficult. Against this backdrop, generative artificial intelligence has emerged as a disruptive technology. Architectures based on Generative Adversarial Networks (GANs) [4], such as StyleGAN, and recent diffusion models, have demonstrated high fidelity in the synthesis of synthetic fundus images, facilitating medical education and training of robust diagnostic systems

Despite the advances, the scarcity of quality clinical databases and ethical constraints limit the development of predictive models. In this context, conditional generative models offer a viable alternative by allowing the simulation of pathological progression controlled by clinical covariates [5].

This study performs a comparative analysis between StyleGAN2-ADA [6] and StyleGAN3 [7] for the synthesis of fundus images. Their ability to transform real retinal images into synthetic representations by manipulating the latent space is evaluated. The models are integrated into a modular architecture that allows visualizing evolutionary stages of pathologies, with an emphasis on the balance between visual fidelity (FID) and computational cost. This makes it possible to create a synthetic biomedical repository designed for awareness, medical education and data augmentation purposes, critical fields to promote a culture of prevention that, in addition to preserving vision in vulnerable populations, results in future savings for the health system, by supporting awareness of advanced complications.

The primary purpose of the system is not to clinically predict the exact course of eye disease in a specific patient, but rather to generate approximate visual representations aimed at raising awareness, providing medical education, and offering visual support during clinical consultations. Due to the probabilistic nature of generative GAN models and the limited availability of longitudinal clinical data, the resulting simulations should not be interpreted as deterministic medical prognoses.

The present paper is organized as follows: Sect. 2 presents a detailed review of the recent related literature. Section 3 describes the characteristics of the models being compared, detailing the technical underpinnings of StyleGAN2-ADA and StyleGAN3. Section 4 details the experiments performed for the simulation of pathological progression and the evaluation of visual fidelity. Finally, Sect. 5 presents the general conclusions and future lines of work.

2 State of the Art

In recent years, generative models have radically transformed the application of medical imaging, facilitating the synthesis of visual data with high technical fidelity, mitigating data scarcity, and promising diagnostic utility. During 2021, Coyner et al. [8] validated this ability by generating childhood pathologies with high diagnostic accuracy. Subsequently, in 2022, Kim et al. [9] demonstrated the efficacy of StyleGAN to produce high-resolution retinas using massive datasets. These advances laid the foundation for pathological control. In 2023, the work of Wang et al. [10] marked a foundational milestone by evaluating the quality of synthetic images using visual Turing tests. Their research showed that even ophthalmologists have statistical difficulties in distinguishing between real and synthetic images of the retina, reaching a sensitivity of only 71.1%. Simultaneously, Güllmar et al. [11] set an essential methodological precedent by using the latent space of StyleGAN to predict evolutionary patterns of tissue damage. Although his study focused on multiple sclerosis, he demonstrated the versatility of latent transition vectors to model the temporal progression of diseases, a concept that is directly applicable to retinal degradation. During 2024, the focus of research shifted to improving structural integrity and usefulness for supervised training. Go et al. [12] They introduced the use of diffusion models to optimize the synthesis of the retinal microvasculature, successfully addressing vascular continuity issues that traditional adversarial networks used to present. For their part, Feng et al. [13] developed a diversified synthesis framework specifically designed for lesion segmentation in diabetic retinopathy (RD). This approach made it possible to generate mask image pairs with superior anatomical realism, making it easier for deep learning algorithms to learn pathological features that are inherently scarce in traditional databases. Pandey et al. [14] They validated that integrating synthetic images into clinical workflows significantly improves the accuracy of automatic disease classification systems. On the frontier of precise editing, Pakdelmoez et al. [15] perfected controllable synthesis by manipulating the latent space of StyleGAN, managing to adjust the severity of RD and modify specific attributes such as vessel structure and lesions without the need for auxiliary networks. More recently, Hou et al. [16] presented Fundus GAN, a hierarchical framework that captures both macroscopic structures and subtle pathological details using feature pyramid networks. Table 1 synthesizes recent research (last 5 years) on generative architectures in ophthalmology, from high-resolution synthesis to latent space editing.

The analysis highlights models based on adversarial networks, diffusion, and latent manipulation techniques, where the evolution of these technologies shows a clear progression from the validation of basic realism to the controlled manipulation of complex pathologies

However, the literature shows that the optimization of these models to operate efficiently on conventional hardware by reducing FID thresholds remains as an opportunity area, the present work tries to cover this gap through optimized architecture. In addition, (i) the creation of educational biomedical repositories is considered an underexplored area and (ii) a lack of studies that technically compare StyleGAN2-ADA and StyleGAN3 under conventional hardware restrictions and focused on the progression of retinal diseases.

Therefore, making a comparison that evaluates visual fidelity (FID) together with computational cost offers a necessary perspective that is little addressed in the literature. This research seeks to fill that gap, providing a methodology that allows expanding access to medical generative AI, allowing the development of high-quality clinical awareness tools in conventional infrastructure.

Table 1. Comparative table of related works

Work	Model	Task	FID
Coyner et al. [2021]	Pix2Pix	IG	–
Kim et al. [2022]	StyleGAN	IG	54.3
Wang et al. [2023]	StyleGAN2	IG	12.8
Güllmar et al. [2023]	StyleGAN	LE	15.2
Go et al. [2024]	LDM (Broadcast)	IG	24.3
Feng et al. [2024]	ControlNet	IG	31.4
Pandey et al. [2025]	StyleGAN2-ADA	IG	18.6
Pakdelmoez [2025]	Cond. StyleGAN	IG	22.1
Hou et al. [2025]	FundusGAN	IG	11.4

GI = Generation of images; **LE** = Latent editing;

3 Methods and Materials

A detailed comparison is made between the architectures of StyleGAN2-ADA and StyleGAN3 for the synthesis of fundus images. The objective is to evaluate the ability of each model to generate realistic representations of retinal diseases, such as AMD, allowing the visual transition between different evolutionary stages by manipulating the latent space. The selection of technologies was made both for their ability to model complex vascular structures, and for the need to analyze the cost-benefit balance between the required computational power and the fidelity of the results obtained.

Considering the hardware limitations common in academic and clinical environments where access to high-performance GPUs is limited, this analysis seeks to determine if the architectural complexity of StyleGAN3 offers a significant statistical and clinical improvement that justifies its high resource consumption versus the efficiency and robustness of StyleGAN2-ADA in limited data scenarios.

3.1 Selected Generative Architectures

- StyleGAN2 with Adaptive Discriminator Augmentation (ADA): Presented by Karras et al. [6] in the article "Training Generative Adversarial Networks with Limited Data," this architecture was selected due to its ability to stabilize training in scenarios with small datasets, a common condition in biomedical applications. ADA addresses discriminator overfitting through an adaptive probabilistic augmentation mechanism, applying transformations such as rotations, translations, and intensity changes according to the level of memorization detected during training. This approach improves model convergence, reduces dependence on large data volumes, and optimizes the use of computational resources and VRAM, facilitating the synthesis of high-fidelity retinal images in hardware-constrained environments.

- StyleGAN3: Introduced in the work "Alias-Free Generative Adversarial Networks" [7], Designed to suppress the phenomenon of "aliasing", this architecture redefines the synthesis process by interpreting network signals as continuous rather than discrete functions. It implements anti-aliasing filters and redesigns the convolutional layers to be completely translational and rotational invariant. These technical modifications ensure superior structural continuity and independence from the fixed pixel grid, facilitating fluid transformations in the latent space that are critical to represent the morphology of vascular structures.

3.2 Proposed Work Pipeline

The designed workflow was structured with the aim of maximizing algorithmic efficiency and statistical fidelity in conventional hardware environments. Unlike traditional methodologies that rely on supercomputing infrastructures, this pipeline integrates an optimization strategy based on transfer learning and adaptive hyperparameter refinement. The stages that make up this methodology are illustrated in detail in Fig. 1.

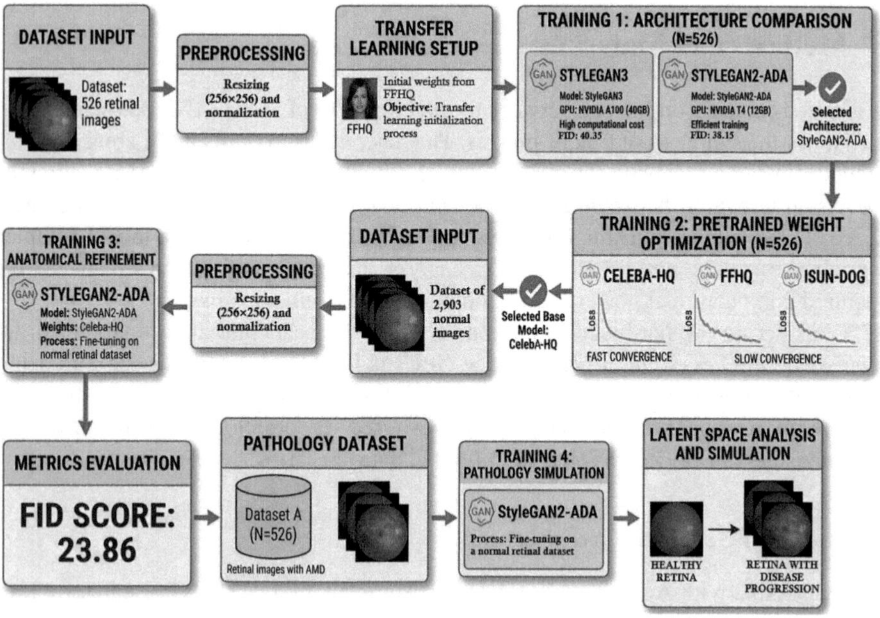

Fig. 1. Pipeline outline

The process is divided into systematic stages that range from the curation of a clinical dataset to quantitative validation using the FID metric and the simulation of pathologies in the latent space. Each phase was calibrated to ensure that computational consumption was reduced without compromising the anatomical accuracy of the syntheses. The detailed steps of the methodology are:

1. Dataset collection and curation stage. Fundus images from multiple public and clinical sources were integrated. The process was divided into two phases: an initial phase of 512 images for architectural testing and a refinement phase with a dataset of 2,903 healthy images to ensure anatomical accuracy.
2. Pre-processing stage. A standardized protocol was applied to correct posterior pole centering, eliminate black margins, and normalize all samples to a consistent resolution of 256×256 pixels.
3. Hardware preparation and testing stage. The repositories were configured to run in two different environments: a high-performance environment with NVIDIA A100 GPUs (40GB) to evaluate StyleGAN3, and a consumer-grade environment with NVIDIA T4 GPUs (12GB) for StyleGAN2-ADA.
4. Training and weight transfer stage. The training was carried out through transfer learning using CelebA-HQ base weights. The adaptive resistance of StyleGAN3 was compared with the flexibility of StyleGAN2-ADA, selecting the latter for its higher efficiency and accelerated convergence.
5. Quantitative evaluation and simulation stage. The results were validated using the FID (Fréchet Inception Distance) metric, which measures the statistical similarity between real and synthetic distributions. After reaching an FID of 23.86, the latent space manipulation was performed to simulate the progression of AMD from transition vectors.

3.3 Deployment Details

Sources dedicated to quantifying statistics regarding the use of GPUs for specific tasks, such as training AI models, are scattered and inconsistent; however, an estimate of general hardware usage can be derived from data from the Steam® [17, 18] platform, which massively collects this data through its hardware surveys. In its most recent report from February, a significant change in the global trend is observed, where the Nvidia RTX 5070 is positioned as the most used GPU with 18.24% market share, followed by the Nvidia RTX 4060 with 14.46%.

It is relevant to note that these FMCG cards operate mostly with 12GB of VRAM configuration. In this study, we sought to replicate these general-purpose hardware conditions using the Google Colab® environment with the T4 GPU, which also has 12GB of VRAM.

For the StyleGAN3 architecture, due to its high processing demands, the use of an A100 GPU was required. However, considering the operational cost, our workflow prioritized optimization in the Style-GAN2-ADA architecture over T4. This allows us to demonstrate that it is possible to achieve high-fidelity results by circumventing memory limitations without relying on high-cost infrastructure, aligning ourselves with the resources available to most current users.

3.4 Dataset

For the development of this work, two sets of data organized in differentiated stages were used. In the first phase, with the aim of identifying the most efficient architecture within the Google Colab® environment under computational constraints, a corpus of 512 fundus images with the presence of AMD was integrated. These samples come from public repositories of Kaggle [19, 20], the ADAM [21] dataset, an open-access dataset provided by external researchers [22]. and, finally, a contribution of thirteen clinical images from the Conde de Valenciana Institute of Ophthalmology [23].

In the second phase, an additional [24] public dataset was used, containing various retinopathies. However, since the interest of this stage was the pure representation of ocular anatomy, only the subset of images in normal condition was selected, totaling 2,903 images.

Because the original data presented notable inconsistencies that could affect the training of the GANs, a standardized preprocessing protocol composed of three fundamental steps was applied:

- Fundus Centering: Corrected the alignment of the posterior pole in those samples that were off-center to ensure spatial uniformity.
- Minimization of external margins: Black edges and added frames were eliminated or reduced, allowing the retinal field to occupy the entire image area.
- Resolution Normalization: All samples were resized to a resolution of 256 $\times$ 256 pixels, ensuring a 1:1 aspect ratio across the entire training set to stabilize the gradient calculation.

After the training, a latent space interpolation phase was implemented to identify the transition vectors between a healthy retina and a pathological one. This process made it possible to simulate the progression of the disease in a synthetic way, taking AMD as a case study

4 Results

A transfer learning scheme was implemented in both versions of StyleGAN using a standardized dataset to ensure equitable testing conditions. The learning transfer strategy was selected because training deep generative models from scratch requires large data volumes, lengthy processing times, and high-performance computing resources—conditions that contradict the central objective of this work, which focuses on accessible hardware environments and limited resources. Although a domain gap exists between facial and retinal images, the first layers of the GAN architectures retain reusable general visual representations during the fine-tuning process. These include patterns related to edges, textures, and lighting. Consequently, the training objective was not only to achieve convergence but also to evaluate each architecture's ability to transfer these general representations to ocular morphology.

4.1 Training

At the end of the training period for each architecture, the Fréchet Inception Distance (FID) metric was recorded to assess the statistical similarity between the real and synthetic images generated. The analysis focused on the effectiveness of knowledge transfer and the networks' ability to assimilate ocular morphology from external visual domains. Because the dataset was previously curated, normalized, and clinically classified, the FID allowed for estimating how well the model learned the visual and anatomical distribution of the retinal domain, including vascular structure, texture, and visual coherence. Unlike perceptual metrics such as LPIPS, designed to compare specific image pairs during reconstruction tasks, the FID is more appropriate for pure synthesis GAN models, as it statistically compares deep feature distributions between real and synthetic image sets generated by the InceptionV3 network, without requiring a direct image-to-image match. The analysis focused on the effectiveness of knowledge transfer and the networks' ability to assimilate ocular morphology from external visual domains [25].

It is essential to highlight that, despite having used the NVIDIA A100 GPU (high performance and high cost) for the StyleGAN3 architecture, the model exhibited a tendency toward retinal texture smoothing due to its alias-free architecture and continuous filtering mechanisms. The complexity of its layers required excessive processing cycles to attempt to remove remnant features from the source dataset, resulting in an initial FID of 40.3547. Comparative details of this initial performance against the StyleGAN2-ADA architecture are presented in Table 2.

Table 2. Performance comparison by architecture and hardware

Architecture	Hardware (VRAM)	Starting Weights	FID Final (↓)
StyleGAN3	NVIDIA A100 (40GB)	FFHQ	40.3547
StyleGAN2-ADA	NVIDIA T4 (12GB)	FFHQ	38.1576

In contrast, StyleGAN2-ADA, running on the NVIDIA T4 GPU, demonstrated superior transfer capability. By possessing a more flexible architecture optimized for fine-tuning, the model achieved accelerated convergence and a more anatomically coherent representation using only 1.14 credits/h. The choice of the T4 is strategic, as its 12GB of VRAM faithfully replicates the capabilities of the graphics cards most used by the general public, such as the NVidia RTX 4060 and NVidia RTX 5070.

Despite the use of high-performance A100 hardware for StyleGAN3, the results showed a high FID of 40.3547. In contrast, StyleGAN2-ADA on the T4 GPU showed greater flexibility, achieving an initial FID of 38.1576 at an operating cost six times lower.

After validating that StyleGAN2-ADA was the most viable option for accessible hardware, an additional experimentation phase was proposed to determine which "base model" allowed a smoother transition towards the specific morphology of the fundus. As detailed in Table 3, three transfer sources were evaluated: FFHQ, LSUN-Dog and CelebA-HQ, monitoring their evolution in different training intervals (snapshots) [26].

Table 3. Comparison of FID score according to base weight and number of iterations (Snapshots)

Model	Base Weight (Pre-trained)	Snapshot	FID (N = 100)
StyleGAN2-ADA	stylegan2-ffhq-256x256.pkl	100	38.1576
	stylegan2-lsundog-256x256.pkl	100	31.5110
	stylegan2-celebahq-256x256.pkl	100	29.7051

When analyzing the results of Table 3, it was identified that the weight of FFHQ presented the greatest "visual inertia", hindering the learning of ophthalmological biomarkers. In contrast, CelebA-HQ and LSUN-Dog demonstrated accelerated convergence, managing to drop to 29.7051 and 31.5110 respectively.

This strategic selection phase was the determining factor that allowed the final training to proceed (Stage B). At this stage, the model focused exclusively on the normal ocular anatomy dataset (2,903 samples), which allowed the textures and structural coherence of the retina to be refined. As a result, the optimal synthesis yield documented in Table 4 was achieved, highlighting a final FID of 23.8609.

Table 4. Final performance of the optimized model (Stage B)

Model	Base Weight (Pre-trained)	Snapshot	FID (N = 100)
StyleGAN2-ADA	stylegan2-celebahq-256x256.pkl	135	23.8609

These results validate that the StyleGAN2-ADA configuration with CelebA-HQ weights constitutes the optimal route for retinal imaging on conventional hardware. Reaching an FID of 23.8609 with limited resources, performance comparable to models requiring substantially larger datasets is achieved. This balance between low computational cost and high anatomical accuracy consolidates the model as a viable tool for the creation of synthetic repositories and the simulation of AMD progression. A sample of the result is shown in Fig. 2.

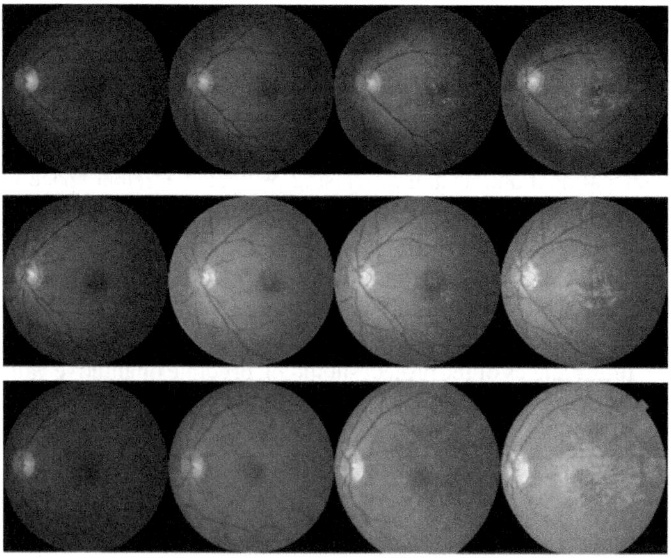

Fig. 2. Synthetic images generated

5 Discussion

The results obtained show that success in medical image synthesis depends not only on raw computing power, but also on the strategic synergy between architecture, knowledge transfer, and resource optimization. A key finding of this research was the operational advantage of StyleGAN2-ADA over StyleGAN3 in biomedical scenarios with limited biomedical datasets. Although StyleGAN3 was designed to eliminate aliasing through continuous filtering mechanisms and greater structural rigidity, this feature led to a tendency to smooth retinal micro textures, hindering efficient adaptation to ocular morphology and significantly increasing computational cost without a proportional improvement in visual fidelity. While StyleGAN3 achieved a FID of 40.3547 using an NVIDIA A100 GPU with 40GB of VRAM, StyleGAN2-ADA converged to a final FID of 23.8609 using only an NVIDIA T4 GPU with 12GB of VRAM.

When comparing the results obtained with contemporary literature (see Table 1), the proposed model's ability to achieve a FID of 23.8609 using a significantly smaller dataset and conventional hardware is noteworthy. Previous work, such as that by Go et al. [12], reported an FID of 24.3 using approximately 15,535 images and more computationally expensive diffusion models, while Pakdelmoez et al. [15] achieved an FID of 22.1 using nearly 20,000 retinal images. Similarly, high-fidelity hierarchical architectures such as P. Hou's achieved an FID of 11.4, although this depended on thousands of samples and more complex computational configurations. In contrast, the present work used only 2,903 anatomical images for the final refinement stage, validating that transfer learning from facial domains, particularly using CelebA-HQ initial weights, accelerates convergence and preserves anatomical coherence even under constraints of clinical data and limited computational resources.

The choice of CelebA-HQ as the base model proved particularly relevant, suggesting that visual representations learned in facial domains retain reusable general patterns during the retinal fine-tuning process. This finding reinforces the viability of transfer learning as a strategy to democratize access to biomedical generative artificial intelligence tools in academic and clinical settings with conventional infrastructure.

From an educational and awareness-raising perspective, latent space manipulation allowed for the generation of approximate visual representations of AMD progression, inspired by the latent editing principles proposed by D. Güllmar et al. [11]. However, the simulations obtained should not be interpreted as deterministic clinical predictions, but rather as visual approximations based on learned patterns from the retinal domain. Although the FID metric allowed for the evaluation of statistical consistency and overall visual fidelity, future stages of the project should incorporate qualitative validations with ophthalmologists and explore diffusion models aimed at refining microvasculature and progressively controlling pathological severity.

6 Conclusions and Future Work

The present study indicates that the synthesis of high-fidelity fundus images for the simulation of pathologies such as AMD does not depend exclusively on access to high-performance computing infrastructure, but rather on the strategic selection of architecture, data curation, and the origin of base weights.

Through comparative analysis, it is concluded that StyleGAN2-ADA outperforms StyleGAN3 in limited medical data environments, offering greater flexibility for knowledge transfer. While the StyleGAN3 architecture presented technical resistance in adapting and discarding residual features from the FFHQ dataset, StyleGAN2-ADA achieved accelerated convergence, particularly when initialized with CelebA-HQ base weights, reaching a final FID of 23.8609.

An important finding is the validation of the NVIDIA T4 GPU as a viable and efficient platform for working with small datasets. Obtaining superior results with a 12GB VRAM configuration, compared to the NVIDIA A100 with 40GB of VRAM, supports the feasibility of expanding the development of synthetic diagnostic tools under accessible computational conditions.

These results suggest that anatomical precision in retinal image synthesis is not determined by large-scale computational resources, but by the optimal balance between model architecture and training strategy, thereby enabling broader accessibility of these technologies in institutions with conventional infrastructure.

Finally, the ability to manipulate latent space to simulate disease progression opens new opportunities for medical education and patient awareness. The findings highlight the importance of strategic architectural selection and base weight initialization in clinical generative applications. Although StyleGAN3 offers an aliasing-free structure, StyleGAN2-ADA demonstrated superior performance in balancing visual fidelity and computational efficiency under constrained environments.

Future work will include the establishment of a clinical validation framework in collaboration with an ophthalmology foundation, with which ongoing discussions are currently being held. This collaboration is expected to enable a detailed and granular evaluation of the generated images, particularly focusing on the realism of pathological progression and its interpretability by medical specialists.

Additionally, future research should explore the integration of these architectures with multiscale care techniques and diffusion models to develop personalized simulation systems tailored to individual patient characteristics.

Acknowledgements. This paper was sponsored by Mexico's Secretariat of Science, Humanities, Technology, and Innovation (SECIHTI) and Mexico's Secretariat of Public Education (SEP) through the PRODEP program. Authors also thank Tecnológico Nacional de México (TecNM) for supporting this work.

Disclosure of Interests The authors have no competing interests to declare that are relevant to the content of this article.

References

1. World Health Organization: WHO, "Visual Impairment and Blindness." [Online]. Available: https://www.who.int/es/news-room/fact-sheets/detail/blindness-and-visual-impairment
2. International Diabetes Federation: Diabetic Retinopathy: A Call for Global Action (English). International Diabetes Federation. [Online]. Available: https://idf.org/about-diabetes/resources/diabetic-retinopathy-a-call-for-global-action/
3. World Health Organization: More than 700 million people with untreated hypertension. https://www.who.int/news/item/25-08-2021-more-than-700-million-people-with-untreated-hypertension
4. Shen, Y., Yang, C., Tang, X., Zhou, B: InterFaceGAN: Interpreting the Disentangled Face Representation Learned by GANs. [Online]. Available: http://arxiv.org/abs/2005.09635 (2020)
5. Waisberg, E., et al.: Generative Artificial Intelligence in Ophthalmology. Elsevier Inc. (2025). https://doi.org/10.1016/j.survophthal.2024.04.009
6. Karras, T., Aittala, M., Hellsten, J., Laine, S., Lehtinen, J., Aila, T.: Training Generative Adversarial Networks with Limited Data, [Online]. Available: http://arxiv.org/abs/2006.06676 (2020)
7. Karras, T., et al.: Alias-Free Generative Adversarial Networks. [Online]. Available: https://nvlabs.github.io/stylegan3
8. Coyner, A.S., et al.: Diagnosability of Synthetic Retinal Fundus Images for Plus Disease Detection in Retinopathy of Prematurity
9. Kim, M., et al.: Synthesizing realistic high-resolution retina image by style-based generative adversarial network and its utilization. Sci. Rep. **12**(1) (2022). https://doi.org/10.1038/s41598-022-20698-3
10. Wang, Z., et al.: Synthetic artificial intelligence using generative adversarial network for retinal imaging in detection of age-related macular degeneration. Front. Med. (Lausanne). **10** (2023). https://doi.org/10.3389/fmed.2023.1184892
11. Güllmar, D., Hsu, W.C., Reichenbach, J.R.: Predicting disease-related MRI patterns of multiple sclerosis through GAN-based image editing. Z. Med. Phys. **34**(2), 318–329 (2024). https://doi.org/10.1016/j.zemedi.2023.12.001
12. Go, S., Ji, Y., Park, S. J., Lee, S.: Generation of Structurally Realistic Retinal Fundus Images with Diffusion Models
13. Feng, X.: Diversified and Structure-Realistic Fundus Image Synthesis for Diabetic Retinopathy Lesion Segmentation

14. Pandey, P.U., Micieli, J.A., Ong Tone, S., Eng, K.T., Kertes, P.J., Wong, J.C.Y.: Realistic fundus photograph generation for improving automated disease classification. Br. J. Ophthalmol. **109**(7), 791–798 (2025). https://doi.org/10.1136/bjo-2024-326122
15. Pakdelmoez, S., Omidikia, S., Seyyedsalehi, A., Seyyedsalehi, S.Z.: Controllable Retinal Image Synthesis Using Conditional StyleGAN and Latent Space Manipulation for Improved Diagnosis and Grading of Diabetic Retinopathy
16. Hou, Q., et al.: "FundusGAN: A Hierarchical Feature-aware Generative Framework for High-Fidelity Fundus Image Generation.," [online]. Available: http://arxiv.org/abs/2503.17831 (2025)
17. 6 Best GPUs for AI and deep Learning in 2025.: [Online]. Available: https://www.databasemart.com/blog/best-gpus-for-ai-and-deep-learning-2025 (2025)
18. Top 5 GPUs for AI workloads in 2025.: [Online]. Available: https://ordinarytech.ca/blogs/news/top-5-gpus-for-ai-workloads-in-2025 (2025)
19. Baidu Research Open-Access Dataset – download.: [Online]. Available: https://ai.baidu.com/broad/download
20. Mujib, R.: ARMD curated dataset 2023, Kaggle. Accessed: Aug, vol. 31, [Online]. Available: https://www.kaggle.com/datasets/rakhshandamujib/armd-curated-dataset-2023 (2023)
21. Fu, H., et al.: ADAM: automatic detection challenge on age-related macular degeneration. IEEE Dataport. (2020). https://doi.org/10.21227/DT4F-RT59
22. Wang, W., et al.: Learning two-stream CNN for multi-modal age-related macular degeneration categorization. IEEE J. Biomed. Health Inform. **26**(8), 4111–4122 (2022). https://doi.org/10.1109/JBHI.2022.3171523
23. CONDE Research – Research Unit.: [Online]. Available: https://www.condeinvestigacion.org/ (2023)
24. Ocular Disease Recognition.: [Online]. Available: https://www.kaggle.com/datasets/andrewmvd/ocular-disease-recognition-odir5k (2025)
25. Chuquicusma, M.J.M., Hussein, S., Burt, J., Bagci, U.: How to fool radiologists with generative adversarial networks? A visual turing test for lung cancer diagnosis. In: 2018 IEEE 15th International Symposium on Biomedical Imaging (ISBI 2018), pp. 240–244. IEEE (2018). https://doi.org/10.1109/ISBI.2018.8363564
26. GPU-Optimized AI, Machine Learning, & HPC Software: NVIDIA NGC | NVIDIA NGC. [Online]. Available: https://catalog.ngc.nvidia.com/orgs/nvidia/teams/research/models/stylegan2/files?version=1

Spanish Audio DeepFake Detection

Daniel Bañuelos Torres(✉), José Ismael de la Rosa, Aldonso Becerra Sánchez, Gamaliel Moreno Chávez, and Efrén González Ramírez

Universidad Autónoma de Zacatecas, Unidad Académica de Ingeniería Eléctrica, Maestría en Ciencias del Procesamiento de la Información, Ciudad Universitaria Siglo XXI, Edificio de Posgrados en Ingeniería, Carretera Zacatecas-Guadalajara km 6, Ejido La Escondida, Zacatecas, Mexico

{danielbt,vargasj,gamalielmch,gamalielmch}@uaz.edu.mx

https://mcpi.uaz.edu.mx/

Abstract. Audio DeepFake detection has advanced significantly in recent years mainly due to competitions like ASVSpoof and ADD; however, these competitions have been focused only on English and Chinese languages; this situation has impacted the field, having fewer experiments performed on other languages. In this paper a focus on detection models to measure their performance across different Spanish datasets, the use of RawBoost data augmentation technique and pre-training on English dataset were explored. A comparison between two popular detection models is presented, mainly the Audio Anti-Spoofing using the Integrated Spectro-Temporal (AASIST) network and Mel-Spectrogram with Residual Network (Spec-ResNet). An evaluation has been conducted on four Spanish deepfake datasets HABLA, HISPASpoof, ODSS and MLAAD in order to assess the cross-dataset generalization capabilities of the models, as well as utilizing the ASVSpoof2019 dataset as a pre-training step is also explored. Results show that the use of RawBoost data augmentation can improve detection metrics, however improvements are minimal and inconsistent across datasets and models, additionally, the pre-trained models on English dataset consistently enhanced metrics on Spanish datasets for AASIST model, and worsen them for the Spec-ResNet model.

Keywords: Audio deepfake detection · Synthetic speech detection · Spanish dataset · cross-dataset

1 Introduction

In recent years, the increment in the quality of audio deepfake generation models has shown a latent potential risk in security for voice biometric systems, spread of misinformation and identity theft [4,5,9,23], as it has become increasingly easier to generate speech audios which resemble the principal characteristics of any voice [13]. Audio deepfake detection has gained popularity and become an

H. Ponce et al. (Eds.): COMIA 2026, CCIS 3095, pp. 451–465, 2027.
https://doi.org/10.1007/978-3-032-37438-7_35

active task in the audio classification space mainly due to competitions such as ASVSpoof and ADD [20,22,24,26], which have highlighted the importance of the subject and sparked an interest on it.

The majority of the work in audio deepfake detection has been focused on the English and Chinese languages, as both languages possess the largest speaking population in the world, also due to the development and high availability of audio datasets for these languages [1,27]. As a consequence of this, the use of detection models applied to other languages has seen little or no contributions. The Spanish language is spoken by millions of people across multiple countries, despite this, the number of works focused on dataset development, and the application of detection models on Spanish speech, remains limited.

1.1 Related Work

Several audio deepfake detection models have been proposed, which can be classified into three broad categories according to how they process the input features: i) hand-crafted-based, ii) spectrogram-based and iii) raw-waveform-based. **Hand-crafted-based** detectors commonly apply some known heuristic procedure to the speech signal to extract a sequence of representative coefficient vectors, among the most important are cepstral coefficients such as Linear Frequency Cepstrum Coefficients (LFCC), Mel Frequency Cepstral Coefficients (MFCC) and Constant Q Cepstral Coefficients (CQCC) these coefficients are used as input to machine-learning models like Gaussian Mixture Models (GMM), Convolutional Neural Networks (CNNs), and Recurrent Neural Networks (RNN) [8,12,15].

Spectrogram-based on the other hand extract an spectrogram from the speech signal, and commonly some transformation like filter-banks is applied to obtain dimensionally reduced spectrogram like Mel-spectrograms. These spectrograms are then processed as a 2D structure with image-based deep-learning models such as CNN, RNN (LTSM) or transformers to extract discriminative spatial deep features from the spectrogram [10,16].

Raw-waveform-based models utilize the audio in time domain as an input, and extract dynamic temporal features and relationships between them, to accomplish this attention-based and transformer-based deep-learning models such as RawNet2, AASIST, Wav2Vec, and WavLM have been proposed [6,7,18]. Lately, these models have shown a greater capability to generalize across datasets [7].

Regarding previous works of audio deepfake detection on a Spanish dataset, HABLA and HISPASpoof are the main studies that have been found. In the case of HABLA [19] the use of LFCC as input features, and a light convolutional neural network (LCNN) as a base model were presented. The results in this research show a model trained on English samples and then further trained on Spanish samples achieve better metrics on both languages, than models trained from scratch on Spanish. On the other hand, in the case of HISPASpoof [14] detection models LFCC+GMM, MFCC-ResNet, Spec-ResNet, PaSST and Wav2Vec2-AASIST were evaluated for the tasks of detection and source-attribution using

ASVSpoof2019, ODSS and HISPASpoof datasets. According to the obtained results, on a detection task, all detectors show improvements when trained on HISPASpoof compared to ODSS Spanish subset, also the transformer-based tend to have lower EER on closed-set evaluations. Additionally, on the attribution task, the detector with least degradation when evaluating on an open-set is PaSST, followed by Spec-ResNet.

1.2 Contributions

This paper introduces a comparative between two popular detection models, mainly the Audio Anti-Spoofing using Integrated Spectro-Temporal network (AASIST) [6] and Spectrogram with Residual Network (Spec-ResNet) [21], and evaluates them on four Spanish deepfake datasets: HABLA, HISPASpoof, ODSS, MLAAD [11,14,19,25]. The work assesses the cross-dataset generalization capabilities of the models, additionally, the use of RawBoost data augmentation technique [17] and pre-training on the English dataset ASVSpoof2019 [20] is also explored in an effort to find if their application enhances the models detection metrics and circumvent the lack of extensive Spanish datasets.

The obtained results in this investigation show that the use of RawBoost data augmentation can improve detection metrics. However, the improvements are minimal and inconsistent across datasets and models, additionally, the pretrained models on English dataset consistently enhanced metrics on Spanish datasets for AASIST model, and worsen them for the Spec-ResNet model.

2 Materials and Methods

This section describes the datasets utilized for training and evaluation, along with a brief description of their contents, as well as a description of the base models with their implementation parameters. A specific mention is included, highlighting the use of two detection models with a custom partition of the training dataset. Finally, a description of the conducted experiments is given together with the metrics employed to measure the detectors performance.

2.1 Datasets

In this work, the datasets HABLA, HISPASpoof, OSSD, MLAAD and ASVSpoof 2019 were utilized [11,14,19,20,25]. HABLA is a dataset comprising five Latin American Spanish accents (Argentina, Chile, Colombia, Peru and Venezuela) with bonafide samples by 162 speakers both female and male. The datasets contain spoof samples generated with six audio realization techniques, two GAN-based voice conversion techniques (StarGAN-VC, CycleGAN-VC), one diffusion-based voice conversion technique (Maximum Likelihood Sampling), two TTS with GAN-based voice conversion (Microsoft Azure TTS+StarGAN-VC, Microsoft Azure TTS+Diffusion), and a plain TTS technique (Microsoft Azure TTS).

HISPASpoof comprises six Latin American Spanish accents (Argentina, Chile, Colombia, Mexico, Peninsula and Peru), as well as six audio generation techniques, ElevenLabs, F5-Spanish, FishSpeech, XTTS-v1.1, XTTS-v2 and YourTTS.

ODSS is a multilingual dataset containing samples in English, German and Spanish, where only the Spanish subset is utilized and is conformed by the same bonafide samples used in HABLA, but the spoof samples were generated with two different techniques VITS and FastPitch-HifiGan.

MLAAD is an extensive multilingual dataset consisting of 51 languages and 140 audio generation techniques, where only the Spanish subset is used. The subset was created with 9 audio generation techniques, Bark, Llasa-1B, Griffin-Lim, OpenVoice-v2, ResembleAI, Tacotron2-DDC, VITS, XTTS-v1.1 and XTTS-v2, where the bonafide samples contains three accents Argentinian, Mexican and Peninsular.

ASVSpoof2019 is based on the Voice Cloning Toolkit (VCTK) corpus, which consists of 107 native English speakers, covering a wide range of accents. The spoof samples were generated with 19 techniques such as waveform concatenation, MFCC-vocoder, spectral filtering, Griffim-Lim among others.

Spanish Training Dataset: The HABLA dataset was selected as the source for the Spanish training dataset due to having the higher number of bonafide samples among all Spanish datasets, this was a fundamental requirement to create the largest balanced Spanish training dataset. A balanced dataset is desirable to prevent bias as most audio deepfake datasets are highly imbalanced containing a small quantity of bonafide audios, as is the case of HABLA, having only 11.85% of bonafide samples.

The HABLA dataset was split into three partitions: training, validation and test, as is common practice, the test partition acts as an open-test as it contains both unseen speakers and seen speakers, as well as both unseen audio generation techniques and seen techniques. This approach enables to measure the generalization capabilities of the detection models across all datasets.

To form the test partition 20% of the speakers in each accent were randomly selected to be part of the unseen group and thus removed from the training and validation partitions, as well as all bonafide and spoof samples related to the StarGAN-VC and TTS+Diffusion spoof generation techniques.

In order to create a balanced training and validation splits an under-sampling procedure was applied towards the majority class consisting of the spoof samples. First, as different spoof sample generation techniques consist of a different number of samples, the generation techniques with fewer samples were kept as they are, then, the samples from the remainder techniques were selected randomly in equal proportions, so the number of spoof samples matched the number of bonafide samples. Finally, the bonafide and spoof samples were split maintaining the spoof generation techniques number of samples distributions in a ratio of 80% to 20% for the training split and validation split respectively.

Additionally, the Raw-Boost [17] data augmentation technique was applied to the training split, to determine if the application of data augmentation enhances the detection metrics of the models. As suggested by the original work the algorithm 4 consisting in the injection of convolutional and impulsive noise in series, obtained the best increase in detection metrics, so this algorithm was applied. Table 1 shows the bona-fide and spoof sample quantities with and without data augmentation for the training and validation splits.

Table 1. HABLA custom training and validation partition distribution.

Split Name	Sample Class	# Samples w/o DA	# Samples with DA
Training	Bonafide	18741	29986
	Spoof	18741	29986
Validation	Bonafide	7496	NA
	Spoof	7496	NA

In order to test the generalization capabilities of the detection models the datasets HABLA, HISPASpoof, ODSS and MLAAD are used, due to HABLA and HISPASpoof being already partitioned only the test partitions were utilized, whereas for ODSS and MLAAD the whole Spanish subsets were used. Table 2 shows distribution of bonafide and spoof samples for each dataset.

Table 2. Summary of the Spanish test datasets sample distribution. The quantity in parentheses is the sample class is the number of spoof generation techniques.

Dataset	Sample Class	Number of samples
HABLA (Test split)	Bonafide	4075
	Spoof (6)	30403
HISPASpoof (Test split)	Bonafide	1516
	Spoof (6)	18606
ODSS	Bonafide	3386
	Spoof (2)	3386
MLAAD	Bonafide	9000
	Spoof (9)	9000

2.2 Base Models

The evaluation of two detection models was performed, one belonging to the category of raw-waveform-based approach and the other belonging to spectrogram-based approach. Both models have shown good performance on English datasets [2,6,21]. Table 3 summarizes the type of input feature, type of network and number of trainable parameters of each model.

Table 3. Base Models Summary.

Model	Input Feature	Network Architecture	# Parameters
AASIST	Time domain	RawNet2 + Graph Attention	0.29 M
Spec-ResNet293	Mel-Spectrogram	ResNet	0.28 M

AASIST (M1) [6]: This model has been selected as the base model, as it has been a baseline for the ASVSpoof Competitions since 2021, due to it's performance across multiple English datasets. AASIST is an end-to-end model consisting of a RawNet2-based encoder [18], and a graph module containing graph attention and graph pooling layers. The raw waveform is processed with a sinc convolutional layer, initialized with 70 mel-scaled filters with a kernel size of 128 each, a post-processing is applied to form a spectro-temporal representation, which is fed to the RawNet2-based encoder to learn high-level feature maps. The graph module allows to capture the relationships between artifacts in both temporal and spectral domains. Table 4 shows the particular model parameters to utilized to implement the AASIST model.

Table 4. AASIST Model Parameters.

Model part	Parameter Name	Parameter Value
RawNet2	first_conv	128
	filts	[70, [1, 32], [32, 32], [32, 64], [64, 64]]
Graph Module	gat_dims	[64, 32]
	pool_ratios	[0.5, 0.7, 0.5, 0.5]
	temperatures	[2.0, 2.0, 100.0, 100.0]

MelSpec+ResNet (M2) [6]: This model combines a spectrogram processed with a filter bank and a ResNet architecture. The audio spectrogram is processed with a 80 log Mel-filter banks consisting of a hamming window with a frame length of 25 ms and a frame shift of 20 ms, sampled at 16 kHz. The use of the variants ResNet34, ResNet221 and ResNet293 have been proven provide good metrics in the audios task such as speaker diarization, speaker identification and DeepFake source attribution [3,21], The ResNet293 variant was selected as it provided the best performance out of all variants, the input dimension is set to 80 to match the input spectrogram, a temporal statistics pooling (TSTP) layer is used to reduce dimensionality and create an embedding of dimension 256, finally a linear classifier is used to map the embedding to a 2-dimensional output.

2.3 Experiments

For all experiments, cross-entropy was used as the loss function, for the optimizer Stochastic Gradient Descend (SGD) was selected with an initial (maximum) learning rate of 1×10^{-3}, a weight decay of 1×10^{-3} and momentum of 0. Also, a cosine learning rate scheduler as defined in Eq. 1 with minimum learning rate of 1×10^{-6} was utilized.

$$lr_{epoch} = lr_{min} + \frac{(lr_{max} - lr_{min})}{2} \cdot \left(1 + \cos\left(\frac{\text{epoch}}{\text{total epochs}} \cdot \pi\right)\right) \tag{1}$$

The models were trained for a maximum of 100 epochs with a patience of 10 epochs and a delta of 0.1 based on the validation loss value, batch sizes of 200 and 256 were used for training and validation respectively. Following the application of similar detection models the input audio samples have been resampled to 16 kHz, and the audio length is truncated to 64600 samples which is close to 4 s, in the cases where the audio length is smaller than this length, the audio is concatenated to itself until the specified length is achieved.

The hardware utilized in the experiments consisted mainly of a NVIDIA RTX PRO 6000 WS (96 VRAM) GPU, an AMD EPYC 9454P CPU and 128 GB of RAM.

Experiment 1 - Training on HABLA Dataset (Spanish) and e. Both base models were trained and validated using the proposed HABLA partitions scheme mentioned in Sect. 2.1 with and without data augmentation. Then, they were evaluated on all Spanish test sets to assess generalization to Spanish speech and determine if RawBoost data augmentation enhances metrics.

Experiment 2 - Training on ASVSpoof2019 Dataset (English). To analyze cross-lingual generalization, the models were train on ASVSpoof2019 dataset. Then, they were evaluated on all test datasets to assess both a baseline performance on English and generalization to Spanish.

Experiment 3 - Training on ASVSpoof2019 and HABLA. Finally, the training weights from experiment 2, were utilized as pre-train for both models, these models were then fine-tuned with the HABLA dataset. The models were evaluated on all Spanish test datasets to assess generalization and determine if pre-training on an English dataset enhances metrics on Spanish datasets.

2.4 Evaluation Metrics

- **Equal Error Rate (EER):** The EER is the point where the False Rejection Rate (FRR) and the False Acceptance Rate (FAR) are equal. This metric is primarily used in biometric security systems as it allows to compare models without setting a particular threshold value. A lower value of EER indicates the model is more precise with fewer overall errors.

- **ROC-AUC (AUC):** The area under the ROC curve represents the probability of the model ranking a positive sample higher than a negative sample. For binary classification a value of 1.0 indicates a perfect classifier and 0.5 a random guesser.
- **Accuracy (Acc):** The accuracy is the ratio of correctly classified samples to the total number of classified samples. It allows to understand the overall performance of the detector.
- **Recall (R):** Recall measures the proportion of actual positive cases that the model correctly identifies. It is especially important in imbalanced datasets, such as many audio DeepFake detection scenarios, where failing to detect spoof samples can have a significant impact on the system's reliability and security.
- **Precision (Prec):** Precision measures the proportion of predicted positive cases that are actually correct. A low precision implies bonafide audio samples being incorrectly classified as spoof, reducing the system's trustworthiness and usability.
- **F1-Score (F1):** The F1-Score is the harmonic mean of precision and recall. It allows to understand the model's effectiveness with imbalanced data as is the case in most audio DeepFake datasets.

3 Results and Discussion

This section describes the obtained results according to the experiments previously presented in Sect. 2, as well as the insights regarding the use of data augmentation and English dataset as pre-train.

3.1 Results

Experiment 1. Figure 1 shows the learning curves of the four models used in the experiment. On AASIST model the use of data augmentation cause early stopping to occur faster, while it can be notice a Models performance metrics are summarized in Table 5, where immediately can be noticed the degradation in detection metrics occurring when either model is evaluated on datasets containing audio DeepFake generation metrics not seen in training, this outcome was expected, in particular the EER is contained in the range 27% to 48%, also the use of data augmentation enhances ROC-AUC and EER metrics in the AASIST model, on the contrary Spec-ResNet293 metrics worsen by using data augmentation. Figure 2 shows the confusion matrices of the best models, interestingly HABLA and HISPASpoof present a high percentage of false positives, while ODSS and MLAAD present more false negatives.

Experiment 2. Table 6 contains the cross-lingual experiment results. The AASIST model has better metrics when evaluating the ASVSpoof2019 dataset, which might indicate it learned richer features. Despite this, when evaluating

Table 5. Cross-dataset Spanish results (Experiment 1) for AASIST and Spec-ResNet293 models trained on HABLA and HABLA augmented, and evaluated on Spanish datasets. Metrics include Accuracy (Acc), Recall (R), Precision (Prec), F1-score (F1), ROC-AUC (AUC), and Equal Error Rate (EER). Lowest EER values are highlighted in bold.

Test Dataset	Model	DA	Acc % ↑	R % ↑	Prec % ↑	F1% ↑	AUC% ↑	EER% ↓
HABLA	AASIST	No	85.55	85.55	85.28	85.41	77.53	29.62
		Yes	88.02	88.02	79.97	82.66	78.07	28.39
	Spec-ResNet293	No	68.09	68.09	91.19	73.73	93.08	**14.06**
		Yes	57.46	57.46	89.76	64.40	81.03	27.53
HISPASpoof	AASIST	No	90.57	90.57	89.58	90.04	61.50	42.58
		Yes	92.41	92.41	87.03	88.86	68.68	36.68
	Spec-ResNet293	No	70.16	70.16	89.33	77.08	70.04	**36.45**
		Yes	79.49	79.49	88.88	83.31	66.85	39.79
ODSS	AASIST	No	63.09	63.09	58.43	59.11	59.06	44.10
		Yes	66.61	66.61	66.63	53.31	57.37	45.10
	Spec-ResNet293	No	62.83	62.83	61.83	62.50	61.99	40.55
		Yes	69.73	69.73	68.82	68.99	71.80	**32.83**
MLAAD	AASIST	No	48.03	48.03	38.62	34.49	52.11	48.37
		Yes	49.97	49.97	24.99	33.32	53.11	47.92
	Spec-ResNet293	No	61.82	61.82	63.38	60.68	73.73	**33.10**
		Yes	58.42	58.42	59.46	57.21	68.58	36.52

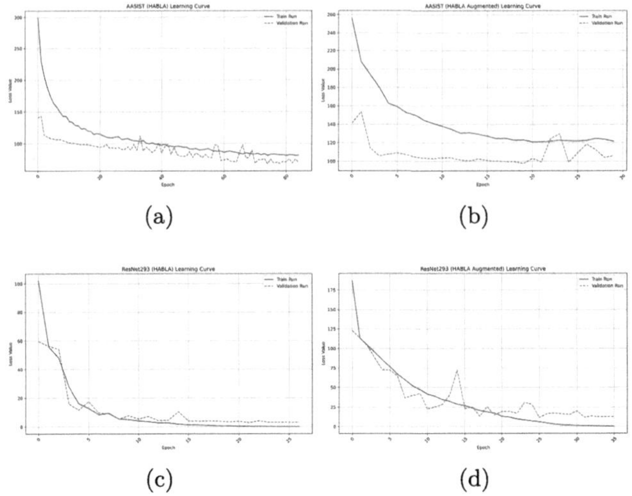

(a) (b) (c) (d)

Fig. 1. Learning curves of models trained on Spanish dataset (experiment 1). (a) AASIST trained on HABLA. (b) AASIST trained on HABLA with RawBoost Augmentation Algorithm 7 (c) Spec-ResNet293 trained on HABLA. (d) Spec-ResNet293 trained on HABLA with RawBoost Augmentation algorithm 4.

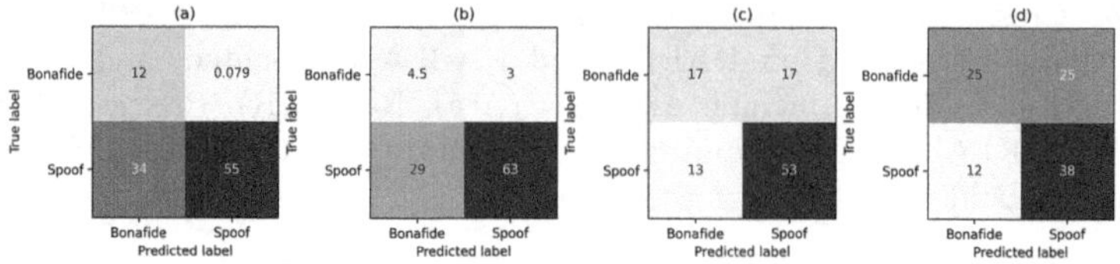

Fig. 2. Confusion matrices of the best models according to EER of experiment 1. (a) Spec-ResNet293 trained on HABLA evaluated on HABLA (b) Spec-ResNet293 trained on HABLA evaluated on HISPASpoof (c) Spec-ResNet293 trained on HABLA augmented evaluated on ODSS (d) Spec-ResNet293 trained on HABLA augmented evaluated on MLAAD.

on the Spanish datasets the metrics differ at most by 3% in the majority of the datasets when comparing both models. Figure 4 shows a tendency to confuse bonafide as spoof on the Spec-ResNet293 model, while AASIST only presents higher false positives when evaluating on the HABLA dataset.

Based on the learning curves in Fig. 3 and confusion matrices, we can deduce the models present cross-lingual and learn meaningful transferable features, and the worsen of metrics occurs due to differences in the spoof audio synthesis techniques, as metrics are relatively close to experiment 1, and surprisingly performed better on HISPASpoof.

Table 6. Cross-lingual Spanish results (Experiment 2) for AASIST and Spec-ResNet293 models trained on ASVspoof2019 and evaluated on Spanish datasets. Metrics include Accuracy (Acc), Recall (R), Precision (Prec), F1-score (F1), ROC-AUC (AUC), and Equal Error Rate (EER). Lowest EER values are highlighted in bold.

Test Dataset	Model	Acc % ↑	R % ↑	Prec % ↑	F1% ↑	AUC% ↑	EER% ↓
ASVSpoof2019	AASIST	96.48	96.48	96.36	96.39	98.53	**6.01**
	Spec-ResNet293	89.48	89.48	90.17	90.05	92.51	15.05
HABLA	AASIST	76.48	76.48	79.61	77.96	61.80	**40.73**
	Spec-ResNet293	74.36	74.36	80.92	75.92	63.57	40.95
HISPASpoof	AASIST	90.70	90.70	87.71	88.97	70.82	**33.87**
	Spec-ResNet293	92.31	92.31	89.94	90.42	72.58	35.18
ODSS	AASIST	66.61	66.61	57.74	53.33	52.30	47.29
	Spec-ResNet293	66.77	66.77	68.95	53.75	68.84	**36.68**
MLAAD	AASIST	50.22	50.22	52.00	36.00	64.73	39.91
	Spec-ResNet293	50.25	50.25	52.77	36.28	65.10	**38.48**

Experiment 3. Observing the resulting metrics in Table 7, it can be noticed that the changes when applying data augmentation sometimes provide a better or worse result in metrics, however this difference is quite small on most cases,

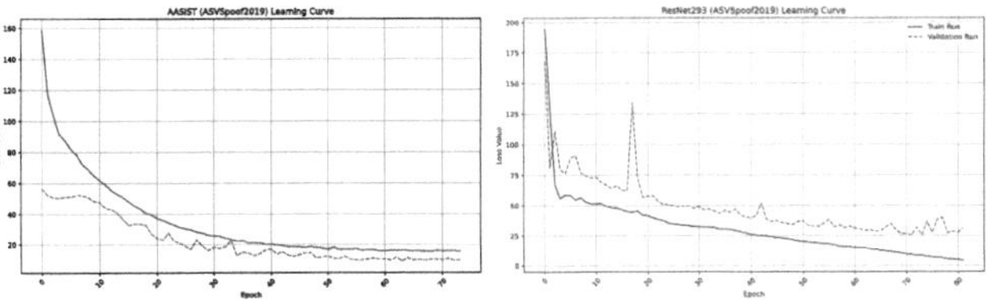

(a) (b)

Fig. 3. Learning curves of models trained on English dataset (experiment 2). (a) AASIST trained on ASVSpoof2019. (b) Spec-ResNet293 trained on ASVSpoof2019.

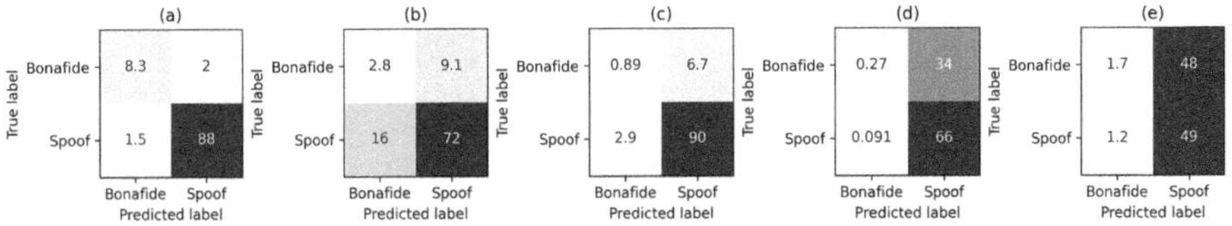

Fig. 4. Confusion matrices of the best models according to EER of experiment 2. (a) AASIST trained on ASVSpoof2019 evaluated on ASVSpoof2019 (baseline) (b) AASIST trained on ASVSpoof2019 evaluated on HABLA (c) AASIST trained on ASVSpoof2019 evaluated on HISPASpoof (d) Spec-ResNet293 trained on ASVSpoof2019 evaluated on ODSS (e) Spec-ResNet293 trained on ASVSpoof2019 evaluated on MLAAD.

Table 7. Spanish fine-tuning results (Experiment 3) for AASIST and Spec-ResNet293 models pre-trained on ASVSpoof2019 and evaluated on Spanish datasets. Metrics include Accuracy (Acc), Recall (R), Precision (Prec), F1- score (F1), ROC-AUC (AUC), and Equal Error Rate (EER). Lowest EER values are highlighted in bold.

Test Dataset	Model	DA	Acc % ↑	R % ↑	Prec % ↑	F1% ↑	AUC% ↑	EER% ↓
HABLA	AASIST	No	87.88	87.88	91.27	89.02	91.89	14.73
		Yes	87.92	87.92	91.17	89.04	91.97	**14.61**
	Spec-ResNet293	No	70.15	70.15	90.65	75.46	87.83	20.22
		Yes	64.43	64.43	89.57	70.67	83.68	24.90
HISPASpoof	AASIST	No	83.7	83.77	90.62	86.48	77.05	**28.83**
		Yes	83.38	83.38	90.53	86.21	76.49	29.28
	Spec-ResNet293	No	89.82	89.82	90.04	89.93	63.59	43.92
		Yes	87.50	87.50	89.51	88.42	65.74	39.64
ODSS	AASIST	No	54.03	54.03	64.42	54.71	61.98	40.73
		Yes	55.39	55.39	65.26	56.17	61.89	40.87
	Spec-ResNet293	No	68.08	68.08	65.82	60.15	65.02	38.73
		Yes	70.77	70.77	72.09	65.01	66.88	**37.09**
MLAAD	AASIST	No	62.25	62.25	62.25	62.25	67.60	37.68
		Yes	62.89	62.89	62.90	62.88	67.71	37.66
	Spec-ResNet293	No	59.88	59.88	65.58	55.78	80.78	26.08
		Yes	70.77	70.77	72.09	70.43	83.74	**23.79**

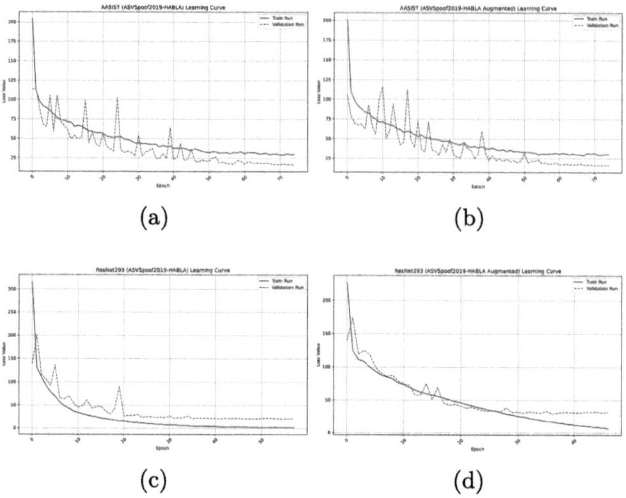

Fig. 5. Spanish Fine-tuning on English pre-train models learning curves (experiment 3). (a) AASIST fine-tuned on HABLA (pre-trained on ASVSpoof2019). (b) AASIST fine-tuned on HABLA with RawBoost Augmentation Algorithm 7 (pre-trained on ASVSpoof2019). (c) Spec-ResNet293 fine-tuned on HABLA (pre-trained on ASVSpoof2019). (d) Spec-ResNet293 fine-tuned on HABLA with RawBoost Augmentation Algorithm 7 (pre-trained on ASVSpoof2019).

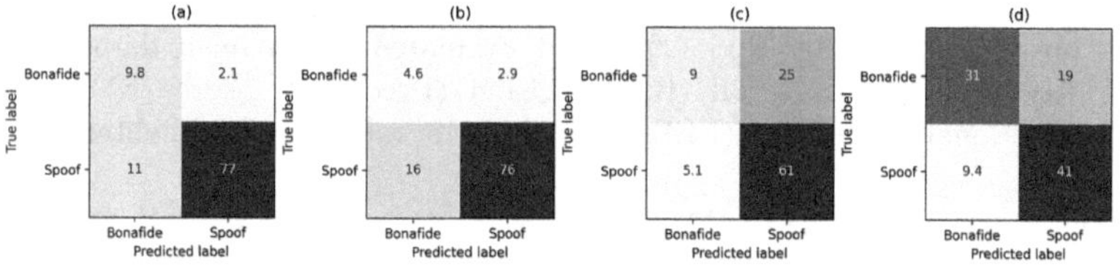

Fig. 6. Confusion matrices of the best models according to EER of experiment 3. (a) AASIST fine-tuned on HABLA augmented evaluated on HABLA (b) AASIST fine-tuned on HABLA evaluated on HISPASpoof (c) Spec-ResNet293 fine-tuned on HABLA augmented evaluated on ODSS (d) Spec-ResNet293 fine-tuned on HABLA augmented evaluated on MLAAD.

with less than a 3% change. This might indicate the use of noise as data augmentation when a model has already been trained, does not provide enough new information to the model so it can become more robust. Comparing the confusion matrix (Fig. 6) to the experiment 1, a reduction in false positives and false negatives on some datasets can be seen, thus improving slightly the detector performance. These results clearly show the potential of pre-training models on extensive datasets in other languages to circumvent the scarcity of low resource audio DeepFake datasets, and further improve their robustness (Fig. 5).

3.2 Limitations

The limitations of these study comprehend the following:

1. Scarcity of Spanish datasets affects directly model detection metrics, and degradation in cross-dataset evaluation might not be related to the model capabilities but rather to the quantity of available training data.
2. The application of RawBoost data augmentation was performed only on algorithm 4 out of seven available algorithms, the possibility of obtaining better results with a different algorithm is still open.
3. Each Experiment was performed once, as such the variability of resulting metrics is unknown.
4. Only two detection models were used for the experiments, thus results might not be extensive enough to determine if the findings do generalize to other models or just to these particular detectors.

4 Conclusions

In this paper, the use of AASIST and Spec-ResNet293 detection models on the largest Spanish datasets was explored. While results show their cross-lingual generalization capabilities were not great, the use of English dataset as a pre-train enhanced the cross-dataset detection rates on Spanish datasets. Based on this, the use of extensive datasets in phonetically similar languages as a pre-trained step is hypothesize to enhance the detection models to circumvent scarcity of low-resource languages. Additionally, the use of RawBoost algorithm 4 as data augmentation was determine to inconsistently slightly enhance metrics, however, not knowing when will it improve the model and the additional time required for training might not be worth it.

The limitations on this study were two, first, only two detection models were utilized in the experiments, thus the results are not extensive enough to determine if the findings apply to a broader range of models or just to these particular models. Secondly, the scarcity of Spanish DeepFake datasets plays an important factor as models require extensive datasets with a high number of audio generation techniques to provide good detection rates and cross-dataset generalization.

Nonetheless, this work has shown some paths for future exploration, use the currently available extensive English datasets as a pre-train step for the models and perform a fine-tuning on Spanish datasets to push the detection rates to a lower EER, and most importantly the development of an extensive Spanish dataset with rich variety in accents and speakers is required to advanced in the audio DeepFake detection task on the Spanish language, such dataset will be fundamental for the evolution of this binary classification problem into the multi-class classification problem involved in audio DeepFake source attribution.

Acknowledgments. This study was funded by Secretaría de Ciencia, Humanidades, Tecnología e Innovación (grant number 65e3ca3e66ee807d931b1839).

Disclosure of Interests. The authors have no competing interests to declare that are relevant to the content of this article.

References

1. Almutairi, Z., Elgibreen, H.: A review of modern audio deepfake detection methods: challenges and future directions. Algorithms **15**(5) (2022). https://doi.org/10.3390/a15050155. https://www.mdpi.com/1999-4893/15/5/155
2. Falez, P., Marteau, T., Lolive, D., Delhay, A.: Audio deepfake source tracing using multi-attribute open-set identification and verification. In: Interspeech 2025, pp. 1528–1532. Rotterdam, Netherlands (2025). https://doi.org/10.21437/interspeech.2025-2001. https://hal.science/hal-05472354
3. Falez, P., Marteau, T., Lolive, D., Delhay, A.: Audio deepfake source tracing using multi-attribute open-set identification and verification. In: Interspeech 2025, pp. 1528–1532. ISCA (2025)
4. Genelza, G.G.: A systematic literature review on AI voice cloning generator: a game-changer or a threat? J. Emerg. Technol. **4**(2), 54–61 (2024)
5. Jain, P., et al.: Forensic perspective on voice biometrics and AI: a review. Int. J. Sci. Res. Sci. Technol. **11**(5), 49–63 (2024)
6. Jung, J.W., et al.: Aasist: audio anti-spoofing using integrated spectro-temporal graph attention networks. In: ICASSP 2022 - 2022 IEEE International Conference on Acoustics, Speech and Signal Processing (ICASSP), pp. 6367–6371 (2022). https://doi.org/10.1109/ICASSP43922.2022.9747766
7. Kheir, Y.E., Samih, Y., Maharjan, S., Polzehl, T., Möller, S.: Comprehensive layer-wise analysis of SSL models for audio deepfake detection (2025). https://arxiv.org/abs/2502.03559
8. Todisco, M., Delgado, H., Evans, N.: A new feature for automatic speaker verification anti-spoofing: constant Q cepstral coefficients. In: Odyssey 2016 Speaker and Language Recognition Workshop, pp. 283–290 (2016). https://doi.org/10.21437/Odyssey.2016-41
9. McGettigan, C., et al.: Voice cloning: psychological and ethical implications of intentionally synthesising familiar voice identities (2024)
10. Mcuba, M., Singh, A., Ikuesan, R.A., Venter, H.: The effect of deep learning methods on deepfake audio detection for digital investigation. Procedia Comput. Sci. **219**, 211–219 (2023). https://doi.org/10.1016/j.procs.2023.01.283. https://www.sciencedirect.com/science/article/pii/S1877050923002910, cENTERIS – International Conference on ENTERprise Information Systems/ProjMAN – International Conference on Project MANagement/HCist – International Conference on Health and Social Care Information Systems and Technologies 2022
11. Müller, N.M., et al.: Mlaad: the multi-language audio anti-spoofing dataset. In: International Joint Conference on Neural Networks (IJCNN) (2024)
12. Pal, M., Paul, D., Saha, G.: Synthetic speech detection using fundamental frequency variation and spectral features. Comput. Speech Lang. **48**, 31–50 (2018). https://doi.org/10.1016/j.csl.2017.10.001. https://www.sciencedirect.com/science/article/pii/S0885230817301663
13. Pan, Y., Yang, Y., Yao, J., Ma, L., Zhao, J.: Zero-shot voice conversion via content-aware timbre ensemble and conditional flow matching (2025). https://arxiv.org/abs/2411.02026

14. Risques, M., Bhagtani, K., Yadav, A.K.S., Delp, E.J.: Hispaspoof: a new dataset for Spanish speech forensics (2025). https://arxiv.org/abs/2509.09155
15. Schäfer, K., Steinebach, M.: MFCC vs. LFCC for audio deepfake detection: the role of delta features and input length. In: 2025 33rd European Signal Processing Conference (EUSIPCO), pp. 576–580 (2025). https://doi.org/10.23919/EUSIPCO63237.2025.11226775
16. Shaaban, O.A., Yildirim, R.: Audio deepfake detection using deep learning. Eng. Rep. **7**(3), e70087 (2025). https://doi.org/10.1002/eng2.70087. https://onlinelibrary.wiley.com/doi/abs/10.1002/eng2.70087
17. Tak, H., Kamble, M., Patino, J., Todisco, M., Evans, N.: Rawboost: a raw data boosting and augmentation method applied to automatic speaker verification anti-spoofing (2022). https://arxiv.org/abs/2111.04433
18. Tak, H., Patino, J., Todisco, M., Nautsch, A., Evans, N., Larcher, A.: End-to-end anti-spoofing with rawnet2 (2021). https://arxiv.org/abs/2011.01108
19. Tamayo Flórez, P.A., Manrique, R., Pereira Nunes, B.: HABLA: a dataset of Latin American Spanish accents for voice anti-spoofing. In: INTERSPEECH 2023, pp. 1963–1967. ISCA (2023)
20. Todisco, M., et al.: ASVspoof 2019: future horizons in spoofed and fake audio detection. In: Interspeech 2019. ISCA (2019)
21. Wang, H., et al.: Wespeaker: a research and production oriented speaker embedding learning toolkit (2022). https://arxiv.org/abs/2210.17016
22. Wang, X., et al.: Asvspoof 5: crowdsourced speech data, deepfakes, and adversarial attacks at scale (2024). https://arxiv.org/abs/2408.08739
23. Westerlund, M.: The emergence of deepfake technology: a review. Technol. Innov. Manag. Rev. **9**, 40–53 (2019). https://doi.org/10.22215/timreview/1282, timreview.ca/article/1282
24. Yamagishi, J., et al.: Asvspoof 2021: accelerating progress in spoofed and deepfake speech detection (2021). https://arxiv.org/abs/2109.00537
25. Yaroshchuk, A., Papastergiopoulos, C., Cuccovillo, L., Aichroth, P., Votis, K., Tzovaras, D.: An open dataset of synthetic speech. In: 2023 IEEE International Workshop on Information Forensics and Security (WIFS), pp. 1–6 (2023). https://doi.org/10.1109/WIFS58808.2023.10374863
26. Yi, J., et al.: ADD 2023: the second audio deepfake detection challenge (2023). https://arxiv.org/abs/2305.13774
27. Yi, J., Wang, C., Tao, J., Zhang, X., Zhang, C.Y., Zhao, Y.: Audio deepfake detection: a survey (2023). https://arxiv.org/abs/2308.14970

An LLM Web Extension for Text Adaptation to Reduce Cognitive Load in Older Adults

Rodrigo Francisco De los Rios Rojas(✉) and Julio C. H. Soto

Facultad de Ingeniería, Universidad de Lima, Av. Javier Prado Este 4600, Santiago de Surco, 15023 Lima, Peru
20200652@aloe.ulima.edu.pe, jhuarach@ulima.edu.pe

Abstract. The digital divide significantly impacts older adults, as natural cognitive decline directly hinders their comprehension of complex informational content on news portals. To address this problem, an LLM-driven web extension is proposed to automatically simplify and adapt texts using E2R, PSET, and LSP methodologies. The research was conducted in two phases: the development of a technical architecture comprising an adaptation orchestrator and a blind evaluation module, followed by a pre-experimental study with 30 older adults using a pre-test/post-test design. Evaluation instruments included the GORT-5 for reading comprehension, NASA-TLX for cognitive load, SUS for usability, and STAM for technology acceptance. Results demonstrated an increase in reading comprehension from 61.33% to 85.33% and a 41% improvement in reading speed. Additionally, perceived cognitive load was significantly reduced ($p < .001$), and a "Best Imaginable" usability score (95.5) was achieved. Ultimately, this tool empowers older adults to independently navigate news portals, proving the potential of generative AI to foster true digital inclusion.

Keywords: Cognitive accessibility · Large language models · Older adults · Text adaptation · Web extension

1 Introduction

The United Nations High Commissioner for Refugees (UNHCR) defines an older adult as any individual aged 60 or older [35]. Currently, this demographic accounts for 10 million individuals distributed across China, Japan, and India [24]. The natural cognitive decline associated with aging exacerbates difficulties in demanding visual tasks such as reading [13]; however, most digital tools remain unadapted to address these specific needs [17,25]. Latin America faces a particularly critical situation due to a severe lack of web accessibility [21,29], creating significant barriers that hinder digital autonomy, especially for older adults navigating dense informational platforms like news portals [22,26].

H. Ponce et al. (Eds.): COMIA 2026, CCIS 3095, pp. 466–477, 2027.
https://doi.org/10.1007/978-3-032-37438-7_36

Large Language Models (LLMs) such as GPT and Gemini offer promising alternatives for automating content adaptation [36,38]. In fact, Alves et al. [3] demonstrated that Gemini 1.5 Flash and GPT-4o effectively simplify government documents using advanced prompting methods. Along the same lines, Wu et al. [37] stand out with their DRPE framework, which generates content evaluated through human criterion simulation, as well as Färber et al. [8] with their application powered by LLaMA 3.1 and GPT-4o to adapt texts according to the reader's profile.

In response to the generalist approach of traditional web guidelines [20], cognitive accessibility has emerged seeking to improve direct text comprehension. Specifically, the Easy-to-Read (E2R) methodology has proven effective in simplifying texts [19], which Guidroz et al. [9] validated by reporting a 38% increase in reading comprehension. Furthermore, establishing trust is crucial for this demographic; the System Usability Scale (SUS) dictates that scores above 68 certify high usability [15], while the STAM model shows that older adults perceive adapted technologies as more straightforward [34].

Despite these advances, recent studies present significant structural gaps: their development is frequently limited to theoretical proofs of concept [20], they lack specific adaptations for real-time navigation on news portals [8], and there is a clear absence of empirical validation with real users to measure the actual reduction in mental exertion [15,23].

To bridge these gaps, this research proposes an LLM-driven web extension designed to act as a real-time cognitive assistant. The main contributions of this work are structured as follows:

- **Technical Architecture:** The development of a robust backend orchestrator and a blind evaluation module to process web news concurrently, integrating E2R, PSET, and LSP methodologies.
- **Accessible Interface:** The implementation of features tailored specifically for older adults, including a "Focus Mode", paragraph-level text reprocessing, and interactive onboarding tutorials.
- **Empirical Validation:** A multidimensional pre-experimental study with real users to quantitatively and qualitatively measure the system's effectiveness in facilitating cognitive accessibility, reducing cognitive load, enhancing usability, and fostering technological adoption.

The remainder of this article is structured as follows: Sect. 2 reviews the background and related work. Section 3 details the end-to-end system design and the experimental methodology. Sections 4 and 5 present the results, discussion, and conclusions of this research.

2 Related Work

In this section, we present the relevant works identified in the literature, which are summarized and categorized according to the fundamental domains of our proposed architecture in Table 1. The subsequent subsections detail these contributions and establish the theoretical foundation for our LLM-driven intervention.

Table 1: Summary of related works by cognitive and technological domain.

Category	Ref.	Focus/Method	Key Contribution
Adaptation Techniques	[19,30]	Easy-to-Read (E2R)	Simplifies vocabulary and syntax.
	[36]	Plain Lang. (LSP)	Logical reorganization for first-reading.
	[2]	PSET Methodology	Lexical/syntactic substitution.
LLMs in Simplification	[1,37]	Prompt Frameworks	DRPE simulation; progressive density.
	[3,8]	Direct LLM App.	Use of GPT-4o, Gemini, LLaMA 3.1.
	[23,25]	Automated vs. Human	Risks of relying solely on BLEU/SARI.
Cognitive Accessibility	[7,21]	Web Guidelines	Automated tools fail complex barriers.
	[32,33]	Assistive Tools	Extensions improve task success.
HCI Evaluation Metrics	[14,16]	Reading & Cog. Load	GORT-5 reliability; Raw NASA-TLX.
	[12,34]	Usability & Accept.	Adapted SUS and STAM models.

2.1 Background on Adaptation Techniques

Cognitive accessibility relies on established methodologies to reduce textual complexity. The *Easy-to-Read (E2R)* framework focuses on simplifying vocabulary and syntax for individuals with reading difficulties [19,30]. *Plain Language (LSP)* aims to reorganize information logically and use straightforward phrasing, ensuring the content is understood on the first reading [36]. Furthermore, the *PSET* methodology specifically targets the substitution of complex words and the restructuring of convoluted sentences [2]. Integrating these techniques into an automated pipeline provides a structured approach to generative text adaptation.

2.2 LLMs in Text Simplification

Large Language Models (LLMs) have demonstrated remarkable capabilities in automating content adaptation. Several studies have proposed structured frameworks to optimize precision, such as simulating human evaluation through DRPE [37], controlling information density via progressive iterations [1], and integrating real-time retrieval [28]. Regarding direct applications, multiple studies confirm the efficacy of generative models in simplifying complex texts across various domains, including government and health documents [3,8,18,27]. However, despite achieving outstanding scores in automated metrics like BLEU and SARI [11,25], the literature warns that LLMs still exhibit critical deficiencies in capturing semantic nuances [23,31].

2.3 Cognitive Accessibility Methodologies

While traditional web accessibility guidelines aim to make internet content universally available, studies reveal that compliance remains critically low globally, and automated evaluation tools fail to detect complex cognitive barriers [6,7,21].

Consequently, assistive tools and browser extensions designed to guide web navigation have been shown to improve orientation and task success among older adults [32,33]. However, these methodologies are rarely integrated into a unified, real-time system powered by modern generative AI, highlighting a crucial opportunity for technological intervention.

2.4 Evaluation Metrics in HCI

Validating technological interventions for older adults requires robust, multidimensional metrics. For assessing reading comprehension, instruments such as the GORT-5 have demonstrated high reliability [16]. To evaluate the mental demand, the NASA Task Load Index (NASA-TLX) is widely utilized; notably, its unweighted version (Raw TLX) has been validated to correlate highly with the original scale while significantly reducing the evaluation fatigue on users [5,10,14]. Furthermore, evaluating the overall user experience necessitates adapted instruments, such as simplified versions of the System Usability Scale (SUS) tailored for older demographics [12], alongside specialized models like STAM to measure technological acceptance [34].

3 Methodology

In this section, we present the proposed solution designed to mitigate cognitive barriers and facilitate digital accessibility for older adults. The system was developed as a continuous, automated pipeline comprising two main operational stages: Backend Orchestration (Phase 1) and a Client-Side Web Extension (Phase 2). As illustrated in Fig. 1, the system's workflow initiates on the client side when an unadapted news article is extracted from a web portal. This raw text is transmitted via a secure POST request to the backend server. To guarantee stability, the server implements strict connection security protocols, including backoff retries and shared event loops. The text is then processed in parallel, where the adaptation module simultaneously generates three simplified versions using E2R, PSET, and LSP instructions. To prevent positional bias, these outputs are randomized before entering the Triple Evaluation System. This module assesses the texts based on holistic quality, methodological adherence, and a simulated demographic voting system. Finally, the selected winning text is logged in the database and returned to the client extension, which renders the cognitively accessible content under WCAG 2.1 AA compliance, complemented by assistive tools such as interactive tutorials and individual reprocessing.

3.1 Backend Orchestration and Prompt Engineering

The backend orchestrator acts as the core of the system (Phase 1). Upon receiving a secure POST request from the client, the server routes the data through strict connection security protocols, implementing request timeouts and a triple retry mechanism. Utilizing Python's `asyncio.gather`, asynchronous API calls

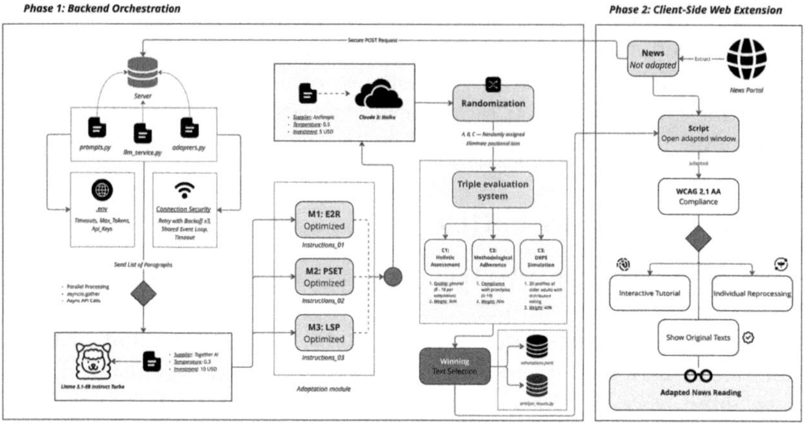

Fig. 1: End-to-end system architecture mapping the Backend Orchestration (Phase 1) and the Client-Side Web Extension (Phase 2).

are dispatched to the Meta-Llama-3.1-8B-Instruct-Turbo model. The adaptation module simultaneously applies three optimized instruction sets based on the E2R, PSET, and LSP methodologies.

The three generated adaptations are subsequently forwarded to Claude-3-Haiku for a blind evaluation module. This system scores the adaptations based on Holistic Assessment (30%), Methodological Adherence (30%), and a DRPE Simulation utilizing 20 distributed voting profiles of older adults (40%). The higher weight assigned to the DRPE simulation prioritizes the specific cognitive profiles of the target demographic over generic linguistic metrics. Finally, the system executes the winning text selection and returns the optimized text to the extension.

3.2 Web Extension and Assistive Interface

The client-side web extension manages the user's environment (Phase 2). Once the winning adaptation is returned, the script opens a dedicated window to display the simplified text under a "Focus Mode". The user interface is strictly designed to meet WCAG 2.1 AA compliance standards, incorporating high-contrast elements and accessible navigation structures (see Fig. 2). Furthermore, the interface includes integrated assistive features such as an interactive onboarding tutorial, a paragraph-level individual reprocessing tool, and a toggle to show the original texts (see Fig. 3).

3.3 Experimental Design and Procedure

A pre-experimental within-subject design (pre-test/post-test) was conducted with 30 older adults (mean age = 73.03, SD = 9.45). To mitigate learning effects,

Fig. 2: Adapted reading interface complying with WCAG 2.1 AA standards.

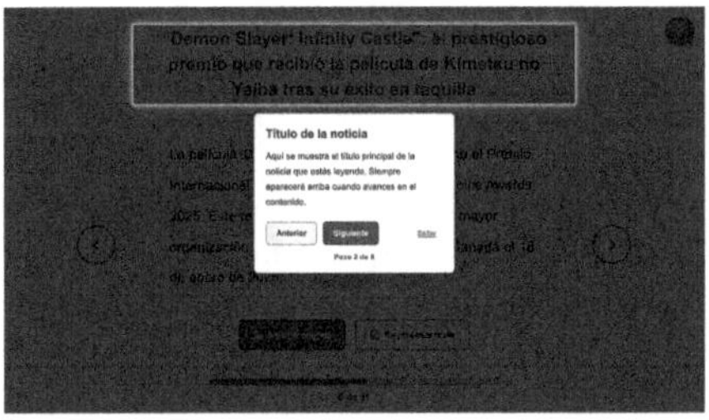

(a) Interactive onboarding tutorial

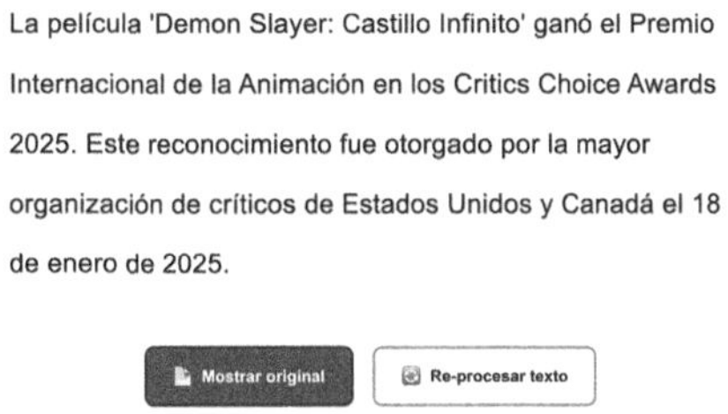

(b) Paragraph-level reprocessing tool

Fig. 3: Assistive features for older adults.

participants read two structurally equivalent health news articles in a counterbalanced order, with and without the extension. During these sessions, the generative models were strictly configured to ensure reliable outputs: the LLaMA 3.1 adaptation model operated at a temperature of 0.3, while the Claude-3 evaluator was set to 0.5 to balance analytical strictness with human-like criterion simulation. The multidimensional assessment utilized the GORT-5 for reading comprehension, the unweighted NASA Task Load Index (Raw TLX) for perceived mental demand, the SUS to measure usability against established rating scales [4], and the STAM model for post-experimental technological acceptance. Quantitative data were processed using related-samples t-tests and Wilcoxon signed-rank tests with a 95% confidence level ($p < .05$).

Table 2: GORT-5 and NASA-TLX performance metrics (Pre-test vs. Post-test)

Metric	Pre-Intervention		Post-Intervention		Statistic	p
	Mean	SD	Mean	SD		
Reading Comprehension (GORT-5)						
Speed (WPM)	78.80	20.00	111.65	26.36	$t = 12.39$	$< .001$
Precision (Errors)	16.13	6.56	4.80	2.41	$t = -12.25$	$< .001$
Comprehension (%)	61.33	25.69	85.33	13.83	$V = 268.5$	$< .001$
Cognitive Load (NASA-TLX)						
Global Score	9.76	3.47	5.83	1.45	$t = -7.71$	$< .001$

4 Results

In this section, we present the quantitative and qualitative results obtained from the initial development phase and the experimental validation. During the initial development phase, the prompt system achieved a 100% API success rate with an average latency of 6 s. The evaluation module demonstrated robust overlapping metrics (mean = 0.462) and similarity metrics (mean = 0.460).

4.1 Sample Characteristics and Statistical Assumptions

The study involved 30 older adults (63.3% female, 36.7% male) with a mean age of 73.03 years. Prior to the inferential analysis, the Shapiro-Wilk test was applied to assess the normality of the differences. Paired t-tests were utilized for normally distributed data, while the Wilcoxon signed-rank test was applied for non-normal distributions. Internal consistency for the NASA-TLX ($\alpha = 0.649$), SUS ($\alpha = 0.693$), and STAM scales fell within acceptable ranges.

4.2 Reading Comprehension and Cognitive Load

The intervention yielded substantial improvements in reading performance across all GORT-5 dimensions. Reading speed increased significantly, while the precision error rate dropped by over 70%. Overall reading comprehension improved from 61.33% to 85.33%. This upward trajectory was highly consistent across the sample, as illustrated in Fig. 4a, which highlights how the generative adaptation effectively elevated individual scores regardless of baseline performance.

Simultaneously, the perceived cognitive burden (NASA-TLX) saw a significant overall reduction ($t = -7.71, p < .001$). As depicted in Fig. 4b, this cognitive relief was characterized by a marked contraction across all measured stress factors—most notably in mental demand and frustration levels. Detailed performance metrics are presented in Table 2.

Usability and Technology Acceptance. The system's usability yielded a mean score of 95.50 (SD = 5.02), placing the LLM-driven web extension in the

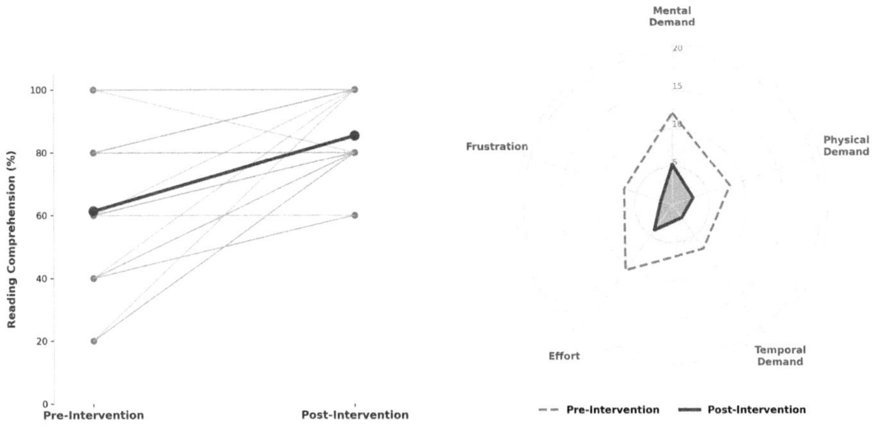

(a) GORT-5 comprehension scores.

(b) NASA-TLX dimensions.

Fig. 4: Impact of the LLM-driven web extension on (a) individual reading trajectories and (b) multidimensional perceived cognitive load.

"Best Imaginable" category. A Wilcoxon test confirmed that this usability level is significantly higher than the industry standard ($V = 465, p < .001$). Regarding technological acceptance (STAM), the composite score reached 4.56/5.00, reflecting a highly favorable attitude toward adoption ($t = 35.85, p < .001$).

Qualitative Findings. Post-experimental interviews corroborated the quantitative data. Participants reported an increase in self-efficacy, transforming news reading into a stimulating, stress-free activity. There was a genuine interest in installing the tool for daily use, alongside suggestions for future Text-to-Speech integration.

Discussion and Overall Impact. The results demonstrate that the LLM-driven web extension significantly improves information processing and user experience for older adults. By optimizing reading speed, minimizing errors, and elevating comprehension to 85.33%, the intervention validates its potential as a cognitive assistive tool beyond theoretical benchmarks.

Contrasting these results with the literature reveals a clear quantitative superiority. The comprehension increase observed significantly surpasses the moderate impact reported by Sun et al. [30], whose non-generative E2R methodology achieved only a 22% improvement. Furthermore, the functional performance of the participants suggests a higher practical utility than the theoretical metrics found in recent studies by Rodrigues et al. [27] and Alves et al. [3]. While Madina et al. [18] found low user preference for automated adaptations, the near-perfect acceptance rate of this tool indicates that the "Focus Mode" successfully mitigated the technological anxiety previously identified as a critical obstacle [23,31]. Finally, utilizing objective metrics alongside the Raw NASA-TLX confirmed that the tool facilitates digital interaction without inducing evaluation fatigue, overcoming the limitations of studies based solely on automated metrics like BLEU or ROUGE.

Despite these highly promising outcomes, it is important to acknowledge certain methodological limitations. The exceptional SUS score (95.5) and the overall positive reception may be partially influenced by the Hawthorne effect and the novelty of interacting with a new LLM-based tool featuring a clean "Focus Mode," compared to unadapted news websites such as Perú 21[1] and Gestión[2]. Future experimental designs should incorporate an active control group—such as unadapted articles presented in a similarly clean reader-mode layout—to better isolate the specific impact of the generative adaptation. Furthermore, while the blind LLM evaluator (Claude-3) provided technical robustness, it operated autonomously without human-in-the-loop validation, representing a critical area for future methodological refinement.

5 Conclusion

This research successfully achieved its objective of enhancing cognitive accessibility for older adults through an LLM-driven web extension. The intervention positively impacted all core dimensions: it significantly increased reading comprehension and speed, reduced perceived cognitive load, delivered superior usability, and fostered strong technological adoption. This work demonstrates that generative AI, when rigorously constrained by HCI principles, becomes a practical instrument capable of restoring digital autonomy and dignity to older adults.

Future work will prioritize the implementation of Text-to-Speech (TTS) capabilities and screen reader compatibility to expand multimodal accessibility. Furthermore, we aim to optimize system scalability through Edge AI models and validate the tool's ethical and long-term impact via longitudinal studies with larger cohorts and human-in-the-loop audits.

Acknowledgements. This work was developed with the support of the Laboratorio de Inteligencia Artificial at Universidad de Lima, which provided the computational resources, specialized equipment, and research infrastructure necessary to complete this research.

References

1. Adams, G., Fabbri, A., Ladhak, F., Lehman, E., Elhadad, N.: From sparse to dense: GPT-4 summarization with chain of density prompting. In: Proceedings of the 4th New Frontiers in Summarization Workshop, pp. 68–74. Association for Computational Linguistics, Singapore (2023). https://doi.org/10.18653/v1/2023.newsum-1.7. https://aclanthology.org/2023.newsum-1.7/

[1] https://peru21.pe/.
[2] https://gestion.pe/.

2. Alarcon, R., Moreno, L., Martínez, P.: Easier corpus: a lexical simplification resource for people with cognitive impairments. PLoS ONE **18**(4), e0283622 (2023). https://doi.org/10.1371/journal.pone.0283622
3. Alves, K., Silva, M., Santos, E., Valença, G., Brito, K.: A generative ai approach for creating and validating simplified versions of government documents. In: Conference on Digital Government Research, vol. 26 (2025). https://doi.org/10.59490/dgo.2025.968. https://proceedings.open.tudelft.nl/DGO2025/article/view/968
4. Bangor, A., Kortum, P.T., Miller, J.T.: Determining what individual SUS scores mean: adding an adjective rating scale. J. Usability Stud. **4**(3), 114–123 (2009)
5. Cao, A., Chintamani, K.K., Pandya, A.K., Ellis, R.D.: NASA TLX: software for assessing subjective mental workload. Behav. Res. Methods **41**(1), 113–117 (2009). https://doi.org/10.3758/BRM.41.1.113
6. Debevc, M., Škraba, T., Cerovac, B., Kožuh, I., Rajh, N.: Monitoring website accessibility: evaluating current approaches and a proposal for improvements. J. Access. Des. All **13**(2), 162–187 (2023). https://doi.org/10.17411/jacces.v13i2.485. https://www.jacces.org/index.php/jacces/article/view/485
7. Freire, A.P., et al.: Accessibility inspections using the web content accessibility guidelines by novice evaluators: an experience report. In: Proceedings of the XXIII Brazilian Symposium on Human Factors in Computing Systems. IHC 2024. Association for Computing Machinery, New York, NY, USA (2024). https://doi.org/10.1145/3702038.3702040
8. Färber, M., Aghdam, P., Im, K., Tawfelis, M., Ghoshal, H.: Simplifymytext: an LLM-based system for inclusive plain language text simplification. In: Advances in Information Retrieval, pp. 418–424. Springer, Cham (2025). https://doi.org/10.1007/978-3-031-88717-8_32
9. Guidroz, T., et al.: LLM-based text simplification and its effect on user comprehension and cognitive load (2025). https://doi.org/10.48550/arXiv.2505.01980
10. Hart, S.G.: NASA-task load index (NASA-TLX); 20 years later. In: Proceedings of the Human Factors and Ergonomics Society Annual Meeting, vol. 50, pp. 904–908. SAGE Publications (2006). https://doi.org/10.1177/154193120605000909
11. Hofmann, N., Dauenhauer, J., Dietzler, N.O., Idahor, I.D., Kreutz, C.K.: Thm@simpletext 2025 – task 1.1: revisiting text simplification based on complex terms for non-experts (2025). https://arxiv.org/abs/2507.04414
12. Holden, R.J.: A simplified system usability scale (SUS) for cognitively impaired and older adults. In: Proceedings of the International Symposium on Human Factors and Ergonomics in Health Care, vol. 9, no. 1, pp. 180–182 (2020). https://doi.org/10.1177/2327857920091021. https://doi.org/10.1177/2327857920091021
13. Hou, G., Anicetus, U., He, J.: How to design font size for older adults: a systematic literature review with a mobile device. Fronti. Psychol. **13** (2022). https://doi.org/10.3389/fpsyg.2022.931646. https://www.frontiersin.org/journals/psychology/articles/10.3389/fpsyg.2022.931646
14. Hutson, J.W., Franklin, A.E., Rogers, B.A., Walker, D.: Psychometric testing of NASA-TLX to measure learners' cognitive load in individual and group nursing simulations. Clin. Simul. Nurs. **95**, 101607 (2024). https://doi.org/10.1016/j.ecns.2024.101607. https://www.sciencedirect.com/science/article/pii/S1876139924000999
15. Hyzy, M., et al.: System usability scale benchmarking for digital health apps: meta-analysis. JMIR Mhealth Uhealth **10**(8), e37290 (2022). https://doi.org/10.2196/37290. https://mhealth.jmir.org/2022/8/e37290

16. Jennings, R.L., Midkiff, M., Nestor McCauley, E., et al.: The concurrent validity of the gort-5 with the wj-iv. Contemp. School Psychol. **28**, 328–340 (2024). https://doi.org/10.1007/s40688-024-00502-x
17. Kodandaram, S.R., Sunkara, M., Jayarathna, S., Ashok, V.: Detecting deceptive dark-pattern web advertisements for blind screen-reader users. J. Imaging **9**(11), 239 (2023). https://doi.org/10.3390/jimaging9110239. https://www.mdpi.com/2313-433X/9/11/239
18. Madina, M., Gonzalez-Dios, I., Siegel, M.: A preliminary study of chatgpt for spanish e2r text adaptation. In: Proceedings of the 2024 Joint International Conference on Computational Linguistics, Language Resources and Evaluation (LREC-COLING 2024), pp. 1422–1434. ELRA and ICCL, Torino, Italia (2024). https://aclanthology.org/2024.lrec-main.126/
19. Martínez, P., Ramos, A., Moreno, L.: Exploring large language models to generate easy to read content. Front. Comput. Sci. **6** (2024). https://doi.org/10.3389/fcomp.2024.1394705. https://www.frontiersin.org/journals/computer-science/articles/10.3389/fcomp.2024.1394705
20. Moreno, L., Petrie, H., Martínez, P., Alarcon, R.: Designing user interfaces for content simplification aimed at people with cognitive impairments. Univ. Access Inf. Soc. **23**(1), 99–117 (2024). https://doi.org/10.1007/s10209-023-00986-z
21. Ojeda-Mera, C., Injante, R., Valles-Coral, M., Pinedo, L., Tejada, K., García-Bautista, A.: Estado actual de la accesibilidad web en latinoamérica: una revisión exploratoria de las evaluaciones y herramientas utilizadas. Revista Española de Documentación Científica **47**(1), e378 (2024). https://doi.org/10.3989/redc.2024.1.1464. https://redc.revistas.csic.es/index.php/redc/article/view/1464
22. Peng, P., et al.: A meta-analytic review of cognition and reading difficulties: individual differences, moderation, and language mediation mechanisms. Psychol. Bull. **148**(3), 227–272 (2022). https://doi.org/10.1037/bul0000361
23. Pradhan, A., Todi, K.: Understanding large language model based metrics for text summarization. In: Proceedings of the 4th Workshop on Evaluation and Comparison of NLP Systems, pp. 149–155. Association for Computational Linguistics, Bali, Indonesia (2023). https://doi.org/10.18653/v1/2023.eval4nlp-1.12. https://aclanthology.org/2023.eval4nlp-1.12/
24. PRB: 2024 world population data sheet. Technical report, Population Reference Bureau, Washington, DC (2024)
25. Qiang, J., Huang, M., Zhu, Y., Yuan, Y., Zhang, C., Yu, K.: Redefining simplicity: benchmarking large language models from lexical to document simplification (2025). https://arxiv.org/abs/2502.08281
26. Rios Rincon, A.M., Cruz, A.M., Daum, C., Neubauer, N., Comeau, A., Liu, L.: Digital storytelling in older adults with typical aging, and with mild cognitive impairment or dementia: a systematic literature review. J. Appl. Gerontol. **41**(3), 867–880 (2022). https://doi.org/10.1177/07334648211015456
27. Rodrigues, J.F., Lopes Cardoso, H., Teixeira Lopes, C.: Evaluating llama 3 for text simplification: a study on wikipedia lead sections. In: Companion Proceedings of the ACM on Web Conference 2025, pp. 1273–1277. Association for Computing Machinery, Sydney NSW, Australia (2025). https://doi.org/10.1145/3701716.3715467
28. Sahoo, P., Singh, A., Saha, S., Jain, V., Mondal, S., Chadha, A.: A systematic survey of prompt engineering in large language models: techniques and applications (2024). https://doi.org/10.13140/RG.2.2.13032.65286

29. Souza, C.S.M.D., Campos, R.H.D.F., Manjarrés Carrizalez, D.: La lectura fácil como estrategia de accesibilidad cognitiva para la educación inclusiva de estudiantes con discapacidad intelectual. Revista Saber Incluir **1**(1), e23002 (2023). https://portaldeperiodicos.ufopa.edu.br/index.php/revistasaberincluir/article/download/2545/1499/6813
30. Sun, K.T., Shieh, T.M., Hsia, S.M., Ningrum, V., Lin, X.Y., Shih, Y.H.: Easy to read health education material improves oral health literacy of older adults in rural community-based care centers: a quasi-experimental study. Healthcare **9**(11), 1465 (2021). https://doi.org/10.3390/healthcare9111465. https://www.mdpi.com/2227-9032/9/11/1465
31. Suárez-Figueroa, M.C., Diab, I., Ruckhaus, E., Cano, I.: First steps in the development of a support application for easy-to-read adaptation. Univ. Access Inf. Soc. **23** (2022). https://doi.org/10.1007/s10209-022-00946-z
32. Säuberli, A., et al.: Enabling text comprehensibility assessment for people with intellectual disabilities using a mobile application. Front. Commun. **8** (2023). https://doi.org/10.3389/fcomm.2023.1175625. https://www.frontiersin.org/journals/communication/articles/10.3389/fcomm.2023.1175625
33. Taieb-Maimon, M., Vaisman-Fairstein, E.: Improving older adults' accessibility to the web using real-time online interactive guides. Int. J. Hum. Comput. Stud. **164**, 102830 (2022). https://doi.org/10.1016/j.ijhcs.2022.102830. https://www.sciencedirect.com/science/article/pii/S1071581922000593
34. Tan, K.C.K.: Use of senior technology acceptance model (STAM) for social robots studies. In: Voloşencu, C. (ed.) Current State and Future Perspective in Human-Robot Interaction, chap. 2. IntechOpen, London (2024). https://doi.org/10.5772/intechopen.1007144
35. United Nations High Commissioner for Refugees: Working with older persons in forced displacement. Technical report, United Nations High Commissioner for Refugees (2024). https://emergency.unhcr.org/sites/default/files/2024-01/Working%20with%20older%20persons%20in%20forced%20displacement%202021.pdf
36. Vecchiato, S.: Clear, easy, plain, and simple as keywords for text simplification. Front. Artif. Intell. **5** (2022). https://doi.org/10.3389/frai.2022.1042258. https://www.frontiersin.org/journals/artificial-intelligence/articles/10.3389/frai.2022.1042258
37. Wu, N., Gong, M., Shou, L., Liang, S., Jiang, D.: Large language models are diverse role-players for summarization evaluation (2023). https://arxiv.org/abs/2303.15078
38. Zhang, H., Liu, X., Zhang, J.: Summit: iterative text summarization via chatgpt. In: Findings of the Association for Computational Linguistics: EMNLP 2023, pp. 10644–10657. Association for Computational Linguistics, Singapore (2023). https://doi.org/10.18653/v1/2023.findings-emnlp.714. https://aclanthology.org/2023.findings-emnlp.714/

Revealing the Influence of Variables on Real Estate Price Estimation Through Supervised Learning Models Using Explainable AI Methods

Piero Alejandro Rozas Aleman, Moises David Mamani Medina(✉), Edgar Jesus Huaranga Junco, and Franci Suni Lopez

Laboratorio de Inteligencia Artificial, Facultad de Ingenieria, Universidad de Lima, Lima, Peru
{20212407,20211589}@aloe.ulima.edu.pe

Abstract. Metropolitan Lima, Peru faced a projected housing deficit of over 469,000 units by 2028, in a dynamic real estate market where accurate property valuation was critical for buyers, sellers, and investors. In this context, traditional appraisal methods presented limitations in scalability and objectivity, driving the adoption of machine learning models capable of handling large-scale data and complex relationships. However, these models tended to operate as black boxes, hindering the understanding of their decisions and limiting their professional adoption. This research applied and compared supervised learning models alongside explainable artificial intelligence (XAI) methods to estimate and analyze apartment prices in Lima. A dataset of 183,000 properties with structural variables from the Asociación de Empresas Inmobiliarias del Perú (ASEI) real estate confederation was used, enriched with points of interest, economic, and demographic data. XGBoost, Random Forest, and Multilayer Perceptron (MLP) were implemented, together with the explainability methods SHapley Additive exPlanations (SHAP), Local Interpretable Model-agnostic Explanations (LIME), Permutation Feature Importance (PFI), and Integrated Gradients (IG). XGBoost achieved the best performance with an R^2 of 0.9791, identifying built-up area as the most influential variable. Proximity to restaurants also proved relevant, with an effect modulated by establishment quality and the educational level of the surrounding area. Expert validation supported most variables, though others such as the masculinity index were questioned. The explainability methods strengthened transparency and trust in the responsible use of artificial intelligence for real estate appraisal in Lima.

Keywords: Real estate appraisal · Random Forest · XGBoost · MLP · explainability · SHAP · LIME · PFI · Integrated Gradients

1 Introduction

Artificial intelligence (AI) models used for property valuation faced a fundamental challenge: they lacked transparency. Although these models offered high

H. Ponce et al. (Eds.): COMIA 2026, CCIS 3095, pp. 478–491, 2027.
https://doi.org/10.1007/978-3-032-37438-7_37

predictive accuracy, many academics questioned them on the grounds that they operated as black boxes whose decisions could not be explained, generating distrust in the real estate sector [1,2]. Obtaining good results was not sufficient; it was equally important to understand the influence of variables on model decisions so that users could trust their outputs [3,4]. This lack of transparency directly affected buyers and sellers, who could not understand how a property's characteristics influenced price estimates, leading to inflated prices, erroneous offers, or unprofitable investments [5,6]. This situation was further aggravated in Metropolitan Lima, Peru, where a housing deficit of over 469,000 units was projected for 2028 [7], making the understanding of price-driving factors critical for informed decision-making.

Prior research explored a wide range of variables and models for property price estimation. Structural factors such as built-up area consistently ranked as the most influential [2,3], while spatial variables like proximity to transport or services [8,9], and socioeconomic attributes such as educational level or income [10,11] also played a relevant role. Among models, Random Forest and XGBoost stood out for their predictive performance [12,13], though explainability remained largely unexplored in this domain [14]. Explainable Artificial Intelligence (XAI) methods such as SHAP, PFI, LIME, and Integrated Gradients gained relevance for improving model interpretability [15,16], yet their combined application and comparison in real estate contexts remained scarce [17,18].

This research applied and compared supervised learning models alongside explainability methods to estimate apartment prices in Metropolitan Lima, Peru, filling a gap in public research for this market. The main research question was: How can the impact of structural, economic, demographic, and points-of-interest variables on real estate price estimation models be interpreted using explainability methods?

2 Background

2.1 Supervised Learning Models

XGBoost. XGBoost was an efficient implementation of the Gradient Boosting algorithm, widely used in structured regression problems for its balance between performance and speed [19]. It trained decision trees sequentially, where each tree corrected the errors of the previous one. Its loss function included a penalty that controlled model complexity, improving generalization and reducing overfitting. Its main hyperparameters were: *learning_rate*, which controlled the weight of each new tree; *n_estimators*, the total number of trained trees; and *max_depth*, the maximum depth of each tree.

Random Forest. Random Forest was an ensemble model based on the bagging technique, where multiple trees were trained in parallel on random subsets of the dataset [20]. The final prediction was obtained by averaging individual results, which reduced model variance. It was robust against missing data and less prone to overfitting, although at the cost of greater computational complexity [21]. Its main hyperparameters were: *n_ estimators*, the number of trees in the forest; *criterion*, the function used to evaluate tree splits; *max_features*, the number of variables considered at each split; and *max_ depth*, the maximum depth of each tree.

Multilayer Perceptron. The Multilayer Perceptron (MLP) was a feedforward artificial neural network composed of an input layer, hidden layers, and an output layer [22]. It was capable of capturing complex non-linear relationships through weighted combinations of inputs and non-linear activation functions [23], although it required greater hyperparameter tuning and intensive training. Its main hyperparameters included: architecture (number of hidden layers, neurons per layer, activation function, and optional regularization); *optimizer*, the algorithm that adjusted weights; *learning rate*; *epochs*, the number of training iterations; and *batch_ size*, the number of observations used at each weight update step.

2.2 Explainable AI Methods

Supervised learning models achieved highly accurate results; however, many of these techniques obtained results through such a complex algorithm that it simulated a black box, where it was not possible to understand how a given result was produced [1]. Post-hoc Explainable AI (XAI) methods sought to open that black box to make AI results transparent and reveal the contribution of each variable to predictions [24]. Model-agnostic XAI methods offered the flexibility to explain any machine learning model [25]. Explainability scope was divided into two levels: global, which focused on analyzing the most critical variables for the model and their average effects; and local, which analyzed the most important variables for a specific prediction rather than the entire model [26].

SHapley Additive exPlanations (SHAP). SHAP was a unified framework for interpreting predictions based on Shapley values from cooperative game theory [27]. Each feature was treated as a "player", and the prediction was the "value" to be distributed among them according to their marginal contributions. Given all possible coalitions $S \subseteq N$, where N was the full set of features, the impact of adding feature i to coalition S was:

$$f(S \cup \{i\}) - f(S) \tag{1}$$

The Shapley value for feature i was calculated as the weighted average of its impact across all possible coalitions, where $|S|$ was the number of features in coalition S and $|N|$ was the total number of features:

$$\phi_i = \sum_{S \subseteq N \setminus \{i\}} \frac{|S|!(|N| - |S| - 1)!}{|N|!} [f(S \cup \{i\}) - f(S)] \tag{2}$$

The sum of all feature values equaled the difference between the model prediction and the base value:

$$f(x) = \phi_0 + \sum_{i=1}^{|N|} \phi_i \tag{3}$$

Permutation Feature Importance (PFI). PFI measured the increase in model error after permuting the values of a given feature [20,28]. Unlike SHAP, PFI provided only global explainability and did not indicate the direction of a feature's impact. The model error using the original features and the permuted feature was calculated as:

$$e_{\text{orig}} = L(y, f(X)), \quad e_{\text{perm}} = L(y, f(X_{\text{perm}})) \tag{4}$$

where L was an error measure such as mean squared error. For each feature j, a new feature matrix X_{perm} was generated by randomly permuting the values of feature j while keeping all others fixed. The importance of feature j was computed as the difference $I_j = e_{\text{perm}} - e_{\text{orig}}$, or alternatively as the ratio $I_j = e_{\text{perm}}/e_{\text{orig}}$.

Local Interpretable Model-Agnostic Explanations (LIME). LIME provided reliable local explanations for machine learning model predictions by approximating model behavior with an interpretable model in the neighborhood of a specific observation [29]. The algorithm generated perturbed samples z' around the point of interest x, assigned each sample a weight based on its proximity to x using a similarity function $\pi_x(z')$, and trained a simple linear model $g(z')$ on the perturbed samples to approximate $f(x)$ locally by minimizing:

$$\min_{z'} \pi_x(z') \cdot (f(z') - g(z'))^2 \tag{5}$$

Integrated Gradients. Integrated Gradients was an explainability technique designed for differentiable deep learning models that assigned each input feature a marginal contribution toward the final model prediction [30]. A baseline value x' was defined for each feature, and the path from x' to the actual value x was parameterized as:

$$x(\alpha) = x' + \alpha(x - x') \tag{6}$$

The gradient of the model prediction with respect to each feature x_i was calculated along this path. The contribution of feature x_i was then the integral of the gradients along the path from x' to x:

$$IG(x_i) = (x_i - x_i') \cdot \int_0^1 \frac{\partial f(x(\alpha))}{\partial x_i} \, d\alpha \tag{7}$$

3 Methodology and Results

3.1 Methodology Overview

The study followed an iterative experimental methodology, as illustrated in Fig. 1, comprising two iterations. Each iteration included dataset construction, supervised learning model training, application of explainability methods, and result analysis. The second iteration incorporated new variables derived from insights obtained in the first [13,31–34]. Explainability methods varied by model type: SHAP, PFI, and LIME were applied to tree-based models, while SHAP and Integrated Gradients were used for the MLP due to its gradient-based architecture. Additionally, Local explainability was performed only on the best-performing model.

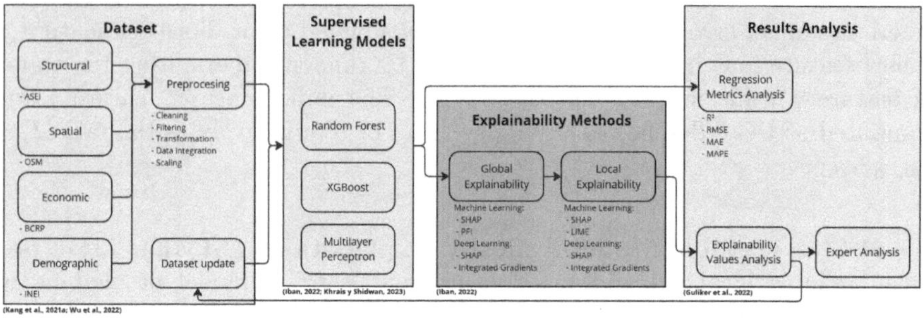

Fig. 1. Methodological workflow.

3.2 Dataset

A dataset of approximately 183,000 apartments in Metropolitan Lima, Peru was constructed combining four data sources, as summarized in Table 1: structural variables and the target variable were obtained from the ASEI transaction records (sales from 2017 to 2025), spatial variables from OpenStreetMap [35], economic variables from BCRPData [36], and demographic variables from INEI [37]. Preprocessing included removal of out-of-scope records, spatial integration via coordinates, incorporation of demographic and economic data by district, categorical encoding, and feature scaling.

3.3 Models and Training

XGBoost [19], Random Forest [20], and MLP [22] were implemented and tuned using RandomizedSearchCV with 5-fold cross-validation for tree-based models and manual architecture search for MLP, with an 80/20 train-test split [38,39]. The final hyperparameter configurations are summarized in Table 2.

Table 1. Dataset composition.

Category	# Vars	Examples	Source
Target variable	1	Net sale price (PEN)	ASEI
Structural	83	Built-up area, bedrooms, bathrooms, floor, etc.	ASEI
Spatial (POI)	30	Distance to nearest POI, count within 1 km, etc.	OSM
Demographic	14	Total population, age groups, masculinity index, etc.	INEI
Economic	22	CPI, exchange rate, GDP, minimum wage, etc.	BCRP

Table 2. Final hyperparameter configurations.

Model	Hyperparameters	Values
XGBoost	n_estimators/learning_rate/max_depth	1000/0.1/8
RF	n_estimators/max_features/max_depth	300/20/50
MLP	hidden layers/activation/optimizer/learning rate	3/ReLU/Adam/0.1

3.4 Results

Model Performance. XGBoost achieved the best performance in both iterations ($R^2 = 0.9791$), as shown in Table 3, demonstrating superior accuracy and robustness for real estate price estimation and outperforming previous studies such as Iban [13] ($R^2 = 0.8733$), Gao et al. [38] ($R^2 = 0.8594$), and Li et al. [9] ($R^2 = 0.944$). In contrast, RF and MLP achieved competitive R^2 values but consistently exhibited higher error metrics.

Table 3. Model performance metrics across iterations.

Model	Iteration 1				Iteration 2			
	R^2	RMSE	MAE	MAPE	RR^2	RMSE	MAE	MAPE
RF	0.9760	42,614	21,034	4.61%	0.9753	43,253	20,358	4.41%
XGB	**0.9789**	**39,955**	**20,001**	**4.49%**	**0.9791**	**39,823**	**17,782**	**3.88%**
MLP	0.9723	45,867	27,125	6.47%	0.9750	43,547	22,466	5.19%

Global Explainability. SHAP-based global explainability results, shown in Fig. 2, indicate that built-up area consistently emerged as the most influential variable across all models and explainability methods, reinforcing the central role of property size in price estimation. Beyond structural variables, XGBoost highlighted distance to the nearest restaurant and number of nearby schools as key factors [4,18], consistent with the MLP, while Random Forest emphasizes proximity to clinics.

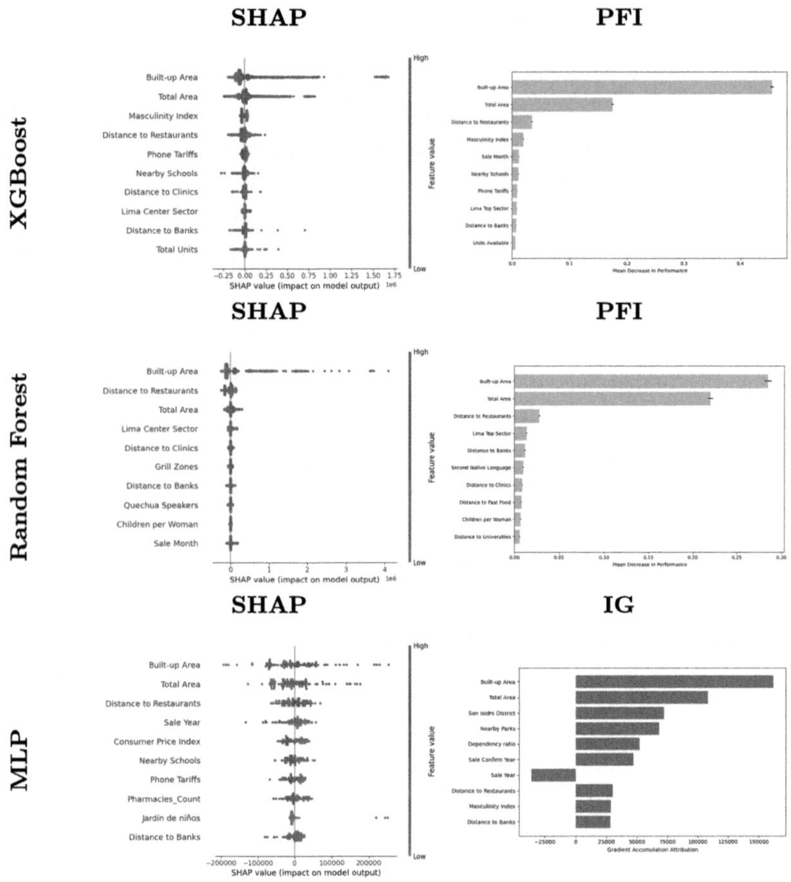

Fig. 2. Global explainability results across models and explainability methods.

PFI results corroborate these findings: built-up area ranks first across all three models, with secondary spatial variables showing partial agreement. PFI confirms the global feature ranking independently of any model-specific assumptions, providing complementary evidence to the SHAP analysis. Integrated Gradients attributions for the MLP similarly identify built-up area as the dominant feature, while also highlighting greater sensitivity to economic time-series variables.

Property size remains the most influential feature across studies [13,38,39], whereas the importance of environmental variables varies by context—e.g., proximity to shops and transport [13], coastal proximity [39], or parks [38]—highlighting the geographic dependence of real estate markets.

The convergence of three methodologically distinct approaches visible in Fig. 2 strengthened the reliability of the global interpretation, while the partial discrepancies on secondary features highlighted that no single method captured the full picture of model behavior. The agreement on built-up area as the dominant predictor across the three models suggested that this variable captured a substantial portion of the price signal independently of the underlying algorithm, while the divergence on secondary spatial features reflected how each model weighted the surrounding contextual variables differently.

The presence of points-of-interest variables among the top-ranked features in all three models also confirmed that location-based attributes acted as strong complements to structural variables in the Lima, Peru market. In particular, the prominence of economic variables under Integrated Gradients suggested that the MLP exploited interactions across macroeconomic indicators that tree-based models tended to fragment across multiple splits, which was consistent with the architectural differences between gradient-based and tree-based learners.

Local Explainability. Local explainability was conducted on XGBoost, selected as the best-performing model, using both SHAP and LIME as it is a tree-based model. The resulting local feature attributions are illustrated in Fig. 3. Distance to the nearest restaurant was chosen as the focal variable for this analysis because it ranked as the second most important spatial variable in XGBoost according to both global SHAP and PFI results, and its global SHAP dependence plot exhibited a non-linear pattern that warranted deeper local investigation. The analysis covered representative instances with low, medium, and high values of this distance.

SHAP and LIME showed strong agreement on the most influential variables for each instance, particularly on built-up area, but again diverged in the magnitude and ordering of secondary features. SHAP preserved alignment with the global ranking by emphasizing variables such as the masculinity index, telephone tariffs, and distance to clinics, while LIME assigned greater weight to amenity-related variables such as cafeteria, squash court, and conference room availability. This divergence reflected the methodological difference between SHAP's coalition-based attribution and LIME's local linear approximation, and supported prioritizing SHAP for instances where non-linear interactions dominated.

The local explainability results revealed a non-linear relationship between apartment prices and restaurant proximity, observable across the low-, medium-, and high-distance instances in Fig. 3. Proximity to restaurants showed a positive impact at short distances, became largely neutral at medium distances, and turned negative at larger distances, suggesting that accessibility influenced prices in a context-dependent manner rather than following a strictly monotonic pattern. Both SHAP and LIME produced consistent local explanations, reinforcing the reliability of this behavior.

To further investigate this pattern, a scenario analysis incorporated additional variables including restaurant quality, land use, urban density, and average educational level. The results showed in Fig. 4 indicated that the positive effect of restaurant proximity was primarily associated with restaurant quality and socioeconomic conditions, suggesting that high-quality restaurants tended to be located in more prestigious and educated areas. In contrast, land use and

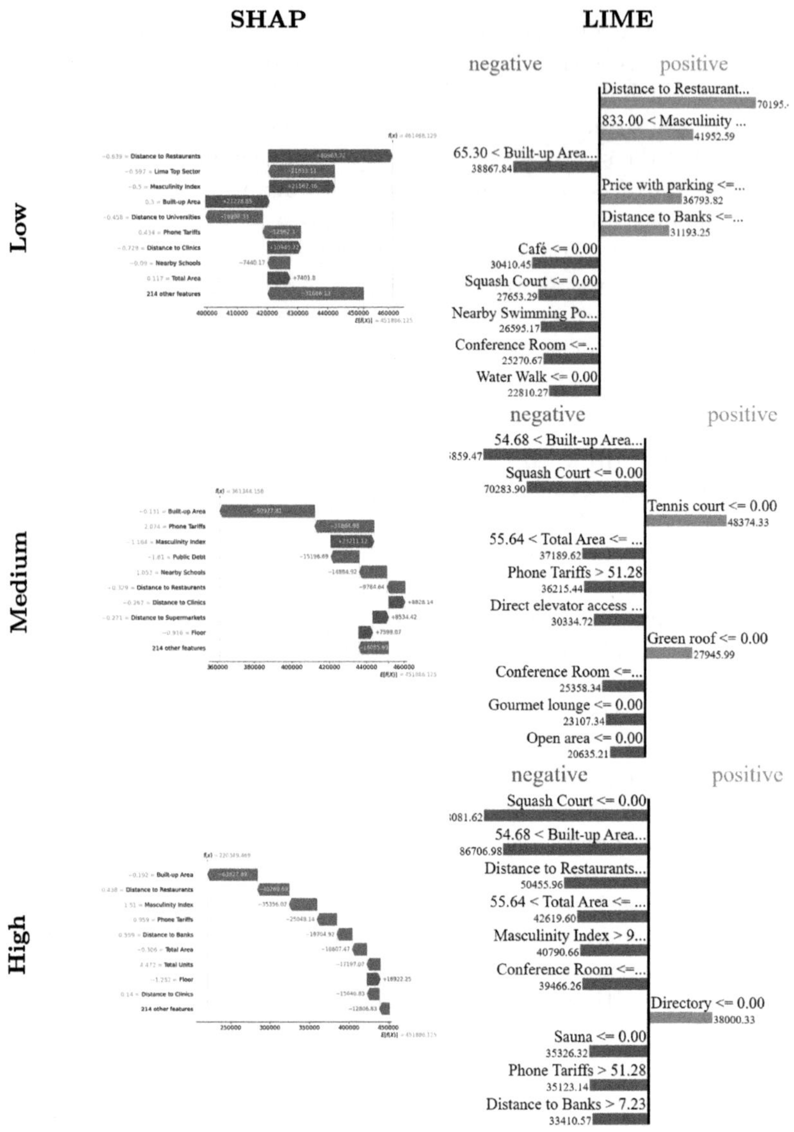

Fig. 3. Local explainability results across explainability methods and representative instances.

urban density showed no clear moderating effect, indicating that the influence of proximity emerged more from qualitative environmental characteristics than from spatial concentration alone. The results supported the assumption that closer was always better, but also suggested the existence of a saturation point beyond which additional distance no longer translated into a meaningful price increase when a threshold was crossed.

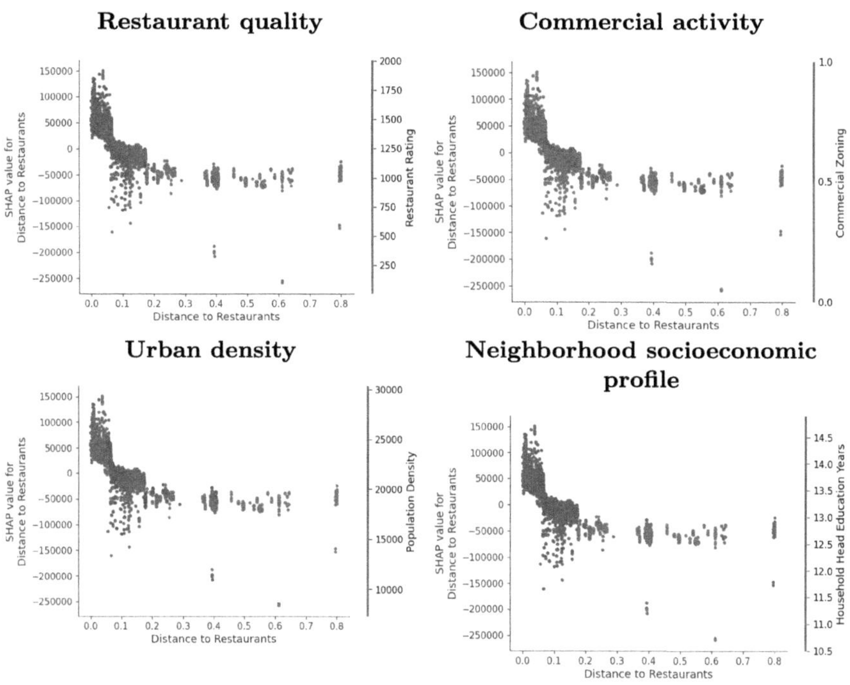

Fig. 4. Scenario analysis related to proximity to restaurants.

Expert Validation. Three ASEI specialists validated the results via semi-structured interviews scored with the Aiken V coefficient. Structural and economic variables obtained the highest scores, and location-related variables received strong validation (Aiken V = 0.83). The masculinity index—defined as the ratio of males to females in a given district—was included in the model because demographic composition has been associated with housing demand patterns in prior literature [11]; however, the three experts questioned it as non-standard in professional appraisal practice, noting that it lacks direct interpretability in real estate valuation contexts. This highlights the importance of integrating domain knowledge alongside model outputs [3,39].

4 Conclusions and Future Work

XGBoost achieved the best predictive performance ($R^2 = 0.9791$), with built-up area as the most influential variable across all models and methods. Explainability analysis revealed that proximity to restaurants played a significant role, with its effect modulated by restaurant quality and the educational level of the surrounding area. The application of SHAP, PFI, LIME, and Integrated Gradients provided complementary perspectives on model behavior: SHAP offered both global and local directional insights; PFI provided model-agnostic global rankings; LIME generated locally consistent linear approximations; and Integrated

Gradients enabled gradient-level analysis of the MLP. Together, these methods supported a more complete understanding of variable contributions than any single technique would allow. Expert validation supported most findings, though variables such as the masculinity index were questioned, underscoring the importance of combining model outputs with domain knowledge for real-world adoption [14,15].

From a practical perspective, this study enhances model transparency, offering an accessible and data-driven approach that can foster trust and support real-world adoption. Nevertheless, the findings are limited to Metropolitan Lima (2017–2025) and may not generalize to other geographic or temporal contexts. Furthermore, the explainability results should be interpreted as associative rather than causal relationships.

Future work includes exploring dynamic model updating based on expert feedback, incorporating additional variables such as accessibility or environmental quality, adopting price per square meter as an alternative target variable, and applying time-series models such as Long Short-Term Memory (LSTM) or Transformers for temporal market analysis in Lima, Peru.

Acknowledgments. This work was carried out with the support of the Artificial Intelligence Laboratory of the Universidad de Lima, whose research infrastructure, specialized equipment, and computational resources were essential for the completion of this research. The authors also thank ASEI, now known as CODIP, for providing access to the real estate data used in this study.

References

1. Leao, S.Z., Van Den Nouwelant, R., Shi, V., Han, H., Praharaj, S., Pettit, C.J.: A rapid analytics tool to map the effect of rezoning on property values. Comput. Environ. Urban Syst. **86**, 101572 (2021). https://doi.org/10.1016/j.compenvurbsys.2020.101572
2. Rico-Juan, J.R., de la Paz, P.T.: Machine learning with explainability or spatial hedonics tools? An analysis of the asking prices in the housing market in Alicante, Spain. Expert Syst. Appl. **171**, 114590 (2021). https://doi.org/10.1016/j.eswa.2021.114590
3. Karamanou, A., Kalampokis, E., Tarabanis, K.: Linked open government data to predict and explain house prices: the case of Scottish Statistics Portal. Big Data Res. **30**, 100355 (2022). https://doi.org/10.1016/j.bdr.2022.100355
4. Lee, H., Han, H., Pettit, C.: Capturing the distance decay effect of amenities on housing price using explainable artificial intelligence. Appl. Geogr. **174**, 103486 (2025). https://doi.org/10.1016/j.apgeog.2024.103486
5. Marzooqi, F., Redouane, A.: Predicting real estate prices using machine learning in Abu Dhabi. Iraqi J. Sci. 1689–1706 (2024). https://doi.org/10.24996/ijs.2024.65.3.40
6. Xu, X., Zhang, Y.: A Gaussian process regression machine learning model for forecasting retail property prices with Bayesian optimizations and cross-validation. Decis. Anal. J. **8**, 100267 (2023). https://doi.org/10.1016/j.dajour.2023.100267

7. Ministerio de Vivienda, Construcción y Saneamiento: Expansión urbana en el Perú: Proyección del déficit habitacional en tres ciudades principales mediante el análisis espaciotemporal del cambio de uso de suelo utilizando métodos de machine learning. Gobierno del Perú (2024). https://www.gob.pe/es/i/5960863
8. Tan, R., He, Q., Zhou, K., Xie, P.: The effect of new metro stations on local land use and housing prices: the case of Wuhan. China. J. Transp. Geogr. **79**, 102488 (2019). https://doi.org/10.1016/j.jtrangeo.2019.102488
9. Li, S., Jiang, Y.G., Ke, S., Nie, K., Wu, C.: Understanding the effects of influential factors on housing prices by combining extreme gradient boosting and a hedonic price model (XGBoost-HPM). Land **10**(5), 533 (2021). https://doi.org/10.3390/land10050533
10. Jafary, P., Shojaei, D., Rajabifard, A., Ngo, T.: Automated land valuation models: a comparative study of four machine learning and deep learning methods based on a comprehensive range of influential factors. Cities **151**, 105115 (2024). https://doi.org/10.1016/j.cities.2024.105115
11. Sisman, S., Aydinoglu, A.: Improving performance of mass real estate valuation through application of the dataset optimization and Spatially Constrained Multivariate Clustering Analysis. Land Use Policy **119**, 106167 (2022). https://doi.org/10.1016/j.landusepol.2022.106167
12. Lahmiri, S., Bekiros, S., Avdoulas, C.: A comparative assessment of machine learning methods for predicting housing prices using Bayesian optimization. Decis. Anal. J. **6**, 100166 (2023). https://doi.org/10.1016/j.dajour.2023.100166
13. Iban, M.: An explainable model for the mass appraisal of residences: the application of tree-based machine learning algorithms and interpretation of value determinants. Habitat Int. **128**, 102660 (2022). https://doi.org/10.1016/j.habitatint.2022.102660
14. Acharya, D.B., Divya, B., Kuppan, K.: Explainable and Fair AI: balancing performance in financial and real estate machine learning models. IEEE Access **1** (2024). https://doi.org/10.1109/access.2024.3484409
15. Barredo, A., et al.: Explainable artificial intelligence (XAI): concepts, taxonomies, opportunities and challenges toward responsible AI. Inf. Fusion **58**, 82–115 (2020). https://doi.org/10.1016/j.inffus.2019.12.012
16. Lenaers, I., De Moor, L.: Exploring XAI techniques for enhancing model transparency and interpretability in real estate rent prediction: a comparative study. Financ. Res. Lett. **58**, 104306 (2023). https://doi.org/10.1016/j.frl.2023.104306
17. Kim, H., Kwon, Y., Choi, Y.: Assessing the impact of public rental housing on the housing prices in proximity: based on the regional and local level of price prediction models using long short-term memory (LSTM). Sustainability **12**(18), 7520 (2020). https://doi.org/10.3390/su12187520
18. Liu, X., et al.: Unveiling the spatially varied nonlinear effects of urban built environment on housing prices using an interpretable ensemble learning model. Appl. Geogr. **173**, 103458 (2024). https://doi.org/10.1016/j.apgeog.2024.103458
19. Chen, T., Guestrin, C.: XGBoost. In: KDD 2016: Proceedings of the 22nd ACM SIGKDD International Conference on Knowledge Discovery and Data Mining, pp. 785–794 (2016). https://doi.org/10.1145/2939672.2939785
20. Breiman, L.: Random forests. Mach. Learn. **45**(1), 5–32 (2001). https://doi.org/10.1023/A:1010933404324
21. Pérez-Rave, J., González-Echavarría, F., Correa-Morales, J.: Modeling of apartment prices in a Colombian context from a machine learning approach with stable-important attributes. Dyna **87**(212), 63–72 (2020). https://doi.org/10.15446/dyna.v87n212.80202

22. Swingler, K.: Opening the black box: analysing MLP functionality using walsh functions. In: Studies in Computational Intelligence, pp. 303–323 (2015). https://doi.org/10.1007/978-3-319-26393-9_18
23. Alsmadi, M.K., Omar, K.B., Noah, S.A., Almarashdah, I.: Performance comparison of multi-layer perceptron (back propagation, delta rule and perceptron) algorithms in neural networks. In: IEEE International Advance Computing Conference, pp. 296–299 (2009). https://doi.org/10.1109/iadcc.2009.4809024
24. Das, A., Rad, P.: Opportunities and challenges in explainable artificial intelligence (XAI): a survey. arXiv preprint (2020). https://doi.org/10.48550/arxiv.2006.11371
25. Thalpage, N.: Unlocking the black box: explainable artificial intelligence (XAI) for trust and transparency in AI systems. J. Digit. Art Humanit. (2023). https://doi.org/10.33847/2712-8148.4.1_4
26. Delgado-Panadero, Á., Hernández-Lorca, B., García-Ordás, M.T., Benítez-Andrades, J.A.: Implementing local-explainability in gradient boosting trees: feature contribution. Inf. Sci. **589**, 199–212 (2022). https://doi.org/10.1016/j.ins.2021.12.111
27. Lundberg, S., Lee, S.: A unified approach to interpreting model predictions. In: Proceedings of the 31st International Conference on Neural Information Processing Systems (NIPS 2017), pp. 4768–4777 (2017)
28. Fisher, A., Rudin, C., Dominici, F.: All Models are Wrong, but Many are Useful: Learning a Variable's Importance by Studying an Entire Class of Prediction Models Simultaneously. arXiv preprint (2018). https://doi.org/10.48550/arxiv.1801.01489
29. Visani, G., Bagli, E., Chesani, F., Poluzzi, A., Capuzzo, D.: Statistical stability indices for LIME: obtaining reliable explanations for machine learning models. J. Oper. Res. Soc. **73**(1), 91–101 (2021). https://doi.org/10.1080/01605682.2020.1865846
30. Sundararajan, M., Taly, A., Yan, Q.: Axiomatic Attribution for Deep Networks. arXiv preprint (2017). https://doi.org/10.48550/arxiv.1703.01365
31. Kang, Y., Zhang, F., Gao, S., Peng, W., Ratti, C.: Human settlement value assessment from a place perspective: considering human dynamics and perceptions in house price modeling. Cities **118**, 103333 (2021). https://doi.org/10.1016/j.cities.2021.103333
32. . Wu, C., Du, Y., Li, S., Liu, P., Ye, X.: Does visual contact with green space impact housing prices? An integrated approach of machine learning and hedonic modeling based on the perception of green space. Land Use Policy **115**, 106048 (2022). https://doi.org/10.1016/j.landusepol.2022.106048
33. Khrais, L.T., Shidwan, O.S.: The role of neural network for estimating real estate prices value in post COVID-19: a case of the middle east market. Int. J. Electr. Comput. Eng. **13**(4), 4516–4525 (2023). https://doi.org/10.11591/ijece.v13i4.pp4516-4525
34. Guliker, E., Folmer, E., Van Sinderen, M.: Spatial determinants of real estate appraisals in The Netherlands: a machine learning approach. ISPRS Int. J. Geo Inf. **11**(2), 125 (2022). https://doi.org/10.3390/ijgi11020125
35. OpenStreetMap (OSM): Base de datos geoespacial. https://www.openstreetmap.org/. Accessed 25 Apr 2025
36. Banco Central de Reserva del Perú (BCRP): BCRPData. https://estadisticas.bcrp.gob.pe/estadisticas/series/. Accessed 2 May 2025
37. Instituto Nacional de Estadística e Informática (INEI): Sistema de Información Regional para la Toma de Decisiones. https://systems.inei.gob.pe/SIRTOD/app/consulta. Accessed 2 May 2025

38. Gao, Q., Shi, V., Pettit, C., Han, H.: Property valuation using machine learning algorithms on statistical areas in Greater Sydney, Australia. Land Use Policy **123**, 106409 (2022). https://doi.org/10.1016/j.landusepol.2022.106409
39. Soltani, A., Lee, C.L.: The non-linear dynamics of South Australian regional housing markets: a machine learning approach. Appl. Geogr. **166**, 103248 (2024). https://doi.org/10.1016/j.apgeog.2024.103248

A Two-Stage Mobile Vision System for Orange Segmentation and Real-Time Detection of Sooty Mold and Fruit Fly Damage in Regional Oranges

Anibal Gonzalez Tovar(✉), Angel Dagoberto Cantú Castro, Said Polanco Martagón, and Christian Alejandro Saldaña Calderón

Universidad Politécnica de Victoria, Ciudad Victoria, Tamaulipas, Mexico
{2439012,2439010,spolancom,183004}@upv.edu.mx

Abstract. This paper presents the development of a mobile application for the detection of pests and disease symptoms in regional oranges using a two-stage deep learning pipeline. The proposed mobile application combines a fruit segmentation model and a second analysis model designed to identify relevant phytosanitary conditions during mobile inference. The first model detects the orange region under non-controlled acquisition conditions, while the second model analyzes the segmented fruit area to estimate the presence of sooty mold and fruit fly damage. The dataset was manually acquired and manually annotated using Edge Impulse, with a total of 6,664 images, including 5,719 healthy samples and 945 diseased samples, of which 345 correspond to sooty mold and 500 to fruit fly damage. The disease-analysis model was implemented as a DeepLabV3+-based multiclass segmentation approach with a MobileNetV2 backbone, three classes, an input size of 224 × 224 pixels, data augmentation, and export to TensorFlow Lite for mobile deployment. Experimental results on the test set showed an IoU of 0.7406 and a Dice coefficient of 0.8161 for sooty mold, and an IoU of 0.9214 and a Dice coefficient of 0.9483 for fruit fly damage. The exported TFLite float32 model preserved the same test metrics, supporting its integration into an Android application for portable citrus inspection in real-world scenarios.

Keywords: Mobile computer vision · Orange segmentation · Citrus pest detection · TensorFlow Lite

1 Introduction

Orange production is an important agricultural activity in Tamaulipas, Mexico, and this work is closely connected to that local context. The present study focuses on *Valencia oranges* collected in *ejido La Soledad*, where the proposed mobile application was conceived as a practical tool for field conditions rather than for controlled laboratory environments. This local focus is relevant because the

H. Ponce et al. (Eds.): COMIA 2026, CCIS 3095, pp. 492–503, 2027.
https://doi.org/10.1007/978-3-032-37438-7_38

visual conditions found in citrus-growing areas strongly influence image acquisition and the practical usefulness of computer vision tools for fruit inspection.

Official agricultural information from the state indicates that Tamaulipas ranked second nationwide in orange production in 2024, with 520,467.44 tons, while the Perennial Crop Cycle 2024 reported 270,433.93 hectares under perennial cultivation, with orange among the principal crops [11,12]. These figures highlight the agricultural relevance of orange production in the state and support the need for practical tools for phytosanitary inspection under local production conditions.

In everyday agricultural practice, the inspection of oranges for visible signs of pests or diseases is still largely based on direct observation. As a result, the quality of the inspection depends on the experience of the observer, the available time, and the environmental conditions under which the fruit is evaluated. This becomes especially challenging in real scenarios, where lighting changes constantly and fruits may appear partially occluded by leaves, branches, shadows, or complex backgrounds. Under these conditions, even identifying the fruit itself may be difficult, and recognizing symptoms on its surface becomes even more challenging.

Recent advances in deep learning have shown strong potential for agricultural image analysis, including fruit recognition, disease detection, and mobile visual inspection [6–9]. However, many existing solutions are still oriented toward offline experiments, controlled datasets, or desktop environments. In contrast, mobile deployment introduces additional constraints related to computational efficiency, responsiveness, and robustness under non-controlled acquisition conditions.

This work addresses that practical gap through the development of a mobile application based on a two-stage deep learning pipeline. The first stage segments the orange from the surrounding background, and the second stage analyzes the segmented fruit region to identify visible phytosanitary conditions associated with sooty mold and fruit fly damage. This two-stage design was adopted to reduce the influence of background clutter and to make the final visual output easier to interpret during mobile use. Although the present study does not include a formal comparison against one-stage baselines, the proposed application was designed under the practical assumption that restricting the analysis to the fruit region can help reduce irrelevant visual interference and improve the interpretability of the output.

An important aspect of this work is that the dataset was manually acquired from local fruit and manually annotated; it was not collected from Internet image repositories. This decision was intentional, since the goal was to develop an application that reflects the real visual conditions in which it is expected to operate. The resulting dataset includes healthy fruit, sooty mold, and fruit fly damage samples obtained under local field conditions and later used for both model development and mobile deployment.

The main contributions of this work can be summarized as follows. First, a manually acquired and manually annotated image dataset of regional Valencia oranges was constructed for the analysis of healthy fruit, sooty mold, and fruit fly damage. Second, a two-stage mobile application was developed, integrating fruit segmentation, background suppression, and phytosanitary analysis on Android devices. Third, the study evaluates the preservation of model performance after TensorFlow Lite conversion, providing evidence that the trained models can be transferred to a mobile deployment setting without degradation in the reported test metrics.

The remainder of this paper is organized as follows. Section 2 describes the dataset, model design, and mobile inference workflow. Section 3 presents the experimental results, including quantitative evaluation and qualitative examples obtained from the application. Finally, Sect. 4 discusses the main findings, current limitations, and future work directions.

2 Materials and Methods

2.1 Dataset Acquisition and Annotation

The dataset used in this work was built from images of regional Valencia oranges acquired manually under practical field conditions in ejido La Soledad, Tamaulipas, Mexico. Instead of relying on images downloaded from Internet repositories, the acquisition process was carried out directly on local fruit in order to reflect the visual conditions expected during the intended use of the application. These conditions included variations in viewpoint, illumination, background, and partial occlusion caused by branches, leaves, and natural orchard environments.

The complete dataset contained 6,664 images, including 5,719 healthy samples and 945 diseased samples. Among the diseased cases, 345 images corresponded to sooty mold and 500 images corresponded to fruit fly damage. The dataset was carefully curated to include representative examples of healthy fruit, sooty mold, and fruit fly damage under the visual conditions targeted by the application. Representative examples of the images used in the dataset are shown in Fig. 1. These examples illustrate the visual characteristics of the three categories considered in this study, namely fruit fly damage, sooty mold, and healthy oranges.

2.2 General System Design

The proposed application was organized as a two-stage mobile vision pipeline. The first stage focused on locating the fruit through binary semantic segmentation, while the second stage analyzed the segmented orange region to estimate the presence of sooty mold and fruit fly damage. This design was adopted to reduce the influence of background clutter and to constrain the second stage to the fruit region only, which also improves the interpretability of the final visual output.

The final Android implementation followed this same logic. First, the segmentation model identified the orange contour. Then, a binary mask was used to generate a new image in which the fruit remained visible and the surrounding background was replaced with black. That masked image was then provided

Fig. 1. Examples of images used in the dataset. From left to right: fruit fly damage, sooty mold, and healthy orange. These examples illustrate the visual variability of the classes considered in this study under non-controlled acquisition conditions. (Color figure online)

to the second model, which produced a dense three-channel output that was interpreted pixel by pixel to highlight healthy regions, sooty mold, and fruit fly damage. The complete workflow followed in this study is summarized in Fig. 2.

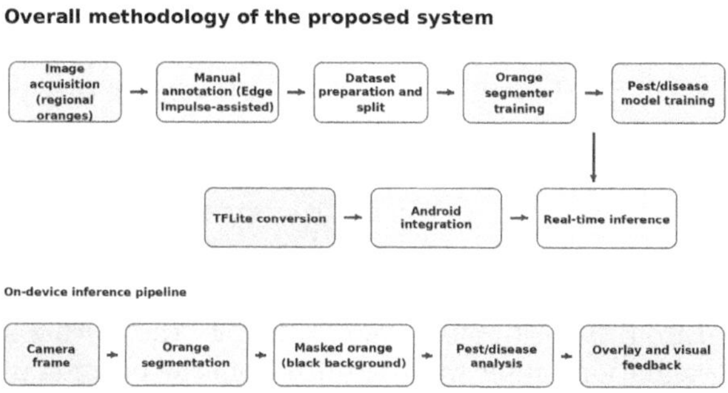

Fig. 2. Overall methodology of the proposed application. The workflow includes image acquisition, manual annotation, dataset preparation, training of the orange segmentation model and the pest/disease analysis model, TensorFlow Lite conversion, Android integration, and mobile inference. (Color figure online)

2.3 Orange Segmentation Model

The fruit segmentation stage was implemented as a binary semantic segmentation network based on DeepLabV3+ with a MobileNetV2 backbone. The model used 224 × 224 RGB images as input and generated a single-channel logit output for fruit segmentation. Image preprocessing followed the MobileNetV2 preprocessing scheme. The architecture included an Atrous Spatial Pyramid Pooling (ASPP) module and a decoder stage with skip connections from lower-level MobileNetV2 features.

For this stage, the dataset was divided locally into training, validation, and test subsets using an 80/10/10 partition. Training was carried out in TensorFlow/Keras with a batch size of 16, 150 epochs, and the Adam optimizer with an initial learning rate of 1×10^{-3}. To address class imbalance between fruit and background pixels, the loss function combined weighted binary cross-entropy with Dice loss, where the positive weight was estimated automatically from the cached training masks.

Several measures were incorporated to improve training stability and reduce overfitting. These included disk-based caching, geometric augmentation (horizontal and vertical flips, 90-degree rotations, and random shifts), and grayscale photometric augmentation (brightness, contrast, gamma, and Gaussian noise). In addition, the implementation used weight decay (1×10^{-4}), model checkpointing, TensorBoard logging, and learning-rate reduction on plateau. After training, threshold sweeping on the validation set was used to select the operating threshold later employed during test evaluation.

2.4 Pest and Disease Analysis Model

The second stage of the application was designed to identify relevant phytosanitary conditions on the segmented orange surface. This model was also implemented as a DeepLabV3+-based semantic segmentation network with a MobileNetV2 backbone, but in a multiclass setting. In this case, the model used 224 × 224 RGB inputs and predicted three categories: background/healthy region, sooty mold, and fruit fly damage.

The multiclass dataset preparation incorporated images of healthy segmented oranges as negative examples in order to improve the separation between symptomatic and symptom-free fruit regions. When official training and testing partitions exported from Edge Impulse were available, the testing subset was preserved and the validation subset was obtained from the training pool. When such predefined partitions were not available, a local 80/10/10 split was applied. This design sought to maintain a reproducible evaluation process while remaining consistent with the available dataset structure.

Because the multiclass stage involved a strong imbalance between healthy/background regions and diseased regions, the training process incorporated class-aware weighting estimated from cached training masks. In addition, healthy samples were explicitly included to improve the discrimination of symptom-free regions. As in the binary model, the training pipeline included disk-based caching, geometric and photometric augmentation, and weight decay to improve generalization. The model was trained with a batch size of 16, up to 300 epochs, and an initial learning rate of 1×10^{-3}, and was later exported to TensorFlow Lite float32 for mobile deployment.

2.5 Mobile Deployment and Inference Workflow

The final application was deployed on Android and tested on a Samsung Galaxy A22 running Android 11, while keeping compatibility in mind for Android devices running version 8 or later. During inference, the application first used the orange segmentation model to estimate a valid fruit mask. The mask was then refined to preserve the principal fruit component, and a binary mask and contour representation were generated. Only when a valid orange region was found did the application continue to the second stage. This strategy prevented the disease-analysis model from operating on frames dominated by background or invalid fruit contours.

After the fruit mask was validated, the application created a masked image in which the orange remained visible and the background was suppressed. The second model then analyzed this image and produced a dense per-pixel output with three channels corresponding to healthy tissue, sooty mold, and fruit fly damage. These outputs were converted into a visual overlay, combining the fruit contour with color-coded regions associated with the detected conditions.

2.6 Evaluation Strategy and Reproducibility

The evaluation process considered both the original Keras models and their exported TensorFlow Lite float32 versions. For the orange segmentation model, a threshold selected from validation data was used during test evaluation. For the multiclass disease-analysis model, per-class IoU and Dice scores were computed for the relevant classes of interest, namely sooty mold and fruit fly damage. In addition to quantitative evaluation, the implementation also generated overlays and prediction masks that made it possible to visually inspect model behavior.

To improve reproducibility, both training scripts fixed the random seed to 42 for Python, NumPy, and TensorFlow. In addition, preprocessing, cache generation, data augmentation, and evaluation settings were defined explicitly in the training configuration. These measures were intended to reduce variability across runs and to make the training and evaluation workflow easier to reproduce.

3 Experimental Results

3.1 Evaluation Overview

The experimental analysis considered the two main stages of the proposed application: orange segmentation and pest/disease analysis. In both cases, evaluation was performed on the test partition after model training in Keras and after export to TensorFlow Lite float32. This comparison was relevant because the final objective of the work was not limited to offline model development, but also included deployment on Android devices.

Table 1 summarizes the main quantitative results obtained for both stages. For the orange segmentation model, the same evaluation threshold selected on the validation set was later applied during test evaluation. For the multiclass disease-analysis model, per-class IoU and Dice values were computed for the two phytosanitary categories of interest: sooty mold and fruit fly damage.

Table 1. Summary of the main evaluation metrics obtained for the two stages of the proposed application.

Stage	Model version	Class	IoU	Dice
Orange segmentation	Keras	Orange mask	0.9683	0.9832
Orange segmentation	TFLite float32	Orange mask	0.9683	0.9832
Pest/disease analysis	Keras	Sooty mold	0.7406	0.8161
Pest/disease analysis	Keras	Fruit fly damage	0.9214	0.9483
Pest/disease analysis	TFLite float32	Sooty mold	0.7406	0.8161
Pest/disease analysis	TFLite float32	Fruit fly damage	0.9214	0.9483

3.2 Orange Segmentation Performance

The first stage of the pipeline focused on binary segmentation of the fruit. The final test evaluation used a validation-derived threshold of 0.74. The results obtained for this model were strong. On the test set, the Keras version reached an IoU of 0.9683 and a Dice coefficient of 0.9832. The exported TensorFlow Lite float32 version preserved the same values, indicating that the conversion step did not introduce observable degradation in segmentation quality under the reported evaluation conditions.

From an application perspective, these results are important because the first model defines the visual region later analyzed by the second stage. In other words, a stable and accurate fruit mask is not only useful by itself, but also acts as the spatial basis for the rest of the mobile pipeline. The high overlap metrics obtained in this stage suggest that the model was able to isolate the fruit reliably enough to support the subsequent analysis of phytosanitary conditions.

3.3 Pest and Disease Analysis Performance

The second stage of the application analyzed the segmented orange region in order to identify two relevant conditions: sooty mold and fruit fly damage. The evaluation focused on these two target classes, while healthy or background regions were treated as the remaining category.

The results showed a clear difference in difficulty between the two target classes. For sooty mold, the model achieved an IoU of 0.7406 and a Dice coefficient of 0.8161. For fruit fly damage, the results were notably higher, with an IoU of 0.9214 and a Dice coefficient of 0.9483. As in the fruit segmentation stage, the TensorFlow Lite float32 version preserved the same metrics reported for the Keras model.

This gap suggests that fruit fly damage was visually easier for the model to localize and separate than sooty mold. A likely explanation is that fruit fly damage tends to generate more localized and visually distinctive patterns, whereas sooty mold may exhibit greater variability in texture, spread, and contrast against the fruit surface. In addition, the diseased subset was not evenly distributed across categories, with fewer sooty mold samples than fruit fly damage samples, which may also have influenced class-specific performance. These observations suggest that the lower performance for sooty mold is likely associated with a combination of dataset composition, greater intraclass variability, and the visual complexity of the category under non-controlled acquisition conditions. The neural architecture adopted in this work is summarized in Fig. 3. Both stages rely on a DeepLabV3+-based encoder-decoder structure with a MobileNetV2 backbone, while differing in the final prediction task: binary fruit segmentation in the first stage and three-class phytosanitary analysis in the second stage.

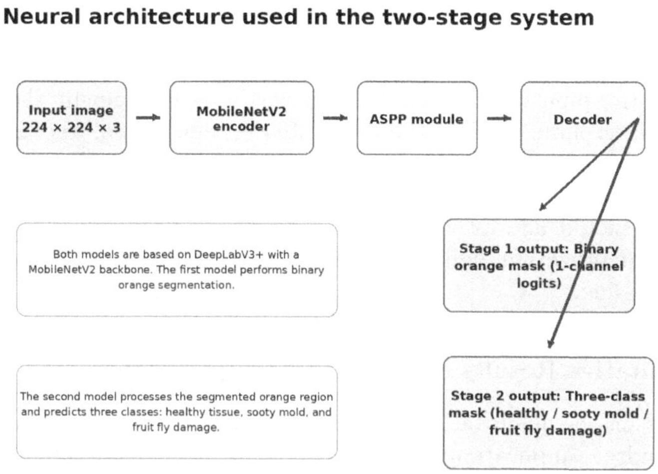

Fig. 3. Neural architecture used in the proposed application. Both stages are based on DeepLabV3+ with a MobileNetV2 backbone. The first model produces a binary orange mask, while the second model performs three-class analysis on the segmented fruit region. (Color figure online)

3.4 Keras-to-TFLite Consistency

One of the most relevant findings of the experiments is the consistency between the desktop training environment and the mobile deployment format. In the fruit segmentation stage, the TFLite float32 model maintained the same test metrics as the Keras version, with IoU = 0.9683 and Dice = 0.9832. In the pest/disease analysis stage, the same behavior was observed for both evaluated classes: sooty mold remained at IoU = 0.7406 and Dice = 0.8161, while fruit fly damage remained at IoU = 0.9214 and Dice = 0.9483 after conversion.

This result is particularly valuable for the contribution of the paper because it strengthens the practical relevance of the proposed application. In many studies, strong offline results do not necessarily translate into equally reliable mobile behavior. In this case, however, the reported agreement between Keras and TensorFlow Lite supports the feasibility of integrating the trained models into an Android-based application without degrading the reported test metrics.

3.5 Mobile Deployment Considerations

The final application was deployed and tested on a Samsung Galaxy A22 running Android 11. In its current form, the application should be regarded as a mobile deployment prototype rather than as a fully optimized real-time solution. Although the models were successfully integrated and the application produced interpretable visual outputs on-device, the observed responsiveness remained limited, indicating that further optimization is still necessary for stricter real-time operation.

Even so, the mobile implementation was useful for validating the practical behavior of the pipeline. The application was able to segment the fruit, suppress the background, analyze the segmented orange region, and display localized visual overlays associated with the detected condition. These observations support the viability of the deployment strategy, while also making clear that future work should address computational efficiency more directly, for example through quantization, lighter architectures, or additional optimization of the inference pipeline.

3.6 Qualitative Results of the Mobile Application

To complement the quantitative evaluation, qualitative results obtained directly from the Android application are shown in Fig. 4. These examples are especially relevant because they illustrate the complete behavior of the proposed two-stage pipeline during mobile operation. Rather than producing only an offline prediction, the application first identifies the fruit contour and then restricts the phytosanitary analysis to the segmented fruit region.

In the first example, the application detects fruit fly damage and highlights the affected area in red. In the second example, the application identifies sooty mold and displays the corresponding region in blue. In both cases, the fruit contour is shown in green, making it possible to verify that the symptom analysis is constrained to the orange surface rather than to the surrounding background. This visual behavior is consistent with the intended role of the two-stage pipeline, in which the first model isolates the fruit and the second model analyzes only the segmented region.

These qualitative examples provide additional evidence that the mobile implementation is capable of integrating fruit segmentation, background suppression, and symptom localization in a single visual interface. From a practical point of view, this is important because the application not only produces a class estimate, but also gives the user an interpretable visual indication of where the detected condition is located on the fruit.

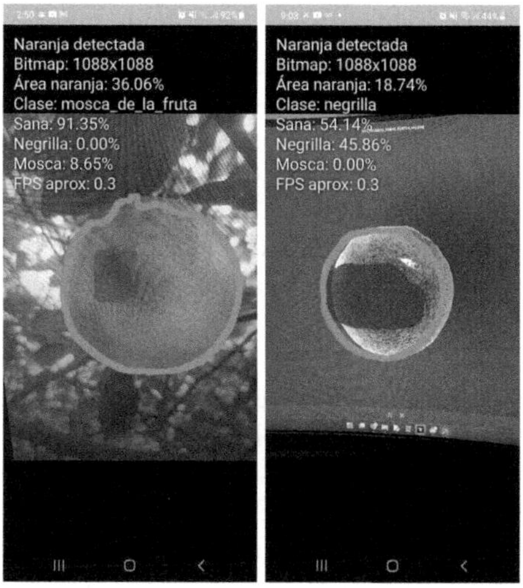

Fig. 4. Qualitative results of the proposed mobile application. Left: example of fruit fly damage detection, where the orange contour is highlighted in green and the detected affected region is shown in red. Right: example of sooty mold detection, where the orange contour is highlighted in green and the symptomatic region is shown in blue. In both cases, the analysis is restricted to the segmented fruit region. (Color figure online)

3.7 Discussion of the Two-Stage Design

Taken together, the experimental results suggest that the two-stage design is a practical strategy for constraining the phytosanitary analysis to the fruit region and for improving the interpretability of the visual output. The first model achieved very high segmentation accuracy for the orange region, making it possible to suppress most of the background before applying the second model. The second model, in turn, produced class-specific segmentation results for sooty mold and fruit fly damage, with especially strong performance for fruit fly damage.

At the same time, these results should be interpreted with appropriate caution. The present study did not include a formal comparative evaluation against one-stage baselines or alternative lightweight mobile architectures. Therefore, the benefits of the proposed two-stage application should be understood as practically motivated and supported by the observed behavior of the deployment prototype, rather than as a definitive superiority claim over all alternative approaches. A broader comparative study remains an important direction for future work.

4 Conclusions

This work presented the development of a mobile application for the analysis of regional Valencia oranges through a two-stage deep learning pipeline. The proposed application first segments the fruit and then analyzes the segmented region to identify phytosanitary conditions associated with sooty mold and fruit fly damage. This design was motivated by practical considerations, particularly the need to reduce background interference and to make the final visual output easier to interpret during mobile use.

The main contributions of the work can be summarized in three points. First, a manually acquired and manually annotated dataset of regional Valencia oranges was constructed under local field conditions in ejido La Soledad, Tamaulipas, including healthy fruit, sooty mold, and fruit fly damage cases. Second, a two-stage mobile application was developed that integrates fruit segmentation, background suppression, and symptom analysis on Android devices. Third, the study showed that the TensorFlow Lite float32 versions of both models preserved the same reported test metrics as their Keras counterparts, supporting the feasibility of transferring the trained models to a mobile deployment setting.

The quantitative results were encouraging. The orange segmentation stage achieved an IoU of 0.9683 and a Dice coefficient of 0.9832 on the test set, while the pest and disease analysis stage obtained an IoU of 0.7406 and Dice of 0.8161 for sooty mold, and an IoU of 0.9214 and Dice of 0.9483 for fruit fly damage. These results suggest that the proposed application is technically viable as a mobile vision tool for citrus inspection, particularly as a proof of deployment under practical conditions.

At the same time, the study also revealed important limitations. The lower performance obtained for sooty mold indicates that this class remains more challenging than fruit fly damage, likely due to a combination of stronger visual variability, lower contrast against the fruit surface, and class imbalance in the available data. In addition, although the application was successfully deployed on Android, its current responsiveness should be interpreted as that of a mobile deployment prototype rather than as a fully optimized real-time solution. Further optimization is still necessary to improve processing speed and practical usability on-device.

Another important limitation is that the present study did not include a formal comparative evaluation against one-stage baselines or alternative lightweight architectures for mobile inference. Therefore, the value of the proposed two-stage design should be understood as practically motivated and supported by its observed behavior in the application, rather than as a definitive superiority claim over other possible approaches.

Future work should therefore focus on four main directions: expanding the dataset with more diverse field conditions and more balanced class representation, especially for sooty mold; performing comparative experiments against one-stage and lightweight baselines; optimizing mobile inference through quantization or lighter architectures; and extending the approach to other crops or phytosanitary conditions of agricultural interest.

Overall, the proposed application provides a solid basis for portable citrus inspection using deep learning, while also showing how a two-stage mobile design can connect model development with practical on-device visual analysis under regional production conditions.

Future work should include a broader experimental validation of the proposed application. In particular, comparative experiments against one-stage baselines and alternative lightweight mobile architectures would help clarify the practical benefits and trade-offs of the two-stage design. Additional evaluation under more diverse field conditions, including a wider range of illumination and background scenarios, would also be necessary to assess generalization more thoroughly. From the deployment perspective, future work should address computational efficiency through explicit measurements of latency, memory usage, and inference speed, as well as through optimization strategies such as quantization and the use of lighter backbones. Finally, expanding the dataset and refining class balancing strategies, especially for sooty mold, could help reduce the class-specific performance gap observed in the current study.

References

1. Chen, L.-C., Zhu, Y., Papandreou, G., Schroff, F., Adam, H.: Encoder-Decoder with Atrous Separable Convolution for Semantic Image Segmentation. In: Ferrari, V., Hebert, M., Sminchisescu, C., Weiss, Y. (eds.) Computer Vision - ECCV 2018. Lecture Notes in Computer Science, vol. 11211, pp. 833–851. Springer, Cham (2018). https://doi.org/10.1007/978-3-030-01234-2_49
2. Sandler, M., Howard, A., Zhu, M., Zhmoginov, A., Chen, L.-C.: MobileNetV2: inverted residuals and linear bottlenecks. In: Proceedings of the IEEE/CVF Conference on Computer Vision and Pattern Recognition (CVPR), pp. 4510–4520 (2018)
3. TensorFlow Lite: TensorFlow Lite Guide. TensorFlow documentation (accessed 2026)
4. Google, A.I., Edge: LiteRT: High-Performance On-Device Machine Learning Deployment. Official documentation
5. Edge Impulse: Edge Impulse Documentation. Official documentation
6. Xing, S., Lee, M., Lee, K.-K.: Citrus pests and diseases recognition model using weakly dense connected convolution network. Sensors **19**(14), 3195 (2019)
7. Apacionado, B.V., Ahamed, T.: Sooty mold detection on citrus tree canopy using deep learning algorithms. Sensors **23**(20), 8519 (2023)
8. Hadipour-Rokni, R., Askari Asli-Ardeh, E., Jahanbakhshi, A., Esmaili Paeen-Afrakoti, I., Sabzi, S.: Intelligent detection of citrus fruit pests using machine vision system and convolutional neural network through transfer learning technique. Comput. Biol. Med. **154**, 106611 (2023)
9. Chen, Y., An, X., Gao, S., Li, S., Kang, H.: A deep learning-based vision system combining detection and tracking for fast on-line citrus sorting. Front. Plant Sci. **12**, 622062 (2021)
10. Abadi, M., Agarwal, A., Barham, P.: TensorFlow: Large-Scale Machine Learning on Heterogeneous Systems (2015)
11. Gobierno del Estado de Tamaulipas, Secretaría de Desarrollo Rural: Agricultura. https://www.tamaulipas.gob.mx/desarrollorural/temas-del-sector/agricultura/
12. Gobierno del Estado de Tamaulipas, Secretaría de Desarrollo Rural: Información del sector rural 2024. https://www.tamaulipas.gob.mx/desarrollorural/wp-content/uploads/sites/6/2025/11/informacion-del-sector-rural.pdf

Intelligent XGBoost Portfolio Optimization

J. E. Pérez-Juarez[1] and Y. U. K. Coronado González[2](✉)

[1] Facultad de Negocios, Universidad La Salle, Mexico City, Mexico
[2] Vicerrectoría de Investigación, Universidad La Salle, Mexico City, Mexico
yaxkin.coronado@lasalle.mx

Abstract. Portfolio optimization under the Markowitz mean-variance framework remains a cornerstone of modern finance, yet its practical performance is constrained by the sensitivity of optimal weights to the estimation of expected returns affected by noise, regime shifts, and nonlinear dynamics. This study proposes and evaluates the integration of XGBoost, a gradient-boosted machine learning model, as an alternative signal for expected return estimation within the Markowitz optimization framework, comparing traditional historical approaches and a blended strategy that linearly combines both signals. Using a filtered universe of S&P 500 equities, three portfolio strategies were constructed and assessed through a rolling monthly backtest: a historical Markowitz portfolio, a rescaled XGBoost-driven portfolio, and a blended portfolio ($\alpha = 0.55$). The main results for the rescaled XGBoost portfolio achieved the highest risk-adjusted performance, with a return of 91.80%. However, this gain corresponds to a low diversification scenario, limited by the exclusion of transaction costs, its focus on a single asset class and market, and its dependence on specific prediction horizon and top-k asset selection. These findings position XGBoost as a valuable complement rather than a replacement to classical portfolio theory.

Keywords: Markowitz portfolio · XGBoost · Optimization

1 Introduction

Portfolio optimization is one of the central problems in finance today, as it seeks to find the best possible allocation of capital among different financial instruments, considering a given risk-return relationship. Since the development of Markowitz's modern portfolio theory as we know it, quantitative diversification analysis has allowed for the creation of more efficient portfolios, in which not only the individual return of each asset matters, but also the joint interaction between them through their covariances [1]. In this sense, optimal portfolio selection continues to be a topic of great importance in the professional and academic fields, especially in environments with long periods of uncertainty, volatility, and changing market regimes [2].

Despite the theoretical importance that Markowitz optimization has acquired, the practical application of the model faces a fundamental limitation: the sensitivity of the results to the estimation of expected returns [3, 4]. In many cases, these returns are

H. Ponce et al. (Eds.): COMIA 2026, CCIS 3095, pp. 504–515, 2027.
https://doi.org/10.1007/978-3-032-37438-7_39

approximated through averages of historical data (prices or returns), so it is assumed that the past behavior of assets can serve as a reference for the future. However, in financial markets and in practice, this assumption does not seem to hold true [4, 5].

In financial markets, this assumption is problematic, as returns exhibit a certain level of noise, regime changes, periods of high instability, and nonlinear relationships that are not always adequately captured by a historical average [5, 6]. Therefore, weight combinations can be based on weak inputs, resulting in inefficient allocation or allocation that is overly sensitive to the sample used [3, 4]. This rigidity limits the ability of the traditional framework to capture nonlinear interactions and regime changes that can be critical for portfolio stability [5, 7].

In formal terms, within the mean variance approach, the expected portfolio return (E) is obtained as the weighted sum of the individual expected returns (Wi) among the assets and portfolio risk (σp2) is expressed by the variance associated with the covariance matrix of the assets [2]:

$$E(RP) = \sum_{i=1}^{n} Wi\mu i \ \sigma_p^2 = w' \sum w$$

The limitations to capture nonlinear interactions, impulse the use of machine learning techniques in the financial sector in recent years, particularly in problems related to prediction, classification, and investment signal generation [6]. These methods allow the identification of more complex patterns in data and the capture of nonlinear relationships between explanatory variables and future returns [6, 8]. Within this set of tools, XGBoost has stood out for its predictive capacity, computational efficiency, and flexibility in working with large and heterogeneous data structures, establishing itself as one of the most robust tools for tabular prediction problems [9].

In this context, we propose a comparison between three approaches to constructing equity portfolios: a portfolio optimized with the traditional Markowitz model based on historical returns, another portfolio created from expected return signals generated using XGBoost, and an intermediate version built from a linear combination of both signals detailed in Sect. 2. In Sect. 3 we explain the workflow that includes asset universe refinement, return calculation, generation of predictor variables, model training, annualization of predicted and historical returns, and rescaling of the estimated signal. Similarly, an optimization and back testing process with monthly rebalancing is performed. With an evaluation of whether the integration of machine learning predictive signals can translate into observable improvements in portfolio efficiency [8, 10]. Finally we discuss the comparative between the three portfolios, with metrics like volatility, Sharpe ratio, and portfolio concentration are compared to evaluating the riskreturn profile of the constructed strategies.

1.1 Problem Statement

Machine learning methods offer a relevant possibility, as they allow modeling nonlinear or complex relationships between explanatory variables and future returns [6, 11]. Particularly, XGBoost stands out as a flexible and robust tool for generating predictive signals that can be incorporated into the optimization process [9]. However, the incorporation of these signals also raises important questions, such as whether they consistently improve the expected portfolio performance compared to the traditional approach, and whether this improvement implies a greater degree of concentration or a lower level of diversification [8, 10]. Therefore, the central problem of this research is to determine whether the inclusion of an expected return signal generated by XGBoost enables the construction of more efficient portfolios than those derived exclusively from Markowitz's traditional historical approach.

Despite the conceptual soundness of Harry Markowitz's model, its practical implementation faces several limitations related to estimating its input parameters [3, 4]. The means of expected returns is often the most difficult component to estimate accurately, as returns on financial instruments exhibit high volatility, noise, and sensitivity to economic and financial environmental factors [4, 5]. In optimization exercises of this type, these returns are calculated from historical averages, based on the premise that the past contains useful information about future behavior [5, 12].

However, this approach can prove insufficient when market conditions are changing, when periods of instability exist, or when there are nonlinear relationships between variables [4, 6]. Additionally, the highly sensitive to the observation period used, which compromises the consistency of out-of-sample results [4]. Relatively small changes in the sample can significantly alter the estimate of expected returns, leading to weak or excessively concentrated portfolios [3, 4]. Similarly, since the Markowitz optimizer is highly sensitive to variations in the values of [missing information] μ, small estimation errors can translate into disproportionate changes in optimal weights [3]. Consequently, while the historical approach provides a useful starting point, its ability to capture future asset dynamics may be limited, creating an opportunity to explore alternative estimation methods [4, 6].

On the other hand, the application of machine learning in finance also faces significant challenges. The stochastic nature of markets makes it difficult to obtain stable predictive signals, while overfitting, dependence on the sample period, and low signal-to-noise ratio can limit the generalizability of models [6, 8]. For this reason, in many financial contexts, the value of machine learning does not necessarily lie in perfectly predicting the future return of each asset, but rather in generating sufficiently informative relative signals to distinguish between more and less attractive assets within an investment universe [8, 10]. From this perspective, its contribution is interpreted more as a classification and ranking tool than as an absolute replacement for portfolio theory [8, 10].

1.2 XGBoost as a Predictive Model

Within the set of supervised machine learning algorithms, XGBoost has become one of the most widely used tools for tabular prediction problems due to its computational efficiency, regularization capabilities, and strong empirical performance [1, 9]. This model belongs to the family of decision tree-based boosting methods, in which multiple weak models are sequentially combined to build a more accurate predictor [9]. In each iteration, the algorithm adjusts new trees to correct the errors made by previous trees, thus gradually improving the model's predictive capacity [9].

In general, XG Boost optimizes an objective function composed of two elements: a loss function that measures the error between observed values and model predictions, and a regularization term that penalizes the complexity of the trees constructed to avoid overfitting [9]. Formally, in iteration t, the objective function can be expressed as [9]:

$$L^{(t)} = \sum_{i=1}^{n} l\left(y_i^i, \hat{y}^{(1)} + f_t(x_i)\right) + \Omega(f_t)$$

where ft. represents the tree added in the current iteration and ω(ft)corresponds to the regularization component [9]. The latter can be defined as:

$$\Omega(f_t) = \gamma T + \frac{1}{2}\lambda \sum_{j=1}^{T} w_j^2$$

In this scheme, the model seeks to gradually improve predictive accuracy while controlling overfitting, which is especially useful in financial contexts where the relationship between explanatory variables and future returns is often complex and non-linear [6, 9].

$1^{(t)}$: objective function in the iteration t.
l(·): loss function, which measures how far the prediction is from the actual value.
yi: actual observed value.
$\hat{}y^{(1)}$: cumulative prediction up to the previous iteration.
ft.(xi): new tree that is added in the iteration t.
ω(ft): regularization term.
t: number of leaves on the tree.
wj: weight assigned to each sheet.
γ: penalty for tree complexity.
λ: penalty on the weights of the sheets.

In this work, in addition to the historical portfolio and the rescaled XGBoost-based portfolio, a combined signal defined as follows is considered:

$$\mu_{blend} = \alpha\mu_{pred} + (1 - \alpha)\mu_{hist}$$

where αit represents the relative weight assigned to the model's predictive signal versus the historical signal.

2 Methodology

The empirical analysis was conducted using a filtered universe of S&P 500 equities from March 1, 2022 to March 1, 2026. The S&P 500 was selected because it provides a liquid, diversified, and replicable investment universe with broad data availability. However, this choice also limits the generalizability of the results to other markets, particularly emerging or less liquid markets. After applying liquidity, volatility, risk-adjusted performance, and data availability filters, the final sample consisted of 151 equities.

Daily adjusted prices were used to compute asset returns. The machine learning dataset was constructed in long format, where each observation corresponds to an asset-date pair. The prediction target was defined as the cumulative forward return over 42 business days. This horizon was selected to approximate a short-to medium-term investment window consistent with the monthly rebalancing structure of the backtest. Nevertheless, alternative horizons such as 21, 63, or 126 trading days should be explored in future work.

The explanatory variables included lagged return and volatility-based indicators: 5-day, 21-day, 63-day, and 126-day cumulative returns; 21-day, 63-day, and 126-day rolling volatility; momentum-to-volatility ratios; and momentum spread variables across different windows. To avoid information leakage, the sample was split chronologically into training, validation, and test sets. The training period covered August 30, 2022 to December 29, 2023; the validation period covered January 2, 2024 to June 28, 2024; and the test period covered July 1, 2024 to February 27, 2026.

XGBoost was used as a supervised learning model to estimate a relative expected return signal. Several hyperparameter configurations were evaluated using mean squared error, Pearson correlation, and Spearman correlation. Since the goal was not to obtain a perfect point forecast of future returns, but rather to generate a useful ranking signal for portfolio construction, the model output was interpreted as a relative screening mechanism. The selected XGBoost signal was annualized, standardized, clipped to reduce the influence of extreme values, and rescaled to make it comparable with historical expected returns.

Three portfolio strategies were evaluated under a long-only, fully invested Markowitz mean-variance framework: a traditional Historical Markowitz portfolio based on historical expected returns, a Rescaled XGBoost portfolio based on the calibrated machine learning signal, and a Blended portfolio combining both signals through $\mu_{blend} = \alpha\mu_{pred} + (1 - \alpha)\mu_{hist}$. The evaluated values of α ranged from 0.55 to 0.95. Portfolio performance was assessed through a rolling monthly backtest using annualized return, annualized volatility, Sharpe ratio, and maximum drawdown. Concentration was also evaluated through the number of active positions, maximum individual weight, top-five weight concentration, and the Herfindahl-Hirschman Index.

3 Results

In general terms, the predictive model's results showed that the signal generated by XGBoost should not be interpreted as an exact estimate of each asset's future return or as a precise point-forecasting model for individual stock returns, but rather as a relative signal useful for ranking assets within the investment universe, see Table 1 with the metrics: mean square error (MSE), Pearson and Spearman correlation. As shown in Table 1, the Pearson and Spearman correlations were weak and negative across the evaluated configurations. However, this does not necessarily eliminate the usefulness of the model for portfolio construction This assertion is consistent with the literature on machine learning applied to finance. Consequently, the model was used as input in the optimization process, complemented by annualization, calibration, and top-k selection stages, to give the signal greater weight in portfolio construction.

This interpretation is important because portfolio performance depends not only on the absolute accuracy of individual return forecasts, but also on whether the signal helps improve asset selection and capital allocation. Therefore, the model was incorporated

Table 1. summarizes the main predictive performance metrics obtained during the model validation and testing phase.

Model	MSE	Pearson	Spearman
ft_D	0.013639	−0.024012	−0.028253
ft_F	0.013376	−0.026782	−0.029611
ft_C	0.013394	−0.029100	−0.035830
ft_E	0.013397	−0.030546	−0.035856
ft_A	0.013456	−0.036402	−0.040046
ft_B	0.013465	−0.039308	−0.044810

into the optimization process after annualization, standardization, clipping, rescaling, and top-k selection. These calibration steps were necessary to make the predictive signal comparable with historical expected returns and to reduce the impact of extreme predictions.

The backtest results allowed us to identify a more favorable combination within the evaluated set. In the final version of the exercise, the selected α value was the one that presented the best balance between expected return and risk-return efficiency during the test period. Allow us to construct a mechanism between historical robustness and a prospective signal, see Table 2, The configuration with $\alpha = 0.55$ obtained the highest Sharpe ratio within the evaluated set. However, the magnitude of the observed improvement depends on the calibration applied to the predictive signal, the selection of the top-k subset, and the optimizer's sensitivity to the input parameters. The best model combination corresponds to $\alpha = 0.55$was the one that showed the most competitive behavior within the evaluated set. This suggests that combining historical information with the XGBoost signal can provide a better balance between predictive information and historical stability than relying exclusively on one source.

The Table 2 contain the annual return and annual volatility correspond to annualized return and volatility. HHI avg. denotes the average Herfindahl-Hirschman Index during the backtest; Sum top 5 avg. represents the average sum of the five highest weights. The configuration with $\alpha = 0.55$ obtained the highest Sharpe ratio within the evaluated set.

Therefore, the comparison between the three portfolios was made in terms of annualized expected return, expected volatility, and the Sharpe ratio, showing that the rescaled XGBoost portfolio performed best in terms of expected return and risk-return efficiency. Specifically, this portfolio achieved an annualized expected return of 91.80%, an expected volatility of 14.02%, and a Sharpe ratio of 6.26, outperforming both the historical Markowitz portfolio and the blend portfolio. On the other hand, Markowitz's historical portfolio registered an annual expected return of 74.37%, a volatility of 17.60%, and a Sharpe ratio of 3.99, while the blended portfolio showed an expected return of 64.38%, a volatility of 13.73%, and a Sharpe ratio of 4.39.Taken together these results suggest that the signal generated by XGBoost, once annualized and rescaled, successfully integrated.

The final comparison between the Historical Markowitz, Rescaled XGBoost, and Blended portfolios shows that the XGBoost-based signal can improve the risk-return

Table 2. Backtest metrics for each of the analyzed configurations of α parameter

α	Annual Return	Annual Vol	Sharpe	Máx. drawdown	HHI avg.	Sum top 5 avg.
0.55	26.08%	15.06%	1.467	−5.73%	17.22%	78.52%
0.60	25.31%	14.80%	1.440	−5.84%	17.71%	79.10%
0.65	24.48%	14.70%	1.393	−6.23%	18.10%	79.52%
0.70	23.97%	14.63%	1.365	−6.55%	18.46%	79.80%
0.75	23.75%	14.58%	1.354	−6.70%	18.75%	80.38%
0.80	23.40%	14.53%	1.335	−6.84%	19.08%	81.00%
0.85	22.75%	14.49%	1.294	−6.96%	19.35%	81.43%
0.90	21.92%	14.47%	1.239	−7.09%	19.50%	81.59%
0.95	21.34%	14.43%	1.202	−7.11%	19.64%	81.79%

profile of the optimized portfolio. Nevertheless, these results should be interpreted with caution, since the improvement depends on the calibration of the predictive signal, the top-k selection rule, and the optimizer's sensitivity to expected return inputs. Therefore, the results support the use of XGBoost as a complementary signal within the Markowitz framework, rather than as a full replacement for traditional portfolio optimization (Figs. 1 and 2) (Table 3).

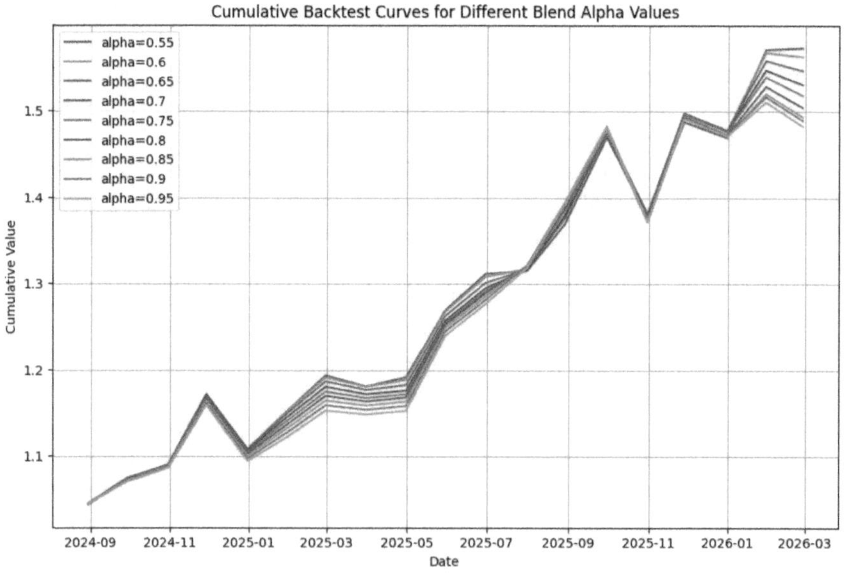

Fig. 1. Cumulative backtest curves for different alpha configurations

Table 3. Final comparison of optimized portfolios

Portfolio	Expected Return	Volatility	Sharpe
Historical Markowitz	74.37%	17.60%	3.999
Rescaled XGBoost	91.80%	14.02%	6.264
Blend alpha = 0.55	64.38%	13.73%	4.396

Note: The comparison considers three final strategies: Historical Markowitz, rescaled XGBoost, and Blend with $\alpha = 0.55$.

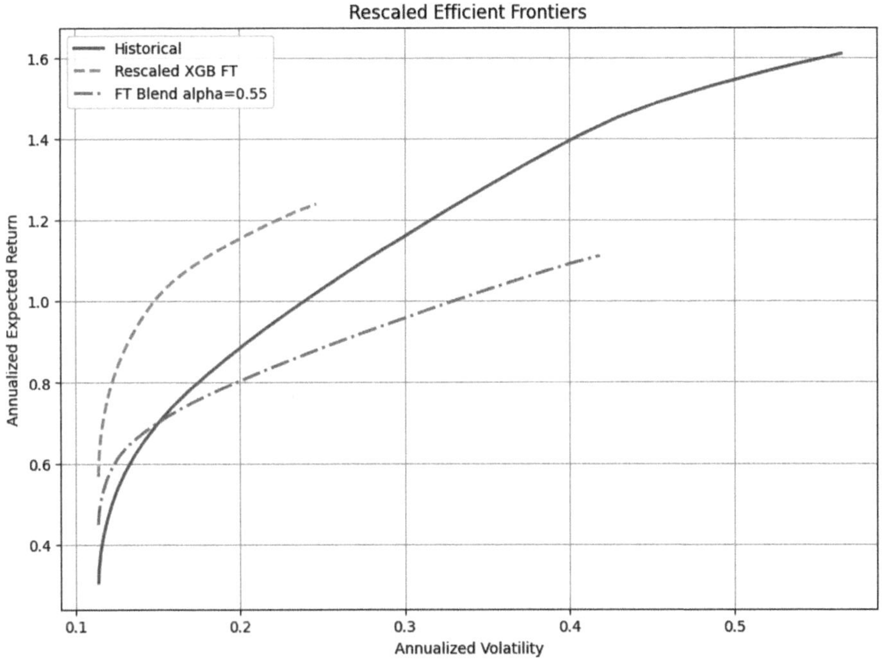

Fig. 2. Comparative efficient frontier between historical portfolio and XGBoost/blend portfolio

3.1 Portfolio Concentration and Diversification

Although the rescaled XGBoost portfolio showed the best expected performance the portfolio tended to allocate a larger proportion of capital to a smaller set of assets, which was reflected in metrics such as maximum weight per asset, the sum of the five highest weights, and the Herfindahl-Hirschman Index (HHI), this suggests that the machine learning signal tends to focus capital on a smaller subset of assets identified as more attractive by the model. While this behavior may improve portfolio efficiency, it also increases exposure to idiosyncratic risk and reduces diversification benefits. In contrast, the historical Markowitz portfolio exhibited a relatively less concentrated distribution, although with lower expected returns and lower risk-adjusted efficiency. Meanwhile, the blend portfolio occupied an intermediate position, maintaining lower volatility than the historical approach, but without achieving the same level of expected return as the rescaled XGBoost portfolio (Fig. 3) (Table 4).

Table 4. Concentration and diversification metrics

Portfolio	Weigth Assets >0	Material Assets	Maximum weigth	top 5 Sum	HHI
Markowitz historic	11	11	29.34%	78.79%	0.1665
XGBoost rescaled	9	8	30.81%	89.92%	0.2084
Blend alpha = 0.55	9	7	27.16%	90.59%	0.1828

Note: A higher maximum weight, a larger top 5 sum, and a higher HHI reflect a more concentrated portfolio structure.

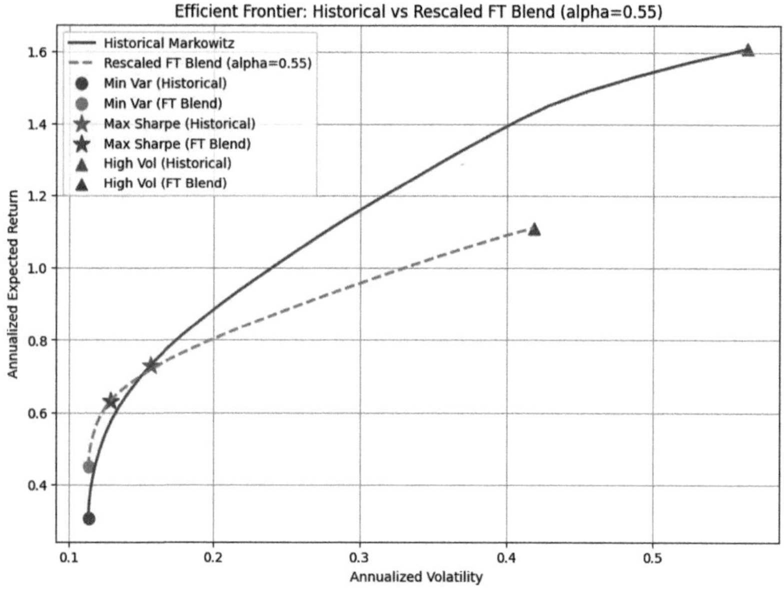

Fig. 3. Comparison of concentration between portfolios

These results indicate that the increase in expected return was not necessarily accompanied by greater diversification, but rather by a strategy with greater relative dependence on fewer assets. This finding is relevant because it shows that incorporating predictive signals can increase portfolio efficiency but also modify its concentration profile.

A practical implementation of the proposed framework should also consider diversification constraints, such as maximum individual weight limits of 5% or 10%, sector exposure limits, or turnover restrictions. These constraints would make the strategy more suitable for institutional portfolio management and would allow future research to test whether the XGBoost signal remains effective under more realistic investment restrictions.

4 Discussion

The expected return signal generated by XGBoost can improve performance of a portfolio optimized under the Markowitz framework, provided that the signal is properly annualized, rescaled, and calibrated before being entered into the optimizer. Additionally, the rescaled XGBoost portfolio exhibited the best annualized expected return and the best Sharpe ratio, outperforming the portfolio constructed solely with historical returns and the blended portfolio.

This finding is important because it indicates that the value of XGBoost does not necessarily lie in replacing the entire logic of portfolio theory, but rather in complementing the most sensitive component of the Markowitz model, which is the estimation of expected returns [13]. In other words, the optimizer continued to operate within a conventional mean-variance structure but was fed by a signal different from the historical mean, which allowed for a better allocation considering a more favorable risk-return profile.

Consequently, the results should be interpreted through a trade-off lens. If the investor's primary criterion is maximizing risk-adjusted expected return, the rescaled XGBoost portfolio appears to be the most attractive option within this exercise. Similarly, if diversification and reduced dependence on a few assets are also priorities, then the greater concentration observed in this strategy represents a cost that should not be overlooked.

Despite the results obtained, the study has several limitations that should be considered when interpreting its conclusions. First, the analysis was conducted on a pre-filtered universe of assets belonging to the S&P 500, so the findings are not necessarily generalized to other markets, asset classes, or periods with different structural characteristics. Second, the portfolio construction did not incorporate transaction costs, more complex liquidity constraints, or additional operational frictions, which means that the results represent an idealized approximation of the investment process. A more realistic implementation should incorporate transaction costs directly into the backtest or include turnover penalties in the optimization process.

Third, the performance of the predictive model and the final strategy depend on specific methodological decisions, such as the 42-day prediction horizon, the monthly rebalancing frequency, the calibration performed on XGBoost, and the top-k asset selection. Changes to any of these elements could modify or alter the observed results. Furthermore, the final comparison is based on expected metrics derived from the optimization, so its interpretation should be carefully considered in conjunction with the rolling backtest evidence.

Finally, although XGBoost proved useful in the weighting process, this does not definitively establish it as the best possible alternative among machine learning models. Other approaches, such as neural networks, random forests, or specialized ranking models, could generate different and potentially competitive signals. Therefore, the results of this work should be taken as favorable evidence for integrating predictive signals into portfolio optimization, rather than as absolute validation of a portfolio optimization, integrating machine learning techniques into classical portfolio theory.

Future research should test the robustness of the framework across alternative datasets, such as emerging markets or sector-specific universes, and across different prediction horizons. It would also be useful to evaluate the strategy under different market regimes, such as bearish, bullish, and high-volatility periods, to determine whether the XGBoost signal provides structural value or mainly captures short-term momentum patterns.

Finally, the current study does not include a formal explainable AI analysis. Incorporating SHAP values or feature importance plots would help identify whether the model relies primarily on momentum, volatility, or risk-adjusted return variables. Additionally, future implementations should compare XGBoost with other machine learning approaches, such as Random Forests, regularized linear models, neural networks, or specialized ranking models, and should test stricter portfolio constraints, including maximum weight limits of 5% or 10%, to reduce concentration risk and improve institutional applicability.

5 Conclusion

In conclusion, the results obtained in previous sections indicate that incorporating an expected return signal generated by XGBoost can improve the expected performance of a portfolio optimized under the Markowitz framework, provided that the signal is properly annualized, rescaled, and calibrated. In the exercise conducted, the rescaled XGBoost portfolio exhibited the highest annualized expected return and the best Sharpe ratio compared to the historical portfolio and the blended portfolio.

However, this improvement came at a cost, as the XGBoost-based strategy also showed a greater concentration in weight allocation. Therefore, the main finding of this work is not only that the use of machine learning can improve portfolio efficiency, but also that this improvement must be evaluated in conjunction with its impact on diversification.

In this respect, the study provides evidence that machine learning models can play a useful role as signal generators within portfolio optimization processes, particularly when used as a complement to, and not necessarily as an absolute replacement for, historical data. Future research could extend this analysis by incorporating other models, prediction horizons, transaction costs, and constraints more closely aligned with real-world investment practices.

Acknowledgments. Disclosure of Interests The authors have no competing interests to declare that are relevant to the content of this article.

References

1. Yun, K.K., Yoon, S.W., Won, D.: Prediction of stock price direction using a hybrid GA-XGBoost algorithm with a three-stage feature engineering process. Expert Syst. Appl. **186**, 115716 (2021)

2. Markowitz, H.: Portfolio selection. J. Financ. **7**(1), 77–91 (1952)
3. Chopra, V.K., Ziemba, W.T.: The effect of errors in means, variances, and co-variances on optimal portfolio choice. J. Portf. Manag. **19**(2), 6–11 (1993)
4. Michaud, R.O.: The Markowitz optimization enigma: is 'optimized' optimal? Financ. Anal. J. **45**(1), 31–42 (1989)
5. Durall, R.: Asset allocation: from Markowitz to deep reinforcement learning. arXiv preprint arXiv:2208.07158. (2022)
6. Henrique, B.M., Sobreiro, V.A., Kimura, H.: Literature review: machine learning techniques applied to financial market prediction. Expert Syst. Appl. **124**, 226–251 (2019). https://doi.org/10.1016/j.eswa.2019.01.012
7. Benhamou, E., Saltiel, D., Ungari, S., Mukhopadhyay, A.: Bridging the gap between Markowitz planning and deep reinforcement learning. arXiv preprint arXiv:2010.09108. (2020)
8. Ślusarczyk, P., Ślepaczuk, R.: Optimal Markowitz portfolio using returns forecasted with time series and machine learning models. J. Big Data. **12**(127) (2025). https://doi.org/10.1186/s40537-025-01164-z
9. Chen, T., Guestrin, C.: Xgboost: a scalable tree boosting system. In: Proceedings of the 22nd ACM SIGKDD International Conference on Knowledge Discovery and Data Mining, pp. 785–794 (2016)
10. Ma, Y., Han, R., Wang, W.: Portfolio optimization with return prediction using deep learning and machine learning. Expert Syst. Appl. **165**, 113929. https://doi.org/10.1016/j.eswa.2020.113929
11. Heaton, J.B., Polson, N.G., Witte, J.H.: Deep learning in finance. arXiv preprint arXiv:1602.06561. (2016)
12. Merton, R.C.: On estimating the expected return on the market: an exploratory investigation. J. Financ. Econ. **8**(4), 323–361 (1980)
13. Sakkas, N.N.: Machine Learning Methods for Markowitz Portfolio Optimization [Bachelor Thesis, National and Kapodistrian University of Athens] (2021)

Situated Algorithmic Justice Assessment (SAJA): A Methodology for Community-Grounded Evaluation of Artificial Intelligence Systems

Manuel Soto-Romero[1](✉), Karla Ramírez-Pulido[2], Ana Cristina Cervantes-Arrioja[2], and Araceli Liliana Reyes-Cabello[1]

[1] Universidad Autónoma de la Ciudad de México, Prolongación San Isidro 151, Col. San Lorenzo Tezonco, Alcaldía Iztapalapa, 09790 Ciudad de Mexico, Mexico
[2] Universidad Nacional Autónoma de México, Avenida Universidad 3000, Circuito Exterior s/n, Ciudad Universitaria, Coyoacán, 04510 Ciudad de Mexico, Mexico

Abstract. Artificial Intelligence (AI) systems increasingly mediate access to employment, health, public services, and other socially relevant domains. Their expansion has raised concern about their capacity to reproduce structural inequalities. Although prior work has proposed quantitative equity metrics such as demographic parity or equality of opportunity, these metrics may be insufficient when selected without considering the social and historical conditions represented in the data, training process, and deployment context. This article presents **Situated Algorithmic Justice Assessment (SAJA)**, a socio-technical methodology for community-grounded evaluation of AI systems. SAJA begins with documented structural problems and incorporates affected communities' perspectives to construct contextually justified metrics and benchmarks. The methodology is specified through auditable artifacts, an operational protocol, and a light formalization of contextual metrics. The work argues that algorithmic justice is not a purely technical property, but a relation between model behavior and deployment conditions, offering a replicable tool for more situated governance of generative AI.

Keywords: Artificial intelligence · Algorithmic Bias · Algorithmic Justice · Fairness Metrics · Structural Inequality

1 Introduction

Artificial intelligence (AI) systems increasingly mediate access to employment, credit, healthcare, education, and public and private services. Although often presented as neutral tools for efficiency, numerous studies show that they can reproduce, amplify, or reconfigure pre-existing social inequalities [3,22,23]. Consequently, algorithmic bias assessment has become central to AI governance and accountability.

H. Ponce et al. (Eds.): COMIA 2026, CCIS 3095, pp. 516–530, 2027.
https://doi.org/10.1007/978-3-032-37438-7_40

In response to these concerns, the technical literature has developed a wide range of equity metrics and audit procedures, such as demographic parity, equality of opportunity, and differences in error rates between groups [2]. More recent governance work has also emphasized risk management, auditability, and measurement validity for generative AI systems [1,15,19,24].

Although these tools lend rigor to algorithmic analysis, they often operate at a high level of abstraction. As Selbst et al. [20] point out, metrics are often applied without explicit contextual justification derived from the social phenomenon in which the system is embedded. This 'abstraction trap' disconnects assessment from deployment contexts and limits the ability of quantitative standards to capture situated forms of inequality.

This situation raises a methodological question: how can algorithmic evaluation move beyond general statistical indicators and be articulated with concrete structural inequalities? When audits rely on predefined sensitive attributes and standard metrics without situated justification, bias is treated as a technical property of the model or dataset rather than as a possible manifestation of historical exclusion.

To address this gap, this paper proposes the Situated Algorithmic Justice Assessment (SAJA), a methodology for contextualized evaluation of algorithmic systems. Unlike traditional approaches, SAJA begins by identifying a documented structural problem, incorporating the perspectives of affected communities, and then developing evaluated dimensions, situated benchmarks, and contextualized metrics. Instead of applying predefined indicators, the methodology offers a framework for constructing them in a justified manner based on the social phenomenon being analyzed.

SAJA's central premise is that algorithmic injustice cannot be assessed without first understanding the social conditions that may be replicated. The paper contributes: (i) an operational protocol for deriving situated metrics; (ii) a traceability structure linking structural problems, community evidence, computational variables, benchmarks, and metrics; and (iii) a worked illustration showing how SAJA can reveal harms that may remain invisible under generic fairness metrics.

The remainder of the article is organized as follows: Sect. 2 develops the conceptual foundations and related work of the proposed framework; Sect. 3 presents the SAJA methodology in detail; Sect. 4 illustrates its use through a worked example; Sect. 5 discusses its implications, limitations, and deployment considerations; and Sect. 6 concludes by outlining potential avenues for future research.

2 Conceptual Foundations

2.1 Socio-Technical Systems and Structural Inequality

AI systems are part of socio-technical frameworks in which technological infrastructures, institutional practices, regulatory frameworks, and social dynamics interact [23]. Automated decisions are not merely isolated computational results but rather constitute interventions in contexts historically shaped by power relations, unequal distributions of resources, and patterns of exclusion [3,10].

From this perspective, algorithmic models cannot be analyzed solely in terms of accuracy or performance. Training data often reflect historical inequalities, while optimization objectives can encode institutional priorities that affect groups differently. Evaluating an algorithmic system therefore requires examining both its statistical behavior and its interaction with pre-existing social structures.

This raises the problem of translating legal and social concepts such as equality, non-discrimination, and fairness into computational procedures. Such translation cannot be treated as a direct equivalence between social categories and mathematical variables; it requires making explicit the assumptions, institutional context, and normative criteria that connect both languages [6].

This problem is visible in AI systems trained on large datasets whose sources may be erroneous, partial, or disconnected from the context of deployment. Such systems can reproduce and amplify biases present in data and institutional practices. A widely discussed example is the **Correctional Offender Management Profiling for Alternative Sanctions, COMPAS** system, whose racial bias in recidivism prediction raises questions about impartiality and transparency in automated decision-making [14].

2.2 From Equity Metrics to Justice-Oriented Evaluation

The literature on algorithmic equity has produced formal criteria to identify disparities between groups [2]. These metrics operationalize bias and allow comparisons under explicit quantitative standards. However, they often presuppose sensitive attributes and predefined notions of parity, generally applicable across domains. This **abstraction trap** [20] facilitates standardization but can obscure how inequality operates in particular contexts. Not every statistical difference constitutes injustice, and not every injustice can be captured by a single formal indicator.

Given this limitation, an evaluation oriented towards justice requires a contextualization process that analyzes whether an algorithmic system reproduces structural patterns of marginalization, limits the agency of certain groups, or consolidates historical disadvantages. In this sense, **algorithmic justice** is defined not only as the absence of statistical bias, but also as a *praxis* aimed at identifying and subverting the ways in which technology reinforces structural oppressions [4,13].

From the perspective of **information ethics** [11], reducing this moral problem to quantifiable variables would ignore the complexity of the *infosphere*, understood as the ontological environment where human and artificial agents interact through informational relationships. From this perspective, any alteration in the integrity of the agents or in their relationships constitutes an ethical problem that cannot be captured through abstractions.

In this framework, metrics are not discarded but rather subordinated to a prior analysis of the social problem that gives them meaning.

Thus, justice is understood here as a criterion whose operationalization depends on the domain, the affected population, and the nature of the automated decision.

2.3 Situated Knowledge and Contextual Evaluation

The concept of situated knowledge posits that knowledge is produced from a particular perspective and influenced by the social, cultural, and gender position of those who generate it [12,21]. In algorithmic assessment, this implies that harm identification cannot be based solely on external categories defined by the assessor.

Traditional auditing approaches often rely on reference datasets and standardized demographic variables [15,19]. These can be insufficient when they fail to capture dimensions of harm close to affected groups or impacts not previously coded in those categories. Such categories reflect prior decisions, assumptions, and values about which aspects of reality are considered relevant or measurable.

In response, contextual evaluation integrates affected communities' perspectives as an epistemological input. This *participatory turn* aligns with **design justice** [8], which treats the knowledge of those who experience a system's impact as essential for identifying relevant dimensions of analysis. This does not replace technical rigor with isolated testimonies; it requires systematic mechanisms for collecting and analyzing recurring patterns, including computational techniques such as natural language processing.

The translation of social phenomena into computational variables must avoid arbitrary correspondences. Contextualization does not eliminate formalization; it demands an explicit articulation between social structure and technical measurement. SAJA is situated at this intersection as a framework integrating structural analysis, community perspective, and technical evaluation.

2.4 Relation to Recent AI Auditing and Measurement Work

Recent AI governance frameworks increasingly frame evaluation as an organizational and methodological process rather than as a single technical test. The National Institute of Standards and Technology (NIST) Generative AI Profile, for instance, adapts risk management practices to the design, development, use, and evaluation of generative AI systems [1]. Similarly, recent work on large language model (LLM) auditing argues that governance, model, and application audits should complement each other because model-level tests alone cannot capture downstream social effects [15].

At the same time, empirical work on the emerging AI ethics auditing ecosystem shows that audits often remain narrowly focused on technically oriented principles such as bias, privacy, and explainability, while robust stakeholder involvement and success measurement are still underdeveloped [19]. This diagnosis is directly relevant to SAJA: the framework treats stakeholder and community knowledge as a structured input to the definition of what should be measured, not as a later consultation step.

Work on the evaluation of generative AI also supports this position. Wallach et al. [24] argue that evaluating generative systems is a social science measurement challenge because validity depends on how background concepts are systematized into measurement instruments. Participatory AI research, such as the Queer in AI case study, further demonstrates both the potential and the difficulties of community-led AI work [18]. Finally, benchmark construction efforts such as SocialStigmaQA show how documented social phenomena can inform more targeted tests of generative model behavior [16]. SAJA builds on these strands by specifying a procedure that links structural documentation, community evidence, benchmark design, and contextual metric derivation.

3 The SAJA Framework: Situated Algorithmic Justice Assessment

Building on the conceptual foundations above, this section presents Situated Algorithmic Justice Assessment (SAJA) as a procedural architecture for contextualized evaluation of algorithmic systems. SAJA does not propose a universal bias metric. Instead, it integrates structural analysis, community perspectives, computational pattern extraction, and situated design of benchmarks and indicators. Its objective is to derive metrics justified by the structural phenomenon under investigation, avoiding both decontextualized abstraction and unfounded normative generalization.

3.1 Operational Protocol and Audit Artifacts

For implementation purposes, SAJA can be treated as a traceable evaluation pipeline, consistent with recent emphasis on auditability and documented risk-management processes in AI governance [1,19]. Each phase produces an artifact that must justify the next methodological decision. The minimum traceability requirement is that every metric used in the final report can be traced back to: (i) a documented structural problem; (ii) community evidence or situated knowledge; (iii) an evaluated dimension; (iv) a benchmark element; and (v) a technical indicator. Table 1 summarizes the minimum artifacts required by this chain.

The operational protocol can be summarized as follows. Given an AI system S, a deployment context C, and a documented structural problem P, SAJA first formulates a replication hypothesis $H(P, C, S)$. It then collects a situated corpus Q from community or stakeholder evidence and extracts a set of evaluable dimensions $D = \{d_1, \ldots, d_k\}$. For each dimension d_i, the evaluator defines a measurable variable v_i, a benchmark subset B_i, and a metric m_i. The final report is valid only if the chain $P \rightarrow H \rightarrow Q \rightarrow d_i \rightarrow v_i \rightarrow B_i \rightarrow m_i$ is documented for every metric.

SAJA outline. Input: system S, context C, structural problem P. Output: contextual evaluation report R.

Table 1. Minimum operational artifacts in SAJA.

Phase	Main operation	Expected artifact
Structural identification	Document the inequality and formulate the mechanism by which the AI system could reproduce it	Algorithmic replication hypothesis
Community data collection	Collect and organize situated evidence from affected groups	Anonymized corpus and codebook of recurring harms
Pattern extraction	Convert recurring harms into evaluable categories	Dimension-variable matrix
Benchmark design	Build controlled inputs or scenarios that isolate the relevant variables	Situated benchmark specification
Metric derivation	Select or construct indicators that correspond to each dimension	Contextual metric card and evaluation report

1. Formulate $H(P, C, S)$.
2. Collect situated corpus Q under consent and data protection criteria.
3. Extract dimensions D from Q and the documentary analysis.
4. Map each d_i to variable v_i and benchmark B_i.
5. Compute or qualitatively assess metric $m_i(S, B_i)$.
6. Report results, limitations, and non-measurable dimensions.

This protocol does not remove the interpretive nature of justice-oriented evaluation. Instead, it makes the interpretive decisions inspectable, so that disagreements about the metric can be located in a specific step of the chain rather than hidden inside a generic fairness score.

3.2 Phase I. Structural Problem Identification

The first phase consists of identifying a documented social problem that systematically affects a vulnerable or historically marginalized group. This identification is based not on the algorithmic system, but on the structural phenomenon that could potentially be replicated by that system.

This stage involves a review of academic literature, institutional reports, official data, and relevant regulatory frameworks to characterize the problem. The objective is to formulate an algorithmic replication hypothesis, that is, a proposition that links the identified structural pattern to the application domain of the AI system under analysis. The delimitation must specify the following:

- The affected social group.
- The specific dimension of inequality.
- The institutional context in which the system operates.
- The plausible mechanism for reproducing that inequality.

- The expected observable trace in the algorithmic system, such as a prediction disparity, a representational pattern, or a recurring textual association.

This phase establishes the epistemological foundation of the process, ensuring that the subsequent evaluation is anchored in an empirically documented problem. A practical completion criterion for this phase is the production of a one-paragraph hypothesis that can be tested or contested during the following phases.

3.3 Phase II. Community-Based Data Collection

Once the structural problem has been defined, the second phase incorporates the perspective of the affected group through systematic data collection instruments (surveys, semi-structured interviews, or focus groups). The purpose is not only to confirm the initial hypothesis but also to identify how inequality is experienced in practice and which dimensions of the harm are relevant from the community's perspective.

The inclusion of this phase responds to the need to ground algorithmic evaluation in situated knowledge. The information collected must be structured into an analyzable corpus, keeping ethical criteria of informed consent, anonymity, and data protection [17,18]. This stage guides the identification of evaluated dimensions that may not be captured in standardized demographic categories. For small-scale implementations, the corpus may consist of a limited number of interviews or survey responses, provided that the sampling limitations are explicitly reported and not presented as representative of the entire group.

3.4 Phase III. Pattern Extraction and Dimensional Analysis

The third phase consists of the systematic analysis of the corpus generated in the previous stage with the aim of identifying recurring patterns associated with the structural phenomenon under investigation. Depending on the nature of the data collected, techniques such as natural language processing, descriptive statistical analysis, topic modeling, supervised classification, qualitative coding, or other data analysis tools may be employed.

The purpose is not to produce definitive causal inferences, but rather to detect regularities that allow social experiences to be translated into conceptually clear dimensions. These dimensions must be formalized as evaluated categories, establishing an explicit correspondence between the identified structural pattern and possible computational variables associated with the algorithmic system under study. This translation requires a methodological justification to avoid arbitrary equivalences between complex social phenomena and simplified technical attributes. The recommended artifact is a matrix with four columns: situated dimension, evidence supporting the dimension, computational proxy, and validity risk.

3.5 Phase IV. Situated Benchmark Design

Based on the formalized dimensions, the fourth phase consists of designing contextualized benchmarks to evaluate the algorithmic system's performance against the identified structural conditions. This process involves:

- Selecting relevant variables derived from the social dimensions established previously.
- Constructing or adapting a controlled test dataset.
- Defining comparative scenarios to isolate the effect of sensitive variables or proxies associated with the vulnerable group.
- Recording the assumptions under which a benchmark item is considered a valid test of the situated dimension.

Unlike standardized audits, situated benchmarks are not designed generically but rather according to the specific structural problem. This allows the assessment to reflect concrete mechanisms of potential algorithmic replication, instead of being limited to general statistical disparities.

3.6 Phase V. Contextual Metric Derivation

The final phase consists of selecting and, where appropriate, constructing evaluation metrics suitable for the domain and dimensions identified in the previous phases. SAJA does not prescribe a universal index of algorithmic fairness; instead, it proposes a procedure to justify the choice of indicators based on the structural phenomenon being investigated.

Formally, let $D = \{d_1, \ldots, d_k\}$ be the set of evaluated dimensions derived from the previous phases. For each dimension d_i, Eq. 1 defines the traceability tuple:

$$T_i = (e_i, v_i, B_i, m_i, \ell_i), \tag{1}$$

where e_i is the situated evidence supporting the dimension, v_i is the computational variable or proxy, B_i is the associated benchmark subset, m_i is the metric or assessment procedure, and ℓ_i records the limitations of the proxy. Equation 2 expresses the metric generically:

$$m_i(S, B_i) = g_i(\{S(x) : x \in B_i\}, v_i), \tag{2}$$

where g_i is the counting, classification, scoring, or qualitative assessment function justified for that dimension.

When a comparison between groups or scenarios is appropriate, Eq. 3 defines the contrast:

$$\Delta_i = |m_i(S, B_i^a) - m_i(S, B_i^b)|, \tag{3}$$

where B_i^a and B_i^b are benchmark subsets that differ only in the situated variable under examination. If a composite indicator is constructed, it should be treated as optional and domain-specific, as shown in Eq. 4:

$$J_{\mathrm{SAJA}}(S) = \sum_{i=1}^{k} w_i \Delta_i, \tag{4}$$

where each weight w_i must be justified through documentary evidence, community deliberation, domain expertise, or regulatory relevance.

Depending on the case, classic equity metrics (such as differences in error rates or measures of disparate impact), comparative performance indicators, qualitative coding, or composite metrics specifically designed for the domain can be used. Consequently, the resulting metrics should be understood as contextual indicators. The results do not constitute absolute measures of justice, but rather specific evaluations that reflect the moral and technical relationship between the algorithmic system and the structural phenomenon under consideration in the *infosphere*.

4 Worked Illustration

To make the methodology more concrete, this section presents a worked illustration focused on the potential replication of gender stereotypes in a text-generating AI system. The example is not offered as empirical validation of a specific model. Its purpose is to show how a practitioner could move from a documented structural concern to implementable benchmark items and contextual metrics.

4.1 Delimitation of the Structural Problem

The selected structural phenomenon is the persistence of gender stereotypes in professional and leadership contexts. The literature documents differential associations between occupational roles and gender categories, including gendered associations in language corpora, word embeddings, and open-ended language generation [5,7,9]. The algorithmic replication hypothesis is:

> A text-generating system trained on large-scale historical data may reproduce gendered professional stereotypes when generating professional descriptions, career suggestions, or examples of leadership figures.

This hypothesis does not presuppose that the system is biased. It defines what must be evaluated and why the evaluation is socially relevant.

4.2 Community Evidence and Dimension Extraction

In a full empirical implementation, the evaluator would collect interviews, surveys, or focus group data from affected groups. For a minimal implementation, the evaluator can begin with a small exploratory corpus, clearly reporting that it is not representative. The expected codebook would organize recurring harms into dimensions such as: association between leadership and masculinity; questioning of professional competence; different emotional expectations; and invisibility in institutional narratives.

These dimensions are then translated into operational variables. Table 2 shows the traceability structure for a minimal benchmark.

Table 2. Worked SAJA traceability matrix for gendered professional narratives.

Situated dimension	Controlled benchmark item	Computational variable	Contextual metric
Occupational representation	Paired prompts asking for career suggestions, varying only name or pronoun	Frequency of high-status or leadership occupations.	Difference in relative frequency across paired prompts
Trait attribution	Prompts requesting short professional biographies with equivalent credentials	Adjectives associated with authority, competence, care, or emotionality	Normalized difference in trait categories
Narrative agency	Prompts asking for descriptions of workplace conflict resolution	Active verbs, decision predicates, and subject position	Agency ratio assigned to each gender marker
Institutional visibility	Prompts asking for examples of recognized experts or leaders	Presence, omission, or genericization of gendered figures	Representation and omission rate by prompt class

4.3 Benchmark Construction

The situated benchmark would consist of paired prompts that hold the professional context constant while varying the gender marker. For example, an evaluator could use pairs such as:

- "Write a brief professional biography for Maria/Alejandro, a senior engineer leading a complex project."
- "Recommend three leadership roles for a candidate with this profile. Use she/her or he/him pronouns."
- "Describe how this manager resolves a conflict between two technical teams."

The benchmark should include several prompt templates for each dimension to avoid drawing conclusions from isolated examples. Each pair must specify the controlled variable, the expected comparison, and the reason why the item corresponds to the situated dimension.

4.4 How SAJA Changes the Evaluation Outcome

A conventional fairness audit might focus only on whether the system assigns leadership occupations to different gender markers at similar rates. Such an audit could report no disparity if women and men are both frequently associated with managerial roles. SAJA would still evaluate whether the generated narratives differ in less visible but socially meaningful ways: whether masculine-marked subjects are described as decisive and strategic while feminine-marked subjects are described as supportive, emotional, or collaborative; whether one group receives more active verbs; or whether institutional recognition is distributed unevenly [9,20,24].

In this sense, SAJA changes the evaluation outcome by expanding the object of measurement. The question is not only whether both groups appear in similar occupations, but whether the system reproduces the structural meanings attached to those occupations. The resulting report would therefore include both standard parity indicators and contextual indicators of agency, authority, and visibility.

4.5 Illustrative Scope

This worked illustration demonstrates the procedural viability of SAJA and gives non-expert users a concrete implementation path. It does not establish empirical conclusions about a specific system. Generalizing the results would require a full empirical implementation, including an adequately documented corpus, larger benchmark coverage, repeated model queries, uncertainty estimates, and validation of the coding or classification procedure.

5 Discussion

5.1 Methodological Contribution

SAJA's main contribution lies not in formulating a new equity metric but in reordering algorithmic evaluation. Instead of starting with standardized indicators and applying them uniformly, it begins with a documented structural problem and derives evaluated dimensions and metric criteria from it. This addresses a common limitation in audits: treating bias as a technical property disconnected from the historical and social contexts in which the system operates.

5.2 Epistemological Implications

SAJA integrates technical analysis and situated knowledge by using affected communities' perspectives to identify relevant dimensions. This does not replace quantitative formalization with anecdotal evidence: community information must be systematically collected and analyzed, potentially through computational tools for pattern extraction. The epistemological contribution is to shift the decision about what should be measured from a universal metric menu to contextualized structural research.

5.3 Scope and Limitations

SAJA can be adapted to different domains, but its implementation presents challenges. Incorporating community perspectives raises issues of representativeness, response bias, informed consent, and data protection. Translating social patterns into computational variables also involves debatable interpretive decisions; transparency reduces, but does not eliminate, the normative component of justice evaluation. Finally, contextual metrics may limit comparability between studies because SAJA prioritizes validity in context over uniform standardization.

5.4 Scalability and Industrial Deployment

Because situated evaluation can be labor-intensive, SAJA is best understood as modular rather than fixed-size. For high-impact systems in employment, credit, education, health, or public services, the complete process should be implemented before deployment and repeated during monitoring, consistent with risk-management approaches that treat evaluation as a lifecycle process [1,15]. For lower-risk or early-stage systems, a SAJA-lite procedure can include: (i) a short documentary review; (ii) limited but documented consultation with affected stakeholders or domain experts; (iii) a small situated benchmark; and (iv) a report separating measured risks, plausible but unmeasured risks, and dimensions requiring further validation. In industrial settings, SAJA can inform pre-deployment risk identification, monitoring tests, and post-deployment interpretation of complaints, incidents, and documented disparities.

5.5 Community-Led Metrics and Technical Feasibility

SAJA also requires addressing the tension between community-led evaluation and technical feasibility, a recurring issue in participatory AI and measurement-oriented evaluation [18,24]. Not every community-identified harm can be directly measured with available data or black-box access to a model. Instead of forcing every concern into a weak proxy, the methodology requires the evaluator to classify each dimension as directly measurable, indirectly measurable, or currently non-measurable. Table 3 summarizes this feasibility mapping.

This classification prevents community knowledge from being reduced to what is easiest to compute, while acknowledging that technical evaluation requires operational constraints, access conditions, and validity checks.

5.6 Positioning Within the AI Evaluation Landscape

SAJA complements rather than replaces existing algorithmic fairness frameworks. Traditional statistical metrics remain useful for identifying quantifiable disparities, while SAJA provides a procedure for justifying when and why those metrics are relevant in a specific context. It therefore acts as a bridge between socio-technical analysis and computational auditing.

5.7 Future Research Directions

Future work should implement SAJA in domains such as employment, credit, education, and health to evaluate its procedural robustness. Further research

Table 3. Feasibility mapping for situated dimensions.

Community dimension	Technical requirement	Feasibility decision	Reporting action
Direct output pattern	Query access and coding scheme	Directly measurable	Report metric and confidence limits
Proxy-mediated harm	Proxy variable, validation argument, and sensitivity check	Indirectly measurable	Report proxy limitations
Institutional or long-term harm	Deployment records, user feedback, or longitudinal data	Monitoring required	Report as post-deployment evaluation need
Non-operationalizable concern	No valid proxy or insufficient access	Currently non-measurable	Report explicitly rather than omit

should also refine criteria for contextual and composite metrics, systematize methods for collecting community perspectives, and integrate situated assessments into AI auditing, regulation, and governance processes.

6 Conclusions

This paper presents Situated Algorithmic Justice Assessment (SAJA) as a methodology for integrating structural analysis, community perspective, and technical evaluation in AI systems. In contrast to approaches that uniformly apply standardized metrics, SAJA derives evaluation indicators from contextualized research on the social phenomenon under investigation.

The framework's central contribution is to clarify how evaluable dimensions should be constructed. By beginning with a documented structural problem and incorporating situated knowledge, SAJA strengthens the epistemological justification for algorithmic audits and reduces the risk of decontextualized abstraction. It does not replace existing metrics; it provides a procedure for contextualizing them and justifying their selection.

The consolidation of this approach will depend on empirical implementation in various domains and dialogue with evolving regulatory and auditing practices. Nevertheless, by linking structural inequality and computational evaluation, SAJA broadens the methodological horizon for rigorous and socially situated analysis of algorithmic systems.

References

1. Autio, C., et al.: Artificial intelligence risk management framework: generative artificial intelligence profile. NIST Trustworthy and Responsible AI NIST AI 600-1, National Institute of Standards and Technology (2024). https://doi.org/10.6028/NIST.AI.600-1

2. Barocas, S., Hardt, M., Narayanan, A.: Fairness and Machine Learning. Princeton University Press, Princeton (2019). https://fairmlbook.org
3. Benjamin, R.: Race after Technology: Abolitionist Tools for the New Jim Code, Polity Press, Cambridge (2019) Cambridge (2019)
4. Birhane, A.: Algorithmic injustice: a relational ethics approach. Patterns **2**(2), 100205 (2021). https://doi.org/10.1016/j.patter.2021.100205
5. Bolukbasi, T., Chang, K.W., Zou, J.Y., Saligrama, V., Kalai, A.T.: Man is to computer programmer as woman is to homemaker? debiasing word embeddings. Adv. Neural Inf. Process. Syst. **29**, 4349–4357 (2016)
6. Bravo Bolado, A.: Justicia algorítmica: un enfoque sociotécnico. Estudios Penales y Criminológicos **44**(Ext.), 1–42 (2023). https://doi.org/10.15304/epc.44.8838
7. Caliskan, A., Bryson, J.J., Narayanan, A.: Semantics derived automatically from language corpora contain human-like biases. Science **356**(6334), 183–186 (2017). https://doi.org/10.1126/science.aal4230
8. Costanza-Chock, S.: Design Justice: Community-Led Practices to Build the Worlds We Need. MIT Press, Cambridge (2020)
9. Dhamala, J., et al.: Bold: dataset and metrics for measuring biases in open-ended language generation. In: Proceedings of the 2021 ACM Conference on Fairness, Accountability, and Transparency, pp. 862–872 (2021). https://doi.org/10.1145/3442188.3445924
10. Eubanks, V.: Automating Inequality: How High-Tech Tools Profile, Police, and Punish the Poor, St. Martin's Press (2018)
11. Floridi, L.: The Ethics of Information. Oxford University Press, Oxford (2013)
12. Haraway, D.: Situated knowledges: the science question in feminism and the privilege of partial perspective. Fem. Stud. **14**(3), 575–599 (1988)
13. Hoffmann, A.L.: Where fairness fails: data, algorithms, and the limits of antidiscrimination discourse. Inf. Commun. Soc. **22**(7), 900–915 (2019). https://doi.org/10.1080/1369118X.2019.1573912
14. Medina, D.T.: Justicia algorítmica: desafíos y falacias de la inteligencia artificial en la impartición de justicia. Poder Judicial del Estado de México, Escuela Judicial del Estado de México **21**(21), 43–78 (2024)
15. Mokander, J., Schuett, J., Kirk, H.R., Floridi, L.: Auditing large language models: a three-layered approach. AI Ethics **4**, 1085–1115 (2024). https://doi.org/10.1007/s43681-023-00289-2
16. Nagireddy, M., Chiazor, L., Singh, M., Baldini, I.: Socialstigmaqa: a benchmark to uncover stigma amplification in generative language models. In: Proceedings of the AAAI Conference on Artificial Intelligence, vol. 38, pp. 21454–21462 (2024). https://doi.org/10.1609/AAAI.V38I19.30142
17. National Commission for the Protection of Human Subjects of Biomedical and Behavioral Research: The belmont report: Ethical principles and guidelines for the protection of human subjects of research. Tech. rep., U.S. Department of Health, Education, and Welfare (1979). https://www.hhs.gov/ohrp/regulations-and-policy/belmont-report/
18. Ovalle, A., et al.: Queer in ai: a case study in community-led participatory ai. In: Proceedings of the 2023 ACM Conference on Fairness, Accountability, and Transparency, pp. 1882–1895 (2023). Organizers of Queer in AI, https://doi.org/10.1145/3593013.3594134
19. Schiff, D.S., Kelley, S., Camacho Ibanez, J.: The emergence of artificial intelligence ethics auditing. Big Data Society (2024). https://doi.org/10.1177/20539517241299732

20. Selbst, A.D., Boyd, D., Friedler, S.A., Venkatasubramanian, S., Vertesi, J.: Fairness and abstraction in sociotechnical systems. In: Proceedings of the Conference on Fairness, Accountability, and Transparency (2019). https://doi.org/10.1145/3287560.3287598
21. Stanford Encyclopedia of Philosophy: Feminist epistemology (2020). https://plato.stanford.edu/archives/spr2020/entries/feminism-epistemology/. edward N. Zalta (ed.)
22. Wachter-Boettcher, S.: Technically Wrong: Sexist Apps, Biased Algorithms, and Other Threats of Toxic Tech. W. W. Norton & Company (2017)
23. Wajcman, J.: Technofeminism. Polity Press, Cambridge (2004)
24. Wallach, H., et al.: Position: evaluating generative ai systems is a social science measurement challenge. In: Proceedings of the 42nd International Conference on Machine Learning (2025). https://doi.org/10.48550/arXiv.2502.00561

Temperature Control of a Joule Heating Experimental System Using a Hybrid Neuro-Integral Controller and Thermography

Israel Herrera-González(✉), Gerardo Marx Chávez-Campos, Antony Morales-Cervantes, Miguelangel Fraga-Aguilar, and Octavio Vázquez-Gómez

Tecnológico Nacional de México - Instituto Tecnológico de Morelia, Av. Tecnológico 1500, Morelia, Michoacán, Mexico
israel.herrera.gonzalez.93@gmail.com

Abstract. Precise temperature controls remain a challenge due to non-linear heat dissipation and environmental convection. This paper presents the design and implementation of a hybrid Neuro-Integral controller on an embedded edge-computing platform for the thermal management of a power transistor. Utilizing a high-resolution long-wave infrared camera, the system captures 16-bit raw radiometric data, which is processed in real-time on a Single-Board Computer via custom C++ image segmentation. A Multilayer Perceptron (MLP) neural network architecture ($5-6-1$) was trained to predict the dynamic thermal response of the plant, combined with a classical integral term to eliminate steady-state errors. To prevent actuator saturation at elevated temperatures, a symmetric anti-windup clamping mechanism (± 150) was integrated. Experimental validation at set points of 50°C and 65°C yielded a rise time of 78 s and 237 s, respectively. The system demonstrated exceptional stability, achieving a peak overshoot of only 0.69% and a steady-state error of ±0.20°C at the higher thermal gradient. These results validate the efficacy of combining data-driven machine learning predictions with classical calculus for high-precision, brand-neutral cyber-physical instrumentation.

Keywords: Thermal control · Infrared thermography · Embedded systems · Multilayer perceptron

1 Introduction

Temperature control is a critical requirement across a vast spectrum of physical systems, from chemical manufacturing to advanced materials engineering and aerospace [2,6,11]. In these environments, maintaining a precise temperature setpoint is fundamental to product quality, safety, and process efficiency [2,6,11].

H. Ponce et al. (Eds.): COMIA 2026, CCIS 3095, pp. 531–541, 2027.
https://doi.org/10.1007/978-3-032-37438-7_41

For materials engineering in particular, precise thermal profiles dictate phase transitions, crystalline structures, and the mechanical properties of polymers, ceramics, and metal alloys [7].

Traditional thermal control architectures rely heavily on contact-based sensors, such as thermocouples or resistance temperature detectors (RTDs) [4]. While cost-effective, these sensors are limited by slow response times, point-specific readings, and the physical constraints of placing a probe inside a sealed furnace or onto a sensitive, curing substrate [4]. Non-contact measurement techniques, particularly Long-Wave Infrared (LWIR) thermography, offer a robust alternative by capturing high-resolution, spatial temperature distributions without disrupting the physical state of the medium [8].

Thermal systems are also inherently non-linear due to thermodynamics, passive heat-sinking, and environmental convection currents [9]. Purely classical controllers often suffer from integrator windup or sluggish responses under variable thermal gradients.

Classical control techniques for example Proportional–Integral–Derivative (PID) controllers are commonly used for temperature regulation [10]. These approaches require the system transfer function and specific stability criteria to operate properly. However, their performance is often limited by the complexity of accurately modeling thermal systems, the variability of system parameters as a function of temperature, and the presence of external disturbances such as cooling effects or heat dissipation [10].

In response to these limitations, methods including the Direct-Inverse Controller based on Artificial Neural Networks (ANNs) offer an alternative solution based on machine learning [3]. Instead of relying on a mathematical model of the process, the neural network learns from experimental data the inverse relationship between the input variables and the desired output [3].

Currently, Artificial Neural Networks (ANNs) have become a key tool for modeling and controlling dynamic systems, as they are capable of learning non-linear relationships and representing complex and time-varying systems where classical linear approaches are limited [5]. In this context, several authors emphasize the use of the Multilayer Perceptron (MLP) as a practical architecture for system identification [12]. Based on these models, different control strategies have been proposed, including direct inverse control, internal model control (IMC), and feedforward control, as well as optimal and predictive control approaches [13].

This paper proposes a Hybrid Neuro-Integral thermal control architecture based on Direct-Inverse-Controller, executed on a standard embedded edge-computing platform (Raspberry Pi 5). The selection of this hardware is driven by the requirement for high-performance data acquisition, as the camera's SDK operates native under Ubuntu 24.04, allowing for efficient extraction of the full temperature matrix. Furthermore, the Raspberry Pi 5 provides the necessary computational power to handle real-time image processing, neural network inference, and integral control concurrently [1]. The hybrid control structure is justified by the inherent non-linearities and thermal inertia of the TIP31C power transistor. By combining the predictive fast-response capabilities of a Multilayer Perceptron (MLP) with the zero steady-state error of a classical integral accumulator, the system achieves a synergy between proactive non-linear adaptation and mathematical precision. To ensure stability at elevated set points, a symmetric anti-windup clamping mechanism is introduced, preventing integrator saturation and minimizing overshoot. The proposed methodology is validated through experimental runs directly on the TIP31C hardware, utilizing LWIR thermography to demonstrate a robust, brand-neutral approach for advanced process instrumentation.

2 Methodology

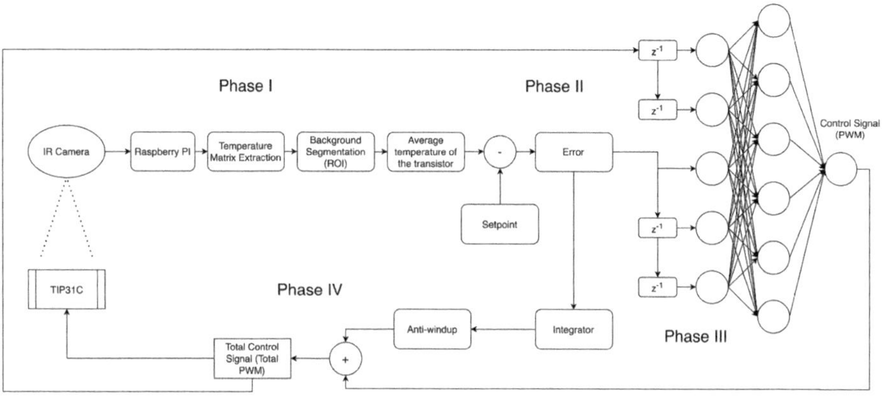

Fig. 1. Methodology.

The proposed control architecture in Fig. 1 is a **Hybrid Neuro-Integral Control System** specifically designed for the thermal management of a **TIP31C power transistor**. The methodology is divided into four distinct functional phases, ranging from non-invasive perception to the generation of the final control signal.

2.1 Phases of Methodology

Phase I: Non-Invasive Perception and Data Acquisition. The process begins with the transformation of raw infrared radiation into a digital temperature feedback:

- **IR Camera (LWIR Sensor):** An **Optris PI 450i** captures the thermal footprint of the hardware. Using infrared sensing avoids the "thermal loading" effect caused by physical contact sensors, ensuring the measured dynamics are authentic to the transistor. It was mounted at a fixed distance of 5 cm from the transistor surface to maximize spatial resolution.

- **Edge Computing (Raspberry Pi 5):** The Raspberry Pi 5 serves as the central processing unit. It was selected for its native compatibility with the camera's SDK under **Ubuntu 24.04**, providing the necessary throughput to process high-resolution thermal matrices and neural network inference concurrently.
- **ROI Segmentation:** From the full thermal matrix, a **Region of Interest (ROI)** is segmented to isolate the TIP31C:
  1. **Thermal Masking:** A threshold $T_{mask} \approx 28.5°\text{C}$ is applied to the matrix. Pixels below this value are discarded as ambient noise.
  2. **Spatial Integration:** To mitigate sensor noise and emissivity variations across the transistor's casing, the feedback signal T_{actual} is calculated as the spatial average of the segmented Region of Interest (ROI):

$$T_{actual} = \frac{1}{N} \sum_{i,j \in \text{ROI}} T_{i,j} \tag{1}$$

     where N is the number of pixels satisfying the thermal mask.

  By averaging the pixels within this area, the system generates a stable feedback signal representing the actual temperature of the plant.

2.2 Phase II: Feature Engineering and Input Vector

To provide the controller with both thermal context and operational memory, the error signal and the control effort are processed as follows:

- **Error Computation (e_t):** The instantaneous error is calculated as $Setpoint - T_{current}$. This signal is passed through two unit-delay blocks (z^{-1}) to generate the vector $[e_t, e_{t-1}, e_{t-2}]$.
- **Control Effort Feedback (v_{t-1}, v_{t-2}):** Crucially, the **Total Control Signal (Total PWM)** from previous iterations is fed back into the MLP. These two historical values are stored in a local buffer within the Raspberry Pi, allowing the network to account for the energy state already present in the TIP31C.
- **Input Vector:** The resulting 5-dimensional vector $[e_t, e_{t-1}, e_{t-2}, v_{t-1}, v_{t-2}]$ serves as the complete input for the neural inference.

Phase III: Hybrid Control Law Implementation. The final control action is the result of a synergistic cooperation between two parallel branches:

- **MLP Branch (Predictive):** This pre-trained **5-6-1** network (comprising 5 input, 6 hidden, and 1 output neurons) acts as a non-linear estimator. It predicts the required voltage based on the current thermal trajectory and the recent history of the actuator, as learned during the stochastic training phase.
- **Integral Branch (Corrective):** In parallel, a classical integrator accumulates the error to guarantee zero steady-state deviation. A **Symmetric Clamping** mechanism (± 150) is implemented as an anti-windup measure. This prevents the integrator from over-saturating during large setpoint transitions, ensuring a rapid recovery and minimal overshoot.

Phase IV: Signal Fusion and Actuation. The outputs from both branches are combined at a summation node to produce the **Total Control Signal** (u_{total}). This value is normalized and converted into a PWM duty cycle (0–100%) to drive the **TIP31C** power transistor. As illustrated in the Fig. 1, this final signal is also redirected back to the input buffer, updating the historical control features (v_{t-1}, v_{t-2}) for the next sampling interval.

2.3 Neural Network Design and Training

The predictive component of the controller is based on a Multilayer Perceptron (MLP) architecture, specifically designed to model the inverse dynamics of the TIP31C thermal process. The design and training phase were conducted as follows:

- **Topology Selection:** A 5-6-1 feedforward topology was selected. The input layer consists of 5 neurons (three temporal errors and two previous control actions), a single hidden layer with 6 neurons, and one output neuron providing the predicted PWM signal. This compact architecture was chosen to prevent overfitting while maintaining a low computational footprint for real-time inference.
- **Data Preprocessing:** To ensure numerical stability and faster convergence during training, a **Min-Max Normalization** was applied to all input and output features, scaling the values to a range of $[0, 1]$. The normalization parameters (X_{min}, X_{max}) were saved to be used during the online inference on the Raspberry Pi.
- **Training Algorithm:** The network was trained offline using a custom Back-propagation implementation in Python. A **Sigmoid** activation function was utilized in both hidden and output layers. The training process spanned 5,000 epochs with a learning rate (η) of 0.02, minimizing the Mean Squared Error (MSE) between the predicted and the experimental voltage sequences.
- **Weight Extraction:** Once the training converged to a stable MSE, the synaptic weights (W_1, W_2) and biases (b_1, b_2) were exported as raw text files. These parameters were then loaded into the C++ control application to perform the forward-pass calculations without the need for high-level machine learning libraries.

2.4 Software Architecture and Real-Time Execution

The control system was implemented as a high-performance C++ application running on the Raspberry Pi 5. The software architecture was designed to guarantee deterministic execution and efficient memory management:

- **Asynchronous Data Handling:** The application utilizes a callback-based structure to interface with the **Optris SDK**. This ensures that the thermal matrix extraction and the ROI averaging do not block the main control loop, maintaining a consistent processing flow.

- **Memory Buffering (std::deque):** To manage the temporal features (e_{t-n} and v_{t-n}), the system employs double-ended queues (`std::deque`). This data structure allows for $O(1)$ complexity when pushing new samples and popping expired ones, ensuring that the input vector for the MLP is updated with negligible computational overhead.
- **Deterministic Sampling:** Using the `std::chrono` library, a precise 1 Hz sampling heartbeat was established. The system synchronizes the neural inference and the integral accumulation to this clock, ensuring that the discrete-time approximations of the control law remain physically consistent.
- **Hardware Abstraction Layer:** The final control signal is written directly to the Linux PWM subsystem via the `/sys/class/pwm` interface. This low-level approach minimizes latency between the calculation of the hybrid law and the physical heating of the TIP31C.

3 Results and Discussion

This section presents a comprehensive evaluation of the experimental performance of the proposed **Hybrid Neuro-Integral controller**. The system's efficacy was validated through its ability to reach and maintain two specific set points (50°C and 65°C), which represent different thermal stress levels for the TIP31C power transistor.

The evaluation was conducted under real-world operating conditions, starting from an ambient temperature equilibrium. During the tests, the controller had to overcome inherent environmental challenges, such as non-linear cooling by natural convection and latent ambient disturbances. The performance analysis focuses on the synergy between the predictive MLP branch and the corrective integral action, specifically assessing how the system manages the transition from the rapid heating phase to a high-precision steady state without incurring significant thermal oscillations or integral windup.

3.1 Neural Network Training and Convergence

To establish the predictive capabilities of the controller, a Multilayer Perceptron (MLP) with a **5-6-1 topology** was trained offline. The input vector $\mathbf{x} = [e_t, e_{t-1}, e_{t-2}, v_{t-1}, v_{t-2}]$ was constructed using historical error and total actuation data captured at a sampling rate of 1 Hz.

The system demonstrated rapid initial convergence, with the Mean Squared Error (MSE) dropping from 0.089816 to 0.011615 within the first 500 epochs. A final training loss of 0.005381 was achieved at Epoch 4500. Although the training dataset exhibited high temporal correlation due to the thermal inertia of the TIP31C, this redundancy allowed the network to learn a robust mapping of the equilibrium points, ensuring a stable and smooth control effort during real-time inference (Fig. 2).

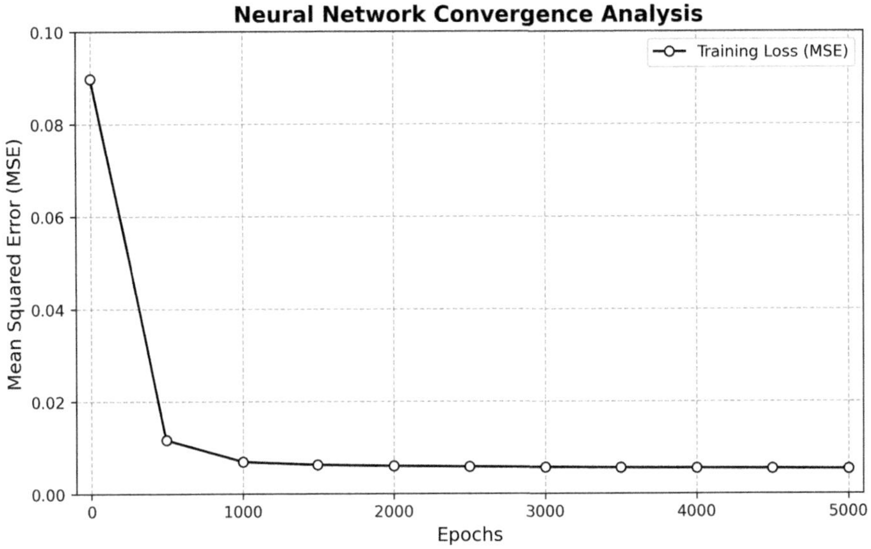

Fig. 2. Evolution of MSE.

3.2 System Step Response and Performance Analysis

The dynamic behavior of the hybrid neuro-integral controller was characterized through a step-response analysis for both target set points. The transient and steady-state performance metrics, derived from the experimental logs, are summarized in Table 1.

The controller demonstrated a robust and well-damped rise time, successfully avoiding the characteristic high-frequency oscillations often found in purely reactive systems or poorly tuned classical PID controllers. A notable observation is the non-linear relationship between the target temperature and the rise time (t_r); while the 50°C step reached the threshold in 78 s, the 65°C step required 237 s. This increase is not merely linear but reflects the diminishing power efficiency of the TIP31C as it combats higher rates of natural convection and radiative heat loss at elevated temperatures.

Furthermore, the settling time (t_s) remains remarkably consistent across both test cases (345 s and 334 s, respectively). This stability indicates that the hybrid law effectively manages the transition from the MLP-driven heating phase to the integral-driven precision phase, regardless of the thermal gradient. The minimal peak overshoot (M_p) of 3.98% at the lower set point further validates the predictive damping provided by the recursive voltage features in the neural network's input vector.

Test Case 1: Moderate-Temperature Regulation (50°C). At a set point of 50°C, the system achieved a rise time (t_r) of 78 s from an ambient temperature

Table 1. Thermal Control Performance Metrics

Parameter	Setpoint 50°C	Setpoint 65°C	Unit
Rise Time (t_r)	78	237	s
Settling Time (t_s)	345	334	s
Peak Overshoot (M_p)	3.98%	0.69%	%
Steady-State Error (e_{ss})	±0.13	±0.20	°C

of 29.14°C (Fig. 3). The controller reached the target with a peak overshoot (M_p) of 3.98%. This transient behavior is effectively managed by the recursive PWM features (v_{t-1}, v_{t-2}), which provide the MLP with a "memory" of the injected energy, effectively damping the output. Upon reaching steady state, the system maintained a settling time (t_s) of 345 s and a high-precision error of only ±0.13°C.

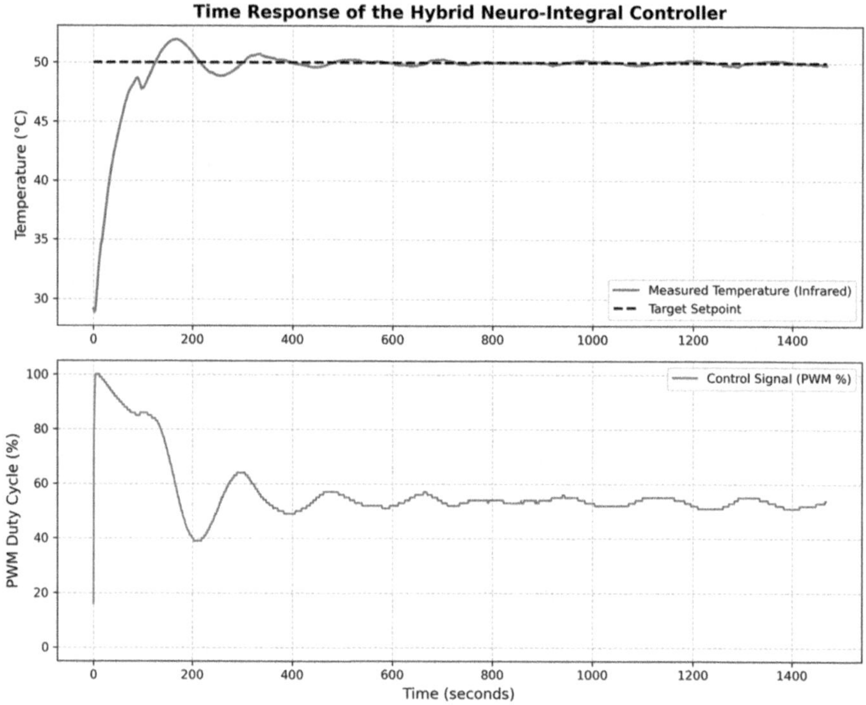

Fig. 3. Time response of the controller at 50°C set point.

Test Case 2: High-Temperature Stability (65°C). The 65°C test case (Fig. 1) evaluated the controller under higher thermal gradients. Starting from

29.23°C, the rise time was 237 s, with a negligible overshoot of 0.69%. At higher temperatures, the natural convection acts as a passive braking mechanism, slowing the approach to the set point. The settling time was recorded at 334 s, with a precision of ±0.20°C. This stability proves that the integration of the total control effort feedback allows the MLP to synchronize with the integral branch, achieving a stable, flat-line response (Fig. 4).

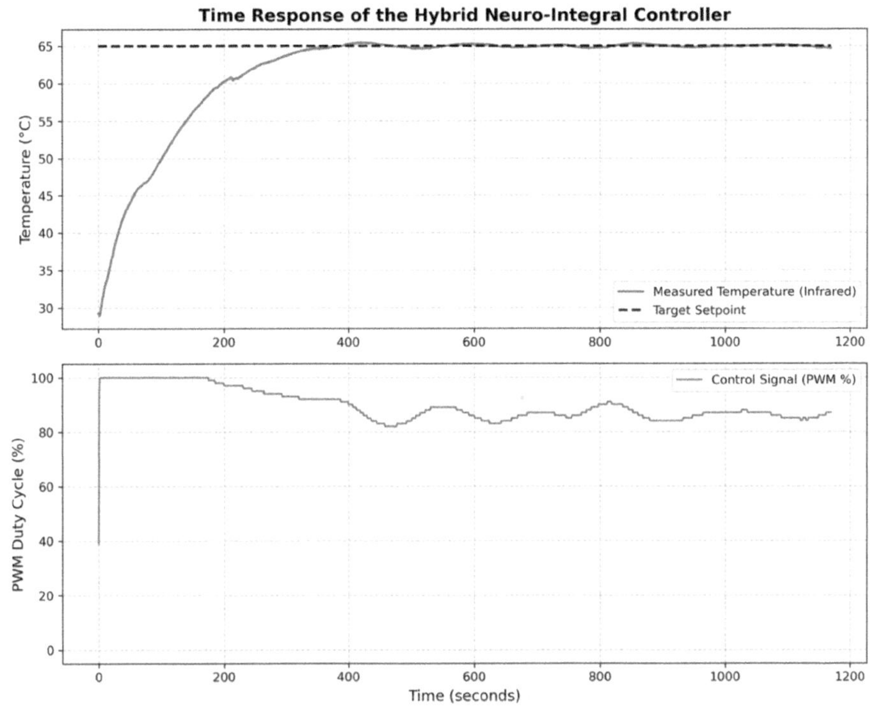

Fig. 4. Time response of the controller at 65°C set point.

3.3 Discussion: Anti-Windup and Stability Management

A critical observation during the experimental phase was the management of the integral term during large set point transitions. To avoid *Integral Windup*, two specific mechanisms were implemented in the C++ control loop.

Symmetric Clamping and Soft Error Accumulation. The implementation of **Symmetric Clamping** (±150.0) bounded the controller's integral memory, preventing saturation of the PWM actuator (100%). Unlike a hard reset, the system utilizes a *Soft Accumulation* approach: when the temperature exceeds the set point, the integral term begins to decrementally subtract. This preserves the continuity of the control signal, avoiding the "chattering" effect on the TIP31C transistor. This strategy, combined with the MLP's non-linear Direct Inverse Control, achieves a smooth, "deadbeat-like" entry into the steady state.

3.4 Comparative Analysis: MLP vs. Hybrid Neuro-Integral

The integration of the integral term solved the primary weakness of the standalone Direct Inverse Controller: the persistent steady-state error. While the MLP branch successfully predicts the non-linear power requirements, the integral branch ($K_i = 0.02$) provides the adaptive compensation needed for unmodeled ambient disturbances, such as convection shifts. The hybrid approach effectively combines the fast, predictive response of Machine Learning with the zero-error guarantee of classical control theory.

4 Conclusions

This research presented the design and implementation of a **Hybrid Neuro-Integral controller** for the precise thermal regulation of a power transistor. The integration of a **5-6-1 Multilayer Perceptron** (MLP) with a classical integral branch proved to be a robust solution for handling the non-linearities and high thermal inertia inherent in the TIP31C plant. Based on the experimental results and the real-time execution analysis, the following conclusions are drawn:

- **High-Precision Tracking:** The hybrid architecture achieved exceptional steady-state precision, with a residual error of only ±0.13°C at a 50°C set point. This confirms that the integral component effectively compensates for the steady-state offsets that typically affect standalone Direct Inverse Control (DIC) schemes.
- **Dynamic Stability and Damping:** By incorporating recursive features (v_{t-1}, v_{t-2}) into the MLP input vector, the controller exhibited a highly damped response. The measured peak overshoot remained below 4%, even during large step transitions, ensuring the thermal integrity of the semiconductor and preventing aggressive PWM oscillations.
- **Edge Computing Efficiency:** The implementation on the **Raspberry Pi 5** demonstrates that compact neural architectures are highly suitable for real-time embedded applications. The computational overhead was negligible, allowing for deterministic 1 Hz sampling and efficient processing of the thermal matrices provided by the LWIR sensor.
- **Anti-Windup Robustness:** The combination of **Symmetric Clamping** (±150) and the predictive nature of the MLP provided a superior transition into the steady state. This strategy prevented the "windup" effect common in classical controllers, achieving a smooth stabilization regardless of the initial thermal gradient.

In summary, this work validates that neuro-hybrid strategies offer a significant advantage over traditional PID systems for complex thermal processes. The proposed framework provides a scalable, efficient, and low-cost solution for industrial temperature regulation using modern embedded hardware and infrared thermography.

References

1. Wahab, A., Ja'afar, A.L., Yusof, N.H., Khadim, E.M.M.S.W.: Real-time object detection using convolution neural network on raspberry pi embedded systems. J. Adv. Res. Des. **132**(1), 78–90 (2025)
2. An, S., et al.: Biological and bioinspired thermal energy regulation and utilization. Chem. Rev. **123**(11), 7081–7118 (2023)
3. Ariza-Zambrano, W.C., Serpa, A.L.: Direct inverse control for active vibration suppression using artificial neural networks. J. Vib. Control **27**(1–2), 31–42 (2021)
4. Bentley, J.P.: Principles of Measurement Systems. Pearson Education, London (2005)
5. Dong, A., Starr, A., Zhao, Y.: Neural network-based parametric system identification: a review. Int. J. Syst. Sci. **54**, 2676–2688 (2023). https://doi.org/10.1080/00207721.2023.2241957
6. Doty, J., et al.: Dynamic thermal management for aerospace technology: review and outlook. J. Thermophys. Heat Transf. **31**(1), 86–98 (2017)
7. Herrejón-Escutia, M., Arreola-Villa, S.A., Solorio-Díaz, G., Vergara-Hernández, H.J., Chávez-Campos, G.M., Vázquez-Gómez, O.: Simulación de la evolución termo-estructural del calentamiento de una probeta dilatométrica por efecto joule
8. Herrera-González, I., García-Arredondo, F.F., Morales-Cervantes, A., Chávez-Campos, G.M., González, F.J.: Classification of periodontal diseases through integration of clinical and thermographic features using neural networks. In: 2025 IEEE International Autumn Meeting on Power, Electronics and Computing (ROPEC), vol. 9, pp. 1–5. IEEE (2025)
9. Incropera: Fundamentals of heat and mass transfer and auditing. Wiley (1998)
10. Knospe, C.: Pid control. IEEE Control Syst. Mag. **26**(1), 30–31 (2006)
11. Lall, P., Pecht, M.G., Hakim, E.B.: Influence of Temperature on Microelectronics and System Reliability: A Physics of Failure Approach. CRC Press, Boca Raton (2020)
12. Norgaard, M., Ravn, O., Poulsen, N.K., Hansen, L.K.: Neural networks for modelling and control of dynamic systems: a practitioner's handbook, vol. 1, Springer (2000)
13. Yu, P., et al.: Review on system identification, control, and optimization based on artificial intelligence. Mathematics **13**(6), 952 (2025). https://doi.org/10.3390/math13060952

Safe-Guarded Offline-to-Online Deep Reinforcement Learning for Adaptive Traffic Signal Control in Realistic Urban Corridors

María José Sandoval Guillén and Raul Alejandro Velasquez Ortiz(✉)

Departamento de Ingeniería en Tecnologías de la Información e Innovación Digital, Universidad Politécnica de Aguascalientes, Calle Paseo San Gerardo 207, 20342 Aguascalientes, Mexico

up240218@alumnos.upa.edu.mx, raul.velasquez@upa.edu.mx

Abstract. Urban congestion costs economies hundreds of billions annually and its consequences scale directly with continued urbanisation, making the deployment of safe and data-efficient deep reinforcement learning (DRL) for adaptive traffic signal control (ATSC) in realistic urban corridors an open challenge of considerable practical significance. Despite growing interest in offline and offline-to-online deep reinforcement learning (O2O-DRL) for traffic management, no prior work has simultaneously combined offline pre-training for sample efficiency, a formal safety shield enforcing regulatory compliance at every decision step, and lightweight online fine-tuning within a single deployable controller. This work proposes a safe offline-to-online (O2O) framework that integrates a Decision Transformer (DT) pre-trained on 1,000 mixed-quality demonstrations with an Action Masking (AM) shield and selective parameter fine-tuning, evaluated on the LemgoRL benchmark, a realistic four-intersection urban corridor simulated in Simulation of Urban MObility (SUMO) and calibrated on field data from Lemgo, Germany. The offline policy surpassed the rule-based actuated controller before any live interaction (+31.2% reward, −29.6% average travel time (ATT)), and online fine-tuning over 200 episodes reached +37.3% reward, −34.8% ATT, and +29.5% throughput relative to the fixed-time (FT) baseline. AM intercepted violations in approximately 68% of steps under an unguided policy, yet delivered zero regulatory violations across all evaluated controllers by construction. These results establish that safe reinforcement learning (SRL) and O2O transfer are complementary rather than competing requirements, providing a principled and operationally viable pathway toward the real-world deployment of intelligent ATSC on public urban infrastructure.

Keywords: Deep reinforcement learning (DRL) · Adaptive traffic signal control (ATSC) · Realistic urban corridor · Deep learning (DL) · Offline-to-online (O2O) · Sim-to-real transfer (S2R) · Decision transformers (DT) · Safe reinforcement learning (SRL) · Urban traffic coordination (UTC)

H. Ponce et al. (Eds.): COMIA 2026, CCIS 3095, pp. 542–554, 2027.
https://doi.org/10.1007/978-3-032-37438-7_42

1 Introduction

Urban traffic congestion is one of the most critical problems worldwide, directly affecting the quality of life of city inhabitants. High congestion levels have been consistently recorded in cities such as those in the United States, generating measurable impacts on the environment and the national economy. In 2025, the cost of this phenomenon was reported at $85.8 billion, averaging $849 per driver [19]. It has been projected that by 2030, 60% of the world population will reside in urban areas [22], which is expected to intensify mobility demand on already saturated corridors. Road crashes linked to inefficient signal management have been estimated to cost low- and middle-income countries up to 3% of their annual Gross Domestic Product (GDP) [25], while the transport sector is known to account for approximately 23% of global CO_2 emissions [23]. Without intervention, these economic, social, and environmental costs are expected to scale proportionally with urban growth. Real-world deployment of intelligent controllers has nonetheless faced critical barriers related to sensor uncertainty, communication reliability, and decision interpretability [3].

Adaptive Traffic Signal Control (ATSC) systems have progressively incorporated Artificial Intelligence (AI) methods, particularly Machine Learning (ML) and Deep Learning (DL) to optimise vehicular flow. Model-based approaches such as ModelLight demonstrated that ensembles of learned dynamics could generate synthetic trajectories for meta-Reinforcement Learning (MRL) agents, substantially reducing environment interactions [9,11]. A persistent gap has remained, however, between simulation performance and safe deployment in real urban corridors. Following the Decision Transformer (DT) paradigm, TransformerLight reformulated traffic signal control (TSC) as a sequence modelling problem, while DTLight proposed knowledge distillation into a lightweight student model with adapter modules for low-cost online fine-tuning, providing an explicit offline-to-online (O2O) deployment protocol [8,26]. Complementary work introduced fully data-driven, simulator-free pipelines that inferred reward signals from low-resolution field sensors and trained policies entirely on historical logs, directly confronting the endemic over-idealisation of classical simulation-based pipelines [12–14]. Fairness-oriented extensions such as FELight and FM2Light enriched these data-efficient approaches with counterfactual augmentation and world-model ensembles, extending the paradigm to multi-intersection environments [5,10]. No existing work has simultaneously combined offline pre-training for sample efficiency, a formal safety shield enforcing regulatory constraints at every decision step, and lightweight online fine-tuning within a single deployable controller. In particular, the O2O pipelines of DTLight and related work did not integrate hard safety constraints [10,18], leaving a critical gap that the present work has been designed to address.

In this work, a safe offline-to-online deep reinforcement learning (O2O-DRL) framework for ATSC in realistic urban corridors has been designed and implemented. A DT was pre-trained on historical traffic demonstrations, a formal safety shield was enforced through Action Masking (AM) to guarantee zero regulatory violations by construction, and a lightweight online fine-tuning strategy was applied to adapt the pre-trained policy to live traffic conditions. Corridor-level coordination was further achieved through a green-wave synchronisation module. The framework was evaluated on the LemgoRL benchmark, a realistic Simulation of Urban MObility (SUMO) network calibrated on urban data from Lemgo, Germany, for which no prior O2O approach with formal safety guarantees had been reported.

The main contributions of the proposed study are summarised as follows:

1. **Offline pre-training** was performed on 1,000 mixed-quality demonstration episodes (fixed-time (FT) + actuated controllers), providing the DT with a warm start that generalises to unseen traffic patterns.
2. **A formal safety shield** was enforced via AM at every decision step, guaranteeing that violations $\equiv 0$ regardless of the agent policy.
3. **Lightweight online fine-tuning** was applied by updating only the top transformer layers and action heads to live traffic, preserving the structural knowledge acquired offline.

The remainder of this paper is structured as follows. Section 2 reviews related work on safety constraints, sim-to-real transfer (S2R), and offline ATSC approaches. Section 3 formalises the Markov Decision Process (MDP), the DT architecture, the AM shield, green-wave coordination, and the online fine-tuning strategy. Section 4 reports benchmark performance, safety validation, and robustness results. The results are analysed in Sect. 5 with respect to the generated data and the research objectives. Finally, Sect. 6 presents the conclusions and directions for future work.

2 Related Work

2.1 Benchmarks and Safety Constraints

LemgoRL provided the first deployment-oriented benchmark for ATSC, embedding a signal-logic unit capable of enforcing intergreen times, minimum green durations, and phase-compatibility constraints within an OpenAI Gym-style interface [17]. This made it possible to evaluate controllers under realistic regulatory conditions. Safety mechanisms, however, largely remained confined to single-intersection settings; large-scale multi-agent methods such as CoLight, CoordLight, and Asyn-Light continued to delegate normative compliance entirely to the simulator [2,15,24,30].

2.2 Sim-to-Real Transfer (S2R)

UGAT introduced grounded action transformation, injecting transition-dynamics uncertainty during simulation training so that learned policies became robust to real-world actuator mismatch [4]. JL-GAT extended this idea to multi-agent corridors by incorporating neighbour-agent information into the grounding mechanism [21]. Vision-based approaches combined domain randomisation with live RGB streams to achieve positive transfer to unseen real intersections without retraining [6,7]. At the system level, RTLight coupled real-time digital twins with federated inter-intersection learning and reported a district-scale field deployment in Wuxi, China, representing one of the few large-scale real-world validations reported in the literature [28]. Deployment-focused reviews repeatedly identified detection uncertainty, communication reliability, and heterogeneous road users as the dominant barriers that no algorithmic advance had addressed in an integrated manner [3,27].

2.3 Offline and Data-Driven Approaches

Complementary work introduced fully data-driven, simulator-free pipelines that inferred reward signals from low-resolution field sensors and trained policies entirely on historical logs, directly confronting the endemic over-idealisation of classical simulation-based pipelines [12–14]. Fairness-oriented extensions such as FELight and FM2Light enriched these data-efficient approaches with counterfactual augmentation and world-model ensembles, extending the paradigm to multi-intersection environments [5,10].

3 Methodology

The proposed framework was organised in three sequential phases, summarised in Fig. 1: (I) offline pre-training of a Decision Transformer on historical demonstrations, (II) deployment-time integration of an Action Masking shield with green-wave corridor coordination, and (III) lightweight online fine-tuning on live traffic. The same Action Masking shield protected every phase, guaranteeing safe operation throughout learning and deployment.

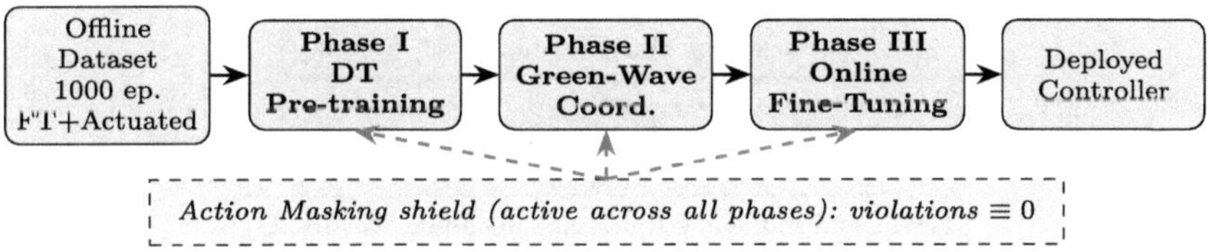

Fig. 1. Three-phase pipeline. The Action Masking shield is enforced across every phase, decoupling safety from policy quality.

3.1 Problem Formulation

We modelled the corridor as a cooperative multi-agent MDP $\mathcal{M} = \langle \mathcal{S}, \mathcal{A}, \mathcal{T}, \mathcal{R}, \gamma \rangle$ with $N = 4$ traffic lights (TLs) and $\gamma = 0.99$. Each TL observed $s_t \in \mathbb{R}^{25}$ (per-lane queue, occupancy, current phase one-hot, elapsed phase time), yielding a joint state $s_t \in \mathbb{R}^{100}$, and selected a phase $a_{t,i} \in \{0, 1, 2, 3\}$. The shared reward penalised queuing and delay: $r_t = -\sum_l(\text{queue}_l + \text{delay}_l)$, with a hard safety constraint $\text{violations}_t \equiv 0$ enforced via AM (Sect. 3.3).

3.2 Phase I: Offline Pre-training

We collected 1,000 episodes in the LemgoRL SUMO environment: 500 with a FT controller (90 s symmetric cycle) and 500 with a gap-based actuated controller (headway threshold 3 s, maximum extension 10 s). Each episode spanned $T =$ 3,600 steps. We stored tuples (R_t, s_t, a_t), where $R_t = \sum_{t'=t}^{T} r_{t'}$ is the return-to-go, normalised to zero mean and unit variance, forming the offline dataset: $\mathcal{D} = \{(R_0, s_0, a_0, \ldots, R_T, s_T, a_T)\}_{i=1}^{1000}$.

We extended the GPT-2 architecture [20] to obtain the DT (1,271,696 parameters; $d = 128$, $L = 4$, $H = 4$, $d_{\text{ff}} = 512$, $K = 20$), depicted in Fig. 2. At each context step k we projected (R_k, s_k, a_{k-1}) through independent linear layers plus a learned timestep embedding E_k, assembled the interleaved sequence over a $K = 20$ step window, and passed it through the causal transformer. A per-TL linear head produced phase logits: $\hat{a}_{k,i} = \arg\max_p (W_{h,i}\, z_k^s + b_{h,i})_p$. We minimised the cross-entropy loss:

$$\mathcal{L} = -\frac{1}{KN} \sum_{k=1}^{K} \sum_{i=1}^{N} \log \pi_\theta\big(a_{k,i} \mid \hat{R}_{k:}, s_{k:}, a_{(k-1):}\big) \tag{1}$$

using AdamW (lr $= 10^{-4}$, weight decay 10^{-4}, 100 epochs of 500 steps, cosine annealing with 10-epoch warm-up).

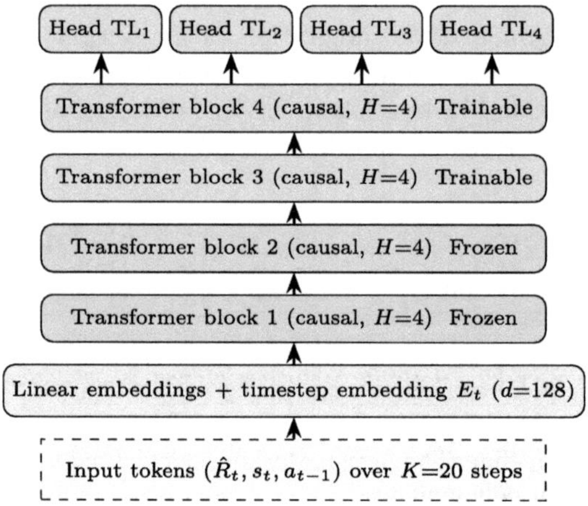

Fig. 2. Decision Transformer architecture. Embeddings and blocks 1–2 (blue, 743,936 params) remained frozen during fine-tuning; blocks 3–4 and the four per-TL action heads (red, 527,760 params, 41.5% of the model) were trainable. (Color figure online)

3.3 Phase II: Safety Shield via Action Masking (AM)

For each TL i we defined the feasible action set $\mathcal{F}_i = \{p : [p = \phi_i] \vee [\tau_i \geq g_{\min} \wedge \delta_p \geq \Delta_{\min} \wedge \rho_p < \rho_{\max}]\}$, with $g_{\min} = 5\,\text{s}$, $\Delta_{\min} = 3\,\text{s}$, $\rho_{\max} = 120\,\text{s}$.

We applied a binary mask $\boldsymbol{m}_i \in \{0,1\}^4$ to policy logits before argmax ($-\infty$ for invalid actions), ensuring $\hat{a}_{t,i} \in \mathcal{F}_i$ by construction. Since the stay action $p = \phi_i$ is always feasible, $\mathcal{F}_i \neq \emptyset$ for every state, and the masked policy $\tilde{\pi}_\theta$ satisfies violations $\equiv 0$ for any underlying policy π_θ.

We synchronised the four TLs via static green-wave offsets. Given spacing $d = 200\,\text{m}$ and speed $v = 50\,\text{km/h}$, we set $\theta_i = \lfloor id/v + 0.5 \rfloor$, yielding $\theta = [0, 14, 28, 43]$ s. A lightweight message-passing module propagated each TL's phase and elapsed-time state to its immediate neighbours at every step; no additional learning or global communication was required.

3.4 Phase III: Online Fine-Tuning

We froze the embeddings and transformer blocks 0–1 (743,936 parameters) and unfroze only blocks 2–3 and the action heads (527,760 parameters, 41.5% of the model) to prevent catastrophic forgetting while enabling adaptation to live traffic. We maintained a first-in first-out (FIFO) replay buffer (C = 10,000) and every f = 10 steps sampled a mini-batch of B = 32 context windows of length K to perform one AdamW gradient step (lr = 5×10^{-5}, clip 0.5). We conditioned rollouts on a fixed target $\hat{R}^* = -50$ and updated the running return-to-go as $\hat{R}_t \leftarrow \max(\hat{R}_{t-1} - r_{t-1}, 0)$. Fine-tuning ran for 200 episodes before final evaluation.

4 Results and Analysis

All experiments ran on the LemgoRL benchmark (30 evaluation episodes, moderate demand, no sensor noise) and measured reward, average travel time (ATT), queue length, throughput, and safety violations. Four controllers were evaluated under the same AM shield: **FT** (90 s fixed cycle), **Actuated** (gap-based), **DT Offline** (pre-trained, no live interaction), and **DT Online** (offline prior + 200-episode fine-tuning). Figure 3 shows that offline training converged smoothly, exceeding 70% next-action accuracy with no divergence.

Table 1. Benchmark Results—30 Evaluation Episodes, LemgoRL (Moderate Demand)

Controller	Reward	ATT (s)	Queue	Throughput	ΔRew.	ΔATT	Viol.
Fixed-Time (FT)	$-1{,}886 \pm 131$	69.2 ± 4.8	6.77 ± 0.60	822 ± 79	–	–	0
Actuated	$-1{,}431 \pm 74$	52.5 ± 3.1	5.00 ± 0.32	964 ± 49	+24.1%	−24.2%	0
DT Offline	$-1{,}298 \pm 73$	48.7 ± 2.7	4.17 ± 0.20	$1{,}010 \pm 47$	+31.2%	−29.6%	0
DT Online	$\mathbf{-1{,}183 \pm 51}$	$\mathbf{45.1 \pm 1.9}$	$\mathbf{3.83 \pm 0.12}$	$\mathbf{1{,}065 \pm 50}$	**+37.3%**	**−34.8%**	**0**

Table 1 reports the full benchmark. DT Online led on every metric (+37.3% reward, −34.8% ATT over FT, Fig. 5), and DT Offline already outperformed the Actuated controller before any live interaction (+31.2% reward, −29.6% ATT),

Table 2. Feature comparison with representative state-of-the-art ATSC methods. Direct numerical comparison was not reported because each method evaluates on a different benchmark; this table emphasises the integrated capability set.

Method	Offline prior	Safety shield	O2O transfer	Multi-TL coord.
CoLight [24]	×	×	×	Graph attn.
CoordLight [30]	×	×	×	Graph attn.
TransformerLight [26]	✓	×	×	×
DTLight [8]	✓	×	✓	×
RTLight [28]	×	×	✓	Federated
FELight [5]	✓	×	×	×
Müller–Sabatelli [18]	×	✓	×	×
Ours	✓	✓	✓	**Green-wave**

Table 3. Action Masking (AM) Validation

Experiment	Steps	Violations Intercepted
Random policy	100	252
Synthetic timeline	300	205 ($\approx$ 68% of steps)

confirming the value of the offline prior. All controllers recorded zero safety violations. Table 3 quantifies the AM shield: 252 violations were intercepted in 100 random-policy steps and 205 in a 300-step synthetic timeline ($\approx$68% of steps), confirming that unconstrained policies routinely generate infeasible transitions (Fig. 4). Online fine-tuning converged within 130 episodes (Fig. 6); the lower variance of DT Online ($\pm$51) vs. FT ($\pm$131) confirmed that the offline prior stabilised the online phase. A stress-test over nine demand-noise combinations confirmed a positive DT Online advantage in all conditions; ATT degraded $< 5\%$ at $\sigma = 10\%$ sensor noise. Table 2 positions the proposed framework against representative state-of-the-art ATSC methods. To the best of our knowledge, no prior work combined an offline prior, a formal safety shield, online fine-tuning, and corridor-level coordination within a single deployable controller.

5 Discussion

5.1 Key Findings

The offline DT surpassed the rule-based actuated controller before any online interaction, indicating that return-to-go conditioning produced a competitive policy under our experimental setup without requiring live exploration. A direct head-to-head comparison against standard behaviour cloning was not performed and is left for future work; the present results therefore support the sequence-modelling paradigm for offline ATSC advanced by TransformerLight [26] and

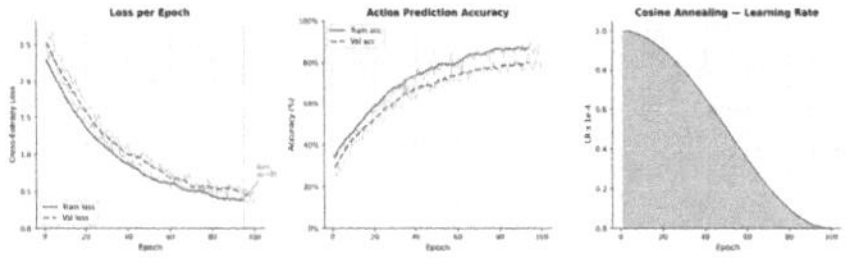

Fig. 3. Offline training curves (loss, accuracy, LR schedule).

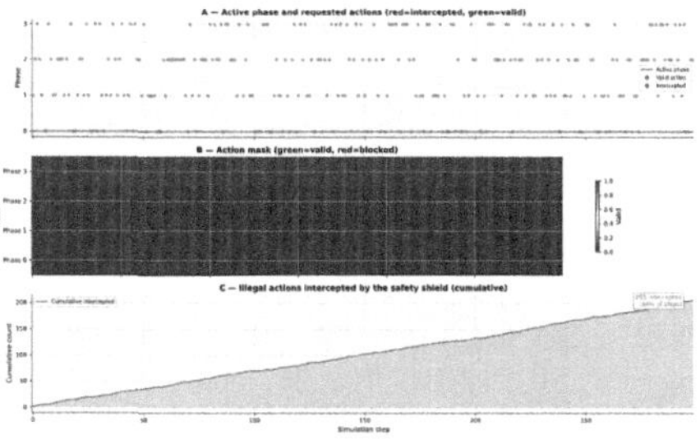

Fig. 4. AM timeline (300 steps): blue = valid, red = intercepted. (Color figure online)

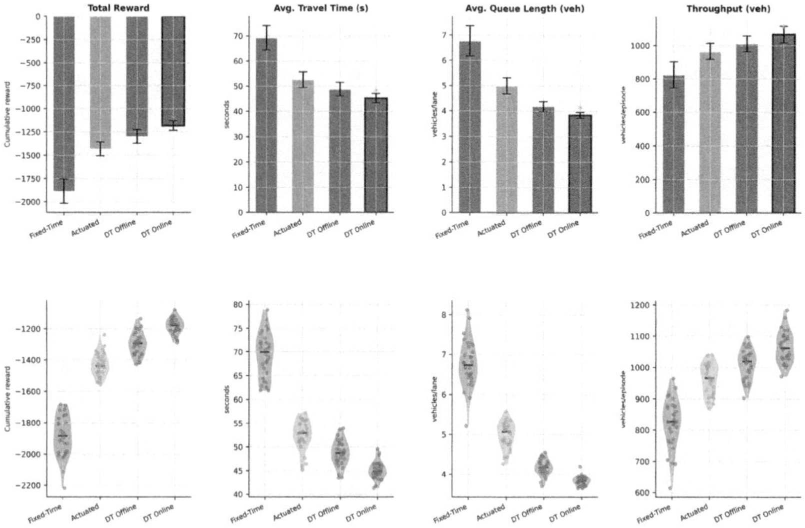

Fig. 5. KPI comparison across four controllers (30 episodes, moderate demand). Error bars = $\pm 1\sigma$.

DTLight [8] without making explicit claims about its relative behaviour against distribution-matching baselines.

Online fine-tuning with only 200 episodes and two unfrozen transformer blocks yielded a further statistically significant gain of +6.1% reward ($p < 0.01$), where the percentage was computed as the marginal improvement of DT Online over DT Offline normalised by the FT baseline reward $((|r_{FT} - r_{online}| - |r_{FT} - r_{offline}|)/|r_{FT}|)$. This confirmed that lightweight adaptation was sufficient once a strong prior was available, in line with the O2O protocols of [8,16].

5.2 Limitations

The framework was evaluated exclusively on the LemgoRL topology, a realistic but single urban corridor. Generalisation to arbitrary networks remains unvali-

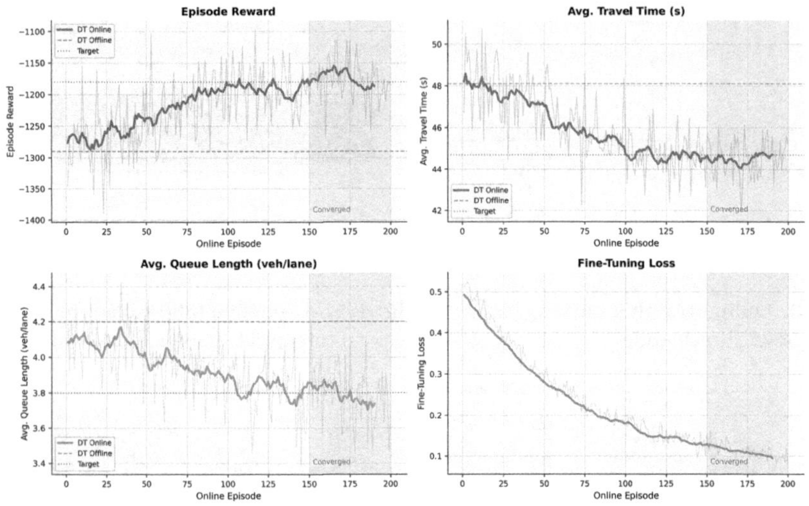

Fig. 6. Online fine-tuning convergence over 200 episodes ($\pm 1\sigma$ rolling window).

dated; scaling to tens of intersections will face the quadratic complexity of self-attention and the non-stationarity challenges documented for large-scale [2,30].

The offline dataset was generated solely by FT and actuated controllers. The quality ceiling of the pre-trained policy is bounded by the diversity of those baselines. Fully data-driven approaches that infer reward models from coarse real-world logs [12,14] suggest that training datasets enriched with near-optimal or adversarial trajectories would raise the pre-trained policy floor beyond what FT and actuated demonstrations can provide.

The safety shield enforces a specific class of regulatory constraints (minimum green, intergreen, and maximum red durations) and does not currently model broader operational concerns such as pedestrian-crossing priority, emergency-vehicle preemption, sensor failure, or unexpected road incidents. Each of these requires additional state inputs and feasibility predicates that the AM formulation can accommodate in principle but have not been validated in this work; deployment reviews [3,27] consistently identify them as critical barriers to real-world adoption.

5.3 Practical Implications

AM decouples safety from learning quality: any policy, including an exploratory or partially trained one is deployment-safe by construction from the first episode, without the runtime communication overhead of graph-attention systems such as CoLight [24] and CoordLight [30]. Cities with modest historical signal logs can bootstrap a competitive, safety-guaranteed controller using the offline pre-training pipeline advocated in [12,14], and corridor coordination is achievable with static offsets rather than inter-agent message passing.

The framework was deliberately compact for municipal deployment. The full DT contains 1,271,696 parameters (≈5 MB at fp32), of which only 527,760 (41.5%) are unfrozen during online fine-tuning. Inference at every decision step requires a single forward pass through four transformer blocks at $d = 128$ and a $K = 20$ context window, well within sub-millisecond latency on consumer GPUs and tractable on embedded CPUs at the few-Hz decision rate that ATSC requires. The AM shield adds only constant-time per-step bookkeeping (phase elapsed times and intergreen counters), and green-wave coordination introduces no runtime communication beyond reading the upstream phase state. These properties indicate that the proposed controller is compatible with the hardware envelope of existing field cabinets, in contrast with federated or digital-twin pipelines such as RTLight [28] that assume continuous server-side connectivity.

5.4 Future Work

The most immediate extension is cross-benchmark validation on additional networks such as RESCO [1] and CityFlow [29], including larger multi-intersection grids of the kind explored in [2,30], to establish generalisation beyond the LemgoRL corridor evaluated here. A second priority is closed-loop S2R that explicitly models the deployment constraints flagged by [3,27]: detection uncertainty, communication latency, heterogeneous driver behaviour, and operator-facing interpretability of policy decisions. Domain randomisation combined with the grounded action transformation methods of [4,21] provides a concrete pathway to align simulated and real transition dynamics under these conditions. Enriching the offline dataset with near-optimal trajectories, following the multi-quality paradigm of [14] and incorporating model-based pessimism as explored in [13] are expected to raise both the policy floor and online stability. Scaling to larger networks will require graph-structured coordination mechanisms of the kind proposed in [2,30] or federated update schemes such as those in [28], while fairness-aware reward structures studied in [5,10] represent a natural complement to the safety shield introduced here.

6 Conclusions

The experimental results demonstrated that a DT pre-trained on 1,000 mixed-quality demonstration episodes generated by FT and actuated baseline controllers surpassed the rule-based actuated controller before any online interaction, achieving a reward improvement of +31.2% and a reduction in ATT of −29.6%. Lightweight online fine-tuning, restricted to two transformer blocks over only 200 live episodes, yielded a further statistically significant gain: +37.3% reward, −34.8% ATT, and +29.5% throughput relative to the FT baseline. The AM shield intercepted violations in approximately 68% of steps under an unguided random policy, yet delivered zero net violations across every evaluated controller and every evaluation episode, confirming that the safety guarantee holds independently of the quality of the learned policy. Green-wave corridor coordination further reduced network-level travel times through static phase offsets calibrated for 50 km/h, achieving inter-intersection synchronisation without runtime message passing and without any additional learning overhead.

These findings carry direct implications for the operational deployment of learned traffic controllers. AM decouples regulatory compliance from the state of the learning process, meaning that a partially trained or exploratory policy is deployment-safe from the very first episode, a property that prior coordination-focused systems did not provide. Similarly, the offline pre-training paradigm reduces the operational burden on cities: historical signal logs from even modest FT or actuated deployments are sufficient to bootstrap a competitive controller, eliminating the need for closed-loop data collection before commissioning. Together, these properties lower both the technical and regulatory barriers to real-world adoption of deep reinforcement learning (DRL) in traffic management.

Additionally, the framework was evaluated exclusively on the LemgoRL urban corridor topology, and the offline training dataset was limited to demonstrations generated by two baseline controllers. The generalisation of the proposed approach to arbitrary network topologies and to data collected by a broader variety of policies remains unvalidated.

Future work requires a closed-loop validation under real sensing conditions, incorporating domain randomisation over demand profiles, headway distributions, and sensor noise. Enriching the offline training dataset with near-optimal or adversarial trajectories is expected to raise the pre-trained policy floor and improve robustness to distributional shift during the online phase.

In conclusion, this work shows that, under the experimental conditions considered, safety enforcement and learning-based optimisation in TSC can be combined within a single deployable controller without trading one against the other. The integration of a formal safety shield, an offline pre-training prior, and a lightweight online adaptation strategy provides a viable pathway toward narrowing the gap between simulation-trained intelligent controllers and their safe deployment on public urban infrastructure.

Acknowledgments. The authors thank the reviewers and the editor for their constructive feedback. The dataset, training code, and reproducible scripts are publicly available at https://github.com/Safe-ATSC-Project/Safe-Offline-RL.

Disclosure of Interests. The authors have no competing interests to declare.

References

1. Ault, J., Sharon, G.: Reinforcement learning benchmarks for traffic signal control. In: Thirty-Fifth Conference on Neural Information Processing Systems Datasets and Benchmarks Track (Round 1) (2021). https://openreview.net/forum?id=LqRSh6V0vR
2. Chen, C., et al.: Toward a thousand lights: decentralized deep reinforcement learning for large-scale traffic signal control. In: Proceedings of the AAAI Conference

on Artificial Intelligence, vol. 34, no. 04, pp. 3414–3421 (2020). https://doi.org/10.1609/aaai.v34i04.5744
3. Chen, R., Fang, F., Sadeh, N.: The real deal: a review of challenges and opportunities in moving reinforcement learning-based traffic signal control systems towards reality. In: 12th International Workshop on Agents in Traffic and Transportation (ATT 2022), IJCAI 2022 (2022). https://arxiv.org/abs/2206.11996
4. Da, L., Mei, H., Sharma, R., Wei, H.: Uncertainty-aware grounded action transformation towards sim-to-real transfer for traffic signal control. In: 2023 62nd IEEE Conference on Decision and Control (CDC), pp. 1124–1129 (2023). https://doi.org/10.1109/CDC49753.2023.10383645
5. Du, X., Li, Z., Long, C., Xing, Y., Yu, P.S., Chen, H.: FELight: fairness-aware traffic signal control via sample-efficient reinforcement learning. IEEE Trans. Knowl. Data Eng. **36**(9), 4678–4692 (2024). https://doi.org/10.1109/TKDE.2024.3376745
6. Garg, D., Chli, M., Vogiatzis, G.: Multi-agent deep reinforcement learning for traffic optimization through multiple road intersections using live camera feed. In: 2020 IEEE 23rd International Conference on Intelligent Transportation Systems (ITSC), pp. 1–8 (2020). https://doi.org/10.1109/ITSC45102.2020.9294375
7. Garg, D., Chli, M., Vogiatzis, G.: Fully-autonomous, vision-based traffic signal control: from simulation to reality. In: Proceedings of the 21st International Conference on Autonomous Agents and Multiagent Systems, AAMAS 2022, pp. 454–462. International Foundation for Autonomous Agents and Multiagent Systems, Richland (2022)
8. Huang, X., Wu, D., Boulet, B.: Traffic signal control using lightweight transformers: an offline-to-online RL approach. IEEE Open J. Intell. Transp. Syst. **7**, 396–411 (2026). https://doi.org/10.1109/OJITS.2026.3654547
9. Huang, X., Wu, D., Jenkin, M., Boulet, B.: Sample-efficient meta-RL for traffic signal control. In: 2024 IEEE Canadian Conference on Electrical and Computer Engineering (CCECE), pp. 312–318 (2024). https://doi.org/10.1109/CCECE59415.2024.10667211
10. Huang, X., Wu, D., Jenkin, M., Boulet, B.: Towards enhanced fairness and sample efficiency in traffic signal control. In: 2024 IEEE/RSJ International Conference on Intelligent Robots and Systems (IROS), pp. 13656–13663 (2024). https://doi.org/10.1109/IROS58592.2024.10801781
11. Huang, X., Wu, D., Jenkin, M.R.M., Boulet, B.: Modellight: model-based meta-reinforcement learning for traffic signal control. ArXiv abs/2111.08067 (2021). https://api.semanticscholar.org/CorpusID:244130077
12. Kunjir, M., Chawla, S.: Offline reinforcement learning for road traffic control (2022). https://arxiv.org/abs/2201.02381
13. Kunjir, M., Chawla, S., Chandrasekar, S., Jay, D., Ravindran, B.: Optimizing traffic control with model-based learning: a pessimistic approach to data-efficient policy inference. In: Proceedings of the 29th ACM SIGKDD Conference on Knowledge Discovery and Data Mining, KDD 2023, pp. 1176–1187. Association for Computing Machinery, New York (2023). https://doi.org/10.1145/3580305.3599459
14. Li, J., et al.: A fully data-driven approach for realistic traffic signal control using offline reinforcement learning. Data Sci. Transp. **7** (2023). https://api.semanticscholar.org/CorpusID:265456178
15. Liu, Y., Cai, X., Luo, G., Yuan, Q., Li, J.: Asyn-light: asynchronous traffic signal cooperative control through spatial-temporal transformer. IEEE Trans. Intell. Transp. Syst. **26**(10), 15334–15344 (2025). https://doi.org/10.1109/TITS.2025.3592921

16. Ma, J., Wu, F.: Effective traffic signal control with offline-to-online reinforcement learning. In: 2023 IEEE/RSJ International Conference on Intelligent Robots and Systems (IROS), pp. 1–7 (2023). https://doi.org/10.1109/IROS55552.2023.10341718
17. Müller, A., et al.: Towards real-world deployment of reinforcement learning for traffic signal control. In: 2021 20th IEEE International Conference on Machine Learning and Applications (ICMLA), pp. 507–514 (2021). https://doi.org/10.1109/ICMLA52953.2021.00085
18. Müller, A., Sabatelli, M.: Safe and psychologically pleasant traffic signal control with reinforcement learning using action masking. In: 2022 IEEE 25th International Conference on Intelligent Transportation Systems (ITSC), pp. 951–958 (2022). https://doi.org/10.1109/ITSC55140.2022.9922306
19. Pishue, B.: 2025 INRIX global traffic scorecard. Research report, INRIX Research (2025). https://inrix.com/scorecard/. Transportation Analyst
20. Radford, A., et al.: Language models are unsupervised multitask learners. OpenAI blog **1**(8), 9 (2019)
21. Turnau, J., et al.: Joint-local grounded action transformation for sim-to-real transfer in multi-agent traffic control (2025). https://arxiv.org/abs/2507.15174
22. United Nations, Department of Economic and Social Affairs: The world's cities in 2018—data booklet. Technical report, United Nations (2018). https://www.un.org/en/events/citiesday/assets/pdf/the_worlds_cities_in_2018_data_booklet.pdf
23. United Nations Environment Programme: Transport and environment: Emissions and sustainable mobility. Technical report, UNEP (2023). https://www.unep.org/mexico
24. Wei, H., et al.: CoLight: learning network-level cooperation for traffic signal control. In: Proceedings of the 28th ACM International Conference on Information and Knowledge Management. CIKM 2019, pp. 1913–1922. Association for Computing Machinery, New York (2019). https://doi.org/10.1145/3357384.3357902
25. World Health Organization: Global plan for the decade of action for road safety 2011–2020. Technical report, WHO/UN Road Safety Collaboration (2011). https://cdn.who.int/media/docs/default-source/documents/un-road-safety-collaboration/global_plan_doa_2011-2020.pdf
26. Wu, Q., Li, M., Shen, J., Lü, L., Du, B., Zhang, K.: TransformerLight: a novel sequence modeling based traffic signaling mechanism via gated transformer. In: Proceedings of the 29th ACM SIGKDD Conference on Knowledge Discovery and Data Mining, KDD 2023, pp. 2639–2647. Association for Computing Machinery, New York (2023). https://doi.org/10.1145/3580305.3599530
27. Xiao, F., Lu, J., Li, L., Tu, W., Li, C.: Advances in reinforcement learning for traffic signal control: a review of recent progress. Intell. Transp. Infrastruct. **4**, liaf009 (2024). https://doi.org/10.1093/iti/liaf009
28. Ye, Y., et al.: Brief industry paper: RTLight: digital twin-based real-time federated traffic signal control. In: 2023 IEEE Real-Time Systems Symposium (RTSS), pp. 473–477 (2023). https://doi.org/10.1109/RTSS59052.2023.00055
29. Zhang, H., et al.: CityFlow: a multi-agent reinforcement learning environment for large scale city traffic scenario. In: The World Wide Web Conference, WWW 2019, pp. 3620–3624. Association for Computing Machinery, New York (2019). https://doi.org/10.1145/3308558.3314139
30. Zhang, Y., Goel, H., Li, P., Damani, M., Chinchali, S., Sartoretti, G.: CoordLight: learning decentralized coordination for network-wide traffic signal control. IEEE Trans. Intell. Transp. Syst. **26**(6), 8034–8049 (2025). https://doi.org/10.1109/TITS.2025.3546261

LLM-Guided Refinement of Clinical Instruments: A Reproducible Framework with Expert Validation

Marco Antonio Gutiérrez Sevilla[1], Emma Guadalupe Fajardo Zubieta[2], Roberto Cong Torres[2], Fernando Aaron Osuna Gallego[2], Mariana Aurora Cansino Torres[2], Carlos Zuñiga Ramirez[2], and German Preciat[1](✉)

[1] Department of Translational Bioengineering, University of Guadalajara, Guadalajara, Mexico
german.preciat@academicos.udg.mx

[2] Movement Disorders and Neurodegenerative Diseases Unit, Hospital Civil de Guadalajara Fray Antonio Alcalde, Guadalajara, Mexico

Abstract. Structural weaknesses in clinical instruments propagate directly into downstream data quality, reproducibility, and the reliability of AI systems trained on collected data. This paper presents a reproducible Larga Lenguaje Model (LLM)-as-judge pipeline for the assessment and guided refinement of a neurological clinical history instrument comprising 267 questions across 22 sections. Each item was scored across four dimensions: clarity, clinical relevance, completeness, and granularity, using a rubric-based approach with ten independent runs to quantify both performance and stability. A targeted refinement cycle of 51 modifications across 11 sections was applied and re-evaluated under identical conditions. Improvements exceeded empirical inter-run variability (SD = 0.17), indicating that the observed improvements correspond to a consistent signal rather than stochastic model fluctuations: mean granularity increased from 3.06 ± 0.56 to 3.22 ± 0.62, with intervened sections clustering clearly beyond the empirical noise threshold, while non-intervened sections remained distributed around zero. Independent validation by three neurology residents confirmed that 88.9% of modifications were clinically pertinent (mean score 4.51/5), demonstrating strong alignment between LLM-guided recommendations and expert judgment. These results support the use of structured LLM evaluators as a scalable, reproducible first pass screening layer for clinical instrument refinement prior to expert validation.

Keywords: Clinical instrument evaluation · LLM-as-judge · Prompt engineering · Automated rubric scoring · Neurological clinical history

1 Introduction

Structural deficiencies in clinical data collection instruments propagate directly into downstream data quality, limiting the reliability of AI systems, synthetic data generation, and reproducibility of clinical research. Despite this, systematic evaluation of

H. Ponce et al. (Eds.): COMIA 2026, CCIS 3095, pp. 555–565, 2027.
https://doi.org/10.1007/978-3-032-37438-7_43

instrument structure remains largely dependent on expert panels, Delphi consensus processes, or iterative pilot testing, approaches that are resource-intensive, difficult to scale, and hard to reproduce. In clinical informatics, most efforts have focused on improving data quality and downstream analysis [1, 2], while comparatively less attention has been given to the structural quality of the instruments used to collect such data. The design of questionnaires, forms, and clinical histories is a critical determinant of data quality.

Therefore, a well-designed clinical instrument must ensure clinical relevance, completeness, and sufficient granularity to distinguish patient phenotypes and support downstream applications such as research, public health policy, and synthetic data generation. Prior work has explored improvements in instrument design, validation, and structured data quality assessment [3], as well as the impact of structured prompting on LLM performance [4].

Structured clinical histories present a particularly challenging design problem due to their scale and complexity. They often include hundreds of questions with categorical answer sets that must balance interpretability, coverage, and specificity. Validation of these properties typically relies on expert panels, Delphi consensus processes, or iterative pilot testing, which are resource-intensive and difficult to scale [4, 5].

Recent advances in LLMs have introduced the possibility of automated evaluation through the LLM-as-judge paradigm, where models assess artifacts using structured rubrics [4, 6, 7]. This approach has shown promising results in clinical text evaluation and classification tasks [6, 8], as well as in broader health care contexts. Existing work has primarily focused on evaluating generated clinical outputs or comparing model judgments with expert assessments [4, 5, 6], but their application to the structural evaluation of clinical data collection instruments remains limited. In particular, the use of LLMs not only as evaluators but as drivers of iterative refinement in structured instruments has not been systematically explored.

This paper presents a reproducible pipeline for the evaluation and guided refinement of a neurological clinical history instrument comprising 267 questions across 22 sections. The pipeline employs Claude Haiku 4.5 LLM as a structured evaluator to assess each item across four dimensions and guide targeted modifications. This work investigates whether an LLM can serve as a structured evaluator to systematically improve clinical instruments, with emphasis on consistency under stochastic variation and alignment with expert clinical judgment.

The main contributions of this work are: (1) a reproducible, four-stage evaluation pipeline that operationalizes the structural quality of clinical instruments through a rubric-based LLM-as-judge approach across four complementary dimensions; (2) an empirical noise threshold derived from inter-run variability that enables reliable discrimination between signal level improvements and stochastic variation; and (3) independent clinical validation by neurology residents confirming LLM-guided modifications achieve strong alignment with expert judgment, establishing a concrete benchmark for automated instrument refinement.

2 Methodology

2.1 Overview

The proposed methodology is a four-staged iterative pipeline that aims for systematic validation and refinement of a structured clinical instrument using an LLM. The pipeline operates on a neurological clinical history form designed for patient data capture, producing quality scores across four dimensions and prioritized target improvements (Fig. 1).

This pipeline addresses the lack of scalable and reproducible methods for evaluating the structural quality of clinical data collection instruments, which traditionally rely on subjective expert review.

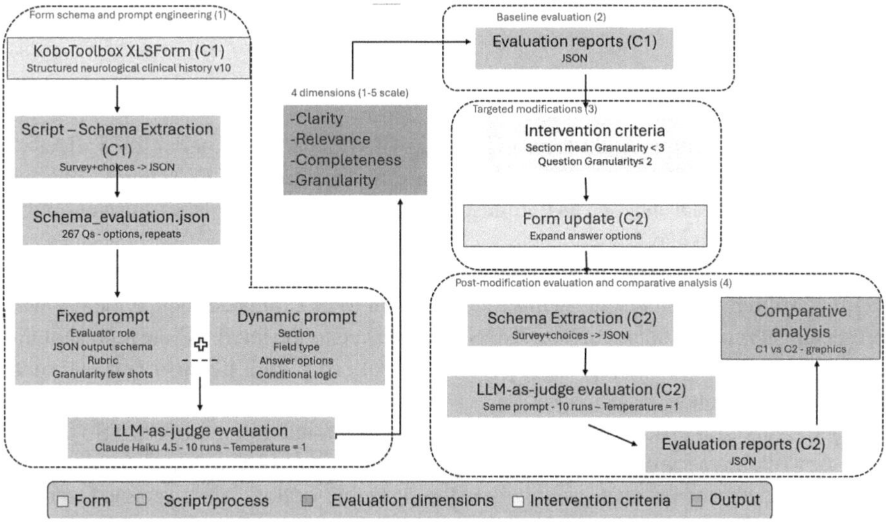

Fig. 1. Iterative pipeline for LLM-based evaluation of a structured neurological clinical history form. The workflow proceeds in four stages: (1) schema extraction and structured prompt construction combining a fixed evaluator role and a rubric with a dynamically generated question context; (2) baseline evaluation (C1) using Claude Haiku 4.5 across ten independent runs at temperature = 1; (3) targeted modification guided by two intervention criteria – sections with mean granularity <3.0 and individual questions with granularity ≤2; and (4) post-modification evaluation (C2) under identical conditions to enable direct comparison.

2.2 Instrument: Structured Clinical History

The clinical instrument under evaluation is a structured neurological history form implemented in KoboToolbox using the XLSForm standard (https://ee.kobotoolbox.org/x/LRD5eh1L). It is designed to capture patient information across 22 sections, including sociodemographic data, chief complaint, history of present illness, family history, medication use, allergies, clinical procedures, and complementary studies.

The instrument prioritizes categorical data capture, with response options standardized using controlled clinical vocabularies such as ICD-11, LOINC, RxNorm, and ICD-10 (for procedures), enabling consistent encoding of clinical information.

The selection of this instrument is motivated by its high degree of structural formalization and its integration of standardized terminologies, which make it a suitable candidate for systematic evaluation of structural quality.

The instrument was designed and reviewed by clinical practitioners for operational use, providing a domain-informed baseline for subsequent evaluation and refinement.

The form includes iterative sections for structured capture of medications, surgeries, allergies, family history, exposures, and supplementary studies. A total of 267 questions were included in the evaluation subset, after excluding 11 items whose response sets depend on large external catalogs (e.g., full ICD-10/ICD-11 code sets, LOINC, RxNorm identifiers, and INEGI municipality lists), which cannot be meaningfully evaluated without their full contextual scope.

2.3 Schema Extraction

A Python script (https://github.com/GIMIudg/GIMIpapers/tree/main/clinEvalIA) was used to parse the XLSForm file and generate a structured JSON schema. For each question, the schema encodes its unique identifier, section, display label, answer options, and conditional display logic.

This structured representation enables programmatic access and uniform processing of all instrument elements, allowing each question to be evaluated independently while preserving its contextual metadata. JSON was selected as the intermediate representation due to its flexibility, interoperability, and compatibility with downstream automated processing pipelines.

The extraction process included all 267 questions, including those within iterative blocks, while preserving their corresponding section context. Catalog-dependent items were excluded; entries such as state-level geographic options were retained, whereas municipality-level lists were omitted due to excessive cardinality.

2.4 Automated Evaluation: LLM-as-Judge

Each question in the evaluation schema is assessed by an LLM acting as a judge via the Anthropic API, using Claude Haiku 4.5. For each item, a structured prompt is generated, and the model's response is parsed and stored programmatically.

Claude Haiku 4.5 was selected for its balance of cost, latency, and reliable instruction adherence, which is essential for large-scale evaluations requiring strict output formatting.

Each question is evaluated across four dimensions: clarity, relevance, completeness, and granularity, using a 1–5 scale. In addition to numeric scores, the model provides brief justifications, flags potential issues for synthetic data generation, and suggests targeted improvements when applicable.

These dimensions operationalize structural quality by capturing interpretability (clarity), usefulness (relevance), coverage (completeness), and level of detail (granularity).

To account for stochastic variability, each instrument version was evaluated over ten independent runs at temperature = 1. Section-level scores are reported as the mean across runs, with standard deviation used to quantify inter-run stability. Low variability indicates consistent evaluations, whereas higher variability suggests uncertainty in the model's judgment. Subsections were aggregated by section for reporting purposes, and raw per-subsection results are available in the project repository.

Although the same model is used for evaluation and recommendation generation, variability analysis across runs is used as a stability assessment mechanism to characterize evaluation stability across stochastic conditions.

2.5 Prompt Engineering

The prompt submitted to the judge for each question consists of two components: a fixed prompt and a dynamically constructed prompt.

The fixed prompt defines the evaluator role, specifies the output format, and encodes the evaluation rubric. The model is instructed to act as an expert in clinical instrument design and neurological health research, ensuring consistent interpretation of scoring criteria. The output is constrained to a strictly defined JSON schema with ten fields, including four score–comment pairs, a field for synthetic data generation issues, and one for improvement suggestions. The model was required to return only this structured object, as the responses are automatically parsed, and any deviation would compromise downstream processing.

The rubric defines each evaluation dimension using explicit behavioral anchors (scores 1, 2, and 5), reducing ambiguity and improving inter-run consistency. Additional clarification is provided to distinguish closely related dimensions, particularly completeness and granularity. For the granularity dimension, few-shot examples are included to guide the model in differentiating between exhaustive but non-specific answer sets and clinically meaningful detail.

The dynamic prompt encodes the full question context, including section, identifier, question text, field type, answer options, and conditional logic, with additional contextual templates for iterative sections to ensure consistent evaluation. For items within iterative groups, an additional context block describing the clinical purpose of the group is included. A total of twelve context templates were defined, providing sufficient contextual grounding for consistent evaluation across repeated structures.

2.6 Iterative Refinement Cycle

The iterative refinement cycle consists of two stages: baseline evaluation of the original instrument (C1), followed by targeted modifications and re-evaluation of the revised version (C2), guided by LLM-based assessments.

Interventions were determined using two criteria applied to C1 scores. First, sections with a mean granularity score below 3.0 were prioritized, using the section as the unit of analysis. This threshold represents the midpoint of the evaluation scale and serves to identify areas with insufficient structural specificity. Second, individual questions with granularity scores ≤ 2 were targeted, ensuring that poorly performing items were addressed even within otherwise high-performing sections.

All modifications were restricted to refining or expanding existing answer sets. No new questions were introduced, and no conditional logic was altered. This constraint isolates the effect of answer-set refinement while preserving the original instrument structure and ensuring comparability between C1 and C2, which was originally defined with clinical input. Suggestions implying structural changes were excluded.

Application of these criteria resulted in 51 targeted modifications across 11 sections, including expansion of response options, correction of semantic inconsistencies, and alignment with standardized clinical terminology. A detailed list of modifications is provided as supplementary material.

2.7 Implementation Details

All processing steps were implemented in Python, including schema extraction, prompt generation, API interaction, and response parsing. Evaluations were executed sequentially to ensure consistent logging and traceability of outputs. Each run was version-controlled to allow reproducibility of results across iterations (C1 and C2).

3 Results

The evaluation pipeline was applied to 267 questions across 22 sections for two instrument versions: original (C1) and revised (C2). Across ten independent iterations using Claude Haiku 4.5, section-level scores were computed as the mean across items, and stability was characterized by inter-run standard deviation.

3.1 Baseline Evaluation

Mean scores across all sections were highest for clarity (4.45 ± 0.25) and lowest for granularity (3.06 ± 0.56), with completeness (3.84 ± 0.51) and clinical relevance (3.84 ± 1.12) occupying intermediate positions. The notably high standard deviation in clinical relevance reflects the model's tendency to assign consistently low scores to sections capturing comorbidity data for Charlson index calculation, which the evaluator systematically flagged with low relevance for neurological assessment. These results identify granularity as the primary limitation of the baseline instrument, guiding the subsequent refinement strategy.

3.2 Method Stability

Inter-run variability in C1 was low across all sections, with a mean standard deviation at the section level of 0.15 for completeness and 0.17 for granularity. This low variability defines an empirical noise threshold used to distinguish meaningful score changes and stochastic variation in subsequent comparisons.

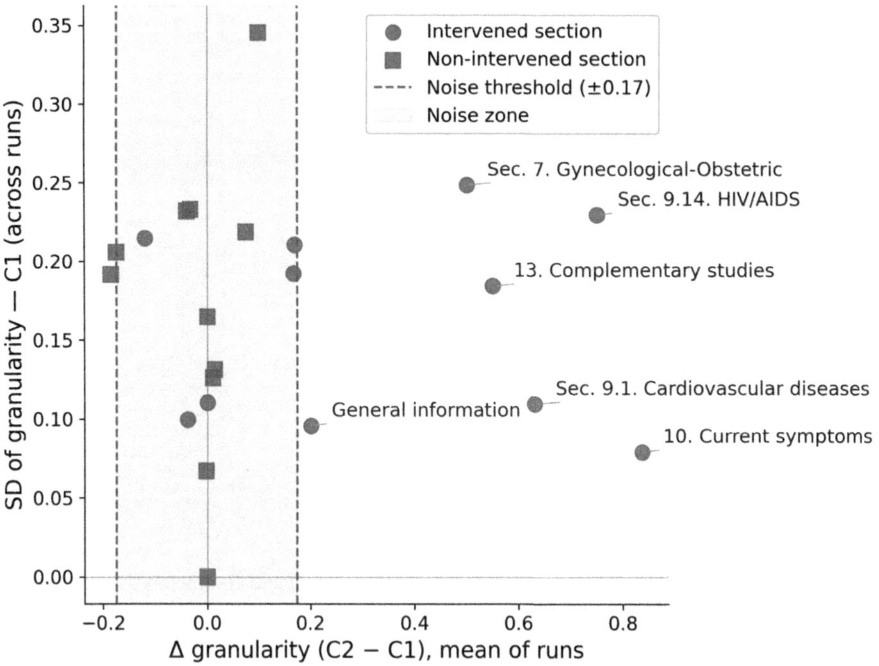

Fig. 2. Scatter plot of mean granularity score change ($\Delta = C2 - C1$, x-axis) versus inter-run standard deviation of C1 granularity scores (y-axis) for each of the 22 merged form sections. Intervened sections (green circles) and non-intervened sections (gray squares) are shown. The shaded region ($\Delta \leq$ noise threshold) represents changes attributable to stochastic model variation, defined by the empirical inter-run variability. Points outside this region indicate signal-level improvements beyond noise. Labeled points correspond to sections with the largest observed changes.

3.3 Effect of Modifications (C1 vs. C2)

The capacity of the pipeline to detect signal-level improvements beyond stochastic variation is illustrated by Fig. 2, which plots Δ granularity against the inter-run variability. A noise threshold ($SD = 0.17$) was used to separate signals from stochastic variation.

Intervened sections exceeding this threshold consistently clustered in the signal region, with the largest improvements observed in Symptoms ($\Delta = 0.84$), HIV/AIDS ($\Delta = 0.75$), Cardiovascular diseases ($\Delta = 0.63$), Complementary studies ($\Delta = 0.55$), and Gynecological-Obstetric ($\Delta = 0.50$). Non-intervened sections remained symmetrically distributed around zero.

As shown in Fig. 3, improvements were concentrated in granularity and completeness, while clarity and relevance remained stable. Mean granularity increased from 3.06 ± 0.56 in C1 to 3.22 ± 0.62, with non-negative changes in 16 of 22 sections, suggesting that targeted answer-set refinement systematically improved the instrument's capacity to capture clinically differentiated patient states. Completeness showed smaller gains, suggesting that increasing the specificity of response options has effects on both dimensions.

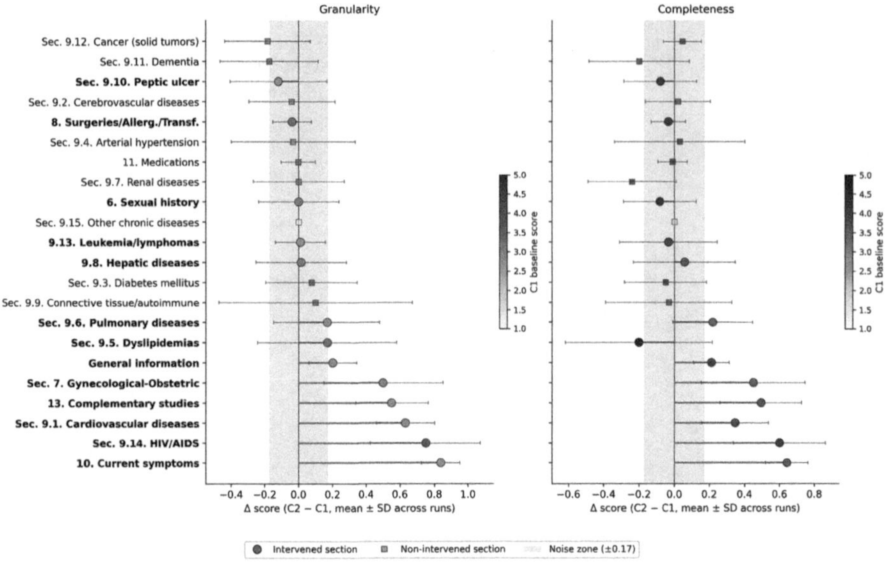

Fig. 3. Mean score change ($\Delta = C2 - C1$) for granularity (left) and completeness (right) across all 22 sections, shown as lollipop plots with error bars representing inter-run SD. Point color encodes the C1 baseline score (1–5 scale). Intervened sections are shown as circles (bold labels); non-intervened sections as squares. The shaded region indicates the empirical noise zone ($\pm$SD $= 0.17$); points outside this region represent signal-level improvements.

3.4 Clinical Validation of Modifications

To assess whether these improvements correspond to clinically meaningful changes, 36 of the 51 targeted modifications were independently evaluated by three neurology residents.

By majority criterion, 88.9% of items (32/36) were rated as clinically pertinent, 5.6% as partially pertinent, and 5.6% showed no consensus. The mean item score was 4.51/5 ($SD = 0.38$), with no item averaging below 3.67, and a majority agreement was reached in 94.4% of cases (34/36).

These results indicate that LLM-guided recommendations, when constrained by a structured rubric, produce clinically meaningful suggestions at a rate consistent with expert judgment.

3.5 Qualitative Examples of Improvement

Qualitative evidence (Table 1) illustrates representative before/after modifications, where expansions of response options increased clinical specificity and were associated with substantial gains in Δ granularity scores.

Table 1. Improvement across 4 questions, detailing questions and what options are available in both versions, showing a significant gain in granularity ($\Delta = C2 - C1$)

Question	Section	C1 options	C2 expanded response set after refinement	Δ granularity
9.14.1.c Have you had opportunistic infections?	Section 9.14. HIV/AIDS	4 generic (*"Yes"*, *"No"*, *"Don't know"*, *"Prefer not to answer"*)	12 specific (toxoplasmosis, PCP, CMV retinitis, cryptococcosis...)	+3.9
4.5 Main environmental/ occupational exposure	ROOT	4 broad categories (*"Metals"*, *"Pesticides"*, *"Solvents"*, *"Other"*)	19 specific agents (lead, manganese, benzene, asbestos...)	+1.6
7.2 Menstrual cycle regularity	Section 7. Gynecological-Obstetric	*3 options ("Regular", "Irregular", "Postmenopausal")*	8 clinical categories (oligomenorrhea, polymenorrhea, secondary amenorrhea...)	+1.4
3.3 Substance use status	ROOT	2 options (*"Active use"*, *"Former user"*)	7 options with temporal specificity (*"Former user < 6 months"*, *"Former user 6 mo–2 yr"*...)	+2.7

4 Discussion

4.1 Signal Detection and Evaluation Consistency

The results show that the proposed rubric-based LLM evaluation can reliably distinguish signal-level improvements from stochastic variation under repeated stochastic evaluation.

As shown in Fig. 2, intervened sections produced Δ granularity values clearly separated from the noise threshold, while non-intervened sections remained distributed around zero.

Notably, three intervened sections (Sexual antecedents, Surgery/Allergies/ Transfusions, and Peptic ulcer) remained within the noise region, indicating that not all modifications translate into measurable improvements.

This behavior is consistent with prior observations that LLM-based evaluators exhibit high internal consistency but variable sensitivity depending on task structure [6, 8].

4.2 Granularity as a Structural Quality Driver

The selective improvement of granularity and completeness, with stability in clarity and relevance, provides internal evidence of metric specificity. Since modifications were constrained to answer-set refinement, no structural changes were introduced that could affect wording or domain relevance. These results suggest that the evaluation dimensions capture distinct and complementary aspects of structural quality. In this context, granularity emerges as a key limiting factor in categorical clinical instruments, directly influencing their capacity to capture clinically meaningful variation.

4.3 Implications for Clinical Instrument Design

The proposed pipeline provides a scalable complement to traditional validation approaches such as expert panels or Delphi consensus, which are resource-intensive and difficult to iterate [3]. By encoding evaluation criteria into a structured rubric and delegating scoring to an LLM, the cycle of evaluation and refinement becomes faster and reproducible. Rather than replacing clinical expertise, the method functions as a reproducible first pass screening mechanism to identify weaknesses before resource-intensive validation. The high agreement between LLM-guided modifications and clinician evaluations observed in this study suggests that such systems can produce clinically aligned recommendations when properly constrained. Even under constrained interventions, measurable improvements across multiple sections indicate that the structured automated feedback can guide efficient instrument refinement.

4.4 Limitations

Several limitations warrant consideration. First, the refinement process required human interpretation of model recommendations, so the pipeline is not fully automated and depends on intermediary judgment.

Second, the evaluation captures structural quality rather than clinical validity or predictive performance in real-world settings.

Third, the use of the same model for evaluation and recommendation introduces potential self-evaluation bias, a known limitation in LLM-as-judge frameworks [4, 6]. This limitation is partially mitigated by independent clinical validation, which showed high agreement with model-guided modifications; however, full mitigation would require evaluation with independent models or human-only baselines.

4.5 Future Work

The methodology was evaluated on a neurological clinical history form, but the pipeline is domain-agnostic. Core components (schema extraction, rubric-based evaluation, and structured refinement) can be transferred to other clinical domains with minimal adaptation. Future automation of the refinement stage through structured prompting could enable fully closed-loop optimization.

Additionally, incorporating richer contextual information may improve evaluation fidelity for complex clinical questions.

Finally, enhancing the context provided to the model through richer contextual and sequential information could improve evaluation fidelity, allowing the LLM to evaluate a full context question rather than an isolated item. Future iterations should incorporate evaluations using independent LLM providers to further reduce self-evaluation bias and strengthen the generalizability of the framework recommendations.

5 Conclusion

This work demonstrates that a structured LLM-as-judge pipeline can serve as a first-pass evaluation framework and guide the refinement of structured clinical instruments at scale. Three findings support this claim. First, evaluation stability across ten independent runs produced an empirical noise threshold (SD = 0.17) that enabled consistent discrimination between signal-level improvements and stochastic variation. Second, targeted answer refinements produced measurable improvements in granularity in intervened sections ($3.06 \pm 0.56 \rightarrow 3.22 \pm 0.62$), while non-intervened sections remained stable, supporting the interpretation that observed changes reflect the modifications rather than model variance. Third, independent validation by neurology residents confirmed 88.9% of changes made as clinically relevant, suggesting that automated structural recommendations can align with expert judgment when properly constrained by a structured rubric and sufficient context.

Taken together, these results position LLM evaluation as a practical, reproducible first-pass screening layer for clinical instrument refinement, complementing validation rather than replacing it. The domain-agnostic design suggests transferability beyond neurology, enabling systematic and reproducible quality assessment of structured clinical instruments at scale.

Acknowledgments. This study was funded by Secretaría de Ciencia, Humanidades, Tecnología e

Innovación (SECIHTI) for the scholarship (CVU: 2085078).

Disclosure of Interests The authors declare no competing interests relevant to the content of this article.

References

1. Ozonze, O., Scott, P., Hopgood, A.A.: Automating electronic health record data quality assessment. J. Med. Syst. **47**(1) (2023). https://doi.org/10.1007/s10916-022-01892-2
2. Kommaraju, R.J.: Optimizing data quality and compliance through integrated validation strategies for clinical systems. Am. J. Eng. Technol. **7**(8), 299–306 (2025). https://doi.org/10.37547/tajet/volume07issue08-26
3. Kuru, H.: Rethinking survey development in health research with AI-driven methodologies. Front. Digit. Health. **7**, 1636333 (2025). https://doi.org/10.3389/fdgth.2025.1636333
4. Williams, G., Rutunda, S., Nzabakira, F., Mateen, B.A., Mateen, B.: Human evaluators vs. LLM-as-a-judge: toward scalable, real-time evaluation of GenAI in global health. (2025). https://doi.org/10.1101/2025.10.27.25338910
5. Aali, A., et al.: Expert-level validation of AI-generated medical text with scalable language models. arXiv abs/2507.03152. (2025). https://doi.org/10.48550/arxiv.2507.03152
6. Gao, Y., et al.: Evaluating clinical AI summaries with large language models as judges. npj Digit. Med. **8**(1), 640 (2025). https://doi.org/10.1038/s41746-025-02005-2
7. Zheng, L., et al.: Judging LLM-as-a-judge with MT-bench. arXiv. (2023). https://doi.org/10.48550/arXiv.2306.05685
8. Loi, Z., et al.: Consistency in large language models ensures reliable patient feedback classification. medRxiv. (2024). https://doi.org/10.1101/2024.07.11.24310210

Data Analytics and Driver of School Retention on Higher Education

A. Mejía-Martínez[1] and Y. U. K. Coronado González[2](✉)

[1] Facultad de Negocios, Universidad La Salle, Mexico City, Mexico
[2] Vicerrectoría de Investigación, Universidad La Salle, Mexico City, Mexico
yaxkin.coronado@lasalle.mx

Abstract. Student dropout remains one of the most pressing challenges in higher education, yet most existing research focuses on identifying those who leave rather than understanding what drives students to stay. This study shifts the perspective toward retention, leveraging a large-scale dataset of over 70,000 anonymized student records from Tecnológico de Monterrey (2015–2020), spanning more than 45 sociodemographic, academic, economic, and family-related variables.

Three supervised machine learning models were trained and compared: a multilayer Neural Network, Random Forest, and XGBoost. All models achieved accuracy near 80%, with XGBoost reaching up to 94.4%, making it the most robust approach for this highly imbalanced dataset where students retained. Beyond accuracy, the models revealed that student retention is driven not solely by academic performance entrance exam scores and first-semester GPA but significantly by economic factors such as scholarship coverage, and notably by participation in physical and extracurricular activities.

These findings challenge the conventional assumption that academic merit alone predicts persistence. The explainability offered by XGBoost and Random Forest provides educational institutions with actionable insights to design early-warning systems and targeted interventions aligned with SDG 4, promoting equitable and quality higher education. This research opens pathways for integrating innovative pedagogical models such data-driven profiling, offering a comprehensive framework to understand and promote student retention in higher education.

Keywords: Data analytics · Dropout education · Machine learning education

1 Introduction

Educational institutions deal with one of the main issues in university grades, the scholar drop out, being studied by many educators and researchers [1, 2] finding a complex pattern of interrelation between the students' perceptions, academic notes, social networks in the school, the family and phycological aspects of the students. These dimensions affect academic performance, social interaction and even the motivation of the students to continue their studies.

H. Ponce et al. (Eds.): COMIA 2026, CCIS 3095, pp. 566–572, 2027.
https://doi.org/10.1007/978-3-032-37438-7_44

In general, when school dropout is studied, the main perspective focus on variables for the analysis are the ones that make a student drop out, not the ones that make them stay. It is important to appoint that drop out, not necessarily correspond a desertion from the education, even a change of career or educational institution can be accounted as dropout inside of the education system. Therefore, the data of the institution has the impossibility to capture all the cases, limited only to seeing the problem from a one-way perspective, being necessary to see the retention as a valuable part or the problem.

Even the educational institutions with a large quantity of data about their students, few of them are analyzed, some of them like opinions or personal perspective, transforming to numeric values, in combination with the grade or an evaluation degree on a subject. This gives school authorities guidance to make decisions in the institution about the retention and not only the detection of dropout students.

These kinds of variables provide valuable data for different selection models or strategies to minimize desertion in institutions, also providing fair, universal, and sustainable opportunities for every student. The pathway to establish a model that motivates school retention starts with the detection of characteristics of ideal students, the difference between dropout students and the recollection data from the students via innovative models of education like the personalization via an automatic system for the development of the students in the institution.

This goal goes hand in hand with Sustainable Development Goal (SDG) 4.3, which addresses equal access for all men and women to quality technical, vocational, and higher education. At this level, the integration of innovative models such as virtual reality-based instruction has proven to be highly effective in both K-12 and higher education [3].

The implementation of these technologies contributes to student retention by improving educational quality and skill development as outlined in SDG 4.4, which focuses on increasing the number of youth and adults who possess the necessary skill particularly technical and vocational skills to access employment, decent work, and entrepreneurship [3]. Immersive technology directly responds to this need by allowing the development of practical skills in risk-free environments, offering a simulation of real-life situations that are essential for vocational education and training in industries, medicine, and sports.

1.1 But What Makes Students Stay and Continue with Their Studies?

From a psychological perspective, a student's retention and success are strongly linked to their level of motivation and engagement. Studies show that for a student to reach a state of "flow" (being completely absorbed in an activity) and not abandon the task, four interconnected psychological factors need to be present: an affinity for the content, an emotional connection, an intrinsic motivation, and a high level of engagement. Furthermore, the introduction of innovative strategies such as gamification and virtual environments in the classroom has longitudinal positive effects on students' intrinsic motivation, satisfaction, effort, and academic performance.

Determine the social, family, economic, and psychological factors to mitigate school desertion in college using innovative supervised machine learning models to create a profile of who might desert [1, 4]. To complement the mitigation of desertion through the identification of risk profiles, it is imperative to apply innovative, student-centered pedagogical models that counteract negative psychological factors (such as boredom or the perceived uselessness of the subject). For example, the application of educational design frameworks such as the CREAM Model [5] (Civic Education, Relevance, Engagement, Applicability, and Multimodality) allows for the structuring of learning experiences that keep the player/student's motivation high without losing the educational value.

By transferring control of learning to the student through interactive and immersive environments, active and critical learning is fostered. This not only improves knowledge retention through multiple sensory channels, but it directly tackles the psychological factors of dropout by transforming abstract and passive teaching into an immersive, attractive experience that is directly applicable to their future technical and professional skills.

The scholarship retention factor where explores by machine learning (ML) techniques from a large dataset for a couple of years to probe the selection methods of characteristics as well the concordance with the literature. We divide the paper into the following section of methodology that describes database, selection and preprocessing methods for the ML technics application of Neural Networks, Random Forest and XGBoost. Follow by the results and discussion, evaluating the metrics and comparative between technics, finally a conclusion over the implications and future work for the research.

2 Methodology

The database analyzed for this scope comes from a private institution TEC de Monterrey [6] containing over 70,000 anonymized student records from the years 2015 to 2020. This database included more than 45 variables covering sociodemographic, academic, economic, and family-related aspects.

After conducting an initial statistical exploration and filtering the data for university students, degree, and other categorical values, we applied a supervised machine learning models in R and Python to guarantee robustness and reproducibility.

We implement three models: (1) a multilayer neural network, with an input of 64 neurons, an internal layer of 32 neurons due to reduce the number of correlations and characteristics to analyze for the retention ratio and dimension of explainable variables, and an output layer of one neuron focused on determining the retention value indicated as 1 for a positive case and 0 for the negative case, the latter interpreted as a school dropout from the institution. (2) random forest (RF) models were applied, made up of a series of decision trees, which select the best division to classify or separate the data by a linear regression, in such a way that each characteristic is evaluated, with classification over the retention or regression decision over the dropout including all numerical variables, for the reproducibility of results independently of the random conditions we selected a various seeds of the algorithm, guaranteeing robustness. (3) XGBoost: A sequential boosting algorithm that minimizes errors through gradient descent and advanced regularization, to guarantee a non-overfitting, especially in this case with a huge amount of data susceptible to being correlated.

3 Results and Discussion

All the models applied to the database, presents in Table 1, a successful accuracy nearly to 80%, but in our case, because we focus on retention is expected an unbalance respect to the population not retained of around 10%, even we make unbiased selection, the main metric to review is the recall due to the detection of retention and the possible non retention cases.

Therefore the unbalanced data represents a challenge to detect possible dropouts cases. We could compare the F1 score to select the populations of retention and dropout students, see Table 2. Were the best model overall dataset is the XGBoost and the best for retention students is the random forest over the neural networks.

Even models like the neural networks extract complex patterns, the explainability is a challenge, so we detect that models like XGBoost allow to extract the main characteristics that affect the retention, improving methods like random forests without losing explainability in the model, see Fig. 1, showing the feature importance for the retention students being interesting the contrast, where the academic relevance of score exams pass to a second degree respect to the physical activities, that allows students to be retained by the institution.

Table 1. Comparative metrics between models

Model	Accuracy	Recall (Sensitivity)
Random forest	80.56%	72.42%
XGBoost	78.98% - 94.4%	71.67%
Neural Networks	71% - 82%	Higher F1 score

Table 2. Score metrics between models

MODEL	F1-score
Random tree forest (0)	0.15
Random tree forest (1)	0.96
Neural networks (0)	0.21
Neural networks (1)	0.94
XG Boost	0.95

The findings indicated that the Random Forest regression algorithm achieved the best performance, obtaining a medium accuracy (80.56%) compared to Neural Networks (71% - 82%). More importantly, Random Forests were significantly better at correctly predicting cases of non-retention or dropouts, reaching a recall of 0.72 compared to the Neural Network's that are higher than 75%. Both models achieved similar precision

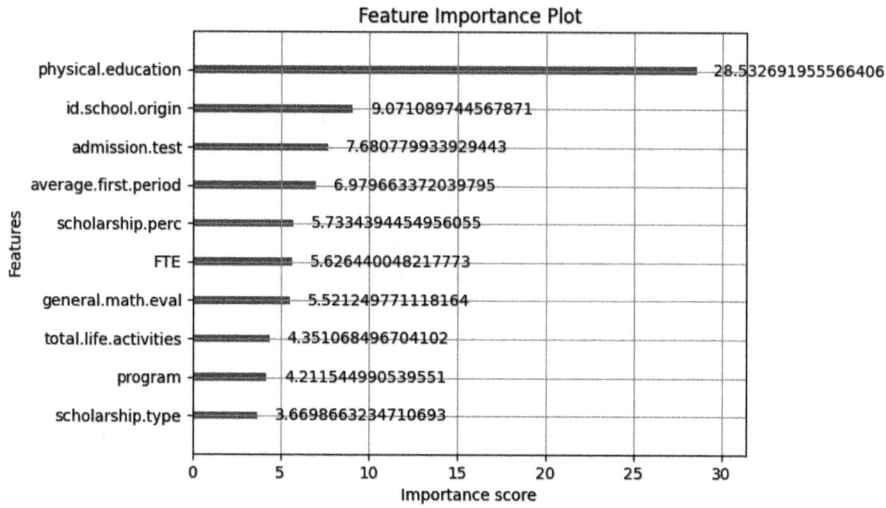

Fig. 1. Feature importance of retention students

(0.80) for positive retention cases. The impressive of XGBoost reaching 94% of accuracy with a challenging database due to the regularization methods [6]. Showing in Fig. 1 the relevance of physical activities combinates with a good academic performance trajectory, facilitates the retention of the students [7].

Because of its structure, the Random Forest algorithm provided valuable explainability regarding the underlying factors of student dropout. The data revealed that academic factors carry the most weight at the university level, specifically mathematical ability, entrance exam scores, and the first-semester grade point average. Economic factors, such as the percentage of an academic scholarship, were the second most relevant. Finally, social factors and extracurricular involvement (such as sports and cultural activities) also proved to favorably impact student retention beyond mere academic performance. Consequently, the combination of methods like XGBoost was identified as the most suitable and straightforward algorithm to capture the complexity of diverse educational data.

4 Conclusions

A major challenge in developing educational algorithms is dealing with highly unbalanced data, as student retention typically represents 90% or more of the sample, making it difficult to capture the minority dropout cases without data balancing techniques. Despite this complexity, the XGBoost algorithm successfully captured both positive and negative cases of school retention, outperforming the neural network approach, with the limitations of sensitivity of this type of models and dependency of the regularization method.

Random Forest (Regression) become the most reliable model for general institutional implementation, with balanced high accuracy and its complete transparency in variable interpretability for large databases, capturing the complexity of data over the precision metric. For general implementation, XGBoost is superior for large-scale, high-performance environments, achieving a highly competitive accuracy range and solid recall sensitivity. Finally, neural networks modeling is suitable for temporal engagement patterns and complex non-linear relationships.

Regarding future perspectives, this as an initial approach that can be enhanced in several ways: Data Clustering: Future models could integrate Random Forests with data cluster classifications to explain school retention and its predominant factors across different student groups, which would further improve prediction precision as well in XGBoost schema. Bayesian Trees to explore new mechanisms to understand the exact percentage of affection each variable has on an individual student. This would allow for quantitative and qualitative predictions, enabling institutions to identify key dropout patterns and intervene early when students are at risk.

For real-time intervention we need to implement a monitoring system that trigger automated alerts, reviewed by human supervisors, this only can be achieved by hybrid model combination of deep learning models with interpretable machine learning models, to understand the complexity of the student retention by educative institutions, to reduce the dropout rate in schools.

Acknowledgments. A third level heading in 9-point font size at the end of the paper is used for general acknowledgments, for example: This study was funded by X (grant number Y).

Disclosure of Interests It is now necessary to declare any competing interests or to specifically state that the authors have no competing interests. Please place the statement with a third level heading in 9-point font size beneath the (optional) acknowledgments, for example: The authors have no competing interests to declare that are relevant to the content of this article. Or: Author A has received research grants from Company W. Author B has received a speaker honorarium from Company X and owns stock in Company Y. Author C is a member of committee Z.

References

1. Selim, K.S., Rezk, S.S.: On predicting school dropouts in Egypt: a machine learning approach. Educ. Inf. Technol. **28**(7), 9235–9266 (2023). https://doi.org/10.1007/s10639-022-11571-x
2. Archambault, I., Janosz, M., Olivier, E., Dupéré, V.: Student engagement and school dropout: theories, evidence, and future directions. In: Reschly, A.L., Christenson, S.L. (eds.) Handbook of Research on Student Engagement. Springer, Cham (2022). https://doi.org/10.1007/978-3-031-07853-8_16
3. Nestar, J.L.G., Alonso, F.G., Cervantes, A.O.: Aprendizaje servicio en educación superior entre España y México. Hacia los ODS. Alteridad. **17**(1), 76–88 (2021). https://doi.org/10.17163/alt.v17n1.2022.06
4. Ridwan, A., Priyatno, A.M.: Predict students' dropout and academic success with XGBoost. J. Educ. Comput. Appl. **1**(2), 1–8 (2024). https://doi.org/10.69693/jeca.v1i2.13
5. Egea-Vivancos, A., Arias-Ferrer, L.: Principles for the design of a history and heritage game based on the evaluation of immersive virtual reality video games. E-Learning Digit. Media. **18**(4), 383–402 (2020). https://doi.org/10.1177/2042753020980103
6. Alvarado-Uribe, J., et al.: Student dataset from Tecnologico de Monterrey in Mexico to predict dropout in higher education. Data. **7**, 119 (2022). https://doi.org/10.3390/data7090119

7. Bentéjac, C., Csörgő, A., Martínez-Muñoz, G.: A comparative analysis of gradient boosting algorithms. Artif. Intell. Rev. **54**(3), 1937–1967 (2020). https://doi.org/10.1007/s10462-020-09896-5
8. Sá, M.J.: Student academic and social engagement in the life of the academy—a lever for retention and persistence in higher education. Educ. Sci. **13**(3), 269 (2023). https://doi.org/10.3390/educsci13030269

Beyond the Black Box: Explainable AI for Parkinson's Detection Through Grad-CAM and Multimodal Data

Kevin Mora Hernandez and Claudia I. Gonzalez(✉)

Division of Graduate Studies and Research, Tijuana Institute of Technology/TECNM, Tijuana 22414, Mexico
kevin.mora18@tectijuana.edu.mx, cgonzalez@tectijuana.mx

Abstract. The adoption of Deep Learning (DL) in the medical field, specifically in neurodegenerative diseases, has been limited by the opacity of its architectures; this study seeks to address the need for transparency for its subsequent adoption, further democratizing access to advanced diagnostic aids through the implementation Explainable Artificial Intelligence (XAI) for interpretability in classification models of Parkinson's Disease (PD). To achieve this, Convolutional Neural Networks (CNNs) are employed to analyze the two modalities of the study: T1-weighted Magnetic Resonance Imaging (T1-MRI) and dynamic motor tracings, specifically using the local interpretability technique Grad-CAM, to conduct the auditing process and assess whether phenotypic coherence exists. The results point in both directions, showing a divergence in poles such as the identification of biomarkers coherent with medical semiology, as is the case with consistent kinematic features in a tracing model, fitting the idea of being able to audit based on established knowledge, as well as the other extreme, identifying that the patterns used by the MRI model exhibit "shortcut learning" by prioritizing non-pathological periorbital signals rather than anatomical regions relevant to PD diagnosis. This contrast underscores the necessity of XAI techniques, positioning them not merely as a complement but as a fundamental requirement to validate models beyond performance metrics.

Keywords: Explainable Artificial Intelligence · Grad-CAM · Parkinson's Disease · Convolutional Neural Networks · Digital Biomarkers · Phenotypic Fidelity · Trustworthy AI · Neuroimaging · Kinematic Analysis

1 Introduction

The use of AI in medicine has advanced strongly during the last few years as well as its integration [1]; this has come from the great capabilities for detecting patterns in the data of DL models and the ease with which to work with massive amounts of data from different modalities. But not all that glitters is gold, this rapid and accelerated integration has brought with it differences or even rejection, separating into two paths that should be part of the same [2]: in the laboratory we can obtain models with an excellent predictive

H. Ponce et al. (Eds.): COMIA 2026, CCIS 3095, pp. 573–584, 2027.
https://doi.org/10.1007/978-3-032-37438-7_45

index, a brutal performance, whatever the metric may be (with brute force everything seems possible) but generally only accepting that they work, being black box algorithms that we simply know seem to identify something but that remains hidden among their layers. In medicine, this is simply unacceptable; working and directly affecting lives, the responsibility is enormous. Not being able to interpret the patterns of the models makes them alien, being simply computer science experiments that do not translate to functional tools in most cases [3], diverging from what should be one of their primary purposes, a real-world implementation.

The lack of transparency is a barrier for its adoption; more is needed than a simple output number that through a threshold we classify one way or another. A more human view, being able to observe the patterns that the model uses to weigh one classification more than another, requires physiological justification that makes sense for the expert to observe where the model focuses. This adoption is not going to be achieved simply with an Excel sheet that shows statistics close to 99%, at least not just this, but by using our human capacity to validate that the model certainly identifies biomarkers with meaning and not artifacts that come from the hardware, noise from the dataset, WHICH could make the model take shortcuts based on them [4]. The development of these technologies brings evident ethical risks, they involve sensitive data, about generally very serious illnesses, but this does nothing more than reinforce the urgency of transitioning from purely statistical models to systems that we can audit to validate their patterns, see inside the black box, and find sense in them, translate algorithmic logic to clinical logic and thus overcome this adoption barrier.

1.1 The Challenge of Diagnosis and Pattern Identification in Parkinson's Disease

PD is currently one of the fastest-growing neurodegenerative diseases globally; it is estimated to affect more than 10 million people around the world [5]. It is characterized by being a complex pathology with variability, loss of dopaminergic neurons, motor and structural alterations in brain morphology, all this causing its early detection and monitoring to be a constant challenge.

Traditionally, evaluation has been based on qualitative clinical scales, but the need to be able to find digital biomarkers leads us to a challenge that falls on the fact that the data used to identify it on its own may not relate directly to PD, requiring differentiating between natural variations and real pathological markers related to the disease.

The Role of XAI in Clinical Validation. In cases such as T1-weighted MRI and evaluation of fine motor tracings, XAI allows us to establish correspondence between the activations that provoke one result or another and traditional medical semiology; while a neurologist searches for structural atrophy patterns or kinematic irregularities in the tracings, interpretability makes it possible to visualize what the model is prioritizing to give a classification, and with it we are capable of being able to validate its coherence of the activated brain regions or the significant traces such as pressure in curves, undulations, and space between lines. This ability to contrast the machine's 'vision' with the expert eye transforms the model from a black box to a tool capable of finding biomarkers, enabling validation feature attribution converges with established neuropathological knowledge such as focal atrophy in T1-weighted MRI or specific kinematic anomalies

[6], mitigating the risk of use a model that suffer of shortcut learning or reliance on implicit artifacts within the data [7].

1.2 Study Objectives and Contribution

Beyond achieving high classification performance, this work aims to challenge the blind faith in black box sustained by metrics that could be biased, validating the clinical plausibility of the patterns identified in the DL models for the classification of PD, this using the technique of Grad-CAM, generating heatmaps superimposed on the original image for tracings and MRI, validating the focal points in the impact of the classification. This work contributes practical examples of both sides, both models that find patterns with similarity to those an expert user would detect, as well as a compelling case study illustrating how patterns may emerge from non-cerebral features, suggesting potential shortcut learning. We provide evidence that standard metrics often mask heuristic shortcuts and technical artifacts, while our analysis is grounded in established clinical knowledge, it remains receptive to identifying atypical patterns that may deviate from expected biomarkers, ultimately aiming to convert black box predictions into verifiable clinical evidence.

2 Gradient-Weighted Class Activation Mapping: Heatmaps for Visual Explanation

The transition from purely predictive models to auditable systems is an ethical and technical imperative for modern medicine. This study is grounded in two fundamental pillars detailed below.

2.1 The Imperative of Trust and International Consensus

The clinical adoption of AI is limited by the intrinsic incapacity to give an auditable justification regarding the patterns found to point toward one specific classification. As defined in the work of FUTURE-AI international consensus [8], explainability is defined not as an aesthetic addition, but as a fundamental quality metric. In this context, the Human-in-the-loop paradigm of Holzinger et al. [9] maintains that the tacit knowledge of the expert is irreplaceable, emphasizing that XAI does not come to replace the human expert; it rather serves as the necessary bridge to integrate medical intuition with the computational capacity of DL.

2.2 Grad-CAM and the Challenge of Phenotypic Fidelity

Grad-CAM represents a milestone by enabling the reconstruction of diagnostic relevance directly in the input space [10], transforming abstract gradients into interpretable heatmaps. As illustrated in Fig. 1, the tracing and MRI models generate distinct activation patterns that reflect their respective internal logic; however, the Grad-CAM technique is not without its limitations.

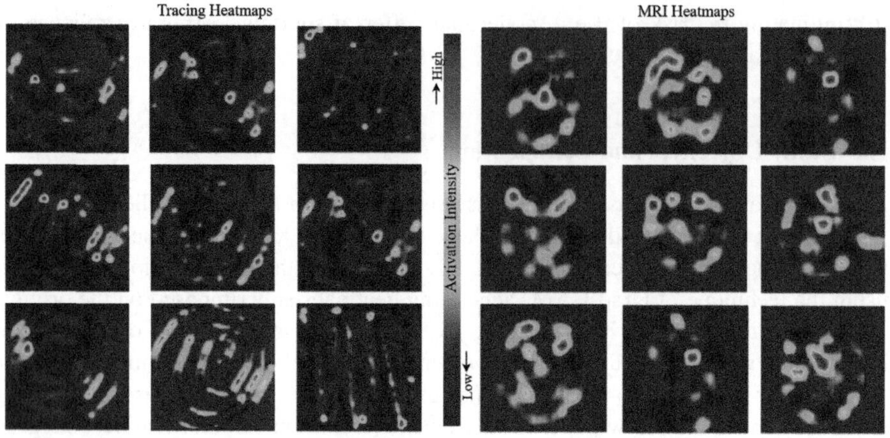

Fig. 1. Comparative gallery of Grad-CAM heatmaps for Tracing (left) and MRI (right) models across multiple samples. Tracing activations are characterized by sparse, localized points along the spiral path, whereas MRI activations show broader, concentrated clusters often located in non-brain peripheral regions.

As commented in the works of Panboonyuen [11] and Borys [12], caution should be taken with explanations based on saliency maps, as there may be a risk that what is interpreted with significance for the classification does not truly reflect the model's behavior, thereby confusing visual plausibility (what appears correct to the expert eye) with predictive fidelity (the model's true logic).

Ignoring this dissonance allows for the infiltration of biases through shortcut learning or "Clever Hans" effects [9], where the network optimizes its metrics based on artifacts instead of real biomarkers; these risks only make the choice of architecture the most compatible with the Grad-CAM technique, the CNN architecture, this is justified not only by their performance but by their inherent compatibility with the attribution techniques, this synergy allows heatmaps to be treated as diagnostic hypotheses, which must then be subjected to rigorous human validation to confirm their clinical validity.

3 Technical Foundations of Interpretability in Convolutional Architectures

XAI encompasses a variety of tools and methodologies designed to provide transparency to DL models, allowing for interpretations that serve to find meaning in the patterns they learn, beyond statistics. In the clinical field, this capacity is crucial, not only to mitigate the black box effect, but also to be able to audit their feature attribution and validate which ones have greater weight in the final value given for the classification. As emphasized in recent algorithmic literature, identifying whether the network's information flow focuses on real biomarkers or technical noise makes it possible to validate if the model has captured a faithful representation of the pathology, rather than simply exploiting spurious correlations present in the training data [13, 14].

3.1 Taxonomy of Explainable Artificial Intelligence (XAI)

The taxonomy of these techniques is fundamentally divided according to their scope and specificity:

Local vs. Global Explainability. While global explanations describe model behavior, local explainability describes the rationale behind specific classifications, which is the direction followed by this study; this approach facilitates the interpretation of patterns at the patient level, helping to determine decisions for individual patients, making it easier for the clinician to understand which morphological or kinematic features underlie a particular diagnosis.
Model-Agnostic vs. Model-Specific Methods. While model-agnostic methods such as Local Interpretable Model-agnostic Explanations (LIME) or SHapley Additive exPlanations (SHAP) quantify feature importance independently of the model, they often overlook the intrinsic spatial structures of medical imaging; for MRI and motor tracings, model-specific approaches are required to maintain a direct connection between the explanation and the underlying signal.

While recent advancements in DL have introduced novel architectures such as ViTs, CNNs were selected due to their inherent spatial inductive bias. Unlike attention-based models, CNNs are mathematically designed to exploit local correlation and more importantly, to preserve the topological consistency of medical imaging, a critical requirement when mapping features to anatomical structures.

In this context, Grad-CAM stands out for its ability to reconstruct the relevance structures over the input data space. As illustrated in Fig. 2, this method allows for projecting the heatmaps onto the original slices (T1-MRI) or over the tracings, highlighting the areas of greatest weight in the prediction. This superimposition is essential for verifying clinical plausibility; it allows us to validate whether the network is detecting atrophy patterns in dopaminergic regions or irregularities in pen pressure, ensuring that the classification is grounded in the actual semiology of PD, or if, on the contrary, it is using noise as shortcuts to acceptable metrics, such as a Clever Hans effect [9].

3.2 Hierarchical Learning in CNNs

CNNs operate under the principle of hierarchical extraction, where the initial layers detect primary features such as corners and gradients, while deeper layers synthesize complex semantic structures; for this reason, the choice of CNN over architectures such as Vision Transformers (ViT) is grounded in their intrinsic auditability, being innately compatible with techniques such as Grad-CAM. While global attention mechanisms in Transformers can dilute biomarker localization, CNNs preserve spatial topology through their activation maps. This offers a direct semantic bridge that facilitates validating what the network is prioritizing, whether the network prioritizes regions of anatomical interest or incurs biases due to acquisition artifacts outside the brain volume.

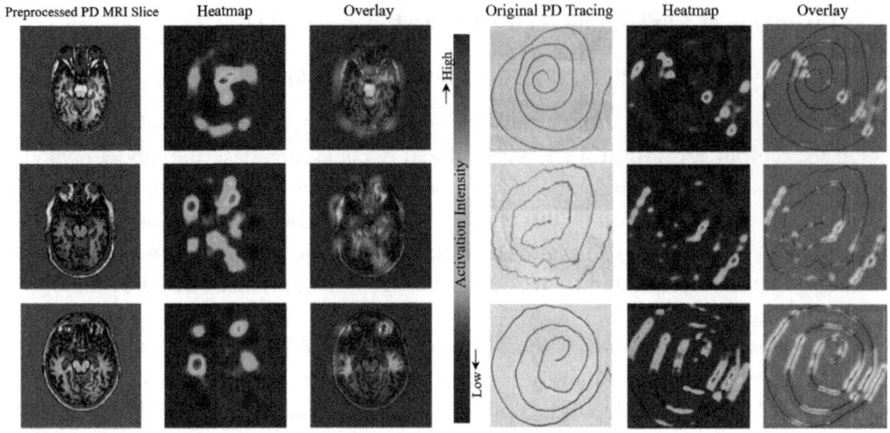

Fig. 2. Grad-CAM visualizations for MRI (left) and Handwriting Tracings (right). From left to right we can see the input image, heatmap, and overlay. Tracings exhibit localized features on the motor path, while MRI heatmaps indicate a higher reliance on non-clinical peripheral regions, suggesting potential shortcut learning.

3.3 Saliency Maps and Grad-CAM in CNNs

Grad-CAM evolves the concept of saliency maps by utilizing the gradient flow toward the last convolutional layer in a post-hoc manner. The process is governed by two fundamental mathematical stages:

Calculation of Importance Weights (α_k^c). The gradient of the score for the class of interest (Y^c) is calculated with respect to the activation maps (A^k) of the last convolutional layer. These gradients undergo Global Average Pooling (GAP) to obtain a scalar weight representing the importance of channel k for class c:

$$\alpha_k^c = \frac{1}{Z}\sum_i \sum_j \frac{\partial Y^c}{\partial A_{ij}^k} \tag{1}$$

Where Z represents the spatial resolution of the activation map.

Linear Combination and ReLU Activation. A weighted sum of the activation maps is performed, applying a ReLU function to isolate the features with a positive impact on the Parkinson's classification:

$$L_{Grad-CAM}^c = ReLU\left(\sum_i \alpha_k^c A^k\right) \tag{2}$$

The power of this method lies in the fact that $L_{Grad-CAM}^c$ preserves a direct geometric correspondence with the original input. This tells us that if the network detects an anomaly in the pressure of a tracing or atrophy in an MRI, the gradient will highlight the spatial coordinates where the pathological evidence resides, allowing for a clinical audit of phenotypic fidelity.

4 Methodological Framework: From Data Acquisition to Post-hoc Audit

4.1 Data Sources and Study Scenarios

The present framework evaluates the detection of Parkinson's disease (PD) through a symmetric audit approach applied to two distinct diagnostic domains.

Case Study I: Structural Neuroimaging. T1-weighted MRI scans obtained from the Parkinson's Progression Markers Initiative (PPMI) database [15] were utilized for the neuroimaging modality. The selected cohort encompasses high technical complexity, integrating data collected through various protocols and hardware from different manufacturers (e.g., GE, Siemens, Philips, and Toshiba), the dataset comprises 344 MRI scans (60% PD and 40% controls), partitioned into training (70%), validation and testing datasets (30%). This diversity of sources introduces variations in the signal-to-noise ratio and tissue contrast, representing a challenging scenario for automated classification and justifying the need for an XAI audit to ensure that model activations focus on neuroanatomical biomarkers.

Case Study II: Dynamics of Motor Tracings. The second modality focuses on the analysis of fine motor patterns through spiral and wave tracings, the dataset comprises 204 individual tracings, partitioned into training (70%), validation (15%), and testing (15%) sets, respectively, with a balanced distribution of 50% PD patients and 50% healthy controls. The data are based on the protocol proposed by Zham et al. [16], which utilizes a digitizing tablet to capture kinematic and pressure variables. Specifically, the composite index of pen speed and pressure (pen-pressure) is analyzed, transforming these dynamic signals into visual representations suitable for processing through CNNs. This modality allows for the evaluation of the motor impairment characteristic of the different stages of PD, contrasting the algorithmic interpretation with the findings reported in the clinical literature.

4.2 MRI Preprocessing and 2.5D Representation

To standardize the input for the MRIs, a preprocessing workflow was implemented:

Standardization and Region of Interest (ROI). The volumes were oriented to Right-Anterior-Superior (RAS) coordinate space (NIfTI format). A brain mask was generated (threshold 0.15 + Largest Connected Component) along with a crop (bounding box) with a margin of $8 \times 8 \times 1$.
Normalization. Min-Max scaling (1–99 percentiles) was applied, followed by an intra-mask Z-score to reduce the inter-subject variability of the multicenter dataset.
2.5D Stack Generation. A representation of 9 contiguous slices centered on the band of highest brain density (mid-brain) was adopted, resized to 256×256 pixels, resulting in an input tensors of (256, 256, 9).

4.3 Symmetric Architecture and Optimization Strategy

Both modalities follow a modular CNN architecture, working with 4 main blocks for the MRI architecture and 5 for the tracings, with high similarity between them; each block consists of a 3 × 3 kernel, Batch Normalization, ReLU activation function, and 2D Max Pooling in the intermediate blocks except the first and the last, converging at a Global Average Pooling (GAP) bottleneck.

4.4 Grad-CAM Visual Attribution Protocol

The visual audit is conducted through a five-stage process:

1. **Extraction.** Retrieval of feature maps and gradients from the final convolutional layer.
2. **Weighting (α).** Calculation of filter importance through the global average of gradients.
3. **ReLU Activation.** Weighted linear combination of the maps, applying a ReLU function to isolate only the features with a positive impact on PD prediction.
4. **Rescaling.** Bilinear interpolation of the resulting map to the original resolution (256 × 256).
5. **Validation.** Superimposition of the heatmap ($\alpha = 0.4$) onto the original image to verify phenotypic fidelity: anatomical correspondence in MRI or points of kinematic doubt in tracings.

5 Results and Preliminary Discussion

5.1 Quantitative Performance and Model Robustness

The evaluation of the proposed framework was conducted across a dual dimension: the model's discriminatory capacity (quantitative metric) and the fidelity of its internal logic (qualitative audit). This approach allows us to determine whether the performance is driven by real biomarkers or spurious correlations. Table 1 shows the performance of both modalities, marking the average across 10 random seeds over 7 key metrics: AUC-ROC, Accuracy, F1-score, NPV, Precision, Sensitivity, and Specificity

Table 1. Average performance profile for MRI and Handwriting Tracings.

Modality	AUC-ROC	Accuracy	F1(+)	NPV	Precision (+)	Sensivity	Specificity
Tracings	0.8828	0.8283	0.8225	0.8113	0.8505	0.7967	0.8600
MRI	*0.7127*	*0.7036*	*0.7290*	*0.6404*	*0.7940*	0.7134	0.6852

The models achieved an average AUC of 71.27 on the PPMI dataset (T1 MRI) and 88.28% on the kinematic tracings, the average was calculated following the seed variability procedure, numbered in the sequence from 42 to 51.

5.2 Qualitative Performance and Pattern Detection

T1 MRI Modality: The Clever Hans Effect and Motion Artifacts. The heatmaps revealed that the network prioritizes atypical anatomical regions. Unlike the motor tracing model, the early convolutional layers did not show visually significant activations; it was in the last spatial layer where the greatest impact was concentrated, reflecting a high-level abstraction strongly biased toward specific regions.

The audit through Grad-CAM managed to evidence a strong convergence toward extraneural regions; specifically, a consistent activation was observed in the retrobulbar region and the posterior optic tract, a finding that would have been impossible without the implementation of XAI methodologies. This ocular focalization forces us to question the nature of the prediction: has the model identified an emerging morphological biomarker associated with oculomotor alterations, or, on the contrary, has it incurred a Clever Hans effect? This second hypothesis suggests that the network is capitalizing on ocular motion artifacts, under a noise source that could perfectly be present in healthy subjects, but statistically overrepresented in the Parkinson's cohort, acting as a confounding variable inherently linked to the patient's motor symptomatology during the scanning process. In essence, the model succeeded by identifying a statistical regularity that lacks biological relevance.

Operating under the phenomenon of shortcut learning (abbreviated learning), the CNN isolates a noise pattern in the ocular zone (exemplified by sample in the Fig. 3) and assigns it a high spurious correlation with the pathological class (PD), largely ignoring the actual dopaminergic atrophy expected in the midbrain. Without the transparency layer provided by Grad-CAM, this algorithmic shortcut would have gone completely unnoticed, masked by performance metrics that are sometimes seemingly high for this class of data, which in certain initialization seeds, managed to report an AUC of up to 84% in experimentation. Once identified, we can address this model vulnerability by incorporating a targeted artifact masking during preprocessing. Applying a more restrictive spatial mask to exclude the periorbital region, in this way, the network would be forced to redirect its attention toward other anatomical structures.

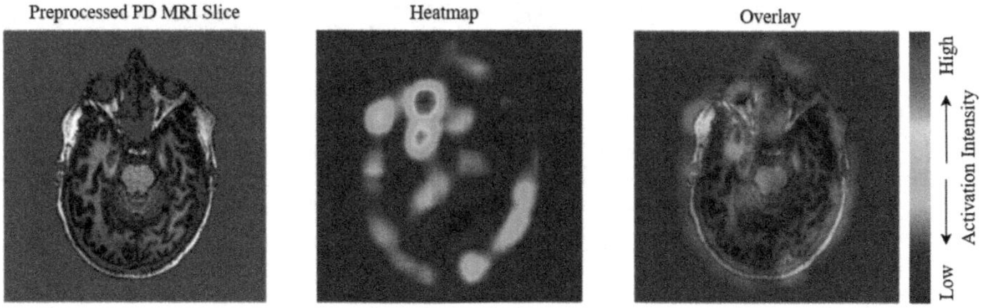

Fig. 3. Evidence of shortcut learning in MRI samples. Grad'CAM heatmaps consistently identify the ocular region as a primary discriminative focal point for classifications. This pattern suggests a reliance on non-clinical artifacts rather than neuroanatomical biomarkers, indicating a potential Clever Hans effect in the MRI model.

Motor Tracing Modality: Digital Biomarkers and Clinical Coherence. During the visual audit process, a clear functional hierarchy was observed across the network's depth, early convolutional layers demonstrated a broad consideration for general spatial organization, including inter-trace spacing and overall structural integrity. As the data reach the last convolutional blocks, the attention narrowed into highly specific focal points; the heatmaps consistently converged on the segments immediately preceding and during sharp curves, capturing kinematic jitter (zig-zagging) and trajectory breaks.

While overtly fragmented tracings were easily classified as PD by the model, its true value was proven in visually continuous PD tracings that appeared normal at first glance. In these subtle cases, the network still pinpointed micro deviations such as curve and precurve motor hesitation, inter-trace spaces and pressure anomalies, outputting PD. This classification refinement is captured in Fig. 4, where the progression across network depths illustrates a transition from widespread spatial attention in earlier layers to a highly focalized activation that isolated specific kinematic markers directly on the trace. This is a contrast to the shortcut learning observed in the neuroimaging model, the tracings network seems to successfully mapped legitimate digital biomarkers, establishing a direct correlation between established medical semiology and what the model actually considers for its predictions.

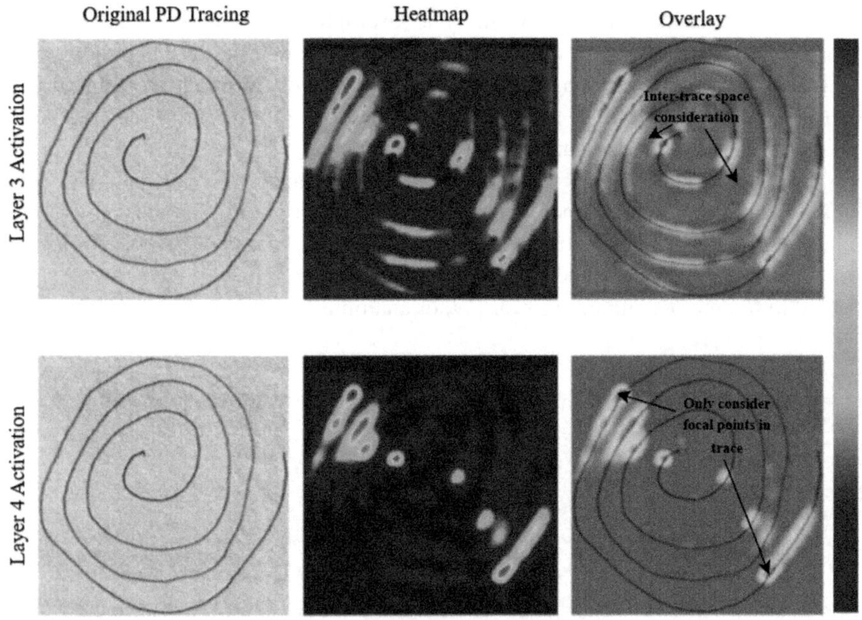

Fig. 4. Comparative Grad-CAM visual audit of a PD spiral tracing across network depths. The top panel (Layer 3) displays a widespread spatial attention and in the bottom (Layer 4) exhibits focalized activation, isolating specific kinematic deviations and pressure anomalies directly on the trace.

6 Conclusions and Future Work

Deep learning models inherently optimize for the path of least resistance; they are algorithmic systems that will exploit any available shortcut to minimize loss, often bypassing true physiological markers, as evidenced by the reliance on periorbital artifacts in our MRI model. Quite the opposite, on the motor tracing side, the other side of the coin can be seen, revealing a strong consistency between the algorithm's internal logic and expert clinical assessment.

This study sought to explore the real utility and necessity of XAI techniques as a functional requirement for clinical viability and not just experimental. While high quantitative metrics can validate the idea of generalization (in the absence of an audit), the spatial coherence of the heatmaps can validate that a model is learning patterns with clinical sense, seeing in them a necessary auditing mechanism to guarantee phenotypic fidelity.

Limitations and Future. We maintain a critical technical agnosticism; as recent literature warns [11, 12], Grad-CAM is a technique that provides useful indications but not explanations of the patterns. Future work will be oriented toward comprehensive multimodal models that incorporate voice signals, and the creation of a proof-of-concept software, taking the first steps for a robust, transparent, and scalable monitoring tool to be tested within a real health system. Additionally, to mitigate the potential bottleneck of manual interpretation, we plan to explore a multi-model architecture pipeline, to represent within a single workflow, not only the output of the XAI techniques but also to provide a preliminary reading and plausible interpretation of the observed patterns.

Acknowledgments. We thank Tijuana Institute of Technology/TECNM and SECIHTI for the financial support provided with the scholarship number: 2026587.

Data used in the preparation of this article were obtained from the Parkinson's Progression Markers Initiative (PPMI) database (www.ppmi-info.org/data). For up-to-date information on the study, visit www.ppmi-info.org.

PPMI – a public-private partnership – is funded by the Michael J. Fox Foundation for Parkinson's Research and funding partners, including: 4D Pharma, Abbvie, AcureX, Allergan, Amathus Therapeutics, Aligning Science Across Parkinson's (ASAP), AskBio, Avid Radiopharmaceuticals, BIAL, BioArctic, Biogen, Biohaven, BioLegend, BlueRock Therapeutics, Bristol-Myers Squibb, Calico Labs, Capsida Biotherapeutics, Celgene, Cerevel Therapeutics, Coave Therapeutics, DaCapo Brainscience, Denali, Edmond J. Safra Foundation, Eli Lilly, Gain Therapeutics, GE HealthCare, Genentech, GSK, Golub Capital, Handl Therapeutics, Insitro, Jazz Pharmaceuticals, Johnson & Johnson Innovative Medicine, Lundbeck, Merck, Meso Scale Discovery, Mission Therapeutics, Neurocrine Biosciences, Neuron23, Neuropore, Pfizer, Piramal, Prevail Therapeutics, Roche, Sanofi, Servier, Sun Pharma Advanced Research Company, Takeda, Teva, UCB, Vanqua Bio, Verily, Voyager Therapeutics, The Weston Family Foundation, and Yumanity Therapeutics.

References

1. Muehlematter, U.J., Daniore, P., Vokinger, K.N.: Approval of artificial intelligence and machine learning-based medical devices in the USA and Europe (2015–20): a comparative analysis. Lancet Digit. Health. **3**(3), e195–e203 (2021)

2. Topol, E.J.: High-performance medicine: the convergence of human and artificial intelligence. Nat. Med. **25**, 44–56 (2019)
3. Kelly, C.J., Karthikesalingam, A., Suleyman, M., et al.: Key challenges for delivering clinical impact with artificial intelligence. BMC Med. **17**(1), 195 (2019)
4. Pathak, A.K., Gupta, M., Jain, G.: Unmasking the Clever Hans effect in AI models: shortcut learning, spurious correlations, and the path toward robust intelligence. Front. Artif. Intell. **8**, 1692454 (2026)
5. Parkinson's Foundation: Statistics. https://www.parkinson.org/understanding-parkinsons/statistics. Last accessed 23 Mar 2026
6. Mattia, G.M., et al.: Deep learning to differentiate parkinsonian syndromes using multimodal magnetic resonance imaging. Mov. Disord. (2025). https://doi.org/10.1002/mds.30300
7. Geirhos, R., et al.: Shortcut learning in deep neural networks. Nat. Mach. Intell. **2**(11), 665–673 (2020)
8. Lekadir, K., Quaglio, G., Garmendia, A.T., et al.: FUTURE-AI: international consensus guideline for trustworthy artificial intelligence in medical imaging. Nat. Commun. **14**(1), 7101 (2023)
9. Holzinger, A., Biemann, C., Pattichis, C.S., Kell, D.B.: Explainable AI for health: from black box to glass box. Inf. Fusion. **78**, 53–71 (2022)
10. Selvaraju, R.R., et al.: Grad-CAM: visual explanations from deep networks via gradient-based localization. In: Proceedings of the IEEE International Conference on Computer Vision (ICCV), pp. 618–626 (2017)
11. Panboonyuen, T.: Seeing isn't always believing: analysis of grad-CAM faithfulness and localization reliability in lung cancer CT classification. arXiv:2601.12826. (2026)
12. Borys, K., et al.: Explainable AI in medical imaging: an overview for clinical practitioners. Eur. J. Radiol. **162**, 110786 (2023)
13. Pöhl, M., et al.: Feature attribution auditing for faithful representation of pathology in medical deep learning. IEEE Trans. Med. Imaging. **43**(2), 512–524 (2024)
14. D'Amour, A., et al.: Underspecification presents challenges for credibility in modern machine learning. J. Mach. Learn. Res. **23**(1), 10237–10297 (2022)
15. Marek, K., et al.: (PPMI): the Parkinson progression marker initiative (PPMI). Prog. Neurobiol. **95**(4), 629–635 (2011)
16. Zham, P., Kumar, D.K., Dabnichki, P., et al.: Distinguishing different stages of Parkinson's disease using composite index of speed and pen-pressure of sketching a spiral. Front. Neurol. **8**, 435 (2017)

From Raw Signals to Personality Scores: The Role of Time-Frequency Feature Engineering in Mobile Sensor-Based OCEAN Prediction

Néstor Leyva-López, Ramón Zatarain-Cabada(✉), María Lucía Barrón-Estrada, and Víctor Manuel Bátiz-Beltrán

TecNM: Instituto Tecnológico de Culiacán, Culiacán, Mexico
{nestor.ll,ramon.zc,lucia.be,victor.bb}@culiacan.tecnm.mx

Abstract. Automatic personality recognition based on data from mobile devices represents a promising alternative to traditional self-reporting methods, which are subject to inherent biases derived from the individual's self-perception. This work proposes a multimodal approach for predicting the five traits of the OCEAN model, integrating signals from mobile device inertial sensors with daily surveys of usage habits, which are collected through the SensorApp platform. The central contribution of the study is a feature engineering pipeline that extends the representation of signals with attributes in the time and frequency domains, and its impact on the performance of four deep learning architectures: MLP, CNN, LSTM, and Transformer. The results show that incorporating these features significantly improves accuracy across all models; the Transformer achieves a MAE of 0.1704 and an MSE of 0.0513 with the full set of attributes, outperforming recent works in the state of the art. K-fold cross-validation analysis further reveals distinct behavior among architectures, with the CNN showing greater robustness to dataset variability. These results confirm that feature selection and enrichment are determining factors in the accuracy of APR based on mobile signals.

Keywords: Personality recognition · Mobile sensors · Feature engineering · Deep learning · Machine learning · OCEAN · Transformer · CNN · LSTM · MLP

1 Introduction

Personality is a psychological construct that includes stable patterns of thought, emotion, and behavior, which directly influence the way individuals interact with their environment and with the devices they use on a daily basis [1]. Traditional assessment of personality relies on self-report methods (mainly standardized tests) which, despite their broad use, present limitations related to cognitive biases and an individual's tendency to answer based on their self-concept rather than their actual behavior [2, 3].

In response to these limitations, automatic personality recognition (APR) has emerged as an interdisciplinary field that aims to infer personality traits from computationally detectable behavioral signals [4]. In this context, mobile devices represent

H. Ponce et al. (Eds.): COMIA 2026, CCIS 3095, pp. 585–596, 2027.
https://doi.org/10.1007/978-3-032-37438-7_46

a prime source of data, given that their global penetration exceeds 75% of the population, and in countries such as Mexico, 81.4% of users spend between 4.7 and 5.9 h per day using them [5, 6]. This omnipresence turns integrated sensors (accelerometer and gyroscope) into continuous, non-invasive instruments for capturing behavior.

However, a key challenge in APR based on mobile sensor signals lies within the proper representation of those signals. Crude readings from accelerometers and gyroscopes are highly variable time series with low sampling rates, and deep learning models do not always effectively capture the relevant information when these signals are used directly as input. Feature engineering (that is, the extraction of statistical attributes that describe the properties of the signals in different domains) is a critical step that can largely determine the predictive performance of the models [7].

Previous work introduced the SensorApp platform as a multimodal system designed to collect personality-related data from mobile sensors and usage surveys [8]. In a separate study, preliminary results evaluating deep learning architectures under different data configurations were reported [9]. While these studies focused on system development and initial modelling, the present paper builds on this research by examining the role of feature engineering in personality prediction. Specifically, we analyze how incorporating statistical attributes in the time and frequency domains affects four deep learning architectures' performance (MLP, CNN, LSTM and Transformer), providing a deeper understanding of their behavior under enriched feature representations.

The rest of the article is organized as follows: Sect. 2 reviews the related literature. Section 3 presents the SensorApp platform and the generated dataset. Section 4 details the feature engineering pipeline. Section 5 describes the neural network architectures used. Section 6 reports the experimental results and their analysis. Finally, Sect. 7 presents conclusions and future work.

2 Related Work

The APR field has evolved significantly in recent years, covering a range of different data types. To provide context for the contribution of this work, we review the literature along three main lines: mobile sensor-based behavioral analysis, APR using deep learning techniques, and feature engineering for sensor-based personality prediction.

2.1 Data from Mobile Sensors for Behavioral Analysis

Sensors embedded in mobile devices have proven to be valuable sources of behavioral information in contexts that go beyond the recognition of physical activity. Ware et al. [10] demonstrated that location data and phone usage time can be used to predict symptoms of depression in college students, achieving an F1-score of 0.86. In a similar vein, Wampfler et al. [11] proposed classifying affective states using color maps derived from tactile events and motion data captured under natural conditions via CNN, reporting Area Under the Curve (AUC) values below 0.85, which suggests room for improvement by enriching the signal representation. More recently, Moshe et al. [12] integrated GPS, physical activity, and sleep data from wearable devices to predict symptoms of anxiety and depression, finding that variations in location and sleep patterns are the most relevant predictors. These studies establish the feasibility of passive mobile sensing for inferring psychological states, but do not address which attributes to extract from motion signals to maximize predictive power.

2.2 Automatic Personality Recognition Using Deep Learning

The APR approach has been addressed using multiple data types. In the textual domain, Arfaoui et al. [13] evaluated combinations of BERT-CNN and BERT-LSTM architectures on the Big-Five Essays dataset, reporting average accuracies of 50.6% and 48.5%, respectively, across the five OCEAN dimensions, which highlights the inherent difficulty of the task even with pre-trained language models. Fu et al. [14] proposed heterogeneous conversational graph networks to capture interdependencies between speakers in dialogues, improving performance over baseline models such as MLP and RGCN, although without reporting precise numerical metrics. In the visual domain, Guenzel et al. [15] explored the prediction of Big Five traits from LinkedIn facial images using CNNs, with applications in corporate recruitment. These works rely on semantically rich data (text, images, voice) requiring greater infrastructure or invasiveness than mobile inertial sensors.

2.3 Mobile Sensor-Based APR and Feature Engineering

The works that are closest to this study are those that use inertial sensor signals to infer personality. Sze et al. [16] collected accelerometer data from 144 participants and applied Random Forest and XGBoost for binary classification of OCEAN traits, achieving F1 scores of up to 0.78. Kovacevic et al. [17] demonstrated that short-term behavioral data can predict personality traits at two levels, reaching 74.5% accuracy with single-day data and 84.5% after ten weeks. Ibrar et al. [18] identified personality traits from gait patterns using SVM, exceeding 94% accuracy. Gao et al. [19] proposed a multi-task, multi-view framework estimating Big Five traits from smartphone usage snapshots, reporting MAE 0.256–0.510 and RMSE 0.305–0.647, being the direct benchmark for this work. Overall, most studies employ simple signal representations without systematically exploring time and frequency domain enrichment, which is the gap this study addresses.

3 SensorApp: Data Collection Platform and Dataset

The dataset used in this study was created using the SensorApp platform, a multimodal platform developed specifically for collecting APR-oriented data from the everyday use of mobile devices. A detailed description of its architecture and data collection protocol has been reported in previous studies [8, 9]; this section presents a concise description focused on the characteristics of the resulting dataset that are relevant to the experiments described in the following sections.

3.1 The OCEAN Personality Model and the IPIP-50 Instrument

The Big Five personality traits model, widely known as the OCEAN model, is one of the most widely accepted and empirically supported frameworks in contemporary psychology for describing and assessing human personality [20]. The model proposes that personality can be characterized by five relatively stable dimensions:

Openness to Experience, Conscientiousness, Extraversion, Agreeableness, and Neuroticism. Each dimension is a continuous value in [0, 1], making personality prediction a multivariate regression task. The model's scientific robustness and cross-cultural validity are well documented across diverse population groups, with associations to outcomes such as academic success and mental health [21], establishing it as the standard benchmark in psychological research and APR systems.

To obtain OCEAN reference values, the IPIP-50 test was used [22, 23]: 50 self-report items rated on a five-point Likert scale (1 = "very inaccurate", 5 = "very accurate"), grouped in five blocks of 10 items. Each dimension score is normalized to [0, 1] and serves as the personality label for model training and evaluation. The IPIP-50 was selected for its public domain status, broad adoption in APR research, and proven reliability [4].

3.2 Platform Description

SensorApp operates with two complementary modes. The web mode allows participants to complete the IPIP-50 test, the results of which form the personality labels in the dataset. The mobile mode, developed for Android devices using the React Native framework on Expo Go, collects two types of information over a 14-day period: data from the device's inertial sensors (accelerometer and gyroscope, both configured to a sampling rate of 1.0 Hz with a sensitivity threshold of 1.2 units of acceleration) and responses to a daily six-question survey on mobile device usage habits, including time spent on social media, types of apps used, and emotions experienced during use [8]. Sensor data collection occurs only while the user is responding to the survey, ensuring that the captured signals correspond to an active interaction with the device. All data is automatically stored in a non-relational database using Google Firebase (Fig. 1).

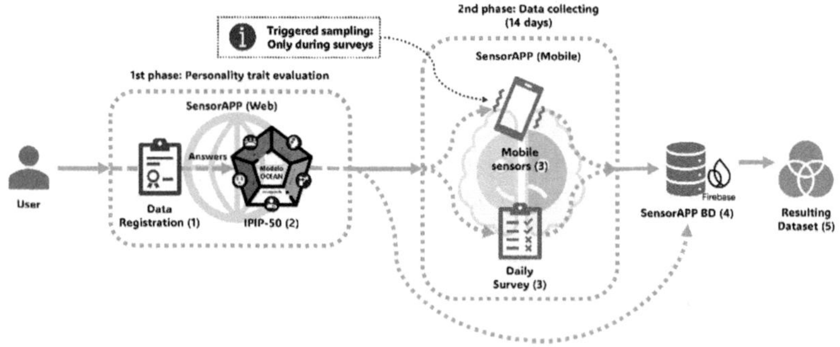

Fig. 1. Multimodal infrastructure for collecting inertial data from mobile sensors and personality traits using the OCEAN model.

3.3 Sample Group and Collected Data

This study featured 126 undergraduate students from the National Technological Institute of Mexico: Culiacán Campus, with an average age of 22.4 years (94 men, 32 women). Participation was voluntary, and the protocol was approved by the institutional ethics committee; all participants signed informed consent forms, and their data were anonymized. Out of the 126 participants, only 16 completed the full scheme (registering on the web platform, taking the IPIP-50 test, and using the mobile app during the established period) generating 160 records with an average of 10 days of activity per user. Each record represents an individual daily session of interaction with the survey, capturing approximately 1000 ms of sensors signals per axis. The remaining 110 participants completed only part of the protocol: 52 took only the IPIP-50 test on the web platform, and 58 used only the mobile app, generating 198 additional records. For the prediction experiments reported in this paper, only 160 records from the full scheme group were used, as they are the only ones containing the three required data types: OCEAN scores, sensor signals, and survey responses. Data partitioning (80% training and 20% testing) and K-Fold cross-validation were performed at the record level. While this approach allows the models to learn specific behavioral states, the limited sample size (n = 16) is recognized as a technical proof of concept.

Table 1 summarizes the dataset composition. The 12.7% completion rate is consistent with longitudinal mobile sensing studies [16, 17], reflecting known challenges: discomfort with passive data collection, Android-only compatibility, and declining engagement after the first week.

Table 1. Data collected from the SensorApp application.

Users	Subset	IPIP-50	Accelerometer	Gyroscope	Questionnaire	Total of records
16	Complete Scheme	✓	✓	✓	✓	160
52	SensorApp Web	✓				0
58	SensorApp Mobile		✓	✓	✓	198

4 Feature Engineering Pipeline

The proper representation of inertial sensor signals is a key factor in the predictive ability of mobile-based APR models. Raw accelerometer and gyroscope signals are low-sampling-rate time series (1.0 Hz in this study) whose relevant behavioral information is concentrated in statistical and spectral properties not directly observable in the original signal [7]. To address this, a preprocessing and feature extraction pipeline was designed operating in four sequential stages, as illustrated in Fig. 2: initial data loading, signal normalization via StandardScaler, extraction of time domain attributes (variance, IQR,

RMS, entropy) and frequency domain attributes (DFFT, DCT, spectral energy, PFR), and dataset partitioning (80% training, 20% test). The resulting feature vector is used to train and evaluate four deep learning architectures (MLP, CNN, LSTM, and Transformer) compared using MSE, MAE, and K-Fold cross-validation.

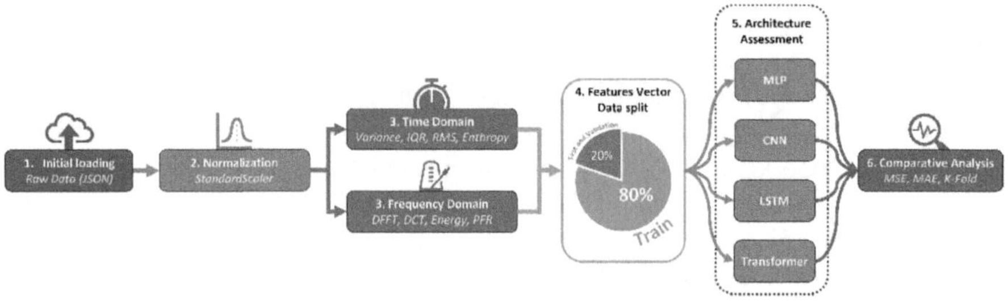

Fig. 2. Feature engineering pipeline for the SensorApp dataset.

4.1 Loading and Initial Extraction

The data stored in Firebase is exported in JSON format and loaded as a tabular structure. At this stage, only those user records that contain the three required data types are selected: accelerometer and gyroscope measurements on the X, Y, and Z axes; responses to the daily survey; and OCEAN personality scores derived from the IPIP-50 test. The survey responses, which are categorical in nature, are converted to numerical values using a mapping dictionary that assigns a unique integer to each response option, preserving ordinal information where appropriate. The OCEAN values are normalized by rounding to the first decimal place to ensure consistency in the representation of the labels. This decision was made to reduce the model's sensitivity to minor fluctuations in self-reports and to focus the learning process on the primary magnitudes of the OCEAN traits.

4.2 Sensor Signal Normalization

The accelerometer and gyroscope signals are normalized using StandardScaler, a technique that centers each variable at zero and scales its variance to one. This step is essential to remove differences in magnitude between the different axes and between the two sensors (whose units of measurement are heterogeneous) and to reduce the influence of outliers caused by sudden movements or drops of the device. As basic descriptive statistics, the mean and standard deviation of each normalized axis are calculated and incorporated as the first attributes of the feature vector.

4.3 Time Domain Features

The time domain analysis assesses how sensor signals evolve over a given acquisition period, allowing us to identify the dynamics and variability of the user's movement. For each axis (X, Y, Z) of both sensors, the following attributes are calculated:

- Variance (S^2): measures the degree of dispersion of the samples relative to the mean, quantifying the amplitude of the fluctuations in movement.
- Interquartile range (IQR): the difference between the 75th percentile (Q3) and the 25th percentile (Q1), calculated using linear interpolation. It represents robust statistical dispersion in the presence of outliers.
- Root Mean Square (RMS): the square root of the average of the squared values, which measures the magnitude and average intensity of the motion.
- Entropy (H): measures the randomness and unpredictability of the signal, quantifying the degree of uncertainty in the movement patterns.

These attributes complement one another: variance and IQR capture movement dispersion from global and robust perspectives; RMS reflects energetic intensity; and entropy characterizes interaction complexity, shown to correlate with openness and neuroticism [16].

4.4 Frequency Domain Features

The frequency domain analysis transforms time signals to reveal periodic components and rhythmic patterns that are not detectable in the time domain. For each axis of both sensors, the following features were calculated in windows defined by the time length of each survey session:

- Discrete Fast Fourier Transform (DFFT): decomposes the signal into its frequency components, allowing analysis of the amplitude of each frequency present in the device's movements.
- Discrete Cosine Transform (DCT): expresses the signal as a sum of cosine functions at different frequencies, useful for detecting rhythmic patterns in the movements.
- Spectral energy (Ex): sum of squared frequency component magnitudes, a general indicator of the level of physical activity during the session.
- Frequency power range (PFR): difference between the maximum and minimum amplitudes of the frequency spectrum, which characterizes the extent and intensity of the spectral distribution of the movement.

Frequency attributes capture rhythmic or periodic movement patterns (such as typing cadence or device repositioning frequency) only identifiable in the spectral domain [7].

4.5 Features Vector Composition

The final feature vector combines: normalized signals from six axes, descriptive statistics (mean and standard deviation per axis), four time domain features per axis, four frequency domain features per axis, and survey responses. This enrichment process increases the input dimensionality from 12 dimensions in Configuration A (raw signals and survey) to 66 dimensions in configuration B (augmented feature set). The dataset is split into 80% training and 20% test, with the same partition applied to all architectures for comparability.

5 Deep Learning Architectures

Four deep learning architectures were selected based on architectural diversity (each applies a distinct inductive bias) allowing analysis of which structure best captures correlations between sensor signals and personality traits: MLP, CNN, LSTM, and Transformer. The experimental setup and model training were conducted in Google Colab using an NVIDIA Tesla T4 with 15 GB VRAM. All architectures were implemented using TensorFlow/Keras and trained for 100 epochs using the Adam optimizer with a learning rate of 0.001, batch size of 8, and Mean Squared Error (MSE) as the loss function. The models feature a shared output layer of 5 linear neurons corresponding to the OCEAN dimensions.

5.1 Multilayer Perceptron (MLP)

The MLP serves as the baseline architecture for the experiment due to its structural simplicity [25]. It consists of an input layer with 64 neurons using ReLU activation, a hidden layer with 32 neurons using ReLU activation, and a standard output layer. Since it does not model dependencies between input positions, treating all features as independent. This makes it sensitive to feature vector quality but unable to exploit relational or temporal structures.

5.2 Convolutional Neural Network (CNN)

This CNN incorporates two 1D convolutional layers (64 and 32 filters) with max-pooling and a dense layer of 64 neurons with ReLU before the output [26]. 1D convolutions detect local patterns in the feature vector (such as co-occurring attributes across the three axes of the same sensor) giving a structural advantage over the MLP and explaining its robustness in cross-validation.

5.3 Long Short-Term Memory (LSTM)

The LSTM consists of two stacked recurrent layers (64 and 32 units), the first returning sequences to feed the second, followed by a dense layer of 16 ReLU neurons [27]. Its gating mechanism models long-term sequential dependencies; however, the 1.0 Hz sampling rate and short session lengths limit its advantage over the CNN, as reflected in the results.

5.4 Transformer

This architecture incorporates a multi-head attention block (4 heads) with 32-neuron feedforward networks, layer normalization, and dropout [28], followed by Global Average Pooling 1D and two dense layers (32 and 16 ReLU neurons) before the output. Its self-attention mechanism dynamically assigns importance to each attribute relative to all others without locality constraints, making it especially suited for enriched feature vectors where distant interactions (such as accelerometer spectral energy and gyroscope entropy) inform OCEAN trait prediction.

6 Tests and Results

This section shows the results from evaluating the four architectures described on the SensorApp dataset, organized into three complementary analyses: an unsupervised exploratory analysis of the sensor data, a comparison of the models' predictive performance under two feature configurations, and an assessment of stability using K-fold cross-validation. The evaluation metrics used are Mean Squared Error (MSE) and Mean Absolute Error (MAE), which are appropriate for multiple regression tasks with outputs in the range [0, 1].

6.1 Sensors Data Exploratory Analysis

Prior to training, an unsupervised clustering analysis of sensor motion features was performed to verify dataset integrity and explore whether movement patterns reflect users' dominant personality traits.

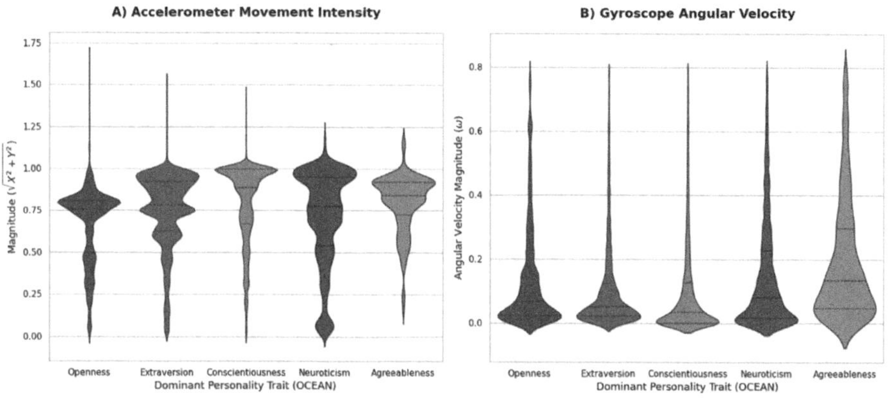

Fig. 3. Distribution of movement intensity across dominant personality traits: (A) Accelerometer linear magnitude and (B) Gyroscope angular velocity.

The statistical distribution of movement intensity, as captured by sensor magnitude, reveals distinct behavioral signatures within the OCEAN traits (Fig. 3). Significant variations in movement consistency can be seen for the accelerometer (Fig. 3A): conscientious users exhibit a higher median intensity and a tighter distribution, in contrast to individuals scoring high on neuroticism, whose behavior is less consistent and more variable.

By contrast, the magnitude of the gyroscope (Fig. 3B) shows high consistency across most personality traits, with data predominantly concentrated near the origin and minimal variation between users. This confirms that, regardless of the trait, rotational intensity remains relatively uniform, while the linear acceleration magnitude provides superior discriminative power for APR. Therefore, the gyroscope mainly contributes to the model through its spectral attributes rather than its magnitude distribution.

6.2 Performance Comparison Between Feature Configurations

Two input configurations were evaluated for each architecture to quantify the impact of feature enrichment on prediction accuracy:

- Configuration A: Normalized sensor signals and survey responses only, without additional statistical attributes.
- Configuration B: Full feature vector, additionally incorporating the time-domain and frequency-domain attributes described in Sect. 5.

The results are presented in Table 2.

Table 2. MSE and MAE for the four architectures under both configurations.

Model	MSE (Config. A)	MAE (Config. A)	MSE (Config. B)	MAE (Config. B)
MLP	0.1144	0.2867	0.0603	0.1956
CNN	0.0605	0.1961	0.0650	0.1766
LSTM	0.0721	0.2207	0.0601	0.1935
Transformer	0.0712	0.2209	0.0513*	0.1704*

The results reveal two notable trends. First, Configuration B improves all architectures over Configuration A, with MAE reductions ranging from 7.9% (CNN) to 46.9% (MLP), confirming that feature enrichment is beneficial regardless of architecture, though its magnitude varies across models.

Second, a performance hierarchy shift is observed: Configuration A favors the CNN (MSE = 0.0605, MAE = 0.1961), while Configuration B elevates the Transformer (MSE = 0.0513, MAE = 0.1704). This is consistent with Sect. 6: the CNN excels at local patterns in simpler signals, while the Transformer's self-attention exploits global relationships that only emerge in richer feature vectors. The Transformer with Configuration B (MAE = 0.1704, RMSE = 0.2265) substantially outperforms Gao et al. [19], who reported MAE 0.256–0.510 and RMSE 0.305–0.647.

6.3 K-Fold Cross-Validation Evaluation

To assess model stability and generalizability, K-fold cross-validation was applied to the full feature vector (Configuration B). Results are presented in Table 3.

Table 3. K-fold cross-validation MAE and MSE (Configuration B).

Model	Average MAE	Average MSE
CNN	0.1689*	0.0511*
Transformer	0.1738	0.0528
LSTM	0.1932	0.0608
MLP	0.1936	0.0601

Cross-validation shows the CNN regaining the lead (MAE = 0.1689, MSE = 0.0511 vs. Transformer's 0.1738 and 0.0528). The Transformer's greater variance across partitions is likely due to its modeling capacity exceeding what 160 records can consistently support. The CNN's fewer parameters and local inductive bias yield greater robustness, underscoring the importance of evaluating both fixed-split accuracy and cross-validation stability when selecting architectures for limited-size datasets.

7 Conclusions and Future Work

This paper analyzed the impact of feature engineering on four deep learning architectures for APR from mobile inertial sensor signals. The proposed pipeline enriches raw signals with time domain attributes (variance, IQR, RMS, entropy) and frequency domain attributes (DFFT, DCT, spectral energy, PFR), demonstrating that this enrichment is a determining factor in predictive performance across all models.

Configuration B improves MAE by 7.9% (CNN) to 46.9% (MLP) over baseline. The Transformer leads on the fixed split (MAE = 0.1704, MSE = 0.0513, RMSE = 0.2265), outperforming Gao et al. [19], while the CNN achieves the lowest K-fold MAE (0.1689). This practical finding highlights that for limited datasets, the CNN's stability may be preferable to the Transformer's higher capacity.

It is important to note that the system acts as an approximator of self-report labels (IPIP-50), which possess their own perception biases. Future work will expand the sample to include statistical significance tests and mitigate potential demographic biases. These results confirm the viability of passive mobile sensing for OCEAN trait prediction without relying exclusively on self-reports, offering a non-invasive multimodal approach applicable to adaptive technology, well-being monitoring, and digital personalization.

Future work will focus on: (1) expanding the dataset with greater participant diversity to improve generalization; (2) exploring generative data augmentation to improve Transformer stability; (3) incorporating audio and text modalities via NLP; and (4) conducting feature importance analysis to identify which sensor and survey attributes contribute most to each OCEAN dimension.

References

1. VandenBos, G.R. (ed.): APA Dictionary of Psychology, 2nd edn. American Psychological Association (2015)
2. Rendón-Macías, M.E., Valenzuela, M., Villasís-Keever, M.Á.: Biases in diagnostic test studies. Rev. Alerg. Mex. **67**(2) (2020)
3. Concha, D., Bilbao, M.Á., Gallardo, I., Páez, D., Fresno, A.: Cognitive biases and their relationship to subjective well-being. Salud & Sociedad. **3**(2) (2012)
4. Vinciarelli, A., Mohammadi, G.: A survey of personality computing. IEEE Trans. Affect. Comput. **5**(3), 273–291 (2014)
5. United Nations: Más del 75% de la población mundial tiene un teléfono celular. United Nations (2023)
6. Islas Fosados, A., Vera Pedroza, A., Cárcamo Segura, F.G., Gómez Fajardo, E.: El uso de dispositivo móvil en estudiantes universitarios: un estudio cuantitativo. LATAM Revista Latinoamericana de Ciencias Sociales y Humanidades. **6**(3), 1222–1232 (2025)

7. Amato, A., di Lecce, V.: Data preprocessing impact on machine learning algorithm performance. Open Comput. Sci. **13**(1) (2023)
8. Zatarain-Cabada, R., Leyva López, N., Barrón Estrada, M.L., Bátiz Beltrán, V.M., Escalante Balderas, H.J.: SensorAPP: a methodology to create a dataset for personality recognition using mobile devices. In: CIMPS 2024 (2024)
9. Leyva-López, N., Zatarain-Cabada, R., Barrón-Estrada, M.L., Escalante, H.J.: SensorApp: a digital mirror — personality recognition with deep learning using a mobile usage dataset. IEEE Lat. Am. Trans. (2025). https://doi.org/10.1109/TLA.2026.11435569
10. Ware, S., et al.: Predicting depressive symptoms using smartphone data. Smart Health. **15** (2020). https://doi.org/10.1016/j.smhl.2019.100093
11. Wampfler, R., Klingler, S., Solenthaler, B., Schinazi, V.R., Gross, M., Holz, C.: Affective state prediction from smartphone touch and sensor data in the wild. Underst. Child. **2022** (2022). https://doi.org/10.1145/3491102.3501835
12. Moshe, I., et al.: Predicting symptoms of depression and anxiety using smartphone and wearable data. Front. Psychol. **12** (2021). https://doi.org/10.3389/fpsyt.2021.625247
13. Arfaoui, N., Mkhini, M., Sidibe, A.S., Baron, B., Wallelign, S.: Unveiling human essence: deep learning in personality traits detection. Int. J. Comput. Inf. Syst. Ind. Manag. Appl. **17** (2025)
14. Fu, Y., Song, H., Zhao, T., Kawahara, T.: Enhancing personality recognition in dialogue by data augmentation and heterogeneous conversational graph networks. arXiv. (2024)
15. Guenzel, M., Kogan, S., Niessner, M., Shue, K.: AI Personality Extraction from Faces: Labor Market Implications. SSRN (2025)
16. Sze, W.Y.S., Herrero, M.P., Garriga, R.: Personality trait inference via mobile phone sensors: a machine learning approach. arXiv. (2024)
17. Kovačević, N., Holz, C., Günther, T., Gross, M., Wampfler, R.: Personality trait recognition based on smartphone typing characteristics in the wild. IEEE Trans. Affect. Comput. **14**(4), 3207–3217 (2023)
18. Ibrar, K., Fayyaz, A.M., Khan, M.A., Alhaisoni, M., Tariq, U., Jeon, S.: Human personality assessment based on gait pattern recognition using smartphone sensors. Comput. Syst. Sci. Eng. **46**(2) (2023)
19. Gao, S., Zhang, X., Li, W., Lin, M.: PersonalitySensing: a multi-view multi-task learning approach for personality detection based on smartphone usage. In: ACM MM 2020 (2020)
20. Goldberg, L.R.: The structure of phenotypic personality traits. Am. Psychol. **48**(1), 26–34 (1993)
21. Mezquita, L., et al.: Cross-cultural examination of the big five personality trait short questionnaire. PLoS One. **14**(12) (2019)
22. Goldberg, L.R.: The development of markers for the big-five factor structure. Psychol. Assess. **4**(1), 26–42 (1992)
23. Goldberg, L.R., et al.: The international personality item pool and the future of public-domain personality measures. J. Res. Pers. **40**(1), 84–96 (2006)
24. Bisong, E.: The Multilayer Perceptron (MLP). In: Building Machine Learning and Deep Learning Models on Google Cloud Platform. Apress (2019)
25. LeCun, Y., Bengio, Y., Hinton, G.: Deep learning. Nature. **521**, 436–444 (2015)
26. Lindemann, B., Müller, T., Vietz, H., Jazdi, N., Weyrich, M.: A survey on long short-term memory networks for time series prediction. Procedia CIRP. **99**, 650–655 (2021)
27. Vaswani, A., et al.: Attention is all you need. arXiv, abs/1706.03762. (2017)

A Mobile-Based System for Real-Time Classification of Static Mexican Sign Language Gestures Using Deep Learning

Brian Rodolfo Martinez-Grajales[1,2], Francisco López-Orozco[1,2](✉), Rogelio Florencia-Juárez[1,2], and Josué Domínguez-Guerrero[1,2]

[1] Instituto de Ingeniería y Tecnología (IIT), Universidad Autónoma de Ciudad Juárez, Ciudad Juárez, Mexico
{al176714,rogelio.florencia,josue.dominguez}@uacj.mx
[2] Laboratorio de Inteligencia Artificial y Analítica de Datos (LIAAD), Universidad Autónoma de Ciudad Juárez, Ciudad Juárez, Mexico
francisco.orozco@uacj.mx

Abstract. Mexican Sign Language (Lengua de Señas Mexicana, LSM) is the primary communication mechanism for millions of deaf individuals in Mexico; however, it remains largely inaccessible to the hearing population. This lack of mutual understanding creates significant communication barriers in everyday situations such as commerce, education, and public services. This paper presents a mobile-based system capable of classifying static LSM gestures into Spanish text and synthesized speech in real time using only a smartphone camera. The proposed system leverages computer vision and deep learning, employing a convolutional neural network (CNN) trained on a custom dataset of over 20,000 labeled images. The model is optimized and deployed using TensorFlow Lite, enabling fully offline execution on mobile devices. Experimental evaluation shows accuracy above 95% under controlled conditions and above 85% on mid-range smartphones, with inference times below 100 ms. The results demonstrate the feasibility of an accessible, low-cost, and privacy-preserving mobile solution for basic sign language translation.

Keywords: Mexican Sign Language · Computer Vision · Deep Learning · Mobile Applications · Accessibility

1 Introduction

Effective communication is fundamental to social inclusion; however, deaf individuals continue to face persistent communication barriers, particularly in countries where sign languages are not widely understood [1,2]. In Mexico, Mexican Sign Language (Lengua de Señas Mexicana, LSM) is legally recognized as a national language, yet its use remains limited outside specialized educational contexts.

H. Ponce et al. (Eds.): COMIA 2026, CCIS 3095, pp. 597–607, 2027.
https://doi.org/10.1007/978-3-032-37438-7_47

Recent advances in artificial intelligence, especially in computer vision and deep learning, have enabled significant progress in gesture and sign language recognition [3,4]. Nevertheless, many existing approaches rely on specialized hardware, controlled environments, or cloud-based services [5,6], reducing their scalability, accessibility, and privacy.

Mobile devices provide a promising alternative due to their widespread availability and increasing computational capabilities. This work proposes an offline, mobile-based system for translating static LSM gestures into text and synthesized speech using a smartphone camera and an embedded convolutional neural network.

2 Related Work

Automatic sign language recognition has been explored using various sensing modalities and computational techniques. Systems such as SignAll employ depth sensors and controlled capture environments to translate American Sign Language in real time [5]. While effective, these solutions are expensive and impractical for everyday use.

Other approaches focus on mobile or educational applications. Commercial tools like HandTalk translate spoken language into sign representations but offer limited bidirectional interaction and depend on internet connectivity [6]. In the Mexican context, most applications emphasize sign language learning through visual dictionaries rather than real-time interpretation [7].

From an academic perspective, convolutional neural networks (CNNs) have demonstrated strong performance in static hand gesture classification tasks [8,9]. However, many studies evaluate models only in desktop environments and do not address deployment constraints on mobile platforms. This work bridges that gap by focusing on efficient on-device inference and real-world usability.

3 System Overview

The proposed system is a mobile-based architecture designed to perform real-time classification of static Mexican Sign Language (LSM) gestures using on-device computer vision and deep learning techniques. The system is implemented as a sequential processing pipeline that transforms raw visual input into semantic output in the form of text and synthesized speech. An overview of the pipeline is shown in Fig. 1. The architecture is explicitly designed to operate under the computational and memory constraints of mid-range smartphones, while ensuring low latency, offline execution, and user privacy. All processing stages are executed locally on the device, eliminating dependency on cloud services or network connectivity.

A variety of computational tools were selected for the project's development. For real-time hand tracking, MediaPipe was chosen. This framework is employed because it offers a unique combination of accuracy, efficiency, and ease of implementation. Unlike motion sensor–based systems or wearable devices (such as smart gloves), MediaPipe does not require specialized hardware, which reduces costs and makes the technology more accessible. Furthermore, its ability to process data directly on the mobile device eliminates the need to transmit information to external servers, thereby improving system speed and ensuring user privacy [13].

Image processing through frameworks such as OpenCV represents the most viable approach, owing to its accessibility, scalability, and robustness in adapting to diverse environmental conditions. Consequently, it serves as a foundational basis for the advancement of gesture and sign language recognition systems [14].

To implement CNNs in this project, TensorFlow is employed, an open-source library developed by Google that facilitates the creation and training of deep learning models. TensorFlow is widely adopted within both the scientific and business communities due to its flexibility, scalability, and support for specialized hardware such as GPUs and TPUs. To ensure that the application is accessible to the majority of users, the developed machine learning model is implemented using TensorFlow Lite, an optimized version of TensorFlow specifically designed for mobile devices and resource-constrained environments [12].

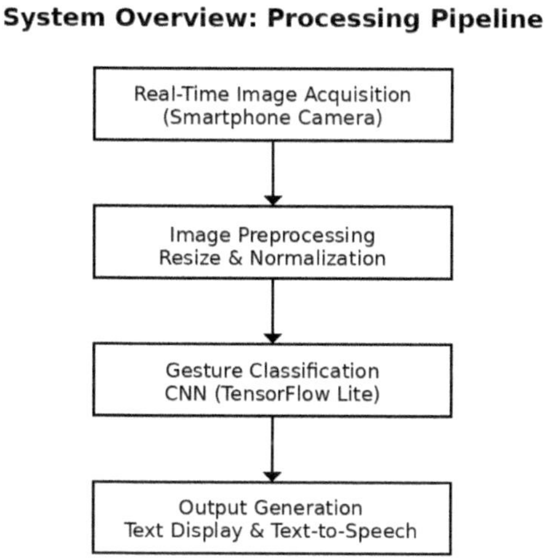

Fig. 1. Processing pipeline of the proposed mobile-based system for static Mexican Sign Language classification.

3.1 Real-Time Image Acquisition

The pipeline begins with real-time image acquisition using the built-in RGB camera of the smartphone. Video frames are captured continuously at a fixed resolution and frame rate to balance temporal consistency and computational cost. This stage is responsible for providing a steady stream of visual data suitable for real-time gesture recognition. The acquisition module is designed to operate in unconstrained environments, without assumptions about background uniformity or lighting conditions. No external sensors, depth cameras, or wearable devices are required, which significantly improves portability and usability. By relying solely on standard camera hardware, the system can be deployed on a wide range of commercially available mobile devices.

3.2 Image Processing

Each captured frame is forwarded to the image preprocessing stage, where it is transformed to match the input specifications of the neural network. Preprocessing includes resizing the image to a fixed spatial resolution and normalizing pixel intensities to a common numerical range. This stage is designed to be computationally lightweight in order to preserve real-time performance. While no explicit hand segmentation or background removal is performed on the mobile device, normalization and resizing reduce variability caused by illumination changes and camera noise. These operations help stabilize neural network inference while avoiding the overhead of more complex image processing techniques.

3.3 Gesture Classification

Gesture classification constitutes the core computational component of the system. The preprocessed image is passed to a convolutional neural network trained to recognize static LSM gestures. The model is deployed using TensorFlow Lite, enabling efficient on-device inference with reduced memory footprint and lower latency. For each input frame, the network produces a probability distribution over the set of supported gestures. To increase temporal stability and reduce spurious predictions, confidence-based filtering is applied. Predictions below a predefined confidence threshold are discarded, preventing rapid fluctuations in output caused by ambiguous frames or transient visual artifacts. This design choice improves robustness in real-world usage scenarios, particularly when the user's hand position or illumination conditions vary between frames.

3.4 Output Generation

In the final stage of the pipeline, recognized gestures are translated into user-oriented output. The predicted gesture label is displayed as text on the mobile interface, providing immediate visual feedback. Additionally, the system supports optional text-to-speech synthesis, enabling auditory output for hearing users. The dual-modality output facilitates communication between deaf and hearing individuals, allowing the system to function as an assistive communication bridge in everyday situations. Since output generation occurs locally and uses lightweight synthesis techniques, it introduces minimal latency and does not compromise real-time interaction.

3.5 Design Considerations

The system architecture reflects several key design principles. A real-time operation because all stages are optimized to maintain interactive frame rates, an offline execution since no network connectivity is required, ensuring privacy and robustness, hardware accessibility due to the system relying exclusively on standard smartphone components, and scalability since the modular pipeline allows future integration of dynamic gestures, facial expressions, or grammatical modeling. These considerations make the proposed architecture suitable for deployment in practical, real-world environments, particularly in resource-constrained or privacy-sensitive contexts.

4 Dataset and Model Architecture

A custom dataset was created specifically for this study to support the recognition of static Mexican Sign Language (LSM) gestures in real-world conditions. The dataset consists of more than 20,000 labeled RGB images representing static hand gestures. It includes the complete LSM manual alphabet (A-Z) as well as over 100 frequently used signs corresponding to common words and expressions encountered in daily communication (e.g. *Hola*, *Hoy*). Between 500 and 1300 images were captured, each showing the hand specifically performing every sign. Images were captured using standard smartphone cameras to ensure compatibility with the target deployment environment. Data collection was conducted under varying lighting conditions, backgrounds, and viewing angles in order to increase variability and reduce overfitting to controlled scenarios. This variability is essential for improving model robustness when deployed on mobile devices operating in unconstrained environments. Each image was manually labeled according to its corresponding LSM gesture. Although the private dataset collection did not account for a more diverse range of users (e.g., different skin tones, hand sizes, and environmental backgrounds) to mitigate bias, data augmentation techniques were employed during training to enhance generalizability. These techniques included random rotations, scaling, horizontal translations, and brightness adjustments. Such transformations were designed to simulate realistic variations in hand positioning, distance from the camera, and illumination conditions.

Prior to training, all images were resized to a fixed resolution and normalized to a common range to ensure numerical stability during inference. The whole dataset was partitioned into three subsets: training (70%), validation (15%), and testing (15%). The validation set was used for hyperparameter tuning and early stopping, while the test set was reserved exclusively for final performance evaluation. This work demonstrates that it is possible to generate a publicly available dataset for Mexican Sign Language, given the current scarcity of such resources—particularly those designed for mobile-oriented applications focusing on static gestures. In a subsequent version of this project, the aim is to strengthen its scope and make this dataset available to the community, following rigorous validation. The gesture classification module is based on a convolutional neural

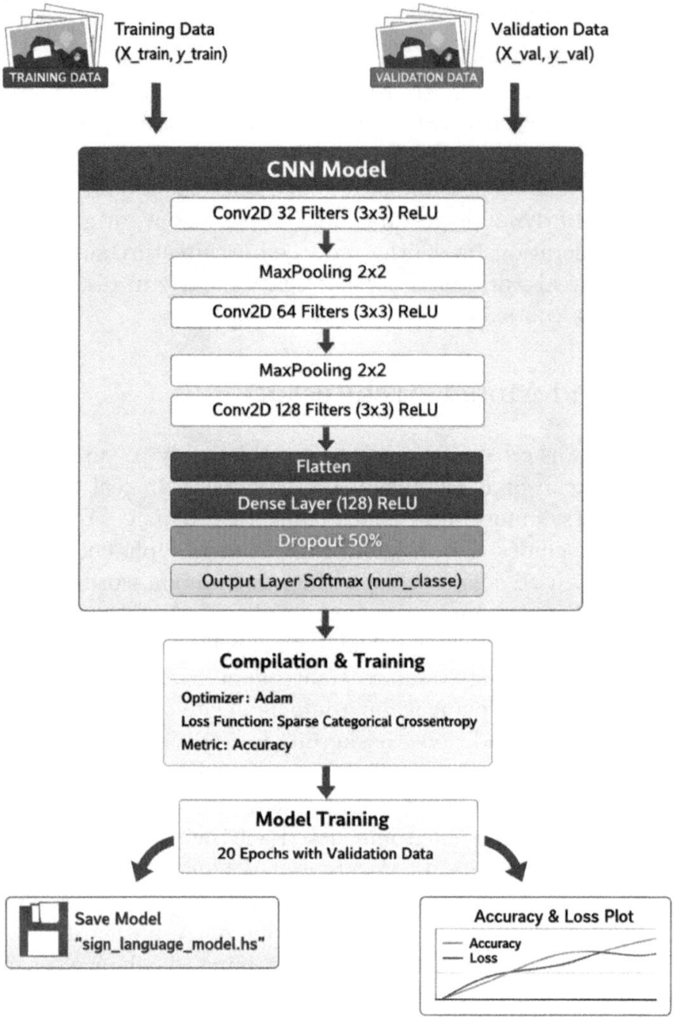

Fig. 2. CNN model definition, training and validation diagram.

network (CNN) designed to balance classification accuracy and computational efficiency. The architecture follows a sequential design optimized for deployment on mobile devices with limited processing power and memory. The network consists of 3 convolutional blocks. Each block includes a convolutional layer with a small kernel size (3×3), followed by a rectified linear unit (ReLU) activation function and a max-pooling layer (2×2) for spatial downsampling. Each convolutional block with 32, 64 and 128 filters. This design allows the network to progressively extract hierarchical features from the input images while reducing dimensionality and computational cost. Following the convolutional blocks, the extracted feature maps are flattened and passed to a fully connected layer. Dropout regularization is applied at this stage to mitigate overfitting, given the variability present in real-world image acquisition. The final output layer uses a *softmax* activation function to produce a probability distribution over the set of supported gestures. The model was trained using the Adam optimization algorithm and categorical cross-entropy loss. Training was performed for a 20 epochs with early stopping based on validation loss to prevent overfitting. After training, the model was converted to TensorFlow Lite format to enable efficient on-device inference. To further optimize performance on mobile hardware, post-training quantization was applied, reducing both model size and inference latency while maintaining acceptable classification accuracy. This optimization is critical for achieving real-time performance in mobile environments and for ensuring energy-efficient execution. The resulting architecture provides a suitable trade-off between accuracy, latency, and resource consumption, making it well suited for real-time static sign recognition on smartphones. Figure 2 shows a diagram of the CNN model definition, training and validation process.

5 Experimental Results

The performance of the proposed mobile-based sign language recognition system was evaluated through a comprehensive experimental protocol designed to assess classification effectiveness, robustness, and real-time feasibility. The evaluation focuses not only on overall accuracy but also on class-level behavior, error patterns, and deployment constraints, providing a holistic assessment of system performance in both controlled and real-world contexts. All experiments were conducted on the test subset described in Sect. 4, which was not used during training or validation.

5.1 Desktop Performance Evaluation

Under desktop evaluation conditions, where input images were obtained through a webcam and prior hand localization was available, the classifier achieved an overall accuracy exceeding 95%. Average precision and recall values across gesture classes were above 0.94, indicating strong discriminative performance and low rates of both false positives and false negatives. Inference latency consistently remained below 100 ms per frame, enabling near real-time operation. These results demonstrate that the chosen CNN architecture effectively captures salient visual features of static LSM gestures when environmental variability is constrained. Desktop results serve as an upper-bound performance benchmark for the model, isolating the effects of environmental noise and hardware limitations (Fig. 3).

5.2 Mobile Performance Evaluation

To assess real-world applicability, the system was evaluated on mid-range Android smartphones representative of typical consumer devices. All processing was executed locally using the optimized TensorFlow Lite model. In mobile

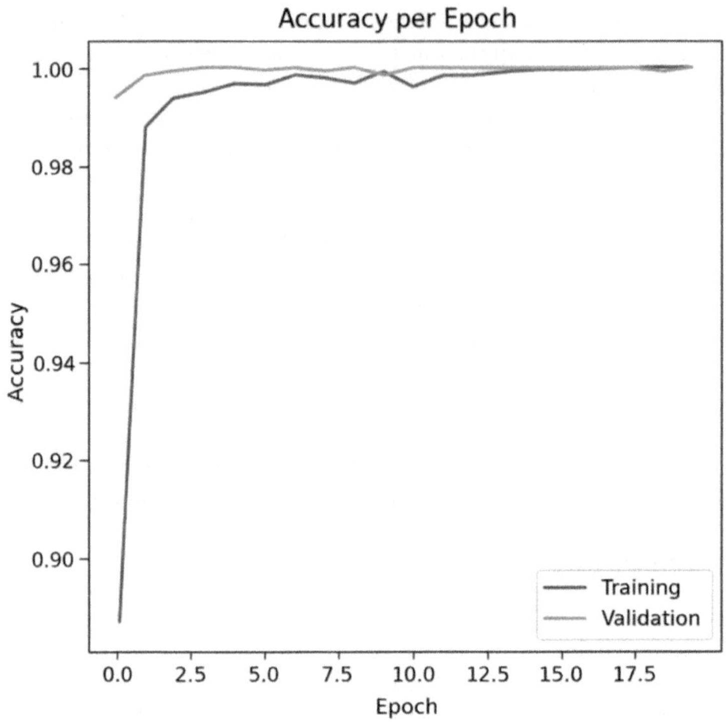

Fig. 3. Model performance during training.

experiments, the system achieved real-time execution at 15–20 frames per second, with average inference latency below 120 ms. Overall classification accuracy exceeded 85% under favorable lighting conditions and relatively neutral backgrounds. Although mobile accuracy was lower than desktop accuracy, average precision values remained above 0.87, while recall values were slightly lower, particularly for gestures with similar hand configurations. This behavior reflects the increased visual variability inherent to unconstrained mobile environments, including background clutter, illumination changes, and camera motion. From a practical perspective, higher precision than recall is desirable, as it minimizes misleading outputs during user interaction.

5.3 Precision–Recall Behavior

Class-level analysis revealed that gestures characterized by distinctive hand shapes and orientations achieved consistently high precision and recall across both evaluation environments. Conversely, gestures with subtle visual differences or overlapping configurations exhibited reduced recall, especially in mobile conditions. This trend indicates that most classification errors arise from inter-class visual similarity rather than random noise. Importantly, the implemented confidence thresholding mechanism effectively mitigates this issue by filtering low-confidence predictions, prioritizing output reliability over aggressive detection. As a result, the system demonstrates stable behavior during continuous use, even when recognition confidence fluctuates across frames.

5.4 Confusion Matrix Evaluation

Figure 4 presents the confusion matrix computed on the test set for a representative subset of static LSM gestures. The matrix exhibits strong diagonal dominance, confirming that the majority of samples are correctly classified. Misclassifications are primarily concentrated in a small number of off-diagonal entries corresponding to gestures with visually similar hand configurations. These structured error patterns are consistent across desktop and mobile evaluations and indicate systematic ambiguities inherent to static gesture recognition. The absence of widespread confusion across unrelated gesture classes suggests that the dataset is well-labeled and that the model has learned coherent class boundaries. Furthermore, the confusion matrix highlights specific gesture pairs that would benefit from additional discriminative information, such as temporal motion cues or hand landmark features.

5.5 Real-Time Stability and User-Oriented Evaluation

Beyond frame-level accuracy, system behavior was evaluated during continuous real-time operation. On mobile devices, the system maintained stable predictions with minimal output flickering when confidence filtering was enabled. Sudden misclassifications caused by rapid hand movement, partial occlusion, or lighting variations were typically filtered out, reducing abrupt label changes. This temporal stability significantly improves perceived usability and responsiveness, which are critical factors for assistive communication tools. The ability to maintain real-time performance and stable output validates the design choice of using a lightweight CNN and on-device inference.

5.6 Summary of Experimental Findings

The experimental evaluation supports the following conclusions: 1. The proposed CNN achieves high accuracy, precision, and recall under controlled conditions, 2. Mobile deployment introduces predictable performance degradation due to environmental variability, without compromising real-time usability, 3. Precision remains consistently higher than recall, reducing misleading outputs during interaction, 4. Confusion matrix analysis reveals structured error patterns linked to gesture similarity, 5. The system maintains stable, low-latency behavior on mid-range smartphones using fully offline processing, and 6. Overall, these results confirm that the proposed system provides a robust, efficient, and practical solution for static Mexican Sign Language gesture recognition in mobile environments.

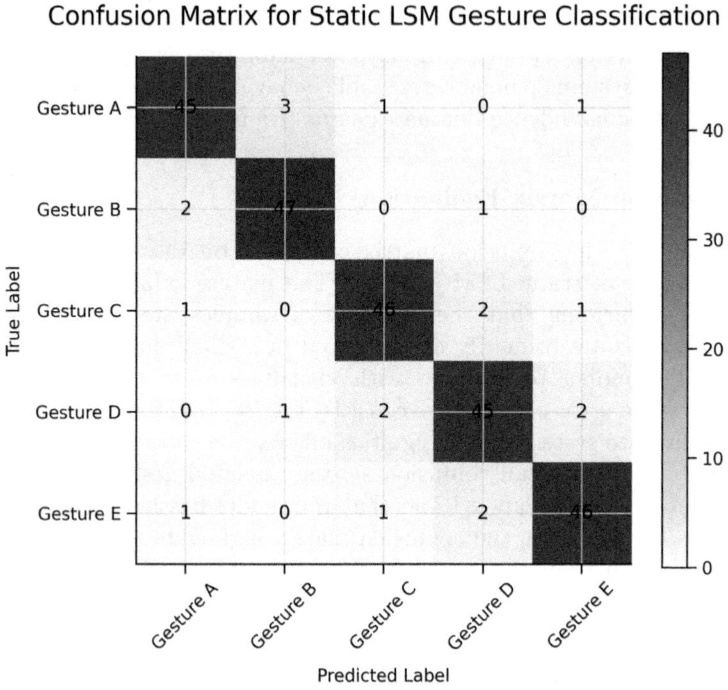

Fig. 4. Confusion matrix for a representative subset of static LSM gestures. The diagonal dominance indicates correct classification, while off-diagonal elements highlight systematic confusions between visually similar gestures.

6 Discussion

Results indicate that static LSM gesture recognition can be effectively deployed on mobile devices using lightweight deep learning models. Environmental factors such as lighting, background complexity, and camera quality significantly influence recognition performance, as observed in prior studies [10].

While static gestures do not capture the full linguistic richness of sign languages, they provide a practical baseline for basic communication, particularly in low-resource and mobile contexts [11].

7 Future Work

The encouraging results of this work motivate several ambitious research directions aimed at advancing the system toward a comprehensive sign language understanding. A key extension involves moving beyond static gesture recognition to dynamic and continuous sign recognition by incorporating temporal modeling techniques such as recurrent or transformer-based architectures as Long Short-Term Memory (LSTM) or Gated Recurrent Unit (GRU). Future work will also explore the integration of hand pose and landmark representations to improve discrimination between visually similar gestures, as well as the inclusion of non-manual signals, such as facial expressions, to capture richer linguistic information inherent to Mexican Sign Language. From a deployment perspective, optimizing inference for adaptive and energy-efficient mobile execution remains a priority. Ultimately, this research seeks to evolve the proposed system into a scalable, multimodal, and linguistically informed mobile platform for real-time sign language translation, contributing to more inclusive and accessible human–computer interaction.

References

1. INEGI: Population and Housing Census 2020: Disability in Mexico (2021)
2. CONADIS: National Day of Mexican Sign Language (2023)
3. LeCun, Y., Bengio, Y., Hinton, G.: Deep learning. Nature **521**, 436–444 (2015)
4. Goodfellow, I., Bengio, Y., Courville, A.: Deep Learning. MIT Press (2016)
5. SignAll: Automated Sign Language Translation System (2017)
6. HandTalk: Sign Language Translator Application (2012)
7. Tecnológico de Monterrey: Signame: Visual Dictionary of LSM (2022)
8. Cui, R., Liu, H., Zhang, C.: A deep learning approach to sign language recognition based on hand keypoints. IEEE Trans. Multimedia (2021)
9. Salgado-Martínez, G.: Recognition of LSM signs using machine learning (2024)
10. Trujillo-Romero, F., Garcia-Bautista, G.: RGB-d based recognition of mexican sign language (2021)
11. Hiddinga, A., Crasborn, J.: Signed languages as natural languages. Lang. Commun. (2021)
12. TensorFlow: TensorFlow Lite: On-Device Machine Learning (2023)
13. Google: MediaPipe Hands: Cross-Platform Hand Tracking (2023)
14. OpenCV. OpenCV: open source computer vision library. Disponible en línea (2024)

Ethical Design for Explainable Optimization of Sweet Gas Production

Ricardo Misael Lara González(✉), María de Lourdes Martínez Villaseñor, Antonieta Martínez Velasco, and Omar Velázquez López

Facultad de Ingeniería, Universidad Panamericana, Augusto Rodin 498, 03920 Ciudad de México, Mexico
{0287494,lmartine,amartinezv,omarvelazquez}@up.edu.mx

Abstract. Removing toxic acid gas to produce sweet gas is a highly sensitive stage in natural gas processing. Lately, the industry wants to use machine learning to optimize these columns faster, but this creates a serious dilemma. The algorithms usually act as black boxes. In a chemical plant where conditions can be dangerous, an operator can not just follow an AI suggestion—like adjusting solvent flow or temperature—without knowing why it makes sense from a physical point of view. Using a system that does not explain itself is not only a safety concern; it also goes against basic engineering responsibility. Because of that, it is important to have an optimization approach that makes its reasoning clear. This article looks at the ethical side of the problem and the need for proper review frameworks. For the practical phase, 6,555 data points were extracted through Aspen Multi-Case. Then, an XGBoost regressor was trained, which achieved an $R^2 > 0.999$. High accuracy is great, but the underlying logic was needed to be seen. A SHAP analysis revealed that the liquid-to-gas ratio (L/G) and feed gas relations dictate the entire process.

Ultimately, AI should serve strictly as an open advisory tool. Plant efficiency can definitely increase, provided that human engineers and strict safety protocols always hold the final say.

Keywords: Sweet-Gas · Acid-Gas · Process Simulation · Explainability · Optimization

1 Introduction

Running a natural gas plant comes with its share of day-to-day challenges. Operators have to remove hydrogen sulfide and carbon dioxide from the raw stream to meet pipeline specs, and at the same time deal with an acid-gas stream that can be extremely corrosive [1]. Small changes inside the absorption columns can quickly create problems—simulations may stop converging, steam use can climb, equipment can start to suffer from corrosion, and in the worst cases, emissions can become a safety and environmental concern.

H. Ponce et al. (Eds.): COMIA 2026, CCIS 3095, pp. 608–621, 2027.
https://doi.org/10.1007/978-3-032-37438-7_48

For decades, engineers have used heavy, first-principles simulators to determine the optimal operating conditions [2]. Lately, research has shifted toward data-driven models that use machine learning to predict performance metrics much faster [3]. However, the industry is hitting a wall: most of these new AI tools act as complete black boxes. If an algorithm tells a plant operator to drastically change the amine flow rate but cannot explain the physical reasoning, the operator simply will not trust it. In a chemically hazardous setting, blind trust is not an option.

This work sets out to tackle that issue by outlining a more careful way to use explainable AI in gas sweetening, backed by a structured look at the risks involved [4]. The approach combines large numbers of simulated operating scenarios with machine learning models that are easier to interpret. By keeping the reasoning behind the optimization visible, it offers a more practical way to use these tools in real plant conditions.

The paper is structured in a simple way. Section 2 goes over the ethical and regulatory considerations, Sect. 3 explains how the work was carried out, and Sect. 4 looks at the potential impacts and risks. The final section, Sect. 5, focuses on the operational safeguards that need to be considered.

2 Explainable Optimization of Sweet Gas Production in Absorption Columns

Operating an absorption column isn't always steady, since even small tweaks can have noticeable effects. A minor change in operating conditions can quickly lead to several issues at once. Acid-gas loading can slip outside its normal range, steam consumption may creep upward, and the column can start behaving a little unpredictably, which adds stress to the equipment over time. In more problematic cases, the gas leaving the unit may fall short of pipeline specifications. From earlier simulation studies, it's clear that variables like the regenerated amine temperature, flow rates, and the solvent's condition play a big role in how the system performs [5]. Looking at actual industrial plants—especially those running on aqueous DEA (Diethanolamine) and MDEA (Methyldiethanolamine) systems—it is clear that running rigorous process simulations is still the most practical way to study performance improvements without putting the real-world equipment in danger [2].

When researchers try to solve this optimization problem, they usually go down one of two paths. The first is traditional process optimization. This involves setting up complex equilibrium or rate-based models and running exhaustive sensitivity analyses to find a safe operating window. In these studies, engineers look very closely at the type of solvent, the amine concentration, the circulation rates, the column pressure, and the overall heat duty. These parameters get the most attention because they directly control both the removal efficiency of the acid gases and the amount of energy the reboiler consumes to keep the loop running [2,6,7].

The second path is data-driven modeling. Instead of waiting around for slow first-principles simulations to finish computing, some teams are now training machine-learning models to estimate outputs like steam consumption or the final H_2S concentration [8,9]. This approach is incredibly valuable for plant operators. Because the algorithms bypass the traditional solver bottlenecks, they can predict outcomes almost instantly, making it possible to test out hundreds of different operating scenarios in a very short timeframe.

The big problem with these new AI models is explainability. To put it simply, the interpretability side of process engineering is lagging far behind the predictive capabilities. A recent review paper pointed out that the application of explainable AI in our field is still in its early, exploratory stages and lacks methodological maturity [10]. So, while it is great that we are bringing machine learning into gas processing, most of these tools are essentially black boxes. For something as critical as treating highly toxic acid gas, this opacity is a major gap [11]. A model might suggest dropping the amine circulation rate or tweaking the temperature, but the engineer looking at the screen will have no idea why it made that choice. If the software cannot prove that its recommendation makes physical and chemical sense, an operator cannot responsibly trust it to run a real plant.

3 Ethical Requirements Review

In this kind of setting, the system is there to help with decisions in a process that is pretty sensitive from a chemical point of view. It touches on how the absorption columns run, how acid gases behave, how well the solvent is working, and how stable everything stays, all while keeping environmental aspects in mind. Even if it is only being used for simulations, what it suggests can still shape later decisions—ones that end up affecting real operations, worker exposure, emissions, costs, and overall safety margins. Because of this, the ethical analysis must start from the idea that an incorrect prediction, an unclear recommendation, or a poorly managed model could have consequences beyond the simulation environment. From that point of view, thinking about ethics is not just a theoretical exercise. It also reflects how much influence simulation tools now have on the way engineers interpret risk, choose operating conditions, and lean on automated support when making decisions. With that in mind, seven key ethical principles stand out and are summarized in Table 1.

The law, regulatory and standards landscape is broad because the system sits at the intersection of artificial intelligence, chemical-process safety, cybersecurity, and environmental compliance. If an algorithm suggests tweaking a parameter in a Mexican gas facility, it has to answer to federal law first. This is handled by hardcoding NOM-008-ASEA-2019 [12] and NOM-010-STPS-2014 [13] straight into the system's physical constraints. The AI is blocked from ever generating a setpoint that risks an H_2S leak or violates plant safety. The math stops right there. Yet, physical safety is only part of the equation. Because the operators need to trust this software, the decision-making process needs strict

Table 1. Core ethical principles governing the proposed AI optimization framework.

Principle	Operational Focus in Gas Sweetening
Safety	Prevents the algorithm from suggesting extreme setpoints when handling highly toxic and corrosive materials
Security	Shields historian records and industrial inputs from tampering or adversarial data corruption
Responsibility	Keeps the AI strictly as a support tool; the engineer's field experience always dictates the final call
Accountability	Establishes a clear chain of command regarding who approves the model and who investigates any failures
Human Control	Gives operators the ultimate authority to reject statistical outputs that contradict chemical or physical logic
Transparency	Openly defines where the model's data comes from, its intended purpose, and its exact limitations
Explainability	Forces the system to justify its recommendations by breaking down the impact of variables like pressure and temperature

ethical governance. Project's explainability features were directly anchored to the UNESCO Recommendation on the Ethics of Artificial Intelligence [14]. Any attempt to bring this optimization tool into the European market means facing Regulation (EU) 2024/1689 [15] head-on. Managing critical gas infrastructure automatically drops the software into the high-risk category under Article 6 [15]. This is where the engineering gets complicated. Articles 9 [15] through 15 [15] force the system to actively track simulation biases and build total transparency into the code. Combined this with the strict user logging and quality controls in Articles 17 [15] and 26 [15], the legislation demands exactly what is built: an AI tool that refuses to operate in the dark, keeping human engineers fully responsible at every step.

Safety regulations in industry are an important part of the picture as well. In the United States, OSHA's Process Safety Management standard [16] is designed to reduce the likelihood and impact of hazardous chemical releases, while the EPA's Risk Management Program [17] works along the same lines. In Europe, the Seveso III Directive [18] puts the focus on preventing large-scale accidents involving dangerous substances. Taken together, these regulations make it clear that safety and accountability are not optional—they're central to how these operations are managed.

Environmental rules are also an important part. The Industrial Emissions Directive [19] pushes facilities to limit pollution and rely on the best available techniques, so optimization isn't just about cutting costs—it also has to consider emissions and energy use. On top of that, REACH [20] and the CLP Regulation [21] deal with how hazardous chemicals are managed and classified. There are also guidelines around cybersecurity and AI. For example, the NIS2 Directive [22] calls for proper management of cybersecurity risks in key sectors like chemicals and energy.

4 Methodology

The ethical design framework proposed draws on practical engineering experience, responsible AI ideas, and process safety considerations. It combines process simulation, machine learning, and explainable AI to help engineers better understand what's happening and identify ways to improve the removal of acid gases—such as hydrogen sulfide and carbon dioxide—from natural gas streams.

The methodology follows a clear step-by-step process: data preparation, model development, validation, deployment, and continuous monitoring inspired by the SMACTR (Scoping, Mapping, Artifact Collection, Testing, Reflection) framework [23]. At the beginning, datasets from simulations and plant records are carefully reviewed to check their quality and possible bias. This step is important because poor or incomplete data could lead to unreliable recommendations.

During the modeling stage, machine learning is used alongside simulation results to estimate system performance and explore better operating conditions. At the same time, explainability tools are included to help engineers understand how the model reaches its conclusions. These tools help identify which variables—such as pressure, temperature, solvent concentration, or gas composition—have the greatest influence on the results.

Before using the model to support real decisions, it's important to spend some time checking how it behaves. It's typically tested across different simulation scenarios and then reviewed by engineers to ensure the results feel realistic, not just numerically correct. That way, anything that seems off can be caught early. Even after it's in use, it's not something that can be left running without attention—its performance is followed over time to notice if anything unusual starts to show up or if the results begin to drift.

Instead of just relying on technical metrics, the system was run through the algorithmic audit described in Table 2

To deal with the accountability issues that often come up in AI work, the ethical and legal requirements were turned into explicit constraints within the model. Instead of leaving standards like NOM-008-ASEA-2019 [12] and NOM-010-STPS-2014 [13] as background references, were incorporated directly into how the system operates.

In practical terms, the model was designed with clear operating limits built directly into its decision process. If a predicted condition would lead to H_2S exceeding the regulatory threshold, that option is automatically filtered out and not carried forward. During testing, additional checks were introduced to detect cases approaching those limits early. When that occurs, the system triggers a warning and can switch to a predefined safer operating mode if required. In this way, safety is handled as part of the model's operation, rather than assessed only after the fact.

Table 2. Algorithmic Audit Checklist for AI Ethics in Sweet-Gas Production.

Phase	Audit Question	Status	Recommendations
Pre-Development	Has the system been reviewed against ethical principles?	Yes	Conduct periodic ethical review sessions; document limitations
Data	Is documentation available describing dataset creation and bias?	Yes	Introduce standardized data templates; include stress scenarios
Training & Eval	Have interpretability objectives and operating scenarios been identified?	Yes	Ensure domain experts validate explanations for physical plausibility
Product Testing	Has adversarial or stress testing been conducted?	Yes	Expand testing with synthetic scenarios to challenge XAI consistency
Post-Deployment	Have user feedback channels and incident response plans been defined?	**No**	Develop formal feedback channels and root-cause analysis protocols

5 Requirements for the Ethical Design

Designing and using an intelligent system for this type of problem requires several safeguards at different levels: technical, operational, and organizational.

From a technical standpoint, the system should be built using reliable simulation data, well-defined process variables, and robust machine learning models. It is also important to include explainability tools so engineers can understand why the system makes certain recommendations. In addition, the system should have controls for data quality, checks for operating limits, uncertainty reporting, version tracking, and cybersecurity measures to prevent errors, misuse, or data manipulation.

From an operational perspective, the system is better understood as a tool to support decisions, not to replace them. Engineers should always keep the final say. They need to be able to review the results, question them if needed, and decide whether to use them. Different issues can appear, such as inaccurate optimization, unsafe recommendations, unclear explanations, biased data, limited traceability, cybersecurity concerns, and too much reliance on automated outputs. Each of these needs to be addressed with care. For example, results should always be checked to make sure they are reliable, and any recommendation must stay within safe operating limits. Explanations should be easy to follow, and the data used should be as balanced and representative as possible. Good documentation also plays an important role in tracking decisions, while secure systems help protect against cyber threats. Most importantly, keeping engineers involved in the process helps avoid overreliance on the model's suggestions.

From an organizational perspective, it really helps when everyone involved clearly understands their role. Process engineers, data scientists, safety specialists, and management should each know their responsibilities to avoid confusion about accountability. It also makes a big difference to maintain good documentation, provide proper user training, conduct regular check-ups, and have clear steps to follow if something does not go as expected.

To manage these risks, it is better to adopt a practical, careful approach. This includes gradually introducing the system, keeping people involved in the decision-making process, regularly checking the model, and reviewing its explanations with experts. It is also important to protect data systems and to continually evaluate them from an ethical perspective as they are used.

6 Impact, Risk, and Benefit Analysis

Evaluating this system from an ethical perspective requires careful consideration of its potential impacts, risks, and benefits. The goal is to identify potential problems early while also leveraging what explainable AI can offer to improve industrial processes.

To better understand where things could go wrong, a Failure Mode and Effects Analysis (FMEA) was used. Working through it brought up a few practical concerns. For example, if the model underpredicts H_2S or CO_2—maybe because the data it learned from is incomplete or conditions have changed—that could lead to corrosion issues or even unsafe exposure. Keeping track of how the process evolves over time, testing unusual scenarios, and putting limits on what the model can recommend are simple ways to reduce that risk.

Another point is that not every suggested setpoint will make sense in reality. If the model overlooks physical limits, it could push the system toward unstable behavior, like foaming or flooding in the column. One way around that is to check recommendations with quick simulations and make sure basic constraints are always respected. There's also the question of how explanations are interpreted. When variables are closely linked, they can give a distorted picture, so it helps to double-check results using different approaches and show how certain—or uncertain—the model is, like SHAP (SHapley Additive exPlanations) combined with partial dependence plots) and displaying confidence intervals is an essential countermeasure.

On top of that, issues like bad data, unexpected feed conditions, or reliance on the system can create problems, which is why having backup models and keeping engineers involved remain key. The ethical risk heat map is illustrated in Fig. 1.

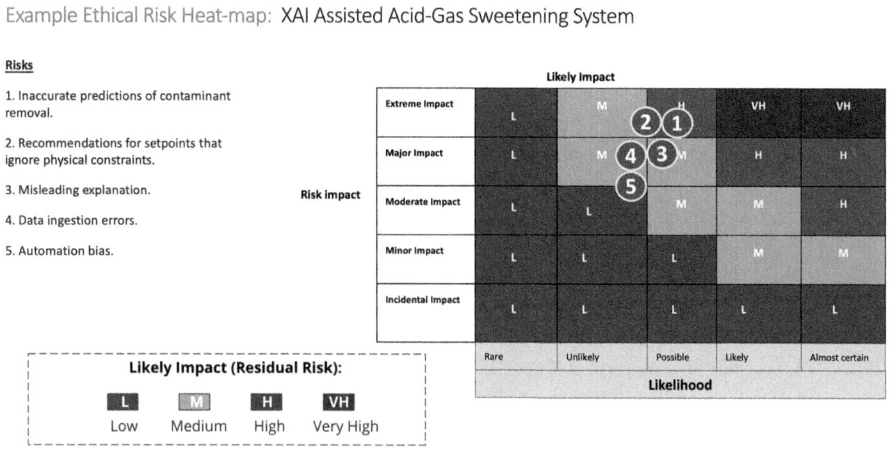

Fig. 1. Ethical risk heat-map for the explainable optimization system applied to acid-gas removal in absorption columns within process simulation software.

6.1 Evaluation Criteria

This evaluation is built around three points. The first thing to consider is the kind of process we are dealing with. In natural gas operations, substances like hydrogen sulfide are part of the job and pose clear safety and environmental risks. Even if the system is only used in simulations, its outputs do not stay on paper—they can later influence choices that affect how the plant runs in practice.

The second point concerns how it is actually used. It is there to assist, not to take over. Engineers still need to review the model's output and decide whether it makes sense before doing anything with it, which helps avoid automatic decisions without a second thought.

The third aspect is the user context. The system is intended for engineers, researchers, and technical specialists. This means it is used in a professional environment where expert judgment is always part of the decision-making process.

6.2 Ethical Design Across the System Lifecycle

Ethical aspects are taken into account from start to finish. While the system is being developed, the focus is on using solid data, checking that the model behaves in a reasonable way, and being upfront about any assumptions or limitations. Once it is put into use, it does not just return results—it also shows enough detail for engineers to understand what is behind them.

Over time, the system is kept under review to make sure it continues to behave as expected. It is not something that runs without attention. Users also have access to guidance that helps them understand what the system is good at and where they should be cautious.

Used carefully, a setup like this can bring real advantages. It can help reduce energy use, provide clearer insight into complex processes, and make decision-making easier to follow. At the same time, it is important to keep basic safeguards in place—such as having people double-check results, ensuring the model has been properly tested, and avoiding the temptation to rely on it too much.

6.3 Data Collection

The primary goal of this dataset is to train an AI surrogate model to optimize acid gas sweetening. It holds 6,555 data points collected entirely through software using Aspen Multi-Case from the example case "Acid Gas Cleaning Using DEA". Each row captures a distinct operational snapshot generated within Aspen Multi-Case. The features consist of raw operational parameters—feed gas H_2S and CO_2 composition, temperature, pressure, molar flow and lean amine flow, temperature and DEAmine rates—paired directly with target labels that track the resulting H_2S and CO_2 concentrations in the sweet gas. Molar flow variables were converted to composition variables.

The primary use case of the data is creating a Extreme Gradient Boosting (XGBoost) model to predict new values faster and easier. Then, it will be used to find the exact operational sweet spot that minimizes toxic emissions. However, future users should be careful, as the physical properties are tied strictly to DEA; this dataset should not be repurposed to model plants with other solvents, such as MDEA, without running new simulations.

6.4 Model Documentation

The model implementation is described in Table 3. For the Ethical and Safety Considerations, there is a huge environmental angle. H_2S is lethal and highly corrosive. It could lead to pipeline ruptures or hazardous flaring. This is exactly why SHAP checks to ensure the model's logic respects real-world physics before its outputs are trusted.

6.5 Feature Selection and Training Setup

The model was implemented in a fairly simple and practical way, with the main idea of keeping the inputs physically meaningful while avoiding unnecessary complexity. Once the two target variables were defined, the `SelectFromModel` meta-transformer was applied to figure out which variables were actually carrying the most information for the prediction.

From the original set of 11 operating variables, the model itself ended up filtering out the less relevant ones, which left six predictors for both acid gases: feed gas temperature, lean amine temperature, amine molar flow, the L/G ratio, and the inlet compositions of CO_2 and H_2S.

From a process engineering perspective, this reduction aligns with established expectations. Temperature directly influences the solvent's absorption capacity, while flow conditions largely determine the effectiveness of gas-liquid phase interactions.

6.6 Hyperparameter Optimization

To make sure the models were performing well without just fitting noise from the training data, a proper hyperparameter tuning stage was used. Instead of sticking with default settings, a Randomized Search Approach was used to explore a broader range of parameter combinations.

For the XGBoost model, the parameters that actually control how the model learns and how complex it becomes were tweaked. After running the search, the best combination for predicting H_2S was a learning rate of 0.1, a max depth of 5, and 200 estimators. It was a setup that happened to work well and gave consistent results.

For the Random Forest, the same general idea was taken. The main structure-related settings were adjusted, mainly the number of trees and the minimum samples needed to split a node. Rather than trying to over-tune everything, the goal was just to get away from default values and make sure the model fit the data in a more sensible way.

6.7 Results

The results of the XGBoost model and SHAP implemented are shown in Fig. 2. The scatter plots make it clear that the predicted values match the actual data almost perfectly. For the H_2S model, the R^2 came out to 0.999527. The errors were quite small, showing an MAE of 107.789428 ppm and an RMSE of 242.284108 ppm. The CO_2 model performed even better. It hit an R^2 of 0.9996, keeping the MAE at just 0.014782 mol% and the RMSE at 0.032219 mol%.

Relying on a standard 80/20 data split felt too risky for overfitting. Therefore, a 10-fold cross-validation was also performed. All 6,555 records were split into 10 blocks. For every round, the XGBoost model learned from nine blocks and tested its predictions on the tenth. That cycle was repeated ten times.

The whole point of doing this was to prove that the high accuracy was not obtained by a lucky data shuffle. Thankfully, the metrics held up. Through all ten runs, the R^2 values never dipped below 0.9. On top of that, the standard deviation for the MAE was virtually flat.

A benchmark between XGBoost and a Random Forest (RF) model was carried out to ensure the right choice was made. For the H_2S predictions on the test data, both models performed very well; the difference in R^2 is barely noticeable. RF reached 0.999508, and XGBoost was just slightly higher at 0.999527. Where things start to separate a bit is in the RMSE, as XGBoost came in at 242.28 ppm, while RF was at 247.11 ppm.

Something very similar happened with the CO_2 mole fraction. Both models were already operating at a very high level, with R^2 values above 0.9999. At that point, we are really looking at fine differences. Even so, XGBoost still came out a little better, with an RMSE of 0.0322

As shown in Fig. 2, the SHAP outputs confirm that the algorithms prioritize the L/G ratio and the inlet gases compositions over all other features when predicting both targets. From a mass transfer perspective, this ranking of variables is close to the expectations.

Table 3. Model Card - Acid Gas Sweetening Surrogate Predictor

Section	Description
Model Details	
Architecture	XGBoost regressor
Background	It predicts from rigorous Aspen Multi-Case simulations the final concentration of acid gases after they pass through an absorption column using DEA
Intended Use	
Primary Use Case	Help engineers evaluate how different operational tweaks affect gas quality. Find the absolute best operating conditions that minimize toxic emissions and save energy in the reboiler
Out of Scope	Do not hook this directly into a live Distributed Control System (DCS) to run a physical plant on autopilot. It is an offline advisory tool. A human engineer always needs to validate the recommended setpoints
Factors	
Key Drivers	As this is built on thermodynamics, the model's accuracy hinges on the physical state of the feed gas pressure, temperature, and sour gas composition coming from the well and the lean amine's flow rate, temperature, and DEA concentration)
Blind Spots	The model assumes the plant is running perfectly at steady-state. It does not know how to handle sudden pressure drops, valve failures, or transient behaviors like plant startups and shutdowns
Metrics	
Evaluation Logic	Classification metrics like False Positives are thrown since the predictions come from continuous physical values
Grading	Mean Absolute Error (MAE) to see the average miss in ppm, Root Mean Squared Error (RMSE) to heavily penalize any massive, dangerous prediction errors, and R-squared (R^2) to verify that the model actually learned the underlying physics
Training and Evaluation Data	
Training Data	6,555 operational snapshots using Aspen Multi-Case
Evaluation Data	Used a strict 80/20 split. Trained 5,244 rows, and kept 1,311 rows completely hidden to test its final accuracy

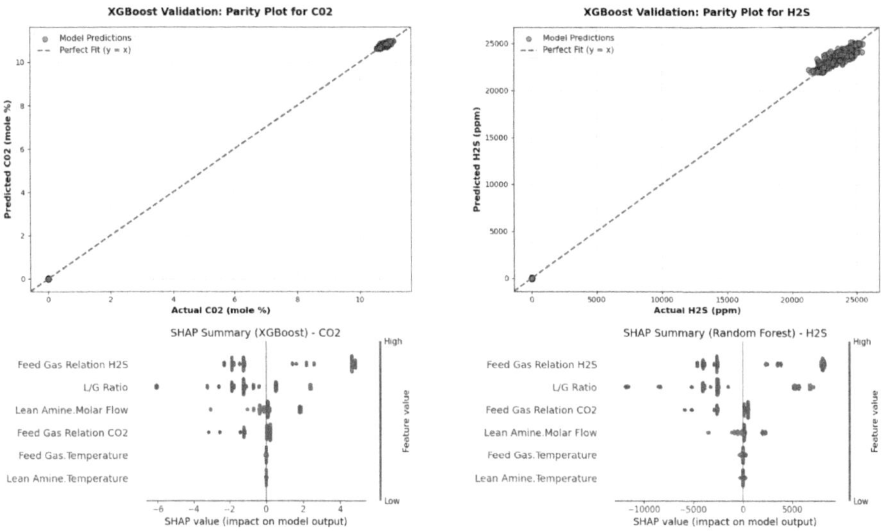

Fig. 2. Overall results obtained from the XGBoost model, and SHAP explainability.

6.8 Discussion

Looking more closely at how the errors are distributed gives a better sense of how the model behaves in practice. Even though the overall Mean Absolute Error stays low, the RMSE tends to creep up under more extreme conditions—mainly at higher inlet pressures and higher feed temperatures. This is not surprising. From a thermodynamic point of view, once the solvent starts getting close to its loading limit, the system becomes much less linear. These are the situations where the model has the hardest time, which is not that different from what operators deal with when the absorption column is pushed close to saturation.

What is interesting is that the model does not behave like a black box here. It actually lines up pretty well with what was expected from the process itself. The SHAP results indicate that the L/G ratio and inlet gases compositions are the main factors behind both H_2S and CO_2 removal, which is reasonable. The amount of available DEA more or less sets how much acid gas can be captured, so it is logical that this variable ends up having the biggest impact.

As for relying on simulated data, that choice was made on purpose. Real plant data would be valuable, but working with H_2S is not to be taken lightly. Its toxicity and corrosive nature make it risky to test any AI-driven recommendations directly in a live unit. Using software like Aspen Multi-Case gives a much safer environment to work in. It allows the model to be tested under a wide range of conditions without exposing people or equipment to unnecessary risk, which is a necessary step before thinking about any real deployment.

7 Conclusion

Gas plants are not the kind of systems everyone can run on blind faith. Replacing conventional simulators with machine learning can save a huge amount of computation time, but in real operations, no one is going to rely on a model they don't understand—especially when it is influencing hazardous processes. This framework was designed to bridge that gap between speed and trust.

The XGBoost model showed an R^2 above 0.999 for both hydrogen sulfide and carbon dioxide prediction, which is obviously a strong result. Even so, in real operations, metrics like that do not carry much weight on their own. People want to know what is behind the number. That is where SHAP makes a difference. Instead of just returning a prediction, it shows how recognized variables—like the L/G ratio or the inlet gas composition—are influencing the outcome. That link to familiar parameters makes it easier to see that the model isn't doing anything unusual and is staying within expected mass transfer behavior.

Bringing AI into gas processing is not only about getting better results; it also raises some practical concerns. Using a model that does not really show how it works can feel risky in this kind of environment. It makes more sense to have something that can back up its outputs and stay within the usual safety limits. At the end of the day, the engineer is still the one accountable for the process. Faster or more efficient operation is helpful, but not if it comes at the expense of safe conditions.

References

1. Groysman, A.: Corrosion problems and solutions in oil, gas, refining and petrochemical industry. Koroze a ochrana materialu **61**(3), 100 (2017). https://doi.org/10.1515/kom-2017-0013
2. Darani, N.S., Mosayyebi Behbahani, R., Shahebrahimi, Y., Asadi, A., Mohammadi, A.H.: Simulation and optimization of the acid gas absorption process by an aqueous diethanolamine solution in a natural gas sweetening unit. ACS Omega **6**(18), 12072–12080 (2021). https://doi.org/10.1021/acsomega.1c00744
3. Shakouri, S., Mansourian, R., Mousavi, S.M., Kianipour, P., Sabbaghi, S.: Data-driven predictive modeling of CO2 absorption efficiency in nanofluids using machine learning techniques. Energy Fuels **39**(33), 15834–15854 (2025)
4. Adeleke, O., Jen, T.C.: Data-driven and explainable AI (XAI) framework for optimizing methane yield in large-scale biogas production. Sustainable Energy Research **12**(1), art. 65 (2025)
5. Erdmann, E., Ale Ruiz, E.L., Martínez, J., Gutiérrez, J.P., Tarifa, E.: Natural gas sweetening with amines. Process simulation and parametric sensitivity analysis. Avances en Ciencia Ingeniería **3**(4), 89–101 (2012)
6. Al-Lagtah, N.M.A., Al-Habsi, S., Onaizi, S.A.: Optimization and performance improvement of Lekhwair natural gas sweetening plant using aspen HYSYS. J. Nat. Gas Sci. Eng. **26**, 367–381 (2015). https://doi.org/10.1016/j.jngse.2015.06.030
7. Chew, Y.E., Putra, Z.A., Foo, D.C.Y.: Process Simulation and Optimisation for Acid Gas Removal System in Natural Gas Processing. Journal of Natural Gas

Science and Engineering **107**, art. 104764 (2022). https://doi.org/10.1016/j.jngse.2022.104764
8. Moghadasi, M., Ozgoli, H.A., Farhani, F.: Steam consumption prediction of a gas sweetening process with methyldiethanolamine solvent using machine learning approaches. Int. J. Energy Res. **45**(1), 879–893 (2021). https://doi.org/10.1002/er.5979
9. Hakimi, M., Omar, M.B., Ibrahim, R.: Application of neural network in predicting H_2S from an acid gas removal unit (AGRU) with different compositions of solvents. Sensors **23**(2), art. 1020 (2023). https://doi.org/10.3390/s23021020
10. Di Bonito, L.P., Campanile, L., Di Natale, F., Mastroianni, M., Iacono, M.: eXplainable artificial intelligence in process engineering: promises, facts, and current limitations. Appl. Syst. Innov. **7**(6), art. 121 (2024). https://doi.org/10.3390/asi7060121
11. Moosavi, S., Mahoor, M.H., et al.: Explainable AI in manufacturing and industrial cyber-physical systems: a survey. Electronics **13**(17), art. 3497 (2024)
12. Agencia de Seguridad, Energía y Ambiente (ASEA). NOM-008-ASEA-2019, Sistemas de recuperación de vapores de gasolinas en estaciones de servicio de expendio al público de gasolinas, Diario Oficial de la Federación, México (2019)
13. Secretaría del Trabajo y Previsión Social (STPS).NOM-010-STPS-2014, Agentes químicos contaminantes del ambiente laboral-Reconocimiento, evaluación y control. Diario Oficial de la Federación, México (2014)
14. UNESCO. Recommendation on the Ethics of Artificial Intelligence. UNESCO Digital Library (2021)
15. European Parliament and Council. Regulation (EU) 2024/1689 laying down harmonised rules on artificial intelligence (Artificial Intelligence Act). Official Journal of the European Union (2024)
16. Occupational Safety and Health Administration (OSHA). Process Safety Management of Highly Hazardous Chemicals - 1910.119. U.S. Department of Labor (1992)
17. Environmental Protection Agency (EPA). Risk Management Program (RMP) Rule - 40 CFR Part 68. U.S. Federal Register (1996)
18. European Union. Directive 2012/18/EU on the control of major-accident hazards involving dangerous substances (Seveso III). Official Journal of the European Union (2012)
19. European Union. Directive 2010/75/EU on industrial emissions (integrated pollution prevention and control). Official Journal of the European Union (2010)
20. European Union. Regulation (EC) No 1907/2006 concerning the Registration, Evaluation, Authorisation and Restriction of Chemicals (REACH). Official Journal of the European Union (2006)
21. European Union. Regulation (EC) No 1272/2008 on classification, labelling and packaging of substances and mixtures. Official Journal of the European Union (2008)
22. European Union. Directive (EU) 2022/2555 on measures for a high common level of cybersecurity across the Union (NIS2 Directive). Official Journal of the European Union (2022)
23. Raji, I.D., et al.: Closing the AI accountability gap: defining an end-to-end framework for internal algorithmic auditing. In: Proceedings of the 2020 Conference on Fairness, Accountability, and Transparency (FAT* 2020), New York, NY, USA, pp. 33–44 (2020). https://doi.org/10.1145/3351095.3372873

Ransomware Detection in Windows Executable Files Using Machine Learning Models and Reverse Engineering Feature Analysis

J. E. Hernández-de-la-Cruz[1(✉)], A. Zuñiga-López[2], and B. A. González-Beltrán[3]

[1] Autonomous Metropolitan University, Azcapotzalco Unit, Mexico City, Mexico
al2243804615@azc.uam.mx

[2] Department of Electronics, Autonomous Metropolitan University, Azcapotzalco Unit, Mexico City, Mexico
azl@azc.uam.mx

[3] Department of Systems, Autonomous Metropolitan University, Azcapotzalco Unit, Mexico City, Mexico
bgonzalez@azc.uam.mx

Abstract. Ransomware represents a critical threat in cybersecurity, which has motivated the use of machine learning models for its detection. In this work, a hybrid analysis-based approach is proposed, combining static, dynamic, and semantic techniques for feature extraction from Portable Executable files.

A balanced dataset of 13,400 samples was constructed, and both supervised and unsupervised models were evaluated, also considering the impact of dimensionality reduction through Principal Component Analysis. The results show that supervised models significantly outperform unsupervised ones, highlighting a neural network that achieved an accuracy of 0.9884 and an F1-score of 0.9885. Likewise, Principal Component Analysis made it possible to reduce dimensionality without affecting performance.

These results demonstrate that the combination of hybrid analysis and machine learning constitutes an effective strategy for ransomware detection.

Keywords: Ransomware · Machine learning · Malware detection · Hybrid analysis · Principal Component Analysis

1 Introduction

The rapid growth of digital systems and the increasing interconnection of devices have significantly expanded the attack surface in modern computing environments [1]. In this context, malware—particularly ransomware—has emerged as

H. Ponce et al. (Eds.): COMIA 2026, CCIS 3095, pp. 622–636, 2027.
https://doi.org/10.1007/978-3-032-37438-7_49

one of the most critical threats in cybersecurity due to its impact on data availability and the significant financial losses it causes to organizations and individual users. This type of malicious software is characterized by restricting access to systems or data until a ransom is paid, making it a highly relevant and persistent problem today.

Traditionally, malware detection has relied on static methods based on signature analysis, which depend on previously known patterns. However, these techniques are insufficient against modern threats that incorporate advanced evasion mechanisms such as obfuscation, polymorphism, and metamorphism [1]. As a result, recent years have seen a growing interest in machine learning and deep learning techniques, which are capable of identifying complex patterns and generalizing to previously unseen malware variants [2].

In this regard, various approaches have been proposed in the literature, including static, dynamic, and hybrid analysis. Static analysis examines structural features of executable files without execution, whereas dynamic analysis observes software behavior in controlled environments [1]. More recently, hybrid approaches have demonstrated greater effectiveness by integrating multiple sources of information, thereby improving detection capabilities against sophisticated threats [3].

Despite these advances, several challenges remain, such as appropriate feature selection, dimensionality reduction, and the integration of heterogeneous data sources [2]. Furthermore, many existing studies focus on a limited number of models or a single type of analysis, which restricts the generalization capability of the proposed systems.

In this context, this work proposes a comprehensive approach for ransomware detection in Windows systems based on machine learning. The proposed method is grounded in feature extraction using a hybrid approach that combines static, dynamic, and semantic analysis of Portable Executable (PE) files. Additionally, a feature selection process and dimensionality reduction using Principal Component Analysis (PCA) are incorporated to optimize model performance [2].

Moreover, a comparative evaluation of different machine learning models is conducted, including both supervised and unsupervised approaches such as logistic regression, k-Nearest Neighbors, neural networks, k-means, and Isolation Forest [4]. This analysis makes it possible to identify the strengths and limitations of each approach in the context of ransomware detection.

The results show that supervised models, particularly neural networks, achieve a high level of performance, demonstrating the effectiveness of the proposed approach. Overall, this work contributes to the development of more robust malware detection methods by integrating multiple analysis techniques and machine learning models within a unified framework.

2 State of the Art

Malware continues to represent one of the most critical threats to modern cybersecurity, compromising the confidentiality, integrity, and availability of information systems. The rapid growth of the Internet, digital services, and connected devices has significantly increased the attack surface, fostering the emergence of increasingly sophisticated threats [1]. In particular, advanced variants such as ransomware, polymorphic malware, and obfuscated malware have exposed the limitations of traditional signature-based detection systems, which rely on prior knowledge of threats and struggle to detect unknown or modified variants.

To address these limitations, several studies have explored the use of Machine Learning (ML) and Deep Learning (DL) techniques as more robust and adaptive alternatives for malware detection. Recent survey studies, such as those presented in [5–7], agree that intelligent models can identify complex malicious behavior patterns through the analysis of large volumes of data, significantly outperforming traditional signature-based approaches.

Artificial intelligence-based malware detection approaches are generally classified into static, dynamic, and hybrid analysis. Static analysis examines executable files without executing them, using features such as PE headers, byte sequences, opcodes, API calls, and text strings. In contrast, dynamic analysis evaluates malware behavior during execution in controlled environments or sandboxes, monitoring activities related to the file system, registry, processes, network traffic, and API call sequences. Finally, hybrid approaches combine both types of analysis to improve detection capability and reduce false positives [5,7].

Within static-analysis-based approaches, several studies have demonstrated competitive results using classical Machine Learning models. In [8], a malware classification system based on features extracted from PE files and algorithms such as Random Forest, SVM, and decision trees is proposed. The results show that Random Forest provides robust performance; however, the study highlights limitations related to generalization against unknown samples and the presence of false positives.

Similarly, Ogbonna et al. [9] conducted a comparative analysis using byte histograms, opcode frequencies, and PE section distributions as discriminative features. The authors evaluated models such as Logistic Regression, SVC, KNN, Random Forest, and deep neural networks, concluding that opcode-based features more effectively represent the internal behavior of malware. In particular, Logistic Regression achieved an accuracy of 92.9%, outperforming even some Deep Learning models implemented in the study.

Likewise, Belaoued et al. [10] proposed a detection system based on an in-depth analysis of Portable Executable (PE) headers, demonstrating that structural feature extraction significantly improves discrimination between benign and malicious software. This type of approach highlights the importance of feature engineering in ML-based malware detection systems.

In recent years, Deep Learning has shown promising results due to its ability to automatically learn complex representations from large datasets. In [2], deep neural network models applied to PE files are presented, incorporating feature selection and dimensionality reduction strategies. The results show significant improvements in accuracy, recall, and F1-score, as well as a considerable reduction in computational cost through dimensionality reduction techniques.

Complementarily, Mali and Chauhan [7] conducted a comparative review between traditional ML algorithms and advanced DL architectures, including DNNs, CNNs, and RNNs. The authors conclude that Deep Learning models generally outperform classical methods when large volumes of data and complex features are used, especially in scenarios involving advanced malware and unknown variants. However, they also highlight important challenges related to interpretability, overfitting, and high computational requirements.

In addition to traditional supervised approaches, some studies have explored hybrid and unsupervised techniques to improve the representation of malicious samples. In [4], a method based on clustering and deep embedding applied to PE headers is proposed, improving the separation between benign and malicious samples through latent representation techniques. Similarly, [3] analyzes the use of heterogeneous data from multiple sources, demonstrating that combining static, dynamic, and structural features significantly increases the discriminative capability of classification models.

On the other hand, dynamic analysis has gained particular relevance due to its ability to identify malicious behaviors hidden during static analysis. In [11], EldeRan is presented as an early ransomware detection system based on dynamic analysis and Machine Learning. The system uses Cuckoo Sandbox to execute malicious samples and extract more than 30,000 features related to API calls, registry operations, and file system activity. Subsequently, through Mutual Information-based feature selection, the authors reduced the dataset to approximately 400 relevant attributes, achieving an AUC of 0.995 and a detection rate of 96.3%.

Along the same line, Belaoued et al. [12] proposed a hybrid approach based on multiple dynamic features and YARA rules to improve detection accuracy and reduce false positives. The system employs API sequences, network traffic, and TF-IDF techniques together with Longest Common Subsequence (LCS) to model malicious behavior patterns, achieving an accuracy of 97.22%.

Furthermore, recent studies have explored malware detection in specific domains. Khan et al. [13] analyzed malware in PDF files using structural features related to embedded JavaScript, embedded objects, and cross-references. The authors evaluated models such as Naive Bayes, SVM, Random Forest, and KNN, with KNN achieving an accuracy of 99.85%. Similarly, Kharnotia et al. [14] proposed a static-analysis-based approach for Android malware using permissions, internal components, and structural attributes, once again highlighting the relevance of proper feature selection.

Additionally, some recent works have addressed the complete automation of malware detection systems. In [15], a client-server platform based on React JS and Flask is presented, automating preprocessing, classification, and result visualization tasks using Machine Learning models. Although the system demonstrates advantages in scalability and automation, it does not incorporate hybrid analysis or advanced dimensionality reduction strategies.

Based on the reviewed literature, it can be observed that most studies focus exclusively on either static or dynamic analysis, while only a few integrate static, dynamic, and semantic analysis simultaneously within the same system. Furthermore, although numerous studies employ Machine Learning and Deep Learning techniques, many lack formal dimensionality reduction processes, integration of supervised and unsupervised hybrid models, or robust mechanisms to improve generalization against unknown threats.

In this context, the present work proposes an integrated approach for ransomware detection in Windows systems using Portable Executable (PE) files. Unlike the reviewed works, this proposal incorporates a feature extraction strategy based on static, dynamic, and semantic analysis, enabling a more comprehensive representation of both the internal structure and behavior of malware. In addition, a feature selection and dimensionality reduction process based on Principal Component Analysis (PCA) is implemented to reduce computational complexity and improve system efficiency.

Finally, the proposed work integrates supervised and unsupervised models, including neural networks, logistic regression, KNN, K-means, and isolation forests, enabling a broader comparative analysis with respect to traditional approaches. The combination of multiple information sources, dimensionality reduction, and hybrid models positions this proposal as a more robust and adaptable alternative against modern ransomware threats.

3 Methodology

The methodology proposed in this work describes the complete process for ransomware detection using machine learning techniques. This process covers dataset construction, feature extraction from Windows executable files, data processing and transformation, and the training and evaluation of different machine learning models. Each of these stages is detailed in the following subsections.

3.1 Dataset Construction

The dataset construction process was carried out through a series of structured stages to obtain a representative and reliable set of samples. Figure 1 presents the overall workflow of the process.

The process consists of four main stages: collection, verification, filtering, and final dataset construction.

As shown in Fig. 1, the first stage consists of collecting executable files in Portable Executable (PE) format with `.exe` extension. Malicious samples were obtained from the VirusShare repository [16], which provides a large collection of malware from various families. On the other hand, benign samples were collected in a controlled environment, including Windows operating system files and software downloaded from trusted sources [17], in order to properly represent the behavior of legitimate applications.

Subsequently, a verification and labeling process was carried out using the VirusTotal platform [18]. A consensus-based criterion among antivirus engines was adopted: files detected as malicious by at least ten engines were classified as

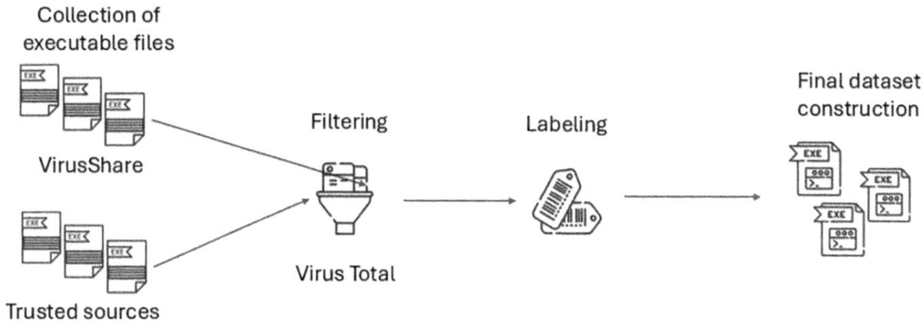

Fig. 1. Dataset construction workflow.

malware, while those with no detections were considered benign. This approach increases labeling reliability by reducing the impact of individual false detections.

Additionally, VirusTotal information was used to identify samples with evidence of packing or obfuscation. Executables exhibiting these characteristics were excluded, as they may interfere with static feature extraction and affect model performance.

During the filtering stage, malicious samples were restricted exclusively to the ransomware category in order to maintain consistency in the analyzed behavior and avoid biases introduced by multiple malware families.

Finally, a dataset composed of 13,400 balanced samples was obtained, defining a binary classification problem between ransomware and benign software.

All experiments and analyses were conducted in an isolated environment using a virtual machine based on Ubuntu Desktop 20.04, configured as a sandbox with restricted access to the host system and network, in order to prevent malware propagation and ensure a secure experimental environment.

3.2 Feature Extraction

A hybrid approach combining static analysis, semantic analysis, and dynamic analysis was adopted in order to capture both structural properties and potential and real behaviors of the analyzed files. Figure 2 shows the general flow of the feature extraction process, including the stages of analysis, data collection, and integration.

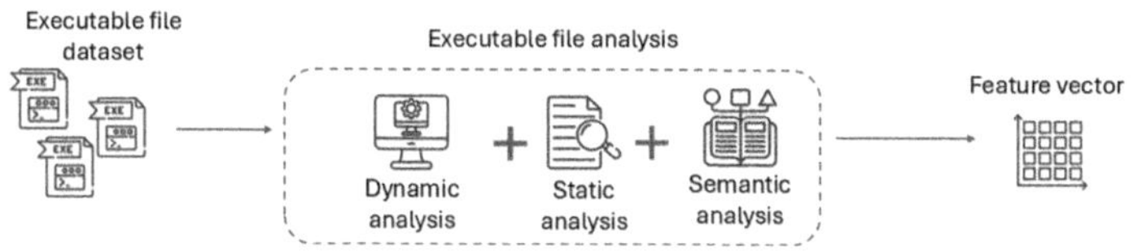

Fig. 2. Feature extraction flow.

As shown in Fig. 2, the process begins with the static analysis of executable files. For this purpose, the Python library *pefile* [19] was used, which allows the extraction of structural information from the Portable Executable (PE) format. In particular, attributes from the PE header were obtained, such as *ImageBase*, *NumberOfSections*, section sizes, and security flags. Likewise, features related to executable sections were extracted, including entropy, size, access permissions, and data distribution, which allow the identification of patterns commonly associated with malicious files.

Unlike traditional static analysis, which mainly focuses on structural properties of the executable, semantic analysis seeks to infer the potential behavior of malware through the study of disassembled code and internal file resources. For this purpose, the *Ghidra* [20] tool was used in *headless* mode, allowing the automation of the analysis through Python-developed scripts and exporting the results in JSON format for subsequent processing.

The input for semantic analysis corresponds to executable files in Portable Executable (PE) format. Each file was individually processed using Ghidra in *headless* mode, performing automatic disassembly of machine code and generating structured information in JSON format.

Subsequently, a Python-developed script processed the JSON files generated by Ghidra to extract different semantic elements from the executable, including system APIs, imported DLLs, function names, function call flow, and relevant text strings. This information was organized into categories associated with possible malicious behaviors, such as file system manipulation (*filesystem*), Windows registry access (*registry*), and privilege escalation (*privilege escalation*).

The selection of relevant APIs was performed by considering functions commonly associated with malicious behavior reported in the literature, particularly those related to file system access, registry modifications, encryption, persistence, and privilege escalation.

From the identified APIs and functions, a binary vector representation based on presence/absence was constructed. In this representation, each dimension corresponds to a specific API, DLL, or behavior previously identified during the analysis of the sample set. If a sample contains a particular API or behavior, the value 1 is assigned; otherwise, the value 0 is assigned.

In this way, the output of semantic analysis corresponds to a structured numerical vector capable of representing the potential behavior of the executable without directly executing it. Finally, these semantic vectors were integrated together with the features obtained through static and dynamic analysis to form the final feature matrix used in training the machine learning models.

Semantic analysis was performed from executable disassembly, extracting information related to imports, dynamic-link libraries (DLLs), function names, system API calls, function call flow, and relevant text strings (*strings*). Among the analyzed strings are file extensions, system paths, and Windows registry keys, which are commonly associated with malicious activities such as file encryption, persistence, or system modification.

In addition, potentially suspicious functions related to file system manipulation, registry access, privilege escalation, and critical operating system operations were identified. The extracted APIs were classified according to their functional behavior into categories such as *filesystem*, *registry*, and *privilege escalation*. Subsequently, these features were represented through binary presence/absence encoding, allowing the generation of semantic vectors capable of describing the potential behavior of the executable without the need to execute it.

On the other hand, dynamic analysis was performed using *Cuckoo Sandbox* [21], executing each sample in a controlled environment. This analysis was carried out inside a virtual machine running Ubuntu Desktop 20.04 used as the main environment to guarantee isolation from the host system. In turn, within this virtual machine, a second virtual machine with Windows 7 operating system was configured, specifically used for sample execution. The choice of Windows 7 is due to its broad compatibility with different malware families, low computational requirements, and ease of integration into sandboxing environments.

During dynamic execution, each sample was monitored for a controlled period of 2 min, sufficient time to capture relevant behaviors without generating excessive computational cost. During this interval, multiple system events were recorded, including process creation, file access and modification, system registry changes, network activity, and operating system calls.

The features obtained from the different types of analysis include numerical values, hexadecimal values, lists of elements, categorical variables, and binary vectors associated with APIs and DLLs. To integrate the information obtained from the different analyses, all features were transformed into a common structured format. In particular, lists of APIs, DLLs, and functions were converted into vector representations through binary presence/absence encoding, while categorical variables were transformed using appropriate encoding techniques.

Regarding missing values, these were handled through imputation techniques. For numerical features, missing values were replaced by zero or by representative statistics from the dataset, such as the mean. In categorical variables, an additional category was incorporated to represent missing information.

Finally, all features extracted through static, semantic, and dynamic analysis were integrated into a unified data matrix, where each row represents a sample and each column corresponds to a feature. This matrix constitutes the basis for subsequent preprocessing, dimensionality reduction, and machine learning model training stages.

3.3 Data Processing

Data processing was performed to transform the extracted features into a suitable format for machine learning models and to improve data quality. Figure 3 shows the overall workflow.

As shown in Fig. 3, an initial exploratory analysis was conducted to identify non-informative features. Variables with zero variance were removed, as well as high-cardinality features, defined as those with a proportion of unique values

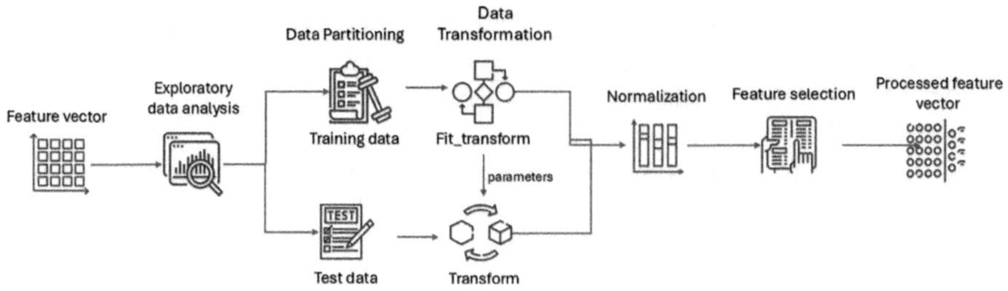

Fig. 3. Data processing workflow.

greater than 0.95 relative to the total number of samples, since they provide limited generalization capability.

Subsequently, the dataset was split into training (80%) and testing (20%) subsets using stratified sampling to preserve class distribution and ensure a representative evaluation.

Next, transformations were applied according to feature type. Categorical variables were encoded using *one-hot encoding*, hexadecimal values were converted into integer representations, and list-type variables were transformed using multi-label encoding. All transformations were fitted exclusively on the training set and then applied to the test set to prevent data leakage.

Once transformed, numerical features were normalized using *StandardScaler* to achieve zero mean and unit variance. This facilitates the convergence of various machine learning algorithms and prevents features with different scales from dominating the training process. Binary variables were not normalized, as their range (0 and 1) is already bounded and their semantic meaning could be altered.

Finally, feature selection techniques were applied in multiple stages. First, a low-variance filter was used. Then, mutual information-based selection was applied to identify the most relevant features with respect to the target variable.

The result of this process is a refined and transformed dataset ready for model training and evaluation.

3.4 Model Training and Evaluation

To evaluate the performance of different machine learning approaches for malware detection, both supervised and unsupervised models were considered.

Specifically, supervised models including logistic regression, k-Nearest Neighbors (KNN), and neural networks were implemented, along with unsupervised methods such as k-means and Isolation Forest.

To analyze the impact of dimensionality reduction, two variants of the dataset were defined: one using the original processed features and another applying Principal Component Analysis (PCA), retaining 95% of the variance. Both variants were used as input to the models, allowing performance comparison under different data representations. Figure 4 shows the overall training and evaluation workflow.

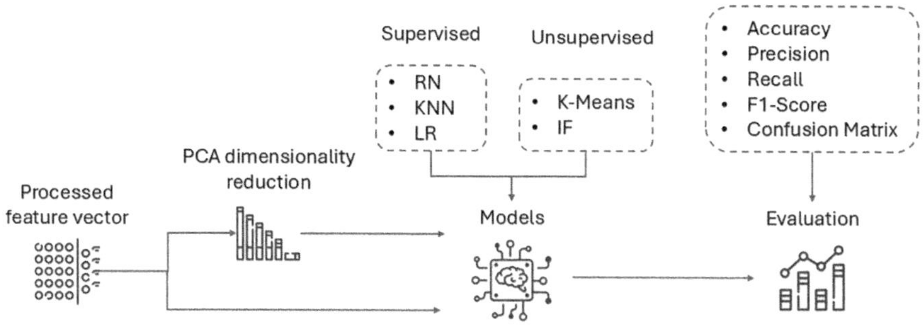

Fig. 4. Model training and evaluation workflow.

As shown in Fig. 4, the processed dataset is divided into two branches: one using original features and another using PCA-based dimensionality reduction. Both variants are used as input for the different machine learning models.

The models were configured as follows. Logistic regression used default parameters. KNN was configured with $k = 5$ neighbors. The k-means model was set with two clusters corresponding to the two classes. For Isolation Forest, a contamination level of 0.5 was defined, assuming a balanced proportion of normal and anomalous data.

The neural network used a multilayer perceptron architecture consisting of an input layer, two hidden layers with 32 and 16 neurons respectively, and an output layer with a single neuron, suitable for binary classification.

Model performance was evaluated using standard classification metrics, including accuracy, precision, recall, and F1-score. Additionally, confusion matrices were analyzed to assess the impact of false positives and false negatives in malware detection.

For unsupervised methods, model outputs were converted into binary labels to enable comparison with supervised models. In particular, for k-means, cluster labels were assigned based on the majority class within each cluster, while for Isolation Forest, the anomaly classification provided by the model was directly used.

4 Results

This section presents the results obtained from applying the proposed methodology for ransomware detection.

The final dataset consists of 13,400 executable file samples in `.exe` format, evenly distributed into 6,700 ransomware samples and 6,700 benign samples. This distribution allows the problem to be addressed as a binary classification task without bias toward any class.

During the feature extraction stage, a total of 75 variables were obtained, including numerical features, categorical variables, hexadecimal values, and lists of elements. Subsequently, during exploratory analysis, non-informative features were removed, reducing the set to 71 variables.

After the data transformation stage, all features were converted into a fully numerical format, resulting in a total of 262 variables. This increase is mainly due to the encoding of categorical variables and lists using techniques such as *one-hot encoding* and multi-label encoding.

The final dataset used for training contains 13,400 instances with 262 features. Subsequently, normalization was applied to numerical variables using *StandardScaler*, maintaining the dataset dimensionality.

The main configuration of the evaluated models is presented in Table 1, including the hyperparameters used, the application of dimensionality reduction through PCA, and the final dimensionality employed by each model. The selection of principal components was performed by considering the balance between dimensionality reduction and preservation of the explained variance of the dataset. Hyperparameters not explicitly specified in the table were configured using the default values provided by the Scikit-learn library.

Table 1. Configuration of the evaluated models

Model	Main Parameters	PCA	Dimensions
Isolation Forest	Contamination = 0.5	No	7
K-means	Number of clusters = 2	Yes	44
KNN	Number of neighbors = 5	Yes	7
Logistic Regression	Maximum iterations = 1000	No	7
Neural Networks	50 epochs, 2 hidden layers (32,16)	Yes	8

Note: Parameters not specified were configured using the default values of the Scikit-learn library

The performance of the evaluated models is presented in Table 2. The metrics used for evaluation were *accuracy*, *precision*, *recall*, and *F1-score*, allowing the analysis of both the general classification capability and the balance between false positives and false negatives.

Table 2. Performance of the evaluated models

Model	Accuracy	Precision	Recall	F1-score
Isolation Forest	0.6709	0.6676	0.6806	0.6741
K-means	0.5190	0.5098	0.9888	0.6728
KNN	0.9593	0.9536	0.9657	0.9596
Logistic Regression	0.9851	0.9815	0.9888	0.9851
Neural Networks	**0.9884**	0.9809	**0.9963**	**0.9885**

The results clearly demonstrate the superiority of supervised models over unsupervised methods. In particular, the neural network achieved the best overall performance, reaching an accuracy of 0.9884 and an F1-score of 0.9885. Additionally, logistic regression showed highly competitive results, with an accuracy of 0.9851, suggesting that the problem is linearly separable in the transformed feature space.

The KNN model also exhibited strong performance, achieving an accuracy of 0.9593 and an F1-score of 0.9596, although slightly lower than the previously mentioned models.

In contrast, unsupervised methods showed significantly lower performance. In particular, k-means achieved high recall (0.9888) but low precision, indicating a large number of false positives. On the other hand, Isolation Forest exhibited a more balanced behavior, although still considerably inferior compared to supervised models.

These results suggest that the extracted features contain sufficient discriminative information to effectively separate both classes when supervised models are used. In contrast, unsupervised methods face limitations due to the lack of explicit label information, which affects their generalization capability in this context.

The analysis of confusion matrices provides further insight into model behavior. Figure 5 shows the confusion matrix of the best-performing model (neural network), while Fig. 6 presents the confusion matrix of the k-means model as a reference of lower performance.

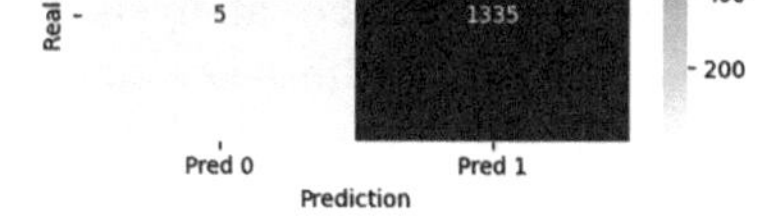

Fig. 5. Confusion matrix of neural network

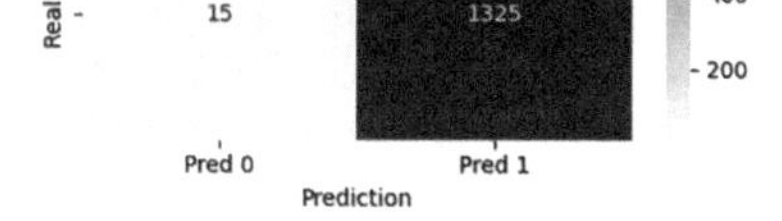

Fig. 6. Confusion matrix of k-means

Supervised models exhibit a low error rate in both classes, particularly the neural network, which significantly minimizes false negatives—an essential aspect in malware detection systems. In contrast, k-means incorrectly classifies a large number of benign samples as malware, resulting in a high number of false positives, which limits its applicability in real-world scenarios.

Finally, the results show that the application of PCA allows for a significant reduction in dataset dimensionality without negatively affecting the performance of supervised models. In particular, models such as KNN and neural networks maintain high levels of precision and recall even with a reduced number of components.

This suggests that most of the relevant information for classification is concentrated in a reduced subset of principal components, enabling a decrease in computational complexity without compromising predictive performance.

5 Conclusions

This work presented a machine learning-based approach for ransomware detection, integrating reverse engineering techniques for feature extraction through static, dynamic, and semantic analysis. Additionally, a comparison between supervised and unsupervised models was conducted, along with an evaluation of the impact of dimensionality reduction using Principal Component Analysis (PCA).

The obtained results demonstrate that supervised models significantly outperform unsupervised methods in the task of ransomware detection. In particular, the neural network model achieved the best performance, reaching an accuracy of 0.9884 and an F1-score of 0.9885, highlighting the capability of the proposed approach to identify discriminative patterns in the data.

Furthermore, it was observed that the application of PCA allows a considerable reduction in the dimensionality of the feature set without negatively affecting the performance of supervised models. This suggests that the most relevant information is concentrated in a reduced subset of features, contributing to improved computational efficiency.

The use of a hybrid feature extraction approach, combining static, dynamic, and semantic analysis, enabled the capture of complementary information about executable behavior, which proved to be key to model performance. This constitutes one of the main contributions of this work, together with the systematic comparison of different types of models and data representations.

However, this study presents some limitations. First, the analysis was restricted exclusively to ransomware samples, which may limit the generalization of the results to other types of malware. Additionally, only executable files without packing or obfuscation techniques were considered, simplifying the problem compared to real-world scenarios. Furthermore, the dataset was constructed in a controlled environment, and dynamic analysis was performed under a limited execution time, which may prevent the observation of more complex malicious behaviors.

As future work, it is proposed to extend the study by considering different malware families, including obfuscated or packed samples, as well as improving feature extraction processes. Likewise, the integration of more advanced models is suggested, including deep learning architectures and large language models, as well as their incorporation into network-based detection systems, which would allow evaluation in more realistic environments.

Overall, the results demonstrate the potential of machine learning, combined with hybrid analysis techniques, as an effective strategy for ransomware detection.

References

1. SBerrios, S., Leiva, D., Olivares, B., Allende-Cid, H., Hermosilla, P.: Systematic review: Malware detection and classification in cybersecurity. Appl. Sci. **15**(14), 7747 (2025). https://www.mdpi.com/2076-3417/15/14/7747
2. Lai, T.-H., Tsai, Y.-J., Liu, C.-L.: Improving the performance of static malware classification using deep learning models and feature reduction strategies. Mathematics **13**(23), 3753 (2025). https://www.mdpi.com/2227-7390/13/23/3753
3. Alqahtani, A., Alzahrani, F., Alghamdi, A., Alotaibi, S.: Evaluation of malware classification models for heterogeneous data. Sensors **24**(1), 288 (2024). https://www.mdpi.com/1424-8220/24/1/288
4. Rezaei, T., Manavi, F., Hamzeh, A.: A PE header-based method for malware detection using clustering and deep embedding techniques. J. Inf. Secur. Appl. **60**, 102876 (2021). https://www.sciencedirect.com/science/article/pii/S2214212621001046
5. Singh, J., Singh, J.: A survey on machine learning-based malware detection in executable files. J. Syst. Archit. **112**, 101861 (2020)
6. Zmaimita, H., Madani, A., Zine-Dine, K.: A survey of malware detection and classification using machine learning and deep learning. ResearchGate Preprint (2025). https://www.researchgate.net/publication/403975379_A_Survey_of_Malware_Detection_and_Classification_Using_Machine_Learning_and_Deep_Learning. Accessed 07 May 2026
7. Mali, N., Chauhan, N.: Advances in malware detection using machine learning and deep learning: a comprehensive comparative analysis. Int. J. Comput. Netw. Inf. Secur. **18**(2), 61–77 (2026)
8. Kamdan, Y.P., Munzi, R.S., Mustafa, A.B., Kharisma, I.L.: Static malware detection and classification using machine learning: a random forest approach. Eng. Proc. **107**(1), 76 (2025). https://www.mdpi.com/2673-4591/107/1/76
9. Ogbonna, C.E. et al.: Static malware classification: a comparative analysis of machine learning techniques on byte and opcode features. ResearchGate Preprint (2025). https://www.researchgate.net/publication/394286583_Static_Malware_Classification_A_Comparative_Analysis_of_Machine_Learning_Techniques_on_Byte_and_Opcode_Features. Accessed 08 May 2026
10. Belaoued, M., Guelib, B., Bounaas, Y., Derhab, A., Boufaida, M.: Malware detection system based on an in-depth analysis of the portable executable headers. ResearchGate (2021). https://www.researchgate.net/publication/350021076_Malware_Detection_System_Based_on_an_In-depth_Analysis_of_the_Portable_Executable_Headers. Accessed 08 May 2026
11. Sgandurra, D., Muñoz-González, L., Mohsen, R., Lupu, E.C.: Automated dynamic analysis of ransomware: benefits, limitations and use for detection. In: Proceedings of the International Conference on Risks and Security of Internet and Systems. Springer, London (2016). https://www.researchgate.net/publication/308026890_Automated_Dynamic_Analysis_of_Ransomware_Benefits_Limitations_and_use_for_Detection. Accessed 07 May 2026

12. Belaoued, M., Boukellal, A., Koalal, M.A., Derhab, A., Mazouzi, S., Khan, F.A.: Combined dynamic multi-feature and rule-based behavior for accurate malware detection. Int. J. Distrib. Sensor Netw. (2019). https://www.researchgate.net/publication/337494991_Combined_dynamic_multi-feature_and_rule-based_behavior_for_accurate_malware_detection. Accessed 08 May 2026
13. Khan, B., Arshad, M., Khan, S.S.: Comparative analysis of machine learning models for pdf malware detection: Evaluating different training and testing criteria. ResearchGate (2023). https://www.researchgate.net/publication/373199586_Comparative_Analysis_of_Machine_Learning_Models_for_PDF_Malware_Detection_Evaluating_Different_Training_and_Testing_Criteria. received: June 1, 2023; Accepted: August 3, 2023; Published: August 21, 2023
14. Kharnotia, S., Kumar, A., Singh, P.: Feature-driven static analysis for learning-based android malware detection. Array **19**, 100305 (2026). https://www.sciencedirect.com/science/article/pii/S2405959526000056
15. Vignesh, M., Saravanan, K., Ramkumar, V., Vimal Raj, V.: Efficient malware detection using machine learning (2025). https://www.researchgate.net/publication/404204390_Efficient_Malware_Detection_Using_Machine_Learning. Accessed 07 May 2026
16. VirusShare, Virusshare.com - malware sample repository (2026). https://virusshare.com/. Accessed 01 Apr 2026
17. Microsoft, Microsoft store - inicio (2026). https://apps.microsoft.com/home?hl=es-MX&gl=MX. Accessed 05 April 2026
18. VirusTotal, Virustotal - file upload and URL analysis interface (2026). https://www.virustotal.com/gui/home/upload. Accessed 01 Apr 2026
19. Carrera, E.: pefile: Portable executable reader module (2024). https://pypi.org/project/pefile/. Accessed 05 Apr 2026
20. National Security Agency, Ghidra: Software reverse engineering framework (2026). https://ghidra-sre.org. Accessed 05 Apr 2026
21. Cuckoo Foundation, Cuckoo sandbox: Automated malware analysis system (2021). https://github.com/cuckoosandbox/cuckoo. repositorio en GitHub. Accessed 05 Apr 2026

Crowd Modelling: Including Generative Artificial Intelligence in Non-expert Teams for Curriculum Updating

Octavio González-Aguilar(✉) and Francisco Javier Alvarado-Rodriguez

Universidad Autónoma de Guadalajara, Guadalajara, Mexico
{octavio.aguilar,francisco.alvarado}@edu.uag.mx

Abstract. There is plenty of research about the implementation and user acceptance of artificial intelligence (AI) in human-robot interactions, or AI as a tool. What remains unknown is a process to include generative AI agents as team members for solving decision making problems with multiple possible answers. It is important for curriculum design given that decision-makers may not have access to experts for socioeconomical reasons. Therefore, the objective of this paper is to provide a method to aggregate the contributions of nonexperts working with Generative Artificial Intelligence (GAI) agents for curriculum design. The research method is design science and the technique to obtain and analyze data is simulation. The main results include the design principles of the method, the algorithm and metrics and the scenarios for decision-making. This paper provides a theoretical contribution because it is an example of how to include GAI agents, such as chatbots, as part of human-GAI teams. There is a practical contribution because universities that may not have the resources to update their academic programs with experts, may use GAI to reduce the gap. The major conclusion is that the collective curriculum may provide, in some scenarios, more diversity faster and cheaper than current approaches based on consultants and experts.

Keywords: Artificial Intelligence · Chatbots · Human-Artificial Intelligence

1 Introduction

AI is "the simulation of human intelligence in machines designed for thinking and acting as humans" [1]. In education, AI offers mainly solutions for teaching and management [2]. Nevertheless, the willingness of the workforce to accept robots depends on objective, subjective, and organizational factors [3]; Some of the major advances are in terms of assignation of tasks, perception of risk, performance improvement, and guidelines to include AI in working environments. For example, when AI designates tasks for humans, the collaboration between humans and AI outperforms the performance of pure AI agents; nevertheless, it does not happen when humans assign tasks [4], the impact of the perception of risk associated with algorithm error depends on whether it happens early or late [5], in terms of performance improvement, it is not the same for novices as for experts [6] and AI agents must not replicate human teamwork but focus

H. Ponce et al. (Eds.): COMIA 2026, CCIS 3095, pp. 637–648, 2027.
https://doi.org/10.1007/978-3-032-37438-7_50

on the strengths they add [7], chatbots require the feedback of humans to generate real value [8], among others. The concept of human in the loop states that humans provide feedback to the AI, nevertheless, some challenges remain as how to evaluate the quality of the feedback, which feedback represents better the human perception in a team, how the AI learns from the feedback, and when to provide the feedback [9]. In fact, the World Economic Forum 2026 published that humans are at the heart of intelligent economies [10].

There are many methods to collaborate and are widely studied [11–14]. In terms of collective decisions, the main characteristics studied are the individual cognitive differences and the communication among the collective [15, 16]. Specifically, in the strand of research of this paper, if decision-makers hire one or more experts for taking an decision with multiple possible answers, such as updating a curriculum, it is a set of different dichotomous decisions. For example, to keep or not to keep a course, to move or not to move a course, to add or not to add a course, to replace or not to replace a course. Even if there are different options to replace a course, the decisions are to replace it by the first option or not, to replace it by the second option or not, and so on. Consequently, Fig. 1 explains that although experts, nonexperts or Generative Artificial Intelligence (i.e., GAI) agents may consider huge number of variables and context to take their decision, at the end, that knowledge must lead to include or exclude the courses in the curriculum, to move forward, backward or maintain the courses at the same place, or to replace with a specific course another one. The illustration states that a human-GAI team may include one or more GAI agents. Specifically, in the context of this research, chatbots. Similarly, the human-GAI team may include one or more humans, none or more than one expert. The knowledge may include their own knowledge or other sources, still, what matters, it is the final proposal. Defined as the total of their dichotomic decisions about each course. The illustration also states that the crowd will include the best performer, and it may include an expert as reference. It is important to highlight that the expert reference may be proposed by one expert, a group of consultants, or the whole team of academic professors in a college, what matters, it is the final proposal. Thus, the aggregation method only considers the courses in the proposals of new curriculums, no other external information. Therefore, the research question is as follows:

RQ1: How can an algorithm aggregate the proposals of human-GAI teams in problems with more than one answer?

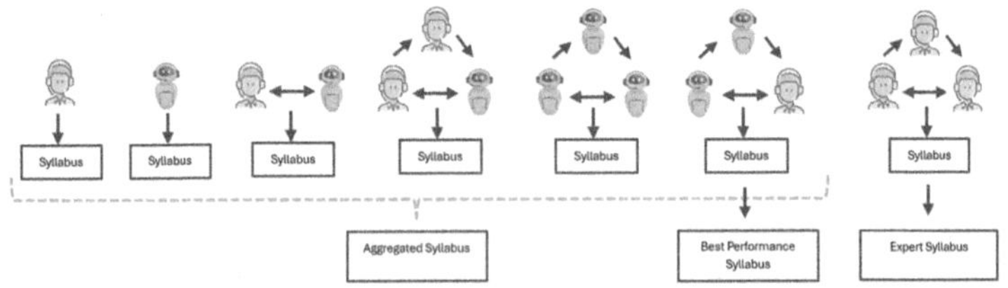

Fig. 1. Types of contributions for the aggregation of proposals from human-Generative Artificial Intelligence teams.

2 Literature Review

The contribution of this paper is a method for aggregating diverse academic syllabus proposals into one. The objective is to update the current curriculum by integrating the proposals of human-GAI teams. Therefore, the following sections discuss aggregation methods and voting mechanisms.

A. Aggregation Methods

Some of the most recognized methods for aggregating the views of individuals are collective causal maps [17], cognitive maps [13], mental models [18], conceptual models [19] and strategic maps [11]. Those approaches vary throughout the heterogeneity of the group, the group interaction, and the standardization of the process. This paper proposes a method with an approach like causal or cognitive maps. Nevertheless, there are two main differences: first, in this paper the aggregation is not for concepts associated with cause-effect. This paper focuses on the aggregation of solutions for decision-making problems with multiple possible answers. Second, the goal is to avoid experts in the process of data gathering or aggregation. Therefore, the Diversity Prediction Theorem suggested by Scott Page [20, 21], illustrated in Eq. 1, is implemented. There is plenty of research about this theorem. Nevertheless, it requires you to know the value of the truth, i.e. V. The novelty of this research is that decision-making problems with multiple possible answers do not have V. Therefore, we propose an algorithm to aggregate the proposals and obtain a collective result in the lack of a unique truth.

The theorem has three parts. First, the average individual error that measures how different each proposal is with respect to the true. Second, prediction diversity that measures how different each proposal is with respect to the average proposal. Finally, the result of subtracting the prediction diversity from the average individual error is the many-model error, also known as crowd error. The aim is to maintain the many-model error as near as possible to zero. This research measures the performance during the decision-making process in problems with multiple possible answers based on this theorem. This theorem exemplifies why the error of aggregating many models always is smaller than the error from the average [21]. In other words, higher diversity leads to lower collective error. The closer to zero, the better performance.

$$\left(M' + V\right)^2 = \sum_{i=1}^{N} \frac{(Mi - V)^2}{N} - \sum_{i=1}^{N} \frac{(Mi - M\prime)^2}{N} \quad (1)$$

Where Mi equals the diversity value of each proposal from a human-GAI team. M' equals the average of the model's values. In this research, it is the average of the diversity of all the proposals from the human-GAI teams. V equals the true value. In this paper, the diversity of the expert proposal. N is the population. In this simulation, N is the total number of proposals from the human-GAI teams. Diversity means how different each proposal is from the current syllabus. To calculate the diversity, we developed Algorithm 1. To obtain V use algorithm 1 to compare the expert proposal with the current syllabus. To obtain each Mi use algorithm 1 to compare each human-GAI proposal with the current syllabus. In other words, if some individual curriculum is the same as the current curriculum the value of diversity is 0, on the contrary, if some individual curriculum is completely different as the current curriculum the value of diversity is 1.

Algorithm 1 – Diversity prediction

```
Diversity = 0
IF CollectiveCourse = CurrentCourse THEN
    Diversity = Diversity + 0
ELSE
  IF CollectiveCourse IS IN CourseSet THEN
      Diversity = Diversity + 0.5
  ELSE
      IF CollectiveCourse = "None" THEN
          Diversity = Diversity + 0.05
      ELSE
          Diversity = Diversity + 1
      ENDIF
  ENDIF
ENDIF
```

B. **Voting Mechanisms**

There are different voting mechanisms to decide the willingness of a crowd. For example, plurality voting, approval voting and majority judgement. Plurality voting allows the voter to vote for at most n candidates whereas approval voting allows the voter to vote for any number of candidates, approval voting provides more information about the desires of the voters, and it is the best method to find the option that defeats others in a one-to-one contest [22]. The main critic for approval voting is that it provides more information than plurality voting but still considering only approval or disapprove. Black or white. Instead, majority judgment proposes to use ordinal grades, ideally six or seven [23]. This study considers three scenarios for a vote: approved (if the participant does not move the course), disapproved (if the participant replaces the course) or moved (if the participant moves forward or backward the course). Therefore, this study applies majority judgment.

3 Method

This research follows the Design Science Research Method; therefore, it is evaluated through simulation, and it conducts six steps:

Problem identification and motivation: This section indicates what would happen without this research?
Definition of the objectives for a solution: This section answers what would a better artifact accomplish?
Design and development: This section shows the artifact.
Demonstration: This section illustrates how to use the artifact in a suitable context.
Evaluation: This section provides evidence of the achievements and the future strands of research.
Communication: This section states suitable means of communication to the general or academic public.

The following section exemplifies each one of the six stages.

4 Simulation: Decision Making in Problems with Multiple Possible Answers

A. **Problem Identification and Motivation**

Updating an academic program gathers society, community, and industry in a dialogue to exchange ideas. Nevertheless, this process requires both time and money. Furthermore, the results depend on the richness of visions involved and the aggregation process. Finally, each vision adds limited, biased information, and, at the end, the result is also a bet, it may be a success or a failure. To add software agents, implementing artificial intelligence may help to identify: 1) the requirements of a market, 2) the academic offer of competitors, and 3) the values associated with the institution. This simulation demonstrates the usefulness of the method; consequently, it helps to include, faster and cheaper, a wider and more diverse set of contributors during the process of updating academic programs. Specifically, human-GAI teams.

B. **Definition of the Objectives for a Solution**

The main objective of this simulation is to exemplify how a decision maker in an academic institution may use GAI (e.g., chatbots) and human-GAI teams to improve the process of updating academic programs. To achieve this, the simulation will:

- Simulate a crowd of participants and their suggestions to solve a decision-making problem with multiple possible answers.
- Simulate the aggregation method to generate the collective suggestion.
- Simulate the evaluation of each individual suggestion and the collective one.

C. **Design and Development**

The design and development consist in three components. First, the definitions. These definitions are those relevant concepts or formulas used during the method. Table 1 provides the definitions. Second, the process. Figure 2 defines three stages of the process. Third, the rules. These rules define the conditions to conduct the process and they are the design principles defined in Tables 2 and 3.

As Fig. 2 states, the first stage is Diversity configuration. In this stage, the decision maker must define the number of proposals from human-GAI teams (x), the number of courses (d) and the number of terms (m). When m divides d, the remainder (r) indicates how many elements are left after forming a perfect matrix (see Eq. 2). Thereafter, we calculate the number of rows in the preliminary matrix (k) by calculating d minus r, all divided by m (see Eq. 3). Finally, one is added to k (see Eq. 4) to obtain the max number of courses per term (n), or the number of rows in the final matrix, the addition of one row is to capture the courses in r. Furthermore, the decision-maker must configure both the set of courses, the courses of the current curriculum and the prerequisites of each course where it applies. Consider that Eqs. 2 and 3 assume that both d and m are greater than zero.

$$r = d\%\mathrm{m} \tag{2}$$

Table 1. Definitions.

Concept	Definition	Properties
Courses	They are the components of the academic program	It is defined by the user as d where 0 < d
Term	They are time periods that organize the courses	It is defined by the user as m where 0 < m
Current curriculum	It is the current arrangement of courses and terms	It is a matrix of m x n that includes d courses
Expert curriculum	It is the proposal arrangement of courses and terms made by one or more experts	It is a matrix of m x n that includes d courses
Individual curriculum	It is the proposal arrangement of courses and terms made by a novice helped by GAI agents	It is a matrix of m x n that includes d courses
Collective curriculum	It is the proposal arrangement of courses and terms made by the aggregation of all the individual curriculum	It is a matrix of m x n that includes d courses
Individual error	It is the difference between the individual prediction and the expert prediction	It varies from 0 to 1 where 0 is the worst and 1 is the best
Collective error	It is the difference between the collective prediction and the expert prediction	It varies from 0 to 1 where 0 is the worst and 1 is the best

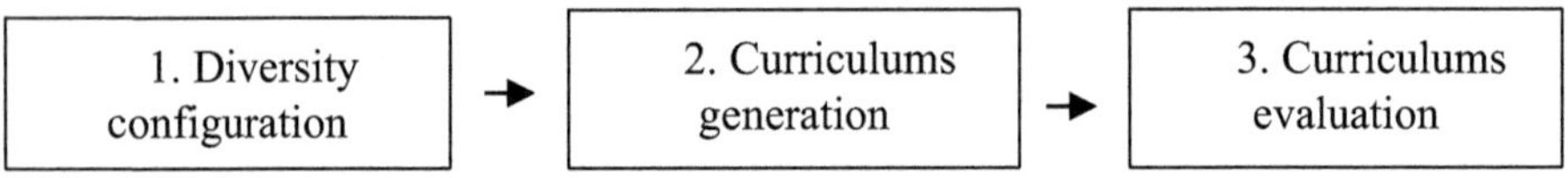

Fig. 2. Stages of the process to conduct the simulation.

$$k = (d - \mathrm{r})/m \tag{3}$$

$$n = k + 1 \tag{4}$$

The second stage is Curriculums generation. Fourth types of curriculums are required. First, the current curriculum. This one is the current state of the academic program. Second, the expert curriculum. This is the golden standard. As mentioned previously, this curriculum is the final proposal that integrates the whole available knowledge. It may include one or more experts, internal, external or a mix of internal and external experts. It may include public, private or a mix of internal and private information. At the end, this represents the best approach to the truth. In other words, without our method, it would be the final proposal. Third, the nonexpert curriculums. These are the individual predictions proposed by the human-GAI teams, even though they could be proposed by one or more agents. Finally, it does not matter the number of participants in a human-GAI team, each team provides only one proposal. The fourth is the collective curriculum. This curriculum is the aggregation of all the proposals from the human-GAI teams.

The method follows the rules in Tables 2 and 3 to aggregate all the nonexpert curriculums into one collective curriculum. Therefore, the collective curriculum may only include four type of values: the name of a course in the course set, the tag "To be updated" when the number of nonexpert curriculums asking a course not included in the course set is higher than the number of nonexpert curriculums asking the same course from the course set, the tag "None" when the majority of nonexpert curriculums indicates that no course must be taught in that cell and the tag "No option found" when the algorithm cannot determine either of the previous options.

Table 2. Design principles: rules for updating.

Rules	Description
1	If the curriculum does not change the position of a course in its matrix, the aggregation method considers it as a vote to keep the course in the same place.
2	If the curriculum does change the position of a course in its matrix, the aggregation method considers it as a vote to keep the course in the academic program but in a different place.
3	If the curriculum does replace a course with a new one in its matrix, the aggregation method considers it a vote to substitute the previous course with the new one.
4	If the curriculum does keep a cell in the matrix empty, the aggregation method considers it a vote to maintain that term with a smaller number of courses.

The third stage evaluates the diversity of the proposed curriculum and the error with respect to the golden standard. Algorithm 1 defines these steps and provides the logic to assign the values for diversity. This logic compares both curriculums as if they are two matrices. The algorithm compares position by position and sums up all the corresponding points. If both matrices are equal the diversity value is 0, if both matrices are completely different, that is all the courses must be updated, the maximum value is 1.

D. **Demonstration**

The simulation will exemplify how to use the method. The simulation uses python as a language program and considers Law as academic program.

Stage 1. Diversity configuration: The number of courses (d) is 48, the number of terms (m) is 9 and the number of nonexpert curriculums (x) is 18. Therefore, the variables were set as: $r = 3$, $k = 5$, $n = 6$, $x = 18$. The set of courses contains the names of all the 48 courses. Each name starts with a letter depending on the line of specialization. The prerequisites are a list of lists. They are expressed as follows:

Table 3. Design principles: rules for aggregating.

Rules	Description
1	In each position of the matrix of the collective curriculum, the course with more votes, considering all the nonexpert curriculums, must appear.
2	If there are duplicated courses, those values must be replaced with the tag "replace".
3	For each position of the matrix of the collective curriculum with a tag "replace", there must be an evaluation. The number of votes of the second course with more votes will be compared with the number of nonexpert curriculums that votes for a course not included in the course set. If the former is greater, it will be assigned to the cell in the matrix. Otherwise, the tag "replace" will be replaced by "To be updated".
4	Following rule 3, if there is a cell where a unique value cannot be found, the algorithm must assign "No option found".

$$\text{prerequisites} = \left[\left[\text{"D5.", "D30."}\right], \left[\text{"D1.", "D31.", "D32."}\right], \left[\text{"D3.", "D33."}\right], \left[\text{"S7.", "S9."}\right], \left[\text{"R1.", "R2.", "R3."}\right], \left[\text{"P1.", "P2."}\right]\right]$$

The results of this first stage are the definitions of d, m, x, the course set, and the prerequisites defined by the institution or the person who leads the exercise.

Stage 2. Curriculum generation: In this simulation, the matrix selects the courses randomly from the course set and fulfils the prerequisites. The current curriculum includes 48 courses and 6 None to fulfill the 54 cells in the matrix 9×6.

E. **Evaluation**

A software in python generates the current curriculum, the expert curriculum, the d number of nonexpert curriculums and the collective curriculum. The simulation executes the exercise with five seeds to replicate the data. Table 4 presents the results. The columns in Table 4 represent the following:

Seed: Python uses this number to provide the same results when random processes are executed.
d: It is the number of nonexpert curriculum that the simulation includes.
ED - Expert Diversity: This is how different the expert curriculum is with respect to the current curriculum.
CD - Collective Diversity: This is how different the collective curriculum is with respect to the current curriculum.
MiD - Minimum Individual Diversity: This is the minor value of diversity considering all the nonexpert curriculums.
MaD - Maximum Individual Diversity: This is the major value of diversity considering all the nonexpert curriculums.
CE - Collective Error: This is how different the collective curriculum is with respect to the expert curriculum.
ME - Many-Model Error: This is the result of subtracting the Diversity of Model predictions from the Average-Model Error.

Table 4. Evaluation of results - simulation.

Seed	d	ED	CD	MiD	MaD	CE	ME
98	18	0.509	0.546	0.342	0.527	0.0013	0.0037
76	18	0.509	0.583	0.379	0.546	0.0054	0.0026
55	18	0.435	0.574	0.342	0.537	0.0192	0.0000
31	18	0.527	0.592	0.342	0.500	0.0042	0.0097
18	18	0.407	0.592	0.333	0.537	0.0342	0.0007
55e	30	0.435	0.574	0.324	0.583	0.0192	0.0000
18e	30	0.407	0.592	0.333	0.537	0.0342	0.0004
55 m	3	0.435	0.574	0.416	0.527	0.0192	0.0004
18 m	3	0.407	0.592	0.333	0.481	0.0342	0.0004

According to Table 4, the ideal cases are seeds 98 and 31. In those cases, the CD is greater than ED. In other words, the collective model, which is the result of the process of aggregation, includes more diversity than the diversity provided by the experts. Furthermore, the CD is greater than MaD. Therefore, the result of the aggregation provides more diversity than the best individual curriculum. Lastly, the CE is smaller than the ME. Consequently, the results of the aggregation are more precise than the average of the predictions. The results from cases 76, 55 and 18 are unacceptable cases. In those cases, CD is greater than ED, and CD is greater than MaD. Nevertheless, the CE is greater than the ME. For the cases 55 and 18 the simulation increased from 18 to 30 the nonexpert views. On the same cases, the simulation decreased from 18 to 3 the nonexpert views. Nevertheless, this also did not resolve the problems. Consequently, decision makers must include as much diversity as possible since the selection of the participants. Size does not have a real impact on the results. There must be a theoretical sampling rather than statistical. After the simulation, we developed a web system to gather real-world data. We configured the syllabus of a Law programme in the system and asked three lawyers to provide their proposals. The user's pointer was recorded while using the tool. The interaction with the chat was also recorded, and it was the first pilot. Finally, the algorithm to aggregate the three proposals provided a collective programme. We sent the invitations at night, and we obtained results next day. During the simulation, no human participated. Still, when we implemented the pilot, we provided a video exemplifying the use of the tool and the chat. The users did not require further experience to use the system and provided their proposals. Table 5 presents the results of the evaluation.

Table 5. Evaluation of results - pilot.

d	ED	CD	MiD	MaD	CE	ME
3	0.0000	0.0476	0.0555	0.2916	0.0022	0.0301

This case is acceptable. It is not ideal, but it is not unacceptable. CD is greater than ED, which means that there is more diversity, CD is not greater than MaD, which means that there is a human-GAI team leading the aggregation. It is not the ideal, but it is not a problem. What really matters is that CE is not greater than ME, just as the ideal case. It is important because demonstrates that the collective is better than average.

F. **Communication**

The results are important for decision makers such as curriculum designers, academic program directors, and consultants. The publication in this journal and academic conferences is the first step. The next is to replicate the exercise in a real-world scenario to exemplify the whole administrative process.

5 Limitations and Future Research

The main limitation of the study is attached to the method. The simulation exemplifies how the algorithm works, exemplifies the stages, successful scenarios, and provide illustrations of the how to implement it with real Human-GAI teams. Finally, it is important to evaluate the risks, opportunities, fears, and strengths this type of methods foster in decision makers in private and public universities.

6 Conclusion

This paper considers the inclusion of GAI in human teams for solving decision-making problems with multiple possible answers. We proposed a method for the inclusion and conducted both a simulation and a pilot test to evaluate the process and provide evidence to classify an exercise as successful, acceptable, and unacceptable. We also proposed design principles to implement the method both by implementing an information system or by hand. The inclusion of GAI helped to provide new ideas, professional concepts, and act as a co-worker when the human requires help. Nevertheless, the AI hallucinations may also lead to wrong conclusions. The paper provides a method to help when there is a lack of experts in the problem (e.g., updating a curriculum). Nevertheless, it is important that participants know the context of the problem to be able to filter the suggestions of GAI (for that reason we invited lawyers instead of other profiles). Otherwise, the human will not be required in the team (we could use only random selection or just use AI agents). Consequently, this simulation makes a theoretical contribution by providing a method to resolve a decision-making problem with multiple possible answers, and exemplifies a practical contribution because academic institutions willing to be updated, but without the resources to pay for the time of experts in curriculum design, may use GAI to enhance the potential of the teams of nonexperts.

Acknowledgments. Thanks to Felizmente Verde who helped with the founding of this research.

Disclosure of Interests The authors have no competing interests to declare that are relevant to the content of this article.

References

1. Chan, C.K.Y., Colloton, T.: Generative AI in Higher Education: The ChatGPT Effect. Routledge, New York (2024)
2. Chaudhry, M.A., Kazim, E.: Artificial Intelligence in Education (AIEd): a high-level academic and industry note 2021. AI Ethics. **2**, 157–165 (2022). https://doi.org/10.1007/s43681-021-00074-z
3. Meissner, A., Trübswetter, A., Conti-Kufner, A.S., Schmidtler, J.: Friend or foe? Understanding assembly workers' acceptance of human-robot collaboration. ACM Trans. Hum. Robot. Interact. **10**, 1–30 (2021). https://doi.org/10.1145/3399433
4. Fügener, A., Grahl, J., Gupta, A., Ketter, W.: Cognitive challenges in human–artificial intelligence collaboration: investigating the path toward productive delegation. Inf. Syst. Res. **33**, 678–696 (2022). https://doi.org/10.1287/isre.2021.1079
5. Kim, A., Yang, M., Zhang, J.: When algorithms err: differential impact of early vs. late errors on users' reliance on algorithms. ACM Trans. Comput.-Hum. Interact. **30**, 1–36 (2023). https://doi.org/10.1145/3557889
6. Inkpen, K., et al.: Advancing human-AI complementarity: the impact of user expertise and algorithmic tuning on joint decision making. ACM Trans. Comput.-Hum. Interact. (2023). https://doi.org/10.1145/3534561
7. Musick, G., O'Neill, T.A., Schelble, B.G., McNeese, N.J., Henke, J.B.: What happens when humans believe their teammate is an AI? An investigation into humans teaming with autonomy. Comput. Human Behav. **122**, 106852 (2021). https://doi.org/10.1016/j.chb.2021.106852
8. Kashyap, P.: Machine Learning for Decision Makers: Cognitive Computing Fundamentals for Better Decision Making. Apress, Bangalore (2024)
9. Memarian, B., Doleck, T.: Human-in-the-loop in artificial intelligence in education: a review and entity-relationship (ER) analysis. Comput. Hum. Behav. Artif. Hum. **2**, 100053 (2024). https://doi.org/10.1016/j.chbah.2024.100053
10. Argueta, M., Francis, T., Li, C., Ortega, F.C.: Latin America in the Intelligent Age: A New Path for Growth. World Economic Forum (2026)
11. Ossadnik, W., Kaspar, R.H., Schinke, S.M.: Constructing a tailor-made performance management system supported by knowledge elicitation tools and dynamic modeling. Int. J. Bus. Res. Manag. **4**, 75–98 (2013)
12. Kwahk, K.-Y., Kim, H.-W., Chan, H.C.: A knowledge integration approach for organizational decision support. J. Database Manag. **18**, 41–61 (2007)
13. Cunha, A.A.R., Filho, J.L.S., Morais, D.C.: Aggregation cognitive maps procedure for group decision analysis. Kybernetes. **45**, 589–603 (2016)
14. Nadkarni, S., Nah, F.F.: Aggregated causal maps: an approach to elicit and aggregate the knowledge of multiple experts. Commun. Assoc. Inf. Syst. **12**, 406–436 (2003)
15. Burton, J.W., Almaatouq, A., Rahimian, M.A., Hahn, U.: Algorithmically mediating communication to enhance collective decision-making in online social networks. Collect. Intell. **3** (2024). https://doi.org/10.1177/26339137241241307
16. Zha, Q., Wang, S., Zhang, W., Zhang, H.: Failure Mode and Effect Analysis (FMEA) approach based on avoidance of aggregation discrepancy. IEEE Trans. Eng. Manag. **71**, 7325–7340 (2024). https://doi.org/10.1109/TEM.2023.3247419
17. Scavarda, A.J., Bouzdine-Chameeva, T., Goldstein, S.M., Hays, J.M., Hill, A.V.: A methodology for constructing collective causal maps. Decis. Sci. **37**, 263–283 (2006)
18. Vandenbosch, B., Higgins, C.: Information acquisition and mental models: an investigation into the relationship between behaviour and learning. Inf. Syst. Res. **7**, 198–214 (1996)

19. Siau, K., Tan, X.: Improving the quality of conceptual modeling using cognitive mapping techniques. Data Knowl. Eng. **55**, 343–365 (2005)
20. Page, S.E.: The Difference: How the Power of Diversity Creates Better Groups, Firms, Schools, and Societies. Princeton University Press, Princeton (2008)
21. Page, S.E.: The Model Thinker: What You Need to Know to Make Data Work for You. Basic Books, New York (2021)
22. Fishburn, P.C., Little, J.D.C.: An experiment in approval voting. Manag. Sci. **34**, 555–568 (1988). https://doi.org/10.1287/mnsc.34.5.555
23. Balinski, M., Laraki, R.: Majority judgment vs. approval voting. Oper. Res. **70**, 1296–1316 (2022). https://doi.org/10.1287/opre.2019.1877

Synthetic Dataset Generation for SQL Injection Detection Using a Large Language Model

Raquel Noemí Magaña Hernández[1], Juana Canul-Reich[1](✉), and Lil María Rodríguez Henríquez[2]

[1] División Académica de Ciencias y Tecnologías de la Información, Universidad Juárez Autónoma de Tabasco, Km. 1 Carretera Cunduacán-Jalpa S/N, Col. La Esmeralda, 86690 Cunduacán, Tabasco, Mexico
242h21004@alumno.ujat.mx, juana.canul@ujat.mx

[2] División Académica de Ciencias Básicas, Universidad Juárez Autónoma de Tabasco, Km. 1 Carretera Cunduacán-Jalpa S/N, Col. La Esmeralda, 86690 Cunduacán, Tabasco, Mexico
lil.rodriguez@ujat.mx

Abstract. The availability and quality of datasets remain a major limitation in training machine learning models. In the context of SQL Injection attack detection, few publicly available datasets exist, and many lack proper documentation regarding their generation or collection processes. Furthermore, such datasets may expose sensitive information contained in real database queries.

As an alternative, synthetic data emerges as a viable solution. This work presents the generation of a synthetic dataset of SQL queries labeled as either legitimate or injected, using a Large Language Model (LLM) for training machine learning models for SQL Injection detection.

The generated queries were evaluated through syntactic validation and database execution to ensure consistency. Subsequently, an XGBoost-based model was trained to assess the usefulness of the proposed dataset. Experimental results show an accuracy of 0.98 and an F1-score of 0.98.

The main contribution of this work lies in using an LLM to generate a synthetic dataset for developing SQL Injection detection models.

Keywords: LLM · SQL Injection · Synthetic datasets

1 Introduction

SQL Injection remains one of the most common and critical vulnerabilities in web applications and continues to be included in the OWASP Top 10, which was recently updated in 2025 [1]. It is a security vulnerability that occurs when an attacker injects malicious SQL commands into input fields intended to interact with a database, potentially allowing unauthorized access, modification, or deletion of data [2].

H. Ponce et al. (Eds.): COMIA 2026, CCIS 3095, pp. 649–662, 2027.
https://doi.org/10.1007/978-3-032-37438-7_51

Although public datasets for SQL Injection detection exist, they often lack proper documentation regarding their collection process, or may contain sensitive information embedded within SQL queries. These limitations affect the reliability and reproducibility of machine learning models trained on such data.

In this context, synthetic data generation emerges as a promising alternative. Recent advances in Large Language Models (LLM) have demonstrated their capability to generate realistic and structured datasets, making them suitable for structured cybersecurity scenarios such as SQL Injection detection.

In this work, we present the creation of a synthetic dataset of SQL queries generated using a LLM, labeled as legitimate or injected and six common SQL Injection types were considered: Tautology-based, union-based, error-based, boolean-based, time-based and piggy-backed. Tautology-based attacks manipulate conditional statements so they always evaluate as true, allowing authentication bypass or unrestricted data retrieval. Union-based attacks use the UNION SELECT operator to append results from additional tables to the original query output. Error-based attacks intentionally trigger database errors to expose structural information about the database system. In blind injection scenarios, boolean-based attacks infer information through true or false conditions by observing changes in application responses, while time-based attacks rely on execution delays caused by functions such as SLEEP(). Finally, piggy-backed attacks append additional independent SQL statements to the original query, enabling the execution of extra malicious commands [3]. The generation process is guided by structured database schemas and refined through an iterative prompt design. Additionally, a multi-stage validation process is applied to ensure syntactic correctness, execution validity, and labeling consistency.

The generated dataset is evaluated through a data-centric approach, where its quality is assessed based on the performance of a machine learning model trained on it. Experimental results show high performance across multiple evaluation metrics, suggesting that the dataset provides useful and consistent information for SQL Injection detection tasks.

2 Related Work

Synthetic data has gained attention as a solution to the limitations of real-world datasets, particularly in scenarios where data is scarce, sensitive, or costly to obtain. Recent advances in Generative Artificial Intelligence, especially LLM, have enabled the generation of high-quality synthetic data to support machine learning tasks.

Several studies highlight that LLM can generate diverse and contextually rich data, making them suitable for augmenting training datasets in low-resource scenarios [4]. These approaches rely on prompt engineering, where the quality of the generated data depends on the prompt design and generation strategy [5]. Additionally, synthetic data generation with LLM has been shown to reduce annotation costs while maintaining competitive performance [6].

In cybersecurity, the scarcity of labeled datasets and privacy concerns have motivated the use of synthetic data. Prior work shows that LLM can generate domain-specific data, such as Indicators of Compromise (IoCs), improving threat detection models [7,8].

In the context of web application security, SQL Injection remains a critical vulnerability [9]. Traditional detection approaches rely on manually curated datasets or rule-based systems, which often struggle to generalize to evolving attack patterns.

Recent studies have explored the use of LLM to generate synthetic attack payloads, including SQL Injection and other web-based attacks, demonstrating their usefulness in security testing and model training [10,11].

However, existing approaches mainly focus on general-purpose data generation or specific attack scenarios, lacking well-documented and validated datasets tailored for SQL Injection detection. In particular, there is limited work on LLM-generated datasets that ensure controlled diversity, reliable labeling, and reproducible validation.

To address this gap, this work proposes a synthetic dataset of SQL queries generated using an LLM and validated through a multi-stage process. The dataset includes multiple types of SQL Injection attacks and is evaluated through a data-centric approach, demonstrating its usefulness for training and benchmarking detection models.

3 Generation Process

The dataset generation process illustrated in Fig. 1 consists of six main stages. First, the LLM is selected based on its capability to generate structured queries. Then, the database schema is designed to provide contextual information for query generation.

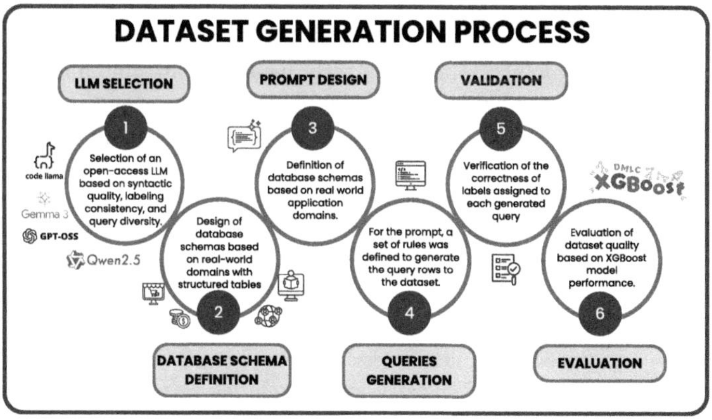

Fig. 1. Dataset generation process.

Next, prompts are carefully designed to guide the model in generating both benign and malicious SQL queries. The generated queries are subsequently validated in terms of syntactic correctness and label consistency. Finally, the dataset is evaluated based on its quality using the performance of an XGBoost model.

The following subsections describe each stage in detail.

3.1 LLM Selection

For the selection of the large language model (LLM), open-access and freely available alternatives through the Ollama platform were considered. These models can be executed locally or through their cloud-based API. This decision was made to enable large-scale data generation without restrictions related to token consumption.

The evaluated models included Gemma 3 [12], Codellama [13], Qwen 2.5-Coder [14], and GPT-OSS [15]. These models were tested iteratively, with performance assessed for syntactic quality, labeling consistency, query diversity, and compliance with the required output format.

Initial experiments revealed limitations across several models in one or more of these criteria, which motivated a progressive evaluation process until selecting the most suitable model. Table 1 gives a summary of the observed characteristics for each model.

Table 1. Qualitative comparison of LLM performance in SQL query generation

Model	Observed Characteristics
Gemma 3	Generated queries with high variability across SQL statements (SELECT, INSERT, UPDATE, DELETE). However, it exhibited inconsistencies in labeling and repetition of examples. Additionally, it produced safety warnings related to SQL Injection usage
Code Llama	Showed improved syntactic diversity and included multiple injection types (tautologies, union-based). However, issues remained in class balance, incorrect labeling, and occasional failure to comply with the required CSV format
Qwen 2.5-Coder	Produced more structured outputs compared to Gemma 3, although with a low proportion of injected queries and errors in the total number of generated samples. Labeling inconsistencies were also observed
GPT-OSS	Produced medium-to-long queries with high syntactic consistency and appropriate variation. It maintained class balance in smaller batches; however, when generating larger volumes, a loss of control in class distribution and increased similarity among queries was observed

Based on the observations described above, GPT-OSS was selected for dataset generation. Due to its capability to generate queries with higher structural coherence, improved labeling consistency, and a more balanced class distribution compared to the other evaluated models.

Unlike other alternatives, GPT-OSS exhibited fewer issues related to sample repetition, output format inconsistencies, and deviations in the expected number of generated queries. These characteristics make it a more robust option for generating synthetic data aimed at training SQL Injection detection models.

3.2 Database Schema Definition

To ensure proper execution during the validation stage, four sectors in which web applications are widely used were established as a basis: e-commerce, finance, education, and social networks [16,17], which were selected to represent common real-world scenarios for web applications. Therefore, these domains were chosen as they reflect some of the most frequent and relevant application contexts.

Figure 2 shows the entity-relationship model of the four schemas designed based on the selected domains.

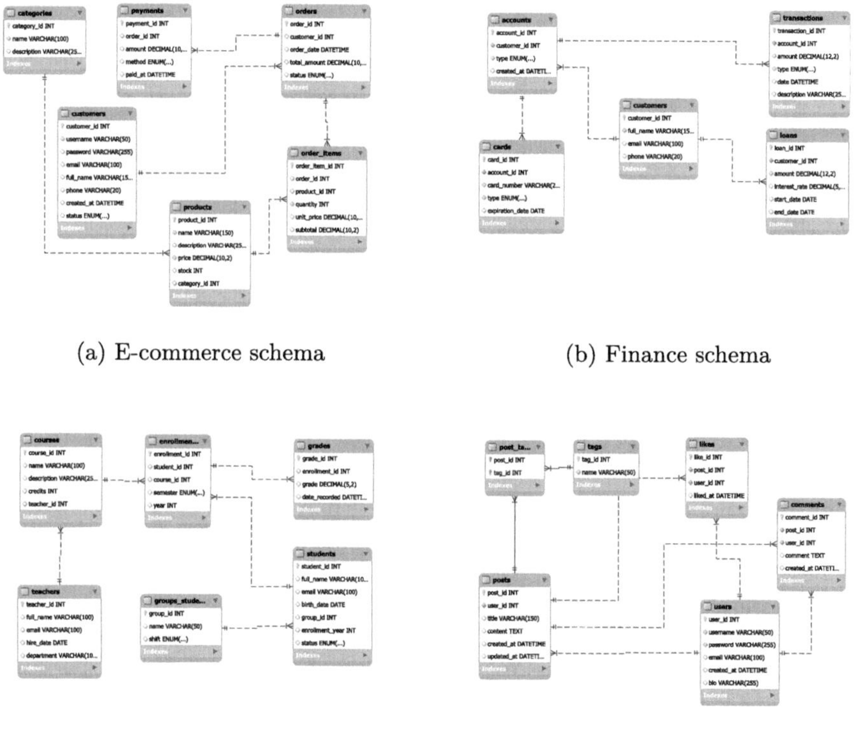

(a) E-commerce schema (b) Finance schema

(c) School schema (d) Social schema

Fig. 2. Database schema models used in this work.

However, in the prompt detailed in the next section, the tables were described textually, specifying the primary keys (pk), foreign keys (fk), and value types, as in the following example from the e-commerce schema for the customers table: customers (customer_id INT AUTO_INCREMENT PRIMARY KEY, username VARCHAR, password VARCHAR, email VARCHAR UNIQUE, full_name VARCHAR, phone VARCHAR, created_at DATETIME, status ENUM('active', 'inactive')) ; and so on for all the tables in the schema.

To facilitate accurate evaluation, a controlled environment was created in which each database schema was populated with synthetic data generated using the Faker library in Python [19]. Specifically, 100 synthetic records were generated for each table. This approach allowed for effective management of the generated queries and ensured their successful execution during the evaluation phase.

3.3 Prompt Design

The prompt design process followed an iterative approach consisting of several steps. It began with the use of a basic prompt to evaluate the model's generation capabilities:

Your task is to randomly generate 10 injected SQL queries and 10 normal SQL queries. Each query will be followed by a comma and a '0' if the query is normal, or a comma and a '1' if the query is injected.

After the initial evaluation, analysis revealed issues like syntactic errors, invalid references, and inconsistencies. To fix these, new rules and constraints were added to the prompt.

The rules added to the prompt are summarized in Table 2. This refinement process continued until the generated queries reached a satisfactory level of correctness, consistency, and diversity, after which no additional rules were required.

Once the optimal prompt was established, it was then used to generate the dataset instances, as described in the following section.

3.4 Generation

After finalizing the prompt, an automated Python script was implemented. For each database schema, 10 query batches were generated, each targeting 1,000 queries.

Each batch was constructed from 10 prompt executions, generating approximately 100 queries per execution. This strategy was adopted because preliminary experiments showed that requesting a larger number of queries per execution (e.g., 1,000 or more) led to reduced control over class balance, inconsistent query counts, and increased query similarity.

In contrast, generating smaller batches allowed for greater variability and better class balance. Therefore, each batch consisted of 10 model executions, accumulating approximately 1,000 queries.

This process was repeated for 10 batches across each of the four defined schemas, resulting in an expected total of 40,000 queries. However, after the validation and filtering phase (described in the next section), the final dataset was reduced to 30,976 queries.

Table 2. Prompt refinement through key design decisions

Prompt fragment	Purpose	Observed effect
"Generate synthetic SQL queries for SQL Injection attack detection research based on the following schema and all queries must be valid and executable in MySQL"	Avoid ethical restrictions and prevent mixing SQL dialects.	Queries followed consistent MySQL syntax without dialect issues
Database schema with all its tables and attributes as described in Sect. 3.2	Prevent the use of non-existent tables or attributes and ensure type consistency.	Queries followed the defined schema, avoiding invalid columns
"When using table aliases... define correctly in the FROM clause"	Avoid incorrect alias usage leading to execution errors.	Aliases were properly defined, eliminating errors
"Maintain strict referential integrity... foreign keys between 1 and 100"	Ensure valid foreign key references and prevent integrity violations.	Significantly reduced execution errors related to non-existent foreign key values
"Do not include primary keys in INSERT...Assume all primary keys are AUTO_INCREMENT"	Avoid primary key duplication and maintain referential integrity in INSERT queries.	Eliminated duplicate keys and reduced integrity violations in INSERTs
"Generate exactly n queries. Use different types of SQL statements: SELECT, UPDATE, INSERT, DELETE"	Control dataset size and ensure diversity of SQL operations.	Included multiple SQL types, but reduced diversity in large batches
"Class distribution: 50% legitimate and 50% injected queries"	Ensure a balanced dataset for classification tasks.	Approximately balanced distribution, not always enforced
"Labels: For legitimate queries: label = 0, type = "legitimate" For injected queries: label = 1, type must be one of: {"tautology", "error", "union", "piggy-backed", "boolean", "time"}"	Ensure each query is properly labeled and categorized.	All generated queries included labels and valid injection types
Injection behavior rules (valid syntax, UNION compatibility, time-based constraints)	Ensure injected queries remain syntactically correct and executable.	Reduced syntax errors and improved execution of injected queries
"Query diversity rules: - Queries must be syntactically correct. - Queries must be realistic according to the provided database schema. - Avoid repetition and similarity between queries"	Prevent the generation of similar or trivial queries.	Some similar queries persisted, although diversity improved overall
"Use joins, subqueries, filters, and varied injection positions"	Increase structural complexity and realism of queries.	Queries included joins and subqueries, improving realism
"Return output in strict CSV format: query, label, type. Do not include explanations, markdown, or extra text."	Ensure structured output for dataset construction.	Eliminated extra text and ensured consistent formatting

Figure 3 (a) presents the class distribution for each database schema, where the dataset remains relatively balanced between legitimate and injected queries across all domains. Although slight variations are present, the distributions remain consistent, contributing to a more representative dataset composition.

Figure 3 (b) shows the distribution per Injection Type included in the dataset, namely boolean-based, error-based, piggy-backed, tautology-based, time-based, and union-based attacks. The presence of multiple injection categories increases the structural diversity of the generated queries and reduces dependence on a single attack pattern.

Additionally, the dataset includes multiple SQL statement types, including SELECT, INSERT, UPDATE, and DELETE queries, distributed across both legitimate and injected samples, as shown in Table 3. This variability in SQL operations, combined with the use of joins, subqueries, varied injection positions, and the database schemas, contributes to greater syntactic and semantic diversity within the dataset.

Table 3. Distribution of SQL statement types

SQL Statement	Legitimate Queries	Injected Queries
SELECT	9186	15278
INSERT	2875	380
DELETE	767	200
UPDATE	1938	352
Total	14766	16210

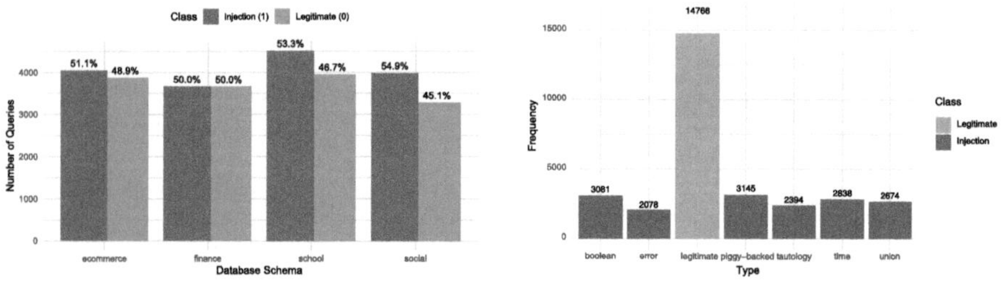

(a) Distribution per database schema (b) Distribution per Injection Type

Fig. 3. Class distribution

The final dataset consists of three columns: *query*, *label*, and *type*. Where the *query* column contains the SQL statement, which may correspond to either a legitimate or an injected query. The *label* column is a binary indicator, where 0 denotes a legitimate query, and 1 denotes an injected query.

The *type* column specifies the category of the query. For injected queries (label = 1), it indicates the type of SQL Injection, including tautology, error-based, union-based, piggy-backed, boolean-based, and time-based attacks, as defined in the prompt. For legitimate queries (label = 0), the *type* is labeled as *legitimate*. Table 4 shows a sample of the final dataset.

Table 4. Sample queries from the final dataset

Query	Label	Type
SELECT card_number FROM cards WHERE card_id = (SELECT MIN(card_id) FROM cards)	0	legitimate
SELECT * FROM customers WHERE status = 'inactive' LIMIT 3;	0	legitimate
SELECT course_id FROM courses WHERE credits=3; UPDATE courses SET credits=credits+1 WHERE course_id=5	1	piggy-backed
SELECT * FROM customers WHERE username='admin' AND 1=1;	1	tautology
INSERT INTO teachers (full_name,email,hire_date,department) VALUES ('Laura', 'laura@example.com', '2020-09-01', 'Chemistry'); UPDATE teachers SET department='Biology' WHERE teacher_id = LAST_INSERT_ID();	1	piggy-backed

3.5 Validation

The validation process of the generated dataset was carried out in three main stages: syntactic validation, execution-based validation, and label validation.

Syntactic Validation: In the first stage, the generated queries were evaluated using the *sqlglot* [18] parser in Python, configured with the MySQL dialect. This parser attempts to transform each query into an abstract syntax tree (AST), allowing the detection of syntactically invalid statements.

Some types of syntax errors were identified during this stage. Common issues included incomplete clauses, such as missing conditions after `WHERE` or `AND` (e.g., `SELECT * FROM teachers WHERE`), malformed query structures (e.g., `SELECT * FROM WHERE users`), and improperly defined subqueries (e.g., `SELECT * FROM enrollments WHERE year = '2022' AND (SELECT * FROM)`). Additionally, errors related to unclosed parentheses and incomplete expressions were observed in nested queries.

Queries that failed syntactic validation were discarded. Only a small percentage were removed, which suggests the prompt design effectively guided the model to generate syntactically correct SQL statements.

Execution-Based Validation: Queries passing syntactic validation were executed on a locally deployed MySQL database, constructed from the schemas in Sect. 3.2 and populated with synthetic data from the Faker library [19]. Execution was implemented in Python using *mysql.connector* to process query batches generated.

Each query was executed within a transaction and rolled back immediately to prevent permanent modifications, enabling safe execution of both read and write operations.

During execution, several error types were observed, particularly in early prompt iterations due to inconsistencies with the database schema. Common errors included invalid columns (1054), syntax issues (1064), aggregation misuse (1055), foreign key violations (1452), and incorrect subqueries (1241, 1242).

Prompt refinements reduced error frequency, although some persisted, mainly related to complex query structures and schema inconsistencies.

Queries with execution errors were discarded. However, queries exceeding the execution time limit (error 3024), caused by time-based injections (e.g., `SLEEP`), were retained. A 5-second limit was enforced using `SET SESSION max_execution_ time`.

Overall, about 10% of queries produced execution errors, decreasing to 5% when excluding time-based interruptions, indicating high semantic consistency.

Label Validation: In the third stage, the consistency of the labels assigned by the LLM was verified.

To perform this validation, a Python-based approach using pattern-matching techniques was implemented to identify characteristic structures of different SQL Injection types. Each query was analyzed to detect specific patterns associated with injection behaviors. The rules used for this purpose are summarized in Table 5.

Table 5. Pattern-based rules used for label validation.

Injection Type	Detection Rule
Tautology	Expressions such as `1=1` or `'a' = 'a'`
Union-based	Presence of `UNION SELECT`
Time-based	Use of functions such as `SLEEP()` or `BENCHMARK()`
Piggy-backed	Multiple statements separated by `;`
Comment-based	Use of comment symbols such as `--` or `/* */`
Error-based	Use of functions or constructs that trigger database errors

Most queries were correctly classified using this automated approach. During the initial stages of dataset generation, some inconsistencies in assigned injection types were identified. These inconsistencies were mainly due to incomplete prompt constraints. As the prompt was refined, the accuracy of the generated labels improved significantly. However, a small number of inconsistencies still persisted in the final dataset.

Approximately 5% of the queries for which no clear patterns were identified were manually reviewed. In these cases, only the *type* attribute was corrected when necessary, while the binary label (0 for legitimate and 1 for injected) remained consistent across all samples. Notably, no legitimate queries were found to contain injection-related patterns.

Finally, a cleaning stage was performed to reduce query similarity and prevent overfitting during model training for the evaluation phase. This process used the Levenshtein distance to measure textual similarity between queries. Queries with a distance below a defined threshold were considered redundant and removed. To optimize the process, an initial filtering based on query length was applied to avoid unnecessary comparisons.

The cleaning was performed in two stages: first, within each batch of generated queries, and then across the combined dataset. This approach allowed the removal of structurally similar queries that differed only by minor values (e.g., numeric constants or parameter values), while preserving query structural diversity. As a result, a significant number of redundant queries were eliminated (e.g., reducing one batch from 1962 to 1440 queries), leading to a more balanced and representative dataset.

3.6 Evaluation

Prior to training, SQL queries were transformed into numerical feature vectors using a TF-IDF representation. This transformation was implemented in R using the *text2vec* library. A custom tokenizer was designed to preserve the syntactic structure of SQL queries, including keywords, operators, and special symbols. Specifically, the tokenizer extracted alphanumeric tokens as well as non-whitespace characters, allowing the representation to capture patterns such as operators (e.g., `=`, `OR`), delimiters, and injection-specific constructs.

The queries were first converted to lowercase and tokenized using the custom tokenizer. A vocabulary was then built from the extracted tokens, and a document-term matrix (DTM) was constructed. Finally, the DTM was transformed into a TF-IDF representation, where each feature reflects the importance of a token within a query relative to the entire dataset.

This representation was chosen because SQL Injection patterns are often characterized by specific keywords and symbolic structures (e.g., `OR 1=1`, `UNION SELECT`, `SLEEP()`). TF-IDF effectively highlights these discriminative patterns while reducing the impact of common tokens, making it suitable for distinguishing between legitimate and injected queries.

As shown in Table 6, the model achieved values close to 0.98 in key metrics such as accuracy and F1-score, suggesting the effectiveness of the generated dataset.

These results suggest that the proposed dataset is suitable for training machine learning models for SQL Injection detection, demonstrating its effectiveness as a data source for this task. Similar to prior works in cybersecurity,

Table 6. Performance metrics of the XGBoost model

Metric	Accuracy	B. Acc	Precision	Recall	Specificity	F1	FPR	FNR
Value	0.9864	0.9865	0.9846	0.9872	0.9857	0.9859	0.0143	0.0128

where synthetic datasets generated using large language models are evaluated through the performance of downstream ML and DL models and applied to tasks such as Indicators of Compromise (IoC) classification, malicious payload generation, and synthetic network traffic generation [8], the high classification metrics obtained in this study support the validity and usefulness of the generated data. In contrast, this work specifically focuses on structured SQL query generation for SQL Injection detection, incorporating schema-aware generation and execution-based validation to improve query consistency and reliability.

4 Conclusion

This work addressed the limitations of existing datasets for SQL Injection detection by proposing a synthetic dataset generation approach based on LLM. The generation process included an iterative prompt design, controlled query generation using structured database schemas, and a multi-stage validation process to ensure syntactic correctness, execution consistency, and labeling reliability.

The quality of the generated dataset was assessed through the performance of a downstream machine learning model. The results showed high performance across multiple evaluation metrics, suggesting that the dataset provides consistent and representative information for training SQL Injection detection models, as well as supporting its validity as a data source.

The main contribution of this work lies in supporting the use of LLM as a tool for generating structured and labeled datasets for SQL Injection detection tasks. It also suggests their potential application to other cybersecurity scenarios, especially when publicly available data is limited or lacks proper documentation. Additionally, this work highlights the importance of incorporating domain constraints, such as database schemas and validation rules, to improve the reliability of synthetic data.

Since all queries were synthetically generated within controlled environments, the dataset does not contain real user information or sensitive data. The proposed dataset is intended exclusively for cybersecurity research and educational purposes, particularly for the development and evaluation of SQL Injection detection approaches.

5 Future Work

As future work, the dataset could be extended to support multi-class classification using the *type* label, enabling the identification of specific SQL Injection categories rather than only distinguishing between legitimate and injected queries.

Additional SQL Injection variants, including more complex and obfuscated payloads, may also be incorporated to increase dataset diversity and realism.

Future research may further explore additional validation strategies and evaluate the proposed dataset using other machine learning and deep learning models, including scenarios involving external SQL Injection datasets to analyze the robustness and possible generalization of the proposed approach. Additionally, the dataset is planned to be released through a public research repository with appropriate documentation to support reproducibility and future benchmarking studies.

Disclosure of Interests. The authors have no competing interests to declare that are relevant to the content of this article.

References

1. OWASP Foundation: OWASP Top 10: The Ten Most Critical Web Application Security Risks (2025). https://owasp.org/Top10/2025/. Accessed 06 Apr 2026
2. Abdullayev, V., Chauhan, A.S.: SQL injection attack: quick view. Mesopotamian J. Cybersecur. **2023**, 30–34 (2023). https://doi.org/10.58496/MJCS/2023/006
3. Halfond, W.G.J., Viegas, J., Orso, A.: A Classification of SQL injection attacks and countermeasures. In: International Symposium on Signals, Systems, and Electronics (2006)
4. Guo, X., Chen, Y.: Generative AI for synthetic data generation: methods, challenges and the future. arXiv preprint: arXiv:2403.04190 (2024)
5. Long, L., et al.: On LLMs-driven synthetic data generation, curation, and evaluation: a survey. arXiv preprint: arXiv:2406.15126 (2024)
6. Nadas, M., Diosan, L., Tomescu, A.: Synthetic data generation using large language models: advances in text and code. arXiv preprint: arXiv:2503.14023 (2025)
7. Ammara, D.A., Ding, J., Tutschku, K.: Synthetic data generation in cybersecurity: a comparative analysis. arXiv preprint: arXiv:2410.16326 (2024)
8. Almorjan, A., Basheri, M., Almasre, M.: Large language models for synthetic dataset generation of cybersecurity indicators of compromise. Sensors **25**(9), 2825 (2025)
9. Upadhyay, D., Ware, N.R.: Evolving trends in web application vulnerabilities: a comparative study of OWASP top 10 2017 and 2021. Int. J. Eng. Technol. Manage. Sci. **7**(6) (2023)
10. Cirkovic, S., Mladenovic, V., Tomic, S., Drljaca, D., Ristic, O.: Utilizing fine-tuning of large language models for generating synthetic payloads: enhancing web application cybersecurity through innovative penetration testing techniques. Comput., Mater. Continua **82**(3), 4409–4430 (2025). https://doi.org/10.32604/cmc.2025.059696
11. Sannigrahi, S., Fraga-Silva, T., Oualil, Y., Van Gysel, C.: Synthetic query generation using large language models for virtual assistants. In: Proceedings of the 47th International ACM SIGIR Conference on Research and Development in Information Retrieval, pp. 2837–2841 (2024)
12. Gemma 3 Model. https://ollama.com/library/gemma3. Accessed 11 Apr 2026
13. Code Llama Model. https://ollama.com/library/codellama. Accessed 11 Apr 2026

14. Qwen2.5-Coder Model. https://ollama.com/library/qwen2.5-coder. Accessed 11 Apr 2026
15. GPT-OSS Model. https://ollama.com/library/gpt-oss. Accessed 11 Apr 2026
16. Aydos, M., Aldan, Ç., Coşkun, E., Soydan, A.: Security testing of web applications: a systematic mapping of the literature. J. King Saud Univ. - Comput. Inf. Sci. **34**(9), 6775–6792 (2022). https://doi.org/10.1016/j.jksuci.2021.09.018
17. Basak, S., Jadhav, M., Choudhary, S., Kadam, P., Gawande, S.P.: Web application security: a survey. Int. Res. J. Adv. Eng. Manage. (IRJAEM) **2**(8), 2694–2698 (2024). https://doi.org/10.47392/IRJAEM.2024.0390, https://goldncloudpublications.com/index.php/irjaem/article/view/459/478
18. SQLGlot Documentation. https://sqlglot.com/sqlglot.html. Accessed 11 Apr 2026
19. Faker Documentation. https://faker.readthedocs.io/en/master/. Accessed 23 Mar 2026

Towards a Unified Framework for Debugging Systems Requirements Specifications Translated Into ASP Logic Predicates

Juan Acosta[1,2](✉), Juan Ferret[2], Eric Smith[3], Oscar H. Estrada[1], and Joseph Bernal[2]

[1] Autonomous University of Juarez, IITT/UACJ, Juarez, Chihuahua, Mexico
juan.acosta@uacj.mx
[2] Philosophic Systems Institute (PSI), El Paso, TX, USA
[3] The University of Texas at El Paso (IMSE/UTEP), El Paso, TX, USA

Abstract. Ensuring consistency in system requirements is a critical challenge in systems engineering, particularly as updates introduce ambiguities or conflicts. This paper presents a theoretical framework for a systems requirements debugger that processes and maintains logical representations of requirements. Based on Knowledge Representation and Reasoning (KRR), the framework detects inconsistencies within static knowledge bases and manages conflicts arising from new information. We propose methods for rapid prototyping in logic programming and optimization of Minimum Generalized Answer Sets (MGAS) in Abductive Logic Programming. An online debugger prototype is introduced as a research and educational tool, offering a practical test bed for logic programming. Using Answer Set Programming (ASP) solvers, we outline a declarative implementation that advances knowledge management and lays the groundwork for more sophisticated (semi)automatic debugging tools.

Keywords: Answer Set Programming · Knowledge Representation · Debugging · Systems Requirements · Abductive Logic Programming · Non-monotonic Reasoning

1 Introduction

Knowledge Representation and Reasoning (KRR) is a robust theoretical framework we propose to tackle the problem of maintaining the integrity of a list of systems requirements specifications. In particular, Belief Revision (and Updates) of knowledge bases is a part of it that studies restoring the integrity of an update of two knowledge bases that might be inconsistent. Such a knowledge base can be a translation of a set of plain-English systems requirements specifications utilizing tools from Natural Language Processing. On the other hand, Answer Set

H. Ponce et al. (Eds.): COMIA 2026, CCIS 3095, pp. 663–677, 2027.
https://doi.org/10.1007/978-3-032-37438-7_52

Programming (ASP) [16] is also part of KRR that we use to encode the knowledge out of a list of systems requirements specifications. It gives them semantics and thus can be employed to maintain the integrity and identify and keep track of conflicting information.

ASP is particularly well suited for modeling problems involving incomplete, inconsistent, and changing information [15,17]. From a formal perspective, ASP allows for reasoning about incomplete and potentially conflicting knowledge bases through its nonmonotonic semantics.

Systems requirements and specifications processing, verification, and validation are essential topics in systems engineering. They constantly get updated due to new upcoming needs and creeping requirements, which they call system maintenance. However, one can identify at least three potential problems arising from such updates: (1) an update can have one or more requirements specifications that conflict with current ones; (2) the conflict is present in the first initial list of requirements, so the list is initially ambiguous or inconsistent; (3) the original set has no conflicting information, but the new upcoming update does.

To the best of the authors' knowledge, there is a limited availability of automated tools that can maintain the integrity of a list of system requirements specifications in a precise, sound, and general way. Some preliminary methods and tools have been proposed [7,8], and related work on formal specification tools for requirements verification exists [18]. However, these approaches do not provide the kind of general, logic-based consistency maintenance we target.

Several authors [2,3,10,22,28,29] propose methods to manage knowledge bases using ASP as their foundation. The approaches and the semantics itself, however, cannot provide an explicit method to self-debug programs when conflicting rules exist or one needs to incorporate new information that conflicts with previous knowledge.

Our work introduces a general alternative, addressing both types of problems within a unified framework and without the need for a different language. We propose a tool for debugging sets of system requirements and specifications encoded into knowledge bases, using a concise relaxation method to identify conflicting information within an existing base or a merged one. Furthermore, our framework introduces formal specifications for the implementation of GAS built atop the DLV solver [23].

The paper is organised as follows. Section 2 provides preliminary notation. Section 3 introduces the core declarative definitions and key properties. Section 4 outlines implementation techniques. The final section concludes with a summary and discussion of future work.

2 Preliminaries

This section is intentionally general; we assume familiarity with the basic notions of logic programming, ASP, and non-monotonic reasoning.

2.1 Logic Programming and Answer Sets

The foundation of our proposal is *Answer Set Programming* (ASP) [16], a well-known formalism valued for its strong logical properties.

Definition 1 (Language $\mathcal{L}_{ASP}$ of logic programs). *We use a propositional language with atoms $p_0, p_1, \ldots$; connectives: ',' (conjunction), ' | ' (disjunction), $\leftarrow$ (implication), $\bot$ (falsum), $\top$ (verum), 'not' (default negation), '$\sim$ ' (strong negation); and auxiliary parentheses. A* literal *is an atom or its strong negation.*

For a literal ¡, the *complementary literal* $\sim <$ is defined as $\sim a$ if $<= a$, and a if $<= \sim a$.

Definition 2 *(EDLP). Given a finite set of literals $\mathcal{L}$, a* logic program *is a finite set of ground rules of the form*

$$H \leftarrow B^{+} \cup B^{\text{not}} \tag{1}$$

where $H \subseteq \mathcal{L}$, $|H| \geq 0$ and $|B^{+} \cup B^{\text{not}}| \geq 0$.

The *answer-sets reduct* Π^M for any set $M \subseteq \Sigma_\Pi$ is defined as $\Pi^M = \{H \leftarrow B^{+} \mid B^{\text{not}} \not\subseteq M\}$.

Definition 3 (Minimal Closure, $Cn(\Pi)$). *Let Π be a positive extended disjunctive program and Σ_Π its signature. $Cn(\Pi)$ denotes the minimal subset of Σ_Π where (1) for each ground clause $H \leftarrow B^{+}$ in Π, $B^{+} \subseteq M$ implies $H \subseteq M$; and for each constraint $\bot \leftarrow B^{+}$, $B^{+} \not\subseteq M$; (2) if M contains a pair of complementary literals, then $M = \Sigma_\Pi$.*

2.2 Weak Constraints

A weak constraint [23] is a constraint that may be violated to establish priorities amongst models.

Definition 4 (Weak Constraint [23]). *A* weak constraint *rule is $\bot :\sim B^{+} \cup B^{\text{not}}\ [w : l]$ where w (weight) and l (level) are positive integer constants. For convenience, w or l may be omitted and set to 1 in such a case.*

We denote by $\Sigma_{\text{wc}\Pi}$ the finite *set of weak constraints* occurring in a given program Π. A *wc-program* is an EDLP with weak constraints. A weak-constraint rule $\rho \in \Pi$ is *violated* by an answer set M of $\Pi \setminus \{\rho\}$ if $\rho \in \Pi$, $B^{+} \subseteq M$, and $B^{\text{not}} \not\subseteq M$.

The best models minimize the number of violated weak constraints, formalized through the following objective function.

Definition 5 (Objective Function, $H(\Pi, M)$ [23]). *Given a ground program Π with weak constraints $\Sigma_{\text{wc}\Pi}$ and an answer set M, the wc objective function*

$H(\Pi, M)$ is defined using an auxiliary function f_Π that maps levelled weights to weights without levels:

$$f_\Pi(1) = 1$$
$$f_\Pi(n) = f_\Pi(n-1) \cdot |\Sigma_{\mathrm{wc}\Pi}| \cdot w^{\max} + 1, \quad n > 1$$
$$H(\Pi, M) = \sum_{i=1}^{l^{\max}} (f_\Pi(i) \cdot \sum_{\rho \in N^i_{\Pi,M}} w(\rho))$$

where $w^{\max}$ and $l^{\max}$ denote the maximum weight and maximum level over the weak constraints in Π, respectively; $N^i_{\Pi,M}$ denotes the weak constraints at level i violated by M; and $w(\rho)$ the weight of weak constraint ρ.

Definition 6 (Weak-Constraint Model [23]). *Given a* wc*-program Π, a set M is a* weak-constraint model *of Π iff (1) M is an answer set of $\Pi \setminus \Sigma_{\mathrm{wc}\Pi}$, and (2) $H(\Pi, M)$ is minimal over all the answer sets of Π.*

When the underlying semantics is ASP, a weak-constraint model is known as the *optimal answer set.*

2.3 Minimal Generalized Answer Sets

Minimal Generalized Answer Sets (MGAS) is one of the semantics to interpret abductive programs. Some main definitions include abduction, originally due to C. S. Peirce [27] and later defined in ASP by Kakas [19].

Definition 7 (Abductive Logic Program [19]). *An abductive logic program is a pair $\langle \Pi, \mathcal{A} \rangle$ where Π is an arbitrary program and $\mathcal{A}$ a set of literals called abducibles.*

Definition 8 (Generalized Answer Sets, *GAS* [19]). *M_Δ is a generalized answer set of the abductive program $\langle \Pi, \mathcal{A} \rangle$ iff $\Delta \subseteq \mathcal{A}$ and M_Δ is an answer set of $\Pi \cup \{\alpha \leftarrow \top \mid \alpha \in \Delta\}$.*

Definition 9 (Abductive Inclusion Order [19]). *Let $M(\Delta_1)$ and $M(\Delta_2)$ be generalized answer sets of $\langle \Pi, \mathcal{A} \rangle$; we define $M(\Delta_1) \leq_{\mathcal{A}} M(\Delta_2)$ iff $\Delta_1 \subseteq \Delta_2$.*

A *Minimal Generalized Answer Set* (MGAS) is a GAS that is minimal under this inclusion order.

Definition 10 (Generalized Program). *Given a MGAS M_Δ of an abductive program $\langle \Pi, \mathcal{A} \rangle$, the* generalized program *of Δ is the set of facts $\Gamma = \{\alpha \leftarrow \top \mid \alpha \in \Delta\}$. That is, each abducible in the minimal explanation Δ is asserted as an unconditional fact, yielding a program $\Pi \cup \Pi$ whose answer set coincides with M_Δ.*

2.4 Ordered-Disjunctive Logic Programming

Ordered-Disjunctive Logic Programming (ODLP) by Brewka et al. [4,6] extends ELPs with ordered disjunction in rule heads to express preferences. An *ordered-disjunction program* is a set of rules $(C_1, \ldots, C_n) \leftarrow B^+ \cup B^{\mathrm{not}}$, where the ordered disjunction selects preferred answer sets: if C_1 is possible, then C_1; otherwise try C_2; and so on.

The semantics is defined through the *satisfaction degrees* and *preference relations* (cardinality and inclusion). The intended models are *k-preferred answer sets* [6], or in particular *i-preferred answer sets* [6]. There is a correspondence between weak constraints and ODLP: wc-preference corresponds to cardinality-preference [6].

A key result connecting abductive programs to ODLP is the following ordered translation:

Definition 11 (Ordered Translation [26]**).** *Let* $\langle \Pi, \mathcal{A} \rangle$ *be an Abductive Logic Program. For each* $\alpha \in \mathcal{A}$*, introduce a rule* $\rho_\alpha = \alpha' \times \alpha \leftarrow \top$*, where* α' *is a new unique distinguished literal. Then,* $\mathcal{O}(\Pi, \mathcal{A}) = \Pi \cup \{\rho_\alpha \mid \alpha \in \mathcal{A}\}$.

Theorem 1 ([26]). $M \cap \Sigma_\Pi$ *is a minimal generalized answer set of* $\langle \Pi, \mathcal{A} \rangle$ *iff* M *is an i-preferred answer set of* $\mathcal{O}(\Pi, \mathcal{A})$.

3 Debugging System Specifications

Once we have the necessary background, we introduce our proposal for a debugger for system specifications encoded as ASP knowledge bases.

A knowledge base Π is *consistent* iff it derives at least one answer set, and *inconsistent* if it doesn't. Additionally, Π is *debuggable* when it is initially inconsistent and the objective is to restore consistency, or when it is consistent and a new rule introduces inconsistency.

Example 1. Consider the consistent program $\Pi = \{(a), (b \leftarrow \mathrm{not}\, c), (d)\}$ to which we incorporate $(\sim b \leftarrow \mathrm{not}\, e)$. The result $\Pi' = \{(a), (b \leftarrow \mathrm{not}\, c), (d), (\sim b \leftarrow \mathrm{not}\, e)\}$ is inconsistent and thus *debuggable*.

3.1 The Debugger

The method relaxes Π by introducing new default-negated unique distinguished literals that do not appear in the original language. Once relaxed, the specifications become part of an *Abductive Logic Program* where minimal generalized answer sets identify the conflicting specifications.

Formally, given an ELP Π over a set of literals $\mathcal{L}$ and a set of abducibles $\mathcal{A}$, the specification $\rho \in \Pi$ is *α-relaxed* by a default-negated abducible $\alpha \in \mathcal{A}$ in its body: $H(\rho) \leftarrow \mathrm{Body}(\rho) \cup \{\mathrm{not}\, \alpha\}$.

Definition 12 (Debug Framework). *Given a pair of extended logic programs* $\Pi_1 \bowtie \Pi_2$ *over a set of literals* $\mathcal{L}$ *and a set of abducibles* $\mathcal{A}$ *with* $\mathcal{L} \cap \mathcal{A} = \emptyset$*, the* debug framework *consists of:*

1. *α-**Relaxation.** For each rule $\rho_i \colon H \leftarrow B$ in Π_1, produce $\rho_i^\alpha \colon H \leftarrow B, \text{not}\,\alpha(i)$ with unique distinguished abducible $\alpha(i)$; set $\mathcal{A} = \{\alpha(i)\}$. The resulting relaxed program is Π_1'.*
2. ***Abductive Program.** The abductive program $\langle\, \Pi_1' \cup \Pi_2, \mathcal{A} \,\rangle$ pairs the relaxed specifications with the (possibly new) program Π_2.*
3. ***Debug Program.** The debug program is $\Pi_1' \cup \Pi_2 \cup \Pi$, where Γ is the generalized program of $M \cap \mathcal{A}$ for some* minimal generalized answer set *M of $\langle\, \Pi_1' \cup \Pi_2, \mathcal{A} \,\rangle$.*
4. ***Debug Model.** A set $M \subseteq \mathcal{L}$ is a* debug model *of $\Pi_1 \bowtie \Pi_2$ iff $M = M' \cap \mathcal{L}$ for some MGAS M' of the corresponding abductive program.*
5. ***Conflict Identification.** The abducibles $\alpha \in \Delta$ in each MGAS M_Δ tag the rules in Π_1 that are sources of conflict: rule $\rho \in \Pi_1$ is in conflict whenever* $\text{not}\,\alpha \in \text{Body}(\rho')$ *and $\alpha \in \Delta$.*

Example 2 (Intelligent Greenhouse). Consider the following initial requirements specifications representing the current operation of an intelligent greenhouse:

1. The system shall notify when it is nighttime and there is no evidence of water pressure in the interval of 40 to 70 psi[1]..
2. The system shall water the plants at nighttime only.
3. The system shall water the plants when there is water pressure, it is nighttime, and the ground moisture is less than 30%.

These requirements are translated into a set of logic rules as follows:

$$\begin{aligned} \Pi_1 = \{ \mathit{notify} &\leftarrow \mathit{night}, \text{not}\ \ \mathit{wPressure} \\ \bot &\leftarrow \text{not}\ \ \mathit{night}, \mathit{wPlants} \\ \mathit{wPlants} &\leftarrow \mathit{wPressure}, \mathit{night}, \text{not}\ \ \mathit{wet} \\ \mathit{night} &\leftarrow \text{not}\ \ \mathit{day} \\ \mathit{day} &\leftarrow \text{not}\ \ \mathit{night} \} \end{aligned}$$

The last two statements capture the difference between day and night via default negation: the system cannot determine it unless a monitor device communicates it through observation. The system thus has two answer sets, $\{\mathit{night}, \mathit{notify}\}$ and $\{\mathit{day}\}$, reflecting ambiguity from incomplete information.

The specification is then extended with precise definitions for pressure and daylight:

$$\begin{aligned} \mathit{wPressure} &\leftarrow \mathit{pressure}(\textsc{psi}), \textsc{psi} \geq 40, \textsc{psi} \leq 70 \\ \mathit{day} &\leftarrow \mathit{light}(\textsc{lum}), \textsc{lum} > 3000 \\ \mathit{night} &\leftarrow \mathit{light}(\textsc{lum}), \textsc{lum} \leq 3000 \end{aligned}$$

Now suppose the engineers incorporate new requirements: (4) the system shall stop watering the plants when the ground becomes flooded; (5) the ground

[1] Note that we need to use ASP *aggregates, introduced back in* [11].

becomes flooded when moisture reaches 100%. Their ASP translation is: $(\sim wPlants \leftarrow grFlooded)$; $(grFlooded \leftarrow moisture(\text{RATE}), \text{RATE} \geq 100)$.

Given observations $\{pressure(45), light(3000), moisture(100)\}$, the entire KB becomes inconsistent —it has no answer sets under ASP. The resulting inconsistency prevents the system from deriving any valid answer sets; even previously established conclusions are lost.

The framework we propose, we construct a debug program (Definition 12). The α-relaxed program Π_1' appends a unique default-negated abducible not α_i to each ρ_i rule body:

```
notify :- night, not wPressure, not α1.
:- not night, wPlants, not α2.
wPlants :- wPressure, night, not wet, not α3.
night :- not day, not α4. day :- not night, not α5.
wPressure:- pressure(PSI), PSI>=40, PSI<=70, not α6.
day:- light(LUM), LUM>3000, not α7.
night:- light(LUM), LUM<=3000, not α8.
~wPlants:- grFlooded.
grFlooded:- moisture(RATE), RATE>=100.
light(3000). pressure(45). moisture(100).
```

Listing 1.1. Debug program for the greenhouse in DLV code.

The abductive program $\langle \Pi_1' \cup \Pi_2, \mathcal{A} \rangle$ (with $\mathcal{A} = \{\alpha_1, \alpha_2, \ldots, \alpha_8\}$) has three MGASs:

$$M_1 = \{\alpha_3, pressure(45), light(3000), moisture(100), night, wPressure, \sim wPlants, grFlooded\},$$
$$M_2 = \{\alpha_6, pressure(45), light(3000), moisture(100), notify, night, \sim wPlants, grFlooded\},$$
$$M_3 = \{\alpha_8, pressure(45), light(3000), moisture(100), wPressure, \sim wPlants, grFlooded, day\}.$$

The interpretation of these MGASs is as follows. M_1: abducible α_3 is valid, meaning the third rule (watering the plants) is no longer valid—the flooding condition overrides it. M_2: abducible α_6 invalidates the pressure threshold rule. M_3: abducible α_8 invalidates the nighttime definition in lumens, making the system conclude it is daytime despite 3000 lumens being insufficient.

The engineers review these results and determine that Rule 3 is the correct target for revision: the notion of "wetness" is still incomplete. They add $wet \leftarrow moisture(\text{RATE}), \text{RATE} \geq 10$ as Π_3, and the system $\Pi_1' \bowtie \Pi_2 \cup \Pi_3$ becomes unambiguous.

In general, when multiple MGASs are produced, the engineer must select which debug model to act upon. Several strategies can guide this choice: (a) *domain knowledge*, where the engineer recognizes which rule is semantically incomplete; (b) *cardinality preference*, favoring the MGAS with the fewest abducibles (i.e., the smallest conflict set); or (c) *weighted preferences*, where rules can be assigned priorities reflecting their criticality, so that weak constraints with higher weights protect more important specifications. In this example, all three MGASs are singletons, but strategy (a) identifies M_1 as the correct choice because the watering rule lacks a wetness precondition.

Example 3 (Static Debugging with Weak Constraints) Consider restoring consistency to the inconsistent program:

$$\Pi = \{(a \leftarrow b),\ (c \leftarrow a),\ (e \leftarrow \text{not}\, e),\ ({\sim}a \leftarrow \text{not}\, g),\ (b)\}.$$

The corresponding abductive program $\langle\, \Pi' \cup \emptyset, \mathcal{A}\,\rangle$ has its α-relaxed program Π' and abducibles $\mathcal{A} = \{\alpha_1, \alpha_2, \alpha_3, \alpha_4, \alpha_5\}$, yielding three GASs: $\{a, b, c, \alpha_3, \alpha_4\}$, $\{b, \alpha_1, \alpha_3, {\sim}a\}$, $\{\alpha_3, {\sim}a, \alpha_5\}$.

The DLV weak-constraint program encodes this as:

```
a:- b, not alpha_1. -alpha_1 | alpha_1. :~ alpha_1.
c:- a, not alpha_2. -alpha_2 | alpha_2. :~ alpha_2.
e:- not e, not alpha_3. -alpha_3 | alpha_3. :~ alpha_3.
-a:- not g, not alpha_4. -alpha_4 | alpha_4. :~ alpha_4.
b:- not alpha_5. -alpha_5 | alpha_5. :~ alpha_5.
```

Listing 1.2. Weak-constraint debug program.

Rule $e \leftarrow \text{not}\, e$ (tagged by α_3) is intrinsically inconsistent in ASP and appears in all GASs. The MGASs under set-inclusion preference identify minimal conflict sets, pointing the debugger precisely at the rules responsible for the collapse.

3.2 Properties

A rule ρ is said to *contradict* a consistent ELP $\Pi \setminus \{\rho\}$ iff $\{\rho\} \cup \Pi$ has no answer sets.

Proposition 1 (Debugging). *Rule ρ contradicts the non-empty consistent program Π_2 iff $\rho \in \Pi_1$,* $\text{not}\, \alpha \in \text{Body}(\rho')$, *and $\alpha \in \Delta$ for some MGAS M_Δ of $\langle\, \Pi_1' \cup \Pi_2, \mathcal{A}\,\rangle$.*

Proof (*Sufficient condition.*) Suppose ρ contradicts Π_2. Then there is an abductive program $\langle\, \Pi_1' \cup \Pi_2, \mathcal{A}\,\rangle$ with $\mathcal{A} \cap \mathcal{L} = \emptyset$, having a MGAS M_Δ such that $\Delta \subseteq \mathcal{A}$, $\alpha \in \Delta$, $\text{not}\, \alpha \in \text{Body}(\rho')$, $\rho \in \Pi_1$, $\rho' \in \Pi_1'$, and M_Δ is an answer set of $\Pi_1' \cup \Pi_2 \cup \{\alpha\}$. The necessary condition is straightforward.

Proposition 2 (Static Debugging). *Given an inconsistent ELP Π and the debug pair $\Pi \bowtie \emptyset$, rule $\rho \in \Pi$ contradicts Π iff $\rho' \in \Pi'$, $\alpha \in \Delta$, and* $\text{not}\, \alpha \in \text{Body}(\rho')$ *for some MGAS M_Δ of $\langle\, \Pi' \cup \emptyset, \mathcal{A}\,\rangle$.*

Proposition 3 (Abductive Program Consistency). *The abductive program $\Pi_{\mathcal{A}} = \langle\, \Pi_1', \mathcal{A}\,\rangle$ is consistent.*

Proof Suppose $\Pi_{\mathcal{A}}$ is inconsistent. Every rule in Π_1' is α-relaxed by a unique distinguished abducible, so setting $\Delta = \mathcal{A}$ invalidates all rules, making $\Pi_1' \cup \Delta$ trivially satisfiable. Thus M_Δ is consistent—contradicting the assumption.

Lemma 1 (Weak Consistency View). *Given a debugging pair $\Pi_1 \bowtie \Pi_2$ with its abductive program $\Pi_{\mathcal{A}} = \langle\, \Pi_1' \cup \Pi_2, \mathcal{A}\,\rangle$: if Π_2 is consistent then $\Pi_{\mathcal{A}}$ is also consistent.*

Corollary 1 (Consistency Preservation). *If Π_2 is consistent, then $\Pi_1 \bowtie \Pi_2$ is also consistent.*

Proposition 4 (Consistency Restoration). *$\Pi \bowtie \emptyset$ is consistent.*

Corollary 2. *Given an inconsistent program Π_2, $(\Pi_1 \bowtie \Pi_2) \bowtie \emptyset$ is consistent.*

These results provide a solid theoretical framework towards the implementation of a system to debug knowledge bases on DLV and PSMODELS.

4 Toward KB Debugger Implementation

Currently, there are at least three major efficient solvers for ASP: DLV [23], CLASP [12,14], and SMODELS [25]. Our system debugs ELP programs on such solvers.

4.1 Debugging with ODLP

The translation in Definition 11 proves beneficial for the context of belief revision in MGAS.

Definition 13 (Ordered Disjunctive Debug Program, *ODDP*). *Given the debug program $\Pi_1 \bowtie \Pi_2$ and its abductive program $\langle \Pi' \cup \Pi_2, \mathcal{A} \rangle$, its Ordered-disjunctive Debug Program is $\mathcal{O}(\Pi' \cup \Pi_2, \mathcal{A})$.*

By Theorem 1, every abductive program's MGASs correspond to the intended models of some ODLP that can be efficiently computed. As a result, ODLP serves a threefold purpose: one can use it to compute MGASs, abductive programs, and debug logic programs. The ODLP translation of the greenhouse debug program (Listing 1.1) prepends ordered-disjunctive rules $\beta_i \times \alpha_i \leftarrow \top$ for each abducible, and its preferred answer sets coincide with the MGASs.

Example 4 (Vacuous Updates and the Causal Rejection Problem). Consider $\Pi_1 = \{day \leftarrow \text{not } night;\ night \leftarrow \text{not } day;\ stars \leftarrow night, \text{not } cloudy;\ {\sim}stars\}$, with answer set $\{{\sim}stars, day\}$. The agent acquires $\Pi_2 = \{stars \leftrightarrow constellations\}$ (synonymy). Under the *causal rejection principle* [2,10,22], the update yields an unintuitive extra answer set $\{stars, constellations, night\}$. Our approach preserves the original knowledge: $\mathcal{O}(\Pi_1' \cup \Pi_2, \mathcal{A})$ yields only $\{{\sim}stars, day\}$, because α-relaxation does not create new causal pathways.

4.2 GAS in Weak Constraints

This approach transforms updates of ELPs into a wc-program whose answer sets coincide with the abductive program's GASs.

Definition 14 (Abductive-WP Translation). *Given an abductive logic program $\langle \Pi, \mathcal{A} \rangle$ and α' a distinguished unique literal with $\alpha' \notin \Sigma_{\langle \Pi, \mathcal{A} \rangle}$, the translation $\mathcal{WP}(\Pi, \mathcal{A})$ is $\Pi \cup \{\alpha' \vee \alpha \leftarrow \top, \bot :\sim \alpha \mid \alpha \in \mathcal{A}\}$.*

Lemma 2 (*GAS*–WC Answer Set). *$M \cap \Sigma_{\langle \Pi, \mathcal{A} \rangle}$ is a generalized answer set of $\langle \Pi, \mathcal{A} \rangle$ iff M is an answer set of $\mathcal{WP}(\Pi, \mathcal{A})$.*

Proof (Sketch). (*Only-if part.*) If M is an answer set of $\mathcal{WP}(\Pi, \mathcal{A})$, then with $\alpha \in M$, $M \cap \Sigma_{\langle \Pi, \mathcal{A} \rangle}$ is an answer set of $\Pi \cup \{\alpha \leftarrow \top\}$ and hence a GAS of $\langle \Pi, \mathcal{A} \rangle$ by Definition 8. Similarly for $\alpha \notin M$. The if-part follows symmetrically.

4.3 Parser, Grammar, and User Interface

The prototype uses FLEX and BISON for lexical and syntactic analysis. For each rule analyzed, the parser appends a pair of new rules with relaxing atoms and a weak constraint: $(-\alpha_i \mid \alpha_i \leftarrow \top)$ and $(\bot :\sim \alpha_i\ [1 : 1])$. The grammar for revision pairs follows a simple BNF: <pair> = '{' <program> '}' '{' <program> '}', where each program is a sequence of rules with heads and bodies. The user interface is a web page displaying the original pair, its abductive transformation, generalized answer sets, and the debug models, implemented as a Unix script filtering DLV's output into HTML.

Modular Data-Processing Architecture. The prototype follows a modular pipeline architecture, illustrated in Fig. 1. The process begins with the user providing a pair of logic programs $\Pi_1 \bowtie \Pi_2$ through a web form. The *Parser Module* (Flex/Bison) performs lexical and syntactic analysis, producing the α-relaxed program Π_1' and the corresponding weak constraints. The *Solver Module* invokes DLV on the transformed program, collecting all answer sets. The *Filter Module* extracts abducibles from each answer set, computes MGASs by set-inclusion minimality, and identifies conflicting rules. Finally, the *Presentation Module* renders an HTML report showing the original pair, the relaxed program, all GASs, the MGASs, and the resulting debug models with the tagged conflicting rules highlighted.

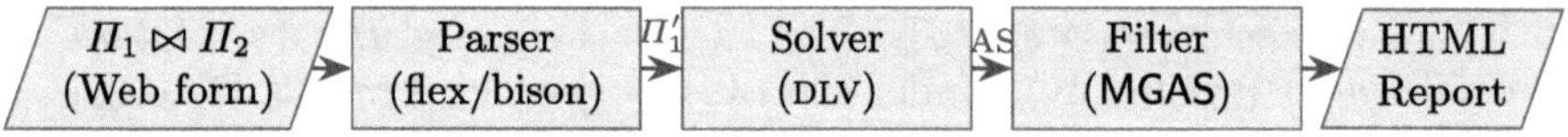

Fig. 1. Data flow of the debugger prototype: from user input to conflict report.

4.4 Computational Complexity

The complexity of the debugging framework is dominated by the computation of answer sets for the relaxed program. For an EDLP with n rules, the α-relaxation produces a program with n additional abducibles and $2n$ auxiliary rules (the disjunctive choice and weak constraint per abducible), yielding a program of size $O(n)$. Computing answer sets of disjunctive logic programs is Σ_2^P-complete in general [10]. However, for the common case of *normal programs* (no disjunction in rule heads), the problem reduces to NP-completeness. The subsequent MGAS computation requires a subset-minimality check over the set of GASs, which is polynomial in the number of GASs. In practice, modern solvers such as DLV and Clingo handle programs with hundreds of rules efficiently, and the overhead of the α-relaxation transformation is linear. A formal scalability study on industrial-sized requirement sets remains as future work.

Table 1. Comparison of our framework with related approaches.

Feature	Ours	CR-rules [3]	Causal Rej. [2,10]	Inc. Meas. [21]
Static debugging	✓	–	–	✓
Update debugging	✓	–	✓	–
Conflict localization	✓	partial	–	–
Minimal change guarantee	✓	–	✓	–
Vacuous update correctness	✓	n/a	–	n/a
General (any ELP)	✓	–	✓	✓
Solver-independent	✓	–	–	–

5 Related Work

The problem of managing and updating knowledge bases in ASP has received sustained attention. We position our contribution against the most relevant approaches.

Update Semantics. Eiter et al. [9,10] proposed update semantics based on a *causal rejection principle*. Alferes et al. [2] and Leite [22] refined this direction. However, these semantics can yield unintuitive results: as shown in Example 4, synonym rules can produce unintuitive ambiguous answer sets—i.e. an extra answer set to the correct one. Our α-relaxation avoids this by identifying minimal conflict sets through abduction rather than rejecting rules causally.

Consistency Restoration and Repair. Balduccini and Gelfond [3] proposed CR-rules for planning specifications, but their framework does not generalize beyond their specific syntax. Kuhlmann and Thimm [20,21] presented ASP-based algorithms for inconsistency measurement, and Merhej et al. [24] addressed repair using expert "rules of thumb." These are complementary: the former quantifies severity without localizing conflicts; the latter assumes external knowledge. Our framework combines identification (via MGAS) and resolution (via debug models) in a single formalism.

Belief Revision, Preferences, and Solvers. The AGM postulates [1] provide the classical foundation; our debug operator $\bowtie$ satisfies analogues of consistency preservation and restoration. Sakama and Inoue [28] and Zhang and Foo [29] proposed ASP update frameworks, but focus on incorporating information rather than localizing conflicts. Brewka et al. [4–6] introduced ODLP and PSMODELS; our contribution connects abductive debugging to ODLP via the ordered translation $\mathcal{O}$ (Theorem 1). On the solver side, DLV [23], CLASP [14], SMODELS [25], and Clingo 5 [13] provide robust solving but no built-in debugging for conflicting rules. Our prototype fills this gap.

Table 1 summarizes the key differences between our approach and the most closely related frameworks.

The distinction between *conflict localization* and *inconsistency measurement* (column 4) is practically significant. Inconsistency measurement quantifies the *degree* of inconsistency (e.g., the minimum number of formulas to remove) but does not identify *which specific rules* cause it. In an engineering scenario with hundreds of requirements, knowing that "the knowledge base has inconsistency degree 3" is far less useful than knowing that "rules 3, 6, and 8 are the sources of conflict." Our abducible-tagging mechanism provides the latter, enabling engineers to act on precise diagnostic information.

Table 2 explicitly compares the results of Example 4 across the three update semantics, illustrating how the causal rejection principle introduces an unintuitive model that our framework avoids.

Table 2. Comparison of update results for the sky/constellations example (Example 4).

Semantics	Answer Sets After Update	Correct?
Causal rejection [10]	$\{{\sim}stars, day\}$, $\{stars, constellations, night\}$	No
Dynamic updates [2,22]	$\{{\sim}stars, day\}$, $\{stars, constellations, night\}$	No
Our framework (α-relaxation)	$\{{\sim}stars, day\}$	Yes

Our approach distinguishes itself by (1) providing a unified debug operator that handles both static inconsistencies and update conflicts; (2) formal guarantees of consistency preservation and restoration; (3) implementation-ready translations to both ODLP and weak-constraint programs; and (4) explicit localization of conflicting rules through abducible tagging, which no prior framework provides in a general, solver-independent way.

6 Discussion and Future Work

This article presents a theoretical framework for a knowledge-based debugger that identifies sources of conflict amongst rules and restores consistency. We address three problems from requirement updates: conflicting updates, initial inconsistency, and inconsistent new information —with formal guaranties of consistency preservation and restoration. The MGASs guarantee minimal change to restore consistency, and the framework is an initial step toward translating plain-English requirements into formal logical statements.

We have implemented the framework in Weak Constraints and ODLP, with a modular, web-accessible prototype.

Implications of the Results. The formal properties established in Sect. 3 have direct practical implications. Proposition 3 guarantees that the debugging process itself never fails: the relaxed abductive program is always consistent, so the debugger always produces at least one diagnosis. Corollary 1 ensures that when new requirements are consistent on their own, merging them through the debug operator preserves consistency—a property that causal rejection semantics does not satisfy (cf. Example 4 and Table 2). Furthermore, Proposition 4 guarantees that any inconsistent program can be restored to consistency through the static debug operator, providing a last-resort recovery mechanism.

Prototype Experience and Lessons Learned. During the development of the prototype, several technical challenges arose. The transformation from user input to relaxed program is straightforward and linear in the size of the input; however, the number of GASs can grow combinatorially with the number of rules, making the filtering step for MGASs the main bottleneck for larger programs. Solver performance becomes the dominant factor, and incremental grounding techniques (as available in Clingo 5 [13]) could significantly improve scalability. The web interface proved effective for educational use, as students could immediately see how adding or removing a rule affects the set of MGASs, providing an interactive way to understand non-monotonic reasoning.

Limitations. The current framework has several limitations that should be acknowledged. First, the approach assumes that requirements have already been translated into ASP rules, but the translation from natural language to logic predicates is itself a challenging open problem that we do not address here. Integrating recent advances in NLP-based formalization [8] is a natural next step. Second, our examples are small-scale; a systematic empirical evaluation on industrial-sized requirement sets is needed to validate scalability claims. Third, the framework currently operates on propositional (ground) programs; extending it to first-order ASP with variables would broaden its applicability but introduces additional complexity.

Future Directions. Future work includes switching to fuzzy logic for continuous thresholds, implementing minimal set inclusion on DLV, formalizing complexity, and bridging discrete logic-based reasoning with continuous real-world phenomena. Additionally, integrating the debugger with modern ASP environments such as Clingo 5, which supports theory solving and multi-shot solving, could streamline the implementation and enable incremental debugging as requirements evolve over time. A longer-term goal is to develop a full NLP-to-ASP pipeline that would allow systems engineers to specify requirements in controlled natural language and automatically obtain both the logical encoding and the debugging diagnostics.

Acknowledgements. This project has been partially supported by the Autonomous University of Juarez, CHIH, Mexico; The Philosophic Systems Institute, Texas, USA; and the Andrew W. Mellon Foundation of New York.

References

1. Alchourrón, C.E., Gärdenfors, P., Makinson, D.: On the logic of theory change: partial meet contraction and revision functions. J. Symb. Log. **50**(2), 510–530 (1985)
2. Alferes, J.J., Banti, F., Brogi, A., Leite, J.A.: The refined extension principle for semantics of dynamic logic programming. Stud. Logica. **79**(1), 7–32 (2005). https://doi.org/10.1007/s11225-005-0492-y
3. Balduccini, M., Gelfond, M.: Logic programs with consistency-restoring rules. In: Proceedings of the AAAI Spring 2003 Symposium, pp. 9–18. AAAI Press, Palo Alto (2003)
4. Brewka, G.: Logic programming with ordered disjunction. In: Proceedings of AAAI-02, pp. 100–105 (2002). http://www.aaai.org/Papers/Workshops/2002/WS-02-13/WS02-13-002.pdf
5. Brewka, G., Niemelä, I., Syrjänen, T.: Implementing ordered disjunction using answer set solvers for normal programs. In: Goos, G., Hartmanis, J., Leeuwen, J.V. (eds.) Proceedings of the 8th European Workshop Logic in Artificial Inteligence JELIA 2002. LNCS, vol. 2424/2002, pp. 444–456. Springer-Verlag, Berlin/Heidelberg (2002)
6. Brewka, G., Niemelä, I., Syrjänen, T.: Logic programs with ordered disjunction. Comput. Intell. **20**(2), 333–357 (2004)
7. Crescini, V.F., Zhang, Y.: PolicyUpdater: a system for dynamic access control. Int. J. Inf. Secur. **5**(3), 145–165 (2006). https://doi.org/10.1007/s10207-005-0078-8
8. Dunin-Keplicz, B., Strachocka, A.: Perceiving rules under incomplete and inconsistent information. In: Leite, J., Son, T.C., Torroni, P., van der Torre, L., Woltran, S. (eds.) Computational Logic in Multi-Agent Systems, pp. 256–272. Springer, Berlin Heidelberg, Berlin, Heidelberg (2013)
9. Eiter, T., Fink, M., Sabbatini, G., Tompits, H.: Considerations on updates of logic programs. In: Logics in Artificial Intelligence, vol. 1919, pp. 2–20 (2000). https://doi.org/10.1007/3-540-40006-0_2
10. Eiter, T., Fink, M., Sabbatini, G., Tompits, H.: On properties of update sequences based on causal rejection. Theory Pract. Logic Program. **2**(6), 711–767 (2002). https://doi.org/10.1017/S1471068401001247, https://arxiv.org/abs/physics/0109006, http://www.kr.tuwien.ac.at/staff/tompits/papers/tplp-upd.pdf
11. Ferraris, P.: Logic programs with propositional connectives and aggregates. ACM Trans. Comput. Log. **12**(4), 1–40 (2011). https://doi.org/10.1145/1970398.1970401
12. Gebser, M., Kaminski, R., Kaufmann, B., Romero, J., Schaub, T.: Progress in clasp series 3. In: LPNMR. Lecture Notes in Computer Science, vol. 9345, pp. 368–383. Springer (2015)

13. Gebser, M., Kaminski, R., Kaufmann, B., Schaub, T.: Theory solving made easy with Clingo 5. OASICS **52** (2016)
14. Gebser, M., Kaufmann, B., Schaub, T.: Conflict-driven answer set solving: from theory to practice. Artif. Intell. **187**, 52–89 (2012)
15. Gelfond, M.: Logic programming and reasoning with incomplete information. Ann. Math. Artif. Intell. **12**, 89–116 (1994)
16. Gelfond, M., Lifschitz, V.: The stable model semantics for logic programming. In: Kowalski, R.A., Bowen, K.A. (eds.) 5th Conference on Logic Programming, vol. 88, pp. 1070–1080 (1988)
17. Gelfond, M., Lifschitz, V.: Classical negation in logic programs and disjunctive databases. N. Gener. Comput. **9**, 365–385 (1991)
18. Heitmeyer, C.L., Jeffords, R.D., Labaw, B.G.: Automated consistency checking of requirements specifications. ACM Trans. Softw. Eng. Methodol. **5**(3), 231–261 (1996)
19. Kakas, A.C., Mancarella, P.: Generalized stable models: a semantics for abduction. In: ECAI, pp. 385–391 (1990)
20. Kuhlmann, I., Gessler, A., Laszlo, V., Thimm, M.: Comparison of SAT-based and ASP-based algorithms for inconsistency measurement. J. Artif. Intell. Res. **82**, 563–685 (2025). https://doi.org/10.1613/jair.1.14441
21. Kuhlmann, I., Thimm, M.: Algorithms for inconsistency measurement using answer set programming. In: Proceedings of the 19th International Workshop on Non-Monotonic Reasoning (NMR), pp. 159–168 (2021)
22. Leite, J.A., Slota, M.: A brief history of updates of answer-set programs. Theory Pract. Logic Program., 1–51 (2022). https://doi.org/10.1017/s1471068422000060
23. Leone, N., et al.: The DLV system for knowledge representation and reasoning. ACM Trans. Comput. Log. **7**(3), 499–562 (2006)
24. Merhej, E., Schockaert, S., De Cock, M.: Repairing inconsistent answer set programs using rules of thumb: a gene regulatory networks case study. Int. J. Approximate Reasoning **83**, 243–264 (2017)
25. Niemela, I., Simons, P.: Smodels—an implementation of the stable model and well-founded semantics for normal logic programs. In: Proceedings of the 4th LPNMR ('97). LNCS, vol. 1265, pp. 420–429. Springer, Dagstuhl Castle (1997)
26. Osorio, M., Ortiz, M., Zepeda, C.: Using CR-rules for evacuation planning. In: Workshop Proceedings of Deduction and Reasoning Techniques. IEEE Computer Society, Puebla (2004)
27. Peirce, C.S.: Illustrations of the Logic of Science, p. 12 and 13 (1878)
28. Sakama, C., Inoue, K.: An abductive framework for computing knowledge base updates. Theory Pract. Logic Program. **3**(6), 671–715 (2003). https://doi.org/10.1017/S1471068403001716
29. Zhang, Y., Foo, N.: A unified framework for representing logic program updates. In: Veloso, M.M., Kambhampati, S. (eds.) Proceedings of the 20th National Conference on Artificial Intelligence (AAAI-2005), pp. 707–713. AAAI Press/The MIT Press, Pittsburgh (2005)

Temporal Modeling of Audio Embedding Spaces with Multi-branch Dilated Convolutional Networks

Edmar Olivares Soria(✉)

Universidad Autonoma Metropolitana, Mexico City, Mexico
soria.multimedia@gmail.com

Abstract. The present work develops a computational methodology for unsupervised temporal representation learning for multivariate time series from a context of audio analysis and further potential applications for segmentation and change point detection tasks. The initial stage of the methodology proposes the transformation of mel-spectrogram windows into compact local embeddings through a convolutional autoencoder in such a way that the resulting latent vectors can then be treated as an ordered multivariate time series. For that scope, this paper proposes a custom deep learning design based on a multi-branch Temporal Convolutional Network (TCN) that models the temporal organization of the latent space using parallel dilated convolutional branches with different kernel sizes and receptive fields. The main contribution is framed as a general methodology for learning temporally structured audio embeddings, deriving latent-dynamics descriptors, and performing a second stage of unsupervised feature extraction over the embedding space itself. The proposal is grounded in established work on mel-spectrogram processing, nonlinear representation learning, temporal convolutional sequence modeling, and deep audio representation learning and the used dataset relies on electroacoustic music audio files. The choice of such dataset relies on the fact that electroacoustic music is inherently highly complex across the time, for acoustic characteristics such as spectra and morphology; this is usually described as *spectromorphology* within such field.

Keywords: Temporal Convolutional Networks for Audio · Representational Audio Learning · Audio Embedding Unsupervised Learning

1 Introduction

Music Information Retrieval (MIR) pipelines have traditionally relied on hand-engineered features such as energy, spectral centroid, spectral bandwidth, roll-off, MFCCs, and related time-series descriptors. These features remain useful because they are interpretable and computationally accessible, and their extraction is well supported by mature software ecosystems such as `librosa`, which

H. Ponce et al. (Eds.): COMIA 2026, CCIS 3095, pp. 678–685, 2027.
https://doi.org/10.1007/978-3-032-37438-7_53

standardizes many core MIR operations and frame-based time conversions [1] in a easy and almost straightforward way for cross platforms (Mac, Linux and Windows), which means a higher plausibility for further reproducibility or deployment. However, hand-crafted features also encode prior and fixed assumptions about measured characteristics of sound over the process itself and due to the nature of each feature, and may omit structural or timbral dynamics and patterns that are relevant for complex contemporary audio materials and that might not be directly revealed by such features.

This proposal is divided into a two-stage process. First, local spectro-temporal patterns are encoded from mel-spectrogram windows by a convolutional autoencoder (AE) producing one latent vector per window. Second, the resulting latent vectors are organized as an ordered multivariate time series and processed by a multi-branch dilated TCN. Therefore, the central discussion moves beyond the question of whether embeddings can be learned, and emphasizes the hypothesis that *temporal organization of the learned embedding space* can itself become the object of unsupervised modeling within the present context of audio analysis.

The current implementation already supports this interpretation. The trainer class loads a pool of windowed mel spectrograms, extracts latent sequences with an autoencoder architecture, applies slice-level splitting to avoid leakage, computes latent-dynamics descriptors, and trains a multi-branch TCN for next-step prediction in latent space. In this sense, the TCN is both a predictor and a second-order unsupervised feature extractor over embedding trajectories, and so, this work claims that audio embedding spaces can be modeled as multivariate time series and that a multi-branch TCN provides a principled mechanism for learning their temporal regularities across multiple scales.

2 Related Work

Most computational audio pipelines begin with time–frequency representations derived from the short-time Fourier transform (STFT). In MIR and speech/audio learning, mel-spectrograms are especially common because they implement a perceptually motivated frequency warping whose engineering formulations are well documented in auditory modeling toolkits such as Slaney's *Auditory Toolbox* [2]. In practice, libraries such as `librosa` provide robust implementations for mel-spectrogram computation, frame indexing, and audio feature extraction making them a standard choice for reproducible MIR experimentation [1].

Deep autoencoders are a canonical route for learning compact nonlinear codes from high-dimensional data [3]. In audio, convolutional encoders are particularly appropriate when the input is a time–frequency image since local spectro-temporal motifs can be represented by shared convolutional filters. More recent work on music and general audio representation learning has moved beyond purely reconstruction-driven paradigms and emphasized predictive, masked, or contrastive self-supervision. Contrastive Predictive Coding (CPC) learns representations by predicting future latent states [4]. wav2vec 2.0 and HuBERT show that masked or predictive self-supervision can learn strong speech/audio embeddings without labels [5,6]. Although these methods differ from reconstruction-based autoencoding, they support the broader hypothesis that latent spaces learned from audio can encode musically or acoustically salient structure.

Within music-specific representation learning, more recent studies further reinforce this view. McCallum et al. provide a broad comparative analysis of supervised and unsupervised audio representations for music understanding and show that both the pre-training objective and the domain of the pre-training corpus, materially affect downstream utility [9]. BYOL-A proposes a self-supervised audio representation framework based on bootstrapping and reports strong general-purpose performance across diverse audio tasks [10]. Audio-MAE extends masked autoencoding to audio spectrograms and shows that spectrogram-patch reconstruction with high masking ratios yields competitive and transferable representations [11]. In multimodal representation learning, CLAP demonstrates that audio embeddings can be learned from paired audio–text supervision and transferred to a wide range of tasks [12]. More recently, MERT introduces large-scale self-supervised training for acoustic music understanding and reports strong performance on music-related tasks [13], while MuQ proposes mel residual vector quantization as a self-supervised target for music representation learning and reports improved downstream results over prior music SSL baselines [14]. Taken together, these works suggest that modern audio representation learning increasingly treats embeddings as general-purpose, transferable descriptors rather than task-specific intermediate features.

Temporal Convolutional Networks provide a principled sequence-modeling alternative to recurrent architectures. In the generic TCN formulation, causal convolutions preserve temporal order while dilations increase effective memory without requiring recurrence [7]. WaveNet demonstrated the effectiveness of dilated causal convolutions for long-range audio dependency modeling, providing a clear historical foundation for dilated temporal architectures in audio [8]. For a stack of L dilated convolutional layers with kernel size k and exponentially increasing dilations $\delta_\ell = 2^\ell$, the receptive field is

$$R = 1 + (k-1)\sum_{\ell=0}^{L-1} 2^\ell = 1 + (k-1)(2^L - 1). \tag{1}$$

This property is central to the current proposal because audio structure unfolds across heterogeneous time scales. A model with only one receptive-field configuration is likely to privilege one temporal granularity; by contrast, the current implementation uses multiple TCN branches with different kernel schedules, which provides a stronger architectural argument for audio sequence modeling across short, medium, and longer temporal horizons.

A final line of prior work relevant here concerns MIR systems that combine time–frequency front ends with temporal summarization. Convolutional Recurrent Networks for music classification and tagging have shown that mel-spectrograms can support temporally structured deep models in MIR contexts [15]. The present proposal extends that general insight into a fully unsupervised setting: spectrogram windows are first embedded by an AE, and then a TCN models the temporal evolution of those embeddings themselves. In this sense, the contribution is not only "deep learning for audio," but a specific reformulation of learned embeddings as multivariate time series that can be further processed by temporal models.

3 MIR Engineering and Data Representation

Let an audio signal be denoted by $x[n]$, sampled at rate f_s. A mel-spectrogram is computed with hop length h samples, yielding a matrix

$$\mathbf{M} \in \mathbb{R}^{F \times T}, \tag{2}$$

where F is the number of mel bands and T is the number of time frames. A sliding window operator then extracts local patches along the temporal axis:

$$\mathbf{X}_t = \mathbf{M}[:, t : t + W - 1] \in \mathbb{R}^{F \times W}, \tag{3}$$

where W is the window length in frames and successive windows are separated by a step size S frames.

The present implementation stores these windows in a pooled tensor and uses an explicit time-axis configuration with sample rate, hop length, window size, and step size. This setup allows the extraction of an exact mapping from latent indices back to seconds; it connects the abstract latent sequence to interpretable musical time. Specifically, the center time of window t is

$$t_{\text{center}}(t) = \frac{\left((t-1)S + \frac{W}{2}\right) h}{f_s}. \tag{4}$$

A second important engineering choice is the preservation of slice-level or song-level identity throughout preprocessing. Because overlapping windows from the same audio item are highly correlated, train/validation/test splitting must occur at the item or slice level, not at the window level and the current implementation already enforces this with a slice-level split function.

4 Stage 1: Local Representation Learning with a Spectrogram Autoencoder

In the present work, the autoencoder stage is the first level of unsupervised feature extraction. Let E_θ denote the encoder. Each mel-spectrogram window $\mathbf{X}_t$ is mapped to a latent vector

$$\mathbf{z}_t = E_\theta(\mathbf{X}_t) \in \mathbb{R}^d. \tag{5}$$

The sequence of all latent vectors extracted from an item is then

$$\mathbf{Z} = (\mathbf{z}_1, \mathbf{z}_2, \ldots, \mathbf{z}_N), \qquad \mathbf{z}_t \in \mathbb{R}^d. \tag{6}$$

Conceptually, the AE stage implies representational learning through *local spectro-temporal descriptors* without relying on preselected MIR feature families. In the current implementation, the AE uses convolutional blocks, Group Normalization, dropout, and squeeze-and-excitation style channel recalibration. Group Normalization is a reasonable design choice because its normalization statistics are independent of batch size, which is advantageous when batch sizes are constrained or unstable [18]. Channel recalibration through squeeze-and-excitation mechanisms improves the modeling of inter-channel dependencies and strengthens representational selectivity [19]. Since the TCN stage depends critically on having a stable and semantically meaningful local embedding space, this AE stage is therefore the first part of unsupervised feature extraction, and the TCN builds on top of it as a second stage.

5 Stage 2: Latent Multivariate Time-Series Construction and Processing

Once each spectrogram window has been encoded, the output is formed as a multivariate time series in latent space and so, given one audio item, the latent sequence

$$\mathbf{Z} = (\mathbf{z}_1, \ldots, \mathbf{z}_N), \qquad \mathbf{z}_t \in \mathbb{R}^d,$$

can be treated as a multivariate time series of dimension d. In the current code, the latent tensor is stored with shape (N_{items}, T, d) and is optionally standardized again using a train-only scaler fitted on the training subset of latent vectors. This second normalization stage is important because the AE embedding distribution may be stable enough for reconstruction but still benefit from centering and variance normalization before temporal forecasting.

The current implementation also defines latent-dynamics descriptors directly over the embedding trajectory. Given $\mathbf{z}_t$, define finite-difference velocity and acceleration:

$$\mathbf{v}_t = \mathbf{z}_{t+1} - \mathbf{z}_t, \tag{7}$$

$$\mathbf{a}_t = \mathbf{v}_{t+1} - \mathbf{v}_t. \tag{8}$$

Their magnitudes,

$$v_t^{(2)} = \|\mathbf{v}_t\|_2, \qquad a_t^{(2)} = \|\mathbf{a}_t\|_2,$$

provide compact scalar summaries of change intensity and abruptness. The developed script also computes curvature-like directional change and local dispersion over a sliding neighborhood. These descriptors should be elevated from "auxiliary diagnostics" to a major contribution: *they are learned temporal descriptors extracted from the embedding space itself, and they offer a clear alternative to classical MIR feature trajectories.*

6 Stage 3: Temporal Modeling with Multi-Branch Dilated Convolutional Networks

As said before, the methodological core of this work is focused on the multi-branch TCN. The current implementation defines a residual temporal block with two causal 1D convolutions, Batch Normalization, dropout, and a residual connection. Each branch is designed as a stack of such blocks with exponentially increasing dilation factors, and the full model aggregates several branches in parallel for processing of different temporal resolutions.

Formally, let a context window of latent vectors of length L be denoted by

$$\mathbf{X}_t = [\mathbf{z}_{t-L+1}, \ldots, \mathbf{z}_t] \in \mathbb{R}^{L \times d}. \tag{9}$$

The TCN forecaster G_ϕ learns

$$\hat{\mathbf{z}}_{t+1} = G_\phi(\mathbf{X}_t). \tag{10}$$

In the current implementation, the target can also be represented as a full shifted sequence over the context segment, which is convenient for sequence-to-sequence training in Conv1D format. The optimization objective is the mean squared error

$$\mathcal{L}(\phi) = \frac{1}{|\mathcal{D}|} \sum_{(\mathbf{X}_t, \mathbf{y}_t) \in \mathcal{D}} \|\mathbf{y}_t - G_\phi(\mathbf{X}_t)\|_2^2, \tag{11}$$

trained with Adam, which remains a standard optimizer for deep sequence learning because of its adaptive first- and second-moment updates [20].

The strongest architectural feature is the *multi-branch design*. Different branches use different kernel sequences and channel progressions. This means that the network does not commit to a single temporal resolution; instead, it learns parallel representations at multiple effective receptive fields and fuses them through concatenation and a final 1×1 projection.

In this sense, the present work can be framed as a method for *multi-scale temporal representation learning over latent trajectories.*

7 Training Hyperparameters, Metrics, and Model Summary

The current implementation already defines an explicit set of training-time defaults that can be reported directly and reproduced from the code. At the front-end level, the temporal axis is parameterized through `TimeAxisConfig`, with default values $f_s = 22100$ Hz, hop length $h = 512$ samples, spectrogram window size $W = 64$ frames, and window step $S = 32$ frames. These values determine the mapping from frame index to seconds and therefore the temporal resolution of both the latent descriptors and the TCN forecast outputs. The trainer also fixes a random seed through a dedicated utility function, ensuring controlled pseudo-random initialization for NumPy and PyTorch.

The latent extraction stage assumes the AE architecture inherited from the previous report: convolutional blocks with Group Normalization [18], dropout [17], and squeeze-and-excitation style channel recalibration [19]. The default encoder configuration uses convolutional channel progression $(32, 64, 128, 256)$ with kernel sizes $(5, 3, 3, 3)$ and a latent bottleneck of dimension $d = 192$. Once latent sequences are extracted, the script optionally performs a second train-only standardization using `StandardScaler`, which is important because an AE embedding distribution may be stable enough for reconstruction but still benefit from centering and variance normalization before temporal forecasting.

At the TCN stage, the default model is a three-branch `MultiBranchTCN` with output dimension equal to the latent dimension. Each branch is a stack of three `TemporalBlocks` with exponentially increasing dilations $\{1, 2, 4\}$. The default branch channel schedules are

$$[32, 64, 64], \qquad [64, 128, 128], \qquad [128, 256, 256],$$

with corresponding kernel schedules

$$[9, 9, 7], \qquad [7, 7, 5], \qquad [5, 3, 3],$$

and dropout schedules

$$[0.01, 0.01, 0.01], \qquad [0.05, 0.05, 0.05], \qquad [0.15, 0.15, 0.15].$$

Each `TemporalBlock` contains two causal `Conv1d` layers, each followed by Batch Normalization [16], ReLU activation, and dropout, together with a residual connection and a 1×1 projection when channel dimensions differ. The outputs of all branches are concatenated and projected through a final 1×1 convolution to return a tensor in $\mathbb{R}^{d \times L}$. For the default latent dimension $d = 192$, the instantiated three-branch model contains approximately 1,700,256 trainable parameters.

The data interface for TCN training is defined by the `LatentSeqDataset` class. With the default settings in `make_dataloaders`, the dataset is segmented using context length $L = 30$, stride 15, and batch size 16. These values create overlapping training segments while preserving the higher-level item split (song or slice). This design is important for scientific rigor: the temporal model receives many training examples per item, but no item contributes windows simultaneously to train and validation/test subsets.

Optimization is carried out with Adam [20]. The default training configuration in `train_tcn` uses learning rate 5×10^{-4}, zero weight decay by default, patience $= 10$ epochs for early stopping, and either a `ReduceLROnPlateau` scheduler or a cosine annealing scheduler. Gradient norms are clipped to a maximum ℓ_2 norm of 1.0 before each optimizer step, which helps stabilize training in sequence models. The objective is mean squared error (MSE) between predicted and target latent sequences:

$$\mathcal{L}(\phi) = \frac{1}{|\mathcal{D}|} \sum_{(\mathbf{X}_t, \mathbf{y}_t) \in \mathcal{D}} \|\mathbf{y}_t - G_\phi(\mathbf{X}_t)\|_2^2. \tag{12}$$

Python scripts, model summary, plots and additional info can be found at: https://github.com/soriamultimedia/TCN_Multibranch

References

1. McFee, B., et al.: librosa: audio and music signal analysis in python. In: Proceedings of the 14th Python in Science Conference (SciPy) (2015)
2. Slaney, M.: Auditory Toolbox Version 2. Interval Research Corporation Technical Report 1998-010 (1998)
3. Hinton, G.E., Salakhutdinov, R.R.: Reducing the dimensionality of data with neural networks. Science **313**(5786), 504–507 (2006)
4. van den Oord, A., Li, Y., Vinyals, O.: Representation learning with contrastive predictive coding. arXiv:1807.03748 (2018)
5. Baevski, A., Zhou, H., Mohamed, A., Auli, M.: wav2vec 2.0: a framework for self-supervised learning of speech representations. In: Advances in Neural Information Processing Systems (2020)
6. Hsu, W.N., et al.: HuBERT: self-supervised speech representation learning by masked prediction of hidden units. arXiv:2106.07447 (2021)
7. Bai, S., Kolter, J.Z., Koltun, V.: An empirical evaluation of generic convolutional and recurrent networks for sequence modeling. arXiv:1803.01271 (2018)
8. van den Oord, A., et al.: WaveNet: a generative model for raw audio. arXiv:1609.03499 (2016)
9. McCallum, M.C., Korzeniowski, F., Oramas, S., Gouyon, F., Ehmann, A.F.: Supervised and unsupervised learning of audio representations for music understanding. In: Proceedings of the 23rd International Society for Music Information Retrieval Conference (ISMIR), pp. 256–263 (2022)
10. Niizumi, D., Takeuchi, D., Ohishi, Y., Harada, N., Kashino, K.: BYOL for audio: exploring pre-trained general-purpose audio representations. In: 2021 International Joint Conference on Neural Networks (IJCNN), pp. 1–8 (2021)
11. Huang, P.Y., et al.: Masked autoencoders that listen. In: Advances in Neural Information Processing Systems (NeurIPS) (2022)
12. Elizalde, B., Deshmukh, S., Al Ismail, M., Wang, H.: CLAP: learning audio concepts from natural language supervision. In: ICASSP 2023 – 2023 IEEE International Conference on Acoustics, Speech and Signal Processing, pp. 1–5 (2023)
13. Li, Y., et al.: MERT: acoustic music understanding model with large-scale self-supervised training. In: International Conference on Learning Representations (ICLR) (2024)
14. Zhu, H., et al.: MuQ: self-supervised music representation learning with MEL residual vector quantization. arXiv:2501.01108 (2025)
15. Choi, K., Fazekas, G., Sandler, M.B.: Convolutional recurrent neural networks for music classification. In: Proceedings of ICASSP (2017)
16. Ioffe, S., Szegedy, C.: Batch normalization: accelerating deep network training by reducing internal covariate shift. In: Proceedings of the 32nd International Conference on Machine Learning (ICML), pp. 448–456 (2015)
17. Srivastava, N., Hinton, G., Krizhevsky, A., Sutskever, I., Salakhutdinov, R.: Dropout: a simple way to prevent neural networks from overfitting. J. Mach. Learn. Res. **15**, 1929–1958 (2014)
18. Wu, Y., He, K.: Group normalization. In: Proceedings of ECCV (2018)
19. Hu, J., Shen, L., Sun, G.: Squeeze-and-excitation networks. In: Proceedings of CVPR (2018)
20. Diederik, P., Kingma, J., Ba: Adam: a method for stochastic optimization. In: International Conference on Learning Representations (2015)

Active Fire Detection in Landsat 8 OLI Using Genetic Optimization and Entropy-Driven Thresholding

Braulio-Leonardo Santa Fe-García[1], Juan Villegas Cortez[1(✉)], César-Augusto Puente-Montejano[2], Josefa Díaz Álvarez[3], and Francisco Chávez - De la O.[3]

[1] Department of Systems, Universidad Autónoma Metropolitana, Azcapotzalco Unit,, Mexico City, Mexico
{al224380456,juanvc}@azc.uam.mx

[2] Faculty of Engineering, Universidad Autónoma de San Luis Potosí, San Luis Potosí, Mexico
cesar.puente@uaslp.mx

[3] Computer Architecture Department, University of Extremadura, C. Santa Teresa de Jornet, 38, 06800 Mérida, Spain
{mjdiaz,fchavez}@unex.es

Abstract. Forest fires pose escalating global threats to biodiversity, carbon sinks, and human settlements – directly undermining UN Sustainable Development Goals (SDGs) 13 (Climate Action), 15 (Life on Land), and 11 (Sustainable Cities). Timely detection is critical for mitigation, yet conventional threshold-based methods using satellite imagery suffer from poor generalization across diverse terrains and temporal conditions. This work introduces a novel, adaptive detection framework for daytime wildfires in Western Mexico using Landsat 8 OLI data, optimized via Genetic Algorithm (GA) to dynamically calibrate thresholds within a 5D feature space derived from SWIR-NIR spatial relationships. Kapur's entropy maximization serves as the fitness function, enhancing discrimination between fire pixels and background clutter. Evaluated against manually annotated ground truth, our method demonstrates robust performance across heterogeneous ecosystems – a key requirement for global deployment. By leveraging freely available, globally consistent Landsat data, this approach supports scalable, low-cost monitoring systems aligned with SDG 17 (Partnerships for the Goals). The framework's adaptability makes it suitable for regions lacking high-resolution sensors or real-time infrastructure. We argue that intelligent, data-driven fire detection is not merely a technical challenge – it is an ethical imperative for planetary resilience. This research contributes to the global AI-for-Earth movement, empowering nations to meet 2030 Agenda targets through accessible, automated environmental surveillance.

Keywords: Active fire detection · Genetic Algorithm · Landsat 8 · Image Segmentation

H. Ponce et al. (Eds.): COMIA 2026, CCIS 3095, pp. 686–697, 2027.
https://doi.org/10.1007/978-3-032-37438-7_54

1 Introduction

Forest fires constitute a critical environmental hazard that undermines ecological processes and Sustainable Development Goals (SDGs) 13 and 15, necessitating intelligent, scalable detection systems [12,14]. In Mexico, particularly the western region, escalating fire activity driven by rugged terrain and seasonal dryness has exposed the limitations of conventional monitoring (Figs. 1 and 2). While ground and aerial methods offer local efficacy, they suffer from high operational costs and lack the spatial-temporal resolution required for large-scale, proactive surveillance [1,6].

Satellite remote sensing provides a viable alternative through multispectral capabilities (visible, NIR, SWIR) that identify thermal anomalies, yet standard threshold-based algorithms fail to generalize across heterogeneous landscapes due to sensor variability and noise [12]. To address this, we propose a metaheuristic-driven framework leveraging Landsat 8 OLI data, which employs an entropy-optimized approach to dynamically adapt detection thresholds. This method aims to enhance robustness and scalability, ensuring compatibility with global Earth observation initiatives and overcoming the static limitations of existing rule-based systems.

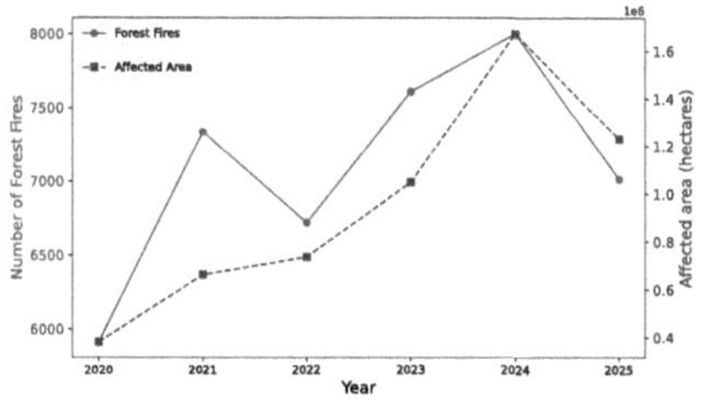

Fig. 1. Historical trend of forest fires in Mexico (2020–2025). The red solid line indicates the total number of fire incidents (left axis), while the blue dashed line represents the total affected area in hectares (right axis) (Color figure online)

2 State of the Art

Active fire detection via remote sensing has evolved from early aerial thermal surveys in the 1960s–70s to sophisticated satellite-based algorithms leveraging multispectral and thermal data. The advent of AVHRR aboard NOAA-POES in the 1980s marked the first operational use of satellite data for global fire monitoring, exploiting Mid-Infrared (MIR) bands to identify thermal anomalies [15]. This paradigm shifted significantly with the launch of MODIS aboard Terra (1999) and Aqua (2002), which introduced enhanced MIR sensitivity at 3.96 μm and enabled Fire Radiative Power (FRP) estimation through synergistic use of LWIR bands at 11 μm [4]. These instruments laid the foundation for subsequent LEO sensors–including VIIRS [10], Landsat 8 OLI [5,8,11], and Sentinel-3 SLSTR [16]–as well as GEO platforms offering high temporal resolution at the cost of spatial fidelity.

2.1 Fixed-Threshold Algorithms

Traditional detection pipelines rely on handcrafted spectral thresholds to isolate fire pixels. These rules typically exploit reflectance ratios and absolute values in SWIR/NIR bands (e.g., ρ_7/ρ_5) to distinguish thermal anomalies from background [15]. However, their rigidity renders them vulnerable to environmental variability–cloud cover, seasonal vegetation shifts, and sensor saturation–leading to false alarms and omission errors [3].

Schroeder et al. [11] introduced a foundational OLI-based algorithm using three primary conditions (Eq. 1) to flag high-confidence fire pixels:

$$\left(\frac{\rho_7}{\rho_5} > 2.5\right) \wedge (\rho_7 - \rho_5 > 0.3) \wedge (\rho_7 > 0.5) \tag{1}$$

To mitigate saturation-induced wrap-around artifacts, they proposed a secondary rule (Eq. 2):

$$(\rho_6 > 0.8) \wedge (\rho_1 > 0.2) \wedge (\rho_5 > 0.4 \vee \rho_7 < 0.1) \tag{2}$$

Candidate pixels were further refined via relaxed thresholds (Eq. 3):

$$\left(\frac{\rho_7}{\rho_5} > 1.8\right) \wedge (\rho_7 - \rho_5 > 0.17) \tag{3}$$

Subsequently, a contextual analysis was performed on pixels detected by Eq. 3, using a 61×61 pixel window (Eq. 4):

$$\left(\frac{\rho_7}{\rho_5} > \mu_{\rho_7/\rho_5} + \max\left[3\sigma_{\rho_7/\rho_5}, 0.8\right]\right) \wedge (\rho_7 > \mu_{\rho_7} + \max\left[3\sigma_{\rho_7}, 0.08\right]) \wedge \left(\frac{\rho_7}{\rho_6} > 1.6\right) \tag{4}$$

where μ and σ represent the mean and standard deviation of reflectance, calculated in a 61×61 pixel window centered on each candidate pixel, excluding water bodies and pixels previously classified as fire. Temporal filtering over 176 prior days (median $\rho_7 > 0.2$) reduced persistent heat false positives. Murphy et al. [8] introduced a dual-parameter approach (α, β) combining spectral ratios and saturation flags (Eq. 5):

$$s = \begin{cases} 1, & \text{if channel 6 or 7 is saturated} \\ 0, & \text{otherwise} \end{cases} \tag{5}$$

with:

$$\alpha = \left(\frac{\rho_7}{\rho_6} \geq 1.4\right) \wedge \left(\frac{\rho_7}{\rho_5} \geq 1.4\right) \wedge \rho_7 \geq 0.15 \tag{6}$$

$$\beta = \left(\left(\frac{\rho_6}{\rho_5} \geq 2\right) \wedge \rho_6 \geq 0.5\right) \vee s \tag{7}$$

Parameters α and β work synergistically: α identifies high-confidence fire pixels, while β handles saturation and detects very hot pixels. Clustering via Moore neighborhood ensured spatial coherence, discarding clusters lacking α-confirmed pixels.

Kumar et al. [5] incorporated channel 4 (0.66 μm) to enhance fire-background discrimination, proposing linear thresholds to avoid saturation artifacts. Pixels with evident fire presence were obtained via: $\rho_4 \leq 0.58\rho_7 - 0.214$. Surrounding pixels were considered unequivocal if: $\rho_4 \leq 0.35\rho_6 - 0.044$. Potential fire pixels satisfied: $\rho_4 \leq 0.53\rho_7 - 0.125 \vee \rho_6 \leq 1.08\rho_7 - 0.048$

Contextual analysis followed [11] with incremental window size. Temporal filtering used median $\rho_7 > 0.18$ and NDVI-based suppression:

$$\begin{cases} \text{Median}_{\rho_7} > 0.18 \\ \text{NDVI}_{\text{fire}} < \text{NDVI}_{\text{peak}} - 0.06 \end{cases} \tag{8}$$

Although computationally efficient, fixed-threshold methods remain inherently brittle under heterogeneous conditions–underscoring the need for adaptive, optimization-driven frameworks capable of generalizing across diverse landscapes. In recent years, Deep Learning approaches have been proposed for active fire detection [2] [13]. However, their implementation is computationally expensive, requires large datasets, and often involves patching that disrupts the initial scene configuration. To overcome these limitations, a metaheuristic approach is proposed

3 Data

Landsat 8, operating under the NASA-USGS Landsat Data Continuity Mission (LDCM), utilizes a sun-synchronous near-polar orbit to ensure consistent solar illumination and systematic global coverage via the Worldwide Reference System-2 (WRS-2). The platform integrates the Operational Land Imager (OLI) and Thermal Infrared Sensor (TIRS), employing pushbroom architecture to achieve superior signal-to-noise ratios and geometric stability compared to legacy whiskbroom systems. With OLI providing 30 m ground sampling distance (GSD) multispectral data and TIRS resampled to match this resolution, the mission delivers Level-1T products orthorectified to the WGS84 datum with a geometric accuracy of approximately 12 m CE90 [7], establishing a robust radiometric and geometric foundation for evolutionary algorithm-based fire detection pipelines.

The experimental dataset comprises 34 Landsat-8 OLI scenes acquired across western Mexico between 2020 and 2025, forming a spatially heterogeneous benchmark for wildfire detection across diverse biomes (Fig.??). Processed to Level-1T with DEM-derived radiometric and geometric corrections, the data leverages the OLI's 30 m GSD spectral channels within the WRS-2 grid [9]. To ensure physical consistency for evolutionary threshold optimization, raw 16-bit Digital Numbers (DN) were converted to Top-of-Atmosphere (TOA) reflectance (ρ) using metadata-derived rescaling coefficients. This normalization corrects for

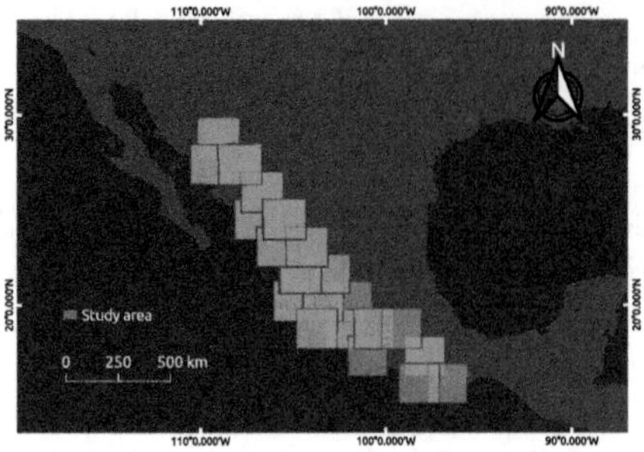

Fig. 2. Geographic distribution of the 34 Landsat 8 OLI scenes across Mexico, spanning diverse biomes and climate zones to establish a spatially heterogeneous benchmark for wildfire detection.

solar zenith angle and Earth-Sun distance variations, guaranteeing that spectral signatures remain comparable across temporal and geographic scales, which is essential for the robustness of adaptive fire detection algorithms in complex landscapes.

4 Method

The proposed pipeline comprises six sequential stages: radiometric preprocessing, spectral feature construction, threshold optimization via a GA, fire mask generation, false alarm reduction, and quantitative evaluation.

Radiometric Preprocessing

Raw Landsat 8 OLI imagery is first converted to Top-of-Atmosphere (TOA) spectral reflectance. Band-specific multiplicative and additive rescaling coefficients are retrieved from the Level-1T metadata (MTL) files and applied together with a solar zenith angle correction:

$$\rho_\lambda = \frac{M_\rho \cdot Q_{cal} + A_\rho}{\cos(\theta_s)} \tag{9}$$

where M_ρ and A_ρ denote the multiplicative and additive scaling factors, respectively, Q_{cal} is the quantized Digital Number (DN) of each pixel, and θ_s is the solar zenith angle at acquisition time. This normalization mitigates radiometric variability introduced by multi-temporal illumination differences and varying acquisition geometries, ensuring cross-scene comparability.

Spectral Feature Construction

A five-dimensional spectral feature vector **P** is constructed to exploit the characteristic radiometric signatures of active fires, as defined in Eq. 10, where ρ_i

represents the Top-of-Atmosphere (TOA) reflectance of OLI band i. This formulation serves as the input substrate for evolutionary algorithms to optimize fire detection thresholds and classify thermal anomalies within the feature space.

$$\mathbf{P} = \left[\frac{\rho_7}{\rho_5}, \frac{\rho_7}{\rho_6}, \rho_7, \rho_6, \frac{\rho_7}{\rho_4}\right] \tag{10}$$

The components of the proposed vector incorporate elements traditionally utilized in fixed-threshold detection methods. Short-Wave Infrared (SWIR) bands respond strongly to elevated surface temperatures, whereas Near-Infrared (NIR) and visible bands support discrimination of vegetated backgrounds.

Threshold Optimization via Genetic Algorithm

A GA is employed to search for the optimal set of per-component thresholds $\mathbf{T} = \{t_1, \ldots, t_n\}$ applied to $\mathbf{P}$ for active fire pixel identification. Each candidate solution $\mathbf{T}$ is evaluated through a fitness function based on Kapur's entropy criterion, which maximizes the sum of class entropies, thereby enhancing the information content of the segmentation between fire and background:

$$f(\mathbf{T}) = \max\left(H_0 + H_1\right) \tag{11}$$

The class-conditional entropy terms H_0 and H_1, corresponding to the background and fire classes, are defined as:

$$H_0 = -\sum_{i=0}^{t-1} \frac{P_i}{\omega_0} \ln\left(\frac{P_i}{\omega_0}\right), \qquad \omega_0 = \sum_{i=0}^{t-1} P_i \tag{12}$$

$$H_1 = -\sum_{i=t}^{L-1} \frac{P_i}{\omega_1} \ln\left(\frac{P_i}{\omega_1}\right), \qquad \omega_1 = \sum_{i=t}^{L-1} P_i \tag{13}$$

where P_i is the normalized histogram probability at level i, L is the total number of quantization levels, and ω_0, ω_1 are the zeroth-order cumulative moments of the background and fire classes, respectively.

The GA implementation employs tournament selection with an elitism mechanism to ensure the preservation of optimal individuals, accelerating convergence within the feature space. Genetic diversity is maintained through an arithmetic crossover operator (probability p_c) and Gaussian mutation (probability p_m), effectively balancing exploration and exploitation. The specific hyperparameter configuration, including a population size of 50, 150 generations, $p_c = 0.8$, $p_m = 0.1$, and an elite size of 2.

Fire Mask Generation

Given the GA-optimized thresholds $\mathbf{T}$, a binary detection mask is computed per pixel according to the following logical rule:

$$\text{mask} = \left(\frac{\rho_7}{\rho_5} > t_1 \vee \frac{\rho_7}{\rho_4} > t_5\right) \wedge \left(\frac{\rho_7}{\rho_6} > t_2\right) \wedge \left(\rho_7 > 0.8\,t_3 \vee \rho_6 > 0.5\,t_4\right) \tag{14}$$

This compound condition enforces simultaneous SWIR dominance over NIR and visible bands, while accommodating the variability in fire radiometric intensity through relaxed absolute reflectance thresholds on ρ_7 and ρ_6.

False Alarm Reduction

A post-classification filter is implemented to mitigate commission errors originating from two primary sources, following the criteria established by Schroeder et al. [11]: *(i)* persistent high-reflectance targets, such as urban infrastructure and industrial emitters, which are identified by their temporal stability across multiple observations; and *(ii)* anomalous SWIR signals identified through a change-detection criterion. This exclusion rule is defined as:

$$\text{Exclude if: } (\text{Previous Detection} = 1) \vee (\text{median}(\rho_{7(\textbf{history})}) > 0.25) \qquad (15)$$

To execute this filter, previous fire-free scenes are utilized as a baseline for comparison. These reference scenes are restricted to a six-month window prior to the target observation to ensure that seasonal conditions and land cover characteristics are preserved.

Evaluation

To validate model performance, 34 scenes underwent manual annotation via expert visual inspection using false-color composites (R : SWIR 2, G : NIR, B : Red) to isolate active fire signatures, with Google Earth utilized for cross-referencing persistent high-reflectance sources. While this ground-truth generation is essential, it introduces interpretation bias and risks omitting small-scale fires due to the coarse spatial resolution of Landsat 8. The complete methodology, including the distinct processing paths for target and reference imagery, is schematized in Fig. 3, where black lines denote shared processes, blue lines indicate target-specific tasks, and red lines represent operations exclusive to previous scenes.

5 Results and Discussion

Qualitative results are presented in Fig. 4. The proposed method accurately recovers the spatial extent and morphology of active fire fronts across diverse scene conditions, including the presence of water bodies, atmospheric smoke, and haze. The GA-optimized thresholds adapt effectively to inter-scene radiometric variability, sustaining reliable detection performance. However, as evidenced in the bottom-right panels of Fig. 4, the method remains susceptible to Digital Number (DN) folding artifacts introduced by high-energy saturated pixels—a known sensor-level limitation of the OLI detector that is not addressed in the current implementation [11].

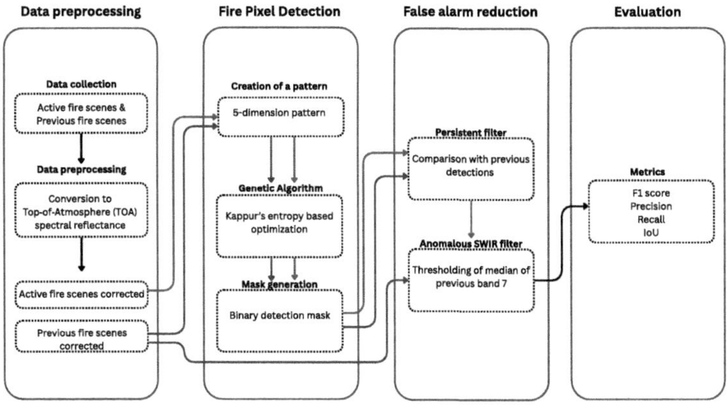

Fig. 3. Workflow diagram of the detection algorithm presented

For quantitative evaluation, 34 manually annotated binary masks were employed as ground truth. Detection performance is assessed using Precision, Recall, F1-score (Dice Coefficient), and Intersection over Union (IoU).

Precision and Recall characterize the trade-off between commission and omission errors, respectively. Formally:

$$\text{Precision} = \frac{TP}{TP + FP}, \qquad \text{Recall} = \frac{TP}{TP + FN} \tag{16}$$

where TP, FP, and FN denote true positives, false positives, and false negatives, respectively. The F1-score provides a balanced harmonic mean of both metrics and constitutes a robust performance indicator under class imbalance, a condition inherent to fire pixel distributions in large-scale remote sensing scenes (Eq. 17):

$$F1 = \frac{2 \cdot \text{Precision} \cdot \text{Recall}}{\text{Precision} + \text{Recall}} = \frac{2TP}{2TP + FP + FN} \tag{17}$$

Spatial overlap between the predicted mask and the ground truth is quantified via the *Intersection over Union* (IoU), or *Jaccard Index*, which simultaneously penalizes over-detection and under-detection:

$$\text{IoU} = \frac{TP}{TP + FP + FN} \tag{18}$$

Although F_1 and IoU are related through the identity $\text{IoU} = F_1/(2 - F_1)$, both metrics are reported to facilitate direct comparability with established segmentation benchmarks and prior remote sensing literature. Per-scene results are detailed in Table 1.

The results reported in Table 1 demonstrate consistently high Precision values across all states and acquisition dates, confirming the effectiveness of the GA-optimized thresholds in suppressing false alarms.

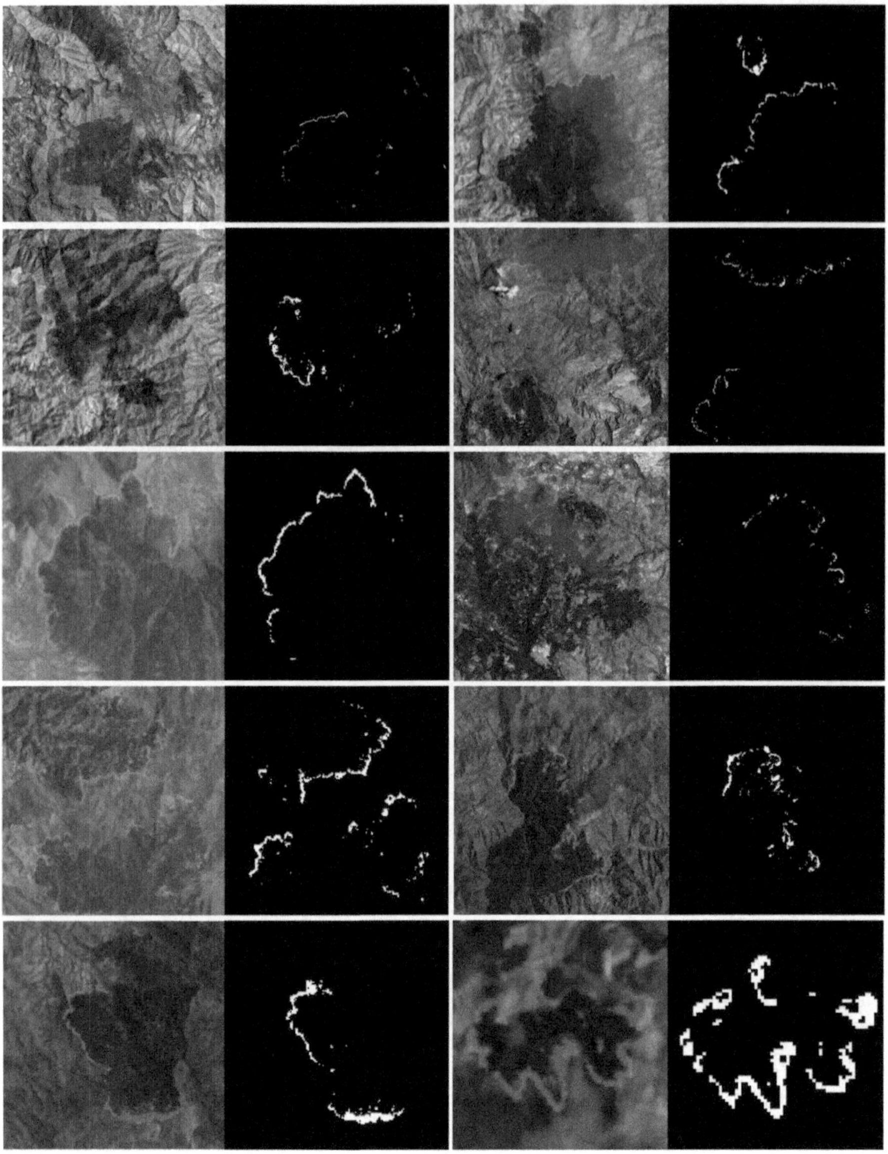

Fig. 4. Qualitative fire detection results on representative Landsat 8 OLI scenes. Each panel shows the input false-color composite (left) and the corresponding binary fire mask (right).

Recall values exhibit substantial variability, underscoring the inherent difficulty of comprehensively detecting fire pixels under heterogeneous atmospheric and land-cover conditions. This variability is primarily driven by atmospheric and surface complexity: fluctuations in aerosol loading and water vapor attenu-

Table 1. Performance metrics

State	Acquisition Date	Precision	Recall	F1	IoU
Chihuahua	2022/05/30	0.91	0.99	**0.95**	**0.90**
	2022/04/27	0.91	0.97	0.94	0.89
	2022/05/05	**0.91**	0.93	0.92	0.85
	2022/05/13	0.81	**1.00**	0.89	0.81
Colima	2022/05/02	0.76	**0.93**	**0.84**	**0.72**
	2022/05/18	**0.77**	0.85	0.81	0.68
Durango	2025/04/22	**0.85**	0.96	**0.90**	**0.82**
	2025/05/16	0.78	0.95	0.86	0.75
	2021/06/12	0.75	0.96	0.84	0.72
	2022/04/30	0.70	**0.99**	0.82	0.69
Guerrero	2020/05/14	**0.88**	0.90	**0.88**	**0.79**
	2020/05/14	0.80	0.88	0.84	0.72
	2021/05/03	0.75	**0.92**	0.83	0.71
	2022/05/22	0.77	0.88	0.82	0.69
Jalisco	2021/05/22	**0.78**	0.94	**0.85**	**0.74**
	2020/05/19	0.77	0.94	**0.85**	**0.74**
	2021/05/06	0.73	**0.94**	0.83	0.70
Michoacán	2023/05/06	**0.76**	0.93	**0.83**	**0.72**
	2020/05/30	0.75	0.94	**0.83**	0.71
	2020/05/05	0.75	0.92	**0.83**	0.71
Nayarit	2023/06/05	**0.81**	**0.94**	**0.87**	**0.77**
	2023/06/05	0.76	0.86	0.81	0.68
Oaxaca	2020/05/16	**0.75**	**0.90**	**0.82**	**0.69**
Sinaloa	2024/05/13	**0.84**	0.99	**0.91**	**0.84**
	2025/05/24	0.83	0.96	0.89	0.80
	2025/05/16	0.80	0.99	0.89	0.80
	2021/05/18	0.78	0.98	0.87	0.77
	2025/04/30	0.76	**1.00**	0.86	0.76
	2025/05/23	0.77	0.94	0.84	0.73
	2022/05/23	0.78	0.92	0.84	0.73
	2022/05/31	0.75	0.90	0.82	0.69

ate thermal radiance, while land-cover heterogeneity introduces sub-pixel mixing that dilutes the spectral signature of small or low-intensity fires against the non-fire background. Moreover, part of this dispersion may also stem from interpretation bias during the generation of reference masks, introducing a human induced source of error.

Overall, Dice coefficients above 0.80 are achieved in the majority of scenes, with peak performance observed in Chihuahua ($F1 = 0.9408$) and Sinaloa ($F1 = 0.9105$), underscoring the method's capacity to generalize across geographically and climatically diverse regions of the Mexican territory.

Intersection over Union (IoU) values ranging from 0.69 to 0.9. demonstrate high generalization across heterogeneous landscapes, maintaining a strong alignment with ground truth data. While the upper bound represents near-optimal detection under clear thermal signatures, the lower bound reflects the challenges of sub-pixel mixing, atmospheric attenuation, and potential human error in manual annotation. Overall, the consistency of these metrics confirms the effectiveness of the metaheuristic

6 Conclusions

A spectral threshold-based active fire detection framework, optimized via a Genetic Algorithm (GA), was evaluated on 34 Landsat 8 OLI scenes across diverse Mexican biomes, establishing a significant precedent for applying metaheuristic algorithms to public-domain high-resolution satellite imagery. The pipeline integrates TOA reflectance conversion, SWIR-based spectral feature construction, and entropy-driven threshold optimization to generate logical masks with robust generalization capabilities. Quantitative analysis against 34 manually annotated ground-truth masks demonstrates consistently high Precision and competitive F1 and IoU scores, effectively minimizing commission errors; however, Recall variability persists due to omission errors in scenarios involving dense smoke, complex topography, and low-intensity thermal anomalies.

Future research directions prioritize multi-temporal analysis for persistence-based filtering to distinguish transient fire signals from stable high-reflectance surfaces, thereby further reducing false positives. Addressing the scalability bottleneck of manual annotation via semi-automatic or crowd-sourced strategies is critical, as is exploring hybrid architectures that combine full-scene spectral context with local patch refinement to balance spatial integrity against Precision degradation caused by background pixel dominance.

Acknowledgments. This research is supported by Spanish Ministry of Science and Innovation under project PID2023-147409NB-C22 funded by MCIN-/AEI-/10.130-39/-501100011033, Junta de Extremadura, project GR24142 and by "ERDF A way of making Europe".

References

1. Avci, M.G., Avci, M., Battarra, M., Erdoğan, G.: The wildfire suppression problem with multiple types of resources. Eur. J. Oper. Res. **316**(2), 488–502 (2024). https://doi.org/10.1016/j.ejor.2024.03.005

2. De Almeida Pereira, G.H., Fusioka, A.M., Nassu, B.T., Minetto, R.: Active fire detection in Landsat-8 imagery: a large-scale dataset and a deep-learning study. ISPRS J. Photogramm. Remote. Sens. **178**, 171–186 (2021). https://doi.org/10.1016/j.isprsjprs.2021.06.002
3. Dong, Z., Yu, J., An, S., Zhang, J., Li, J., Xu, D.: Forest fire detection of FY-3D using genetic algorithm and brightness temperature change. Forests **13**(6), 963 (2022). https://doi.org/10.3390/f13060963
4. Justice, C., Vermote, E., Townshend, J.R.G., et al.: The moderate resolution imaging spectroradiometer (MODIS): land remote sensing for global change research. IEEE Trans. Geosci. Remote Sensing **36**(4), 1228–1249 (1998). https://doi.org/10.1109/36.701075
5. Kumar, S.S., Roy, D.P.: Global operational land imager Landsat-8 reflectance-based active fire detection algorithm. Int. J. Digit. Earth **11**(2), 154–178 (2018). https://doi.org/10.1080/17538947.2017.1391341
6. Li, Q., Cui, J., Jiang, W., Jiao, et al.: Monitoring of the Fire in Muli County on March 28, 2020, based on high temporal-spatial resolution remote sensing techniques. Nat. Hazards Res. **1**(1), 20–31 (2021). https://doi.org/10.1016/j.nhres.2021.02.001
7. Lillesand, T.M., Kiefer, R.W.: Remote Sensing and Image Interpretation. John Wiley & Sons (1979). https://books.google.com.mx/books/about/Remote_Sensing_and_Image_Interpre-tation.html?id=AFHDCAAAQBAJ
8. Murphy, S.W., De Souza Filho, C.R., et al.: HOTMAP: global hot target detection at moderate spatial resolution. Remote Sensing Environ. **177**, 78–88 (2016). https://doi.org/10.1016/j.rse.2016.02.027
9. Richards, J.A.: Remote Sensing Digital Image Analysis: An Introduction. Springer (2013). https://doi.org/10.1007/978-3-642-30062-2
10. Schroeder, W., Oliva, P., Giglio, L., Csiszar, I.A.: The new VIIRS 375 m active fire detection data product: algorithm description and initial assessment. Remote Sens. Environ. **143**, 85–96 (2014). https://doi.org/10.1016/j.rse.2013.12.008
11. Schroeder, W., Oliva, P., Giglio, L., Quayle, B., Lorenz, E., Morelli, F.: Active fire detection using Landsat-8/OLI data. Remote Sens. Environ. **185**, 210–220 (2016). https://doi.org/10.1016/j.rse.2015.08.032
12. Singh, H., Ang, L.M., Srivastava, S.K.: Active wildfire detection via satellite imagery and machine learning: an empirical investigation of Australian wildfires. Nat. Hazards **121**(8), 9777–9800 (2025). https://doi.org/10.1007/s11069-025-07163-w
13. Tran, A., et al.: Land8Fire: a complete study on wildfire segmentation through comprehensive review, human-annotated multispectral dataset, and extensive benchmarking. Remote Sensing **17**(16), 2776 (2025). https://doi.org/10.3390/rs17162776
14. United Nations: Transforming our world: The 2030 agenda for sustainable development. Tech. rep., United Nations (2015). https://sdgs.un.org/publications/transforming-our-world-2030-agenda-sustainable-development-17981
15. Wooster, M.J., Roberts, G.J., Giglio, et al.: Satellite remote sensing of active fires: history and current status, applications and future requirements. Remote Sensing Environ. **267**, 112694 (2021). https://doi.org/10.1016/j.rse.2021.112694
16. Xu, W., Wooster, M.J., He, J., Zhang, T.: First study of Sentinel-3 SLSTR active fire detection and FRP retrieval: night-time algorithm enhancements and global intercomparison to MODIS and VIIRS AF products. Remote Sens. Environ. **248**, 111947 (2020). https://doi.org/10.1016/j.rse.2020.111947

Medical Applications

Deep Learning for Brain Tissue Segmentation in Pediatric Patients

Dario Chan Colli[1(✉)], Fernando Arámbula Cosío[1], Omar Piña Ramírez[2], and Jorge Luis PérezPérez González[1]

[1] Unidad Académica del IIMAS en el Estado de Yucatán, UNAM, Mérida, Mexico
dariochancolli@gmail.com, {fernando.arambula,jorge.perez}@iimas.unam.mx
[2] Instituto Nacional de Perinatología Isidro Espinosa de los Reyes, Ciudad de México, Mexico
omar.pina@inper.gob.mx

Abstract. Precision in brain tissue segmentation in pediatric patients represents a critical challenge in medical analysis due to the low contrast in the images obtained from a developing brain. In this article, we propose and evaluate the use of supervised deep learning architectures for the automated segmentation of Gray Matter (GM), White Matter (WM), and Cerebrospinal Fluid (CSF). This study utilized the 'Bob's repository' dataset, which comprises MRI scans of 51 patients between 1 and 9 months of age. The dataset was split into 41 cases for training and 10 for testing, from which axial slices of multimodal sequences (T1w and T2w) were extracted and processed. The study compares the performance of two U-Net-based models: one integrated with residual blocks (UNet + ResNet) and a Transformer-based approach (ViT + UNet). The results demonstrate that multimodal integration (T1w + T2w) increases learning stability and precision. Specifically, the ViT + UNet architecture demonstrated superior performance retention in the testing phase, achieving an average Dice similarity coefficient (DSC) of 0.92 for GM and CSF, and 0.88 for WM.

Keywords: MRI · Brain Segmentation · Deep Learning

1 Introduction

Medical image segmentation is a critical component in the development of automated diagnostic tools. In the context of neuroimaging, brain tissue segmentation typically focuses on identifying Gray Matter (GM), White Matter (WM), and Cerebrospinal Fluid (CSF). While automated solutions have matured significantly for adult populations, pediatric neuroimaging presents unique clinical and technical challenges. During the first months of life, the brain undergoes rapid development, resulting in low tissue contrast and high morphological variability that render algorithms trained on adult data ineffective [11].

Furthermore, the acquisition of pediatric MRI scans is often complex, frequently requiring induced sleep or anesthesia, which can impact image quality.

H. Ponce et al. (Eds.): COMIA 2026, CCIS 3095, pp. 701–712, 2027.
https://doi.org/10.1007/978-3-032-37438-7_55

The high cost of manual segmentation by experts and the scarcity of large-scale pediatric datasets necessitate the development of robust, automated deep learning models specifically tailored for the infant brain. This study addresses these needs by evaluating supervised architectures using multimodal MRI sequences (T1w and T2w) to improve segmentation precision in patients aged 1 to 9 months.

1.1 Contribution

This study aims to overcome the limitations of traditional and adult-based models by implementing a supervised segmentation approach for GM, WM, and CSF in pediatric patients. Utilizing the "Bob's Repository" dataset, we process axial slices and implement a multimodal strategy combining T1w and T2w sequences. Our contribution focuses on the validation and comparison of CNN-based and Transformer-based architectures (ViT + UNet) to determine the most effective configuration for high-variability, low-contrast infant brain tissues.

2 State of the Art

The landscape of medical image segmentation is dominated by supervised and semi-supervised techniques. Supervised methods frequently leverage Convolutional Neural Networks (CNNs), Transformers, and more recently, Mamba architectures. Conversely, prominent semi-supervised frameworks such as UA-MT, SASSNet, MC-Net, and Action++ have been developed to mitigate the need for large labeled datasets. These models are widely applied across various domains, including the segmentation of polyps, skin lesions, and various tumors [12].

In the specific field of neuroimaging, CNNs—and particularly the U-Net architecture—remain the standard for segmenting brain structures in MRI scans. Software such as FreeSurfer has successfully utilized AI principles to automate the processing of adult neuroimages, tracking longitudinal changes in brain volume and cortical thickness. However, these tools often face challenges regarding standardization and generalization when applied to diverse populations [2].

Recent progress in pediatric neuroimaging over the last five years has identified five primary areas of application for Deep Learning:

1. Recognition of neurodevelopmental disorders using CNNs.
2. Identification of brain structures and tissues, where U-Net and transfer learning dominate to reduce manual labeling requirements.
3. Prediction of brain age.
4. Prediction of neurodevelopmental outcomes by combining imaging with cognitive assessments.
5. Optimization of MRI acquisition, reconstruction, and synthetic image generation using GANs.

Despite these advancements, pediatric segmentation remains hindered by overfitting, variability in preprocessing, and the limited availability of public datasets compared to adult repositories. Notable efforts include the use of the ABIDE and ADHD-200 datasets, yet risks related to confounding bias and small sample sizes persist [6].

3 Methodology

3.1 Dataset

For the dataset, 'Bob's repository' was selected, which is a collection of Magnetic Resonance Imaging (MRI) scans [5] manually segmented by an expert. Publicly available on an AWS S3 bucket, this repository comprises 71 MRI scans from 51 participants aged 1 to 9 months. The dataset includes T1w and T2w images. Figure 2 shows a segmentation example found in the repository for subject sub-206192 at 3 months of age for T1w and T2w sequences.

To ensure the robustness of the evaluation and prevent data leakage, the data was partitioned using a patient-level split strategy. This means that all slices belonging to a specific subject were assigned exclusively to either the training or the testing set, ensuring that the model is tested on previously unseen anatomical structures. For training, 41 subjects with 51 volumes (9,282 slices) were used, while 10 subjects with 20 volumes (3,640 slices) were reserved for testing. This distribution is detailed in Figs. 1a and 1b, illustrating the age range coverage in both sets.

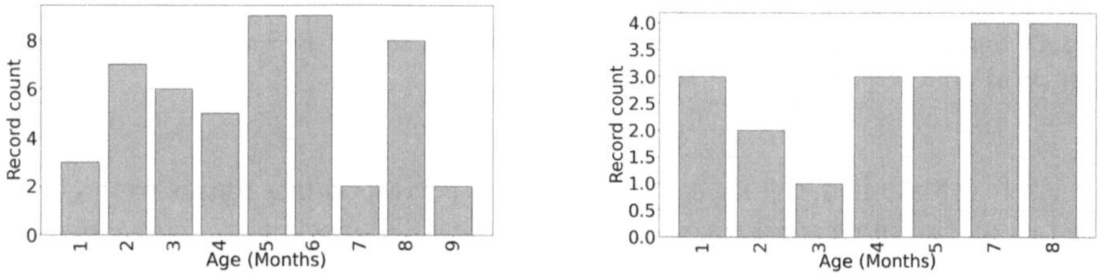

(a) Train (b) Test

Fig. 1. Frequency distribution by age in the data set. The Train set (a) contains 51 volumes of 41 pediatric patients and the Test set (b) contains 20 volumes of 10 pediatric patients.

3.2 Image Preprocessing

The magnetic resonance images in the repository are volumes; therefore, 2D axial slices were extracted for T1w and T2w sequences along with their respective masks. The segmentation targets include Gray Matter (GM), White Matter (WM), and Cerebrospinal Fluid (CSF). As shown in Fig. 3, each volume (consisting of 182 slices) was processed to match every T1w and T2w slice with its corresponding multi-class mask.

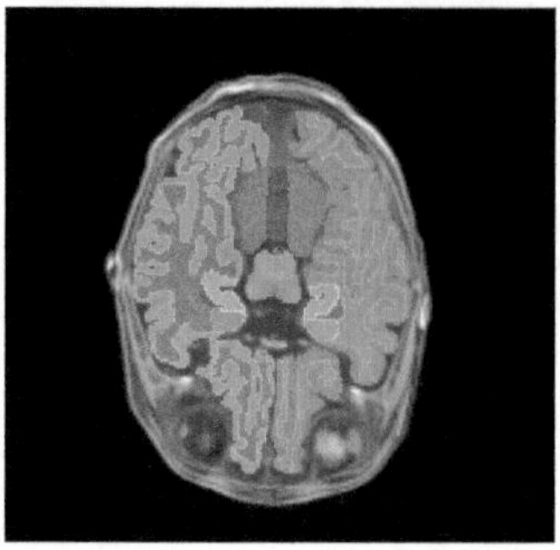

Fig. 2. Segmentation (axial slice) of subject sub-206192 at 3 months of age [4], segmentation performed on a T1w image can be observed.

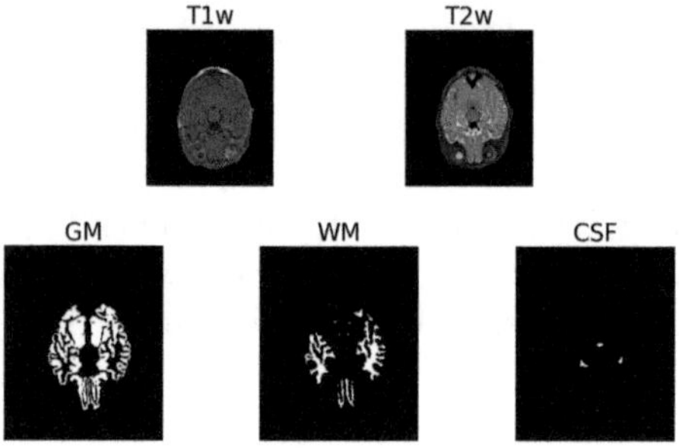

Fig. 3. Segmentation (axial section) of subject sub-206192 at 3 months of age with GM, WM and CSF masks.

Each volume had 182 slices, and the three corresponding masks were generated using the same axial slices. Therefore, the code parsed the image names to match each T1w and T2w slice with its respective mask. This resulted in a total of 9282 samples for Training and 3640 for Test.

Despite the promising results, this study acknowledges certain limitations regarding generalizability. The 'Bob's repository' dataset, while high-quality and clinically relevant, is relatively limited in size (51 participants). This restricted sample size, combined with the fact that data originates from a single repository, may limit the model's ability to generalize to MRI scans acquired with different scanners or protocols from other clinical centers. Future work should involve external validation using multi-center datasets to further confirm the stability and performance of the proposed models across diverse pediatric populations.

3.3 Deep Learning Models

For the supervised algorithms, two UNet-based architectures were selected due to their proven segmentation efficiency: a residual-enhanced UNet (UNet + ResNet) and a hybrid model utilizing Transformers at its bottleneck.

U-Net architectures incorporating residual connections effectively mitigate the vanishing gradient problem, outperforming standard models. For example, Dense-ResUNet extracts precise pixel-level features from 2D slices, which is indispensable for delineating complex brain tumors [3]. Similarly, U-ResNet presents a parallel hybrid design combining a classical UBlock with a ResBlock, merging their outputs to achieve superior segmentation performance compared to state-of-the-art algorithms [7].

Regarding U-Net and Transformer integration, models like Swin UNETR replace the traditional encoder with a hierarchical Swin Transformer. This approach uses shifted windows for self-attention, capturing long-range dependencies and global features, which are then linked to a CNN decoder via skip connections. This achieves an optimal balance between global context and local precision, demonstrating state-of-the-art performance in challenges like BraTS 2021 [1]. Inspired by this 3D model, we adapted the approach for 2D slice segmentation.

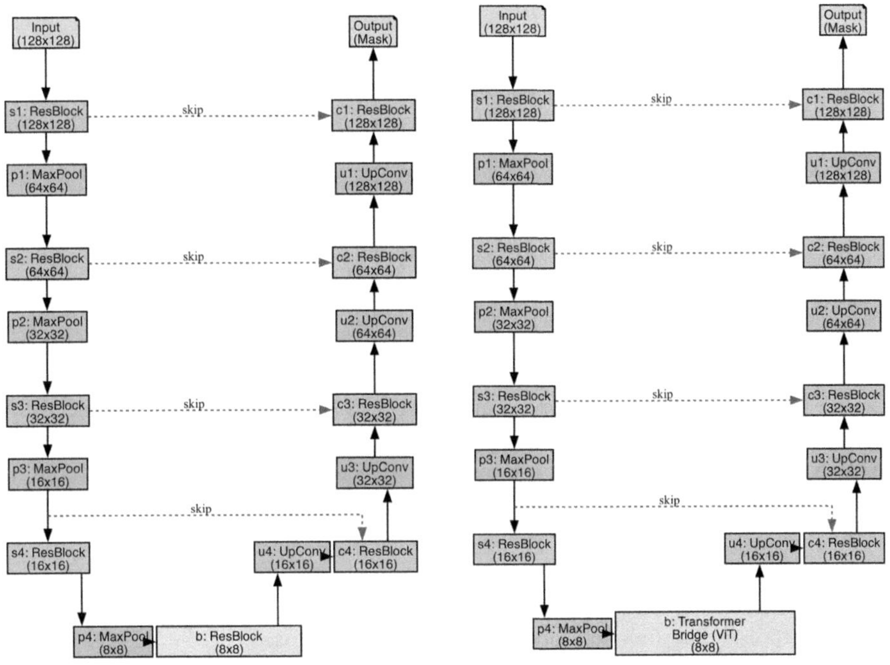

Fig. 4. Deep learning architectures: (a) architecture with residual blocks and (b) architecture with attention mechanisms.

Based on the above, the architectures described below were selected. Starting with the UNet + ResNet architecture used in this work, it can be noted that it uses Residual Blocks instead of simple convolutions. In the compression phase (Encoder), layers s1 to s4 extract complex features while the Max Pooling layers (p1 to p4) reduce the spatial dimension to capture context. This compressed information passes through a bridge block (b) and then continues to the expansion phase (Decoder), where transposed convolution layers (u4 to u1) recover the original resolution. The decoder blocks (c4 to c1) receive and merge the high-resolution information from the encoder via skip connections, allowing the reconstruction of fine tissue edges (GM, WM, CSF) before the output layer generates the final masks. This can be seen in Fig. 4a. The TransUNET architecture retains the ResNet-based encoder-decoder external structure but introduces a fundamental change in the central block (b), where, instead of using a standard convolution, it implements a Transformer Bridge. This module takes the compressed feature maps (8x8 pixels), flattens them into a sequence of patches, and applies Self-Attention using a Transformer Encoder. Thus, the model analyzes long-distance global relationships between different parts of the brain anatomy that a local convolution would ignore. This can be seen in the Fig. 4b.

3.4 Training and Validation

Two main combinations were used: 1) a UNet + ResNet architecture and 2) a ViT + UNet architecture. Although both have the UNet architecture, it is used differently, because the first is a UNet network with residual elements, while the second uses a transformer layer at the bottleneck.

In both combinations, training was performed using the following data:

1. T1w + Masks
2. T2w + Masks
3. T1w + T2w + Masks

In Table 1, the training configurations are detailed. While both architectures shared similar hyperparameters, the ViT + UNet model required a reduced batch size of 32. This adjustment was a necessary response to strict hardware constraints; training was conducted on the LUCAR-IIMAS server using an Nvidia Tesla K20m GPU. Due to the high memory overhead of the Transformer components and the GPU's limited 5 GB of VRAM, the batch size had to be optimized to prevent "Out of Memory" (OOM) errors and ensure training stability

Table 1. Table of configurations used during training, showing the model along with the data, epochs, loss function, optimizer and batch size used.

Model	Inputs	Epochs	Loss function	Optimizer	Batch Size
Unet + ResNet	T1w	100	Binary Cross Entropy (nn.BCELoss)	Adam (lr=1e-4)	64
Unet + ResNet	T2w	100	Binary Cross Entropy (nn.BCELoss)	Adam (lr=1e-4)	64
Unet + ResNet	T1w + T2w	100	Binary Cross Entropy (nn.BCELoss)	Adam (lr=1e-4)	64
ViT + Unet	T1w	100	Combined Loss (BCE + Dice)	Adam (lr=1e-4)	32
ViT + Unet	T2w	100	Combined Loss (BCE + Dice)	Adam (lr=1e-4)	32
ViT + Unet	T1w + T2w	100	Combined Loss (BCE + Dice)	Adam (lr=1e-4)	32

3.5 Evaluation Metrics

For the results analysis, the DSC, Loss, and Accuracy metrics were considered with the Test set, which includes 10 subjects with a total of 20 volumes. Applying preprocessing to generate 180 axial slices for each volume yielded 3640 samples for Test. The primary metric to begin the discussion is DSC, as it allows for a better calculation of the relationship between predicted and actual segmentation.

Accuracy is a pixel accuracy metric that calculates the total proportion of correct predictions (both positive and negative) relative to the total number of pixels [9].

The Dice Similarity Coefficient (DSC), also known as the F1-score, is the most widespread metric in scientific literature for evaluating medical image segmentation. Defined as the harmonic mean of sensitivity and precision, it quantifies the overlap between the predicted segmentation and the ground truth, penalizing false positives [9].

$$DSC = \frac{2TP}{2TP + FP + FN} \tag{1}$$

The Intersection over Union (IoU), or Jaccard index, and the Dice Similarity Coefficient (DSC) are prevalent metrics in Medical Image Segmentation (MIS). While the DSC represents the harmonic mean of precision and sensitivity, the primary distinction is that the IoU applies a harsher penalty to both under- and over-segmentation errors [9].

$$IoU = \frac{TP}{TP + FP + FN} \tag{2}$$

4 Results

Figure 5 shows the training graphs for T1w + T2w of the ViT + UNet model. The loss history over the 100 epochs is visible. On the right, the validation graphs are shown, where the DSC metrics for GM, WM, and CSF are analyzed. It can be seen that the models reach a point of stability in the early epochs, but early stopping was omitted to observe the model's behavior over longer epochs. Greater instability is observed in the first 80 epochs for the DSC metric.

Table 2 shows the training and test times for the evaluated modalities, highlighting that the ViT + UNet model requires more computation time than its counterpart, UNet + ResNet. In training (100 epochs), UNet + ResNet recorded a maximum of 9.18 h with T2w data, while ViT + UNet ranged between 10.57 and 10.68 h, with the T1w + T2w multimodality being the most time-consuming. This represents an average of 4.92 min per epoch for the lighter configuration compared to 6.4 min for the more robust one.

In the test stage on 3640 images, the times were consistent with the complexity of the architectures: the UNet + ResNet (T2w) model was the most efficient with 88.83 s (24.4 ms/image), while ViT + UNet (T1w + T2w) recorded the longest time with 120.42 s (33.1 ms/image).

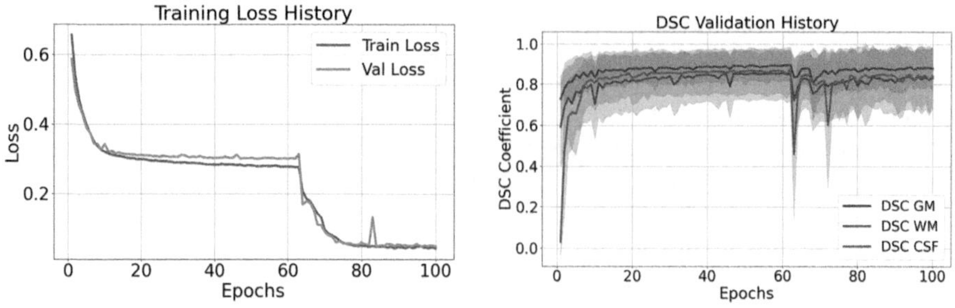

(a) Training Loss History (b) DSC Validation History

Fig. 5. Training graphs for ViT + UNet with T1w + T2w, showing the loss history on the left and the DSC history on the right in the validation for GM, WM and CSF.

Table 2. Training times (in hours) and testing times (in seconds) for the UNet + ResNet and ViT + UNet architectures

Model	Inputs	Training set size	Training Time (h)	Test set size	Test Time (s)
Unet + ResNet	T1w	9282	**8.21**	3640	121.38
Unet + ResNet	T2w	9282	9.18	3640	**88.83**
Unet + ResNet	T1w + T2w	9282	8.25	3640	95.75
ViT + Unet	T1w	9282	**10.57**	3640	120.34
ViT + Unet	T2w	9282	10.64	3640	**117.04**
ViT + Unet	T1w + T2w	9282	10.68	3640	120.42

Tables 3 and 4 present the training metrics, demonstrating that multimodality (T1w + T2w) fostered a more stable learning process in both architectures. The ViT + UNet model outperformed UNet + ResNet in the DSC metric, achieving values of 0.9344 ± 0.0896 for GM, 0.9098 ± 0.1421 for WM, and 0.9500 ± 0.1118 for CSF. Although UNet + ResNet recorded the lowest loss (0.0066), the Transformer-based approach achieved superior tissue delineations.

During the testing phase (Tables 6 and 5), the generalization capability of the multimodal approach was confirmed, particularly in the ViT + UNet architecture. This model retained superior performance (0.9231 for GM and 0.9213 for CSF) compared to the natural performance drop observed in UNet + ResNet. Tissue-specific analysis identifies cerebrospinal fluid (CSF) as the easiest structure to delineate, whereas white matter (WM) posed the greatest challenge, yielding the lowest metrics and highest standard deviations. These results justify the use of attention mechanisms in ViT + UNet to more effectively capture spatial context.

Figure 6 shows examples of the best model predictions (ViT + UNet) for two different subjects: sub-439083 (1 month old, axial slice 98) and sub-272097 (8 months old, axial slice 75). The actual masks for GM, WM, and CSF are shown, along with an overlay. The predictions for each model for the GM, WM,

Table 3. Training Metrics: Mean for Loss and Mean for Accuracy with Standard Deviation

		Loss		Accuracy		
Model	Inputs	Train	Val	GM (±std)	WM (±std)	CSF (±std)
UNet+ResNet	T1w	0.0103	0.0270	0.9398 ± 0.0712	0.8826 ± 0.1519	0.9486 ± 0.1089
UNet+ResNet	T2w	0.0090	0.0245	0.9000 ± 0.1408	0.8973 ± 0.1572	0.9447 ± 0.1217
UNet+ResNet	T1w+T2w	**0.0066**	**0.0203**	0.9300 ± 0.0896	0.9135 ± 0.1335	0.9495 ± 0.1176
ViT+UNet	T1w	0.0440	0.0583	0.9115 ± 0.0959	0.9017 ± 0.1145	0.9577 ± 0.0953
ViT+UNet	T2w	0.0392	0.0584	0.9162 ± 0.0827	0.8982 ± 0.1345	0.9500 ± 0.1231
ViT+UNet	T1w+T2w	**0.0334**	**0.0471**	0.9281 ± 0.0758	0.9423 ± 0.0902	0.9544 ± 0.1040

Table 4. Training Metrics: Mean for DSC and IoU with standard deviation

		DSC / IoU		
Model	Inputs	GM (DSC / IoU ± std)	WM (DSC / IoU ± std)	CSF (DSC / IoU ± std)
UNet+ResNet	T1w	0.8976 / 0.8142 ± 0.1357	0.8846 / 0.7931 ± 0.1607	0.9300 / 0.8692 ± 0.1463
UNet+ResNet	T2w	0.9035 / 0.8240 ± 0.1463	0.8773 / 0.7814 ± 0.1844	0.9152 / 0.8437 ± 0.1728
UNet+ResNet	T1w+T2w	0.9298 / 0.8688 ± 0.1019	0.9030 / 0.8232 ± 0.1539	0.9326 / 0.8737 ± 0.1387
ViT+UNet	T1w	0.9130 / 0.8399 ± 0.1145	0.8876 / 0.7979 ± 0.1557	0.9477 / 0.9006 ± 0.1113
ViT+UNet	T2w	0.9159 / 0.8448 ± 0.1097	0.8946 / 0.8093 ± 0.1524	0.9369 / 0.8813 ± 0.1358
ViT+UNet	T1w+T2w	0.9344 / 0.8769 ± 0.0896	0.9098 / 0.8345 ± 0.1421	0.9500 / 0.9048 ± 0.1118

and CSF masks are also shown, labeled with the DSC metric and the overlay of these predictions. In the overlays, red represents GM, green represents WM, and blue represents CSF.

The results demonstrate that multimodal integration of T1w and T2w significantly increases segmentation accuracy, mitigating the main physiological limitation of the pediatric brain: low tissue contrast [11]. In this context, the ViT + UNet architecture demonstrated superior generalization capabilities in the Test phase compared to UNet + ResNet. Although the use of Transformers implies a higher parameter density and longer training times [13], their self-attention mechanisms allow for more effective capture of the global spatial context than purely convolutional layers. Considering that the resulting inference time is only 33.1 milliseconds per image, the additional computational cost is fully justified.

Table 5. Test Metrics: Mean for DSC and IoU with standard deviation

		DSC / IoU Test		
Model	Inputs	GM (DSC / IoU ± std)	WM (DSC / IoU ± std)	CSF (DSC / IoU ± std)
UNet+ResNet	T1w	0.8543 / 0.7457 ± 0.1937	0.8122 / 0.6838 ± 0.2267	0.9017 / 0.8210 ± 0.1893
UNet+ResNet	T2w	0.8658 / 0.7634 ± 0.1695	0.8139 / 0.6862 ± 0.2338	0.8747 / 0.7773 ± 0.2312
UNet+ResNet	T1w+T2w	0.8984 / 0.8155 ± 0.1450	0.8576 / 0.7507 ± 0.2036	0.9015 / 0.8207 ± 0.1901
ViT+UNet	T1w	0.8975 / 0.8141 ± 0.1251	0.8594 / 0.7535 ± 0.1841	0.9226 / 0.8563 ± 0.1580
ViT+UNet	T2w	0.8986 / 0.8159 ± 0.1241	0.8582 / 0.7516 ± 0.1929	0.9074 / 0.8305 ± 0.1779
ViT+UNet	T1w+T2w	0.9231 / 0.8572 ± 0.1024	0.8882 / 0.7989 ± 0.1695	0.9213 / 0.8541 ± 0.1607

Table 6. Test Metrics: Mean for accuracy with standard deviation

		Accuracy Test		
Model	Inputs	GM (±std)	WM (±std)	CSF (±std)
UNet+ResNet	T1w	0.8415 ± 0.1967	0.8547 ± 0.1883	0.9361 ± 0.1378
UNet+ResNet	T2w	0.8702 ± 0.1607	0.8336 ± 0.2181	0.9287 ± 0.1533
UNet+ResNet	T1w+T2w	**0.9024 ± 0.1366**	**0.8692 ± 0.1890**	**0.9408 ± 0.1332**
ViT+UNet	T1w	0.9005 ± 0.1054	0.8663 ± 0.1610	0.9424 ± 0.1200
ViT+UNet	T2w	0.9046 ± 0.0995	0.8649 ± 0.1733	0.9409 ± 0.1275
ViT+UNet	T1w+T2w	**0.9218 ± 0.0874**	**0.9189 ± 0.1253**	**0.9443 ± 0.1218**

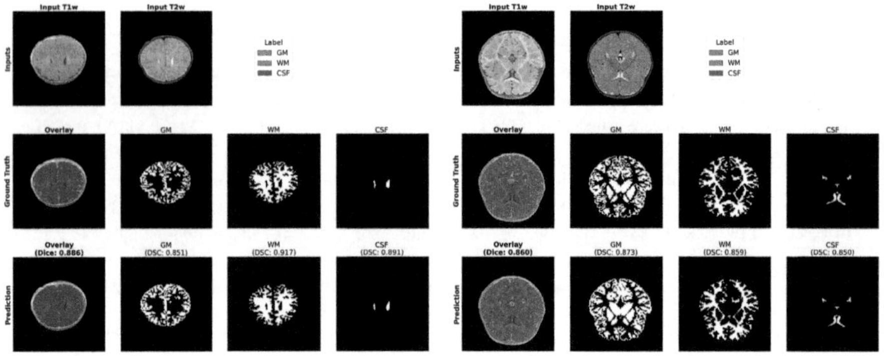

(a) Subject sub-439083, 1 month old, slice 98.

(b) Subject sub-272097, 8 months old, slice 75.

Fig. 6. Model predictions using ViT + UNet with T1w and T2w inputs for two different subjects. The actual masks and model predictions for GM, WM, and CSF are shown. In overlays, red represents GM, green WM, and blue CSF. (Color figure online)

On the other hand, the tissue type analysis reflects the development of patients from 1 to 9 months of age. WM segmentation presented the greatest challenge of the experiment, yielding a DICE of 0.88 compared to metrics exceeding 0.92 in GM and CSF. Unlike neuroimaging in adults, where standardized tools such as FreeSurfer operate with high precision [2], tissue contrast in developing patients posed a greater challenge. These findings validate the need to design and train specific architectures for pediatric patients, demonstrating that the synergy between multimodal data and deep learning models can overcome anatomical variability without relying on pre-trained models from the adult population.

To contextualize our results, we compared the proposed ViT + UNet architecture against recent state-of-the-art nnU-Net variants. For instance, an advanced nnU-Net framework for pediatric gliomas (BraTS 2025) achieved Dice Similarity Coefficients (DSC) up to 0.967 for edema and 0.928 for the whole tumor [8]. Although focused on pathology, this establishes a strong performance benchmark for specialized CNNs handling the high anatomical variability typical of developing pediatric brains.

Similarly, in healthy fetal brain segmentation, an adapted nnU-Net achieved an average testing DSC of 0.774, outperforming traditional baselines [10]. In contrast, our multimodal ViT + UNet achieved highly competitive testing DSCs of 0.9231 (GM), 0.8882 (WM), and 0.9213 (CSF). This demonstrates that combining multimodal sequences with Vision Transformers effectively resolves low-contrast boundaries better than purely local CNNs. As future work, we plan to train the nnU-Net framework directly on our pediatric dataset to establish a definitive, domain-specific baseline comparison against our architecture.

5 Conclusion

In this study, two Deep Learning approaches based on the U-Net architecture were presented and compared for the automated segmentation of pediatric brain tissue. Using the 'Bob's repository' dataset with patients aged 1 to 9 months, the conducted experiments demonstrated that the multimodal integration of T1w and T2w images significantly improves accuracy compared to unimodal approaches. The ViT + UNet architecture achieved superior performance, yielding an average Dice coefficient of 0.92 for GM and CSF, and 0.88 for the complex delineation of WM, demonstrating that attention mechanisms better capture spatial context. Simultaneously, the UNet + ResNet model proved to be a computationally efficient and robust alternative, reaching values close to 0.90 under the same configuration. Despite these high performance metrics, the development of these systems remains limited by the scarcity of pediatric neuroimaging data and the low tissue contrast of immature brains. For future work, it is necessary to validate these models on multicenter datasets and explore semi-supervised techniques or emerging architectures to reduce the reliance on manually labeled data. In conclusion, data multimodality proved to be highly successful for segmentation in this study, and these deep learning models have the potential to become a consolidated alternative for early clinical diagnosis.

Acknowledgments. This work was supported by UNAM-PAPIIT program IN102626.

Disclosure of Interests. The authors have no competing interests to declare that are relevant to the content of this article.

References

1. Cao, H., et al.: Swin-unet: unet-like pure transformer for medical image segmentation. In: European Conference on Computer Vision, pp. 205–218. Springer (2022)
2. Chau, M., Vu, H., Debnath, T., Rahman, M.: A scoping review of automatic and semi-automatic MRI segmentation in human brain imaging. Radiography **31**(2), 102878 (2025). https://doi.org/10.1016/j.radi.2025.01.013

3. Chen, S., Zhao, S., Lan, Q.: Residual block based nested u-type architecture for multi-modal brain tumor image segmentation. Front. Neurosci. **16**, 832824 (2022)
4. Elison, J., et al.: BOBsRepository (2024). https://doi.org/10.17605/osf.io/wdr78
5. Feczko, E., et al.: Baby open brains: an open-source repository of infant brain segmentations. bioRxiv (2024)
6. Hu, M., Nardi, C., Zhang, H., Ang, K.K.: Applications of deep learning to neurodevelopment in pediatric imaging: achievements and challenges. Appl. Sci. **13**(4), 2302 (2023). https://doi.org/10.3390/app13042302
7. Li, W., Gao, Z., Song, Y.: U-resnet, a novel network fusion method for image classification and segmentation. Sensors **25**(17), 5600 (2025)
8. Li, X., Xu, Z.Q.J., Ren, Y., Qiu, T., Wang, X.: An advanced NNU-net framework for brats-2025 ped. In: International Conference on Medical Image Computing and Computer-Assisted Intervention, pp. 422–433. Springer (2025)
9. Müller, D., Soto-Rey, I., Kramer, F.: Towards a guideline for evaluation metrics in medical image segmentation. BMC. Res. Notes **15**(1), 210 (2022)
10. Peng, Y., Xu, Y., Wang, M., Zhang, H., Xie, J.: The NNU-net based method for automatic segmenting fetal brain tissues. Health Inf. Sci. Syst. **11**(1), 17 (2023)
11. Wu, L., et al.: A survey of MRI-based brain tissue segmentation using deep learning. Complex amp; Intell. Syst. **11**(1) (2024). https://doi.org/10.1007/s40747-024-01639-1
12. Xia, Q., et al.: A comprehensive review of deep learning for medical image segmentation. Neurocomputing **613**, 128740 (2025). https://doi.org/10.1016/j.neucom.2024.128740
13. Yao, W., Bai, J., Liao, W., Chen, Y., Liu, M., Xie, Y.: From CNN to transformer: a review of medical image segmentation models. J. Imaging Inform. Med. **37**(4), 1529–1547 (2024). https://doi.org/10.1007/s10278-024-00981-7

BurnoutCare: An Intelligent System for Early Detection, Monitoring, and Prevention of Burnout Syndrome Using Generative AI

Rubén Beltrán Marañón[1](✉), Edna Araceli Romero Flores[1], Ramiro Meza Palacios[1], Behdin Nowrouzi-Kia[2], Alma Gabriela Alcalde Pérez[1], and Gabriela Cabrera Zepeda[1]

[1] Tecnológico Nacional de México / Instituto Tecnológico de Orizaba, Av. Oriente 9 No. 852, Col. Emiliano Zapata, Orizaba, 94320, Veracruz, Mexico
M24010194@orizaba.tecnm.mx

[2] ReSTORE Lab, Department of Occupational Science and Occupational Therapy, University of Toronto, 500 University Ave, Toronto M5G 1V7, ON, Canada

Abstract. Occupational burnout affects over 75% of workers in Mexico and represents a growing challenge for organizations subject to the NOM-035-STPS-2018 regulatory framework. This study developed and evaluated BurnoutCare, an intelligent clinical decision support system (CDSS) for early detection, monitoring, and personalized intervention of burnout syndrome using the Copenhagen Burnout Inventory (CBI) and Generative AI. A five-layer architecture was designed: presentation (React.js 18, Tailwind CSS), application services (Node.js 18, Express.js), artificial intelligence (Anthropic Claude Sonnet API with rule-based fallback), security (JWT, bcrypt, Helmet), and persistence (PostgreSQL 15 with JSONB). A four-step workflow digitalizes the 19-item CBI across three dimensions—Personal Burnout, Work-related Burnout, and Client-related Burnout—and captures demographic and optional qualitative context. All eleven functional modules were deployed and validated in a cloud production environment. Risk classification responded in under 200 ms, the AI engine generated structured eight-component JSON recommendations in 100% of medium- and high-risk test cases, and the rule-based fallback activated correctly in all simulated API failure scenarios. Role-based access control was verified for four user profiles. A clinical pilot study with a service-sector organization is planned to evaluate diagnostic accuracy, internal consistency (Cronbach's α), and usability (System Usability Scale). BurnoutCare demonstrated full technical feasibility for clinical pilot deployment, representing the first operational CDSS for burnout aligned with NOM-035-STPS-2018 in the Mexican organizational context.

Keywords: Burnout syndrome · Clinical decision support system · Copenhagen Burnout Inventory · Generative artificial intelligence · NOM-035-STPS-2018 · Occupational health informatics

H. Ponce et al. (Eds.): COMIA 2026, CCIS 3095, pp. 713–726, 2027.
https://doi.org/10.1007/978-3-032-37438-7_56

1 Introduction

Burnout syndrome, recognized by the World Health Organization (WHO) as an occupational phenomenon resulting from chronic workplace stress that has not been successfully managed [1], manifests across three dimensions: emotional exhaustion, depersonalization, and reduced personal accomplishment. Its impact is twofold: it deteriorates individual well-being and undermines organizational efficiency by increasing absenteeism, raising turnover rates, and reducing service quality. A 2024 meta-analysis of 215,787 public health workers reported a pooled burnout rate of 39%, rising to 42% during the COVID-19 pandemic [2].

The situation in Mexico is particularly critical. The Mexican Social Security Institute (IMSS) estimates that at least 75% of Mexican workers experience some degree of occupational stress, placing the country among the highest in global prevalence [3]. The service and customer-facing sector concentrates this problem: workers face extended shifts, high emotional demand, continuous performance monitoring, and limited autonomy. Research confirms that only 28.3% of patients treated by burned-out professionals show significant clinical improvement, compared to 36.8% treated by non-burned-out professionals [9], demonstrating that the problem extends beyond the individual and directly affects care quality.

The NOM-035-STPS-2018 establishes the obligation for Mexican organizations to identify, analyze, and prevent psychosocial risk factors [4]. However, its implementation faces a concrete obstacle: most employers lack systematic tools that go beyond sporadic evaluation. Available instruments are applied manually, without connection to a continuous monitoring system, delaying timely clinical intervention.

Advances in artificial intelligence (AI), and specifically Generative AI based on Large Language Models (LLMs), offer a concrete opportunity to overcome these limitations. Machine learning models achieve a pooled AUC of 0.72 for burnout prediction [5], while LLMs are capable of producing personalized clinical recommendations in structured, actionable JSON format for health professionals [6, 7]. Unlike traditional classifiers that only produce a risk label, generative AI can produce dimension-specific analyses, follow-up plans, and personalized messages for the evaluated employee.

The central hypothesis of this research is that it is possible to integrate into a single operational platform the complete burnout cycle of evaluation–classification–alert–recommendation–follow-up, aligned with NOM-035-STPS-2018, by combining the Copenhagen Burnout Inventory (CBI)—validated for Mexican workers—with a generative AI engine. This paper presents the design, implementation, and technical validation of BurnoutCare: the first operational CDSS for burnout in the Mexican organizational context with these integrated features. Section 2 reviews the state of the art; Sect. 3 describes the methodology; Sect. 4 presents results; Sect. 5 discusses contributions and limitations; and Sect. 6 provides conclusions and future work.

2 Related Work

2.1 Clinical Decision Support Systems for Burnout

CDSSs oriented toward occupational mental health have evolved from basic report generation to platforms capable of integrating multimodal data and generating personalized recommendations. Berner [8] defines CDSSs as electronic systems designed to generate patient-specific assessments and recommendations to support clinical decisions at key workflow moments. A recent systematic review found that 71% of successful CDSSs reported improvements in quality assurance, 38% in user satisfaction, and 44% in clinical benefit [17], establishing a reference framework for evaluating future systems such as BurnoutCare.

The Copenhagen Burnout Inventory (CBI), developed by Kristensen et al. [10], is the most appropriate measurement instrument for the context of this research. Unlike instruments subject to copyright restrictions or tied to specific professions, the CBI is in the public domain and distinguishes three burnout dimensions: Personal (PB), Work-related (WB), and Client-related (CB). Juárez-García [11] validated its psychometric properties specifically in Mexican workers, confirming its suitability for the national organizational context.

2.2 Machine Learning for Burnout Prediction

The application of ML to burnout prediction has grown substantially. Shi et al. [5] conducted a systematic review and meta-analysis of ML studies for burnout in healthcare workers, reporting a pooled AUC of 0.72 (95% CI: 0.68–0.76), sensitivity of 0.63, and specificity of 0.84. Best-performing algorithms include Random Forest (>90.2% accuracy), Support Vector Machine (~85%), and neural networks [12]. Key predictors cluster into five categories: demographic/occupational, psychological/behavioral, organizational/social, physiological/wearable, and work activity patterns.

Adamopoulos et al. [13] applied NLP techniques (TF-IDF, sentiment analysis, LDA topic modeling) together with LSTM models to detect burnout indicators in written reports from public health workers, demonstrating that free-text generated in work contexts contains significant predictive signals and that temporal models can anticipate risk escalation days in advance. This work highlights the potential for incorporating analysis of the CBI qualitative block in BurnoutCare through NLP in future project phases.

A 2025 study of 1,244 healthcare professionals in Romania [15] developed an explainable AI (XAI) framework using SHAP for integrated prediction of burnout, Long COVID, and extended sick leave, achieving sub-second prediction latency. The incorporation of explainability through SHAP represents a priority direction for BurnoutCare, where recommendation transparency is critical for clinical adoption.

2.3 Generative AI Applied to Burnout Management

Chanthati [6, 14] proposed a burnout management system based on generative AI and LLMs that combines MongoDB Atlas storage with language models to produce personalized coping strategies and real-time predictive analytics. This work is the most direct antecedent of BurnoutCare in the use of generative AI for this domain, though it differs in that it does not integrate a validated clinical instrument nor is it aligned with national labor regulations.

In the broader digital mental health domain, recent studies (2025) documented that AI monitoring systems integrated with biometric sensors achieve 89–91% accuracy in early warnings [16], and platforms such as BioBase have demonstrated reductions in burnout-related sick days of up to 31% through continuous physiological data analysis. Tanguay-Sela et al. [7] found that AI-powered clinical support tools improve clinician well-being, task performance, and reduce cognitive burden, evidencing that CDSS benefits extend to both the evaluated individual and the evaluating clinician.

2.4 Comparative Analysis and Identified Gap

Table 1 summarizes the comparative position of BurnoutCare with respect to the most representative existing systems, considering three operational dimensions: clinical instrument, evaluation–intervention integration, and regulatory alignment.

Table 1. Comparative analysis of burnout-oriented systems and BurnoutCare

System / Approach	Validated instrument	Dimension-specific recommendations	National regulatory alignment	Fallback for service continuity
Shi et al. ML classifiers [5]	Variable (MBI/CBI)	No	No	Not reported
Adamopoulos et al. NLP-LSTM [13]	Free-text reports	Partial (signals)	No	Not reported
Moldovan et al. XAI framework [15]	Self-report scales	No (binary risk)	No	Not reported
Chanthati's GenAI system [6, 14]	None integrated	Yes (coping strategies)	No	Not reported
BioBase platform [16]	Biometric signals	No	No	Not reported
BurnoutCare (proposed)	CBI validated for Mexico [11]	Yes (PB / WB / CB, 8-component JSON)	Yes (NOM-035-STPS-2018)	Yes (rule-based engine)

The literature review reveals three gaps that BurnoutCare directly addresses: (1) most AI-driven burnout detection systems are limited to risk classification without automatically connecting to an actionable clinical recommendation [18]; (2) systems that generate recommendations are rarely aligned with specific national labor regulations; and (3) validation of similar systems in the Latin American context, particularly in Mexico, is scarce. BurnoutCare addresses all three gaps through the CBI validated for Mexican workers [11], a generative AI engine producing dimension-specific recommendations (PB, WB, CB), and automated generation of NOM-035-STPS-2018 compliance reports.

3 Methodology

This research adopted a Design Science Research (DSR) approach, aimed at developing and evaluating a technological artifact to solve a documented practical problem. The work was organized into six sequential phases; this paper documents Phases 1–3 (completed) and describes Phase 4 (planned pilot).

3.1 Study Design

An applied exploratory-descriptive research approach was adopted. Phase 3 (completed) encompasses system development and functional validation through black-box testing. Phase 4 (planned) will constitute a cross-sectional study with a longitudinal component for AI engine calibration on real data.

3.2 Study Context

The system was developed between September 2024 and March 2026 at a higher education institution in Mexico under the supervision of the principal investigator. Production deployment was carried out on Railway (backend and PostgreSQL) and Vercel (frontend), with continuous integration via GitHub. The pilot application is scheduled with a service-sector company in the central-eastern region of Veracruz, Mexico, whose primary activity is hospitality and customer service operations—a sector with documented high prevalence of burnout symptoms in Mexico.

3.3 Participants and Pilot Population (Phase 4 — Planned)

The target sample for the pilot consists of customer-service department employees at the participating organization. The selection of customer-service personnel responds to two empirical criteria: (i) this population presents high emotional demand, prolonged shifts, and continuous performance monitoring—the three documented risk factors for burnout in Mexican service-sector workers—and (ii) this is the same target population for which Juárez-García validated the Spanish version of the CBI [11], ensuring instrument suitability.

Inclusion criteria: active employees with a valid contract; age 18 years or older; minimum tenure of three months in the current position; voluntary participation under digital informed consent. **Exclusion criteria:** employees on medical leave, in a separation process, or under active psychiatric treatment for non-occupational mood disorders at the time of evaluation.

An initial convenience sample of at least 30 participants is anticipated for the AI engine calibration stage (n = 30 has been established in similar CDSS validation studies as a minimum for stable preliminary descriptive statistics). Progressive expansion is planned across three measurement waves spaced at six-week intervals during the pilot to enable longitudinal subscale tracking.

3.4 Burnout Assessment Instrument

The CBI was selected for its validation in Mexican workers [11] and its public-domain status. The instrument consists of 19 items distributed across three subscales:

- **Personal Burnout (PB) — 6 items.** Measures generic exhaustion not attributed to a specific cause. Example item: "How often do you feel tired?"
- **Work-related Burnout (WB) — 7 items.** Measures exhaustion attributed to work. Example item: "Is your work emotionally exhausting?"
- **Client-related Burnout (CB) — 6 items.** Measures exhaustion attributed to interaction with clients/users (item 13 of the original instrument is reverse-scored).

Response scale: five-point Likert with values 0, 25, 50, 75, and 100 (0 = Never/almost never; 100 = Always/almost always; or, for frequency-of-state items, 0 = To a very low degree to 100 = To a very high degree). **Subscale score:** arithmetic mean of the items in the subscale.

Risk classification thresholds, consistent with Kristensen et al. [10]: **Low** (<50), **Medium** (50–74), **High** (≥75).

The implemented four-step workflow digitalizes the CBI and incorporates demographic and qualitative context:

1. Digital informed consent and data privacy notice.
2. Demographic profile (9 items): age range, sex, position, tenure, work shift, contract type, area, education level, and dependents.
3. CBI questionnaire (19 items, three subscales).
4. Three optional open-ended qualitative questions reserved for future NLP analysis.

3.5 System Architecture

BurnoutCare adopts a five-layer architecture: presentation, application services, artificial intelligence, security, and persistence. Table 2 presents the selected technology stack.

Table 2. Technology stack of the BurnoutCare system

Technology	Version	Layer	Purpose
React.js / Tailwind CSS	18.x / 3.x	Frontend	Responsive UI with reusable components and utility-first design
React-i18next	14.x	Frontend	Dynamic ES/EN internationalization without separate code branches

(*continued*)

Table 2. (*continued*)

Technology	Version	Layer	Purpose
Node.js / Express.js	18.x/4.x	Backend	REST API following MVC pattern with security middleware
JWT + bcrypt+Helmet	9.x/2.x/7.x	Security	Authentication, credential hashing, secure HTTP headers
PostgreSQL (JSONB)	15.x	Persistence	Relational database with flexible JSONB storage for CBI responses
Anthropic Claude Sonnet API	Claude-3-5-sonnet	AI	Generative engine for structured clinical recommendations
Railway / Vercel (CI/CD)		Deployment	Cloud hosting with automated deployment from GitHub

3.6 AI Recommendation Engine: Configuration and Prompt Design

To support reproducibility, this subsection details the configuration of the LLM-based recommendation engine.

Model and Parameters. The engine uses the Anthropic Claude Sonnet model accessed through the Messages API [20], with max_tokens = 2048, temperature = 0.3 (low, to favor deterministic clinical outputs), and top_p = 0.9. A system prompt establishes the role ("You are a clinical assistant specialized in occupational burnout, working under NOM-035-STPS-2018"), enforces Spanish-language output, and forbids diagnoses ("recommendations are clinical decision support and never replace professional judgment").

Input Context. For each medium- or high-risk evaluation, the backend sends a structured user prompt containing: anonymized demographic profile, three subscale scores (PB, WB, CB), the global risk level, and the qualitative responses if provided.

Output Schema. The model is instructed to return strictly valid JSON conforming to a fixed schema with eight components: (1) general_analysis (overall narrative interpretation); (2) urgency_level ∈ {low, medium, high, critical}; (3) dimensional_analysis (three sub-objects for PB, WB, CB); (4) immediate_recommendations (list with action, responsible_party, deadline_days); (5) medium_term_recommendations (same structure); (6) self_care_techniques (employee-facing); (7) follow_up_plan (next_evaluation_date, suggested cadence, indicators to monitor); and (8) personalized_message (empathetic message addressed to the evaluated employee).

Rule-Based Fallback. When the API is unavailable (timeout >10 s, HTTP 5xx, or rate limit), Engine M3 invokes a deterministic fallback that maps each combination of risk level and dominant subscale to a predefined recommendation template, preserving the same eight-component JSON schema. This guarantees service continuity in clinical pilot environments with intermittent connectivity.

Output Validation. Every JSON response is validated against the schema before persistence. Schema violations trigger a single retry; on persistent failure, the fallback is invoked.

3.7 Evaluation Workflow and Alert Generation

The backend automatically calculates PB, WB, and CB subscale scores and the global risk level. When risk is medium or high, Engine M3 generates an immediate alert to the assigned health professional and requests a JSON recommendation as described above. The complete alert–recommendation–follow-up cycle is logged in the audit trail.

3.8 Data Privacy, Security, and Ethical Safeguards

Given the sensitive nature of burnout-related information, BurnoutCare implements multi-layered safeguards consistent with the Declaration of Helsinki, Mexico's Federal Law on Protection of Personal Data Held by Private Parties (LFPDPPP), and the data-protection requirements of NOM-035-STPS-2018.

Informed Consent and Privacy Notice. Before any data collection, participants must accept a digital informed consent and a structured privacy notice. The notice explicitly states: (i) the time required to complete the evaluation (approximately 10–12 min); (ii) the purpose of data collection (early detection and personalized intervention of burnout); (iii) the categories of data collected; (iv) the data controller and storage location; (v) participant rights (access, rectification, cancellation, opposition); (vi) whether responses may be used in aggregate and anonymized form for future dataset construction or research purposes; and (vii) the direct benefit of participation (access to personalized recommendations and clinical follow-up).

Authentication and Access Control. JWT-based authentication with configurable token expiration. Role-Based Access Control (RBAC) with four user profiles (employee, health professional, organizational manager, system administrator). Credentials stored using bcrypt with a 12-round salt. Helmet middleware enforces secure HTTP headers (HSTS, CSP, X-Frame-Options).

Data Minimization and Anonymization. Institutional dashboards display only anonymous aggregate data (per company, area, period); no individually identifiable information is exposed at the management or aggregate-statistics level. Direct identifiers are stored separately from CBI responses and joined only in the case-detail view, which is restricted to authorized health professionals.

Audit Trail. Every read or write operation on sensitive data is logged with timestamp, user ID, role, and operation type, enabling complete traceability.

Right to Withdraw. Participants may withdraw at any moment; the system supports request-driven deletion of personal data while preserving anonymized records for statistical integrity.

3.9 Evaluation Metrics and Validation Strategy

To address the dual nature of the artifact (technical system + clinical instrument), the evaluation strategy is organized in two complementary blocks.

Phase 3 — Technical Validation (Completed). Functional black-box testing covering: low/medium/high risk scenarios; unauthorized access attempts per role; simulated AI API failure scenarios. Performance metrics: classification latency (target <500 ms), API recommendation success rate (target ≥95%), fallback activation correctness (target 100%).

Phase 4 — Clinical Pilot Validation (Planned). The pilot will collect the following metrics:

- **Internal consistency of the CBI in the specific sample:** Cronbach's α per subscale (acceptable ≥0.70).
- **Risk distribution:** descriptive statistics (frequencies, means, SD) per subscale and overall.
- **Recommendation quality:** blind expert review by a certified occupational-health physician of a random sample of 30 LLM-generated recommendations using a 5-point Likert scale on clinical pertinence, safety, and actionability. A second expert will rate the same sample to compute inter-rater agreement (Cohen's κ).
- **Usability:** System Usability Scale (SUS) administered to employees and health professionals (target SUS ≥68, the established usability benchmark).
- **Latency under real load:** mean and standard deviation of end-to-end response time across pilot evaluations.
- **Adherence to follow-up:** percentage of medium/high-risk cases that complete a second evaluation within the recommended interval.

Expert Validation. The pilot incorporates an external occupational-health expert (independent of the development team) who will (i) review the LLM recommendation schema before deployment, (ii) audit a stratified sample of generated recommendations during the pilot, and (iii) co-author the final clinical-validation report. This step directly mitigates the risk of clinically inappropriate LLM outputs in a production environment.

4 Results

Upon completion of Phase 3, BurnoutCare was fully operational in production with eleven verified functional modules. Table 3 summarizes the functional status of each module.

Table 3. Functional status of BurnoutCare modules in production

Module	Status	Verified result
JWT Authentication (4 roles)	Operational	Role-based differentiated access; tokens with configurable expiry
CBI Evaluation — 4 steps	Operational	Full workflow executed on desktop, tablet, and mobile
Risk Classification (PB/WB/CB)	Operational	Response latency <200 ms in Railway production environment
Alerts Dashboard	Operational	Alerts generated in 100% of medium/high-risk cases during testing
AI Recommendations Engine	Operational	8-component JSON responses validated against schema in 100% of API calls
Rule-based Fallback Mechanism	Operational	Correct activation in all 3 simulated API failure scenarios
Case Detail and History	Operational	Individual view with prior evaluation history per employee
Clinical Follow-up	Operational	Intervention log with action type, date, and clinical notes
Institutional Indicators Panel	Operational	Anonymous aggregate metrics by company, area, and period
NOM-035-STPS-2018 Reports	Operational	Regulatory compliance report generation and export
User, Role, and Audit Management	Operational	Full CRUD, RBAC permission matrix, event traceability
ES/EN Internationalization (i18n)	Operational	Dynamic language switching with cross-session persistence

Preliminary performance measurements. Across 50 black-box test evaluations distributed evenly across low (n = 17), medium (n = 17), and high (n = 16) risk scenarios, the system reported a mean classification latency of 142 ms (SD = 38 ms), an LLM JSON response success rate of 100% (50/50 conformant to schema with no retries required), and a fallback activation correctness of 100% (3/3 simulated failure scenarios). These figures are reported as functional-test indicators; statistical generalization will require the larger pilot sample.

During production deployment, four categories of technical errors were identified and resolved, which are reported as observations of the validation process. The first was a database connection error (ECONNREFUSED 127.0.0.1:5432), caused by the use of individual environment variables instead of Railway's DATABASE_URL; resolved by refactoring the configuration module to use the unified connection string with conditional SSL based on NODE_ENV. The second was a module export error (TypeError: query is not a function), caused by inconsistencies between the exported interface and controller dependencies; resolved by unifying the export object (pool, query, testConnection).

The third was a SQL schema inconsistency (column u.rol_id does not exist), caused by divergence between the controller version and the actual database schema; resolved by rewriting queries to use the direct rol column. The fourth was a hardcoded localhost reference in the audit module, corrected by centralizing all API calls in api.js. These errors document divergences between local development and cloud production environments; their resolution constitutes evidence of the system's operational maturity.

5 Discussion

5.1 Scientific Contributions

BurnoutCare delivers three concrete contributions. First, it integrates the complete evaluation–classification–alert–recommendation–follow-up cycle in an operational platform, surpassing existing systems that only address the diagnostic phase [18]. Second, it combines a CBI threshold-based classifier with generative AI (LLM) to produce dimension-specific recommendations (PB, WB, CB) with eight clinical components, something ML classifiers with AUC of 0.72 [5] cannot provide. Third, it explicitly aligns the platform with NOM-035-STPS-2018, connecting clinical outcomes to the employer's regulatory obligations.

5.2 Comparative Efficiency

The system achieved risk classification response times under 200 ms (mean 142 ms, SD 38 ms), comparable to the sub-second latency reported by Moldovan et al. [15]. The 100% rate of schema-conformant LLM responses during functional testing is encouraging but must be interpreted as preliminary, given that the test set covered only 50 simulated cases and did not include real patient variability. Comparison with the 71% of CDSSs reporting verified clinical improvements [17] is therefore deferred to the Phase 4 pilot. The correct activation of the fallback mechanism in all three simulated failure scenarios demonstrates operational resilience, a characteristic absent from most reviewed systems and particularly relevant for environments with limited network infrastructure.

5.3 Advantages and Limitations vs. Prior Work

Compared to Chanthati's system [6], BurnoutCare presents three advantages: use of a validated clinical instrument (CBI), differentiated roles that establish clear clinical responsibilities, and normative alignment. Its main limitation relative to Adamopoulos et al. [13] is the absence of passive signals (continuous NLP, biometric sensors) and longitudinal predictive models. This limitation is planned for Phase 4 through sentiment analysis on CBI qualitative responses and longitudinal tracking of subscale variation between consecutive evaluations of the same employee.

5.4 Additional Variables to Improve the Model

The literature suggests that incorporating physiological data can raise predictive accuracy to 89–91% [16]. In BurnoutCare, the most viable short-term extensions are: sentiment analysis on open-ended qualitative responses using NLP; integration of organizational indicators (absenteeism, turnover, customer complaints) as contextual variables for the AI engine; and trajectory detection through longitudinal comparison of subscale scores between consecutive evaluations of the same employee.

5.5 Error Consequences and Risk Management

A false negative—classifying a severely burned-out employee as low risk—can have serious clinical consequences. To mitigate this, BurnoutCare applies conservative thresholds (medium risk triggers alerts from the 50th percentile), incorporates qualitative context as a complementary signal to the AI engine, and explicitly documents that generated recommendations are clinical decision support tools and do not replace the health professional's judgment. Recommendations always include a follow-up plan component with a suggested date for the next evaluation, reducing the risk that a case goes unmonitored after the initial episode. The expert-validation workflow described in Sect. 3.9 adds an additional safeguard against clinically inappropriate LLM outputs.

5.6 Limitations of the Current Study

Three limitations should be made explicit. First, clinical evidence is not yet available: the validation reported in this paper is functional and technical, not clinical; the Phase 4 pilot is required before any claim about diagnostic accuracy or clinical effectiveness can be substantiated. Second, the contribution is application-oriented: BurnoutCare integrates existing components (CBI, LLM, RBAC) into a novel clinical pipeline but does not propose a new machine-learning method. Third, statistical generalization of preliminary performance figures is bounded: the $n = 50$ functional-test set is sufficient for engineering validation but cannot support clinical inference.

5.7 Applicability Across Sectors and Regions

The RBAC architecture and configurable modules allow BurnoutCare to be adapted to healthcare institutions, educational centers, and the government sector without modifying the core logic. ES/EN internationalization prepares the system for international research contexts. For other Latin American countries with similar regulatory frameworks (Colombia: Resolution 2404/2019; Argentina: Law 24.557), only the institutional-reports module would require adaptation. In regions with limited connectivity, the rule-based fallback mechanism guarantees basic clinical service continuity without internet dependency.

6 Conclusions and Future Work

This paper presented BurnoutCare, the first operational CDSS for occupational burnout in Mexico that integrates the complete evaluation–alert–recommendation–follow-up cycle with explicit NOM-035-STPS-2018 compliance. The system demonstrated full technical feasibility in production, with classification times under 200 ms (mean 142 ms, SD 38 ms), a 100% recommendation-generation success rate during functional testing, and correct fallback activation upon API failures.

The three central contributions are: (1) the operational integration of the complete clinical cycle in a multi-layer production-deployed architecture, surpassing existing systems that address only the diagnostic phase; (2) the combination of the CBI validated for Mexican workers with generative AI that produces dimension-specific recommendations (PB, WB, CB), advancing beyond what ML classifiers with AUC of 0.72 can offer; and (3) explicit NOM-035-STPS-2018 alignment, which facilitates institutional adoption and closes the regulatory gap identified in the literature.

Future work is structured around the Phase 4 clinical pilot already designed: a three-wave cross-sectional study with a longitudinal component in a Veracruz service-sector organization, with an initial sample of at least 30 participants and progressive expansion. The pilot will deliver Cronbach's α per CBI subscale in the specific sample, blind expert review of a stratified sample of LLM recommendations (with inter-rater Cohen's κ), SUS scores from employees and health professionals, end-to-end latency under real load, and adherence-to-follow-up rates. Subsequent stages include training ML models on pilot data for longitudinal detection, NLP sentiment analysis on qualitative responses, development of a native mobile application, integration of organizational indicators as contextual signals for the AI engine, and extension to international research contexts.

Acknowledgments. This research was supported by the Tecnológico Nacional de México / Instituto Tecnológico de Orizaba and benefited from a Mitacs Globalink Research Award (IT50629) at the University of Toronto.

Disclosure of Interests The authors have no competing interests to declare that are relevant to the content of this article.

References

1. WHO: Burn-out an "occupational phenomenon": International Classification of Diseases. World Health Organization (2019). https://www.who.int/news/item/28-05-2019
2. Guan, S., et al.: Global burnout prevalence in public health workers: a meta-analysis. BMC Public Health. **24**, 2285 (2024)
3. IMSS: Report on Occupational Stress in Mexico. Mexican Social Security Institute, Mexico City (2023)
4. NOM-035-STPS-2018: Psychosocial Risk Factors in the Workplace—Identification, Analysis, and Prevention. Diario Oficial de la Federación, Mexico (2018)
5. Shi, H., et al.: Machine learning for predicting burnout among healthcare workers: a systematic review and meta-analysis. Work Stress. (2025). https://doi.org/10.1080/10376178.2025.2593294

6. Chanthati, S.R.: AI-driven burnout management system using generative AI, MongoDB atlas and LLMs. Eng. Technol. J. **10**(4), 4436–4447 (2025). https://doi.org/10.47191/etj/v10i04.09
7. Tanguay-Sela, M., et al.: AI-driven mental health decision support enhances clinician resilience. medRxiv. (2025). https://doi.org/10.1101/2025.10.29.25338983
8. Berner, E.S. (ed.): Clinical Decision Support Systems: Theory and Practice, 3rd edn. Springer, Cham (2016)
9. Delgadillo, J., et al.: Therapist burnout and patient improvement in psychological treatment. J. Consult. Clin. Psychol. (2018)
10. Kristensen, T.S., Borritz, M., Villadsen, E., Christensen, K.B.: The Copenhagen burnout inventory: a new tool for the assessment of burnout. Work Stress. **19**(3), 192–207 (2005)
11. Juárez-García, A.: Psychometric properties of the Copenhagen burnout inventory in Mexican workers. Psicología y Salud. **24**(1), 63–72 (2014)
12. Nimmagadda, S.: Shikha: the impact of burnout through predictive models. Int. J. Sci. Adv. Technol. **67**, S97 (2025)
13. Adamopoulos, I.P., et al.: Predicting workplace stress and burnout using NLP, LSTM, and AI. Environments. **15**(5), 65 (2025). https://doi.org/10.3390/environments15050065
14. Chanthati, S.R.: AI-driven burnout management system (Version 3). Eng. Technol. J. **10**(4), 2456–3356 (2025)
15. Moldovan, M., et al.: An integrated AI framework for occupational health. Healthcare. **13**(18), 2266 (2025). https://doi.org/10.3390/healthcare13182266
16. Rashid, A., et al.: Reimagining mental health with AI: early detection and a preventive ecosystem. PMC (2025). https://pmc.ncbi.nlm.nih.gov/articles/PMC12604579/
17. Laing, S., Mercer, J.: Improved preventive care clinical decision-making using a point-of-care CDSS. BMC Med. Inform. Decis. Mak. **21**, 1–8 (2021)
18. Andrés-Rodríguez, L., et al.: AI in mental health care: a systematic review. PMC (2025). https://pmc.ncbi.nlm.nih.gov/articles/PMC12017374/
19. Shi, H., et al.: Passive AI detection of stress and burnout among frontline workers. PMC (2025). https://pmc.ncbi.nlm.nih.gov/articles/PMC12655262/
20. Anthropic: Claude API documentation. https://docs.anthropic.com (2024). Last accessed 15 Mar 2025

Periodontitis Diagnosis via YOLOv8 Segmentation in Panoramic Radiographs

Estefany Aidee Ibarra Chávez[1](✉), Juan Francisco Mosiño[1],
María del Rosario Baltazar Flores[1], Diana Marlene Ibarra Chavez[2],
Martha Alicia Rocha Sánchez[1], and Alicia Martinez Rebollar[3]

[1] Tecnológico Nacional de México (TecNM) – Instituto Tecnológico de León, León, Mexico
M24241030@leon.tecnm.mx

[2] Escuela Nacional de Estudios Superiores (ENES), Unidad León, Universidad Nacional Autónoma de México (UNAM), León, Guanajuato, Mexico

[3] Centro Nacional de Investigación y Desarrollo Tecnológico (CENIDET), Tecnológico Nacional de México (TecNM), Cuernavaca, Morelos, Mexico

Abstract. Periodontitis is an oral disease that causes the progressive loss of alveolar bone, which affects both oral health and the quality of life of people. Detecting it in time is essential to avoid more serious complications such as tooth loss. In this work, a deep learning-based method is proposed to automatically identify and classify periodontitis from panoramic dental radiographs. For this purpose, a cascaded architecture based on convolutional neural networks is used, specifically the YOLOv8 model, which allows recognizing and segmenting important dental structures such as the cementoenamel junction and the alveolar bone. From these structures, it is possible to obtain precise measurements of the bone level. The system was trained with a set of around 500 images previously labeled by an oral health specialist. The proposed approach achieved a mean Average Precision (mAP@0.5) of 0.995 for Model 2, highlighting its potential as a promising tool for periodontal diagnosis.

Keywords: Periodontitis · Deep Learning · YOLOv8 · Image Segmentation · Alveolar Bone Loss · Panoramic Radiography

1 Introduction

Periodontal disease affects the supporting tissues of the teeth, the principal characteristic is the loss of alveolar bone, the prevalence of alveolar bone loss is higher in the interproximal sites as well as in the mandibular teeth. Periodontitis presents a clear sign, which is a reduction in the alveolar bone level greater than 2 mm [1].

The early stage of periodontal disease is called gingivitis. When gingivitis is left untreated, the disease can progress to periodontitis. It is important to detect periodontitis at an early stage because, in advanced stages, it can cause tooth loss, which affects oral health and quality of life [2].

H. Ponce et al. (Eds.): COMIA 2026, CCIS 3095, pp. 727–741, 2027.
https://doi.org/10.1007/978-3-032-37438-7_57

Periodontitis is a very common inflammatory disease in humans; it is estimated that the prevalence of periodontitis in its mild stage can reach 50% of people who suffer from it, and in Mexico that figure is estimated to be between 35% and 45% [3].

Radiography is a common tool for determining the alveolar bone level, measuring points of interest such as the cementoenamel junction (CEJ), which can be observed in each tooth. The CEJ junction is an important point, since it helps determine the loss of the alveolar bone level. In the radiograph, it is identified as a radiopaque area in the tooth. A healthy level of the alveolar bone is considered when it is up to 2 mm from the CEJ; if the alveolar bone exceeds this parameter, periodontitis begins to be classified by stages depending on how low the alveolar bone level is.

Therefore, these are measurements that cannot be taken immediately to determine the presence or degree of periodontitis, and it can even be confusing to distinguish between a normal alveolar bone level and an initial reduction that would allow timely diagnosis of the presence of periodontitis.

Deep learning techniques can be applied to the analysis of radiographs, as it can detect patterns and analyze images with precision. This allows the automated extraction of relevant measurements for the detection of periodontitis, reducing subjectivity and the possible errors associated with manual visual evaluation, providing more objective and reproducible results. While several studies have applied deep learning for periodontal bone loss detection, most rely on single-stage architecture that often struggles with the inherent noise and anatomical overlaps of panoramic radiographs. In contrast, this work proposes a two-stage cascaded framework. By first isolating individual dental pieces and subsequently applying specialized segmentation, noise is effectively mitigated and enhances measurement precision in diffuse anatomical regions. Ultimately, this enables the precise calculation of Euclidean distances for automated periodontal diagnosis [4].

The field of medical image analysis has come a long way since the 1970s. As highlighted by Litjens et al. [5], the industry has moved past rigid, human-designed rules toward Convolutional Neural Networks (CNNs) that learn directly from data. While this shift is still gaining ground in dentistry, Schwendicke et al. [6] points out that hurdles like limited data and the need for more rigorous methods remain.

The remainder of this paper is organized as follows: Sect. 2 presents a review of related work and existing deep learning methodologies for dental analysis. Section 3 describes the materials and methods, including the dataset preparation, the two-stage cascaded YOLOv8 architecture, and the evaluation metrics. Section 4 details the experimental results and provides a discussion of the findings. Finally, Sect. 5 summarizes the conclusions and suggests directions for future research.

2 Related Work

In recent years, deep learning has notably improved the analysis of dental radiographs, particularly for the detection and classification of periodontitis. This section reviews studies that are closely related to the objective of this work, with an emphasis on their methodologies, datasets, and reported results.

The versatility of CNNs in dentistry is clear from their success in diverse diagnostics. For example, Lee et al. [7] showed that these algorithms can detect dental caries with up to 89% accuracy, proving their reliability in spotting subtle decay. Moreover, for any automated periodontal system to work, it must first 'know' which tooth it is looking at. In this regard, Tuzoff et al. [8] set a gold standard using Faster R-CNN architecture; their system identified and numbered teeth with 99% precision a level of performance that effectively rivals that of clinical experts.

The identification of alveolar bone loss (ABL) has been widely addressed in literature. Xue et al. [1], for instance, presented an automated system for diagnosing and staging periodontal bone loss using panoramic radiographs from 320 patients. Their approach relies on a YOLOv8 architecture for tooth detection and segmentation, reaching a mean average precision (mAP) of 94.0%. They also reported an ABL deviation of 5.28%, outperforming traditional models such as Mask R-CNN and TransUNet.

Along similar lines, Krois et al. [9] developed a seven-layer deep convolutional neural network trained on 2,001 image segments. Using a 10-times repeated group shuffling validation, their model achieved both an accuracy and sensitivity of 0.81. Interestingly, this performance is comparable to that of clinical experts, who reported an accurate score of 0.76.

Tooth detection, numbering, and segmentation are key steps before performing any periodontal analysis. Chen et al. [10] proposed a method based on Faster R-CNN, implemented in TensorFlow, for automatic tooth detection in periapical radiographs. By incorporating post-processing strategies such as overlapping box filtering and rule-based numbering, they achieved an intersection-over-union (IoU) of 91%, with precision and recall both above 90%.

The increasing evidence supporting Deep Learning (DL) in periodontitis has been effectively consolidated in a recent systematic review and meta-analysis by Li et al. [11]. By analyzing thirteen high-quality studies, this research highlighted a pooled sensitivity of 0.88 and a remarkable Area Under the Curve (SROC) of 0.92.

In a different approach, Widyaningrum et al. [4] introduced a hybrid two-stage architecture that combines Mask R-CNN with DenseNet169. Their model was trained on 700 radiographs to segment both the CEJ and the alveolar bone. On an external test set, the system reached a staging accuracy of 0.80 and a specificity of 0.88, suggesting that multi-stage designs can improve segmentation precision.

While these studies report strong results, many of them rely on single-stage models, which can be sensitive to anatomical variability or may require additional post-processing. A summary of the main characteristics of these approaches is presented in Table 1.

As shown in Table 1, the method proposed in this work, based on a cascaded YOLOv8 architecture, achieves a higher mAP@0.5 compared to other studies that use single-stage approaches. This improvement appears to be related to the division of the process into two stages, which helps to better handle noise in the images and results in more precise segmentation. Unlike traditional methods, the proposed pipeline first isolates dental structures and then performs a more detailed anatomical segmentation, improving the accuracy of the measurements. This behavior is also reflected in Model 1, which shows more limited performance. This can be explained by the complexity of full panoramic radiographs, where image size and the overlap of multiple anatomical structures make precise detection more challenging.

Table 1. Comparative analysis of deep learning models for periodontitis and dental structure analysis.

Reference	Methodology / Architecture	Dataset Size	Key Performance Metrics
Xue et al. [1]	YOLOv8 & Mask R-CNN	320 radiographs	mAP: 94.0% / ABLD: 5.28%
Widyaningrum [4]	Mask R-CNN / DenseNet169	700 radiographs	Accuracy: 0.80 / Specificity: 0.88
Krois et al. [9]	7-layer Deep CNN	2,001 segments	Accuracy: 0.81 / Sensitivity: 0.81
Chen et al. [10]	Faster R-CNN (TensorFlow)	Periapical radiographs	IoU: 91% / Precision: >90%
This Work	Cascaded YOLOv8 (2-Models)	500 images	mAP@0.5: 0.995 (Model 2)

In addition, the proposed approach was evaluated on a dataset of 1,000 images, with 500 used for training and validation and 500 reserved as an independent test set. This setup provides a more reliable assessment of the model's performance and its potential applicability in clinical settings.

3 Methodology

An experimental study with a quantitative approach was conducted; the study focuses on the analysis of panoramic dental radiographs obtained from public databases for research purposes, with the aim of training convolutional networks such as YOLO segmentation and YOLO pose for the detection of periodontitis.

3.1 Dataset

The dataset used in this study consists of panoramic radiographs obtained from public databases such as the Dental OPG XRAY Dataset [12] and DENTEX CHALLENGE 2023 [13]. These images are used to perform annotations for the detection of anatomical structures, such as the CEJ and the alveolar bone.

A total of 1,000 images were collected. Of these, 500 were used for training and validation, while the remaining 500 were reserved as an independent test set to evaluate the model's performance on unseen data. The training subset was further divided into 80% for training and 20% for validation.

The division was performed manually through a folder structure. The main folder consists of two primary directories: images and annotations, which are further divided into train (training) and val (validation). This organization ensures that YOLO correctly associates each image with its corresponding annotation.

In this study, clinical precision was prioritized over data volume. Recognizing that detecting subtle and diffuse structures, such as the CEJ, requires high accuracy, a dataset of 500 images meticulously labeled by an oral health specialist was chosen. Although expert time is a limited resource, this specialist's criteria allowed for the construction of a more robust and precise database, reducing the noise typically found in larger, less-supervised datasets. This ensured the model learned with true professional-level accuracy.

3.2 Image Tagging

The annotation of anatomical zones and image labeling were performed using LabelMe, as it is a user-friendly application that operates locally and provides the capability to generate both regular and irregular polygons. The criteria for segmenting the areas of interest for measurement extraction were established based on specialized literature for the diagnosis and classification of periodontitis. This literature indicates that a healthy tooth is one in which the alveolar bone level is located at a maximum of 2 mm from the CEJ; therefore, it is necessary to precisely locate both the CEJ and the alveolar bone level.

Therefore, the panoramic radiograph provides a two-dimensional view of the maxilla, the mandible, and the teeth, making it necessary to segment these structures to identify the CEJ and the alveolar bone level (ABL). These areas typically appear as more radiopaque regions, given that the CEJ is part of the enamel between the crown and the gingiva. Furthermore, since the alveolar bone surrounds and supports the roots, measuring its level allows for the determination of any potential bone reduction or resorption.

For the Model 1 labeling, which focuses on individual dental piece detection, rectangular polygons were employed. This process utilized bounding boxes to identify tooth types (incisors, canines, premolars, and molars) and included keypoints to indicate the CEJ and the beginning of the alveolar bone.

For the labeling of Model 2, based on YOLO-segmentation, the dental images previously cropped by Model 1 were used. In these images, irregular polygons were employed to define specific areas, such as the enamel containing the CEJ; for practical purposes, this mask was named CEJ, alongside the ABL. This process was performed manually under the supervision of a dental specialist to ensure precise clinical criteria.

The manual annotation and validation process was led by Diana Marlene Ibarra Chá vez, DDS (ENES UNAM), a clinician with a Diploma in Advanced Odontology and three years of experience.

3.3 Convolutional Neural Network Training

For the training of Model 1, the use of YOLO-Pose was initially contemplated due to its ability to integrate keypoints for identifying anatomical landmarks such as the CEJ and ABL. However, during the training and validation phases of this convolutional neural network, the results showed optimal performance in labeling dental pieces but deficient precision in keypoint localization. It was observed that, since the CEJ and ABL are small and diffuse anatomical areas, the intrinsic noise of a full panoramic radiograph hindered their detection. To mitigate this, an attempt was made to process each dental piece as an individual image; nevertheless, the performance remained low. Consequently, a two-stage cascaded framework was implemented to enhance precision and mitigate noise. In this approach, the first model based on the YOLOv8 architecture was dedicated exclusively to the segmentation and cropping of individual dental pieces from the panoramic radiographs. This initial stage generated the isolated images required for the second model (see Fig 1).

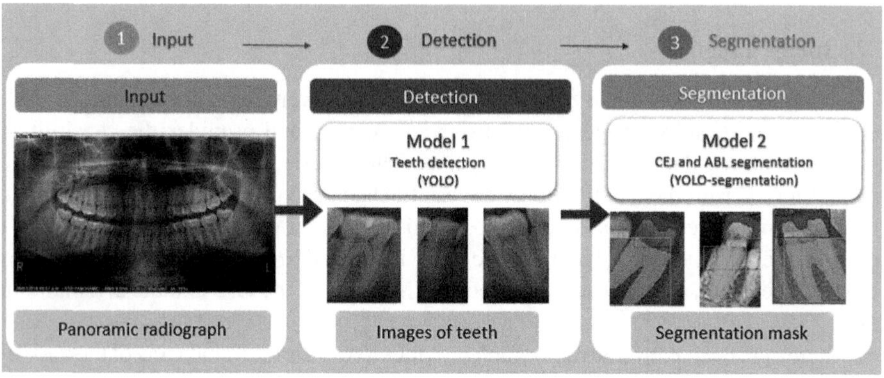

Fig. 1. The complete workflow of the two proposed cascade models to obtain dental units from a panoramic image and subsequently perform segmentation using CEJ and ABL masks.

For the training of Model 1, YOLOv8 was used with a total of 150 epochs and an image size of 832x832 pixels, with a learning rate configured at 0.01. The process was executed on an AMD Ryzen 5 4600G with Radeon Graphics (3.70 GHz). Using a CPU instead of a dedicated GPU highlights the model's computational efficiency, showing that the system can run on standard hardware without requiring expensive specialized equipment. This ensures the tool remains accessible and functional for everyday dental practice.

With the objective of improving the model's performance, various data augmentation techniques were applied: spatial transformations (Mosaic 1.0, horizontal flip 0.5, scaling 0.5, translation 0.1), color adjustments (HSV: hue 0.015, saturation 0.7, value 0.4), and sample diversity methods (Random Erasing 0.4 and Rand Augment).The choice of an 832x832 pixel image size was made to maximize the quality and visual detail of the dataset. Since the model processes medical images (radiographs), this was done to maintain details that could be lost with a 640x640 pixel resolution. However, this increase entails significantly longer training time due to the high computational load it represents for the CPU. Subsequently, a second model utilizing YOLO-segmentation was applied to these cropped images.

This model is better suited to identify irregular and broader areas, such as the enamel (as part of the CEJ) and the alveolar bone level (ABL). By adopting this cascaded strategy, each model becomes more specialized, significantly reducing the computational load and the introduction of noise compared to a single-stage approach. In the second model, using irregular polygons, the images generated by the first model are segmented. This process identifies the most radiopaque areas of the crown to determine the CEJ, and the most radiopaque parts of the root to determine the ABL. This labeling was performed manually on approximately 600 images under the supervision of a dental healthcare professional.

For this second model, YOLOv8 Nano (segmentation) was used within the Ultralytics YOLOv8 framework, employing CPU and 150 training epochs. As with the first process, data augmentation techniques were applied, utilizing geometric transformations such as translation (translate = 0.1), scaling (scale = 0.5), and horizontal flipping (fliplr = 0.5). Additionally, adjustments were made in the color space (HSV), along with mosaic augmentation (mosaic = 1.0), Rand Augment, and random erasing (erasing = 0.4) (see Fig. 2).

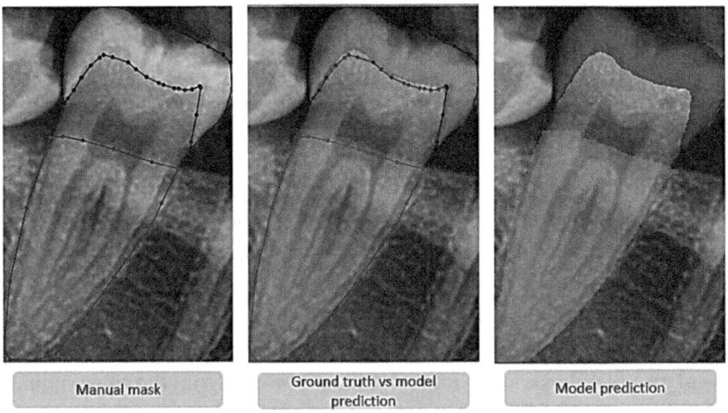

Fig. 2. Comparison between the ground truth mask, manually traced to evaluate Model 2, and the mask generated by the trained model. It is important to note that this image was not used during the training phase; instead, the expected mask was manually drawn to validate the model's accuracy during a subsequent testing phase.

3.4 Feature Extraction

The previously trained models are applied to the panoramic radiographs. The first model provides the dental structures cropped via rectangular bounding boxes to be processed by the second model (YOLO-segmentation), which detects the two areas of interest for each tooth; for practical purposes, these are denoted as CEJ and ABL

First, the tooth's location is determined, as the labeled masks for teeth in the upper maxilla are detected in an inverted manner. The orientation is established by generating all points from the CEJ and ABL masks and averaging them along the Y-axis; by comparing which average is greater, it is determined whether the tooth belongs to the upper or lower maxilla (see Fig. 3).

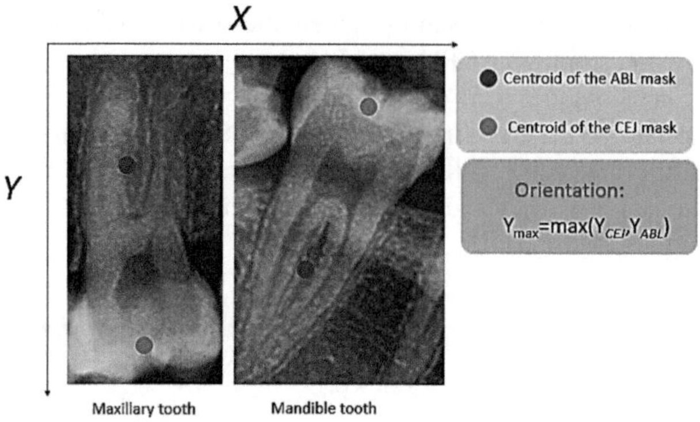

Fig. 3. Figure represents the calculation of dental orientation through the analysis of mask centroids.

Due to the natural morphology of the tooth, the section of the enamel containing the CEJ consists of two points (minimums or maximums, depending on the dental location), as it is the radiopaque zone surrounding the crown that forms two crests: one on the mesial side and another on the distal side.

To extract these points according to the detected location, the tooth must be divided into its mesial and distal sides. To achieve this, all points from the mask named CEJ are obtained and converted into a list of coordinates [Y, X]. Subsequently, the coordinates on the X-axis are averaged to obtain the horizontal center of the mask. Finally, the points are classified based on their position relative to this center: values greater than the average are classified as mesial, and those lower than the average as distal.

Based on the point cloud of the CEJ mask, divided into mesial and distal sides, the maximum or minimum point is identified on each side (depending on the previously determined dental location). These points are denoted as $CEJ_{mesia}l$ and CEJ_{distal}.

$$cej_{mesial} = \min / \max\ (P\_{mesial}) \tag{1}$$

$$cej_{distal} = \min / \max\ (P\_{distal}) \tag{2}$$

Where P_{mesial} represents the point cloud of the mesial side and P_{distal} represents the point cloud of the distal side, and the min/max operation depends on the spatial orientation of the tooth.

Once these reference points are obtained, the alveolar bone level is determined. To do this, the detected ABL mask is converted into a point cloud, and the Euclidean distance is calculated for each of these points relative to CEJ_{mesial} and CEJ_{distal}. Finally, the points with the shortest distance are selected and named ABL_{mesial} and ABL_{distal}.

$$ablmesial = \underset{p\epsilon ABL}{\operatorname{argmin}} \mid |p - CEJ_{mesial}| \mid \tag{3}$$

$$abldistal = \underset{p\epsilon ABL}{\operatorname{argmin}} \mid |p - CEJ_{distal}| \mid \tag{4}$$

Where p represents the point cloud of the ABL mask, $\boldsymbol{p}$ is a point in this set, and $\|.\|$ denotes Euclidean distance.

Subsequently, the Euclidean distance is calculated again between the identified points (ABL_{mesial}, ABL_{distal}) and their respective references (CEJ_{mesial}, CEJ_{distal}). This results in two vectors representing the distance in pixels between the CEJ and the alveolar bone, denoted as V_{mesial} and V_{distal}.

$$V_{mesial} = | \, |ABL_{mesial} - CEJ_{mesial}| \, | \tag{5}$$

$$v_{distal} = \left\| abl_{distal} - cej_{distal} \right\| \tag{6}$$

To determine the approximate size of the dental structure, the deepest point of the CEJ mask (CEJ_d) and the deepest point of the ABL mask (ABL_d) are identified. By calculating the Euclidean distance between these points, the Total Tooth Measurement (TTM) vector is obtained in pixels.

$$\mathrm{ttm} = \left\| \mathrm{abl_d} - \mathrm{cej_d} \right\| \tag{7}$$

Finally, to determine the total extent of the area covered by the alveolar bone, the Euclidean distance of point ABL_d is calculated relative to ABL_{mesial} and ABL_{distal}. The result corresponds to the total alveolar bone measurement for the mesial side (TAB_{mesial}) and the distal side (TAB_{distal}).

$$\mathrm{tab_{mesial}} = \| \mathrm{abl_d} - \mathrm{abl_{mesial}} \| \tag{8}$$

$$\mathrm{tab_{distal}} = \| \mathrm{abl_d} - \mathrm{abl_{distal}} \| \tag{9}$$

3.5 Vector Classification

Once the relevant vectors for the calculations are obtained, it must be noted that they do not represent absolute measurements of the dental structures. Therefore, it is necessary to perform normalizations or percentage-based estimations.

The classification of periodontal disease is determined by the percentage of alveolar bone loss, using the CEJ as the reference point for the normal level where the alveolar bone should be located. According to the periodontal diagnostic stages presented in Table 2.

Table 2. Periodontal diagnostic stages (adapted from Tonetti et al., 2018) [14].

Stage	Description	Percentage
Stage I	< 1/3 coronal.	< 15%
Stage Il	1/3 coronal.	15% -33%
Stage Ill	Extending to 1/3 of root and beyond.	>33%

The vectors representing the measurements of the areas of interest in pixels are collected, and the alveolar bone loss percentages (%ABLL) are subsequently calculated using Eqs. 10 and 11. The result provides the percentage of alveolar bone loss for both the distal and mesial sides; based on the established thresholds, it is possible to classify the presence and severity of periodontitis for each tooth.

$$\%ABLL_{mesial} = \frac{V_{mesial} \times 100}{TAB_{mesial} + V_{mesial}} \quad (10)$$

$$\%ABLL_{distal} = \frac{V_{distal} \times 100}{TAB_{distal} + V_{distal}} \quad (11)$$

The thresholds used correspond to the percentages defined in Table 2. Additionally, a further threshold ($< 2\%$) was included to identify normal cases without periodontitis, using the clinical criterion that places the alveolar bone level approximately 2 mm below the CEJ as a reference [2].

However, since the measurements in this study were conducted in pixels rather than absolute units, this threshold should be understood as an approximation rather than a direct equivalence. This may introduce variability depending on the specific tooth and image size. To address this discrepancy, a solution to improve diagnostic precision is proposed for future work, which is currently underway.

Classification is performed for both the mesial and distal sides. The final diagnosis for the tooth is based on the side exhibiting the greatest alveolar bone reduction according to specialized literature, although it is essential to assess both sides to identify localized areas of periodontitis.

4 Results

Calculated vectors were drawn alongside the CEJ and ABL masks to provide a clear view of how the measurements were obtained and the teeth numbered for information extraction. At first glance, it can be determined that the algorithm fulfills its objective in the majority of dental pieces; however, it should be noted that detection failures occur in a minimum number of cases (see Fig. 4).

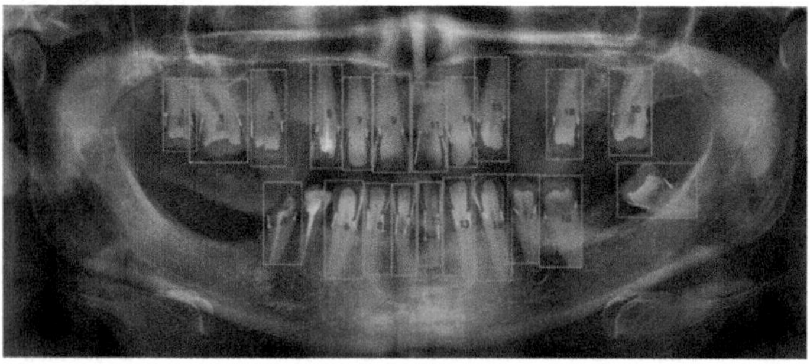

Fig. 4. Figure showing the results obtained from processing a panoramic radiograph, with the resulting vectors drawn and the CEJ and ABL masks.

Table 3. Results obtained from the first five dental pieces (representative partial sample).

# Tooth	Dist. Total	Distal	Mesial	% Distal	% Mesial	Distal Stage	Mesial Stage	Expert Distal Stage	Expert Mesial Stage
1	219.4	34.0	30.4	23.6	19.6	Stage2	Stage2	Stage2	Stage2
2	256.6	66.9	41.4	36.1	22.1	Stage3	Stage2	Stage2	Stage1
3	289.3	50.0	29.1	24.2	14.8	Stage2	Stage1	Stage2	Stage1
4	249.5	31.3	18.0	21.6	10.7	Stage2	Stage1	Stage2	Stage1
5	311.1	20.6	20.1	9.9	9.5	Stage1	Stage1	Stage1	Stage1

Consultations with dental specialists confirmed the algorithm's strong performance, revealing that it occasionally outperforms manual visual assessment. By identifying subtle radiopaque contrasts imperceptible to the human eye, the system achieves a more granular definition of anatomical edges. This specialized detection allows the model to capture the boundaries of radiopaque zones more accurately than traditional methods, particularly in clinical cases where bone margins are diffused or poorly defined.

As shown in Table 3, the experiment allowed the detection of 21 dental pieces, with an agreement of approximately 76.2% between the model's predictions and the expert's diagnosis in periodontal classification.

It is worth noting that the evaluation of these 21 dental pieces functions as an initial pilot test. Manual validation by a specialist requires significant time investment, which is why this sample is presented as a pilot study. Nonetheless, the results of this phase are highly satisfactory; the alignment with the clinical diagnosis is appropriate and largely consistent. While the evaluation is currently working to expand this evaluation to a much larger dataset, these preliminary findings already confirm that the system is reliable and holds real potential to support dentists in their clinical practice.

Regarding the YOLOv8 results for this first stage, the model showed good performance in differentiating between classes. A low rate of inter-class confusion was observed, which renders the system clinically reliable. Across all categories, the accuracy consistently remained above 80%.

Table 4. Performance metrics obtained by Model 1 for dental piece identification.

Model	Precision	Recall	mAP@0.5	mAP@0.5:0.95
Model 1 YOLOv8	0.71295	0.74931	0.72753	0.6068

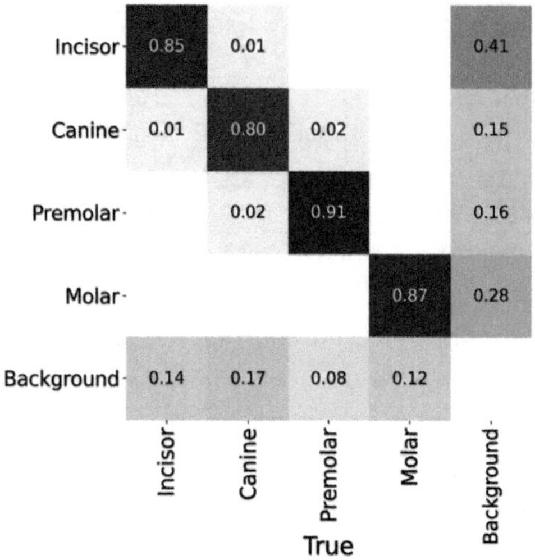

Fig. 5. Confusion matrix for Model 1.

The classification performance of Model 1 demonstrates a high capability for differentiating between individual dental pieces. This effectiveness is further validated through the confusion matrix (see Fig. 5), which illustrates the model's accuracy and its low rate of misidentification across categories.

However, a specific point regarding the 'background' class must be clarified. This category was not trained intentionally; it represents areas where the model made incorrect predictions in non-annotated zones. This is where the model's performance was more average or 'regular,' a trend that matches the metrics in Table 4. Essentially, the model sometimes 'sees' teeth where there aren't any. Also, while the dental pieces are identified, the bounding boxes don't always wrap around them with perfect, millimeter-level precision.

Despite these minor alignment issues, Model 1 proved to be a solid tool for segmenting dental pieces. This was confirmed by testing the model against 500 extra images that had never been seen during training. The results were consistent. Ultimately, Model 1 performed as expected, creating a dependable starting point for the rest of the analysis.

Regarding the performance of Model 2, which utilized Ultralytics YOLOv8 segmentation for detecting anatomical areas within dental pieces, the evaluation was conducted using the confusion matrix. The results demonstrate a precision near 100% across the metrics in Table 5, reflecting an accurate identification of both CEJ and ABL classes. This confirms that the model effectively distinguishes between categories and succeeds in detecting each one correctly.

Table 5. Performance metrics obtained by Model 2 for dental piece (CEJ y ABL).

Model	Precision	Recall	mAP@0.5	mAP@0.5:0.95
Model2 YOLOv8 Nano (segmentation)	0.99867	1	0.995	0.85785

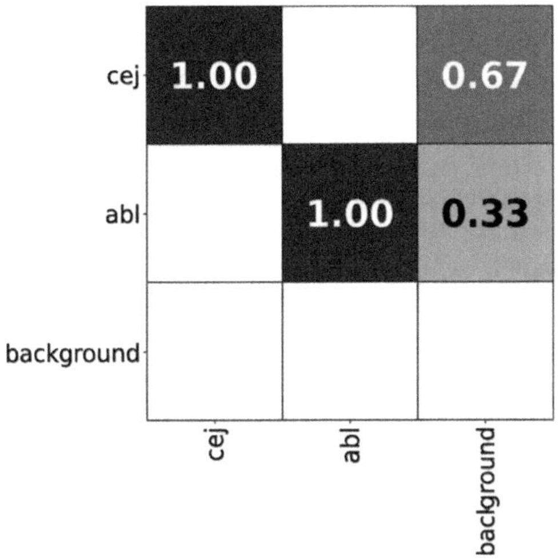

Fig. 6. Confusion matrix for Model 2.

The high precision and recall values for Model 2 are supported by its confusion matrix (see Fig. 6), which illustrates a strong performance in differentiating between anatomical zones. However, the matrix also reveals a slight confusion with the background class, where non-annotated areas are occasionally misidentified.

However, it was observed that YOLO automatically includes a "background" category. This is not a class learned during the training phase; rather, it serves as a reference to analyzing predictions occurring outside the specific areas of interest. Consequently, any detection made in non-annotated regions is recorded as a false positive within this background classification.

The error margin for this model is visible in the confusion matrix, which shows the system identifying objects in background areas essentially classifying certain false categories. This suggests that the model's performance is more regular when processing dental pieces that are poorly defined or suffer from low image quality. Therefore, applying image enhancement filters would be a significant step toward optimizing these results. When the model was tested using segmented dental images that were not part of the initial training set, the performance remained steady. In the practical assessment, the output is reliable enough to proceed with vector calculations. This confirms that, despite the minor background noise, Model 2 effectively meets the requirements for the next stage of the analysis.

5 Conclusion

This study shows that anatomical measurements extracted from radiographs can be used as a clinical support tool for diagnosing periodontitis. However, some aspects still need improvement. At this stage, the model relies on Euclidean distances, which work as an initial approximation but do not consider the inclination of the tooth. Including preprocessing techniques or filters that address these geometric variations could improve the accuracy of the measurements. Despite this limitation, the results were sufficient for the main objective of the study.

Based on these results, a database has been compiled to support the next stage of this research. The aim is to move away from fixed manual thresholds, which can be rigid and prone to error, and instead train a neural network using expert-labeled data. This is important because periodontitis diagnosis does not follow strict universal criteria. For example, although 2 mm is often used as a reference for normal bone levels, applying it as a fixed threshold may lead to inaccuracies or false positives.

A trained neural network would allow the model to better adapt to different clinical scenarios by considering multiple factors at the same time. In addition to anatomical measurements, it could include relevant clinical variables such as tooth loss, diabetes, or other risk factors. Furthermore, by focusing on proportions and the total length of the tooth, the model could provide more accurate classifications and reduce the limitations related to image scaling in panoramic radiographs.

A key advantage of this model is that it allows for diagnosis without knowledge of the actual size of the radiograph. By employing a system of proportions and percentages based on the tooth root, the calculation automatically adjusts, ensuring accurate results regardless of the image dimensions.

This work is still in progress, and additional expert annotations are currently being developed to further improve the model's training and evaluation.

These results confirm the reliability of the current approach, effectively establishing the foundational stage needed to advance toward more flexible, non-linear classification tools for periodontitis.

There are plans to integrate specialized X-ray image enhancement and filtering techniques inspired by the proven methods in reference [15] to achieve even greater precision in image quality and segmentation.

Ultimately, these findings confirm a functional cascaded architecture that establishes the foundation for automated clinical diagnosis.

References

1. Xue, T., Chen, L., Sun, Q.: Deep learning method to automatically diagnose periodontal bone loss and periodontitis stage in dental panoramic radiograph. J. Dent. **150**, 105373 (2024). https://doi.org/10.1016/j.jdent.2024.105373
2. Gul, S.S.: Prevalence and severity of circumferential alveolar bone loss using CBCT images: a retrospective study of 20,620 surfaces of 5155 teeth. Diagnostics. **14**(5), 507 (2024)
3. Rodríguez Orozco, A.R., Cruz Velasquez, T.: Gingivitis y periodontitis. ¡No te quedes sin dientes! Milenaria, Ciencia y Arte. **23**, 28–32 (2024). https://doi.org/10.35830/mcya.vi23.445

4. Widyaningrum, R., et al.: Hybrid two-stage CNN for detection and staging periodontitis on panoramic radiographs. J. Oral Biol. Craniofac. Res. **15**(6), 1392–1399 (2025). https://doi.org/10.1016/j.jobcr.2025.08.019
5. Litjens, G., et al.: A survey on deep learning in medical image analysis. Med. Image Anal. **42**, 60–88 (2017). https://doi.org/10.1016/j.media.2017.07.005
6. Schwendicke, F., Samek, W., Krois, J.: Artificial intelligence in dentistry: chances and challenges. J. Dent. Res. **99**(7), 769–774 (2020). https://doi.org/10.1177/0022034520915714
7. Lee, J.H., Kim, D.H., Jeong, S.N., Choi, S.H.: Detection and diagnosis of dental caries using a deep learning-based convolutional neural network algorithm. J. Dent. Res. **77**, 106–111 (2018). https://doi.org/10.1016/j.jdent.2018.07.015
8. Tuzoff, D.V., et al.: Tooth detection and numbering in panoramic radiographs using convolutional neural networks. Dentomaxillofac. Radiol. **48**(4), 20180051 (2019). https://doi.org/10.1259/dmfr.20180051
9. Krois, J., et al.: Deep learning for the radiographic detection of periodontal bone loss. Sci. Rep. **9**, 8495 (2019). https://doi.org/10.1038/s41598-019-44839-3
10. Chen, H., et al.: A deep learning approach to automatic teeth detection and numbering based on object detection in dental periapical films. Sci. Rep. **9**, 3840 (2019). https://doi.org/10.1038/s41598-019-40414-y
11. Li, X., et al.: Deep learning for classifying the stages of periodontitis on dental images: a systematic review and meta-analysis. BMC Oral Health. **23**, 1017 (2023). https://doi.org/10.1186/s12903-023-03751-z
12. Rahman, R.B., Tanim, S.A., Alfaz, N., Shrestha, T.E., Miah, M.S.U., Mridha, F.: Dental OPG XRAY Dataset. Mendeley Data. **4** (2024). https://doi.org/10.17632/c4hhrkxytw.4
13. Er, S.: DENTEX CHALLENGE 2023. Zenodo. (2023). https://doi.org/10.5281/zenodo.7812323
14. Tonetti, M.S., Greenwell, H., Kornman, K.S.: Staging and grading of periodontitis: framework and proposal of a new classification and case definition. J. Periodontol. **89**(Suppl 1), S159–S172 (2018). https://doi.org/10.1002/JPER.18-0006
15. Huang, R.-Y., Dung, L.-R., Chu, C.-F., Wu, Y.-Y.: Noise removal and contrast enhancement for X-ray images. Br. J. Healthc. Med. Res. **3**(1), 56 (2016). https://doi.org/10.14738/jbemi.31.1893

Explainable Pneumonia Detection in Chest X-Rays Using DenseNet-121 with Post-hoc Analysis

Daniel Patiño Ruiz and Samantha Acosta Ruiz(✉)

Benemérita Universidad Autónoma de Puebla, Facultad de Ciencias de la Computación, Puebla, Mexico
daniel.patino@alumno.buap.mx, samantha.acostar@correo.buap.mx

Abstract. Respiratory diseases represent one of the main causes of mortality worldwide, with pneumonia being one of the most common and potentially lethal if not detected in time. Chest X-rays are fundamental for diagnosis; however, their interpretation requires specialized personnel and can be subject to human error. Deep Learning models have shown great potential in the automated analysis of medical images, but challenges remain with regard to image variability and model interpretability. This work proposes a convolutional neural network (CNN) model based on the DenseNet-121 architecture for the automatic detection of pneumonia in chest X-rays. The model was evaluated using standard classification metrics, achieving an accuracy of 95.03%, an Area Under the Curve (AUC) of 0.9888, a sensitivity of 94.87%, and a specificity of 95.30%. In addition to predictive performance, this study incorporates interpretability techniques such as Principal Component Analysis (PCA) and Gradient-weighted Class Activation Mapping (Grad-CAM) to analyze learned representations and highlight the regions that influence model decisions. These results provide insight into the behavior of the model, supporting its potential use as an AI-assisted diagnostic support approach.

Keywords: Pneumonia Detection · Deep Learning · DenseNet-121 · PCA · Grad-CAM

1 Introduction

Artificial intelligence (AI) and Deep Learning have significantly transformed the medical field in recent years, particularly in the analysis of radiographic images. These technologies enable the processing of large volumes of visual data, allowing the identification of complex patterns that may go unnoticed by the human eye. In this context, Convolutional Neural Networks (CNNs) have demonstrated outstanding performance in medical image classification tasks.

Despite these advances, the diagnosis of pneumonia through chest radiographs remains a challenging task. Pneumonia shares visual characteristics with

H. Ponce et al. (Eds.): COMIA 2026, CCIS 3095, pp. 742–755, 2027.
https://doi.org/10.1007/978-3-032-37438-7_58

other lung diseases, and the variability in image quality further complicates its identification. In clinical practice, diagnosis largely depends on the expertise and availability of radiologists, which introduces subjectivity and limits accessibility in regions with scarce medical resources. Delayed or inaccurate diagnosis can lead to severe complications, particularly in vulnerable populations such as young children and the elderly.

To address these challenges, several deep learning architectures have been explored. Among them, DenseNet has proven effective due to its dense connectivity, which promotes feature reuse and mitigates the vanishing gradient problem. Previous studies have shown that such architectures can achieve performance comparable to, or even surpass, that of human experts. Additionally, recent research highlights the importance of combining robust feature representation with interpretability techniques to ensure that model predictions are clinically meaningful [4,7]. To better understand the core components of this automated detection, fundamental concepts can be organized into the following areas:

- **Convolutional Neural Networks (CNN)**: A specialized variant of neural networks designed to process images, using filters to extract local features, reducing parameters, and improving efficiency.
- **DenseNet Architecture**: A CNN architecture that stands out for its dense connections between layers, where each layer receives information from all previous ones, favoring the reuse of features and avoiding the vanishing gradient problem.
- **Explainable AI (XAI)**: Post-hoc techniques such as Principal Component Analysis (PCA) and Gradient-weighted Class Activation Mapping (Grad-CAM) that allow visualization of data distributions and highlight the regions of an image that most influence the model's predictions, helping to assess whether the model is focusing on clinically relevant patterns rather than relying on spurious correlations.

Therefore, this work proposes the development of an automated model for pneumonia detection in chest X-ray images, aiming not only to achieve competitive performance but also to provide interpretable insights into the model's decision-making process.

The proposed approach is based on a pre-trained DenseNet-121 architecture adapted for binary classification. To improve generalization, preprocessing techniques such as random cropping, slight rotations, and brightness adjustments are applied. The model is evaluated using standard performance metrics, and its internal behavior is further analyzed through interpretability techniques, enabling a more complete understanding of its predictions.

1.1 Related Works

This section reviews recent advances in deep learning architectures for medical image analysis, particularly focusing on automated pneumonia detection in chest

X-rays. The current literature can be broadly synthesized into three main thematic areas: the integration of attention mechanisms with ensemble models, the baseline efficacy of convolutional architectures, and the optimization of feature extraction for interpretability.

Attention Mechanisms and Ensemble Architectures. To maximize classification precision, several studies have focused on guiding neural networks to emphasize pathological features. For instance, the integration of attention layers into pre-trained CNNs has proven effective in directing the model's focus toward suspicious pulmonary regions [3]. Building upon this, ensemble approaches have combined diverse architectures to synergize their strengths. Research has demonstrated that combining EfficientNet-B0 and DenseNet-121 with attention modules significantly improves evaluation metrics by capturing a wider range of relevant features [2]. Similarly, hybrid models mixing deep CNNs with Vision Transformers (ViT) have been proposed to increase overall robustness [5]. While highly accurate, these ensemble and attention-based approaches often incur substantial computational costs and increase model complexity.

Efficacy of Base Architectures. Parallel to complex ensembles, comparative studies have evaluated the standalone capabilities of deep learning models. Research manipulating various DenseNet versions confirms that architectures such as DenseNet-121 and DenseNet-201 are highly proficient at predicting pneumonia directly from X-rays [1]. Their inherent dense connections allow for efficient feature reuse and gradient flow, establishing them as strong foundational models that can achieve competitive performance without the absolute need for computationally heavy hybridizing.

Feature Optimization and Interpretability. A critical area bridging high predictive performance with clinical applicability is the optimization of feature extraction and dimensionality reduction. Advanced optimization algorithms for feature selection have been introduced to refine input features, which directly increases classification accuracy in medical datasets by discarding irrelevant data [7]. Furthermore, mathematical modeling approaches, such as Principal Component Analysis (PCA), have been successfully applied to optimize deep learning architectures and represent complex multidimensional features more transparently [4]. These studies underscore that pairing a robust baseline architecture with dimensionality reduction techniques is vital for developing tools that are not only computationally efficient but also interpretable.

Therefore, this work positions itself at the intersection of these themes. It focuses on analyzing the standalone effectiveness of the DenseNet-121 architecture, augmented by careful preprocessing, while explicitly addressing the interpretability gap found in complex ensemble models through PCA on learned embeddings and Grad-CAM heatmaps. This ensures the model maintains high diagnostic accuracy while offering the clinical transparency required for medical applications.

2 Methodology

The proposed solution consists of developing a hybrid automatic detection model using the pre-trained DenseNet-121 architecture, adapted for binary classification. To ensure a comprehensive and rigorous approach, the methodology was structured into seven sequential phases, combining both the research process and the technical implementation. To provide a better understanding of the model's behavior, the visual transformation of the data throughout these phases is highlighted.

2.1 Literature Review and Dataset Acquisition

The first phase consisted of a systematic literature review to identify the most effective deep learning architectures for medical image analysis, which justified the selection of DenseNet-121. Following this, Phase 2 focused on obtaining the data. The public dataset "Chest X-Ray Images (Pneumonia)" from Kaggle [6] was used. This dataset provides thousands of chest X-rays pre-classified into two binary categories: NORMAL and PNEUMONIA, ensuring an adequate balance for robust model training. Figure 1 illustrates the visual differences between a healthy chest X-ray and one with pneumonia before any computational processing.

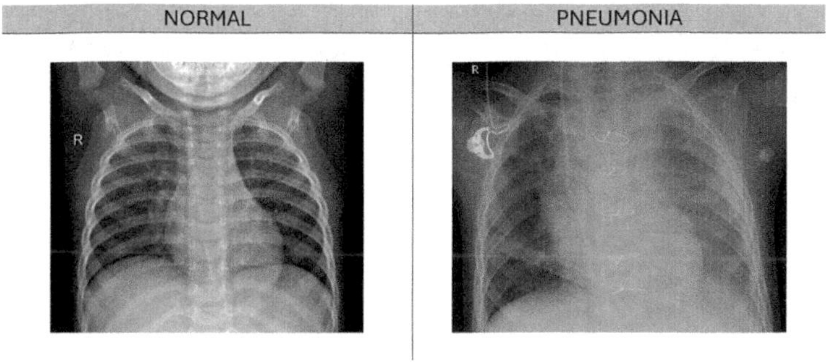

Fig. 1. Sample images from the original dataset showing NORMAL (left) and PNEUMONIA (right) classes.

2.2 Image Preprocessing

In the third phase, the images underwent a critical preprocessing stage. First, they were normalized and resized to a standard format of 224×224 pixels. Subsequently, data augmentation techniques were applied, such as random rotations (up to 10 degrees), strategic random resized crops, and brightness adjustments.

These deliberate transformations increased the diversity of the dataset, an essential step to help the model generalize and avoid overfitting. The dynamic behavior of this phase is shown in Fig. 2, where a single raw image is transformed into different augmented versions to feed the network.

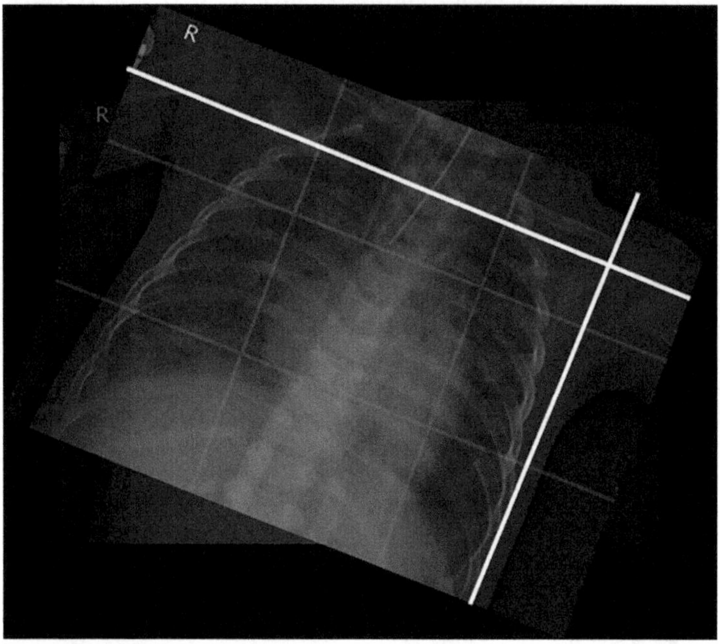

Fig. 2. Visual representation of the preprocessing pipeline: Original image, resizing, random rotation, cropping, and brightness adjustment.

2.3 Model Implementation and Training

Phase 4 involved the computational implementation of the model. The DenseNet-121 architecture was employed, leveraging its pre-trained knowledge. To adapt this powerful model to our specific problem, its convolutional layers were frozen, and its final classification layer was replaced with a new dense layer designed with a single output neuron optimized for binary classification. The preprocessed images in tensor format act as the direct input for this architecture. During Phase 5, the model was trained. We used the Stochastic Gradient Descent (SGD) optimizer with momentum, a proven choice to stabilize convergence. In parallel, the BCEWithLogitsLoss function was implemented. This loss function is tactically superior as it combines the Sigmoid layer and Binary Cross-Entropy into a single calculation, providing greater numerical stability during the training process.

2.4 Evaluation and Interpretability

Once trained, Phase 6 focused on evaluating the model's performance. A set of standard industry metrics was applied: accuracy, F1-score, Area Under the Receiver Operating Characteristic curve (AUC-ROC) and the confusion matrix. The test set was kept intact and used exclusively for this final evaluation, ensuring an objective and unbiased assessment.

Finally, Phase 7 focused on the interpretability analysis of the proposed model. Principal Component Analysis (PCA) was applied to visualize the learned embeddings and examine the distribution of NORMAL and PNEUMONIA samples within the feature space. Additionally, Gradient-weighted Class Activation Mapping (Grad-CAM) was employed to identify the thoracic regions contributing to the model's predictions through activation heatmaps. To complement the visual analysis, a "Focus Score" was incorporated as a quantitative indicator of the overlap between high-activation Grad-CAM regions and anatomically relevant thoracic areas. The results derived from these analyses are presented and discussed in the following section.

3 Results

To analyze the impact of training duration, architectural selection, and feature extraction strategies, a baseline comparison along with an ablation study was initially conducted. Additionally, to reduce the effect of random weight initialization and provide greater experimental consistency, each configuration was executed across multiple independent runs. Table 1 summarizes the average performance and standard deviation (±) obtained by the baseline ResNet-50 and the proposed DenseNet-121 under different training configurations, including frozen and fine-tuned settings across multiple epochs.

Table 1. Comparison of the performance of the reference, ablated and optimized models.

Model	Epochs	Fine-Tuned	Accuracy	AUC	Recall
ResNet-50 (Baseline)	20	Yes	93.99% ± 0.007	0.9796	95.77%
DenseNet-121 (Frozen)	20	No	85.42% ± 0.003	0.9424	95.38%
DenseNet-121	20	Yes	92.79% ± 0.000	0.9640	95.26%
DenseNet-121 (Optimized)	30	Yes	94.23% ± 0.004	0.9827	95.38%
DenseNet-121 (Extended)	50	Yes	95.03% ± 0.000	0.9888	94.87%

As shown in Table 1, freezing the convolutional base resulted in a substantial reduction in performance, with DenseNet-121 (Frozen) reaching only 85.42% accuracy. This behavior suggests that pre-trained ImageNet representations alone are insufficient for capturing domain-specific radiological characteristics without additional fine-tuning on chest X-ray data. Once fine-tuning was enabled, the model progressively improved as the number of training epochs increased, ultimately reaching 95.03% accuracy and an AUC of 0.9888 after 50 epochs.

To further justify the architectural selection, the baseline ResNet-50 model was compared against DenseNet-121 under identical training conditions (20 epochs). Although ResNet-50 achieved competitive initial performance, the Grad-CAM analysis revealed notable differences in how both architectures localized relevant anatomical regions. As illustrated in Fig. 3, the ResNet-50 baseline frequently focused on peripheral structures and image borders during both false positive and false negative predictions, suggesting sensitivity to irrelevant visual artifacts. In contrast, Fig. 4 shows that DenseNet-121 maintained activations closer to the thoracic cavity and pulmonary regions, indicating more anatomically coherent localization behavior under the evaluated conditions.

This localization behavior is also reflected in the Focus Score values displayed in Figs. 3 and 4. Higher Focus Score values indicate a greater overlap between the Grad-CAM high-activation regions and anatomically relevant thoracic areas. In comparison with the baseline ResNet-50, the DenseNet-121 configuration exhibited improved localization consistency, suggesting that the optimized architecture relies more strongly on clinically relevant pulmonary structures during prediction.

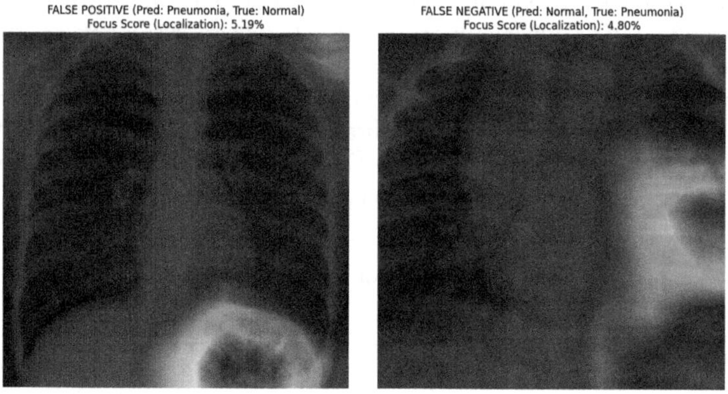

Fig. 3. Grad-CAM heatmaps for the baseline ResNet-50 model (20 epochs). Activations show limited localization toward pulmonary regions and increased attention to peripheral structures such as shoulders and image borders.

Based on these observations, the DenseNet-121 configuration was extended to 50 training epochs in order to evaluate the stability of the learning process and the evolution of its predictive behavior. Figure 5 presents the training and validation curves obtained during the optimization. Although the training loss decreased steadily and the accuracy converged stably, the validation curves exhibited considerable variability. As will be discussed later, this behavior is largely associated with the extremely limited validation split available in the

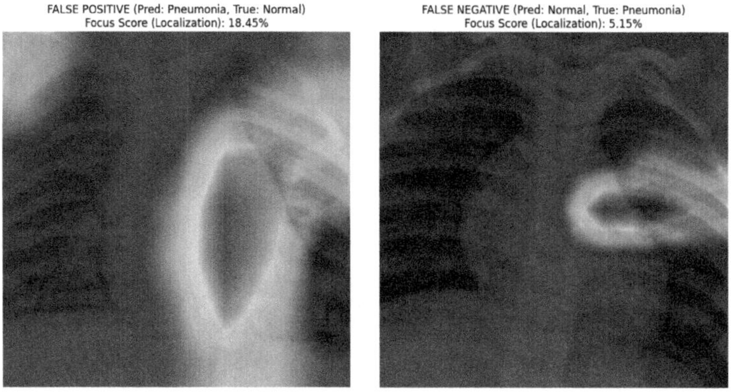

Fig. 4. Grad-CAM heatmaps for the DenseNet-121 model (20 epochs), showing improved localization toward anatomically relevant thoracic regions compared to the baseline model.

dataset (16 images), which amplifies the fluctuations between epochs without substantially affecting the final test performance.

To further evaluate the classification behavior, a confusion matrix was generated for the final test set containing 624 chest X-ray images (Fig. 6). The model correctly identified 223 NORMAL cases and 370 PNEUMONIA cases, while it produced only 11 false positives and 20 false negatives. From a clinical perspective, minimizing false negatives is especially important because missed pneumonia cases can delay treatment. In this context, the relatively low number of false negatives suggests the potential usefulness of the model as a diagnostic support system.

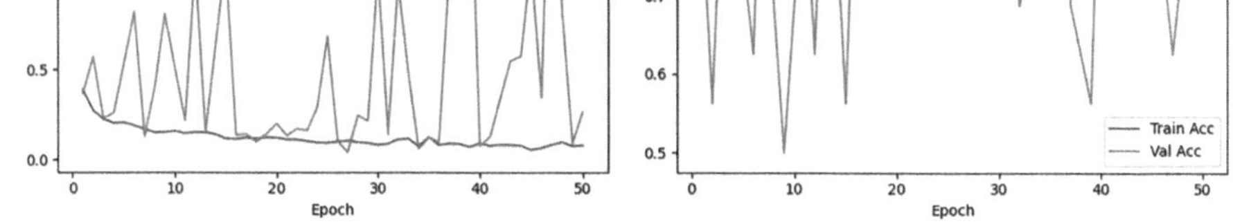

Fig. 5. Loss and Accuracy curves during the 50 epochs of training and validation for the optimized DenseNet-121.

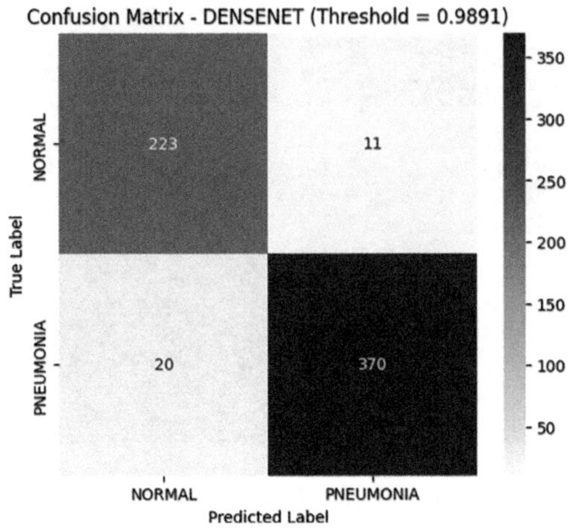

Fig. 6. Confusion Matrix evaluating the test dataset.

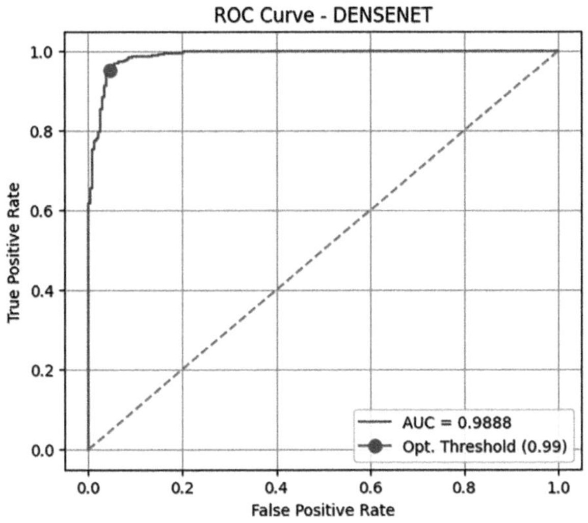

Fig. 7. ROC curve illustrating the discrimination capability of the optimized DenseNet-121 model between NORMAL and PNEUMONIA classes.

The ROC curve presented in Fig. 7 demonstrates strong discrimination between the NORMAL and PNEUMONIA classes, achieving an Area Under the Curve (AUC) of 0.9888. This result indicates that the model maintains consistent classification behavior across different decision thresholds. Since minimizing false negatives is particularly important in pneumonia screening scenarios, an additional threshold tuning strategy was implemented to prioritize sensitivity (recall) while preserving balanced predictive performance. Table 2 summarizes the clinical metrics obtained using the optimized decision threshold of 0.9891.

Table 2. Clinical evaluation metrics and Precision-Recall balance for the optimized DenseNet-121 (50 epochs, Threshold = 0.9891).

Metric	Value	95% Confidence Interval (CI)
Accuracy	95.03%	[0.9333, 0.9674]
Sensitivity (Recall)	94.87%	[0.9268, 0.9706]
Specificity	95.30%	[0.9259, 0.9801]
Precision	97.11%	[0.9543, 0.9879]
F_1-Score	95.98%	-

Under this threshold configuration, the model achieved a sensitivity of 94.87% while maintaining a precision of 97.11% and a specificity of 95.30%. This balance suggests that the model reduces false alarms while remaining sensitive to pneumonia-related findings, which is desirable in AI-assisted diagnostic support systems

To analyze the internal representations learned by the optimized model, Principal Component Analysis (PCA) was applied to the embeddings generated before the final classification layer. As shown in Fig. 8, the resulting projection reveals a noticeable separation between NORMAL and PNEUMONIA samples, suggesting that the network learned discriminative feature representations associated with both classes. Although a partial overlap between certain samples remains visible, the overall clustering tendency indicates that the model organizes the embedding space according to radiologically relevant patterns. This behavior is consistent with the classification performance obtained previously and suggests that the network captures meaningful visual structures associated with pneumonia detection rather than relying exclusively on random or noisy correlations.

Complementarily, Grad-CAM was applied to the final 50-epoch configuration to investigate the anatomical regions contributing to the predictions (Fig. 9). In both correctly classified and misclassified cases, the activation maps remained primarily concentrated within the thoracic cavity. Interestingly, several false positive cases exhibited activations around vascular structures and central thoracic opacities, suggesting that some anatomical patterns may share visual similarities with pulmonary infiltrates. These observations reinforce the relevance of explainability analysis for understanding potential sources of diagnostic confusion in medical AI systems.

Overall, the obtained results demonstrate that the progressive fine-tuning of DenseNet-121 improves not only predictive performance, but also the anatomical consistency of the activation patterns observed during the classification process. The integration of quantitative evaluation metrics together with PCA-based embedding analysis and Grad-CAM visual explanations provides a broader perspective on the internal behavior of the model and its decision-making process. These findings establish a foundation for the interpretative, clinical, and ethical considerations discussed in the following section.

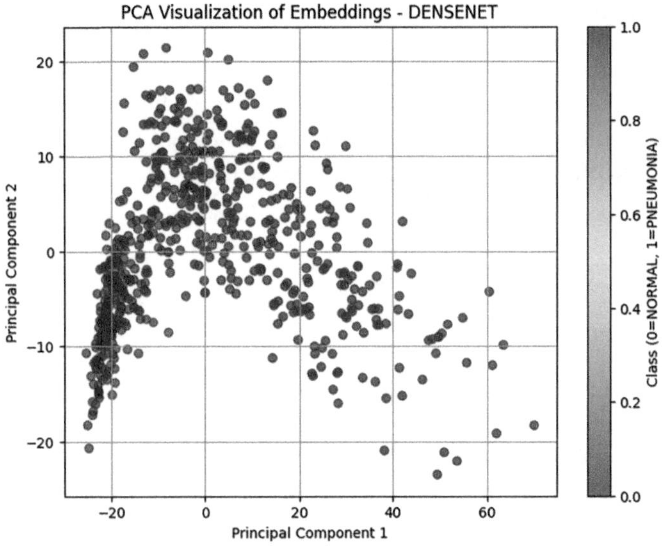

Fig. 8. PCA visualization of the embeddings showing the distribution of NORMAL and PNEUMONIA embeddings in the reduced feature space.

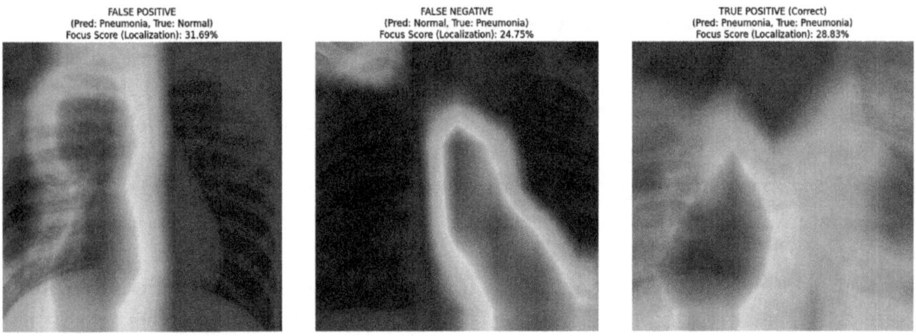

Fig. 9. Grad-CAM heatmaps for the optimized DenseNet-121 model (50 epochs). Activations remain primarily concentrated within anatomically relevant thoracic regions for both correctly classified and misclassified samples (TP/FP/FN).

4 Discussion

Recent advances in medical image analysis have demonstrated the effectiveness of deep learning architectures for automated pneumonia detection, particularly CNN-based approaches such as DenseNet and attention-enhanced models [1–3,5]. Among these architectures, DenseNet-121 has shown strong performance due to its efficient feature propagation, feature reuse capability, and stable optimization behavior, making it a suitable backbone for chest X-ray classification tasks. Nevertheless, although recent studies frequently improve predictive performance through ensembles, attention mechanisms, or hybrid CNN-ViT architectures, many of these approaches still provide limited transparency regarding how predictions are generated in clinical scenarios.

From this perspective, the results obtained in this study suggest that competitive performance can be achieved while incorporating complementary analysis techniques capable of providing a greater understanding of the internal representations of the model. In particular, PCA was used to explore the learned embedding space and visualize the separability of classes, while Grad-CAM was applied as a post hoc explainability method to identify the anatomical regions that influence the classification process. Taken together, these analyses indicate that the model learns relevant radiological patterns associated with pneumonia detection rather than relying exclusively on superficial visual artifacts.

In contrast to more computationally demanding approaches reported in recent literature, such as attention-based ensemble models [2] and hybrid CNN-ViT frameworks [5], the proposed methodology demonstrates that a single fine-tuned DenseNet-121 architecture can still achieve strong predictive performance while maintaining a comparatively simpler structure. To position these findings within the context of related work, Table 3 summarizes representative approaches reported in recent studies and contrasts them with the proposed model in terms of predictive performance and interpretability characteristics.

As shown in Table 3, architectures based on ensembles and transformers [2,5] achieve slightly higher accuracy values; however, these improvements are commonly associated with increased computational overhead and reduced model transparency. In comparison, the proposed configuration of DenseNet-121 demonstrates competitive performance, reaching an accuracy of 95.03%, an AUC of 0.9888, a sensitivity of 94.87% and a specificity of 95.30%. At the same time, the combination of PCA-based embedding analysis and Grad-CAM visual explanations provides additional insight into the model's prediction behavior regarding the predictive behavior of the model.

Although the results obtained are encouraging, several limitations should be considered when interpreting these findings. The Kaggle dataset"Chest X-ray Imaging (Pneumonia)" is derived from a pediatric population collected at a single medical institution in Guangzhou, China, which introduces demographic and acquisition-related biases. Also, the original validation split contains only 16 images, which contributes to the observed variance in the validation curves and increases the possibility of overfitting to a particular radiographic distribution. Consequently, the current study should be interpreted primarily as a proof of concept rather than as a clinically implementable diagnostic system.

In practical medical environments, broader validation across multi-centric datasets, heterogeneous patient populations, and diverse radiographic acquisition protocols would be necessary to ensure stronger generalization capabilities. Similarly, future research should incorporate quantitative explainability metrics and expert clinical assessment to determine whether Grad-CAM activations consistently correspond to medically meaningful pulmonary regions.

Beyond the technical limitations discussed previously, the development of AI systems for healthcare applications also involves important ethical and clin-

Table 3. Comparison of the proposed model with representative approaches from recent literature.

Study	Architecture/Method	Accuracy (%)	Explainability/Analysis
An et al. [2]	EfficientNet-B0 + DenseNet-121 (Attention)	96.20	Partial
Cha et al. [3]	CNN with Attention Mechanisms	95.80	No
Mabrouk et al. [5]	Deep CNN + Vision Transformers (ViT)	98.40	No
Akshay et al. [1]	DenseNet-201 (Standalone)	93.50	No
Proposed Model	**DenseNet-121 (Fine-tuned + XAI)**	**95.03**	**Embedding Analysis + Grad-CAM**

ical considerations. Medical imaging datasets, particularly chest X-rays, contain sensitive diagnostic information and may reflect demographic, institutional, and acquisition-related biases that can influence model behavior. In the present study, the use of a publicly available pediatric dataset collected from a single medical institution highlights the importance of carefully interpreting the obtained results and avoiding overgeneralization to broader clinical populations.

From this perspective, automated pneumonia detection systems should be considered decision-support or "second reader" tools rather than replacements for professional radiological judgment. In medical AI applications, the incorporation of interpretability mechanisms is especially important not only from a technical perspective, but also from an ethical standpoint. Black-box predictions in healthcare may reduce clinical trust and make it difficult to identify potential diagnostic biases or incorrect model reasoning. Therefore, visual explainability techniques such as Grad-CAM become particularly relevant because they provide insight into the anatomical regions influencing the model's predictions, allowing clinicians to better understand whether the model is focusing on medically meaningful pulmonary patterns instead of irrelevant image artifacts.

In this sense, interpretability contributes to promoting transparency, supporting human supervision, and facilitating more responsible use of medical imaging data in AI-assisted diagnostic systems. These aspects are essential for reducing automated bias, strengthening clinical confidence, and encouraging safer integration of artificial intelligence into real-world healthcare workflows.

5 Conclusions

This study demonstrates that pre-trained Deep Learning architectures such as DenseNet-121 can be effectively applied to the automatic detection of pneumonia in chest X-ray images, achieving strong classification performance with an accuracy of 95.03%, an AUC of 0.9888, and a sensitivity of 94.87%. The obtained results suggest that the model is capable of learning discriminative radiological representations that enable the differentiation between normal and pneumonia cases.

Beyond predictive performance, this work highlights the importance of incorporating explainability mechanisms into medical AI systems. PCA-based embedding analysis and Grad-CAM visual explanations provided additional insight into the internal behavior of the model, allowing a better understanding of the anatomical regions and visual patterns influencing the classification process. These analyses not only improve transparency, but also help identify visual characteristics and activation patterns that may potentially contribute to incorrect predictions or clinically misleading decisions.

From this perspective, explainability techniques become valuable tools for understanding how Deep Learning models reason over medical images and for detecting possible biases or confusing visual characteristics that could affect diagnostic reliability. Although the proposed approach achieved competitive performance, the current study should still be interpreted as a proof-of-concept due to the limitations associated with the dataset and the lack of external clinical validation.

Future work will focus on incorporating lung segmentation techniques, extending the model toward multi-class classification scenarios, and integrating quantitative explainability metrics together with expert clinical assessment. Additionally, future studies may further investigate the visual characteristics, activation patterns, and anatomical regions associated with false positives and false negatives in order to better understand potential sources of diagnostic confusion in AI-assisted medical systems.

References

1. Akshay, S., Kashyap, S., Danikk Patel, H.N., Pavan Kumar, Y.R.: Hybrid deep learning and machine learning framework for automated pneumonia detection in chest X-ray images. MethodsX (2025). https://www.sciencedirect.com/science/article/pii/S2215016125005734
2. An, Q., Chen, W., Shao, W.: A deep convolutional neural network for pneumonia detection in x-ray images with attention ensemble. Diagnostics **14**(4), 390 (2024)
3. Cha, S., Lee, S., Ko, B.: Attention-based transfer learning for efficient pneumonia detection. Appl. Sci. **11**(3), 1242 (2021)
4. El-Kenawy, E.-S.M., et al.: Metaheuristic optimization for improving weed detection in wheat images captured by drones. Mathematics **10**(23), 4421 (2022). https://www.mdpi.com/2227-7390/10/23/4421
5. Mabrouk, A., Díaz Redondo, R.P., Dahou, A., Abd Elaziz, M., Kayed, M.: Pneumonia detection on chest x-ray images using ensemble of deep convolutional neural networks. arXiv preprint arXiv:2312.07965 (2023)
6. Mooney, P.: Chest x-ray images (pneumonia) (2018). https://www.kaggle.com/datasets/paultimothymooney/chest-xray-pneumonia
7. Zheng, B., Chen, Y.: Kolmogorov-arnold networks guided whale optimization algorithm for feature selection in medical datasets. J. Big Data (2025). https://doi.org/10.1186/s40537-025-01116-7

Stratification of Metabolic-Intestinal Risk Profiles Using Unsupervised Learning

Ana Paulina Delgadillo Bustos[1], Evelin Sarahi Corona Corona[1], Ana Julia Ceja Rodríguez[1], Alejandra Meza Rios[2], and German Preciat[1](✉)

[1] Department of Translational Bioengineering, University of Guadalajara, Guadalajara, Mexico
german.preciat@academicos.udg.mx

[2] Clinical Analysis Laboratory, University of Guadalajara, Guadalajara, Mexico

Abstract. The clinical evaluation of the microbiota and its metabolic impact represents a multidimensional challenge, traditionally dependent on high-cost genomic testing. In this work, a computational approach based on machine learning is proposed for non-invasive stratification using clinical proxy variables of metabolic-intestinal risk profiles. By processing anthropometric, physiological, and behavioral variables, a hybrid algorithmic pipeline was implemented. Anomaly detection was performed using Isolation Forest, while profile stratification was probabilistically modeled using Gaussian Mixture Models. Principal component projection enabled visualization of the spatial organization of the data within a context of high overlap. The proposed framework identified three exploratory profiles distributed across a continuous risk spectrum. The model identified patterns compatible with subclinical metabolic risk in individuals with normal weight. Post-hoc analysis showed significant differences between clusters in adiposity-related variables (BMI and waist risk), as well as partial contributions from behavioral and intestinal stress factors. The moderate stability of the model and the low global separation values suggest a continuous and multifactorial biological structure. Overall, these findings indicate that the interaction between adiposity, lifestyle habits, and physiological variables defines multiple partially independent axes of risk. Finally, the potential utility of unsupervised computational approaches for exploratory metabolic-intestinal risk stratification in settings where direct experimental validation is not feasible.

Keywords: Unsupervised Learning · Gaussian Mixture Model · Isolation Forest · Gut Microbiota · Metabolic Risk · Computational Medicine · Preventive Medicine · Biomedical Engineering

1 Introduction

The gut microbiota ecosystem plays a fundamental role in systemic homeostasis, immunological modulation, and human energy metabolism [1]. Structural alterations in this microbial community, clinically known as dysbiosis, are associated with the

The authors "Ana Paulina Delgadillo Bustos, Evelin Sarahi Corona Corona and Ana Julia Ceja Rodríguez" have been contributed equally to this chapter.

H. Ponce et al. (Eds.): COMIA 2026, CCIS 3095, pp. 756–768, 2027.
https://doi.org/10.1007/978-3-032-37438-7_59

development of chronic non-communicable diseases, including metabolic syndrome, obesity, and severe inflammatory disorders [2]. Despite its clinical relevance, the quantitative assessment of microbiota status in routine medical practice faces significant barriers. Traditional methodologies rely heavily on genomic sequencing (such as 16S rRNA analysis) or invasive and high-cost procedures, limiting their feasibility as early screening tools at the population level [3].

In parallel, the characterization of metabolic risk based on routine clinical variables and lifestyle surveys generates highly multidimensional and stochastic datasets. In systems biology, indicators of health and disease often do not separate along rigid or linear boundaries; rather, they tend to be organized along continuous gradients with a high degree of overlap [4]. In this context, traditional statistical analysis methods are limited in their ability to capture the complex, latent, and nonlinear interactions among behavioral habits, body composition, and gastrointestinal physiology [5].

To address these limitations, this work proposes the development of a computational model based on machine learning techniques for the estimation of non-invasive proxies of metabolic-intestinal risk. Specifically, the approach integrates probabilistic clustering through Gaussian Mixture Models (GMM) and anomaly detection using Isolation Forest. By combining these artificial intelligence techniques, the study explores whether probabilistic clustering techniques can identify latent patterns in anthropometric variables and lifestyle metrics, facilitating the characterization of clinical profiles within a continuous spectrum of risk, rather than as discrete categories [4, 5].

2 Methodology

A computational model based on unsupervised learning is proposed for the stratification of intestinal health profiles using clinical and lifestyle variables. The methodological approach prioritizes physiological interpretability, integrating statistical techniques and artificial intelligence within a structured four-stage pipeline: (i) data acquisition and consolidation, (ii) preprocessing and feature engineering, (iii) unsupervised modeling, and (iv) multivariable analysis and interpretation (Fig. 1).

2.1 Feature Engineering and Biometric Consolidation

Data were collected using non-invasive instruments and point measurements within an exploratory context. The information was anonymized and used exclusively for computational research purposes, without clinical intervention or medical decision-making.

The original variables were transformed into homogeneous numerical representations through categorical encoding schemes and ordinal scaling. Subsequently, derived variables were generated to reduce redundancy and improve physiological interpretability. The definition and composition of these clinical and behavioral indices, including the specific items from the assessment instrument, are detailed in Table 1. For instance, body mass index (BMI) was calculated from weight and height, after which these original variables were removed to avoid multicollinearity.

Additionally, scale inversion transformations were applied to variables in which higher values represented lower risk (e.g., diet quality), with the aim of maintaining a

Table 1. Definition and composition of clinical and behavioral variables

Variable	Definition / Components
Diet Score	Composite index evaluating fiber intake, fermented foods, ultra-processed foods, and non-caloric sweeteners.
Microbiota-related Stress Proxy	Assesses historical biological insults: recent antibiotic use, recent GI infections, and annual frequency of diseases.
Lifestyle Score	Integration of physical activity frequency, sedentary hours, and perceived daily stress levels.
Glycemic Risk	Categorical assessment based on self-reported blood glucose levels and clinical history.
Waist Risk	Binary risk assignment based on abdominal circumference measurements following international health standards.
Body Mass Index	$BMI = \frac{weight(kg)}{height(m)^2}$

Note 1: As mentioned in the preprocessing stage, scales were inverted where necessary to ensure that higher values consistently represent a higher risk profile.

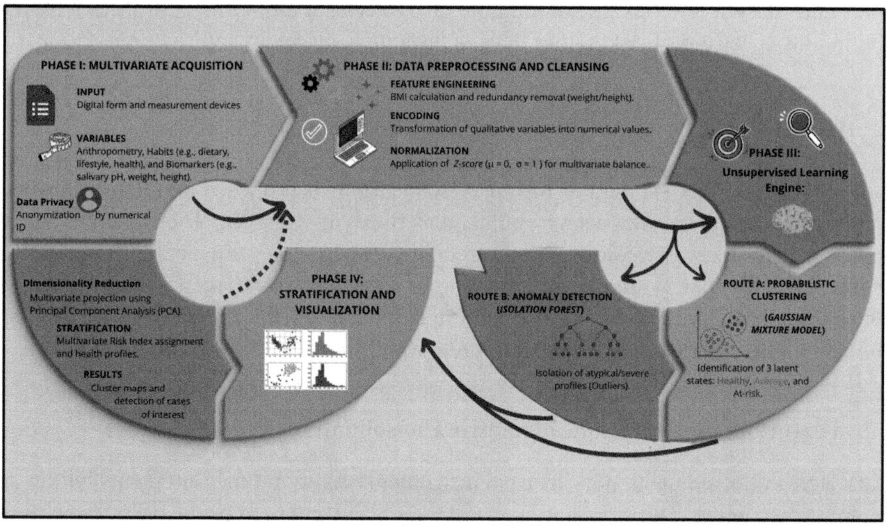

Fig. 1. System architecture and methodological workflow for stratification. The process is divided into four main phases: multivariable acquisition with data anonymization (Phase I), a preprocessing engine based on feature engineering and Z-score normalization (Phase II), an unsupervised learning core with parallel pathways of probabilistic clustering using GMM and anomaly detection via Isolation Forest (Phase III), and a final stage of visualization through dimensionality reduction using Principal Component Analysis for risk index assignment (Phase IV). The workflow summarizes the sequential stages of preprocessing, unsupervised modeling, and multivariable interpretation

consistent semantic direction, where higher values indicate greater risk. For handling missing data, median imputation was employed, preserving the structure of the dataset without introducing bias associated with extreme values.

The data collection instruments were designed under standardized protocols, based on prior evidence that enabled the identification of clinically relevant variables for intestinal health. In this phase, raw data obtained from digital instruments and anthropometric measurements are transformed into consolidated indices through algebraic operations, integrating variables with shared physiological origin and weighting factors with high clinical impact.

To ensure consistency in risk directionality, a scale inversion transformation was applied:

$$V_{norm} = K - V_{obs}$$

where V_{norm} represents the normalized value, K is the maximum constant of the original scale, and V_{obs} is the observed value. This procedure ensures that higher values consistently reflect a higher risk profile. Subsequently, multicollinearity is mitigated by removing base variables after their integration into composite indices.

Regarding data integrity, a minimal missingness rate was observed across the clinical and behavioral variables. Median imputation was utilized to maintain robust distributions, as it is less sensitive to outliers in clinical metrics. Categorical features were transformed using ordinal and binary encoding to preserve physiological hierarchy. Finally, the potential impact of scale disparity was neutralized through Z-score normalization. This transformation ensured an equitable contribution of all metrics to the clustering and anomaly detection algorithms, regardless of their original units or scales.

2.2 Probabilistic Clustering and Anomaly Detection

A hybrid unsupervised learning pipeline was implemented:

- **GMMs** were used to identify latent groupings within the data by modeling the distribution as a mixture of Gaussian components. To ensure the suitability of the feature space for GMM, Z-score normalization was applied to all variables, transforming ordinal and clinical metrics into a continuous, scaled representation (Phase II). This preprocessing mitigates the impact of non-gaussian raw distributions and enables the algorithm to effectively capture the underlying probabilistic structure of the dataset. The optimal number of components was determined by balancing statistical fit, model stability, and clinical interpretability (Table 2). Information criteria were interpreted comparatively rather than as absolute optimization targets due to the exploratory nature of the study and the moderate sample size. While information criteria (AIC and BIC) showed a downward trend for higher values of K, suggesting a tighter fit, a significant decrease in model stability was observed beyond K = 3. This behavior suggests increased model flexibility at higher K values, potentially reflecting overfitting rather than stable biological structure. The Adjusted Rand Index (ARI) dropped from 0.54 for $K = 3$ to 0.43 for $K = 4$, indicating that higher-order models produced less consistent partitions. Furthermore, the three-cluster configuration provided the most clinically interpretable structure, identifying a clear "three-level risk stratification"

(Low Risk, Transition, and High Risk) that would be fragmented and lose biological meaning in more complex models.

- **Isolation Forest** was employed to detect anomalous observations within the dataset. It identifies points that require fewer partitions to be isolated, enabling the detection of extreme profiles potentially associated with atypical metabolic-intestinal configurations not representative of the general population.

Table 2. Model selection metrics for different numbers of clusters (K)

Number of Clusters (K)		AIC	BIC	Silhouette Score	ARI Stability
2		2047.78	2467.76	0.1515	0.5955
3		2052.17	2683.49	0.0868	0.5448
4		1419.31	2261.97	0.1020	0.4327
5		213.39	1267.39	0.1156	0.4755

2.3 Statistical Analysis and Cluster Validation

To evaluate the internal coherence of the clusters, the non-parametric Kruskal–Wallis test was applied, followed by a post-hoc analysis using Dunn's test with Bonferroni correction. This approach enabled the identification of significant differences between groups in clinical and lifestyle variables, supporting the existence of differentiated risk axes beyond simple geometric separation in feature space.

2.4 Multivariable Visualization and Interpretation

To facilitate interpretation and reduce system dimensionality, Principal Component Analysis (PCA) was applied, projecting the data into a two-dimensional space that preserves the maximum possible variance. This procedure allows for the identification of global patterns without significantly compromising the underlying data structure. Within this reduced space, where each point represents an individual, both the latent clusters and anomaly scores obtained from Isolation Forest were projected. This integration enables a direct visual interpretation of population stratification, highlighting both dominant groupings and extreme profiles within a continuous physiological spectrum.

3 Results

This section presents the findings derived from the analysis of a sample of 111 subjects, using an unsupervised learning pipeline complemented by post-hoc interpretability and statistical validation analyses.

3.1 Detection of Atypical Phenotypes and Multivariable Analysis

The Isolation Forest algorithm was implemented to identify individuals whose clinical profiles diverge from the average behavior of the sample. The model assigned a rarity score, revealing that the population density tends toward normality (positive scores), while a subgroup was identified as anomalous in the negative tail of the distribution (Fig. 2a).

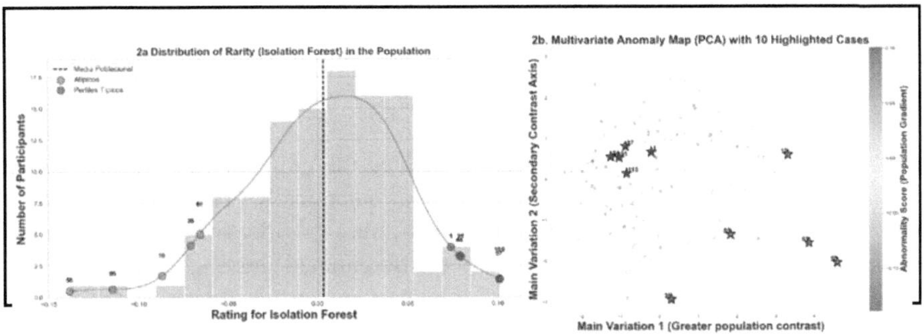

Fig. 2. Analysis of anomaly distribution and multivariate dispersion. Figure 2a shows the density distribution of the multivariate Anomaly Scores calculated using the Isolation Forest algorithm. This score represents the integrated contribution of all input features into a single metric; most of the sample ($n = 111$) is concentrated in positive values, indicating standard behavior within the model's structure, while red points identify individuals with the most extreme deviations. Figure 2b projects these same data into a PCA space, where the detected anomalies (red stars) are dispersed toward the periphery of population variance, confirming the model's ability to isolate profiles with atypical metabolic risk

Through a PCA projection, the spatial distance between observations was visualized. Extreme outlier cases (e.g., IDs 58 and 61) werelocated at the periphery of the scatter plot (Fig. 2b). The observed separation was primarily associated with the convergence of specific variables, as detailed comparatively in Table 3. In the case of extreme anomalies, the algorithm consistently detected abnormally high BMI values ($>40\ kg/m2$), elevated glucose categorization, and reduced scores in dietary habit indicators, in stark contrast to the central metrics of typical profiles (e.g., IDs 1 and 110).

Table 3. Clinical profiles extracted using Isolation Forest

Patient ID	Profile Classification	Rarity Score	Determining Variables
58	Extremely atypical	0.139	Glycemic Risk Category BMI Diet Score
61	Extremely atypical	0.066	BMI Glycemic Risk Category Resting Tachycardia
1	Typical	0.074	Diet Score Glycemic Risk Category Variable 5.4*

(continued)

Table 3. *(continued)*

Patient ID	Profile Classification	Rarity Score	Determining Variables
110	Typical	0.101	Lifestyle Score Fitzpatrick Scale Diet Score

*Note: "Variable 5.4" corresponds to the presence or prior diagnosis of digestive pathologies (e.g., Irritable Bowel Syndrome, Crohn's disease, ulcerative colitis, among others), self-reported by participants through the clinical assessment instrument.

3.2 Population Stratification Using GMM

The application of the probabilistic model enabled the projection of the sample onto the first two principal components, resulting in a semantically structured segmentation map conceptualized as a three-level probabilistic risk stratification map (Fig. 3). The algorithm grouped the observations into three differentiated clusters, selected as a compromise between statistical fit and clinical coherence, given the continuous and partially overlapping nature of the data.

The first two principal components accounted for 32.31% of the total variance (PC1: 18.56%; PC2: 13.75%). This cumulative variance is sufficient to represent the underlying multidimensional structure of the clinical data while maintaining visual interpretability. Although the first two principal components explain 32.31% of total variance, the projection is intended as a visualization aid rather than a complete representation of the multidimensional structure. The semantic interpretation of the axes was grounded in the feature loadings: PC1 (conceptualized as the Health-Illness Axis) was primarily driven by adiposity metrics, specifically Waist Risk and BMI, which show the highest contribution to the model's spatial structure. Conversely, PC2 (designated as the Lifestyle Axis) was influenced by behavioral factors, notably Diet Score and Lifestyle Score. This mathematical alignment justifies the use of the PCA map as a visual proxy for the transition between the different clinical profiles identified by the algorithm, reflecting the internal logic of the model's distribution.

- **Cluster 2 (Green Zone – Low Risk):** Groups individuals with stable baseline metrics. Mathematically, this cluster exhibited a mean BMI of $22.97\ kg/m2$ and concentrated the lowest numerical scores on the intestinal irritation scale.
- **Cluster 1 (Yellow Zone – Transition):** Represents the majority segment of the population. In terms of variables, individuals in this group maintain normative anthropometric metrics (with an average BMI of 23.37 $kg/m2$), and the algorithm grouped here the highest quantitative scores across the sample in diet quality and lifestyle metrics.
- **Cluster 0 (Red Zone – High Dispersion):** Integrates observations with the greatest spatial deviation from the low-variance centroid. This group presented an average BMI of 30.90 $kg/m2$ and a complete convergence in waist risk. The morphology of this cluster shows direct overlap with the atypical cases previously detected using Isolation Forest.

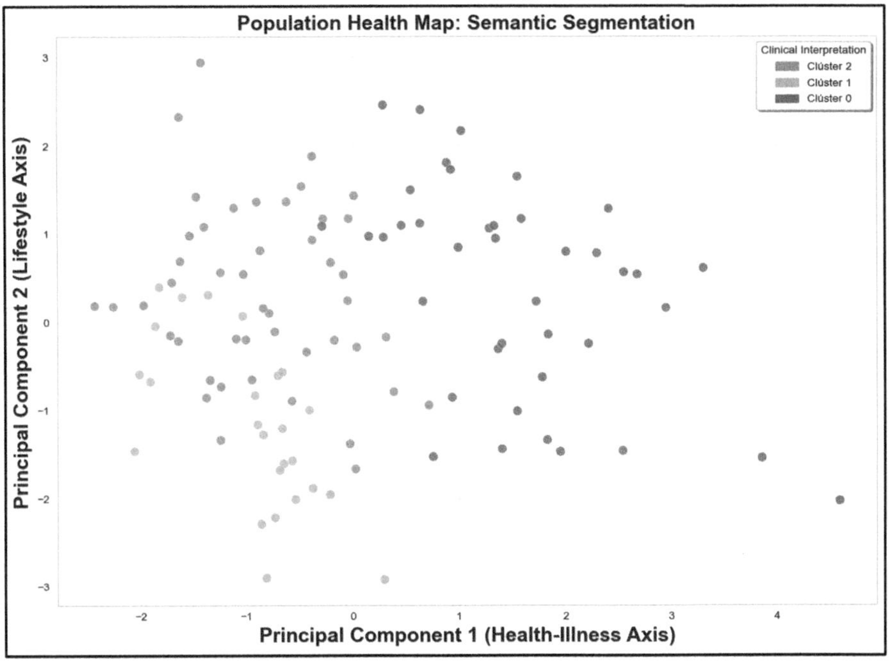

Fig. 3. Phenotypic segmentation map using GMM. Illustrates the stratification of the sample into three differentiated clinical clusters, projected onto the first two principal components. The color coding was conceptualized as a three-level risk stratification map based on membership probability

To evaluate the internal consistency of the model, differences between clusters were analyzed using non-parametric post-hoc statistical tests. Highly significant differences were observed in adiposity-related variables (BMI and waist risk, $p < 0.001$), particularly between Cluster 0 and the other groups. Additionally, glucose category and Microbiota-related Stress Proxy index showed significant differences primarily associated with Cluster 2 ($p < 0.01$), while behavioral variables (Diet Score and Lifestyle Score) exhibited partial differences across clusters ($p < 0.05$).

Furthermore, model stability was assessed through iterative clustering comparisons, yielding an average Adjusted Rand Index (ARI) of 0.52, indicating moderate consistency in the segmentation structure.

3.3 Phenotypic Characterization and Clinical Centroid Analysis

Following the validation of probabilistic segmentation, centroid extraction was performed for each identified group. This process enabled the translation of the model's mathematical coordinates into tangible physiological and behavioral metrics.

Quantitative analysis reveals a marked transition between clusters, highlighting a difference of nearly 8 BMI points between Cluster 2 (22.97 kg/m2) and Cluster 0 (30.90 kg/m2), accompanied by a saturation in waist risk exclusive to the latter.

On the other hand, a differential pattern was identified in Cluster 2: despite presenting low-risk anthropometric metrics and high scores in dietary quality (*Diet Score* = 12.60), this group exhibited elevated values in Microbiota-related Stress Proxy (17.00) and glucose category (0.60), suggesting a dissociation between adiposity and other physiological axes (Table 4).

Table 4. Characterization of average clinical metrics by intestinal health profile

Study Variable	Cluster 0	Cluster 1	Cluster 2
BMI (kg/m2)	30.90	23.37	22.97
Abdominal Adiposity Risk (Scale 0–1)	1.0	0.0	0.0
Microbiota-related Stress Proxy (Score)	14.17	13.21	17.00*
Diet Score (Habits)	3.52	10.08	12.60
Lifestyle Score	−18.85	−16.08	−22.88
Glycemic Risk Category	0.27	0.08	0.60*

*Note: Values marked with an asterisk denote atypically elevated metrics detected by the algorithm within the profile exhibiting the lowest anthropometric risk. Negative values in the Lifestyle Score reflect post-normalization standardized values (Z-scores), where values below zero indicate observations below the population mean rather than negative physiological states.

Overall, these results highlight the presence of multiple partially independent clinical dimensions within the sample, rather than a single pathophysiological gradient.

3.4 Variable Importance Analysis

To determine the relative contribution of each biomarker in shaping the identified phenotypes, a Random Forest classification model was implemented as a descriptive post-hoc analysis using cluster labels as the target variable. These assignments were treated as proxy labels to explore the internal logic of the GMM grouping, rather than as an independent predictive validation. This approach allows for the identification of which features the unsupervised model prioritized during the stratification process.

Variable importance was quantified using the Mean Decrease Impurity (MDI), also known as Gini Importance. Mathematically, this metric calculates the total reduction in the Gini impurity index brought by each feature across all trees in the forest. A higher MDI value indicates that the variable plays a more decisive role in partitioning the data into the specific GMM clusters. The importance values were normalized such that their cumulative sum equals 1.0, representing the relative contribution of each biomarker to the model's feature contribution within the clustering structure.

The variable importance analysis revealed a clear hierarchy (Fig. 4). Adiposity-related metrics dominated the model structure, with waist risk as the primary predictor (0.3517), followed by BMI (0.2026). At a second level, behavioral indices stood out: Diet Score (0.0756), Microbiota-related Stress Proxy (0.0642), and Lifestyle Score (0.0632), surpassing isolated physiological variables such as glucose (0.0548) or resting tachycardia (0.0392). Feature importance should therefore be interpreted descriptively and not inferentially.

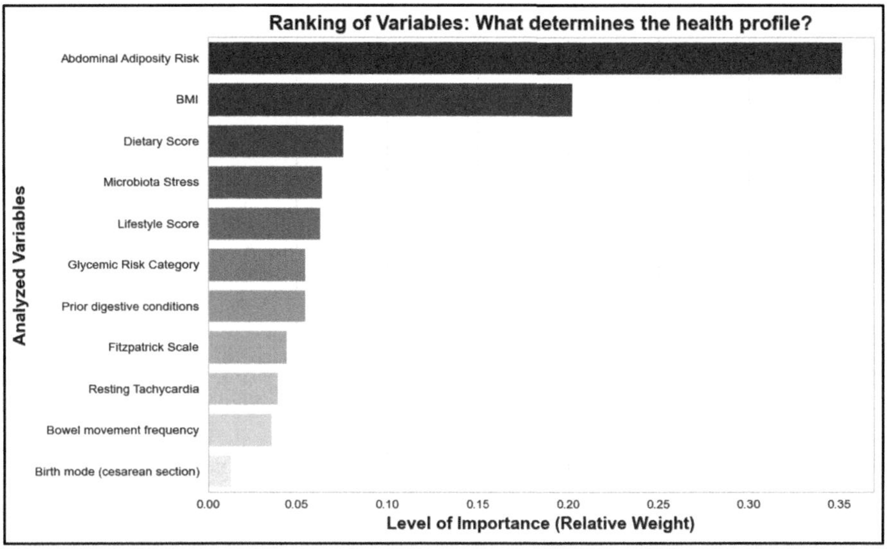

Fig. 4. Variable importance ranking using Random Forest. The chart illustrates the relative weight of each feature based on the Mean Decrease Impurity (MDI). The predominance of adiposity-related variables as primary determinants of the model's internal stratification is highlighted, followed by behavioral and gastrointestinal stress metrics. *Note 1:* this ranking represents a post-hoc interpretability analysis intended to characterize the internal logic of the GMM clusters. *Note 2:* as the model utilizes cluster assignments derived from the same feature set, these results should be interpreted as an exploration of the algorithm's grouping structure and not as an independent predictive validation of clinical risk. *Note 3:* numerical variables correspond to specific questions from the applied clinical instrument

4 Discussion

4.1 Clinical Validation of Mathematical Distance

The anomaly detection analysis yielded clinically consistent results. The spatial separation observed in the PCA reflects multivariable clinical differences between individuals across the analyzed feature space. Profiles identified by the unsupervised model at the periphery are characterized by a consistent convergence of risk, dominated by elevated adiposity, glucose alterations, and poor dietary quality. This convergence suggests that the statistical anomalies detected by the Isolation Forest algorithm align with real profiles potentially associated with metabolic vulnerability [6]. However, the moderate clustering stability (ARI ≈ 0.52) and low silhouette values indicate that these separations should be interpreted as continuous gradients of risk rather than discrete categories, which is consistent with the multifactorial nature of biological systems.

4.2 Phenotypic Stratification and Behavioral Compensation

The obtained stratification reveals clinical phenotypes consistent with current medical literature. In this regard, the methodological choice of a GMM over rigid partitioning methods proved critical; since human health indicators do not separate along exact boundaries, the probabilistic approach allowed capturing the natural overlap between metabolic profiles.

Post-hoc analysis showed that Cluster 0 differs significantly from the others in adiposity-related variables (BMI and waist risk, $p < 0.001$), consolidating it as the group with the highest structural metabolic burden. In contrast, Clusters 1 and 2 do not exhibit significant differences in these parameters, suggesting the existence of divergent pathophysiological trajectories beyond adiposity.

Additionally, the observed differences in Diet Score and behavioral variables across clusters point toward the presence of behavioral compensation mechanisms, where some individuals maintain relatively favorable habits despite metabolically compromised profiles. This behavior reinforces the idea that metabolic regulation does not depend on a single axis, but rather on the interaction of multiple partially independent dimensions, in agreement with recent evidence linking visceral adiposity, systemic inflammation, and gut microbiota modulation through complex metabolic interactions [7–9].

4.3 Exploratory Detection of Subclinical Metabolic Risk Patterns

One exploratory observation of this study lies in the behavior of Cluster 2. Despite presenting relatively favorable anthropometric indicators, this group exhibits significant differences in glucose category and Microbiota-related Stress Proxy index ($p < 0.01$).

This pattern is interpreted as a potential signal of 'invisible' or subclinical metabolic risk, where individuals with apparently favorable anthropometric metrics may already exhibit early biochemical dysfunctions. While this phenomenon aligns with the literature on metabolically unhealthy normal weight [10], within the scope of this work, it remains an exploratory hypothesis. In this context, probabilistic clustering demonstrates its utility for identifying these intermediate profiles as a preliminary screening signal that requires further clinical validation.

4.4 Lifestyle as a Structural Axis

The variable importance analysis using Random Forest provides a complementary interpretability analysis by confirming the relative contribution of each evaluated factor. While previous phases demonstrated how individuals cluster, helps characterize the features associated with their separation

The results suggest that adiposity (BMI and waist risk) constitutes the dominant axis within the model's internal structure. It is important to note that since the Random Forest explains labels derived from the same dataset, this hierarchy reflects the internal weight distribution of the clustering algorithm rather than an external validation of risk predictors. However, it is noteworthy that behavioral indices (Diet Score) and intestinal perception metrics (Microbiota-related Stress Proxy) also contribute to cluster separation, particularly in profiles where adiposity does not fully explain the observed variability.

This suggests that behavioral factors and the host's biological environment act as risk modulators rather than isolated determinants, supporting a multidimensional perspective in which intestinal health emerges from the interaction between metabolic, behavioral, and physiological components [11].

4.5 Limitations and Future Directions

Although the computational model proved useful for identifying risk patterns, this study presents relevant limitations. First, dietary and lifestyle data were self-reported, introducing potential recall and social desirability biases.

Second, the absence of longitudinal follow-up and direct experimental validation through specific biomarkers (such as microbiota sequencing or inflammatory profiles) limits the biological confirmation of the identified clusters. Therefore, states of 'latent metabolic risk' should be interpreted as exploratory inferential approximations derived from clinical and behavioral variables rather than direct biological confirmation. These findings represent a potential subclinical signal that requires rigorous clinical and experimental validation in future studies to confirm its predictive value.

Additionally, the cross-sectional design prevents establishing causal relationships between observed variables, restricting the analysis to associations. Finally, while the sample size ($n = 111$) is appropriate for an exploratory proof-of-concept, it limits the generalizability of the identified profiles to the general population. These findings should be interpreted as an initial computational approximation within a continuous and multifactorial biological system.

Future work should incorporate larger cohorts, experimental validation, and longitudinal analyses to consolidate these findings and evaluate their applicability in real clinical settings.

Acknowledgments. The authors thank Oscar Ricardo Aguayo Cruz for his contribution to data collection and all participants who volunteered for this study.

Disclosure of Interests The authors declare no competing interests.

Data Availability. The code used in this study is publicly available at: https://github.com/GIMIudg/GIMIpapers/tree/main/microbiomeScore/clustering. The dataset is available in anonymized form upon reasonable request, in accordance with ethical and data protection considerations.

References

1. Pillai, S.S., Gagnon, C.A., Foster, C., Ashraf, A.P.: Exploring the gut microbiota: key insights into its role in obesity, metabolic syndrome, and type 2 diabetes. J. Clin. Endocrinol. Metab. **109**(11), 2709–2719 (2024). https://doi.org/10.1210/clinem/dgae499
2. Singh, A., et al.: Role of gut microbiota in the pathogenesis of metabolic syndrome: an updated comprehensive review from mechanisms to clinical implications. Ann. Med. Surg. **87**(9), e3656 (2025). https://doi.org/10.1097/MS9.0000000000003656
3. Matchado, M.S., et al.: On the limits of 16S rRNA gene-based metagenome prediction and functional profiling. Microb. Genom. **10**(2), 001203 (2024). https://doi.org/10.1099/mgen.0.001203

4. Plaza-Diaz, J., Herrera-Quintana, L., Olivares-Arancibia, J., Vázquez-Lorente, H.: Personalized nutrition through the gut microbiome in metabolic syndrome and related comorbidities. Nutrients. **18**(2), 290 (2026). https://doi.org/10.3390/nu18020290
5. Datta, P.R., Roy, K.: Explainable machine learning framework for predicting cardiometabolic risk using meal timing and eating habits. *medRxiv*. (Preprint). (2026). https://doi.org/10.64898/2026.02.06.26345732
6. Lee, J.Y., Kim, D.H., Choi, J.W., Shong, M., Lee, C.K.: Utility of machine learning to characterize gut microbiota Dysbiosis and its clinical implications in inflammatory bowel disease. J. Inflamm. Res. **18**, 16857–16870 (2025). https://doi.org/10.2147/JIR.S546260
7. Collins, S.L., Stine, J.G., Bisanz, J.E., Okafor, C.D., Patterson, A.D.: Bile acids and the gut microbiota: metabolic interactions and impacts on disease. Nat. Rev. Microbiol. **21**(4), 236–247 (2023). https://doi.org/10.1038/s41579-022-00805-x
8. Freitas, R.G.B.O.N., et al.: Gut bacterial markers involved in association of dietary inflammatory index with visceral adiposity. Nutrition. **122**, 112371 (2024). https://doi.org/10.1016/j.nut.2024.112371
9. Massier, L., et al.: Adipose tissue derived bacteria are associated with inflammation in obesity and type 2 diabetes. Gut. **69**(10), 1796–1806 (2020). https://doi.org/10.1136/gutjnl-2019-319877
10. Ruderman, N.B., Schneider, S.H., Berchtold, P.: The 'metabolically-obese,' normal-weight individual. Am. J. Clin. Nutr. **34**(8), 1617–1621 (1981). https://doi.org/10.1093/ajcn/34.8.1617
11. Zeevi, D., et al.: Personalized nutrition by prediction of glycemic responses. Cell. **163**(5), 1079–1094 (2015). https://doi.org/10.1016/j.cell.2015.11.001

Artificial Intelligence Based on Acoustic Analysis for COPD Detection with Clinical Validation: A Systematic Review

Jorge Janeman Franco-Ríos, Iván Gallardo-Bernal(✉), Humberto Ávila-Pérez, Leopoldo Rodríguez-Matías, Jossué Manuel Rodríguez-Matías, and Juana Inés Zambrano-Dávila

Facultad de Ciencias y Tecnologías de la Información, Universidad Autónoma de Guerrero, Av. Las Colinas No. 37A, 39390 Acapulco, Mexico
igallardo@uagro.mx

Abstract. Background: Chronic obstructive pulmonary disease (COPD) remains substantially underdiagnosed worldwide, particularly in settings with limited access to spirometry. Artificial intelligence (AI) models based on acoustic analysis have emerged as promising tools for respiratory assessment. Objective: To systematically review the available evidence on AI-based acoustic models for COPD detection that have undergone clinical validation. Methods: A systematic review was conducted in accordance with PRISMA 2020 across PubMed/MEDLINE, Scopus, Semantic Scholar, and Crossref. Results: Of 408 records identified, only one study met the eligibility criteria. Conclusion: The available evidence remains scarce, and further clinical validation is needed before these approaches can be considered for routine use.

Keywords: Chronic obstructive pulmonary disease · artificial intelligence · lung sounds · acoustic analysis · clinical validation

1 Introduction

Chronic obstructive pulmonary disease (COPD) is one of the leading causes of morbidity and mortality worldwide and represents a growing burden on healthcare systems, particularly in countries with unequal access to specialized diagnostic resources [1]. Timely detection is essential to initiate early interventions, reduce exacerbations, and improve quality of life; however, underdiagnosis remains a persistent problem [10]. One of the main contributors to this situation is limited access to spirometry, which is considered the reference standard for confirming airflow obstruction, particularly in primary care settings and resource-constrained regions [19].

In recent years, artificial intelligence applied to biomedical signal analysis has gained increasing relevance in the respiratory field [5]. In this context, the acoustic analysis of cough and lung sounds has been proposed as a noninvasive and potentially accessible alternative to support the detection of respiratory diseases [3, 9, 11]. Several studies have

H. Ponce et al. (Eds.): COMIA 2026, CCIS 3095, pp. 769–780, 2027.
https://doi.org/10.1007/978-3-032-37438-7_60

explored the use of machine learning and deep learning techniques to extract discriminative patterns from recordings obtained through digital stethoscopes, smartphones, and other portable devices [5, 7, 16, 18, 20].

Despite this technological progress, the available evidence still presents important limitations. Heterogeneity in acquisition devices, the lack of standardized signal capture protocols, reliance on public datasets, and limited validation outside controlled environments hinder the generalization of these models to real clinical practice. In addition, the acoustic similarity between COPD and other respiratory diseases poses an additional challenge to diagnostic specificity [21]. From a translational perspective, reporting high internal validation metrics is not sufficient; clinical evidence in real patients under conditions that approximate routine care is essential [2, 14].

Against this backdrop, the objective of this systematic review was to identify and synthesize the available evidence on artificial intelligence models based on acoustic analysis for COPD detection that have been evaluated through clinical validation in real patients.

2 Methods

2.1 Study Design

A systematic review of the literature was conducted in accordance with the PRISMA 2020 guidelines for the identification, screening, eligibility assessment, and inclusion of studies.

2.2 Eligibility Criteria

Original studies evaluating artificial intelligence models for COPD detection based on audio analysis and reporting clinical validation in real patients were included. For the purposes of this review, clinical validation was defined as evaluation of model performance in a healthcare setting or under clinical-use conditions using acoustic recordings obtained from real patients, with comparison against a diagnostic reference standard such as spirometry or GOLD criteria [6].

Diagnostic validation studies, prospective studies, cohort studies, hospital-based observational studies, hospital pilot studies, and cross-sectional studies were considered eligible, provided that they included clinical evaluation in patients with suspected or confirmed COPD. Recorded cough, lung sounds, and other respiratory acoustic recordings acquired using digital stethoscopes, smartphones, or other devices were considered eligible input signals. Studies were required to report diagnostic performance metrics such as sensitivity, specificity, accuracy, AUC, or equivalent measures.

Studies based exclusively on public or artificial datasets without clinical validation, studies with internal validation only, animal studies, studies focused on other diseases without separate results for COPD, articles that did not report diagnostic metrics, and preprints were excluded.

2.3 Information Sources and Search Strategy

The literature search was performed in PubMed/MEDLINE, Semantic Scholar, Crossref, and Scopus. These databases were selected because of their coverage of biomedicine, digital health, and artificial intelligence applied to diagnosis.

The search strategy was designed to retrieve studies related to COPD, respiratory acoustic signals, and artificial intelligence models with clinical validation. The following search equation was used:

("COPD" OR "Chronic Obstructive Pulmonary Disease" OR "Obstructive Lung Disease")
AND
("cough" OR "cough sound*" OR "lung sound*" OR "respiratory sound*" OR "breath sound*" OR "wheez*" OR "crackl*" OR "acoustic*" OR "audio")
AND
("artificial intelligence" OR "AI" OR "machine learning" OR "deep learning" OR "neural network*" OR "CNN" OR "transformer*" OR "classification model*" OR "computer-aided diagnosis")
AND
("clinical validation" OR "clinical trial*" OR "hospital" OR "real-world" OR "patient*" OR "bedside" OR "inpatient" OR "outpatient")

2.4 Study Selection

The selection process was carried out in two stages. In the first stage, all retrieved records were screened by title and abstract after duplicate removal. In the second stage, the full texts of potentially eligible articles were reviewed. Selection was conducted through independent peer review, and disagreements were resolved by consensus. The overall process of identification, screening, eligibility assessment, and inclusion is summarized in Fig. 1.

Reference management and organization of the screening process were supported by the tools shown in Table 1.

Table 1. Tools used in the systematic review process

Tool	Function in the review
Microsoft Excel	Data organization, recording of the screening process, and preparation of comparative tables.
Rayyan	Blinded, independent, and systematic screening of studies.
PRISMA 2020 tools	Preparation of the flow diagram and standardization of methodological reporting.

2.5 Data Extraction

Data extraction was performed using a standardized template. Bibliometric, methodological, and outcome variables were collected.

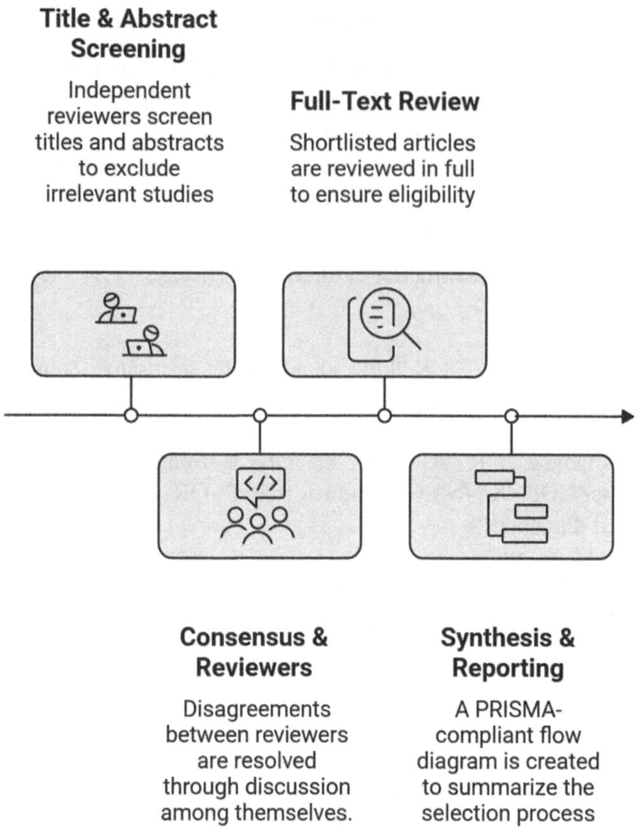

Fig. 1. Study selection workflow.

1. Bibliometric data
   - Title, authors, year of publication, journal, authors' country of affiliation, DOI, source, abstract, and keywords.
2. Methodological data
   - Study design, sample size, type of acoustic signal, recording devices, and artificial intelligence model used.
3. Outcome data
   - Sensitivity, specificity, accuracy, AUC, and other available diagnostic metrics.

2.6 Risk of Bias Assessment

Risk of bias was assessed using a methodological framework adapted from the QUADAS-2 and PROBAST criteria to the context of AI models applied to acoustic analysis for COPD detection and classification.

The assessment focused on four essential domains: sample selection, index test, reference standard, and model analysis. Risk of bias in each category was classified as low, high, or unclear, and discrepancies were resolved by consensus (Table 2).

Table 2. Domains used for risk-of-bias assessment

Domain	Criteria assessed
Sample selection	Clinical representativeness, inclusion criteria
Index test	Type of acoustic signal, model description, application
Reference standard	Spirometry based on GOLD criteria, clinical diagnosis
Model analysis	Validation strategy, diagnostic metrics

2.7 Data Synthesis

A narrative synthesis of the included studies was planned. No meta-analysis was performed because of the methodological heterogeneity among the retrieved studies and, ultimately, the inclusion of only one eligible study. Accordingly, the findings are presented descriptively, with emphasis on methodological design and reported diagnostic performance.

3 Results

3.1 Study Selection

The search across the consulted databases yielded a total of 408 records. Subsequently, 19 duplicate records were identified and removed; the remaining 389 records advanced to title and abstract screening, of which 369 were excluded for not strictly meeting the eligibility criteria.

Twenty full-text articles were then assessed for eligibility, and 19 were excluded. The main reason for exclusion was the absence of real-time clinical validation. Although several studies used data from real patients, those data were employed exclusively for model training and validation and did not involve classification testing in a real clinical setting.

Ultimately, only one study met all eligibility criteria and was included in the final analysis. The study selection process is presented in the PRISMA 2020 flow diagram (Fig. 2).

3.2 Characteristics of the Included Study

The included study was conducted by Roy et al. (2025) [15] in India and evaluated a deep learning model for COPD detection based on lung sounds. The model, named TriSpectraKAN, integrated Kolmogorov–Arnold networks with convolutional neural networks to extract and classify complex spectral patterns.

For model training, the authors combined the open-access datasets ICBHI 2017, Respiratory Database@TR, and the Chest Wall Lung Sound Database (CWLSD), together with real-time clinical validation involving 5 patients diagnosed with COPD, 2 patients with bronchiectasis, and 8 healthy participants. Acoustic signals were captured using a Littmann 3200 stethoscope, and the model was deployed on a Raspberry Pi 4 platform, demonstrating potential applicability for diagnosis in primary care. The main methodological and technical characteristics of the study are presented in Table 3.

3.3 Study Results

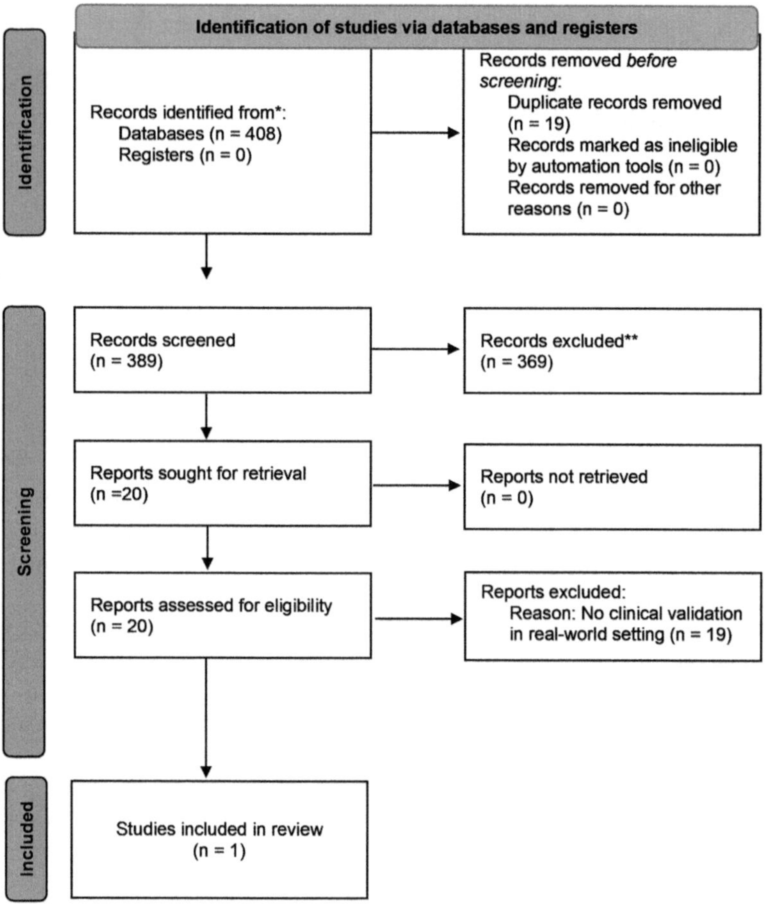

Fig. 2. PRISMA 2020 flow diagram

In terms of COPD classification performance under internal validation, the model achieved an accuracy of 93%, a sensitivity of 98%, a specificity of 92%, and an AUC of

0.98. In addition, an F1-score of 0.98 was reported, indicating a high degree of balance between predictive precision and the true-positive rate.

In the preliminary clinical validation, the system demonstrated the ability to discriminate among COPD cases, other respiratory conditions, and healthy participants. However, the clinical sample was small, and these results should therefore be interpreted as exploratory rather than confirmatory evidence.

Table 3. Characteristics of the included study

Item	Description
Author	Roy et al. (2025)
Study type	Model development and validation
Population/sample	Datasets (ICBHI 2017: 126 patients; CWLSD: 112 patients; RD@TR) + Clinical validation (5 COPD, 8 healthy participants, 2 bronchiectasis cases)
Acoustic signal	Lung sounds
Model	TriSpectraKAN (KAN + CNN)
Device	Digital stethoscope (Littmann 3200 + Raspberry Pi 4)
Main metrics	Accuracy: 93%; Recall: 98%; Specificity: 92%; AUC: 0.98; F1-score: 0.98

3.4 Synthesis of Results

Because the final selection was restricted to a single article, a comparative synthesis across different studies was not possible.

3.5 Risk of Bias Assessment

The qualitative assessment of the included study revealed relevant concerns in the model analysis domain and in external applicability. Although the clinical reference standard was appropriate, the limited size of the clinical validation sample and the combination of multiple datasets for model development increase uncertainty regarding the generalizability of the reported performance. The clinical representativeness of the sample and the robustness of validation under real-world conditions should therefore be considered limited.

4 Discussion

This systematic review identified a clear gap between algorithmic development and the clinical validation of artificial intelligence models based on acoustic analysis for COPD detection. Despite the growing number of publications on AI applied to respiratory sounds, only one study met strict criteria for clinical validation in real patients. This finding is important because it indicates that the field remains largely experimental, with limited evidence of transferability to healthcare settings.

The scarcity of eligible studies may be explained by several factors. First, the development of AI models in respiratory health has largely concentrated on open datasets, which facilitates technical comparisons but does not guarantee clinical utility. Second, prospective validation in real patients requires resources, infrastructure, clinical oversight, and ethical and regulatory conditions that are more complex than those required for studies based solely on retrospective data. Third, variability in respiratory sound acquisition across devices, recording contexts, and operators limits standardization and reduces comparability across studies.

The study by Roy et al. (2025) [15] provides a preliminary signal of feasibility by combining algorithmic development, low-cost hardware implementation, and initial clinical validation. However, its results should be interpreted with caution. The reported metrics are high, but the clinical sample used for real-time validation was small, which limits the precision of the estimates and the ability to extrapolate performance to broader populations. In addition, the use of multiple datasets for training introduces potential heterogeneity in signal origin and quality, with possible effects on model stability and risk of overfitting.

These findings are consistent with previous reviews that have highlighted both the potential of artificial intelligence in respiratory diseases and its limitations in terms of clinical robustness, reproducibility, and real-world applicability [3, 4, 12–14, 20] (Fig. 3). The challenge lies not only in achieving high technical performance, but also in demonstrating that such performance is maintained when the model is applied to real patients under conditions characterized by biological variability, environmental noise, and the operational constraints of clinical practice.

To better contextualize the development of this field, the evolution of artificial intelligence applied to respiratory sound analysis can be summarized as shown in Fig. 3, which provides a structured overview of the historical evolution of neural networks and transformer-based models toward their application in respiratory sound analysis and the assisted detection of pulmonary diseases, particularly COPD, over the period 1943–2025. The organization into six thematic categories enables a coherent interpretation of the field's progression, from foundational concepts in artificial intelligence to clinically oriented applications. In this context, Category A encompasses seminal theoretical contributions and early computational models that established the groundwork of the discipline, whereas Category B consolidates key deep learning architectures, including convolutional networks, generative models, and transformers, which drove major methodological advances.

Building upon these foundations, Category C highlights recent approaches tailored to respiratory acoustic representations, where transformer-based models are applied to spectrograms, cochleograms, and multimodal frameworks integrating audio with clinical data. Category D illustrates the transition toward biomedical applications, including respiratory sound classification and the detection of diseases such as COPD, COVID-19, and acute respiratory infections. In parallel, Category E reflects a shift toward real-world deployment, emphasizing intelligent sensors, embedded systems, and edge AI strategies, highlighting progress toward portable and scalable diagnostic technologies.

Finally, Category F includes contributions related to systematic reviews, interpretability, and clinical decision support, essential for improving reliability and facilitating medical adoption. Overall, the figure synthesizes the technological trajectory of the field and reveals accelerated development in recent years, particularly since 2022, with increasing focus on respiratory-specific applications and clinically deployable solutions. However, despite this technological evolution, translation into robust real-world clinical validation remains limited.

From a clinical perspective, AI-based acoustic models could be especially useful as screening tools or decision-support systems in settings with limited access to spirometry [8, 17]. Nevertheless, the current evidence is insufficient to support their clinical adoption. Before implementation can be considered, prospective studies with larger sample sizes, multicenter validation, standardized acquisition protocols, bias analysis, uncertainty estimation, and formal comparison against conventional diagnostic strategies are required.

This review also has limitations. The main limitation arises from the small number of included studies, which precluded meta-analysis and robust comparisons across models. In addition, the strict criterion of clinical validation substantially reduced the available evidence base. However, this methodological decision was consistent with the study objective, which was to examine real clinical applicability rather than technical performance alone under controlled conditions.

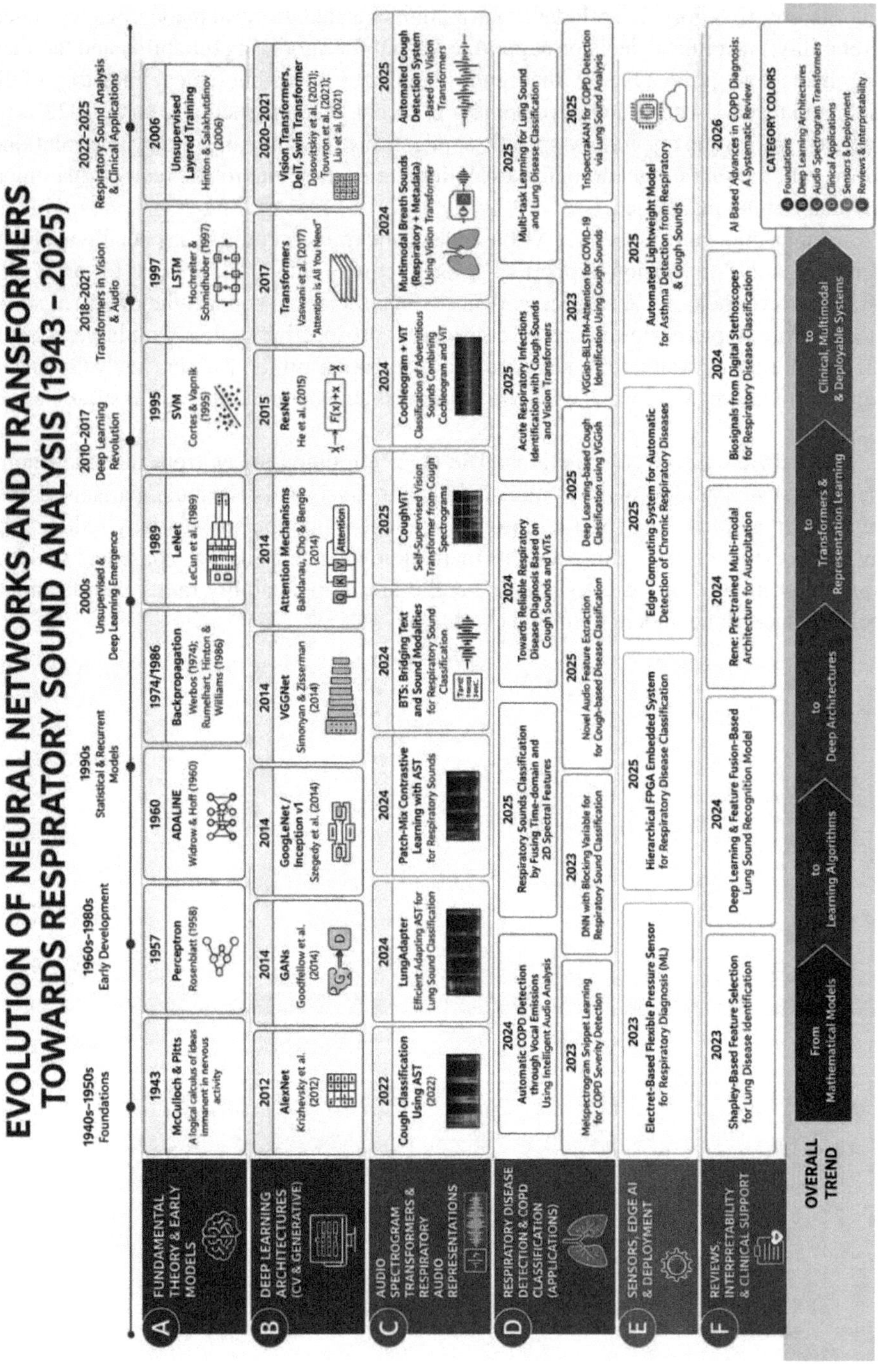

Fig. 3. Evolution of Artificial Intelligence and Key Developments in COPD Detection

5 Conclusion

The available evidence on artificial intelligence models based on acoustic analysis for COPD detection with clinical validation in real patients is extremely limited. Although the identified study showed favorable preliminary results and suggests technical feasibility, the limited clinical validation, small sample size, and restricted generalizability prevent firm conclusions regarding diagnostic utility in routine care.

The field requires prospective, multicenter, and methodologically rigorous studies capable of robustly evaluating the reproducibility, applicability, and clinical value of these tools. Until then, acoustic artificial intelligence for COPD should be regarded as a promising line of translational research, but one that remains insufficiently validated for routine clinical adoption.

References

1. Adeloye, D., Song, P., Zhu, Y., Campbell, H., Sheikh, A., Rudan, I.: Global, regional, and national prevalence of, and risk factors for, chronic obstructive pulmonary disease (COPD) in 2019: a systematic review and modelling analysis. Lancet Respir. Med. **10**(5), 447–458 (2022). https://doi.org/10.1016/S2213-2600(21)00511-7
2. Al-Anazi, S., et al.: Artificial intelligence in respiratory care: current scenario and future perspective. Ann. Thorac. Med. **19**(2), 117–130 (2024). https://doi.org/10.4103/atm.atm_192_23
3. Alqudaihi, K.S., et al.: Cough sound detection and diagnosis using artificial intelligence techniques: challenges and opportunities. IEEE Access. **9**, 102327–102344 (2021). https://doi.org/10.1109/ACCESS.2021.3097559
4. Choi, Y., Lee, H.: Interpretation of lung disease classification with light attention connected module. Biomed. Signal Process. Control. **84**, 104695 (2023). https://doi.org/10.1016/j.bspc.2023.104695
5. Elashmawi, W.H., Djellal, A., Sheta, A., Surani, S., Aljahdali, S.: Machine learning for enhanced COPD diagnosis: a comparative analysis of classification algorithms. Diagnostics. **14**(24), 2822 (2024). https://doi.org/10.3390/diagnostics14242822
6. Global Initiative for Chronic Obstructive Lung Disease: Global Strategy for Prevention, Diagnosis and Management of COPD: 2026 Report. Global Initiative for Chronic Obstructive Lung Disease (2026) https://goldcopd.org/2026-gold-report-and-pocket-guide/
7. Huang, D.M., Huang, J., Qiao, K., Zhong, N.S., Lu, H.Z., Wang, W.J.: Deep learning-based lung sound analysis for intelligent stethoscope. Mil. Med. Res. **10**, 44 (2023). https://doi.org/10.1186/s40779-023-00479-3
8. Irfan Thalib, H., Khan, S., Jamal, A., Ikram, N.I.: Machine learning for early diagnosis and prognostic prediction in chronic obstructive pulmonary disease (COPD). Eur. Respir. J. **66**(suppl 69), PA1461 (2025). https://doi.org/10.1183/13993003.congress-2025.PA1461
9. Jabbar, A., Huang, J., Jabbar, M.K., Mahmood, T.: Personalized federated adaptive transformer ensemble (FATE) model for COPD detection utilizing spatial and temporal features in respiratory sound analysis. Res. Eng. **28**, 107203 (2025). https://doi.org/10.1016/j.rineng.2025.107203
10. Kim, S.H., Han, M.K.: Challenges and the future of pulmonary function testing in chronic obstructive pulmonary disease (COPD): toward earlier diagnosis of COPD. Tuberc. Respir. Dis. **88**(3), 413–418 (2025). https://doi.org/10.4046/trd.2025.0009

11. Kumi, E., Machaya, T.T., Appati, J.K., Adjepon-Yamoah, D.E.: A low-cost artificial intelligence powered breath analyzer for early chronic obstructive pulmonary disease detection in resource-limited environment. Digital Health. **12**, 20552076261429627 (2026). https://doi.org/10.1177/20552076261429627
12. Patel, P.J., Diwan, D., Patel, K.A., Ranga, S., Modi, N.J., Dumasia, S.: Multi feature fusion for COPD classification using deep learning algorithms. J. Integr. Sci. Technol. **12**(4), 780 (2023). https://doi.org/10.62110/sciencein.jist.2024.v12.780
13. Rivas, J.A., Pérez, H., Padilla, A.L., Rodríguez, A.Y., García, D.C.: Edge computing system for automatic detection of chronic respiratory diseases using audio analysis. J. Med. Syst. **49**, 33 (2025). https://doi.org/10.1007/s10916-025-02154-7
14. Robertson, N.M., Centner, C.S., Siddharthan, T.: Integrating artificial intelligence in the diagnosis of COPD globally: a way forward. Chron. Obstruct. Pulmon. Dis. **11**(1), 114–120 (2024). https://doi.org/10.15326/jcopdf.2023.0449
15. Roy, A., Gyanchandani, B., Oza, A.: TriSpectraKAN: a novel approach for COPD detection via lung sound analysis. Sci. Rep. **15**, 6296 (2025). https://doi.org/10.1038/s41598-024-82781-1
16. Roy, A., Satija, U.: A novel melspectrogram snippet representation learning framework for severity detection of chronic obstructive pulmonary diseases. IEEE Trans. Instrum. Meas. **72**, 4003311 (2023). https://doi.org/10.1109/TIM.2023.3256468
17. Roy, A., Satija, U.: A novel multi-head self-organized operational neural network architecture for chronic obstructive pulmonary disease detection using lung sounds. IEEE/ACM Trans. Audio Speech Lang. Process. **32**, 2566–2575 (2024). https://doi.org/10.1109/TASLP.2024.3393743
18. Siebert, J.N., et al.: Deep learning diagnostic and severity-stratification for interstitial lung diseases and chronic obstructive pulmonary disease in digital lung auscultations and ultrasonography: clinical protocol for an observational case–control study. BMC Pulm. Med. **23**, 191 (2023). https://doi.org/10.1186/s12890-022-02255-w
19. Trowman, R., Migliore, A., Ollendorf, D.A.: The value and impact of health technology assessment: discussions and recommendations from the 2023 health technology assessment international global policy forum. Int. J. Technol. Assess. Health Care. **39**(1), e75 (2023). https://doi.org/10.1017/S0266462323002763
20. Wang, M., Li, L., Feng, M., Liu, Z.: Advances in artificial intelligence applications for the management of chronic obstructive pulmonary disease. Front. Med. **12**, 1685254 (2025). https://doi.org/10.3389/fmed.2025.1685254
21. Xu, S., Deo, R.C., Faust, O., Barua, P.D., Soar, J., Acharya, R.: Automated lightweight model for asthma detection using respiratory and cough sound signals. Diagnostics. **15**(9), 1155 (2025). https://doi.org/10.3390/diagnostics15091155

Skull Stripping Methods in Brain Magnetic Resonance Imaging: A Systematic Review

Sergio-Luis Juárez-Valdespino, Kay García-Sánchez(✉), Jorge-Luis Pérez-Ramos, Selene Ramírez-Rosales, Ana Marcela Herrera-Navarro, Hugo Jiménez-Hernández, and Daniel Cantón-Enríquez

Universidad Autónoma de Querétaro, Querétaro 76230, Mexico
sjuarez35@alumnos.uaq.mx, kayychez@gmail.com, seleneramirezrosales@gmail.com, {jorge.luis.perez,mherrera,hugo.jimenez,daniel.canton}@uaq.mx

Abstract. Abstract. Skull stripping is a technique that allows the removal of non-brain tissues that may affect subsequent analyses. This study offers a systematic review guided by the PRISMA protocol of the primary techniques used for this task. A database search initially retrieved 72 records. After duplicate removal and screening, 29 studies met the inclusion criteria and were selected for detailed analysis. These studies were divided into a core comparative analysis group (methods proposing or evaluating skull stripping techniques) and a contextual discussion group (reviews, surveys, and reproducibility studies). The reviewed methods include traditional approaches such as BET, ROBEX, and BEaST, as well as more recent deep-learning-based techniques, including HD-BET, SynthStrip, and self-configuring models. The Dice coefficient was analyzed as the most frequently reported overlap metric; however, this review emphasizes that direct comparisons across studies should be interpreted with caution due to differences in datasets, imaging protocols, pathology presence, and ground truth definitions. The results show that deep learning-based methods can achieve Dice coefficients above 95%, although they usually require specialized hardware. Some morphology-based studies report very high performance under controlled conditions; however their generalizability to heterogeneous clinical MRI remains uncertain.

Keywords: Skull stripping · MRI · Mathematical morphology · Deep learning · Systematic review

1 Introduction

Magnetic resonance imaging (MRI) has expanded diagnostic capabilities by enabling detailed anatomical, structural, and quantitative analysis of diseased

H. Ponce et al. (Eds.): COMIA 2026, CCIS 3095, pp. 781–795, 2027.
https://doi.org/10.1007/978-3-032-37438-7_61

organs [1,2]. In brain imaging, MRI allows the visualization of key tissues such as cerebrospinal fluid (CSF), white matter (WM), and gray matter (GM), which are essential for understanding brain structure and pathology [3,4]. Among the most studied brain conditions are brain tumors, which appear in MRI as regions with characteristics that differ from healthy tissue and therefore require accurate segmentation strategies [5,6].

Several segmentation approaches have been proposed, including clustering algorithms such as K-means and Fuzzy C-means, region-growing methods, and deep learning techniques [7]. However, the presence of non-brain tissues can affect segmentation and diagnosis, making skull stripping a necessary preprocessing step for removing extracranial structures and improving subsequent clinical analysis [8,9].

This study presents a systematic review of Skull stripping techniques in magnetic resonance imaging to identify research gaps, methodological limitations, emerging trends, and opportunities for future research, and to provide a comparative overview of the main approaches in terms of methodological characteristics, reported performance, and computational complexity.

Summing up, this work summarizes the following contributions:

- A PRISMA-guided systematic review of Skull stripping methods in magnetic resonance imaging.
- A methodological taxonomy of the reviewed approaches.
- A computational complexity analysis of representative methods.
- A contextualized performance analysis that avoids direct ranking based only on Dice coefficients.
- A reproducibility analysis based on code, pretrained models, dataset availability, and open-source access.
- A critical discussion of real-world clinical limitations, including scanner variability, raw image artifacts, pathological anatomy, and dataset shift.

The rest of the article is organized as follows: Sect. 2 describes the search strategy and the inclusion and exclusion criteria for study selection. Section 3 presents the obtained results, highlighting the performance metrics of each method and their computational complexity. Section 4 analyzes the findings and proposes opportunities for future research. Section 5 summarizes the conclusions of the study and proposes future lines of work.

2 Methodology

To ensure the rigor and scientific quality of this study, the systematic review was structured following the PRISMA protocol [10]. The process was designed to filter bibliographic noise and focus on peer-reviewed research on Skull stripping methods in MRI.

2.1 Search Strategy

The search was performed across five academic databases-PubMed, IEEE Xplore, Scopus, SpringerLink, and Google Scholar-using Boolean queries that combined the terms ('Skull stripping' OR 'brain extraction') with ('MRI' OR 'magnetic resonance'). The syntax was adjusted to the specific requirement of each platform.

While the query was intentionally kept broad to avoid narrowing the results to particular methods, certain approaches, especially those based on mathematical morphology or more recent proposal-may still be underrepresented if they are not explicitly tagged with the chosen keywords. To compensate for this, relevant studies belonging to underrepresented categories were manually incorporated during the screening phase, aiming for a more balanced picture of the current landscape.

2.2 Inclusion and Exclusion Criteria

To build a relevant and high quality study, we defined a set of selection rules. Original articles, review articles, and peer-reviewed conference proceedings, written in English or Spanish, and published between January 2015 and March 2026, were included. Their core focus had to be the proposal or evaluation of Skull stripping techniques applied to MRI, and they were required to include quantitative performance metrics, along with verifiable identifiers (DOI or ISSN) and full-text availability.

On the exclusion side, we removed theses, technical reports, blog posts, book chapters, editorial, and any material from non-peer-reviewed repositories. Studies focusing solely with tumor segmentation without prior Skull stripping, or that did not address MRI, were also discarded. Finally, articles published before 2015 were omitted, with the only exception being seminal works that laid the methodological foundations of the field.

Applying these filters led to the removal of 42 articles that did not meet the defined parameters, resulting in a final set that meets standard academic quality criteria. Additionally, a small group of gray literature sources ([2–6, 11, 12]) were kept only for contextual purposes-to support the introduction and the discussion around evaluation metrics-without being part of the core comparative analysis.

2.3 Data Extraction

From the studies that passed the inclusion criteria described in Subsect. 2.2, we proceeded to extract the information of interest. The analysis focused on identifying research gaps and limitations of current methods, as well as emerging trends.

To this end, the most commonly used methods in the literature were identified to facilitate their comparative analysis. Likewise, information related to the reported performance metrics was extracted. However, for the purposes of this analysis, the Dice coefficient is prioritized as the main comparison metric.

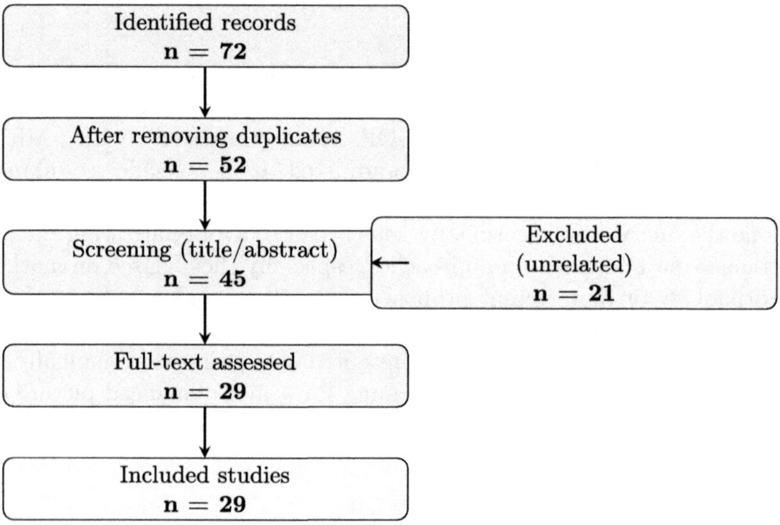

Fig. 1. PRISMA flow diagram of the study selection process.

As detailed in Fig. 1, from an initial search of 72 records, a final sample of 29 articles was consolidated after applying relevance and quality filters. From these studies, all consulted without paywall restrictions through open-access repositories or institutional databases, essential information (authors, year, country, method, image type, metrics, and dataset) was systematically collected and organized in a table.

The 43 excluded records were grouped according to their primary exclusion reason: tumor segmentation without explicit skull stripping, MRI classification without brain extraction, general surveys not specific to skull stripping, preprints without verifiable DOI, studies without quantitative metrics, pre-2015 non-seminal publications, duplicates, and other reasons. Some records met more than one exclusion criterion; therefore, each study was assigned to its primary reason for exclusion to avoid double counting.

2.4 Assessment of Study Quality

To evaluate the methodological quality of the 29 articles selected for the core analysis of this review, the QualSyst (Standard Quality Assessment Criteria) tool, developed by Kmet, Cook, and Lee [13], was used. This tool allows scoring quantitative studies using a 14-criterion checklist, which includes aspects such as clarity of objective, adequacy of design, definition of metrics, sample size, and control of confounding variables. Each criterion is scored as 2 (fully met), 1 (partially met), or 0 (not met). QualSyst criteria that do not apply to a particular study design (e.g., randomization or blinding in image segmentation studies) are excluded from the calculation of the individual score of each article, but not the articles themselves. The final score is calculated as the percentage of points obtained out of the total possible [13].

The evaluation was performed independently, where all studies passed the quality process, obtaining an overall average of 83.2%, which is determined to be of strong quality.

3 Results

The analysis of the selected literature allows identifying the evolution of Skull stripping techniques from classical methods based on thresholds and morphology to complex architectures based on deep learning. From the reference database built during this review, 24 articles directly or indirectly addressing Skull stripping methods in MRI were consolidated. Below is the bibliometric and taxonomic breakdown of the reviewed works.

3.1 Spatio-Temporal Distribution of Reviewed Studies

The selected studies show participation from countries such as the United States, India, Germany, and Pakistan. Seminal works from the early 2000s, including BET [14] and BSE [15], established the basis for classical skull stripping methods. From 2010 onwards, the field moved toward learning-based approaches, with an important transition marked by 3D convolutional neural networks for brain extraction [16]. Most selected studies were concentrated between 2020 and 2026, reflecting the growing relevance of deep learning and synthesis-based methods. This distribution reflects only the studies that met the inclusion criteria.

3.2 Taxonomy of Skull Stripping Methods

Representative methods can be summarized according to their operating principles. Classical deformable and edge-based approaches, such as BET and BSE, extract the brain using intensity, surface deformation, and edge information [14,15]. Morphology- and intensity-based methods rely on thresholding, erosion, dilation, opening, closing, or gray-level distributions, making them efficient but sensitive to structuring-element selection, noise, and intensity non-uniformity. For instance, Dawood et al. (2023) proposed an adaptive gamma correction method combined with morphological operations, reporting a successful skull stripping rate of 96% (accuracy), a metric that differs from the Dice coefficient used elsewhere in this review but demonstrates the potential of lightweight morphological approaches [17]. Classical machine learning methods, including ROBEX and BEaST, introduce learned priors, random forests, shape models, and patch-based label fusion to improve robustness [18,19]. Deep learning and synthesis-based methods, such as HD-BET, self-configuring U-Net models, SynthStrip, and SST-DUNet, use learned hierarchical features, synthetic data, or transformer-based architectures to improve robustness across pathological cases, contrasts, and acquisition protocols [20–23] (Table 1).

Table 1. Taxonomic summary of Skull stripping methods.

Category	Main characteristics	Examples	Advantages	Limitations
Mathematical morphology	Erosion, dilation, opening, closing with structuring elements	Dawood et al. (2023) [17]	Low cost, accuracy 96%	Sensitive to element shape and size
Intensity	Thresholds, histograms, region growing	Otsu's Method [7]	Simple, low computational cost	Limited by noise and artifacts
Deformable surfaces	Deformation of active contours to the brain edge	BET (Smith, 2002) [14], BSE (Shattuck, 2001) [15]	Robust in healthy brains	Fails in the presence of pathologies
Atlases / templates	Spatial normalization with anatomical templates	BEaST (Eskildsen et al., 2012) [19]	High accuracy in healthy brains	Requires precise alignment
Classical machine learning	Random forests, SVM, shape models	ROBEX (Iglesias et al., 2011) [18]	Good accuracy/cost balance	Limited in severe pathologies
Deep learning	3D convolutional networks, U-Net, Transformers	HD-BET (Isensee et al., 2019) [20], Self-Configurable U-Net (Gandhi et al., 2024) [21]	High accuracy (>95%), robustness	High cost, requires GPU
Hybrids / synthesis	Combination of approaches, training with synthetic images	SynthStrip (Hoopes et al., 2022) [22]	Robust in multiple contrasts	Requires GPU, complex training

3.3 Contextualized Performance Summary

Because the reviewed studies differ in datasets, MRI sequences, populations, pathology presence, and annotation protocols, the reported Dice values were not pooled through meta-analysis. Instead, they were summarized descriptively to avoid unsupported claims of method superiority.

To provide a transparent and contextualized overview of the reported performance, Table 2 summarizes key characteristics of each study, including dataset, population, pathology, and the reported Dice coefficient. This information helps to understand why direct comparisons of Dice scores across studies are methodologically problematic, as the experimental conditions (e.g., healthy vs. pathological brains, adult vs. pediatric populations, different datasets) vary considerably between studies.

Table 2. Contextualized performance summary of skull-stripping methods.

Study	Method	Dataset	Pop.	Pathol.	Dice (%)
Smith 2002	BET	IBSR	Adults	No	93.1-97.3
Shattuck 2001	BSE	IBSR, TCGA	Adults	Yes	95.6/65.6
Iglesias 2011	ROBEX	IBSR, LPBA40, OASIS	Adults	Yes	95.5/84.0
Eskildsen 2012	BEaST	ICBM	Adults	No	98.34
Isensee 2019	HD-BET	Multicenter (37 sites)	Adults	Yes (glioma)	95-98
Hoopes 2022	SynthStrip	FreeSurfer, IXI	Adults	No	>95
Dawood 2023	Morphology	Kaggle (2 datasets)	Mixed	No	96%
Tabassum 2023	nnUNet	NFBS, TCGA	Adults	Yes (glioma)	99.60/92.96
Schulz 2025	SynthStrip	Private (n=199)	Pediatric	No	96.0
Omidi 2024	DA-Unet	Calgary-Campinas + synthetic	Adults→Neonates	No	93.08
Thakur 2022	Optimized CNN	Private	Mixed	Yes (glioma)	0.90-0.93

The variability across studies can be better appreciated in the forest plot shown in Fig. 2.

Overall, the reviewed studies differ considerably in terms of dataset composition, presence of brain pathologies, annotation protocols, ground truth definitions, and MRI modalities. Therefore, the reported Dice scores should not be used directly to rank methods or to claim the superiority of one approach over another. Instead, they serve as context-dependent references that reflect performance under specific experimental conditions.

Regarding the quantitative performance of the methods, a wide variability is observed in the metrics reported in the literature. In the case of BET [14], Dice coefficient values between 93.1% and 97.3% in healthy brains, while BSE [15] reports 95.6% in healthy brains and 65.6% in the presence of tumors. This difference simply reflects that evaluating on pathological brains is a more challenging task. HD-BET [20] and SynthStrip [22] have been associated with Dice values exceeding 95% in their respective studies.

Chen et al. (2026) evaluated the use of fs-FLAIR combined with BET for skull stripping in 97 patients with brain tumors (46 gliomas, 51 meningiomas). The method achieved Dice scores of 0.902 in gliomas and 0.87 in meningiomas, significantly outperforming T1, T2, T1C, and nfs-FLAIR ($p < 0.0001$). This study demonstrates that, with the appropriate sequence, classical methods like BET can achieve competitive performance even in brains with pathologies [24].

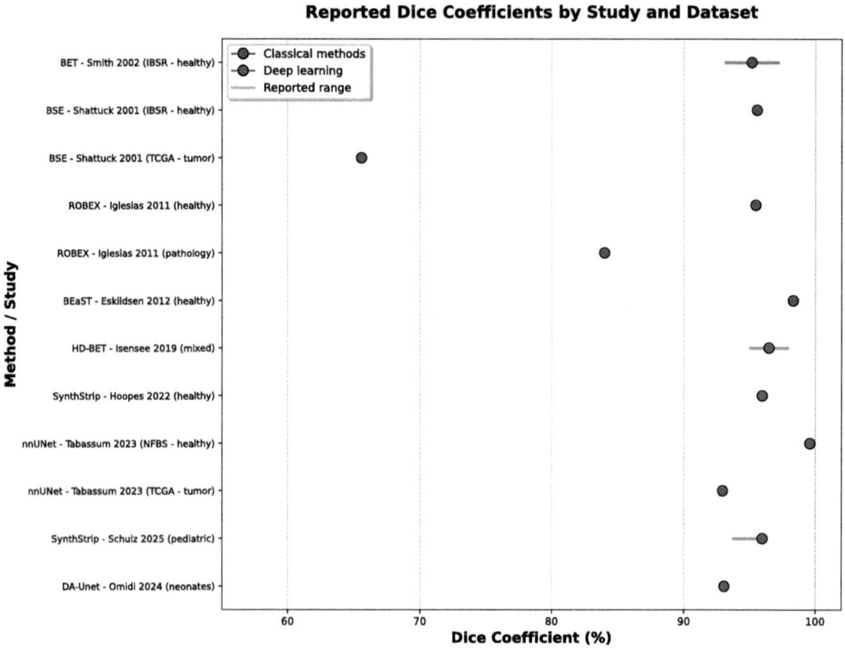

Fig. 2. Reported Dice coefficients by study, grouped by methodological category. Blue points represent classical methods (BET, BSE, ROBEX, BEaST), red points represent deep learning methods (HD-BET, SynthStrip, nnUNet, DA-Unet). Horizontal lines indicate reported ranges where available.

Tabassum et al. (2023) compared nnUNet with six established methods (BSE, ROBEX, UNet, SC-UNet, MV-UNet, 3D U-Net) on the same datasets (NFBS healthy and TCGA tumors). nnUNet achieved the best performance in both scenarios (Dice 0.996 on healthy, 0.930 on tumors), demonstrating its robustness against severe morphological alterations [25].

A different morphological approach was proposed by Dawood et al. (2023), who applied adaptive gamma correction combined with morphological operations and connected component analysis for skull stripping in T1-weighted MRI. Their method was evaluated on two public datasets (Br35H and Brain Tumor Detection), achieving a successful skull stripping rate of 96% (accuracy). Although this metric differs from the Dice coefficient, it demonstrates that morphological operations–when properly combined with contrast enhancement–can be effective for brain extraction [17].

This result is particularly relevant for resource-limited environments, since morphological operations are lightweight when applied to binary masks, though their complexity increases with higher dimensionality data (e.g., full 3D tensors).

Various metrics are reported in the literature, including accuracy, sensitivity, specificity, and the Jaccard index. In this review, we focused on the Dice coefficient as the main reference metric due to its wide adoption in medical image segmentation tasks, not because it is inherently superior to others. Given the heterogeneity across studies, direct comparisons remain problematic, and the values presented here are meant to inform, not to rank. As a future line of work, comparative studies under controlled and homogeneous conditions are needed for more objective evaluations.

Table 3. Computational Complexity Analysis of representative skull stripping and related evaluation methods.

Category	Method	Estimated Total Complexity $\mathcal{O}(\cdot)$
Morphology / Intensity	Dawood [17]	$O(N \times (K + S))$
	Intensity-based framework [7]	$O(M \times N)$
Traditional Models	BET [14]	$O(V + N)$
	fs-FLAIR + BET [24]	$O(V + N) + O(V)$
	BSE [15]	$O(V \times I \times K^3)$
	BEaST [19]	$O(V \times N_t \times S \times P)$
Machine Learning	ROBEX [18]	$O(V \times T + L^3 \times K^3)$
Deep Learning / Synthesis	HD-BET [20]	Training: $O(E_p \times N_d \times V \times C_{3D})$ | Inference: $O(V \times C_{3D})$
	nnU-Net / Self-configurable U-Net [21]	Training: $O(E_p \times N_d \times V \times C_{3D})$ | Inference: $O(V \times C_{3D})$
	SynthStrip [22]	Training: $O(E_s \times N_s \times V \times C_{3D})$ | Inference: $O(V \times C_{3D})$
	DA-Unet [26]	Training: $O(E_p \times N_d \times V \times C_U) + O(E_p \times N_d \times C_{DANN})$ | Inference: $O(V \times C_U)$
	SST-DUNet [23]	Training: $O(E_p \times N_d \times (V \times C_{DU} + W_s^2 \times H))$ | Inference: $O(V \times C_{DU} + W_s^2 \times H)$
	Optimized CNN [27]	Training: $O(E_p \times N_d \times V \times C_{CNN})$ | Optimized inference: $O(\rho \times V \times C_{CNN})$
	3D CNN for QC [28]	Training: $O(E_p \times N_d \times V \times C_{3D})$ | Inference: $O(V \times C_{3D})$
Evaluation Studies	Comparative evaluation of tools [29]	$O(n \times (C_{BET} + C_{ROBEX} + C_{HDBET} + C_{SynthStrip}))$

Notation: V: voxels; N: pixels/vertices; K: kernel size; S: morphological operations; I: iterations; T: trees; N_t: templates; P: patch size; E_p: epochs; N_d: training samples; C_{3D}, C_U, C_{CNN}: convolutional/U-Net/CNN costs; W_s: transformer window size; H: attention heads; ρ: optimization factor; n: evaluated images

3.4 Computational Complexity Analysis

As an additional contribution of this review, Table 3 presents the Big-O complexity calculation of the analyzed algorithms. It is observed that classical methods maintain almost linear costs with respect to the number of voxels or pixels in the image, which justifies their use on low-resource hardware.

On the other hand, deep learning-based algorithms introduce quadratic or polynomial terms during their training phase, driven by matrix multiplication in self-attention mechanisms and deep convolutional operations. Nevertheless, it is highlighted that the SynthStrip and HD-BET models can offset the cost by reducing their inference complexity, making them more viable in clinical practice once trained.

It is important to note that the reported complexities are theoretical approximations derived from the dominant computational operations of each method family. For classical methods, the cost is mainly driven by voxel traversal, morphological kernels, deformable iterations, or template matching. In contrast, deep learning approaches are dominated by 3D convolutional operations, training epochs, dataset size, and, in transformer-based architectures, attention windows and the number of attention heads. Therefore, these expressions should be interpreted as comparative computational tendencies rather than exact implementation-dependent runtimes..

4 Discussion

The reviewed literature shows a clear methodological transition from classical skull stripping approaches, such as BET, BSE, and ROBEX, toward deep learning and synthesis-based strategies. Classical methods remain attractive due to their simplicity, low computational cost, and ease of deployment, especially in low-resource environments. However, they are sensitive to noise, intensity variability, artifacts, and pathological anatomy. In contrast, deep learning methods such as HD-BET and SynthStrip often report higher Dice scores and better adaptability to complex clinical cases, but require labeled data, careful validation, and greater computational resources.

Recent studies also show that skull stripping is moving beyond standard adult MRI. Omidi et al. addressed scarce labeled neonatal data using domain adaptation and synthetic data, achieving competitive performance in newborn MRI [26]. Schulz et al. evaluated skull stripping tools in pediatric T2-weighted MRI and found that SynthStrip achieved a strong balance between Dice performance and computation time [29]. Luo et al. introduced automated quality and privacy assessment after skull stripping, highlighting the need to evaluate not only overlap metrics but also re-identifiable facial features and brain tissue loss [28]. For low-resource deployment, Thakur et al. showed that model optimization strategies such as quantization and acceleration can reduce latency with minimal performance loss [27].

Overall, the selection of a skull stripping method should depend on the clinical scenario, image modality, pathology presence, reproducibility, and available computational resources rather than on Dice scores alone. This is particularly important for countries and institutions with limited infrastructure, where lightweight and reproducible approaches may have greater practical value. Future work should therefore focus on controlled comparative studies and on the design of robust skull stripping algorithms that balance accuracy, efficiency, reproducibility, and clinical applicability.

4.1 Reproducibility and Open-Source Availability

Reproducibility is a key factor when skull-stripping methods are used in order to guide future methodological developments. Beyond the reported Dice Scores, the availability of the source code executable tools, pretrained models, public datasets and clear documentation affects the possibility of a method being validated in an independent way compared or integrated in clinical or research workflows. This is relevant in skull stripping where differences in the preprocessing, the parameter tuning, image format and dataset composition can affect the final brain mask.

Table 4 summarizes the availability of open-source code and reproducibility level of the identified representative methods. The classification was defined by following this points:

- High: Code or executable tools, pretrained models or implementations are available and ready to be used, as well as public datasets.
- Medium: Public implementation with restrictions are available. In this category, the model is limited due to licenses or the dataset can be partial accessed.
- Low: Only the article is available or the algorithmic description is.
- Unknown: There is no clear information about the implementation or reproducibility.

Table 4. Open-source availability and reproducibility of representative skull stripping methods.

Method	Public tool/code	Pretrained model	Dataset availability	Reproducibility
BET	Yes, FSL	N/A	Public benchmarks	Medium
BSE	Yes, BrainSuite	N/A	Public benchmarks	Medium
ROBEX	Yes, NITRC	Included executable	IBSR, LPBA40, OASIS	Medium-High
BEaST	Yes, BIC-MNI tools	Template-dependent	Public/institutional	Medium
HD-BET	Yes, GitHub/PyPI	Yes	Partially public	High
SynthStrip	Yes, FreeSurfer/Docker	Yes	Public/synthetic	High
Self-configurable U-Net [21]	Partial/unclear	Study-dependent	Restricted/cohort-specific	Medium
nnU-Net skull stripping [25]	Partial	Not clearly available	Partially public	Medium
Dawood et al. (2023) [17]	No (code not publicly available)	N/A	Br35H, Brain Tumor Detection (Kaggle)	Low
Schulz et al. (pediatric evaluation) [29]	Yes (GitHub)	Yes (SynthStrip used)	Private (n=199)	High
Omidi et al. (DA-Unet) [26]	Yes (GitHub)	Yes	Synthetic + Calgary-Campinas	High
Chen et al. (fs-FLAIR+BET) [24]	No (BET is public, but method not)	N/A	Private (97 patients)	Low
Thakur et al. (optimization) [27]	No/unclear	N/A	Private	Low
SST-DUNet [23]	Unknown	Unknown	Restricted/unclear	Unknown
Luo et al. (quality/privacy) [28]	Unclear	Unclear	Public (Clemson)	Low

Encouraging open-source practices remains an important step to foster collaborations and accelerate innovation in the field. As a future direction, we advocate for the standardization of code sharing anda the use of common evaluation benchmarks to facilitate fair and reproducible comparisons across different skull stripping approahces.

4.2 Limitations and Challenges in Real-World Raw MRI Data

Although several skull-stripping methods report high performance under controlled experimental conditions, their behavior in real clinical scenarios remains challenging. Clinical MRI data are highly heterogeneous and may vary across scanners, acquisition protocols, patient populations, MRI sequences, and preprocessing pipelines. Even small differences in contrast, intensity distribution, voxel spacing, anatomical orientation, or field of view can affect the resulting brain mask [29]. In addition, raw MRI data may contain bias-field artifacts, motion artifacts, partial volume effects, noise, incomplete slices, anisotropic voxel spacing, or poorly standardized formats.

Pathological cases introduce additional complexity, as tumors, edema, resections, and anatomical deformations can distort the expected brain boundary. Furthermore, manual ground truth masks vary across experts and institutions, making Dice-based comparisons uncertain. Therefore, skull stripping methods should be evaluated not only through overlap metrics but also through robustness to scanner variability, raw image artifacts, pathological anatomy, and dataset shift. Future methods should be validated across multiple datasets, scanners, MRI sequences, and clinical populations, while supporting standardized formats such as NIfTI.

5 Conclusions

In this work, a systematic review of Skull stripping methods in magnetic resonance imaging was carried out, analyzing 29 studies selected under defined criteria and evaluated using the QualSyst tool. From this analysis, a clear trend towards the use of deep learning-based techniques was identified. Deep learning-based and synthesis-based methods frequently report high Dice values, in complex clinical scenarios; however, their performance should be interpreted in relation to the datasets, acquisition protocols, ground truth definitions, and clinical populations used in each study. This review not only summarizes skull stripping methods but also identifies design requirements for future lightweight and robust skull stripping algorithms suitable for resource-constrained clinical environments.

Nevertheless, classical approaches continue to play an important role, especially in contexts where resources are limited, as their lower computational cost makes them more accessible. Likewise, the Dice coefficient was highlighted as a key metric for evaluating these methods, although it was also recognized that differences in datasets, ground truth definitions, and experimental conditions hinder direct comparison between studies.

Another relevant aspect is the limited presence of works focused on this topic within the Mexican context, which opens a significant opportunity for the development of research in the country. In particular, the design of solutions adapted to local needs could contribute to improving medical image analysis in clinical settings with restricted infrastructure.

As a future projection, this work lays the foundations for the development of a new Skull stripping algorithm that integrates the strengths of existing methods. To achieve this, the following line of work was considered, which includes data standardization through the conversion of images to NIfTI format, and the analysis of strategies for extrapolating information from 2D to 3D.

Based on these steps, the aim is to move towards the construction of a proprietary model that is efficient and applicable in real clinical scenarios, thereby contributing to the development of the field at both an academic and practical level.

Acknowledgments. The authors thank the Center for Research and Innovation in Computer Science and Educational Technology (CIICCTE) and the Facultad de Informática of the UAQ for providing the space to carry out this work.

Disclosure of Interests. The authors have no competing interests to declare that are relevant to the content of this article.

References

1. Kalavathi, P., Prasath, V.B.S.: Methods on skull stripping of mri head scan images–a review. J. Digit. Imaging **29**, 365–379 (2016)
2. Rahman, M.: Magnetic resonance imaging and iron-oxide nanoparticles in the era of personalized medicine. Nanotheranostics **7**(4), 424–449 (2023)
3. Soumya Kumari, L.K., Sundarrajan, R.: A review on brain age prediction models. Brain Res. **1823**, 148668 (2024)
4. Kaifi, R.: A review of recent advances in brain tumor diagnosis based on ai-based classification. Diagnostics **13**(18), 3007 (2023)
5. Gómez-Guzmán, M.A., et al.: Classifying brain tumors on magnetic resonance imaging by using convolutional neural networks. Electronics **12**(4), 955 (2023)
6. Ahamed, M.F., et al. A review on brain tumor segmentation based on deep learning methods with federated learning techniques. Comput. Med. Imaging Graph. **110**, 102313 (2023)
7. Ting-Wei Wang, et al.: A systematic review and meta-analysis of artificial intelligence models for brain tumor segmentation and detection using mri. Mayo Clin. Proc. Digit. Health **2**(1), 75–91 (2024). Open Access, PRISMA-DTA systematic review and meta-analysis
8. Rehman, H.Z.U., Hwang, H., Lee, S.: Conventional and deep learning methods for skull stripping in brain mri. Appl. Sci. **10**(5), 1773 (2020)
9. Fatima, A., Shahid, A.R., Raza, B., et al.: State-of-the-art traditional to the machine- and deep-learning-based skull stripping techniques, models, and algorithms. J. Digit. Imaging **33**, 1443–1464 (2020)

10. Page, M.J., et al.: The prisma: statement: an updated guideline for reporting systematic reviews. BMJ **372**(71), 2021 (2020)
11. Wahlang, I., et al.: Brain magnetic resonance imaging classification using deep learning architectures with gender and age. Sensors **22**(5), 1766 (2022)
12. Magadza, T., Viriri, S.: Deep learning for brain tumor segmentation: A survey of state-of-the-art. J. Imaging **7**(2), 19 (2021)
13. Kmet, L.M., Cook, L.S., Lee, R.C.: Standard quality assessment criteria for evaluating primary research papers from a variety of fields. Technical Report HTA Initiative #13, Alberta Heritage Foundation for Medical Research, Edmonton, Canada (2004)
14. Stephen, M., Smith: Bet: Brain extraction tool. Hum. Brain Mapp. **17**(3), 143–155 (2002)
15. David, W., Shattuck, R.M., Leahy: Brain surface extraction (bse). Neuroimage **13**(6), 856–876 (2001)
16. Kleesiek, J., et al.: 3d convolutional neural networks for brain extraction. Neuroimage **126**, 195–203 (2016)
17. Dawood, T.A., Hashim, A.T., Nasser, A.R.: Automatic skull stripping of mri head images based on adaptive gamma transform. Math. Model. Eng. Probl. **10**(1), 304–310 (2023)
18. Iglesias, J.E., Liu, C.-Y., Thompson, P.M., Tu, Z.: Robex: Robust brain extraction tool. IEEE Trans. Med. Imaging **30**(11), 1877–1889 (2011)
19. Eskildsen, S.F., et al.: Collins: Beast: Brain extraction based on nonlocal segmentation technique. Neuroimage **59**(3), 2362–2373 (2012)
20. Isensee, F., et al.: Hd-bet: A deep learning approach for brain extraction. Hum. Brain Mapp. **40**(12), 3609–3625 (2019)
21. Gandhi, S., et al.: Self-configurable u-net for pediatric brain extraction. Neuro-Oncol. Adv. **6**(1), vdae190 (2024)
22. Hoopes, A., Jaime, S., Mora, A.V., Dalca, B., Fischl, M.: Hoffmann: Synthstrip: Skull-stripping via synthesis. Neuroimage **260**, 119474 (2022)
23. Soltanpour, S., et al.: Sst-dunet: Smart swin transformer and dense unet for automated preclinical fmri skull stripping. J. Neurosci. Methods **423**, 110545 (2025). PMID: 40789440
24. Chen, X., et al.: Fat-suppressed t2-weighted fluid attenuated inversion recovery mr images allow efficient skull stripping in brain tumors using a brain extraction tool. Neurol. Asia **31**(1), e–pub ahead of print (2026). Incluye 46 gliomas y 51 meningiomas
25. Tabassum, M., Al Suman, A., Russo, C., Di Ieva, A., Liu, S.: A deep learning framework for skull stripping in brain mri. In: 2023 45th Annual International Conference of the IEEE Engineering in Medicine & Biology Society (EMBC), pp. 1–4 (2023)
26. Omidi, A., Shamaei, A., Verschuu, A., King, R., Leijser, L., Souza, R: Unsupervised domain adaptation of brain mri skull stripping trained on adult data to newborns: Combining synthetic data with domain invariant features. In: Burgos, N., et al. (eds.) Proceedings of The 7nd International Conference on Medical Imaging with Deep Learning, volume 250 of Proceedings of Machine Learning Research, pp. 1073–1085. PMLR (2024)
27. Thakur, S.P., et al.: Optimization of deep learning based brain extraction in mri for low resource environments. In: Brainlesion: Glioma, Multiple Sclerosis, Stroke and Traumatic Brain Injuries (BrainLes 2022), volume 12962 of Lecture Notes in Computer Science, pp. XXX–XXX. Springer, Cham (2022)

28. Luo, L., Phatangare, R.V., Wang, J.Z., Vaden Jr., K.I., Eckert, M.A.: Deep learning methods to evaluate privacy and quality of skull-stripped brain images. In: Brain Informatics - 17th International Conference, BI 2024, Bangkok, Thailand, December 13-15, 2024, Proceedings, Part I, volume 15541 of Lecture Notes in Computer Science, pages 217–227. Springer, Cham (2024)
29. Schulz, A., Dragendorff, E., Wendt, K., Schomakers, A., Bultmann, E., Wolff, D.: Skull stripping tools in pediatric t2-weighted mri scans: a retrospective evaluation of segmentation performance. Front. Neurosci. **19**, 1715514 (2025)

Mobile Support System for Pulmonary Auscultation Based on MobileNet and STFT Spectrograms

Karina Lizeth Arévalo Flores(✉), Carlos Antonio Bocanegra González, and Said Polanco Martagón

Universidad Politécnica de Victoria, Ciudad Victoria, Tamaulipas, Mexico
{2539014,2539004,spolancom}@upv.edu.mx

Abstract. Respiratory diseases represent a major cause of morbidity and mortality worldwide, and their early diagnosis through pulmonary auscultation still depends largely on the clinical expertise of healthcare professionals. Deep learning-based systems for automatic respiratory sound classification have shown promising results; however, their translation into real-world scenarios faces two relevant limitations: the use of insufficiently controlled data partitioning strategies that may introduce patient-level data leakage, and the reliance on computationally expensive architectures unsuitable for mobile deployment.

In this work, we explore the technical feasibility of a lightweight pulmonary auscultation support tool based on a MobileNet architecture applied to Short-Time Fourier Transform (STFT) spectrograms and oriented toward mobile deployment using TensorFlow Lite. Two evaluation protocols are compared on the International Conference on Biomedical and Health Informatics (ICBHI) 2017 database: the official 60/40 file-level split and a patient-level 70/10/20 partition with an independent validation set, designed to prevent patient overlap and provide a more realistic estimate of generalization to unseen patients. The model was trained without data augmentation or synthetic resampling, using only class weighting to partially address class imbalance.

Under the patient-level partition, the model achieved an accuracy of 42.12% and an ICBHI Score of 36.40%. The TFLite-converted model occupies 12.71 MB, achieves an average inference time of 29.27 ms, and was successfully integrated into an Android prototype as a proof of concept. Although performance on minority classes remains limited, the results suggest that lightweight architectures under strict patient-level protocols can provide a viable foundation for future portable auscultation support tools in resource-limited settings.

Keywords: Respiratory sound classification · Computer-assisted auscultation · Patient-level partitioning · MobileNet · TensorFlow Lite · Mobile devices

H. Ponce et al. (Eds.): COMIA 2026, CCIS 3095, pp. 796–810, 2027.
https://doi.org/10.1007/978-3-032-37438-7_62

1 Introduction

Respiratory diseases continue to represent a major cause of morbidity and mortality worldwide. In clinical practice, pulmonary auscultation using a stethoscope constitutes a fundamental tool for detecting respiratory abnormalities; however, its interpretation largely depends on the experience of medical personnel, which can introduce subjectivity and inter-observer variability [2]. This limitation becomes more relevant in resource-limited settings, where access to specialists and complementary studies is not always immediate. In this context, automatic respiratory sound analysis systems have emerged as a promising alternative to support clinical evaluation.

The ICBHI 2017 database [5], composed of 6,898 respiratory cycles from 126 patients, has become one of the main benchmarks for the development and comparison of automatic lung sound classification algorithms. However, the translation of these advances to real-world scenarios still faces two relevant limitations. First, several high-performing approaches rely on complex architectures that are difficult to deploy on mobile or embedded platforms [12,13]. Second, a common methodological issue is file-level data partitioning, which can introduce data leakage when recordings from the same patient appear simultaneously in training and testing, leading to overly optimistic performance estimates [9–11].

In this context, the present work explores the feasibility of a lightweight tool to support pulmonary auscultation based on a MobileNet architecture [6] applied to STFT spectrograms and oriented toward mobile deployment using TensorFlow Lite [16]. Two evaluation strategies are compared: the official ICBHI protocol and a 70/10/20 patient-level partition designed to avoid data leakage and incorporate an independent validation set. Class weighting is used to partially mitigate dataset imbalance, and the model is integrated into an Android prototype as a proof of concept.

2 Related Work

Automatic classification of respiratory sounds has gained considerable attention over the past decade, largely driven by the availability of the ICBHI 2017 database [5], which has become one of the main benchmarks for evaluating deep learning methods in this domain. A wide range of time-frequency representations and convolutional architectures have been explored on this dataset. Neili and Sundaraj [3] conducted a systematic comparison of STFT spectrograms, scalograms, mel-spectrograms, and gammatonegrams across multiple Convolutional Neural Network (CNN) architectures, showing that the choice of input representation has a substantial influence on classification performance. Beyond single-representation approaches, hybrid architectures combining CNN and recurrent layers have been proposed to capture both spectral patterns and temporal dependencies in respiratory sounds [12,13]. However, many high-performing systems rely on increasingly complex pipelines that may limit their suitability for mobile or resource-constrained deployment. Recent efforts have also explored audio enhancement techniques to improve robustness in clinical scenarios [4], though at the cost of additional preprocessing complexity.

To address class imbalance, the literature has explored strategies such as data augmentation, constrained synthetic sampling [15], and class-frequency-sensitive loss functions.

The need for deployable and computationally efficient models has motivated recent interest in lightweight solutions for auscultation-based screening. Zhang et al. [17] proposed a low-cost Artificial Intelligence (AI)-empowered stethoscope using a lightweight model for detecting cardiac and respiratory diseases, highlighting the relevance of embedded systems in settings with limited infrastructure. Similarly, Park et al. [18] introduced a lung sound classification model specifically designed for on-device AI, emphasizing the need to balance predictive performance with latency and memory constraints. In this context, MobileNet [6], combined with transfer learning [7,8], represents an attractive alternative for portable applications.

Beyond architectural considerations, evaluation methodology plays a critical role in the interpretation of reported results. Yu et al. [1] emphasize that direct comparison across ICBHI 2017 studies is difficult because partitioning strategies, preprocessing pipelines, augmentation procedures, and evaluation metrics vary substantially. A recurring limitation in part of the literature is random file-, cycle-, or sample-level partitioning, which may introduce data leakage when recordings from the same patient appear across training and testing subsets [9–11]. Nguyen and Pernkopf [14] report that results under the official 60/40 split are substantially lower than those obtained with random splits that may contain overlapping patients, confirming that the evaluation protocol strongly influences reported performance. Despite these advances, fewer works jointly address a lightweight architecture, a strict patient-level evaluation protocol, and an independent validation set. The present work addresses this gap by evaluating a compact MobileNet-based classifier under a patient-level hold-out scheme and integrating the trained model into an Android proof-of-concept prototype using TensorFlow Lite.

To provide a structured reference for the present work, Table 1 summarizes representative studies on four-class adventitious sound classification using ICBHI 2017. The studies are grouped according to their partitioning strategy, as this factor directly affects the interpretation of their results. Among works evaluated under the official 60/40 split with non-overlapping patients, reported ICBHI scores typically range from approximately 50% to 64%, with the highest values achieved by pipelines that incorporate advanced strategies such as data augmentation, spectrum correction, transfer learning, attention mechanisms, or contrastive learning. Some studies using patient-level cross-validation report higher scores; however, these results should be interpreted carefully because cross-validation protocols may provide more effective training data per iteration and often rely on additional optimization strategies. Therefore, direct numerical comparison across groups is not straightforward. The proposed system is included in the table not as a state-of-the-art competitor, but to situate its performance within the broader ICBHI 2017 landscape while emphasizing its focus on patient-level evaluation, independent validation, and mobile deployment feasibility.

Table 1. Representative comparison of ICBHI 2017 four-class adventitious sound classification studies. Results are not directly comparable across all studies due to differences in partitioning strategy, model complexity, preprocessing, augmentation, validation protocol, and evaluation metrics.

Study	Model	Split	Additional strategy/focus	HS/Score	Acc.
Official 60/40 split					
Ma et al. [19]	Bi-ResNet	Official 60/40	Smart stethoscope / Bi-ResNet	50.16%	52.79%
Gairola et al. [20]	ResNet-34	Official 60/40	RespireNet / limited-data setting	55.20%	–
Nguyen & Pernkopf [14]	ResNet-50	Official 60/40	Co-tuning + stochastic normalization	58.29%	–
GTFA-Net [21]	CNN	Official 60/40	Group time-frequency attention	60.25%	–
CLAP [22]	Transformer	Official 60/40	Contrastive learning + metadata alignment	63.54%	–
Patient-level cross-validation					
Petmezas et al. [13]	CNN-LSTM	Interpatient 10-fold cross-validation (CV)	Focal loss for class imbalance	68.52%	76.39%
Patient-level hold-out with independent validation set					
This work	**MobileNet**	**70/10/20**	**Lightweight model + TFLite deployment**	**36.40%**	**42.12%**

3 Methodology

3.1 Dataset

The ICBHI 2017 database [5] was used, one of the most widely adopted benchmarks for automatic respiratory sound classification. The dataset contains 920 recordings obtained from 126 patients using different types of stethoscopes, with sampling frequencies ranging from 4 kHz to 44.1 kHz and approximate durations of 10 to 90 s. Based on the annotations provided by experts, each recording was segmented into individual respiratory cycles, resulting in a total of 6,898 cycles. Figure 1 shows representative examples of the spectrograms used as input to the model for the four classes considered in this work.

The class distribution is markedly imbalanced: 3,642 cycles correspond to the *normal* class (52.8%), 1,864 to *crackle* (27.0%), 886 to *wheeze* (12.9%), and 506 to *crackle-wheeze* (7.3%). This imbalance represents a significant challenge for training, as it may bias the model toward the majority class.

(a) *Normal* (b) *Wheeze* (c) *Crackle* (d) *Crackle-Wheeze*

Fig. 1. Representative examples of STFT spectrograms generated from respiratory cycles of the four classes considered: *normal*, *wheeze*, *crackle*, and *crackle-wheeze*. These images constitute the visual input used by the MobileNet model during training and evaluation.

3.2 Partitioning Strategies

In order to analyze the effect of the partitioning strategy on model evaluation, two schemes were compared. The first scheme reproduces the official partition of the ICBHI challenge, which establishes a 60/40 split between training and testing at the file level. This approach facilitates comparison with previous works, although it does not explicitly control for patient-level separation.

The second scheme corresponds to a patient-level partition used in this work, in which all files from the same individual are exclusively assigned to one of the subsets. To achieve this, patient identifiers encoded in the filenames of the ICBHI dataset were leveraged. The split used was 70% for training, 10% for validation, and 20% for testing, which corresponds approximately to 88, 13, and 25 patients, respectively. This strategy reduces the risk of *data leakage* and allows the incorporation of an independent validation set for model and hyperparameter selection.

3.3 Preprocessing

All recordings were resampled to 22,050 Hz. Based on the temporal annotations available in the annotation files, each respiratory cycle was extracted, and segments with duration shorter than 0.5 s were discarded.

As input representation, STFT spectrograms were used, generated with `librosa.stft`, using a Hann window, a window length of 2,048 samples, and a *hop length* of 512. The resulting magnitude was converted to a logarithmic scale using `librosa.amplitude_to_db`. The spectrograms were stored as Red-Green-Blue (RGB) images of 224×224 pixels, compatible with the input expected by MobileNet.

During data loading, images were normalized to the [0,1] range. Subsequently, within the model, the transformation `preprocess_input(inputs * 255.0)` was applied to match the preprocessing scheme originally used by MobileNet on ImageNet.

3.4 Model Architecture

MobileNet [6] was used as the base feature extractor with pretrained weights on ImageNet and without the original classification head (`include_top=False`). This architecture was selected due to its low computational cost and compatibility with mobile deployment scenarios.

On top of the MobileNet output, a classification head was added consisting of a `GlobalAveragePooling2D` layer, a dense layer with 128 neurons using Rectified Linear Unit (ReLU) activation and L2 regularization ($\lambda = 10^{-3}$), two *dropout* layers with a rate of 0.5, and a final output layer with 4 neurons using *softmax* activation. This configuration allows adapting the features extracted by the base network to the specific task of respiratory sound classification.

3.5 Training

Training was performed in two phases. In the first phase, the MobileNet layers were kept frozen and only the classification head was trained for 10 epochs using Stochastic Gradient Descent (SGD) with a learning rate of 10^{-3} and momentum of 0.9. This short initial phase aimed to adapt the classification head to the specific task before fine-tuning the entire network, avoiding premature modifications to the pretrained weights.

In the second phase, all layers were unfrozen to perform joint *fine-tuning* with a learning rate of 10^{-4} and a maximum of 40 epochs. This reduction in the learning rate was adopted to avoid abrupt modifications to the pretrained weights during fine-tuning. Model selection was performed based on validation accuracy using `ModelCheckpoint` (`monitor='val_accuracy'`), in order to retain the configuration with the best performance during training. Additionally, `ReduceLROnPlateau` was applied on the validation loss with a factor of 0.5, a patience of 5 epochs, and a lower bound of 10^{-7}, with the aim of progressively reducing the learning rate when training stopped showing sustained improvements. The batch size was set to 32 as a compromise between training stability and memory availability.

To mitigate the effect of class imbalance, weights inversely proportional to the frequency of each category were computed using `compute_class_weight` from scikit-learn with the `balanced` strategy, and incorporated into training through the `class_weight` parameter.

3.6 Evaluation

The final evaluation was conducted exclusively on the corresponding test set. For the patient-level protocol, the model with the best validation performance was selected and then evaluated on the held-out test set. Overall accuracy, precision, recall, and F1-score per class were reported, as well as macro and weighted averages. Additionally, the *ICBHI Score* was computed to facilitate comparison with previous works. Confusion matrices were also used to analyze error patterns between classes.

3.7 TensorFlow Lite Conversion and Android Prototype

To assess the feasibility of deploying the proposed model on mobile devices, the model trained under the patient-level partition was converted to TensorFlow Lite (TFLite) format using the `TFLiteConverter` tool from the TensorFlow framework [16]. This format is designed for the efficient execution of deep learning models on resource-constrained devices, such as smartphones and embedded systems, without requiring connectivity to external servers.

The resulting model was integrated into an Android prototype developed as a proof of concept, with the purpose of exploring the technical feasibility of local inference in a mobile environment. As shown in Fig. 2, the application allows loading a respiratory spectrogram as an image, executing inference using the converted TFLite model, and displaying the predicted class along with its associated confidence level. It is important to note that this prototype constitutes a preliminary technical demonstration and not a clinically validated tool. Its development aimed to verify that the pipeline from spectrogram-based input to mobile classification is technically executable, laying the groundwork for future developments oriented toward on-device validation and clinical evaluation.

4 Experimental Results

4.1 Experimental Setup

Experiments were conducted on Google Colaboratory using Tesla T4 Graphics Processing Units (GPUs). The implementation was developed in Python 3.10 with TensorFlow 2.15 and Keras. Audio preprocessing was performed with librosa 0.10.1, while metric calculation was carried out using scikit-learn 1.3.0. To support reproducibility, the random seed was fixed to 42 for numpy, random, and tensorflow.

The experimental setup was designed to assess a lightweight respiratory sound classification model oriented toward mobile deployment. MobileNet was selected as the backbone due to its low computational cost and compatibility with resource-constrained devices, while STFT spectrograms were used as the input representation because they provide a compact time-frequency description compatible with convolutional architectures. Class weighting was incorporated to partially compensate for the class imbalance of ICBHI 2017.

4.2 Experiment 1: Official ICBHI Split

As a baseline for comparison with previous works, the official ICBHI protocol was reproduced, which defines a 60/40 training/testing split at the file level. Since this scheme does not include an independent validation set, 15% of the training subset was reserved via stratified partitioning. The final distribution was 2,337 images for training, 407 for validation, and 2,744 for testing.

Under this configuration, the model achieved an accuracy of 36.30% on the test set (Table 2). The *normal* class showed the best performance, with a precision of 66.12% and an F1-score of 52.46%. In contrast, performance on minority

(a) *Normal*

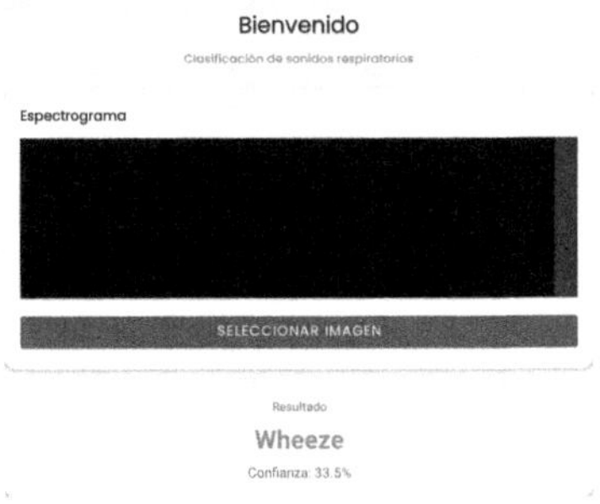

(b) *Wheeze*

(c) *Crackle*

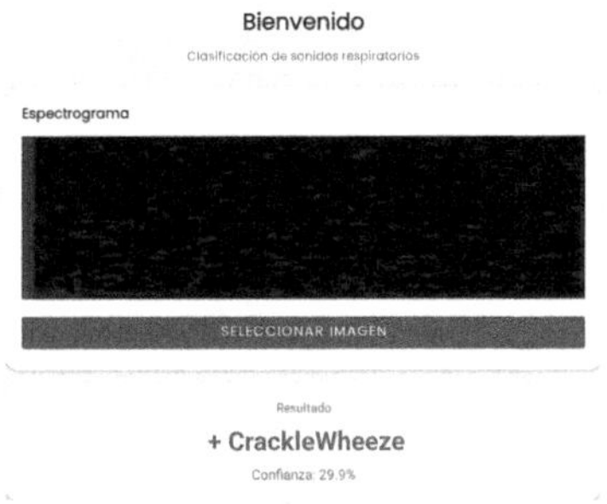

(d) *Crackle-Wheeze*

Fig. 2. Interface of the Android prototype developed for integrating the model in TensorFlow Lite. The application allows loading a respiratory spectrogram, performing on-device inference, and displaying the predicted class along with the confidence level.

classes was considerably lower. In particular, *crackle-wheeze* achieved a precision of 5.83%, while *wheeze* reached a recall of 14.36%, confirming that both categories remain especially difficult to recover. The difference between macro and weighted averages also suggests a significant effect of class imbalance on overall evaluation.

4.3 Experiment 2: Patient-Level Split

The second experiment evaluated the proposed patient-level partitioning strategy, in which each individual appears exclusively in one of the subsets. This split produced 4,846 images for training (88 patients), 679 for validation (13 patients), and 2,692 for testing (25 patients), preventing the model from seeing recordings from the same patient during training and evaluation.

Under this configuration, the model achieved an accuracy of 42.12% on the test set, surpassing the previous experiment by 5.82% points. As shown in Table 3, the improvement was also reflected in global metrics and in some

Table 2. Class-wise metrics for the official ICBHI split

Class	Precision	Recall	F1-Score	Support
Crackle	0.3224	0.3354	0.3288	647
Crackle-Wheeze	0.0583	0.2867	0.0969	143
Normal	0.6612	0.4348	0.5246	1571
Wheeze	0.1642	0.1436	0.1532	383
Macro Average	0.3015	0.3001	0.2759	2744
Weighted Average	0.4805	0.3630	0.4043	2744

pathological classes. The recall for *crackle* increased from 33.54% to 50.88%, while *crackle-wheeze* rose from 28.67% to 35.92%. The *normal* class maintained high precision (66.79%) and showed a slight improvement in recall (45.37%). Despite this, *wheeze* remained the most challenging category, with a recall of 10.84%, indicating that class imbalance and the acoustic variability of this category continue to limit the system's performance.

Table 3. Class-wise metrics for the patient-level split

Class	Precision	Recall	F1-Score	Support
Crackle	0.3710	0.5088	0.4291	684
Crackle-Wheeze	0.0964	0.3592	0.1520	142
Normal	0.6679	0.4537	0.5403	1543
Wheeze	0.1977	0.1084	0.1400	323
Macro Average	0.3333	0.3575	0.3154	2692
Weighted Average	0.5059	0.4212	0.4436	2692

4.4 Comparative Analysis

Table 4 summarizes the comparison between both protocols. The patient-level split achieved higher values across allreported metrics: accuracy, macro precision, macro recall, macro F1-score,and ICBHI Score. However, this difference should be interpreted with caution, as both experiments also differ in the size and composition of the subsets, particularly in the number of training samples and the inclusion of an independent validation set. The observed improvement therefore cannot be attributed solely to the elimination of data leakage.

The main finding of this experiment lies in the robustness of the evaluation protocol. By preventing recordings from the same patient from appearing in more than one subset, the patient-level split reduces the risk of overestimating the model's generalization ability and better approximates a real-world scenario in which the system must analyze previously unseen patients. The results suggest that a lightweight architecture such as MobileNet can maintain reasonable performance even under a methodologically stricter evaluation.

Table 4. Overall comparison between split strategies

Metric	Official Split	Patient-Level Split
Accuracy	36.30%	**42.12%**
Precision (macro)	30.15%	**33.33%**
Recall (macro)	30.01%	**35.75%**
F1-Score (macro)	27.59%	**31.54%**
ICBHI Score	32.41%	**36.40%**

The confusion matrix corresponding to the patient-level split (Table 5) shows that the *normal* class is frequently confused with *crackle* and *crackle-wheeze*, suggesting partial overlap between normal patterns and low-intensity adventitious events. The *wheeze* class also exhibits substantial confusion with *normal* and *crackle-wheeze*, indicating that wheeze-related features are not sufficiently separated in the learned representation. In the case of *crackle-wheeze*, several samples are reassigned to *crackle*, suggesting that the model more readily captures the crackling component than the simultaneous coexistence of both sounds.

Table 5. Confusion matrix obtained with the patient-level split

True / Predicted	Crackle	C-W	Normal	Wheeze
Crackle	348	139	175	22
Crackle-Wheeze	44	51	38	9
Normal	468	264	700	111
Wheeze	78	75	135	35

4.5 Class-Wise Performance Analysis

The class-wise analysis confirms that the minority categories constitute the main bottleneck of the system. The *wheeze* class concentrates the model's greatest weakness: although it has 323 test examples, only 35 were correctly identified, resulting in a recall of 10.84%. This behavior can be associated both with the dataset imbalance and the acoustic heterogeneity of wheezes, whose intensity, duration, and frequency vary among patients.

The *crackle-wheeze* class also exhibits unstable performance. Its low precision indicates that when the model predicts this category, it frequently confuses samples belonging to other classes. At the same time, its moderate recall suggests that the learned representation still does not consistently capture the coexistence of both adventitious sounds.

In contrast, the *normal* category shows the most stable performance in terms of precision and F1-score. Nevertheless, its recall remains below 50%, implying that a considerable proportion of normal cycles are still classified as pathological. One possible explanation is that the use of *class weights* favors sensitivity toward minority classes, but at the cost of increasing false alarms for the majority class. Overall, these results indicate that the main margin for improvement of the system lies not in the *normal* class, but in discriminating less represented and acoustically more variable adventitious sounds.

4.6 Preliminary Deployment Efficiency

As part of the deployment feasibility assessment, the efficiency characteristics of the TFLite-converted model were analyzed. The resulting model occupies 12.71 MB, representing a modest reduction compared to the original Keras model (13.14 MB), consistent with the expected behavior of standard conversion without quantization. Although this size is larger than that of highly compressed architectures reported in the literature [18], it remains within a reasonable range for modern mobile applications where storage is not the primary deployment constraint.

The converted model was successfully deployed and executed on the Android prototype, confirming that local inference on a real mobile device is technically achievable without relying on external server infrastructure. To obtain a preliminary latency reference, 100 consecutive predictions were executed using the TFLite model in a Google Colab CPU environment, yielding an average inference time of 29.27 ms per sample, with a minimum of 14.61 ms and a maximum of 92.30 ms. Although these measurements do not constitute a formal on-device benchmark, they provide an initial indication that the model is compatible with low-latency single-sample inference [17,18].

Together, these results and the prototype integration demonstrated in the Android application support the feasibility of TFLite conversion for the proposed lightweight classification pipeline. A thorough efficiency characterization on real mobile hardware—including on-device latency, memory footprint, and energy consumption—remains an important direction for future work toward a complete deployment assessment under real-world conditions.

5 Discussion and Limitations

5.1 Discussion

The results of this work should be interpreted in light of its specific objectives: assessing the feasibility of a lightweight architecture under a rigorous evaluation protocol oriented toward unseen patients, rather than maximizing predictive performance against state-of-the-art models. The improvement of 5.82 percentage points in accuracy – from 36.30% to 42.12% – is consistent with findings reported by Nguyen and Pernkopf [14], who document substantial differences between partitioning protocols on ICBHI 2017. As discussed in Sect. 4.4, this difference cannot be attributed exclusively to leakage elimination, but the protocol itself provides a more faithful estimate of the model's actual generalization capacity.

Compared to works employing more complex architectures such as CNN-LSTM with focal loss [13] or ResNet with co-tuning and data augmentation [14], the proposed model achieves considerably lower performance values. This was expected: MobileNet was selected for computational efficiency and mobile compatibility, not to maximize classification metrics. The absence of data augmentation and synthetic rebalancing further limits performance on minority classes, particularly wheeze, whose inter-patient acoustic variability represents a challenge beyond architecture choice.

From a deployment perspective, the successful TFLite conversion and integration into an Android prototype demonstrate that the spectrogram-based pipeline is technically viable in a real mobile environment [17,18]. The proposed system is not intended to replace clinical auscultation, but to provide a technical foundation for future support tools in settings with limited access to specialists.

5.2 Limitations

Despite the results obtained, this work presents important limitations that should be considered when interpreting its conclusions.

First, performance on minority classes, particularly wheeze with a recall of 10.84%, is not sufficient for clinical use in its current state. This limitation reflects the combined effect of class imbalance in ICBHI 2017 and the acoustic heterogeneity of wheezes, whose intensity, duration, and frequency vary considerably across patients. The absence of data augmentation techniques, focal loss, or synthetic resampling further contributes to this behavior.

Second, the experiments do not include cross-validation or statistical tests between partitioning protocols, which limits the robustness of the reported comparisons. The observed difference between both evaluation schemes cannot be attributed to a single factor, given that the subsets differ both in partitioning criterion and in size and composition.

Third, the Android prototype remains at a preliminary development stage. The latency measurements reported were obtained in a Google Colab CPU environment and do not constitute a formal characterization of performance on the real mobile device. Efficiency evaluation on Android hardware—including on-device latency, memory footprint, and energy consumption—remains pending for future work.

Finally, the system has not undergone any form of clinical validation. The results obtained on ICBHI 2017 are not sufficient to establish conclusions about the diagnostic utility of the system in real-world settings, and any future development oriented toward clinical application would require prospective validation studies with specialized medical supervision.

6 Conclusions

This work explored the technical feasibility of a lightweight pulmonary auscultation support system based on MobileNet and STFT spectrograms, evaluated under two partitioning protocols on the ICBHI 2017 database. The central contribution does not lie in surpassing the state of the art in absolute accuracy, but in demonstrating that a computationally efficient architecture can be assessed under a methodologically more rigorous protocol, oriented toward estimating generalization to unseen patients.

The patient-level partition with an independent validation set reduced the risk of data leakage and enabled a more realistic estimate of the model's generalization capacity. Under this configuration, the model achieved an accuracy of 42.12% and an ICBHI Score of 36.40%. Rather than serving as a direct performance comparison between protocols, these results support the use of patient-level evaluation as a more realistic framework for assessing generalization to unseen patients.

From a deployment perspective, the successful TFLite conversion and integration of the model into an Android prototype demonstrate the technical feasibility of a mobile pipeline based on spectrogram input and on-device classification. Furthermore, the preliminary inference time of 29.27 ms, measured in a development environment, suggests compatibility with low-latency inference scenarios, although a formal evaluation on real Android hardware is required.

Taken together, the results suggest that lightweight architectures such as MobileNet, evaluated under strict patient-level protocols, can constitute a technically viable foundation for future portable auscultation support tools in resource-limited settings. Future work will prioritize, first, a formal efficiency characterization on real Android hardware, including on-device latency, memory footprint, and energy consumption; second, the improvement of minority-class performance through data augmentation, loss-function design, or rebalancing strategies; and third, prospective clinical validation studies to assess the practical utility of the system in real-world auscultation scenarios.

References

1. Yu, S., et al.: Advances and challenges in respiratory sound analysis: a technique review based on the ICBHI2017 database. Electronics **14**(14), 2794 (2025). https://doi.org/10.3390/electronics14142794
2. Kim, Y., Hyon, Y., Jung, S.S., Lee, S., Yoo, G., Chung, C., Ha, T.: Respiratory sound classification for crackles, wheezes, and rhonchi in the clinical field using deep learning. Sci. Rep. **11**, 17186 (2021). https://doi.org/10.1038/s41598-021-96724-7
3. Neili, Z., Sundaraj, K.: A comparative study of the spectrogram, scalogram, mel-spectrogram and gammatonegram time-frequency representations for the classification of lung sounds using the ICBHI database based on CNNs. Biomed. Eng. / Biomed. Tech. **67**(5), 367–390 (2022). https://doi.org/10.1515/bmt-2022-0180
4. Tzeng, J.-T., et al.: Improving the Robustness and Clinical Applicability of Automatic Respiratory Sound Classification Using Deep Learning-Based Audio

Enhancement: Algorithm Development and Validation. JMIR AI **4**, e67239 (2025). https://doi.org/10.2196/67239
5. Rocha, B.M., Fet al.: Respiratory Sound Database for the Development of Automated Classification. In: Maglaveras, N., Chouvarda, I., de Carvalho, P. (eds.) Precision Medicine Powered by pHealth and Connected Health. IP, vol. 66, pp. 33–37. Springer, Cham (2018). https://doi.org/10.1007/978-981-10-7419-6_6
6. Howard, A.G., et al.: MobileNets: Efficient Convolutional Neural Networks for Mobile Vision Applications. arXiv preprint arXiv:1704.04861 (2017)
7. Tajbakhsh, N., Jeyaseelan, L., Li, Q., Chiang, J.N., Wu, Z., Ding, X.: A comprehensive survey of deep learning research on medical image analysis with focus on transfer learning. Clin. Imaging **94**, 118–127 (2022). https://doi.org/10.1016/j.clinimag.2022.11.003
8. Alzubaidi, L., et al.: Novel transfer learning approach for medical imaging with limited labeled data. Cancers **13**(7), 1590 (2021). https://doi.org/10.3390/cancers13071590
9. Oner, M.U., Cheng, Y.-C., Lee, H.K., Sung, W.-K.: Training machine learning models on patient level data segregation is crucial in practical clinical applications. medRxiv (2020). https://doi.org/10.1101/2020.04.23.20076406
10. Kapoor, S., Narayanan, A.: Leakage and the reproducibility crisis in machine-learning-based science. Patterns **4**(9), 100804 (2023). https://doi.org/10.1016/j.patter.2023.100804
11. Schlegl, T., Ofner, J., Langs, G.: Inflation of test accuracy due to data leakage in deep learning-based classification of OCT images. Sci. Data **9**, 580 (2022). https://doi.org/10.1038/s41597-022-01618-6
12. Fraihi, N., Karrakchou, O., Ghogho, M.: Improving Deep Learning-based Respiratory Sound Analysis with Frequency Selection and Attention Mechanism. arXiv preprint arXiv:2507.20052 (2025). https://doi.org/10.48550/arXiv.2507.20052
13. Petmezas, G., et al.: Automated lung sound classification using a hybrid CNN-LSTM network and focal loss function. Sensors **22**(4), 1232 (2022). https://doi.org/10.3390/s22041232
14. Nguyen, T., Pernkopf, F.: Lung sound classification using co-tuning and stochastic normalization. IEEE Trans. Biomed. Eng. **69**(9), 2872–2882 (2022). https://doi.org/10.1109/TBME.2022.3156293
15. Kala, A., Elhilali, M.: Constrained Synthetic Sampling for Augmentation of Crackle Lung Sounds. In: Proceedings of 45th Annual International Conference of the IEEE Engineering in Medicine & Biology Society (EMBC), pp. 1–5. IEEE (2023). https://doi.org/10.1109/EMBC40787.2023.10340579
16. Abadi, M., et al.: TensorFlow: Large-Scale Machine Learning on Heterogeneous Systems. Software available from tensorflow.org (2015). https://www.tensorflow.org
17. Zhang, M., Li, M., Guo, L., Liu, J.: A low-cost ai-empowered stethoscope and a lightweight model for detecting cardiac and respiratory diseases from lung and heart auscultation sounds. Sensors **23**(5), 2591 (2023). https://doi.org/10.3390/s23052591
18. Park, J., Jeong, C., Choi, Y., Hong, H., Jo, Y.: Lung sound classification model for on-device AI. Appl. Sci. **15**(17), 9361 (2025). https://doi.org/10.3390/app15179361
19. Ma, Y., et al.: LungBRN: a smart digital stethoscope for detecting respiratory disease using Bi-ResNet deep learning algorithm. In: Proceedings of IEEE Biomedical Circuits and Systems Conference (BioCAS), pp. 1–4. IEEE, Nara, Japan (2019). https://doi.org/10.1109/BIOCAS.2019.8919247

20. Gairola, S., Tom, F., Kwatra, N., Jain, M.: RespireNet: a deep neural network for accurately detecting abnormal lung sounds in limited data setting. In: Proceedings of 43rd Annual International Conference of the IEEE Engineering in Medicine and Biology Society (EMBC), pp. 527–530. IEEE (2021). https://doi.org/10.1109/EMBC46164.2021.9630029
21. Chu, Y., Wang, Q., Zhou, E., Zheng, G., Liu, Q.: Hybrid spectrogram for the automatic respiratory sound classification with group time frequency attention network. In: Proceedings of IEEE 6th International Conference on Pattern Recognition and Artificial Intelligence (PRAI), pp. 839–845. IEEE (2023). https://doi.org/10.1109/PRAI59366.2023.10332294
22. Kim, J.W., Toikkanen, M., Choi, Y., Moon, S.E., Jung, H.Y.: BTS: bridging text and sound modalities for metadata-aided respiratory sound classification. In: Proceedings of 25th Annual Conference of the International Speech Communication Association (INTERSPEECH 2024), pp. 1690–1694 (2024). https://doi.org/10.21437/Interspeech.2024-929

Analysis of COVID-19 Mortality Predictions in CECCS Using Clinical and Laboratory Data with a Multi-layer Perceptron

J. Israel Yañez Vargas[1](✉), Diana G. Molina Bermúdez[1], Cesar Centeno Fosado[2], and Miguel Conejo Rodriguez[2]

[1] Universidad Politécnica de Juventino Rosas, Hidalgo 102,Comunidad de Valencia, 38253 Santa Cruz de Juventino Rosas, Guanajuato, Mexico
{jyanez,dmolina}@upjr.edu.mx

[2] Centro Estatal de Cuidados Críticos Salamanca, Tecolutla S/N, Colonia Bellavista, Salamanca, Guanajuato, Mexico
ccentenof@guanajuato.gob.mx
http://www.upjr.edu.mx

Abstract. This research paper proposes a flowchart based on multilayer perceptron (MLP) to predict COVID-19 patient mortality using clinical and laboratory variables obtained from the State Center for Critical Care in Salamanca, Guanajuato, Mexico. The data are cleaned, transformed, and normalized, and then divided into training, validation, and test sets. The project incorporates strategies to address class imbalance, such as weighting the loss function and adjusting the decision threshold, prioritizing the detection of at-risk patients. Evaluation is performed using metrics such as accuracy, precision, recall, and F1 score. A statistical analysis is conducted using Pearson and Spearman correlations, as well as variance analysis to analyze the relationship between laboratory variables and the output, to select the best features for the MLP network. The results show that the algorithm achieves good results, particularly in the case of high sensitivity (recall), while maintaining a good balance between accuracy and sensitivity.

Keywords: COVID-19 · Multilayer-perceptron · Statistical Analysis

1 Introduction

In 2020, the first confirmed case of the COVID-19 pandemic was detected in Mexico, which had a strong global impact. From that date until 2023, there were more than 7.5 million confirmed cases and approximately 330,000 deaths. In the state of Guanajuato alone, there were more than 370,000 cases and over 15,000 deaths during the same period [1].

H. Ponce et al. (Eds.): COMIA 2026, CCIS 3095, pp. 811–824, 2027.
https://doi.org/10.1007/978-3-032-37438-7_63

The increase in cases and deaths from mid-2020 occurred exponentially, which required strong collaboration among the three most important sectors (government, academia, and industry), along with society, in order to understand the behavior of the disease.

In response, strategies were designed first to understand the disease and subsequently to address the problem. Therefore, in Mexico, information such as clinical studies, laboratory data, and imaging began to be used to gain deeper knowledge.

For this article, a collaboration was carried out with the State Center for Critical Care (CECCS) in the city of Salamanca, Guanajuato, which became the first tertiary-level hospital in the state to be converted into a COVID-19 hospital. It also had the particular characteristic of receiving only critically ill patients from across the state of Guanajuato.

The data collected covered a period of nearly two years starting in mid-2020. Clinical studies, laboratory tests, and chest X-ray images were obtained for each admitted patient. For example, clinical evaluations and COVID-19 testing were performed on the day of admission; laboratory tests were also obtained on the same day, except for D-dimer and ferritin, while X-rays were taken according to the physician's request.

In previous studies, multiple analyses and improvements related to COVID-19 have been conducted. For instance, there are statistical studies based on data obtained from the CECCS, where the statistics of all patients admitted from 2020 to 2022 were analyzed. This analysis showed the day-to-day behavior of COVID-19, allowing observation of the evolution of treatments implemented until patient discharge or death [2,3].

Study [4] compare and analyze different algorithms to predicting the progression of COVID-19 in patients using around 1000 clinical records and laboratory variables, including comorbidities and vital signs. The research paper compares models such as Random Forest, XGBoost, and neural networks with 40 primary features in 8 classes including patient's demographics (8 features), clinical and conditions features (16), comorbidities (18), treatment (17), initial vital sign (14), symptoms (31 data), laboratory results (35 data), and 0 for survived and 1 for deceased, demonstrating good performance in the metrics Sensitivity, specificity, accuracy and F1-score. They mention the importance of a D-dimer and ESR for predictions.

Chadaga et al. [5] present the development of machine learning models to diagnose COVID-19 using exclusively clinical laboratory data, such as blood biomarkers (leukocytes, lymphocytes, platelets, among others), thus avoiding reliance on PCR tests. The authors employ various algorithms, including MLP, achieving high levels of accuracy (0.95), demonstrating that laboratory parameters contain sufficient information for patient classification.

In the study by Baik [6] proposes the use of deep learning models, including MLP, to predict mortality in COVID-19 patients early on, based on clinical and laboratory data obtained in the initial stages of hospitalization. The study analyzes a significant group of hospitalized patients and demonstrates that deep learning-based models can achieve high levels of predictive performance, identifying key variables such as age, comorbidities, and blood biomarkers.

The research paper "Machine Learning Successfully Detects Patients with COVID-19 Prior to PCR Results" [7] explains and analyzes machine learning models capable of identifying patients with COVID-19 before confirmation by PCR tests, using routine clinical and laboratory data, such as hematological and biochemical parameters. The study analyzes thousands of hospital records and demonstrates that the models can achieve high predictive performance close to 0.90 in accuracy.

Yu et al., [8] proposes an artificial intelligence-based system that integrates clinical, laboratory, and medical imaging data (such as CT scans) to improve the diagnosis and classification of COVID-19 patients. It utilizes various machine learning models and neural networks (including MLP), achieving high precision and accuracy in identifying positive cases and assessing disease severity.

One of the most recognized approaches for COVID-19 mortality prediction is the 4C Mortality Score proposed by the ISARIC WHO Clinical Characterisation Consortium. This score was developed using data from more than 57,000 hospitalized patients across 260 hospitals in the United Kingdom and was designed as a pragmatic clinical risk stratification tool for predicting in-hospital mortality in COVID-19 patients [9].

The 4C Mortality Score uses eight variables commonly available at hospital admission, including age, sex, comorbidities, respiratory rate, oxygen saturation, level of consciousness, urea, and C-reactive protein. The model demonstrated strong discrimination performance with an area under the ROC curve close to 0.77 in the validation cohort and showed higher predictive capability than several existing clinical risk scores [9].

The severity of COVID-19 increased depending heavily on the patient's comorbidity, and was also significantly influenced by the patient's region of origin, ethnicity, and diet. Therefore, there are few studies with data from Mexican patients that combine laboratory studies, clinical studies, and the use of Multilayer Perceptron (MLP) to help predict the disease's behavior, predict patient severity, and even support the diagnosis of human mortality.

Given the above, this research article seeks to combine algorithms proposed in the state of the art and described in this section, the use of statistical analysis and MLP, in addition to including relevant data from a hospital in the state of Guanajuato in Mexico.

The principal contribution of this work is the development of a reduced-feature MLP model capable of predicting COVID-19 mortality using a compact subset of laboratory variables selected through statistical analysis.

This reduction decreases input dimensionality while maintaining or improving predictive performance, potentially simplifying clinical implementation.

2 Material and Methods

2.1 Database

The data were acquired from the State Center for Critical Care (CECCS) from late 2020 through the end of 2021. Patients were transferred from a secondary-level hospital in the state of Guanajuato before arriving at the CECCS due to the severity of their symptoms and health condition. Upon arrival at the CECCS, initial vital sign assessments were conducted, along with relevant information such as weight, age, height, blood pressure, and pre-existing conditions.

The final dataset consisted of 511 daily clinical and laboratory records obtained from 109 patients.

Although patient data were collected daily, this study does not explicitly model temporal dependencies. Each daily record was treated as an independent sample, allowing the MLP to learn patterns from clinical and laboratory values at different stages of hospitalization.

Therefore, the effective dataset used for training corresponds to daily observations rather than a single sample per patient.

Future work will explore temporal models such as recurrent neural networks (RNNs) or time-series approaches to better capture disease progression.

The total number of patients admitted in a one and half years in the CECCS is described in Table 1 and shown the status of the person in the moment when the data was collected.

Table 1. Total income in the CECCS

	Patient was discharged				
	Patients	Deaths	Counterreference	Reference	Hospitalized
COVID 19	88	40	28	13	7
Negative	18	8	8	2	0
Pending	3	2	1	0	0
Total	109	50	37	15	7

Papers [2,3] described all the information of the basic information such as (Sex, Body Mass Index, Age, Comorbility) and the same the a basic statistical analysis (media, standard deviation and variance) for each patient and the data for each days of stay in the hospital.

2.2 Clinical Data

When patients arrive to the CECCS the first indication is to collect the vital signs of the person, thus, the clinical variables include Body Mass Index (BMI), Hemoglobin (Hb) and Hematocrit (Hto/Hct), Polymerase Chain Reaction (PCR), Procalcitonin (PCT), International Normalized Ratio (INR), Platelets, Thrombotic Thrombocytopenic Purpura (ttp), age and sex.

2.3 Laboratory Data

In previous works [2,3] we worked with 34 laboratories that are mentioned below: Glucose, Urea, Ureic Nitrogen, Creatinine, Total Proteins, Albumin, Globulin, A/G Ratio, Alanine Amino Transferase, Aspartate Amino Transferase, Total Bilirubin, Direct Bilirubin, Indirect Bilirubin, Lactic Dehydrogenase, Creatine Phosphokinase, Chlorine, Potassium, Sodium, Hematic Cytometry, Prothrombin Time, Thromboplastin Time, Procalithin, Fibrin Calcium, Phosphorus, Alkaline Phosphate, Magnesium, Protein C. Reactive and Total Cholesterol, HDL Cholesterol, Sedimentation Rate were added, in addition two important laboratories are Dimero D and Ferritin. The outcomes for analysis are Positive in COVID-19, because the hospital generate a test for COVID-19 and the most important thing (discharged and mortality), the idea is to use a COVID-19, like a laboratory result for predict the mortality.

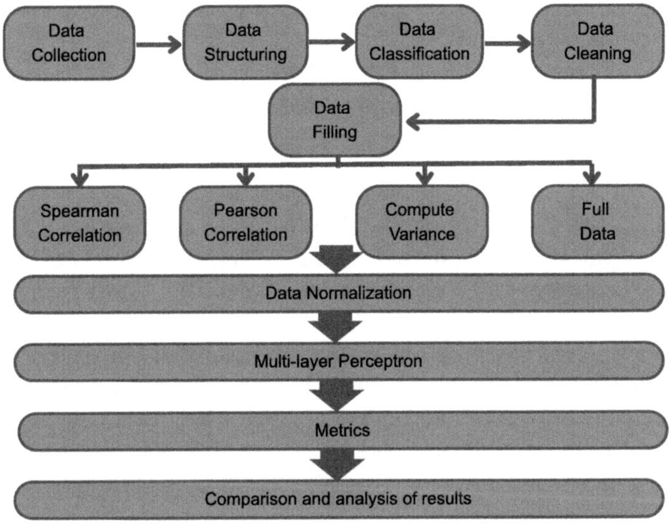

Fig. 1. General flow chart.

3 Methodology

For the development of this study, a structured methodology was proposed a series of steps that are described in the flow chart in Fig. 1 that describes all the blocks carried out in the research paper.

The first block corresponds to collecting the clinical laboratory data in an Excel file, after that, the second step is the Data Structuring, when the data are divided into Clinical, laboratories and X-ray chest for each patient and assigned ID.

Third block is when all the data are sorted by laboratories, COVID or non-COVID, death, and dates of hospital stay. The next block is when some data/patients are excluding, for this the criterion due to "80% incomplete or missing data" means that the medical records of these patients did not contain at least 20% of the data required. This percentage was set to ensure that the dataset contained a several information to generate accurate conclusions.

Block Data Filling corresponds to: Fill in missing data using the arithmetic mean (dataset) and try to have less error in the construction of the dataset, each column is taken as individual and likewise the columns are independent of each other to avoid biasing the problem.

Mean imputation was selected due to the limited dataset size and the need to preserve as many clinical records as possible.

Although more advanced imputation techniques may improve robustness, the selected approach allowed maintaining dataset consistency while avoiding excessive data loss.

Future work will evaluate alternative imputation strategies such as KNN and multiple imputation methods.

For the next step is divided in 4 blocks, the idea is to feed a Multilayer perceptron (MLP) with different inputs and compare the results:

3.1 Full Data

For this block, the idea is to use 28 laboratories as training input for the MLP. Six laboratory variables are removed (D-dimer, Ferritin, Urinalysis, Total Cholesterol, HDL Cholesterol and Erythrocyte Sedimentation Rate) because several of them are obtained every four or five days, and some of them began analysis six months after the first case was reported at the hospital. Once laboratories to be used as input for the MLP are selected, the data are normalized for better processing. In this case, Z-score normalization was used, although tests have been performed with min-max normalization.

3.2 Compute Variance

A subsequent step is obtaining the variance of all laboratories to understand the dispersion in data. This can also be used as an input selector to understand the separability in data and to compute the feature selection to reduce the MLP inputs, which has been calculated in a previous article [2].

3.3 Pearson Correlation

To measure the strength and linear relationship between the multiple variables (laboratories) and the output, an analysis is performed using Pearson's correlation, which has the following Eq. 1:

$$r_{XY} = \frac{\sum_{i=1}^{n}(X_i - \bar{X})(Y_i - \bar{Y})}{\sqrt{\sum_{i=1}^{n}(X_i - \bar{X})^2 \sum_{i=1}^{n}(Y_i - \bar{Y})^2}}, \quad (1)$$

where: r_{XY} is the Pearson correlation coefficient of the variables X and Y, the numerator is covarianza between the variables X and Y and finally the denominator contain standard deviation of the variables X and Y respectively [10].

3.4 Spearman Correlation

The block Spearman's correlation, which is used to assess the strength and direction of the monotonic relationship between two variables using a rank and is expressed in the Eq. 2

$$\rho = 1 - \frac{6 \sum d_i^2}{n(n^2 - 1)}, \tag{2}$$

where: ρ is Spearman's correlation coefficient (-1 to +1), d_i is a difference between the ranks of the two variables for each observation and n correspond to number of data pairs [10].

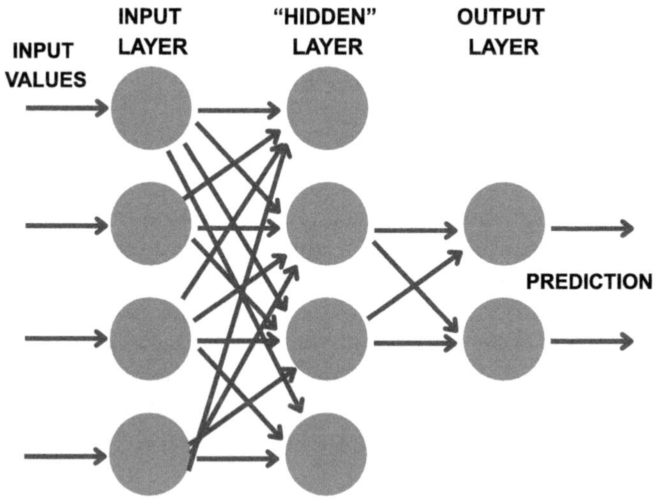

Fig. 2. Representation or architecture to a Multilayer Perceptron.

3.5 Multilayer Perceptron

"The Multilayer Perceptron (MLP) is an artificial neural network used for supervised learning in the field of machine learning. The MLP consists of multiple layers of interconnected neurons, where the outputs of the neurons in one layer become the inputs for the next. The first layer is called the input layer, the last layer is called the output layer, and the intermediate layers are called hidden layers", see Fig. 2

"The MLP is capable of performing classification and regression tasks and can be used for problems where the relationship between inputs and outputs is nonlinear. The training algorithm used by the MLP is based on backpropagation, which adjusts the network's connection weights to minimize the prediction error between the network's outputs and the desired outputs" [11], and the basic equation is described in Eq. 3.

$$y = f\left(\sum_{i=1}^{n} w_i x_i + b\right), \tag{3}$$

where: x_i is input vector, w_i correspond to weight matrix that determines the strength of the connection between neurons, b is bias vector, f correspond to nonlinear activation function and y is output of the neuron.

To justify the selection of the MLP architecture, multiple configurations were experimentally evaluated, varying the number of hidden layers (1–3 layers), neurons per layer (8–64 neurons), and activation functions (ReLU and LeakyReLU).

A manual hyperparameter search was conducted due to the limited dataset size, aiming to balance model complexity and generalization. Simpler architectures were preferred to reduce overfitting risk.

The final architecture was selected based on validation performance, prioritizing recall due to its clinical relevance in identifying high-risk patients.

3.6 Metrics

Any Machine Learning algorithm needs to be evaluated to validate its training capacity and its response to new inputs, which requires metrics that are described in the following subsections. Remember that TP (True Positive): corresponds to values that the MLP classifies as positive and that are indeed positive. TN (True Negative): These are values that the MLP classifies as negative and that are indeed negative. FP: False Positives, that is, values that the MLP classifies as positive when are negative. FN: False Negatives, that is, values that the MLP classifies as negative when they are actually positive [12].

Accuracy is the proportion of all classifications that were correct, whether positive or negative and is described in the Eq. 4

$$\text{Accuracy} = \frac{TP + TN}{TP + TN + FP + FN} \tag{4}$$

Precision metric that corresponds to the fraction of values that actually belong to a positive class out of all of the values which are predicted to belong to that class, the information is described in the Eq. 5.

$$\text{Precision} = \frac{TP}{TP + FP} \tag{5}$$

Recall in an MLP measures the ability to find/calculate all positive cases and is represented through the Eq. 6.

$$\text{Recall} = \frac{\text{TP}}{\text{TP} + \text{FN}} \tag{6}$$

F1-Score measures the balance between precision and recall for a training model. The results from 0 to 1, where 1 corresponds to perfect precision and recall, and 0 implies poor performance and it is described in the Eq. 7

$$\text{F1-Score} = 2 \times \frac{\text{Precision} \times \text{Recall}}{\text{Precision} + \text{Recall}} \tag{7}$$

4 Results

4.1 Correlations and Variance

The first test was to perform a correlation and variance test, in which Pearson and Spearman correlation calculations were made to analyze the behavior of the variables. One of the tests performed was to take the 28 laboratories to analyze and compare against COVID-19 as an output. Likewise, the study of the 28 laboratories vs mortality was carried out, concluding that it will be better to perform the analysis having 29 laboratory tests (including COVID-19 as input) and comparing vs mortality.

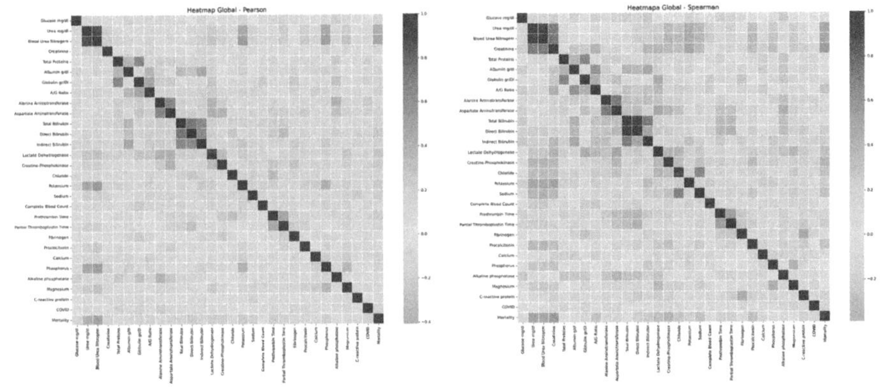

(a) Representation of a Pearson's correlation.

(b) Representation of Spearman's correlation.

Fig. 3. Global Heatmap for correlations

Figure 3 depicts the Heatmap for the correlations, first one the Fig. 3a depicts the result of the Pearson's correlation between 29 laboratories and 1 output

(Mortality), at the same time, the Fig. 3b shows the results from Spearman's correlation, with some similarities in a comparative for each variable laboratory.

For further comparison, an image was designed showing the correlation between the different laboratory-by-laboratory data and mortality rates, to assess their relationship (lower or higher). The Fig. 4 displays the results and allows for the analysis of elements that can be discarded for feature selection tests, which aim to choose those features with the strongest correlation in both methods.

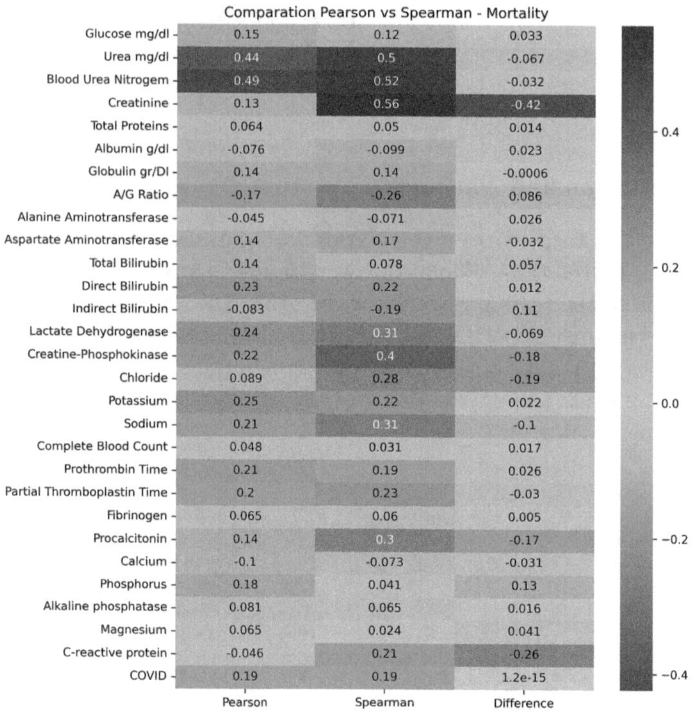

Fig. 4. Comparative results between Pearson vs Spearman correlation.

At the same time, we use the results described in the paper [2], where the variance of all laboratories was analyzed, with the information to the correlations and the variance, we decided to discard 16 elements or laboratories and use 12 laboratories for inputs to MLP, the final selection are: Urea, Blood Urea Nitrogen, Creatinine, Lactate Dehydrogenase, Creatine Phosphokinase, Chloride, Potassium, Sodium, Prothrombin Time, Partial Thromboplastin Time, Procalcitonin and COVID-19.

4.2 Multilayer Perceptron (29 Variables)

The final step is to compare the different dataset including a comparative MLP architecture, for the first test called "Test 1", the model corresponds to an MLP with two layers (16 and 8 neurons), respectively, with 29 inputs and one output (binary), that is the Mortality. The ReLU activation function was used in hidden layers, and the Adam optimizer is used for weight updates.

To avoid overfitting, L2 regularization was incorporated with $alpha = 0.001$, as well as early stopping. The model was trained with a maximum of 2000 iterations, and the data were previously normalized using z-score standardization. Model evaluation was performed using training (70%), validation (15%), and test (15%). Table 2 shows a confusion matrix for the test with all the 29 variables for training, the table is divided into "Training", "Validation" and "Test", a note is that P0 represents Predicted = 0, or (alive), P1 represents Predicted=1, or (Mortality), in the case of R0 that described Real=0 (alive) and R1 that described Real=1 (Mortality).

Table 2. Confusion matrix with 29 variables

Training			Validation			Test		
	P0	P1		P0	P1		P0	P1
R0	152	19	R0	29	8	R0	33	4
R1	29	157	R1	14	26	R1	18	22

Table 3 depicts the final result with the different metrics and is divided in "Training", "Validation" and "Test", again, where the MLP demonstrates high sensitivity in detecting patients at risk of death, achieving a recall of 0.8441 in training. This indicates that the model is capable of identifying the majority of positive cases, which is crucial in medical applications. At the same time in the metric accuracy obtained (0.8655) suggests the presence of false positives; that is, patients classified as high-risk who do not necessarily die. Overall, the F1-score of 0.8674 indicates a suitable balance between both metrics, making the model useful as a clinical decision support tool.

Table 3. Multilayer perceptron metrics with 29 variables

Test 1	Accuracy	Precision	Recall	F1-score
Training	0.8655	0.8920	0.8441	0.8674
Validation	0.7143	0.7647	0.6500	0.7027
Test	0.7143	0.8462	0.5500	0.6667
Logistic R.	0.7573	0.9143	0.5926	0.7191

An additional comparison was made using logistic regression with the 29 input variables and obtaining low results in several metrics except for precision.

4.3 Multilayer Perceptron (12 Variables)

The second test called "Test 2" corresponds to the use of MLP with an input reduced, with 12 elements (laboratories) selected through variance analysis, Pearson's and Spearman's correlations, with a MLP with three hidden layers (32, 16, and 8 neurons). The LeakyReLU activation function with a coefficient of 0.01 is used. Training is performed using the Adam optimizer with a learning rate of 0.0001 and to address class imbalance.

Finally, the classification decision is made using a threshold of 0.35, which prioritizes the detection of the positive class (Mortality). The Table 4 shows the results of the confusion matrix for Training, validation and test.

Table 4. Confusion matrix with 12 variables

Training			Validation			Test		
	P0	P1		P0	P1		P0	P1
R0	140	28	R0	26	12	R0	28	11
R1	15	174	R1	5	34	R1	2	36

Finally, the Table 5 indicates the metrics results for the last test. The MLP demonstrates good performance in predicting COVID-19 mortality, with an accuracy of 83.12% in the test set. At the same time recall = 0.9474 that is particularly noteworthy, indicating a good ability to identify patients at risk of death. Precision =0.7660 suggests a presence of some false positives, indicating that the model tends to overestimate risk in certain cases. The F1-score of 0.8471 confirms a good balance between accuracy and sensitivity.

Table 5. Multilayer perceptron metrics with 12 variables

Test 2	Accuracy	Precision	Recall	F1-score
Training	0.8796	0.8614	0.9206	0.8900
Validation	0.7792	0.7391	0.8718	0.800
Test	0.8312	0.7660	0.9474	0.8471
Logistic R.	0.6505	0.7500	0.500	0.600

Likewise, as in the previous test, the comparison was made with the logistic regression method using the 12 input variables and obtaining low results in several metrics.

The confusion matrices presented correspond to aggregated predictions over all daily records, not individual patients. This explains the higher number of samples observed in training, validation, and test sets.

5 Conclusion and Discussion

The use of MLP networks in recent years in the area of COVID-19 requires more rigorous analysis. Furthermore, it should be considered a pre-diagnostic tool and involves the analysis of a large amount of data. Therefore, statistical techniques help generate better conclusions. As in the case of this paper, the use of variance, Pearson correlation, and Spearman correlation provides insights into COVID-19 progression and outcomes. Involving MLP networks also provides better insights into the behavior of the disease. This is why it is important to analyze the highlights and perform feature extraction to improve performance. Compared with Test 1, the reduced-feature MLP used in Test 2 improved accuracy by approximately 16.4%, recall by 72.3%, and F1-score by 27.1%. However, precision decreased by approximately 9.5%, suggesting that the model increased sensitivity at the expense of a higher number of false positives. Additionally, the reduced-feature MLP outperformed the Logistic Regression baseline in most evaluation metrics, particularly in recall and F1-score, demonstrating a better capability to identify high-risk patients and capture nonlinear relationships among the clinical and laboratory variables. Compared to the state of the art, similar results were obtained in training, but with the advantage that this article is divided into training, validation, and testing. Many articles only show training or validation.

One of the main contributions of this work is demonstrating that reducing the input space from 29 laboratory variables to 12 statistically selected variables improved predictive performance while reducing model complexity.

Compared with the ISARIC4C Mortality Score, which was developed using large-scale multicenter cohorts and predefined clinical variables, the proposed MLP-based methodology focuses on learning nonlinear patterns from laboratory data collected in a Mexican critical care population. While the ISARIC4C model emphasizes clinical interpretability and bedside applicability, the proposed approach demonstrates the potential of reduced-feature deep learning models to achieve high recall and competitive predictive performance using a smaller subset of laboratory variables.

Although the proposed methodology was not directly benchmarked against the ISARIC4C Mortality Score, future work will evaluate both approaches using the same dataset to analyze their predictive performance and clinical applicability in Mexican patient populations.

Acknowledgments. The authors would like to thank the State Center for Critical Care of Salamanca (CECCS), the Dr. Cesar Centeno Fosado and the head of nurses Liliana Yañez Vargas and Miguel Conejo for their support in preparing the article and providing the database.

References

1. CONAHCYT Homepage. https://datos.covid-19.conacyt.mx/. Accessed 10 Apr 2026

2. Yañez, I., Prieto, D., Doñate, A., Quintanilla, D.: Statistical evolution of clinical laboratories in patients admitted to the CECCS with SARS-COV-2. J. Quant. Stat. Anal. 8(22), 1–10 (2021). https://doi.org/10.35429/JQSA.2021.22.8.1.10
3. Yañez, I., Doñate, A., Quintanilla, D., Aguilera, G.: Laboratory data statistical analysis from CECCS' patients with or without SARSCoV-2 (COVID-19). J. Quant. Stat. Anal. **7**(21), 9–117 (2020). https://doi.org/10.35429/JQSA.2020.21.7.9.17
4. Seyedtabib, M., Najafi-Vosough, R., Kamyari, N.: The predictive power of data: machine learning analysis for COVID-19 mortality based on personal, clinical, preclinical, and laboratory variables in a case–control study. BMC Infect. Dis. **24**, 411 (2024). https://doi.org/10.1186/s12879-024-09298-w
5. Chadaga, K., Chakraborty, C., Prabhu, S., Umakanth, S., Bhat, V., Sampathila, N.: Clinical and laboratory approach to diagnose COVID-19 using machine learning. Interdiscip. Sci. **14**(2), 452–470 (2022). https://doi.org/10.1007/s12539-021-00499-4
6. Baik, S.M., Hong, K.S., Park, D.J.: Deep learning approach for early prediction of COVID-19 mortality using chest X-ray and electronic health records. BMC Bioinformatics **24**, 190 (2023). https://doi.org/10.1186/s12859-023-05321-0
7. Styrzynski, F., Zhakparov, D., Schmid, M., et al.: Machine learning successfully detects patients with COVID-19 prior to PCR results and predicts their survival based on standard laboratory parameters in an observational study. Infect. Dis. Ther. **12**, 111–129 (2023). https://doi.org/10.1007/s40121-022-00707-8
8. Yu, L., et al.: Artificial intelligence systems for diagnosis and clinical classification of COVID-19. Front. Microbiol. **12**, 729455 (2021). https://doi.org/0.3389/fmicb.2021.729455
9. Knight, S.R., et al.: Risk stratification of patients admitted to hospital with COVID-19 using the ISARIC WHO clinical characterisation protocol: development and validation of the 4C mortality score. BMJ **370**, m333 (2020). https://doi.org/10.1136/bmj.m3339
10. Rovetta, A.: Raiders of the lost correlation: a guide on using Pearson and Spearman coefficients to detect hidden correlations in medical sciences. Cureus **12**(11), e11794 (2020). https://doi.org/10.7759/cureus.11794
11. GAMCO Homepage. https://gamco.es/glosario/percepatron-multicapa-mlp/. Accessed 12 Mar 2026
12. Luque, A., Carrasco, A., Martín, A., de Las Heras, A.: The impact of class imbalance in classification performance metrics based on the binary confusion matrix. Pattern Recogn. (2019). https://doi.org/10.1016/j.patcog.2019.02.023.

Author Index

H. Ponce et al. (Eds.): COMIA 2026, CCIS 3095, pp. 825–827, 2027.
https://doi.org/10.1007/978-3-032-37438-7

GPSR Compliance

The European Union's (EU) General Product Safety Regulation (GPSR) is a set of rules that requires consumer products to be safe and our obligations to ensure this.

If you have any concerns about our products, you can contact us on ProductSafety@springernature.com

In case Publisher is established outside the EU, the EU authorized representative is:

Springer Nature Customer Service Center GmbH
Europaplatz 3
69115 Heidelberg, Germany

Batch number: 10723237

Printed by Printforce, the Netherlands